# Coherent Raman Scattering Microscopy

# Series in Cellular and Clinical Imaging

*Series Editor*
**Ammasi Periasamy**

PUBLISHED

Coherent Raman Scattering Microscopy
*edited by Ji-Xin Cheng and Xiaoliang Sunney Xie*

FORTHCOMING

The Fluorescent Protein Revolution
*edited by Richard N. Day and Michael W. Davidson*

Second Harmonic Generation Imaging
*edited by Francesco S. Pavone and Paul J. Campagnola*

Imaging in Cellular & Tissue Engineering
*edited by Hanry Yu and N. Aida Abdul Rahim*

Natural Biomarkers for Cellular Metabolism:
Biology, Techniques, and Applications
*edited by Vladimir V. Gukassyan and Ahmed A. Heikal*

Optical Probes in Biology
*edited by Jin Zhang and Carsten Schultz*

SERIES IN CELLULAR AND CLINICAL IMAGING
AMMASI PERIASAMY, SERIES EDITOR

# Coherent Raman Scattering Microscopy

Edited by
Ji-Xin Cheng
Xiaoliang Sunney Xie

CRC Press
Taylor & Francis Group
Boca Raton London New York

CRC Press is an imprint of the
Taylor & Francis Group, an **informa** business

**Front Cover:** Multimodal Imaging of the central nervous system, courtesy of Ji-Xin Cheng. **Back Cover:** The stimulated Raman scattering image of a mouse brain cross-section, showing the distributions of protein (blue) and lipid (green), from X. Sunney Xie's group. Imaging courtesy of Minbiao Ji of Harvard University and Daniel Orringer of Brigham and Women's Hospital.



CRC Press
Taylor & Francis Group
6000 Broken Sound Parkway NW, Suite 300
Boca Raton, FL 33487-2742

First issued in paperback 2018



ISBN 13: 978-1-138-19952-1 (pbk)
ISBN 13: 978-1-4398-6765-5 (hbk)



**Library of Congress Cataloging-in-Publication Data**

Coherent Raman scattering microscopy / edited by Ji-Xin Cheng and Xiaoliang Sunney Xie.
p. ; cm. -- (Series in cellular and clinical imaging ; 1)
Includes bibliographical references and index.
ISBN 978-1-4398-6765-5 (hardcover : alk. paper)
1. Raman spectroscopy. 2. Three-dimensional imaging in medicine. I. Cheng, Ji-Xin. II. Xie, Xiaoliang Sunney. III. Series: Series in cellular and clinical imaging ; 1.
[DNLM: 1. Spectrum Analysis, Raman--methods. 2. Imaging, Three-Dimensional--methods. 3. Microscopy--methods. QC 454.R36]

QC454.R36C64 2013
543'.57--dc23 2012015407

**Visit the Taylor & Francis Web site at**
**http://www.taylorandfrancis.com**

**and the CRC Press Web site at**
**http://www.crcpress.com**

# Contents

# Series Preface

A picture is worth a thousand words.

This proverb says everything. Imaging began in 1021 with use of a pinhole lens in a camera in Iraq; later in 1550, the pinhole was replaced by a biconvex lens developed in Italy. This mechanical imaging technology migrated to chemical-based photography in 1826 with the first successful sunlight-picture made in France. Today, digital technology counts the number of light photons falling directly on a chip to produce an image at the focal plane; this image may then be manipulated in countless ways using additional algorithms and software. The process of taking pictures ("imaging") now includes a multitude of options—it may be either invasive or noninvasive, and the target and details may include monitoring signals in two, three, or four dimensions.

Microscopes are an essential tool in imaging used to observe and describe protozoa, bacteria, spermatozoa, and any kind of cell, tissue, or whole organism. Pioneered by Antoni van Leeuwenhoek in the 1670s and later commercialized by Carl Zeiss in 1846 in Jena, Germany, microscopes have enabled scientists to better grasp the often misunderstood relationship between microscopic and macroscopic behavior, by allowing for study of the development, organization, and function of unicellular and higher organisms, as well as structures and mechanisms at the microscopic level. Further, the imaging function preserves temporal and spatial relationships that are frequently lost in traditional biochemical techniques and gives two- or three-dimensional resolution that other laboratory methods cannot. For example, the inherent specificity and sensitivity of fluorescence, the high temporal, spatial, and three-dimensional resolution that is possible, and the enhancement of contrast resulting from detection of an absolute rather than relative signal (i.e., unlabeled features do not emit) are several advantages of fluorescence techniques. Additionally, the plethora of well-described spectroscopic techniques providing different types of information, and the commercial availability of fluorescent probes such as visible fluorescent proteins (many of which exhibit an environment- or analytic-sensitive response), increase the range of possible applications, such as development of biosensors for basic and clinical research. Recent advancements in optics, light sources, digital imaging systems, data acquisition methods, and image enhancement, analysis and display methods have further broadened the applications in which fluorescence microscopy can be applied successfully.

Another development has been the establishment of multiphoton microscopy as a three-dimensional imaging method of choice for studying biomedical specimens from single cells to whole animals with submicron resolution. Multiphoton microscopy methods utilize naturally available endogenous fluorophores—including NADH, TRP, FAD, and so on—whose autofluorescent properties provide a label-free approach. Researchers may then image various functions and organelles at molecular levels using two-photon and fluorescence lifetime imaging microscopy (FLIM) to distinguish normal from cancerous conditions. Other widely used nonlabeled imaging methods are coherent anti-Stokes Raman scattering spectroscopy (CARS) and stimulated Raman scattering (SRS) microscopy, which allow imaging of molecular function using the molecular vibrations in cells, tissues, and whole organisms. These techniques have been widely used in gene therapy, single molecule imaging, tissue engineering, and stem cell research. Another nonlabeled method is harmonic generation (SHG and THG), which is also widely used in clinical imaging, tissue engineering, and stem cell research. There are many more advanced technologies developed for cellular and clinical imaging including multiphoton tomography, thermal imaging in animals, ion imaging (calcium, pH) in cells, etc.

The goal of this series is to highlight these seminal advances and the wide range of approaches currently used in cellular and clinical imaging. Its purpose is to promote education and new research across a broad spectrum of disciplines. The series emphasizes practical aspects, with each volume focusing on a particular theme that may cross various imaging modalities. Each title covers basic to advanced imaging methods, as well as detailed discussions dealing with interpretations of these studies. The series also provides cohesive, complete state-of-the-art, cross-modality overviews of the most important and timely areas within cellular and clinical imaging.

Since my graduate student days, I have been involved and interested in multimodal imaging techniques applied to cellular and clinical imaging. I have pioneered and developed many imaging modalities throughout my research career. The series manager, Ms. Luna Han, recognized my genuine enthusiasm and interest to develop a new book series on Cellular and Clinical Imaging. This project would not have been possible without the support of Luna. I am sure that all the volume editors, chapter authors, and myself have benefited greatly from her continuous input and guidance to make this series a success.

Equally important, I personally would like to thank the volume editors and the chapter authors. This has been an incredible experience working with colleagues who demonstrate such a high level of interest in educational projects, even though they are all fully occupied with their own academic activities. Their work and intellectual contributions based on their deep knowledge of the subject matter will be appreciated by everyone who reads this book series.

**Ammasi Periasamy, PhD**
*Series Editor*
*Professor and Center Director*
*W.M. Keck Center for Cellular Imaging*
*University of Virginia*
*Charlottesville, Virginia*

# Preface

In 1928, the Indian scientist C.V. Raman and his collaborator K.S. Krishnan reported their discovery of inelastic scattering of light by molecules in a short article in *Nature* [1]. This phenomenon, which later bore Raman's name, was characterized in his own words as "excessive feebleness," due to the low efficiency of inelastic scattering, hence necessitating intense excitation light. This is why it was not until the emergence of lasers in the 1960s that Raman spectroscopy was popularized. Reporting molecular vibrational frequencies, Raman spectroscopy has now become an indispensible tool for chemical analyses in industrial and research laboratories.

This book addresses the question of how Raman spectroscopy can be used to obtain a point-by-point chemical map of a live cell or tissue. Although fluorescence microscopy has been widely applied to biological imaging because of its high sensitivity, it requires fluorophores that might perturb cell functions, and it suffers from the complication of fluorophore photobleaching. Both Raman scattering and infrared absorption are label-free and offer contrast based on vibrational frequencies intrinsic to molecular species. There are, however, drawbacks to both methods. Despite rapid developments, low spatial resolution due to the long wavelengths involved and low sensitivity to absorption measurements hamper infrared microscopy. Raman microscopy has come a long way in recent years, but the image acquisition speed is again limited by the weakness of its signals. Modern Raman spectrometers can acquire a Raman spectrum in a fraction of a second. Although this is fast enough for chemical analysis at a particular location, an image of 200 × 200 pixels would require the significant portion of an hour. Such acquisition speed is insufficient if one needs to follow dynamical processes of biological samples.

Fortunately, coherent Raman scattering (CRS) provides us with a solution. Shortly after the invention of lasers, two nonlinear Raman phenomena were discovered: stimulated Raman scattering (SRS) by Woodbury and Ng of Hughes Aircraft Company [2] and coherent anti-Stokes Raman scattering (CARS) by Maker and Terhune at Ford Motor Company [3]. Interestingly, the latter was not given the acronym CARS until much later [4]. In both of these cases, simultaneous excitation with two laser beams at different frequencies was used. When the difference between the two frequencies matches the frequency of a particular molecular vibration, the interaction of the laser beams and the sample results in coherent vibrational motion of the molecules, hence greatly enhancing the signals.

In many ways, CRS is similar to nuclear magnetic resonance, which is now widely used by chemists to identify molecular species, by biologists to determine macromolecular structures, and by clinicians to image human organs such as brains through magnetic resonance imaging (MRI). Both of these methods are coherent spectroscopy, in contrast to incoherent fluorescence microscopy. Recent advances in CRS microscopy are built upon decades of fundamental studies of nonlinear optics and continuous laser developments. CRS microscopy now allows video imaging of living cells or tissues. This is an increase of several orders of magnitudes, as compared to conventional Raman microscopy, and offers exciting possibilities, not only in scientific laboratories but also in industry and hospitals.

The development of CRS microscopy began in 1982 when Duncan et al. at the U.S. Naval Research Laboratory made the first attempt at CARS microscopy. In a move that was ahead of its time, they used two synchronized picosecond dye laser beams, propagating noncollinearly through the sample [5]. After a long quiescent period during which solid state ultrafast laser technology matured, in 1999 Zumbusch et al. at Pacific Northwest National Laboratory demonstrated a new CARS microscope with a lens that focused on collinear laser beams, which enabled three-dimensional imaging [6]. In 2001, Volkmer et al. at Harvard University demonstrated CARS microscopy with backward scattering light, which permitted imaging of nontransmissive samples [7]. However, CARS contrast was hampered by the so-called nonresonant background, which exists even in the absence of a vibrational resonance and causes distortion of CARS spectra from their corresponding Raman spectra. Many efforts have been made to circumvent these problems. In 2007, Ploetz et al. of Ludwig Maximilian University reported an SRS microscope constructed with a low repetition rate and high peak power laser [8]. In 2008, Freudiger and Min et al. at Harvard University reported a biocompatible SRS microscope with a sensitive detection scheme, which offered improved performance over CARS microscopes [9]. These key developments prompted widespread research activities and broad applications of CRS microscopy to biology and medicine, thanks to the creativity and innovation of the ever-growing community of CRS microscopy practitioners.

In this first book on CRS microscopy, many experts have contributed their perspectives. This book starts by discussing the principles of nonlinear optical spectroscopy, particularly coherent Raman spectroscopy, followed by theories on contrast mechanisms pertinent to CRS microscopy. These two chapters (1 and 2) establish the foundation of CRS microscopy and are followed by the details of CRS microscope construction, various detection schemes, and data analyses. These chapters (3 through 16) provide important technical aspects of CRS microscopy. These chapters (17 through 27) present a survey of applications, including the study of lipid metabolism, the central neuronal system, drug distribution in tissues, cancer detection *in vivo*, and tissue engineering. These chapters demonstrate how CRS microscopy has become a valuable tool for biomedicine.

We would like to thank all the contributors for the tremendous efforts that have made this book possible. We would also like to thank CRC Press, in particular Luna Han, for guiding us through the publication process. We hope this timely collection will provide a comprehensive overview for all those interested in the developments and applications of CRS microscopy.

In 1978, on the occasion of the 50th anniversary of Raman's discovery, Nicolaas Bloembergen, the Nobel laureate in nonlinear optics, stated, "As spontaneous Raman spectroscopy has blossomed and grown during one half-century, it may be predicted with some confidence that coherent nonlinear Raman spectroscopy will yield many new results in the next half-century" [10]. Thirty-four years have passed, and Bloembergen's prediction is indeed being realized through recent developments in CRS microscopy. In the next 16 years, it is our belief that CRS microscopy will deepen our understanding of life processes and will make an impact on human health. It is particularly intriguing to us that Raman has kept, and will continue to keep, us busy for 100 years.

## References

1. C. V. Raman, K. S. Krishnan, *Nature* **121**, 501 (1928).
2. P. D. Maker, R. W. Terhune, *Phys. Rev.* **137**, A801 (1965).
3. E. J. Woodbury, W. K. Ng, *Proc. Inst. Radio Eng.* **50**, 2367 (1962).
4. R. F. Begley, A. B. Harvey, R. L. Byer, *Appl. Phys. Lett.* **25**, 387 (1974).
5. M. D. Duncan, J. Reintjes, T. J. Manuccia, *Opt. Lett.* **7**, 350 (1982).
6. A. Zumbusch, G. Holtom, X. S. Xie, *Phys. Rev. Lett.* **82**, 4142 (1999).
7. A. Volkmer, J. X. Cheng, X. S. Xie, *Phys. Rev. Lett.* **87**, 02390 (2001).
8. E. Ploetz, S. Laimgruber, S. Berner, W. Zinth, P. Gilch, *Appl. Phys. B Lasers Opt.* **87**, 389 (2007).
9. C. W. Freudiger, M. Wei, B. G. Saar, S. Lu, G. R. Holtom, C. He, J. C. Tsai, J. X. Kang, X. S. Xie, *Science* **322**, 1857 (2008).
10. N. Bloembergen, *Proceedings of the Sixth International Conference on Raman Spectroscopy*, E.D. Schmid et al. (eds.), Vol. 1, 335 (Heyden, London, U.K., 1978).



**Ji-Xin Cheng**
**Xiaoliang Sunney Xie**

# Editors

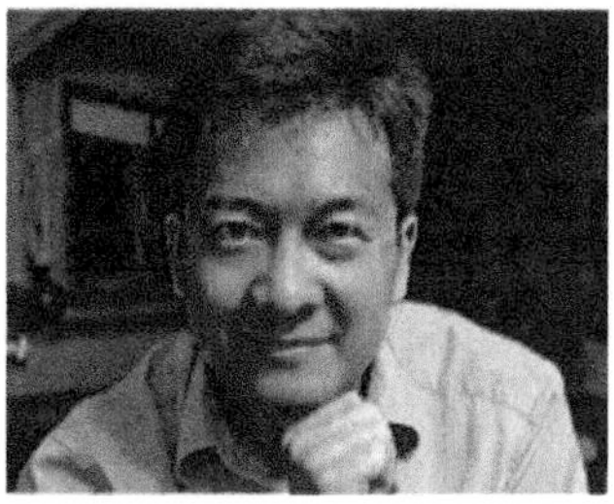

**Ji-Xin Cheng** is professor of biomedical engineering at Purdue University, West Lafayette, Indiana. He received his doctorate from the University of Science and Technology of China, followed by postdoctoral research at the Hong Kong University of Science and Technology and Harvard University. He is a pioneer in the development and biomedical application of molecular vibration–based imaging tools.

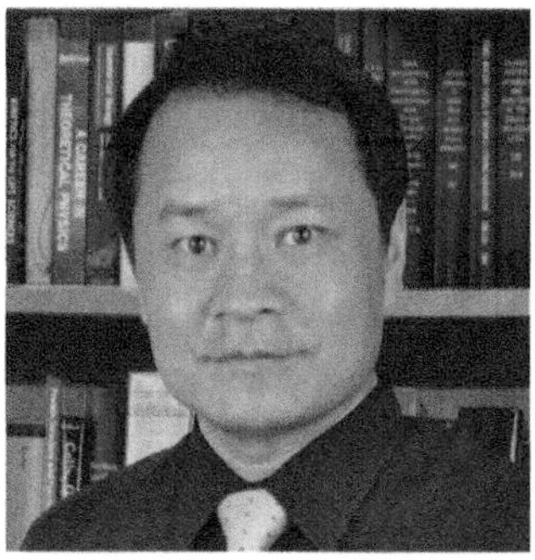

**Xiaoliang Sunney Xie** is Mallinckrodt Professor of Chemistry at Harvard University, Cambridge, Massachusetts. He received his doctorate from the University of California at San Diego, followed by postdoctoral research at the University of Chicago. Xie is known for his innovations in nonlinear Raman microscopy and his pioneering work in the field of single-molecule biophysical chemistry, in particular enzyme dynamics and live cell gene expression.

# Contributors

**Khaled A. Aamer**
National Institute of Standards and Technology
Gaithersburg, Maryland

**Mihaela Balu**
Beckman Laser Institute
University of California, Irvine
Irvine, California

**Erik Bélanger**
Department of Physics
Laval University
Quebec City, Quebec, Canada

**Stefan Bernet**
Department of Physiology and Medical Physics
Innsbruck Medical University
Innsbruck, Austria

**Mischa Bonn**
Department of Molecular Spectroscopy
Max Planck Institute for Polymer Research
Mainz, Germany

**Christian Brackmann**
Department of Chemical and Biological Engineering
Chalmers University of Technology
Gothenburg, Sweden

**Kimberly K. Buhman**
Department of Nutrition Science
Purdue University
West Lafayette, Indiana

**Zhongping Chen**
Beckman Laser Institute
University of California, Irvine
Irvine, California

**Ji-Xin Cheng**
Weldon School of Biomedical Engineering
Purdue University
West Lafayette, Indiana

**Marcus T. Cicerone**
Polymers Division
National Institute of Standards and Technology
Gaithersburg, Maryland

**Daniel Côté**
Department of Physics
Laval University
Quebec City, Quebec, Canada

**James P.R. Day**
Department of Molecular Spectroscopy
Max Planck Institute for Polymer Research
Mainz, Germany

**Katrin F. Domke**
Department of Molecular Spectroscopy
Max Planck Institute for Polymer Research
Mainz, Germany

**Annika Enejder**
Department of Chemical and Biological Engineering
Chalmers University of Technology
Gothenburg, Sweden

**Helen Fink**
Department of Chemical and Biological Engineering
Chalmers University of Technology
Gothenburg, Sweden

**Christian Freudiger**
Department of Chemistry and Chemical Biology
Harvard University
Cambridge, Massachusetts

**Dan Fu**
Department of Chemistry and Chemical Biology
Harvard University
Cambridge, Massachusetts

**Yan Fu**
National Institute of Biomedical Imaging and Bioengineering
National Institutes of Health
Bethesda, Maryland

**F.P. Henry**
Massachusetts General Hospital
Harvard Medical School
Boston, Massachusetts

**Alexander Jesacher**
Department of Physiology and Medical Physics
Innsbruck Medical University
Innsbruck, Austria

**Martin Jurna**
Optical Sciences Group
University of Twente
Enschede, the Netherlands

**Hideaki Kano**
Institute of Applied Physics
University of Tsukuba
Ibaraki, Japan

**Ori Katz**
Department of Physics of Complex Systems
Weizmann Institute of Science
Rehovot, Israel

**Hyunmin Kim**
Surface and Interface Research Group
National Institute of Standards and Technology
Gaithersburg, Maryland

**I.E. Kochevar**
Massachusetts General Hospital
Harvard Medical School
Boston, Massachusetts

**Oleg D. Lavrentovich**
Liquid Crystal Institute
Kent State University
Kent, Ohio

**Thuc T. Le**
Nevada Cancer Institute
Las Vegas, Nevada

**Young Jong Lee**
National Institute of Standards and Technology
Gaithersburg, Maryland

**Jonathan M. Levitt**
Department of Physics of Complex Systems
Weizmann Institute of Science
Rehovot, Israel

**Sang-Hyun Lim**
Department of Chemistry and Biochemistry
University of Texas
Austin, Texas

**Charles P. Lin**
Center for Systems Biology
Massachusetts General Hospital
Harvard Medical School
Cambridge, Massachusetts

**Gangjun Liu**
Beckman Laser Institute
University of California, Irvine
Irvine, California

**Shaul Mukamel**
Department of Chemistry
University of California, Irvine
Irvine, California

**Herman L. Offerhaus**
Optical Science Group
University of Twente
Enschede, the Netherlands

**Jennifer P. Ogilvie**
Department of Physics and Biophysics
University of Michigan
Ann Arbor, Michigan

**Daniel A. Orringer**
Department of Neurosurgery
University of Michigan Health System
Ann Arbor, Michigan

**Cees Otto**
Department of Applied Physics
University of Twente
Enschede, the Netherlands

**Sapun H. Parekh**
Molecular Spectroscopy Department
Max Planck Institute for Polymer Research
Mainz, Germany

**Heung-Shik Park**
Liquid Crystal Institute
Kent State University
Kent, Ohio

**Eric Olaf Potma**
Department of Chemistry
University of California, Irvine
Irvine, California

**Varun Raghunathan**
Department of Chemistry
University of California, Irvine
Irvine, California

**Gianluca Rago**
Department of Molecular Spectroscopy
Max Planck Institute for Polymer Research
Mainz, Germany

**M.A. Randolph**
Massachusetts General Hospital
Harvard Medical School
Boston, Massachusetts

**Monika Ritsch-Marte**
Department of Physiology and Medical Physics
Innsbruck Medical University
Innsbruck, Austria

**Brian G. Saar**
Department of Chemistry and Chemical Biology
Harvard University
Cambridge, Massachusetts

**Yunzhou (Sophia) Shi**
Purdue University
West Lafayette, Indiana

**Yaron Silberberg**
Department of Physics of Complex Systems
Weizmann Institute of Science
Rehovot, Israel

**Mikhail N. Slipchenko**
Weldon School of Biomedical Engineering
Purdue University
West Lafayette, Indiana

**Stephan Stranick**
Surface and Interface Research Group
National Institute of Standards and Technology
Gaithersburg, Maryland

**Michael Sturek**
Department of Cellular and Integrative Physiology
Indiana University School of Medicine
Indianapolis, Indiana

**Gregor Thalhammer**
Department of Physiology and Medical Physics
Innsbruck Medical University
Innsbruck, Austria

**Ling Tong**
School of Medicine
Stanford University
Palo Alto, California

**R. Vallée**
Centre de Recherche Université Laval Robert-Giffard
and
Centre d'Optique, Photonique et Laser
Laval University
Quebec City, Quebec, Canada

**Erik M. Vartiainen**
Department of Mathematics and Physics
Lappeenranta University of Technology
Lappeenranta, Finland

**Andreas Volkmer**
Institute of Physics
University of Stuttgart
Stuttgart, Germany

**Han-Wei Wang**
LI-COR Biosciences
Lincoln, Nebraska

**J.M. Winograd**
Massachusetts General Hospital
Harvard Medical School
Boston, Massachusetts

**Xiaoliang Sunney Xie**
Department of Chemistry and Chemical Biology
Harvard University
Cambridge, Massachusetts

**Delong Zhang**
Department of Chemistry
Purdue University
West Lafayette, Indiana

# Theory

# 1. Theory of Coherent Raman Scattering

## Eric Olaf Potma and Shaul Mukamel



*Coherent Raman Scattering Microscopy*. Edited by Ji-Xin Cheng and X. Sunney Xie © 2013 CRC Press/ Taylor & Francis Group, LLC. ISBN: 978-1-4398-6765-5.

## 1.1 Introduction: The Coherent Raman Interaction

The term "coherent Raman scattering" (CRS) denotes a special class of light-matter interactions. Central to this class of interactions is the particular way in which the material is responding to the incoming light fields: the response contains information about material oscillations at difference frequencies of two incident light fields. Hence, writing the frequencies of the light fields as $\omega_1$ and $\omega_2$, the coherent Raman interaction depends on oscillatory motions in the material at the frequency $\Omega = \omega_1 - \omega_2$. This simple stipulation dresses coherent Raman techniques with many unique capabilities. In particular, since the difference frequency $\Omega$ generally corresponds to a low frequency oscillation which can be tuned into resonance with characteristic vibrational modes $\omega_v$, coherent Raman techniques make it possible to probe the low frequency nuclear vibrations of materials and molecules by using high frequency optical light fields.

Coherent Raman techniques are related to spontaneous Raman scattering. In spontaneous Raman scattering, a single $\omega_1$ mode is used to generate the $\omega_2$ mode, which is emitted spontaneously. Both coherent and spontaneous Raman scattering allow for vibrational spectroscopic examination of molecules with visible and near-infrared radiation.

Compared to spontaneous Raman scattering, CRS techniques can produce much stronger vibrationally sensitive signals. The popularity of CRS techniques in optical microscopy is intimately related to these much improved signal levels, which have enabled the fast scanning capabilities of CRS microscopes. However, beyond stronger vibrational signals, the coherent Raman interaction offers a rich palette of probing mechanisms for examining a wide variety of molecular properties. In general, CRS techniques offer a more detailed control of the Raman response of the medium than what is available through spontaneous Raman techniques. CRS allows a more direct probing of the molecular coherences that govern the Raman vibrational response. When ultrafast pulses are used, CRS methods can resolve the ultrafast evolution of such Raman coherences on the appropriate timescale. CRS techniques also offer more detailed information about molecular orientation than spontaneous Raman techniques. In addition, advanced resonant Raman (coherent or spontaneous) techniques can selectively probe both the electronic and vibrational response of the material, which opens a window to a wealth of molecular information.

In this chapter, we examine the basics of the coherent Raman interaction, which provides a foundation for more advanced topics discussed in subsequent chapters of this book. Here, we focus predominantly on the light-matter interaction itself. We study both the classical and the semi-classical descriptions of the coherent Raman process and discuss strengths and weaknesses of each approach. In addition, we highlight some of the findings obtained with a quantum mechanical model of the CRS process. The propagation of light in the material, which gives rise to several interesting effects in coherent Raman microscopy in the tight focusing limit, is discussed in Chapter 2.

## 1.2 Nonlinear Optical Processes

### 1.2.1 Induced Polarization

Both linear and nonlinear optical effects can be understood as resulting from the interaction of the electric field component of electromagnetic radiation with the charged particles of the material or molecule. Generally, an applied electric field moves positively charged particles in the direction of the field and negative charges in the opposite direction. The electric field associated with the visible and near-infrared range of the electromagnetic spectrum oscillates at frequencies in the $10^3$ THz range. Such driving frequencies are too high for the nuclei to follow adiabatically. The electrons in the material or molecule, however, are light enough to follow the rapid oscillations of the driving field. Consequently, optical resonances in this frequency range are predominantly due to the motions of the electrons in the material.

As a result of the driving fields, the bound electrons are slightly displaced from their equilibrium positions, which induces an electric dipole moment:

$$\mu(t) = -e \cdot r(t) \tag{1.1}$$

where $e$ is the charge of the electron. The magnitude of the dipole depends on the extent of the displacement $r(t)$. The displacement, in turn, is dependent on how strong the electron is bound to the nuclei. The displacement will be more significant for electrons that are weakly bound to the nuclei, and smaller for electrons that are tightly bound. Close to the nuclei, the electron binding potential can generally be approximated by a harmonic potential.

The macroscopic polarization, which is obtained by adding up all $N$ electric dipoles per unit volume, reads:

$$P(t) = N\mu(t) \tag{1.2}$$

In the limit of weak applied electric fields (compared to the field that binds the electrons to the nuclei), the displacement is directly proportional to the electric field. This allows us to write the polarization as:

$$P(t) = \epsilon_0 \chi E(t) \tag{1.3}$$

where

$\epsilon_0$ is the electric permittivity in vacuum

$\chi$ is the susceptibility of the material (we will use SI units unless otherwise stated)

This expression highlights that, in the weak field limit, the induced polarization in the material depends linearly on the magnitude of the applied field. Such linear dependence is the origin of all linear optical phenomena.

### 1.2.2 Nonlinear Polarization

For stronger fields, the electron is farther displaced from its equilibrium position. For larger displacements, the binding potential can no longer be assumed to be harmonic as anharmonic effects become more significant. When the anharmonic shape of the potential becomes important, the dependence between the driving electric field and the induced polarization is not strictly linear, and corrections to the polarization will have to be made. Figure 1.1 illustrates the nonlinearity between the driving field and the induced polarization in the presence of anharmonicity. If the anharmonic contributions to the harmonic potential are relatively small, the displacement $r$ can be expressed as a power series in the field. This implies that the displacement of the electron is no longer linearly dependent on the field as nonlinear corrections grow in importance. In a similar fashion, the polarization can be written as a power series in the field to include the nonlinear electron motions:

$$\begin{aligned} P(t) &= \epsilon_0[\chi^{(1)}E(t)+\chi^{(2)}E^2(t)+\chi^{(3)}E^3(t)+\cdots] \\ &= P^{(1)}(t)+P^{(2)}(t)+P^{(3)}(t)+\cdots \end{aligned} \tag{1.4}$$

where
$\chi^{(n)}$ is the $n$th order susceptibility
$P^{(n)}$ is the $n$th order contribution to the polarization

The coherent Raman effects described in this book can all be understood as resulting from the third-order contribution to the polarization $P^{(3)}$. The magnitude of these effects is thus governed by the strength of the triple product of the incoming fields and the amplitude of the third-order susceptibility $\chi^{(3)}$.

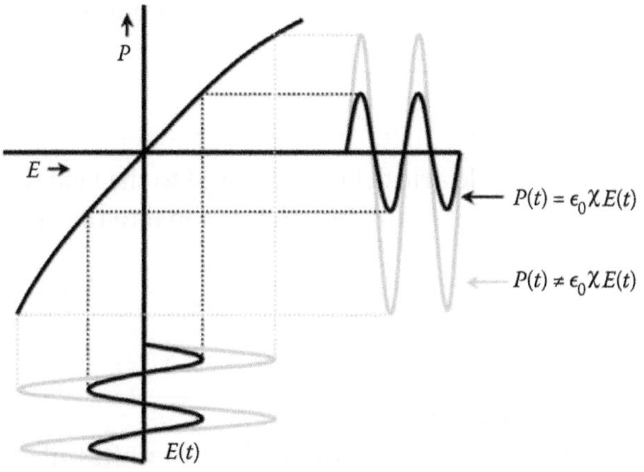


**FIGURE 1.1** Relation between incident electric field and the induced polarization. For weak electric fields, indicated by the black sinusoidal line, only the harmonic part of the potential is relevant and the polarization depends linearly on the field. For strong electric fields, symbolized by the gray sinusoidal line, the anharmonicities of the potential contribute and the polarization depends nonlinearly on the incoming field. In this case, the polarization profile no longer matches the profile of the sinusoidal input modulation.

### 1.2.3 Magnitude of the Optical Nonlinearity

To appreciate the nonlinear origin of coherent Raman effects, it is useful to examine the magnitude of the third-order susceptibility. From the previous discussion it follows that the nonlinearity results predominantly from the electronic anharmonic electron motions. This is indeed the case when ultrafast pulses in the picosecond to the femtosecond range are used, which induce nonlinear optical effects that are typically directly related to the electronic polarizability of the material. We may expect that the nonlinear electron motions become very significant when the applied field is of the order of the field that binds the electron to the atom. This atomic field is $E_a \approx 2 \times 10^7$ esu (in electrostatic units). Hence, in case the applied field is of the order of $E_a$ we expect the nonlinear polarization to be comparable to the linear polarization, i.e., $P^{(1)} \approx P^{(3)}$. Under these (nonresonant) conditions we can write $\chi^{(1)}E_a \approx \chi^{(3)}E_a^3$ and thus estimate that $\chi^{(3)} \approx \chi^{(1)}/E_a^2$. Given that $\chi^{(1)}$ is about unity in the condensed phase, this yields a numerical value for the nonresonant third-order susceptibility of $\chi^{(3)} \simeq 3\times10^{-15}$ [1]. Despite the approximate nature of this estimate, it is surprisingly close to actual measurements of the nonlinear susceptibility. Numerical values of some materials and compounds are given in Table 1.1.

**Table 1.1** Magnitude of $\chi^{(3)}$ as Determined with Third-Harmonic Generation Measurements at the Indicated Excitation Wavelength

| Material | $\chi^{(3)}$ (esu) | $\lambda$ (μm) | Reference |
|---|---|---|---|
| Water | $1.3 \times 10^{-14}$ | 1.06 | [2] |
| Glycine (1 M aqueous) | $1.2 \times 10^{-14}$ | 1.06 | [2] |
| Ethanol | $1.3 \times 10^{-14}$ | 1.06 | [2] |
| Vegetable oil | $1.9 \times 10^{-14}$ | 1.06 | [2] |
| Carbon disulfide | $2.0 \times 10^{-13}$ | 1.91 | [3] |
| Silica | $1.4 \times 10^{-14}$ | 1.06 | [4] |
| BK7 | $2.1 \times 10^{-14}$ | 1.06 | [4] |
| $TiO_2$ (rutile) | $4.0 \times 10^{-12}$ | 1.90 | [5] |

To generate an observable third-order optical signal in practice, applied fields are used that are generally much weaker than $E_a$. This condition is required because otherwise the $\chi^{(3)}$ response cannot be easily isolated from higher order nonlinearities. In addition, fields of the order of $E_a$ would correspond to laser intensities of ~$10^{14}$ W cm$^{-2}$, which is many orders of magnitude too high for applications in microscopy. At the much lower laser intensities relevant to laser scanning optical microscopy (~$10^{10}$ W cm$^{-2}$), the third-order response is orders of magnitude smaller than the linear response, but can nonetheless be detected.

The magnitude of $\chi^{(3)}$ grows larger whenever the electron displacement is enhanced. This is the case under electronically resonant conditions. When the frequency of the driving field is tuned to the frequency of an electronic resonance in the material or molecule, we may expect that the electron displacement is magnified and the third-order nonlinear response is correspondingly stronger. This principle is utilized, for instance, in four-wave mixing microscopy of nanostructures where strong $\chi^{(3)}$-based signals are attained through electronic resonances [6]. In addition to electronic resonances, the presence of nuclear resonances can also affect the electronic nonlinear susceptibility. The coherent Raman effects discussed in this book all derive their chemical sensitivity from these nuclear resonances. In the following sections, we will first introduce a general classification of Raman sensitive techniques, followed by a discussion on the classical Raman effect and the manifestation of the Raman effect in the coherent nonlinear response of the material.

## 1.3 Classification of Raman Sensitive Techniques

Before we discuss the basics of the Raman effect, it is useful to define a couple of terms that will prove useful for interpreting the different types of optical techniques for probing the Raman effect.

### 1.3.1 Coherent versus Incoherent

An important classification is whether the detected signal is coherent or incoherent. The signal is coherent if the optical waves radiated from dipole emitters at different points $\mathbf{r}$ in the sample exhibit a well-defined phase relationship. In this case, the total field, obtained by averaging over all dipole emitters, is non-vanishing and thus $\langle E \rangle \neq 0$. On the other hand, if the phases of the emitted waves are random relative to one another, then the total field averages to zero, i.e., $\langle E \rangle = 0$. This latter case represents an incoherent signal. Note that even though the total field is zero for incoherent signals, the intensity defined by $\langle E^{\dagger} E \rangle$ can be finite.

Conventional spontaneous Raman scattering is an example of an incoherent signal, because the phase of the wave radiated by an individual molecule is uncorrelated with the waves emitted by other molecules in the sample. Rayleigh (elastic) scattering, on the other, is a coherent signal. In Rayleigh scattering, the phase of the scattered waves is not perturbed by a nuclear mode with arbitrary phase, producing scattered radiation with a definite phase relation relative to the incoming waves. The difference between Raman scattered light and Rayleigh scattered light is further addressed in Section 1.4.2. All nonlinear Raman techniques produce coherent signals. Contrary to incoherent Raman, in nonlinear Raman techniques the nuclear oscillators in the sample are correlated by the light fields, producing radiation from different points in the sample with a well-defined phase relationship. All nonlinear Raman techniques discussed in the book are classified as coherent.

### 1.3.2 Linear versus Nonlinear

The linearity of the signal is defined through its dependence on the intensity $I$ of the incident radiation. Optical signals that scale linearly with the average power of the incident radiation are classified as linear techniques. Optical signals that exhibit a quadratic or higher order dependence on the intensity of the input radiation are classified as nonlinear techniques. Incoherent (spontaneous) Raman is linear, whereas CRS techniques

**Table 1.2** Classification of Raman Sensitive Techniques

| | Incoherent | Coh. Homodyne CARS | Coh. Heterodyne CARS | Pump-Probe |
|---|---|---|---|---|
| Common name | Spontaneous Raman | CARS | Heterodyne CARS | SRS |
| Raman resonance | Spontaneous | Stimulated | Stimulated | Stimulated |
| Detection mode | Spontaneous | Spontaneous | Stimulated | Stimulated |
| $N$ scaling | $N$ | $N^2$ | $N$ | $N$ |
| $I$ scaling | $I$ | $I^3$ | $I^2$ | $I^2$ |

$N$ denotes number density of Raman scatterers and $I$ denotes intensity of the incident radiation.

are nonlinear. The intensity dependence of different Raman sensitive techniques is listed in Table 1.2. The linearity of the optical signal with respect to its dependence on $I$ should not be confused with the linearity of the light-matter interaction. For example, although incoherent Raman is a linear technique, it can be described as a nonlinear interaction between photon fields and the material.

## 1.3.3 Homodyne versus Heterodyne Detection

A further classification of the signal is based on the way it is detected. In terms of classical fields, if the sample radiation is detected at an optical frequency different from the incident radiation, the signal intensity is proportional to $|E|^2$. In this case, the signal is classified as homodyne, as the intensity is the square modulus of the emitted field itself. If the emitted field occurs at a frequency that is identical to any of the frequencies contained in the incident radiation $E_{in}$, then the signal intensity is proportional to $|E + E_{in}|^2$. Consequently, the detected intensity contains a mixing term, i.e., $E^* E_{in} + EE^*_{in}$. We define this mixing term as the heterodyne contribution to the signal, as the emitted field is mixed with another field. In terms of quantized fields, the signal is homodyne if detected at a field mode that is initially vacant and heterodyne when detected at a field mode that is already occupied. Note that the current definition, which is commonly used to describe the detection method in molecular spectroscopy, is different from the definition used in the quantum optics and optical engineering literature. In this book, we will use the spectroscopy definition of homodyne and heterodyne signals because it is better suited to classify the different Raman sensitive techniques in a comprehensive fashion.

For instance, conventional coherent anti-Stokes Raman scattering (CARS) is a coherent homodyne technique. It is coherent because the waves emitted from different points in the sample exhibit a definite phase relation, and the detection is homodyne because the detected signal at the anti-Stokes frequency occurs at a field mode different from the input fields. In heterodyne CARS, the emitted field is mixed with another field at the anti-Stokes frequency, usually called local oscillator, and the mutual interference of the fields is detected. The interferometric mixing term is the heterodyne contribution to the signal. In case the one of the incident excitation fields acts as the local oscillator, i.e., detection occurs at a frequency similar to one of the input fields, the signal is self-heterodyned. Raman sensitive pump-probe is an example of a self-heterodyned signal, which is a special case of the heterodyne coherent Raman technique. Raman sensitive pump-probe is commonly called stimulated Raman scattering (SRS), in which the signal is detected at a field mode already occupied by one of the input fields. The designator *stimulated* is, however, somewhat misleading since it is not unique to SRS, as other coherent Raman techniques also have a stimulated component. We address this issue below.

### 1.3.4 Spontaneous versus Stimulated

We can define the stimulated character of Raman sensitive techniques at two levels. The first level pertains to the way the Raman resonance is created. Classically, if the Raman active molecule is driven into resonance by two incident (off-resonance) fields, the Raman resonance is said to be stimulated. The initial phase of the Raman oscillation is determined by the relative phase difference of the input fields. In all nonlinear Raman techniques the Raman resonance is driven in a stimulated fashion. If the molecule is addressed with one (off-resonance) input field, the Raman resonance is established in a spontaneous manner. The phase of the Raman oscillation is determined by the random phases of the nuclear oscillators at equilibrium. This case describes the Raman resonance relevant to incoherent Raman techniques.

The second level relates to the mechanism of detection. This level is best explained in terms of quantized fields. If the field is detected at a field mode that is initially vacant, then the detected signal is spontaneous. This case represents both incoherent Raman techniques and homodyne detected coherent Raman techniques. In both cases, the detection mode is at an optical frequency different from the frequencies carried by the input fields. Note that if a signal is spontaneous in the detection mode, it is not necessarily incoherent. For instance, homodyne CARS is spontaneous in the detection mode, but the detected signal is coherent. If the field is detected at a field mode that is occupied by one of the input fields, then the signal is classified as stimulated. Heterodyne coherent Raman techniques, including Raman sensitive pump-probe, are stimulated in the detection mode. Hence, both heterodyne CARS and SRS are stimulated in the detection mode. In this regard, the term SRS does not exclusively cover the traditional stimulated Raman loss (SRL) or stimulated Raman gain techniques (SRG), as it encompasses more coherent Raman techniques as well. Therefore, a better classification for SRL and SRG techniques would be Raman sensitive *pump-probe*, which more accurately captures the nature of the detected signal. In the remainder of this chapter, we will refer the techniques by their common names (see Table 1.2), but the reader is warned about the existing ambiguities in the current nomenclature.

## 1.4 Classical Description of Matter and Field: The Spontaneous Raman Effect

### 1.4.1 Electronic and Nuclear Motions

Although it is the electrons in the molecule that are set in motion by the visible or near-IR driving fields, their oscillatory motions do contain information about the motions of nuclei. The reason for this is that the adiabatic electronic potential depends on the nuclear coordinates. Since the electrons are bound to the nuclei, nuclear motions will affect the motions of the electrons as well. Hence, the electronic polarizability is perturbed by the presence of nuclear modes. To describe the effect of the nuclear motions, we first connect the electric dipole moment to the polarizability $\alpha(t)$ under the assumption that the driving frequency is far from any electronic resonances of the system:

$$\mu(t) = \alpha(t)E(t) \tag{1.5}$$

In the hypothetical absence of nuclear modes and/or nonlinearities, the polarizability can be approximated as a constant $\alpha_0$. In the presence of nuclear modes, we can express the electronic polarizability in terms of the nuclear coordinate Q, and expand it in a Taylor series [7]:

$$\alpha(t) = \alpha_0 + \left(\frac{\delta\alpha}{\delta Q}\right)_0 Q(t) + \cdots \tag{1.6}$$

The first-order correction to the polarizability has a magnitude of $\delta\alpha/\delta Q$ and can be interpreted as the coupling strength between the nuclear and electronic coordinates. The nuclear motion along Q can be assumed to be that of a classical harmonic oscillator:

$$Q(t) = 2Q_0 \cos(\omega_v t + \phi) = Q_0\left[e^{i\omega_v t + i\phi} + e^{-i\omega_v t - i\phi}\right] \tag{1.7}$$

where

$Q_0$ is the amplitude of the nuclear motion
$\omega_v$ is the nuclear resonance frequency
$\phi$ is the phase of the nuclear mode vibration

When the incoming field is written as $E(t) = Ae^{-i\omega_1 t} + \text{c.c.}$, then the dipole moment is found as:

$$\mu(t) = \alpha_0 A e^{-i\omega_1 t} + A\left(\frac{\delta\alpha}{\delta Q}\right)_0 Q_0\left\{e^{-i(\omega_1 - \omega_v)t + i\phi} + e^{-i(\omega_1 + \omega_v)t - i\phi}\right\} + \text{c.c.} \tag{1.8}$$

The dipole moment oscillates at several frequencies. The first term on the right-hand side of Equation 1.8 describes the process of elastic Rayleigh scattering at the incident frequency. The second term describes the inelastic Raman-shifted frequencies at $\omega_1 - \omega_v$, which is called the Stokes-shifted contribution, and at $\omega_1 + \omega_v$, the anti-Stokes-shifted contribution. The scattering process is illustrated in Figure 1.2. Note that the Raman term is directly proportional to $\delta\alpha/\delta Q$, which describes how the applied field brings about a polarizability change along the nuclear mode. The polarizability change is strongly dependent on the symmetry of the nuclear mode in the molecule, which forms the basis for the selection rules in Raman spectroscopy.

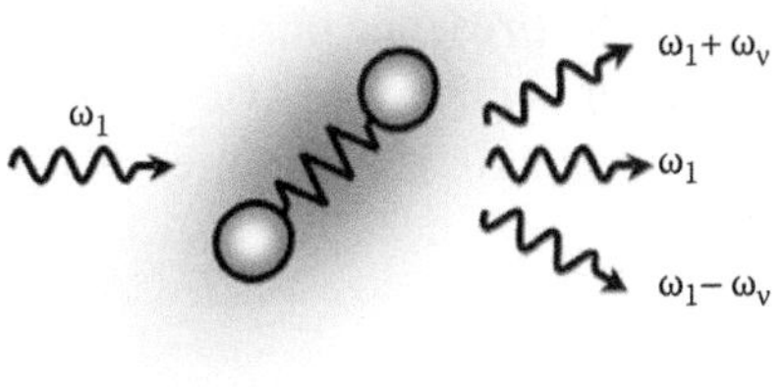


FIGURE 1.2 Schematic of spontaneous Raman scattering. The incoming light is scattered at the molecule into a Rayleigh component at $\omega$ and two Raman-shifted components at $\omega - \omega_v$ and $\omega + \omega_v$, the Stokes and anti-Stokes contributions, respectively.

### 1.4.2 Spontaneous Raman Scattering Signal

Within the classical model for Raman scattering, the harmonic nuclear mode dresses the oscillating dipole with frequency-shifted components. The amplitude of the Stokes and anti-Stokes components are then proportional to the magnitude of the electric field radiated by the dipole at the shifted frequencies. It is instructive to examine the magnitude of the Raman-shifted signal within the framework of the classical model. We will consider the Stokes-shifted component at $\omega_s = \omega_1 - \omega_v$. The derivation of the anti-Stokes component is similar.

The amplitude of the electric field at frequency $\omega_s$, radiated by the oscillating dipole along **r** in the far field, is obtained from electrodynamics in scalar form as:

$$E(\omega_s) = \frac{\omega_s^2}{4\pi\epsilon_0 c^2}|\mu(\omega_s)|\frac{e^{ikr}}{r}\sin\theta \tag{1.9}$$

where

$k$ is the wave vector of the radiated field
$c$ is the speed of light
$\theta$ is the angle relative to the dipole axis
$r$ is the distance from the dipole location to the observation point
$|\mu(\omega_s)|$ is the amplitude of the dipole oscillation at $\omega_s$

The outgoing energy flux along **r** is calculated as the time-averaged Poynting flux $S$:

$$S(\omega_s) = \frac{\epsilon_0 c}{2}|E(\omega_s)|^2 \tag{1.10}$$

The total energy radiated by the (single) dipole is then obtained by integrating the energy flux over the unit sphere. Using $|\mu(\omega_s)|$ from Equation 1.8, the intensity of the Raman-shifted light is:

$$I(\omega_s) = \frac{\omega_s^4}{12\pi\epsilon_0 c^3}Q_0^2|A|^2\left|\frac{\delta\alpha}{\delta Q}\right|^2 \tag{1.11}$$

From Equation 1.11 we see that the classical model predicts a $\omega^4$ dependence of the intensity of the Raman scattered light. In addition, it scales with $|\delta\alpha/\delta Q|^2$ and with the intensity of the incident beam $I_0 = |A|^2$. The phase $\phi$ of the Raman scattered light is dependent on the nuclear mode oscillation. At equilibrium, the nuclear vibrations of different molecules are uncorrelated, i.e., each molecule $i$ carries its own independent phase $\phi_i$. This implies that the phase of the radiated field from one dipole emitter is unrelated to the phase of the radiated field by a second dipole emitter elsewhere in the sample. Consequently, the signal is incoherent and the intensity of the total Raman emission is proportional to Equation 1.11 multiplied by the total number of Raman scatterers in the sample. It is interesting to note that the first term in Equation 1.8, which represents elastic light scattering, is not dependent on the nuclear vibration, and thus does not acquire a random phase $\phi$. This is the reason why Rayleigh scattering is coherent while the Raman-shifted contributions are incoherent.

Experimentally it is useful to define the Raman signal strength in terms of a cross section. The cross section expresses the Raman scattering efficiency of a molecule in a manner analogous to describing light absorption through the absorption cross section (Beer's law). Using the cross section $\sigma$, the total scattered Raman-shifted light from a sample with length $z$ and a molecular number density $N$ is written as:

$$I(\omega_s) = Nz\sigma(\omega_s)I_0 \tag{1.12}$$

Comparing Equations 1.12 and 1.11, it is evident that the Raman cross section is directly proportional to $|\delta\alpha/\delta Q|^2$. This underlines the central importance of the condition of a non-zero polarizability change along the nuclear coordinate.

Unfortunately, the classical model does not offer a correct description of the resonance behavior of the polarizability. In addition, the classical description is unable to predict the ratio between the intensities of the Stokes and anti-Stokes contributions. A quantum mechanical treatment of the molecule is required to account for these effects. Furthermore, because the field is treated classically, the amount of energy exchange between the light fields and the molecule cannot be accurately described, and some corrections to Equation 1.11 are needed. We will address these issues in Section 1.7.2.

Despite these shortcomings, the classical model provides a useful physical picture for interpreting several attributes of the spontaneous Raman scattering process and the coherent Raman scattering process alike. In the next section, we will highlight some of the basic properties of coherent Raman techniques in the context of the classical description.

## 1.5 Classical Description of Matter and Field: Coherent Raman Scattering

The classical description of the coherent Raman effect provides an intuitive interpretation of the light-matter interaction in terms of actively driven nuclear oscillations in the material. For clarity, the following derivation assumes a single harmonic nuclear mode per molecule. Even though this description does no justice to the multitude of vibrational states of actual molecules, it introduces a clear picture in which a driven nuclear mode forms the source for coherent scattering of light. Extending the description to include multiple modes is straightforward.

Briefly, this description divides the coherent Raman process into two steps. First, two incoming fields induce oscillations in the molecular electron cloud. These oscillations form an effective force along the vibrational degree of freedom, which actively drives the nuclear modes. Second, the driven nuclear mode forms the source of a spatially coherent modulation of the material's refractive properties. A third light field, which propagates through the material and experiences this modulation, will develop sidebands that are shifted by the modulation frequency. The amplitude of the field scattered into these sidebands forms the basis of the frequency-shifted coherent Raman signal. Below we will discuss the key elements of the classical model.

### 1.5.1 Driven Raman Mode

In the classical model for the coherent Raman process, we assume that the vibrational motion in the molecule can be described by a damped harmonic oscillator with a resonance frequency $\omega_v$. Similar to the situation encountered in the classical description of the spontaneous Raman process, we can think of the oscillator as the vibrational motion of two nuclei along their internuclear axis $Q$. This system is subject to two incoming light fields $E_1$ and $E_2$, which are modeled as plane waves:

$$E_i(t) = A_i e^{-i\omega_i t} + \text{c.c.} \tag{1.13}$$

where the subscript $i = (1, 2)$ and all propagation factors are included in the amplitude $A_i$. As before, we assume that the frequencies $\omega_1$ and $\omega_2$ are much higher than the resonance frequency $\omega_v$, and that $\omega_1 > \omega_2$. Since the incident frequencies are far from the resonance frequency of the oscillator, the nuclear mode will not be driven efficiently by the fundamental fields. The electrons surrounding the nuclei, however, can follow the incident fields adiabatically. In addition, when the fields are sufficiently intense, nonlinear electron motions can occur at combination frequencies, including the difference frequency $\Omega = \omega_1 - \omega_2$. Under these conditions, the combined optical field exerts a force on the vibrational oscillator:

$$F(t) = \left(\frac{\delta\alpha}{\delta Q}\right)_0 \left[A_1 A_2^* e^{-i(\Omega)t} + \text{c.c.}\right] \tag{1.14}$$

From Equation 1.14 we see that, because the electronic motions are coupled to the nuclear motions through a nonzero $(\delta\alpha/\delta Q)_0$, the modulated electron cloud introduces a time-varying force that oscillates at the difference frequency $\Omega$ and which is felt by the nuclear mode. In the presence of the driving fields, the nuclear displacement $Q$ can then be expressed by the following equation of motion [8]:

$$\frac{d^2Q(t)}{dt^2} + 2\gamma\frac{dQ(t)}{dt} + \omega_v Q(t) = \frac{F(t)}{m} \tag{1.15}$$

where

$\gamma$ is the damping constant

$m$ indicates the reduced mass of the nuclear oscillator

$\omega_v$ is the resonance frequency of the harmonic nuclear mode

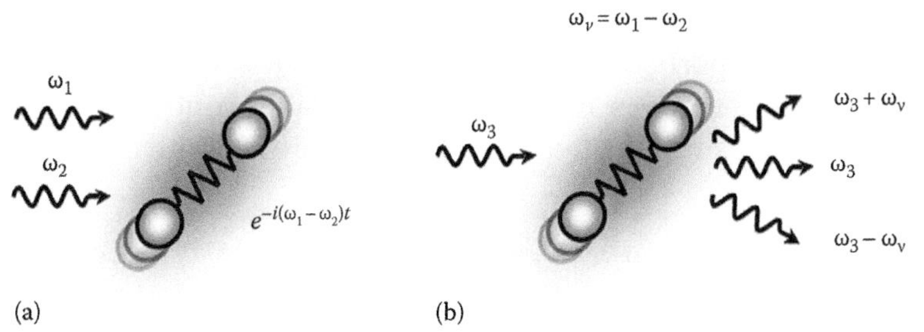


**FIGURE 1.3** Schematic of coherent Raman scattering. (a) Two incoming fields drive a harmonic oscillator at the difference frequency $\Omega = \omega_1 - \omega_2$. (b) The presence of this oscillation causes a fluctuating refractive index for a third field $\omega_3$, which develops sidebands that are shifted $\Omega$ from the fundamental frequency. The amplitude of the sidebands is maximized when $\Omega$ equals the resonance frequency $\omega_v$ of the oscillator, which yields sidebands at $\omega_3 + \omega_v$ and $\omega_3 - \omega_v$.

The time-varying nuclear displacement can be found from Equation 1.15 as:

$$Q(t) = Q(\Omega)e^{-i\Omega t} + \text{c.c.} \tag{1.16}$$

which oscillates at $\Omega$ with the amplitude:

$$Q(\omega_v) = \frac{1}{m}\left(\frac{\delta\alpha}{\delta Q}\right)_0 \frac{A_1 A_2^*}{\omega_v^2 - \Omega^2 - 2i\Omega\gamma} \tag{1.17}$$

The physical interpretation of Equation 1.17 is clear. The nuclear mode is driven by the joint action of the incident fields. The amplitude of the vibrational motion depends on the amplitudes of the applied light fields and the magnitude of the coupling of the nuclear coordinate to the electronic polarizability $(\delta\alpha/\delta Q)_0$. The extent of the vibration also depends on the difference between the effective driving frequency $\Omega$ and the resonance frequency $\omega_v$ of the oscillator. Indeed, the amplitude of the oscillatory motion is largest when the difference frequency $\Omega$ matches the oscillator's resonance frequency (Figure 1.3).

## 1.5.2 Probe Modulation

The presence of the driven nuclear motion affects the optical properties of the material. As a consequence, the applied electric fields $E_1$ and $E_2$ will experience a slightly altered electronic polarizability upon propagating through the material. The effective macroscopic polarization in the material is the sum of the dipole moments as in Equation 1.2.

Using Equations 1.2, 1.5, and 1.6 we can write the polarization as:

$$P(t) = N\left[\alpha_0 + \left(\frac{\delta\alpha}{\delta Q}\right)_0 Q(t)\right]\{E_1(t) + E_2(t)\} \tag{1.18}$$

The terms proportional to $\alpha_0$ correspond to the linear polarization of the material, whereas the terms proportional to $(\delta\alpha/\delta Q)_0$ describe the contribution to the third-order polarization due to the driven Raman mode. This latter contribution is the nonlinear polarization, which, using Equations 1.13 and 1.16, can be written as:

$$P_{NL}(t) = P(\omega_{cs})e^{-i\omega_{cs}t} + P(\omega_2)e^{-i\omega_2 t} + P(\omega_1)e^{-i\omega_1 t} + P(\omega_{as})e^{-i\omega_{as}t} + \text{c.c.} \tag{1.19}$$

where

$\omega_{cs} \equiv 2\omega_2 - \omega_1$ is referred to as the coherent Stokes frequency
$\omega_{as} \equiv 2\omega_1 - \omega_2$ is referred to as the anti-Stokes frequency

The nonlinear polarization thus contains contributions that oscillate at the fundamental frequencies $\omega_1$ and $\omega_2$, as well as contributions that oscillate at the new frequencies $\omega_{cs}$ and $\omega_{as}$. The relation between these frequency components is sketched in Figure 1.4. The amplitude of the polarization at the anti-Stokes frequency is given by:

$$P(\omega_{as}) = \frac{N}{m}\left(\frac{\delta\alpha}{\delta Q}\right)_0^2 \frac{1}{\omega_v^2 - \Omega^2 - 2i\Omega\gamma} A_1^2 A_2^* = 6\epsilon_0 \chi_{NL}(\Omega) A_1^2 A_2^* \tag{1.20}$$

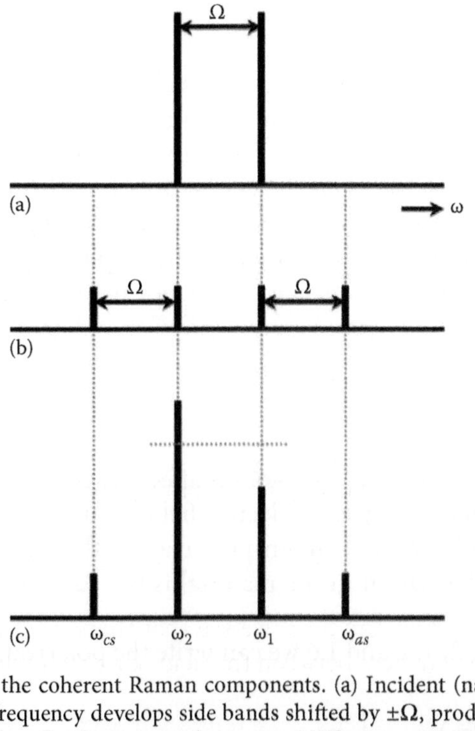


**FIGURE 1.4** Spectrum of the coherent Raman components. (a) Incident (narrow band) frequencies at $\omega_1$ and $\omega_2$. (b) Each input frequency develops side bands shifted by $\pm\Omega$, producing $\omega_{cs}$ and $\omega_1$ for the $\omega_2$ input frequency, and $\omega_2$ and $\omega_{as}$ for the $\omega_1$ input frequency. (c) The intensities of the coherent Raman components after passage through the sample. The $\omega_2$ frequency channel has experienced a gain and the $\omega_1$ frequency channel has experienced a loss.

where the nonlinear susceptibility is defined as:

$$\chi_{NL}(\Omega) = \frac{N}{6m\epsilon_0}\left(\frac{\delta\alpha}{\delta Q}\right)_0^2 \frac{1}{\omega_v^2 - \Omega^2 - 2i\Omega\gamma} \tag{1.21}$$

Similarly, we can write for the other frequency components:

$$P(\omega_{cs}) = 6\epsilon_0 \chi_{NL}^*(\Omega) A_2^2 A_1^* \tag{1.22}$$

$$P(\omega_2) = 6\epsilon_0 \chi_{NL}^*(\Omega) |A_1|^2 A_2 \tag{1.23}$$

$$P(\omega_1) = 6\epsilon_0 \chi_{NL}(\Omega) |A_2|^2 A_1 \tag{1.24}$$

The nonlinear polarizations given in Equations 1.20 through 1.24 describe the four lowest order coherent Raman effects: $P(\omega_{cs})$ is responsible for coherent Stokes Raman scattering (CSRS), $P(\omega_2)$ for stimulated Raman gain (SRG), $P(\omega_1)$ for stimulated Raman loss (SRL), and $P(\omega_{as})$ for coherent anti-Stokes Raman scattering (CARS). We see that the amplitudes of the different nonlinear polarization components all depend on the magnitude of the same $\chi_{NL}$. Therefore, the material polarizations of the four coherent Raman effects are comparable in magnitude: they are all induced by the same nuclear vibration at $\omega_v$. However, this does not imply that the actual detected signals of the four CRS techniques are of similar strength. We will discuss this issue in next section.

### 1.5.3 Energy Flow in Coherent Raman Scattering

As we have seen in the previous section, the induced polarization in the material produces radiation at the fundamental frequencies and at two new frequencies $\omega_{cs}$ and $\omega_{as}$. In the coherent Raman process, energy contained in the fundamental light fields is redirected in two ways. First, there is an energy exchange with the material. In the presence of the driving fields, the material can either gain or lose energy. In case the total energy contained in all the light fields combined is lower after passing through the material, the total energy of the material will be higher. This type of process is called *dissipative*. Second, new light fields can be generated without energy exchange with the material. In this latter process, energy amounts formed by adding and subtracting the incoming light fields are used to generate new light fields while the material acts merely as a mediator. In these so-called *parametric* processes, the total energy of the combined light fields is conserved.

To describe energy flow in the classical model, explicit evaluation of Maxwell's wave equation is required, which connects the induced polarization to a radiating coherent field. All participating waves ($\omega_1$, $\omega_2$, $\omega_{cs}$, $\omega_{as}$) need to be taken into account in a coupled wave equation approach [9]. The coupled equations are then integrated over the (macroscopic) volume that contains the molecules in order to find the energy exchange between the waves and the material as well as the energy exchange among the waves. Such a derivation is beyond the scope of this chapter. Here we wish to highlight the essentials of energy flow in coherent Raman processes without explicitly incorporating wave propagation effects. The following discussion is, therefore, qualitative in nature. Wave propagation is discussed in detail in Chapter 2.

We first discuss the case of homodyne detection. We will start our discussion with considering the fields at the new frequencies $\omega_{as}$ and $\omega_{cs}$. The nonlinear field at the anti-Stokes frequency can be written as:

$$E_{as}^{(3)}(t) = A_{as}e^{-i\omega_{as}t} + \text{c.c.} \tag{1.25}$$

The corresponding intensity associated with this field is given by:

$$I(\omega_{as}) = \frac{\epsilon_0 c}{2}|A_{as}|^2 \tag{1.26}$$

In the lowest order coherent Raman interaction, the only source for the anti-Stokes field is the nonlinear polarization that oscillates at $\omega_{as}$. The magnitude of the anti-Stokes field $A_{as}$ is thus proportional to the magnitude of $P(\omega_{as})$. Using Equation 1.20, we can then write:

$$I(\omega_{as}) \propto |\chi_{NL}|^2 I_1^2 I_2 \tag{1.27}$$

where $I_1$ and $I_2$ are the intensities of the beams at $\omega_1$ and $\omega_2$, respectively. Similarly, we find for the intensity of the coherent Stokes contribution:

$$I(\omega_{cs}) \propto |\chi_{NL}|^2 I_2^2 I_1 \tag{1.28}$$

In the above description, the coherent Stokes and anti-Stokes contributions are detected as homodyne signals, i.e., the signals are directly proportional to the modulus square of the nonlinear polarization. In this limit, the energy contained in the $\omega_{cs}$ and $\omega_{as}$ frequency channels is extracted from the incident fields at $\omega_1$ and $\omega_2$, as can be shown by performing a coupled wave equation analysis [9]. Because the process is parametric, no effective energy exchange with the material has taken place.

The situation changes when an additional field at frequency $\omega_{cs}$ or $\omega_{as}$ is applied to the material. This additional field is commonly referred to as a *local oscillator*, which must exhibit a well-behaved phase relation with the nonlinear polarization in the material. In the presence of a local oscillator, the induced nonlinear polarization is no longer the only source of radiation at the signal frequency. The intensity in the anti-Stokes frequency channel at the detector can now be written as:

$$\begin{aligned} I(\omega_{as}) &= \frac{\epsilon_0 c}{2}\left|E_{as}^{(3)} + E_{as}^{lo}\right|^2 \\ &\propto \left|E_{as}^{(3)}\right|^2 + \left|E_{as}^{lo}\right|^2 + \left[\left\{E_{as}^{(3)}\right\}^* E_{as}^{lo} + \left\{E_{as}^{lo}\right\}^* E_{as}^{(3)}\right] \end{aligned} \tag{1.29}$$

where $E_{as}^{lo}$ is the local oscillator field at the anti-Stokes frequency. The last term on the right hand side of Equation 1.29 represents a heterodyne mixing contribution that depends on both the nonlinear anti-Stokes field $E_{as}^{(3)}$ and the local oscillator field. The heterodyne contribution $I^{het}$ can be recast as:

$$I^{het}(\omega_{as}) = 2A_{as}^{lo}\left[\mathrm{Re}\left\{E_{as}^{(3)}\right\}\cos\phi + \mathrm{Im}\left\{E_{as}^{(3)}\right\}\sin\phi\right]$$
$$= 2\alpha[\mathrm{Re}\{\chi_{NL}\}\cos(\phi-\phi_p) + \mathrm{Im}\{\chi_{NL}\}\sin(\phi-\phi_p)] \quad (1.30)$$

where
$\alpha = \left|A_{as}^{lo}A_1^2A_2\right|$
$A_{as}^{lo}$ is the amplitude of the local oscillator

The phase difference between $E_{as}^{(3)}$ field and the (real) $E_{as}^{lo}$ field is indicated as $\phi$, whereas the phase difference between the radiated field $E_{as}^{(3)}$ and the induced polarization $P(\omega_{as})$ is indicated as $\phi_p$. Let us consider the energy flow of the heterodyne detected signal under the condition of driving the oscillator at the vibrational resonance frequency, i.e., $\Omega = \omega_v$. In this situation, we see from Equation 1.21 that $\chi_{NL}$ is purely imaginary in case nonresonant contributions to the nonlinear susceptibility are ignored. The total detected intensity in the anti-Stokes channel is then:

$$I(\omega_{as}) \propto \left|E_{as}^{(3)}\right|^2 + \left|E_{as}^{lo}\right|^2 + 2\alpha\,\mathrm{Im}\{\chi_{NL}\}\sin(\phi-\phi_p) \quad (1.31)$$

This result indicates that the detected intensity depends on the phase difference $\Delta\phi = \phi - \phi_p$. The actual geometrical phase difference between the induced field and the local oscillator depends on propagation factors that are not included in this simple interference model. In Chapter 2, we will consider a more complete description of the phase different between $E_{as}^{(3)}$ and $E_{as}^{lo}$ at the location of the detector in the context of light propagation. Here, we will use the simple interference model to briefly discuss several values for $\Delta\phi$ that correspond to important cases in the heterodyne detection scheme. For instance, when $\Delta\phi = 0$, the heterodyne term disappears and the total intensity is simply the sum of the (homodyne) anti-Stokes contribution and the local oscillator intensity. However, when $\Delta\phi = -\pi/2$, the heterodyne term is negative and the total energy detected in the anti-Stokes channel is less than the sum of the homodyne contributions ($\mathrm{Im}\{\chi_{NL}\} > 0$; see Equation 1.21). Under these conditions, the CARS process is no longer purely parametric as dissipative interactions, which here scale with $\mathrm{Im}\{\chi_{NL}\}$, also play a role. In case modulation techniques are employed, the heterodyne term can be selectively detected and the resulting signal is directly proportional to $\mathrm{Im}\{\chi_{NL}\}$, the dissipative part of the coherent Raman interaction. The same detection strategy can also be applied to CSRS.

The example in the preceding text illustrates that for a particular coherent Raman process the presence of a phase coherent local oscillator can change the sensitivity of the measurement in terms of probing parametric and dissipative processes. This notion is important when describing the SRL and SRG processes. In SRL, the signal is detected in the $\omega_1$ frequency channel. In this channel, $P(\omega_1)$ is the source of the nonlinear field $E_1^{(3)}$.

Because the frequency of the nonlinear radiation is similar to the frequency of the fundamental light field $E_1$, interference between the two fields will occur. The fundamental $E_1$ field can be interpreted as a local oscillator. The total intensity detected in the $\omega_1$ channel is:

$$I(\omega_1) = \frac{\epsilon_0 c}{2}\left|E_1^{(3)} + E_1\right|^2$$
$$\propto \left|E_1^{(3)}\right|^2 + \left|E_1\right|^2 + 2\beta[\mathrm{Re}\{\chi_{NL}\}\cos\Delta\phi + \mathrm{Im}\{\chi_{NL}\}\sin\Delta\phi] \tag{1.32}$$

with $\beta = I_1 I_2$. At the far field detector, the phase shift $\Delta\phi$ amounts to $-\pi/2$, which implies that the real part of the material response is $\pi/2$ retarded with respect to the $E_1$, while the imaginary part of the material response is out-of-phase with $E_1$ (see Chapter 2). We thus find:

$$I(\omega_1) \propto \left|E_1^{(3)}\right|^2 + \left|E_1\right|^2 - 2\beta\,\mathrm{Im}\{\chi_{NL}\} \tag{1.33}$$

Equation 1.33 thus shows that the total intensity in the $\omega_1$ channel is attenuated because of the presence of the driven oscillator. The loss in the $\omega_1$ channel is the result of destructive interference between the induced field and the fundamental field. Note that the attenuation is mediated by the dissipative part of the interaction as described by the imaginary part of the nonlinear susceptibility. In the $\omega_2$ channel, the $E_2$ excitation field acts as the local oscillator. Using $\Delta\phi = \pi/2$ and $\chi_{NL}^* = -\chi_{NL}$ at the vibrational resonance, we find:

$$I(\omega_2) \propto \left|E_2^{(3)}\right|^2 + \left|E_2\right|^2 + 2\beta\,\mathrm{Im}\{\chi_{NL}\} \tag{1.34}$$

From Equation 1.34 we see that the intensity in the $\omega_2$ channel grows. The gain in the $\omega_2$ channel is due to constructive interference between the induced field and the driving field $E_2$. When modulation techniques are used, the heterodyne portion of the signal can be separately detected and the resulting SRG signal is directly proportional to the dissipative part of the coherent Raman interaction.

The general picture offered by the classical model is that the harmonic oscillator, driven at $\omega_v$, forms a material modulation that affects the amplitude of the fundamental fields $E_1$ and $E_2$. The material modulation gives rise to frequency-shifted radiation at $\omega_1 + \omega_v$ and $\omega_2 - \omega_v$, the CARS and CSRS contributions, respectively. In the homodyne detection mode, the CARS and CSRS signals are sensitive to the parametric part of the interaction. On the other hand, the field contributions at $\omega_1 - \omega_v$ and $\omega_2 + \omega_v$ radiate in the $\omega_2$ and $\omega_1$ frequency channels, respectively, and interference between the nonlinear fields and the fundamental fields will occur. In the SRG channel this interference is constructive, producing a gain of the overall $\omega_2$ field, whereas in the SRL channel the interference is destructive, giving rise to a loss of the amplitude of the $\omega_1$ field. The extend of the loss and gain scales with $\mathrm{Im}\{\chi_{NL}\}$, which describes the dissipative part of the interaction.

# 1.6 Semi-Classical Description: Quantum Matter and Classical Fields

The fully classical model provides a qualitative description of the coherent Raman process in which the nuclear motion is described as a harmonic oscillator. A shortcoming of the classical model is that it does not recognize the quantized nature of the nuclear oscillations. The semi-classical model incorporates the quantum mechanical character of the material into the picture, whereas the description of the field remains classical and hence the name semi-classical. As such, nonlinear susceptibilities can be derived that describe the accessible states of the nuclear mode and the transitions between these states, expressed in material parameters such as transition dipole moments. By including the quantum mechanical material properties, the semi-classical model predicts nonlinear susceptibilities that are quantitatively more meaningful. It also naturally describes the existence of nonresonant contributions to the nonlinear optical response.

## 1.6.1 Wavefunctions of Matter

In the quantum mechanical description, the state of the material is described in terms of molecular wavefunctions. The wavefunctions are a function of space and time and are generally written as a superposition of molecular eigenstates $\psi_n$:

$$\psi(r,t) = \sum_n c_n \psi_n(r,t) \tag{1.35}$$

where the $c_n$ are the projections of $\psi$ along the system's eigenstates. The $r$ coordinate includes both the electronic and nuclear coordinates. The evolution of the wavefunction over time is given by the time-dependent Schrödinger equation:

$$i\hbar \frac{d\psi}{dt} = \hat{H}_0 \psi \tag{1.36}$$

Here $\hat{H}_0$ is the Hamiltonian of the system in the absence of any external field. The hat indicates that $\hat{H}_0$ is an operator. Because $\psi_n$ are eigenstates of the unperturbed Hamiltonian, their evolution can be expressed as:

$$\psi_n(r,t) = a_n(r)e^{-i\omega_n t} \tag{1.37}$$

where

$a_n(r)$ denotes the spatially varying part of the wavefunction
$\omega_n$ is the eigenfrequency associated with eigenstate $\psi_n$

The system's wavefunction is affected by the coupling to an external field. The Hamiltonian is now given by:

$$\hat{H} = \hat{H}_0 + \hat{V}(t) \tag{1.38}$$

where the interaction Hamiltonian is given as:

$$\hat{V}(t) = -\hat{\mu} \cdot E(t) \tag{1.39}$$

The interaction with the electric field happens through the charged particles, electrons and nuclei, of the material, which are set in motion by the optical field applied at time $t$. In the dipole approximation, the extent of the interaction is described by the electric dipole operator:

$$\hat{\mu} = \sum_{\alpha} e_{\alpha} \hat{r}_{\alpha} \tag{1.40}$$

where the sum runs over both nuclei and electrons. Solving the wavefunction for this new Hamiltonian would allow the calculation of several observables. Since we are interested in calculating the optical response of the material, our target is to determine the polarization $P(t)$ of the material in a given volume $V$. Once the wavefunction is known, the polarization can be calculated from the expectation value of the dipole operator:

$$P(t) = N\langle \hat{\mu}(t) \rangle = N\langle \psi(r,t) | \hat{\mu} | \psi(r,t) \rangle \tag{1.41}$$

where
 the bra $\langle \psi |$ and ket $| \psi \rangle$ notation is used
 $N$ is the number density in volume $V$

Finding the driven wavefunction is not trivial, however, and approximate methods have to be used. The most general approach is based on perturbation theory, where $\hat{V}(t)$ is treated as a perturbation and the wavefunction $\psi(r, t)$ is expanded to the $n$th order. Using the perturbation-corrected wavefunction in Equation 1.41 yields contributions to the polarization to various orders in the field. Collecting terms to third-order in the applied field with a coherent Raman resonance at $\omega_1 - \omega_2$ allows for the calculation of the quantum mechanical counterparts to the classical nonlinear susceptibilities given in Equation 1.21. Such a description, however, is rarely used because it is unable to properly account for broadening mechanisms of spectroscopic features due to coupling to other (bath) degrees of freedom. To include such broadening phenomena, a density matrix formalism is commonly employed, as we will briefly describe in the next section.

## 1.6.2 Density Matrix

The density matrix operator is defined as:

$$\hat{\rho}(t) \equiv |\psi(t)\rangle\langle\psi(t)| = \sum_{nm} \rho_{nm}(t) |n\rangle\langle m| \tag{1.42}$$

where $|n\rangle$ is the bra notation of the eigenstates of the unperturbed system. From the definition of the density matrix we see that it depends on the operator $|n\rangle\langle m|$, and the matrix elements $\rho_{nm} = \langle n | \hat{\rho} | m \rangle$. The diagonal elements of the density matrix, $\rho_{nn}$, give the probability that the system is in state $|n\rangle$, while the off-diagonal elements imply that the system is in a coherent superposition of eigenstates $|n\rangle$ and $|m\rangle$. We will call $|n\rangle\langle m|$ with $n \neq m$ the *coherence* and $\rho_{nm}(t)$ the time-dependent amplitude of this coherence. The description of the coherent Raman process in terms of coherences will prove useful for analyzing the different quantum pathway contributions to the overall signal.

The density operator evolves in the Schrödinger picture as:

$$\frac{d\hat{\rho}}{dt} = -\frac{i}{\hbar}[\hat{H},\hat{\rho}] \tag{1.43}$$

where the Hamiltonian is defined as in Equation 1.38 and the brackets indicate the commutator operation of two operators $\hat{A}$ and $\hat{B}$ according to $[\hat{A},\hat{B}] \equiv \hat{A}\hat{B} - \hat{B}\hat{A}$. As in the classical model, we are interested in calculating the polarization of the material in a given volume $V$. The expectation value of the electric dipole operator can be expressed in terms of the density operator as:

$$\langle\hat{\mu}(t)\rangle = \sum_{nm} \mu_{mn}\rho_{nm}(t) \equiv \mathrm{tr}[\hat{\mu}\hat{\rho}(t)] \tag{1.44}$$

The tr symbol denotes the trace over the matrix elements of the operator product between the brackets. Similar to solving for the system's wavefunction, the density matrix of the system is found by a perturbation expansion of $\hat{\rho}(t)$ in powers of the electric field:

$$\hat{\rho}(t) = \rho^{(0)}(t) + \rho^{(1)}(t) + \rho^{(2)}(t) + \rho^{(3)}(t) + \cdots \tag{1.45}$$

here $\rho^{(n)}$ is the $n$th order contribution in the electric field. The zeroth order contribution denotes the unperturbed density matrix at thermal equilibrium and is given as:

$$\rho^{(0)}(t) = \rho(-\infty) = \frac{e^{-\hat{H}/kT}}{\mathrm{tr}\{e^{-\hat{H}/kT}\}} \tag{1.46}$$

where $k$ is Boltzmann's constant. The perturbative expression of each of the components $\rho^{(n)}$ gets increasingly more complex with growing orders of $n$. It is, therefore, helpful to use alternative notation for writing these expressions in a more compact and insightful form. A common tool is the use of Liouville space operators, also known as superoperators. The action of the Liouville space operator $\mathbb{H}$ and $\mathbb{V}(t)$ on an ordinary operator $\hat{A}$ is defined through:

$$\mathbb{H}\hat{A} \equiv [\hat{H},\hat{A}] \tag{1.47}$$

$$\mathbb{V}(t)\hat{A} \equiv [\hat{V}(t),\hat{A}] \tag{1.48}$$

With these definitions, the equation of motion of the density matrix operator can be rewritten as:

$$\frac{d\hat{\rho}}{dt} = -\frac{i}{\hbar}\mathbb{H}\hat{\rho} \tag{1.49}$$

To describe the coherent Raman interaction, we are interested in finding $\rho^{(3)}$, the density matrix contribution that is third order in the electric field. The derivation of $\rho^{(3)}$ is beyond the scope of this chapter, and the reader is referred to the existing literature for details [1,10,11]. Here we merely give the result of $\rho^{(3)}$ as predicted by perturbation theory. The Liouville space notation yields compact expressions for the third-order density matrix contribution:

$$\rho^{(3)}(t) = \left(\frac{-i}{\hbar}\right)^3 \int_0^\infty d\tau_3 \int_0^\infty d\tau_2 \int_0^\infty d\tau_1$$
$$\times\, \mathbb{G}(\tau_3)\mathbb{V}(t-\tau_3)\mathbb{G}(\tau_2)\mathbb{V}(t-\tau_3-\tau_2)\mathbb{G}(\tau_1)\mathbb{V}(t-\tau_3-\tau_2-\tau_1)\rho(-\infty) \tag{1.50}$$

where the time variables $\tau_n$ run over the interval between the application of a light field incident at $t_{n-1}$ and a light field incident at $t_n$, as shown in Figure 1.5. The Liouville space Green's function $\mathbb{G}(\tau)$ describes the propagation of the material system in the absence of the light fields and is given as:

$$\mathbb{G}(\tau) \equiv \theta(\tau)e^{-i\mathbb{H}\tau/\hbar} \tag{1.51}$$

with $\theta(\tau)$ the Heavyside step function. Note that the expression for $\rho^{(3)}$ has an intuitive form: reading from right to left, the system starts out at thermal equilibrium $\rho(-\infty)$ and is subsequently perturbed by successive light fields as described by the $\mathbb{V}$ operator. In between the light-matter interactions, the material system evolves according to the Green's function $\mathbb{G}$. Using the solution of $\rho^{(3)}(t)$ as given in Equation 1.50 we can proceed with the calculation of the expectation value of the third-order polarization. This will be discussed in the next section.

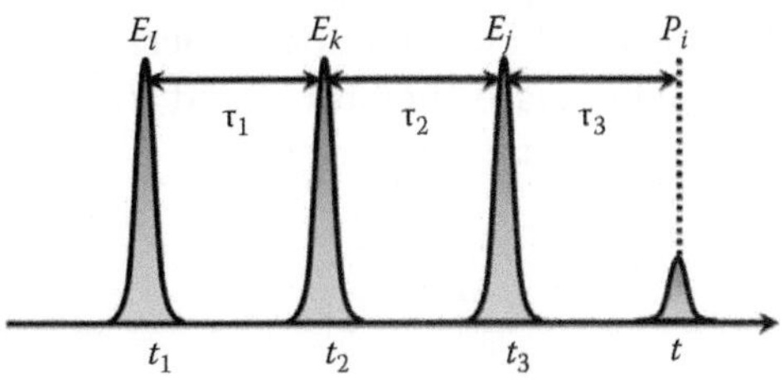


**FIGURE 1.5** Schematic of the time ordering of the incident fields and the induced polarization.

## 1.6.3 Response Functions and Third-Order Susceptibility

### 1.6.3.1 Material Response Function

The polarization can be determined by evaluating the expectation value of the electric dipole operator as given in Equation 1.44. When the polarization is expanded in powers of the electric field, we find for the third-order contribution:

$$P^{(3)}(t) = N\,\mathrm{tr}[\hat{\mu}\rho^{(3)}(t)] \tag{1.52}$$

Using Equation 1.50 we can express the components of the third-order polarization as:

$$P_i^{(3)}(t) = N\sum_{jkl}\int_0^\infty d\tau_3\int_0^\infty d\tau_2\int_0^\infty d\tau_1 R_{ijkl}^{(3)}(\tau_3,\tau_2,\tau_1)$$
$$\times E_j(t-\tau_3)E_k(t-\tau_3-\tau_2)E_l(t-\tau_3-\tau_2-\tau_1) \tag{1.53}$$

where

the indices $\{i, j, k, l\}$ indicate the polarization orientation in cartesian coordinates

$R_{ijkl}^{(3)}$ is the third-order response function, which is given as:

$$R_{ijkl}^{(3)}(\tau_3,\tau_2,\tau_1) = \left(\frac{i}{\hbar}\right)^3 \left\langle \hat{\mu}_i \mathbb{G}(\tau_3)\hat{\mu}_j^L \mathbb{G}(\tau_2)\hat{\mu}_k^L \mathbb{G}(\tau_1)\hat{\mu}_l^L \rho(-\infty)\right\rangle \tag{1.54}$$

where we have used the notation $\hat{\mu}^L$ to indicate the Liouville space version of the electric dipole operator $\hat{\mu}$. The response function describes the time-ordered response of the material to the incoming light fields. The expectation value in Equation 1.54 is to be taken over all the unperturbed eigenstates of the system. The expression of the nonlinear polarization in terms of a time-dependent response function is a natural means to describe time-resolved coherent Raman spectroscopy experiments. For many coherent Raman microscopy applications, however, the time-domain expression is of limited use, as the response is rarely time-resolved in fast imaging applications. Instead, the magnitude of the polarization at different vibrational frequencies is more practically related to imaging experiments. Therefore, we will seek frequency domain expressions of the nonlinear polarization.

To illustrate the form of the nonlinear polarization in the frequency domain, we will initially assume that the light fields are spectrally narrow, a situation directly relevant to picosecond coherent Raman microscopy. In this case we can write for the contribution to the nonlinear polarization that oscillates at the signal frequency $\omega_4 = \omega_1 + \omega_2 + \omega_3$:

$$P_i^{(3)}(t) = P_i(\omega_4)e^{-i\omega_4 t} + \text{c.c.} \tag{1.55}$$

Note that because $P_i^{(3)}(t)$ is a real function of time, the relation $P_i^*(\omega_4) = P_i^{(3)}(-\omega_4)$ must hold. The amplitude of the nonlinear polarization is given by:

$$P_i(\omega_4) = N\sum_{jkl} R_{ijkl}^{(3)}(\omega_4,\omega_1+\omega_2,\omega_1)E_j(\omega_1)E_k(\omega_2)E_l(\omega_3) \tag{1.56}$$

with $R^{(3)}_{ijkl}$ the frequency domain response function defined through:

$$R^{(3)}_{ijkl}(\omega_4,\omega_1+\omega_2,\omega_1)=\left(\frac{-1}{\hbar}\right)^3\left\langle\hat{\mu}_i\mathbb{G}(\omega_4)\hat{\mu}^L_j\mathbb{G}(\omega_1+\omega_2)\hat{\mu}^L_k\mathbb{G}(\omega_1)\hat{\mu}^L_l\rho(-\infty)\right\rangle \tag{1.57}$$

In this expression, we have used the frequency domain Green's function:

$$\mathbb{G}(\omega)=-i\int_0^\infty dt\,\mathbb{G}(t)e^{i\omega t} \tag{1.58}$$

The Green's function describes the frequency content of the density matrix during a given propagation period. The response function provides a detailed account of the evolution of the system in response to the incoming fields in terms of molecular coherences. In particular, the coherence during the second propagator is the material quantity that gives rise to the Raman sensitive signal. In the next section, we will focus on response functions that contain such propagators.

#### 1.6.3.2 Third-Order Susceptibility

The system's response to a particular combination of optical frequencies is conveniently described by the third-order susceptibility $\chi^{(3)}_{ijkl}$. To obtain $\chi^{(3)}_{ijkl}$, we sum over all field permutations of $R^{(3)}_{ijkl}$. For instance, the $\chi^{(3)}_{ijkl}$ for the CARS process is defined through:

$$\chi^{(3)}_{ijkl}(-\omega_4;\omega_1,\omega_2,\omega_3)=-\frac{N}{6\epsilon_0}\sum_p R^{(3)}_{ijkl}(\omega_4,\omega_1+\omega_2,\omega_1) \tag{1.59}$$

The summation indicated by $p$ means that all frequency combinations, both positive and negative, of the applied fields are included that sum up to the final frequency $\omega_4$. The frequency arguments of $\chi^{(3)}_{ijkl}(-\omega_4;\omega_1,\omega_2,\omega_3)$ are organized as follows. Reading from left to right, the first frequency is the detected field. We will use a negative sign when the field is emitted and a positive sign when the field is absorbed. The fields to the right of the semicolon are the applied fields. In the frequency domain, the applied fields are not necessarily time-ordered.

The third-order susceptibility fully describes the response of material following the application of the fields $E_1$, $E_2$, and $E_3$, and forms the link between experimental observations and the underlying material response. The third-order susceptibility contains many terms. Assuming that the material can be described by a four-level system as sketched in Figure 1.6 and all the molecules are initially in the ground state $|a\rangle$, the response function $R^{(3)}_{ijkl}$ consists of eight different quantum pathways [11]. Since there are $p = 3!$ different permutations of the incoming fields, the total number of terms in $\chi^{(3)}_{ijkl}(-\omega_4;\omega_1,\omega_2,\omega_3)$ is $3! \times 8 = 48$. Not all of these terms contribute to the vibrationally resonant coherent Raman response. To illustrate this point, we consider the CARS response where the signal is detected at the frequency $\omega_{as} = 2\omega_1 - \omega_2$. In this case, there are $p = 3$ different permutations of the incoming fields, producing a total of 24 terms to $\chi^{(3)}_{ijkl}(-\omega_{as};\omega_1,-\omega_2,\omega_1)$. Together, these $\chi^{(3)}$ terms describe all the quantum pathways that

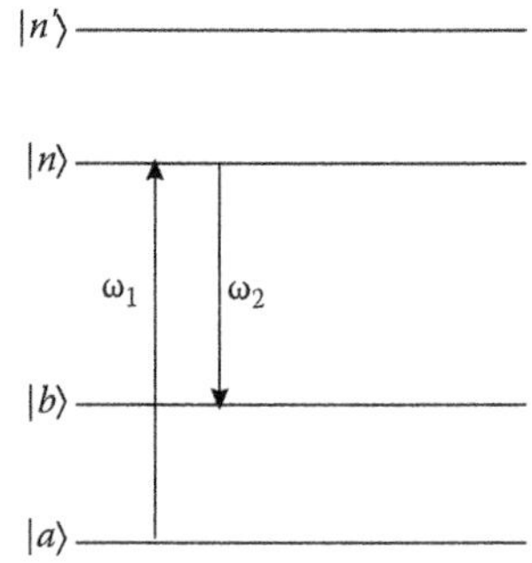


**FIGURE 1.6** Energy diagram of the four-level system discussed in the text.

the system can make due to the perturbations induced by the applied fields $\omega_1$, $-\omega_2$ and $\omega_1$, and producing radiation at $\omega_{as}$.

To gain physical insight into the form of $\chi^{(3)}$, we need to adopt a model for the evolution of the density matrix operator, which in turn determines the functional form of the Green's function propagator. The details of the propagation of the density matrix generally depend on the form of the system's Hamiltonian. Consequently, the evolution of the density matrix can be quite complex. Here we will not consider the complexities associated with elaborate models. Instead, we will focus only on a simple effective relaxation model that assumes that the elements of the time-dependent density matrix $\rho_{nm}$ obey the following equation of motion in the absence of the fields:

$$\frac{d\rho_{nm}}{dt} = -i\omega_{nm}\rho_{nm} - \gamma_{nm}\left(\rho_{nm} - \rho_{nm}^{(0)}\right) \tag{1.60}$$

Here $\gamma_{nm}$ is the dephasing rate associated with the *nm* transition, which depends on both relaxation and pure dephasing contributions. Using this simple model, the matrix elements of the frequency domain Green's function can be written as:

$$\mathbb{G}_{nm,nm}(\omega) = \frac{1}{\omega - \omega_{nm} + i\gamma_{nm}} \tag{1.61}$$

We can now write explicit expressions for the different $\chi_{ijkl}^{(3)}(-\omega_{as};\omega_1,-\omega_2,\omega_1)$ terms that govern the CARS response. These terms are conveniently depicted by Feynman diagrams, some of which are given in Figure 1.7. The diagram in Figure 1.7a, for instance, represents the following term:

$$-\frac{N}{6\hbar^3\epsilon_0}\sum_{ab,nn'}\rho_{aa}^{(0)}\frac{\mu_{an'}^{i}\mu_{n'b}^{l}\mu_{bn}^{k}\mu_{na}^{j}}{\left[\omega_{as}-\omega_{n'a}+i\gamma_{n'a}\right]\left[(\omega_1-\omega_2)-\omega_{ba}+i\gamma_{ba}\right]\left[\omega_1-\omega_{na}+i\gamma_{na}\right]} \tag{1.62}$$

It can be seen from Equation 1.62 that the contributions to $\chi^{(3)}$ become more significant when the denominator terms are minimized. The second denominator, representing

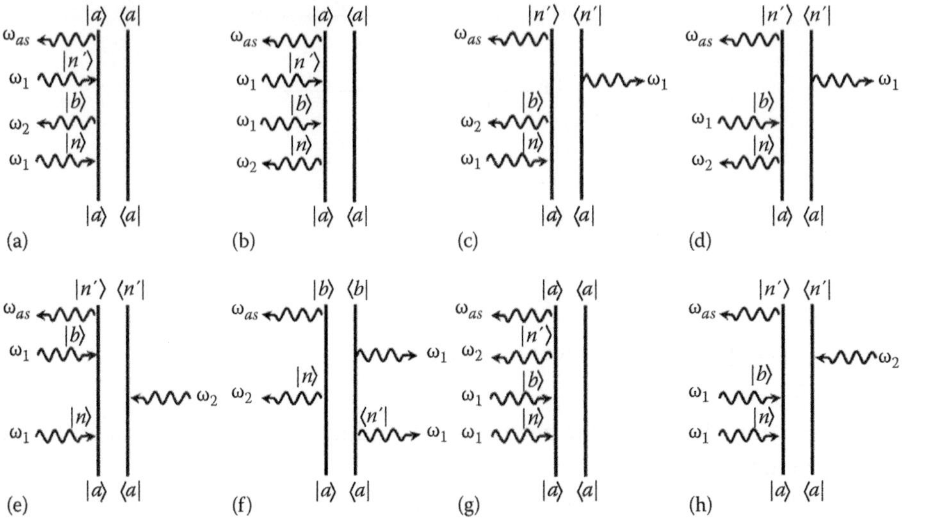

**FIGURE 1.7** Double-sided Feynman diagrams of various contributions to $\chi^{(3)}(-\omega_{as}; \omega_1, -\omega_2, \omega_1)$. (a–d) Diagrams with a two-photon Raman resonance. (e through h) Diagrams without a two-photon Raman resonance. See Ref. [11] for details about Feynman diagrams.

the propagation of the density matrix after two field interactions, is minimized whenever the difference frequency $\omega_1 - \omega_2$ matches a vibrational frequency $\omega_{ba}$. This is the Raman resonance condition. Figure 1.7b through d also contain a two-photon Raman resonance and thus contribute to the vibrationally resonant portion of $\chi^{(3)}(-\omega_{as})$. The contributions represented by Figure 1.7e through h, however, do not exhibit a Raman coherence at $\omega_{ba}$ after two field interactions, and the Raman resonance condition is not fulfilled. In the absence of electronic resonances, the contribution of a nonresonant diagram is generally less than that of a vibrationally resonant diagram. There are 16 more such nonresonant diagrams. The total contribution of the combined nonresonant terms is commonly indicated by $\chi^{(3)}_{NR}$, which is typically not negligible. We thus see that the semi-classical model provides a physical explanation for the existence of the nonresonant background: these are the quantum pathways the system can undergo which contribute to the dipole radiation at $\omega_{as}$ but do not contain propagators at $\omega_1 - \omega_2$ in resonance with the vibrational mode.

Besides the two-photon Raman resonances, $\chi^{(3)}(-\omega_{as})$ can contain additional resonances. Inspection of Equation 1.62 reveals that resonance conditions are achieved when the first and third terms in the denominator are minimized. Such conditions are met if $\omega_1$ and/or $\omega_{as}$ are in resonance with an electronic state of the material. In addition, if the vibrational state $|b\rangle$ is initially populated, electronic resonances with $\omega_2$ can also contribute to $\chi^{(3)}(-\omega_{as})$. These one-photon electronic resonances can boost the magnitude of $\chi^{(3)}(-\omega_{as})$ significantly. When both two-photon Raman resonances and electronic resonances are present, the vibrational information contained in the nonlinear susceptibility is enhanced by the electronic resonance. Resonance enhanced CARS (RCARS), which makes use of this enhancement mechanism, generally has a much higher sensitivity than regular CARS. Vibrationally resonant signals from chromophores down to $\mu$M concentrations have been measured with RCARS [12,13].

Note that electronic resonances do not only enhance the Raman resonant terms in $\chi^{(3)}(-\omega_{as})$, but also the vibrationally nonresonant terms. In addition, vibrationally nonresonant terms containing electronic two-photon resonances can contribute significantly to the overall magnitude of $\chi^{(3)}$. Diagrams (g) and (h) in Figure 1.7 contain such resonances whenever the molecule or medium contains transitions that match the combination frequency $\omega_1 + \omega_1$. The contribution of diagram (h), for instance, is:

$$\frac{N}{6\hbar^3\epsilon_0}\sum_{ab,nn'}\rho_{aa}^{(0)}\frac{\mu_{an'}^{l}\mu_{n'b}^{i}\mu_{bn}^{k}\mu_{na}^{j}}{\left[\omega_{as}-\omega_{bn'}+i\gamma_{bn'}\right]\left[(\omega_1+\omega_1)-\omega_{ba}+i\gamma_{ba}\right]\left[\omega_1-\omega_{na}+i\gamma_{na}\right]} \quad (1.63)$$

which exhibits a two-photon resonance when the system has a two-photon accessible state such that $2\omega_1 = \omega_{ba}$ (note that $b$ is a dummy index that is summed over all states).

In CRS microscopy, we are typically concerned with vibrational resonances of nonabsorbing molecules. In this case, it is not very practical to interpret the experiment in terms of the full structure of $\chi^{(3)}$. For this purpose, the third-order susceptibility is often written in a shorthand notation that highlights only the relevant vibrational resonances contained in the second propagator:

$$\chi^{(3)}(-\omega_{as};\omega_1,-\omega_2,\omega_1)=\chi_{NR}^{(3)}-\sum_{b}\frac{A_b}{(\omega_1-\omega_2)-\omega_{ba}+i\gamma_{ba}} \quad (1.64)$$

where all vibrationally nonresonant terms, including terms with two-photon electronic resonances, are lumped into $\chi_{NR}^{(3)}$. The second term on the right hand side of the equation is the vibrationally resonant contribution $\chi_R^{(3)}$ with $A_b$ the effective amplitude associated with the $\omega_{ba}$ resonance.

#### 1.6.3.3 Frequency Dependence of $\chi^{(3)}$

Many of the CRS imaging properties are directly related to the frequency dependence of $\chi^{(3)}$. The behavior of $\chi^{(3)}(-\omega_{as};\ \omega_1, -\omega_2, \omega_1)$ as a function of the difference frequency $\Omega = \omega_1 - \omega_2$ near the $\omega_{ba}$ resonance is sketched in Figure 1.8. The imaginary part of the resonant portion of $\chi^{(3)}(\Omega)$ shows a maximum around the vibrational resonance frequency, whereas the real part features a dispersive profile. When expressed in terms of amplitude and phase, the nonlinear susceptibility displays the familiar behavior of a driven oscillator, where the phase of the oscillator with respect to the driving field undergoes a $\pi$ step when the driving frequency is transitioning through resonance. This is illustrated in Figure 1.8b. The spectral phase behavior of $\chi^{(3)}$ plays an important role in CRS, and in CARS and CSRS in particular. The resonant nonlinear susceptibility produces a field that is interfering differently with the nonresonant field on each side of the resonance. On the low energy side, the resonant field is in phase with the (spectrally flat) nonresonant field contribution. On the high energy side of the resonance, the resonant field approaches a $\pi$ phase shift relative to the nonresonant field, introducing destructive interference between the two contributions. The destructive interference is reflected in the $|\chi^{(3)}(\Omega)|^2$ spectrum as the dip on the high energy side of the spectral profile, as

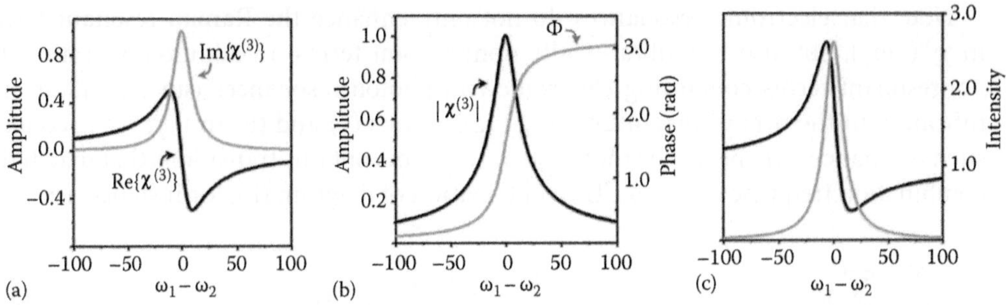


**FIGURE 1.8** Spectral dependence of the nonlinear susceptibility. (a) Real and imaginary parts of the resonant portion of $\chi^{(3)}$. The profiles are normalized to the maximum of Im$\{\chi^{(3)}\}$ and $\gamma_{ba}$ was set to 10 cm$^{-1}$. (b) The nonlinear susceptibility expressed in amplitude and spectral phase $\Phi$. (c) Spectral dependence of $|\chi^{(3)}|^2$ (black line). The ratio $\left|\chi_R^{(3)}/\chi_{NR}^{(3)}\right|$ was set to one and the $|\chi^{(3)}|^2$ profile is normalized to the nonresonant contribution. The gray line indicates the spectral profile of Im$\{\chi^{(3)}\}$, which was scaled to enable a direct comparison with $|\chi^{(3)}|^2$.

sketched in Figure 1.8c. This is the spectrum probed by homodyne-detected CARS and CSRS experiments, which is written in terms of real and imaginary parts as:

$$S(\Omega) \propto \left|\chi^{(3)}(\Omega)\right|^2 = \left|\chi_{NR}^{(3)}\right|^2 + \left|\chi_R^{(3)}(\Omega)\right|^2 + 2\chi_{NR}^{(3)}\,\mathrm{Re}\left\{\chi_R^{(3)}(\Omega)\right\} \tag{1.65}$$

In contrast, the spectrum detected through the heterodyne portion of the SRS signal is directly proportional to the imaginary part of $\chi^{(3)}(\Omega)$:

$$S(\Omega) \propto \pm 2\,\mathrm{Im}\left\{\chi^{(3)}(\Omega)\right\} \tag{1.66}$$

The plus sign pertains to SRG and the minus sign to SRL. The SRG spectrum is shown together with the CARS spectrum for comparison in Figure 1.8c. As discussed in Section 1.7.2, the imaginary part of $\chi^{(3)}(\Omega)$ is proportional to the spontaneous Raman spectrum.

#### 1.6.3.4 Spatial Polarization Properties of $\chi^{(3)}$

Besides the dependence of the third-order susceptibility on the frequencies of the optical fields, the magnitude of $\chi^{(3)}$ is highly dependent on the spatial polarization states of the incident and detected fields. The polarization states of the participating fields are indicated by the indices $\{i, j, k, l\}$ in $\chi_{ijkl}^{(3)}(-\omega_4;\omega_1,\omega_2,\omega_3)$, where the indices run in the order $\{i, j, k, l\} \rightarrow \{\omega_4, \omega_1, \omega_2, \omega_3\}$. The polarization states are considered linear and are expressed in cartesian coordinates. Complex polarization states are obtained by simply taking linear combinations of linear polarizations. The third-order susceptibility is thus a fourth-rank tensor with $3^4 = 81$ separate elements. The number of nonzero and independent elements depends on the spatial symmetry of the material. For low symmetry materials, a large number of elements may be nonzero and independent.

In CRS microscopy of biological materials, the medium is predominantly aqueous and isotropic. For isotropic materials, the number of nonzero elements is reduced to 21. To discriminate these different elements, the notation is commonly simplified by denoting the three orthogonal cartesian coordinates with the numbers {1, 2, 3}. For instance, $\chi^{(3)}_{1111}$ is the element in which all the participating fields have identical polarization orientations. Since the coordinate axes are invariant in an isotropic medium, the following symmetry properties of the 21 elements must apply [1]:

$$\begin{aligned}
&\chi_{1111} = \chi_{2222} = \chi_{3333} \\
&\chi_{1122} = \chi_{1133} = \chi_{2211} = \chi_{2233} = \chi_{3311} = \chi_{3322} \\
&\chi_{1212} = \chi_{1313} = \chi_{2323} = \chi_{2121} = \chi_{3131} = \chi_{3232} \\
&\chi_{1221} = \chi_{1331} = \chi_{2112} = \chi_{2332} = \chi_{3113} = \chi_{3223}
\end{aligned} \quad (1.67)$$

Equation 1.67 shows that the 21 elements can all be categorized into four types of non-zero elements: $\chi_{1111}$, $\chi_{1122}$, $\chi_{1212}$, and $\chi_{1221}$. These element types are not mutually independent, and are related by:

$$\chi_{1111} = \chi_{1122} + \chi_{1212} + \chi_{1221} \quad (1.68)$$

Note that for ordered structures in biological materials, including densely stacked membrane structures and deposited crystalline materials, the symmetry relations listed here for isotropic media do not necessarily hold.

## 1.7 Quantum Description of Field and Matter

In both the classical and semi-classical approaches to the coherent Raman interaction, the optical field is described as a classical electric field. This description does not take into account the quantized nature of the photon. Since the photon field can only exchange energy with the molecule in quantized amounts, the classical and semi-classical descriptions may give inaccurate quantitative estimates of the signal strength in some cases. The quantized nature of light is particularly important when modeling the response from a single or a few Raman scatterers, which produce signals in the single photon regime. A full quantum description where both field and matter are treated quantum mechanically provides a more intuitive and quantitatively more accurate picture of the energy exchange between the molecule and the field. Unlike in models that rely on classical fields, the energy exchange in the quantum filed picture does not explicitly rely on wave propagation. In the following sections, we discuss two quantum-field approaches for describing linear and nonlinear Raman interactions. In a first approach, similar to the semi-classical approach, the material evolution is modeled in terms of the density matrix operator, with the inclusion of quantized field degrees. This description forms a natural connection with the semi-classical model and provides accurate estimates of the Raman cross section. In a second approach, based on a generalized Kramers–Heisenberg model for transition rates, the energy flux from the molecule's perspective is highlighted. In this approach, the material response is no longer described by the nonlinear susceptibility but by more intuitive transition amplitudes instead.

### 1.7.1 Quantum Description of the Field

In the quantum-field description of the linear and nonlinear Raman interactions, the electric field is quantized. Similar to the material degrees, the optical electric field is described by a wavefunction, which we will denote as $|\psi_F\rangle$. The expectation value of the field is given by the expectation value of the optical electric field operator, $\langle \psi_F | \hat{E}(\mathbf{r},t) | \psi_F \rangle$, where the operator is written as:

$$\hat{E}(\mathbf{r},t) = \hat{E}_s(\mathbf{r},t) + \hat{E}_s^{\dagger}(\mathbf{r},t) \tag{1.69}$$

with

$$\hat{E}_s(\mathbf{r},t) = \left( \frac{\hbar \omega_s}{2\epsilon_0 V} \right)^{1/2} \hat{a}_s \, e^{-i(\omega_s t - \mathbf{k}_s \cdot \mathbf{r})} \tag{1.70}$$

$$\hat{E}_s^{\dagger}(\mathbf{r},t) = \left( \frac{\hbar \omega_s}{2\epsilon_0 V} \right)^{1/2} \hat{a}_s^{\dagger} \, e^{i(\omega_s t - \mathbf{k}_s \cdot \mathbf{r})} \tag{1.71}$$

where

$\hat{a}_s^{\dagger}$ and $\hat{a}_s$ are the boson creation and annihilation operators for the mode $s$, respectively

$V$ is the quantization volume of the photon mode $s$ [14,15]

The annihilation operator annihilates a photon from the mode $s$, while the creation operator creates a photon in the mode $s$:

$$\hat{a}_s^{\dagger} \left| \psi_F^s(n) \right\rangle = n_s^{1/2} \left| \psi_F^s(n+1) \right\rangle \tag{1.72}$$

$$\hat{a}_s \left| \psi_F^s(n) \right\rangle = (n_s - 1)^{1/2} \left| \psi_F^s(n-1) \right\rangle \tag{1.73}$$

where $n_s$, an integer, is the photon occupation number of mode $s$. The system's Hamiltonian now includes the contributions from the field in addition to the material degrees of freedom:

$$\hat{H} = \hat{H}_0 + \hat{H}_F + \hat{H}_{int} \tag{1.74}$$

where

$\hat{H}_0$ is the unperturbed Hamiltonian of the material as before

$\hat{H}_F$ is the contribution from the field degrees

$\hat{H}_{int}$ constitutes the interaction between the field and material

The latter two contributions are written in the quantum-field model as:

$$\hat{H}_F = \sum_s \hbar \omega_s \, \hat{a}_s^{\dagger} \hat{a}_s \tag{1.75}$$

$$\hat{H}_{int} = \hat{E}_s(\mathbf{r},t) \hat{V}^{\dagger}(\mathbf{r}) + \hat{E}_s^{\dagger}(\mathbf{r},t) \hat{V}(\mathbf{r}) \tag{1.76}$$

where the dipole operators are of the form:

$$\hat{V}(\mathbf{r}) = \sum_{\alpha=1}^{N} \delta(\mathbf{r} - \mathbf{r}_\alpha) \sum_{a,b>a} \mu_{ab} |a\rangle\langle b| \tag{1.77}$$

The index $\alpha$ runs over all molecules, which are assumed to be identical.

An important difference between the semi-classical approach and the quantum-field description is captured by the expression for $\hat{H}_{int}$ in Equation 1.76: a field-matter interaction involves a change in both the material and the field degrees of freedom. In the classical and semi-classical approach, the signal is obtained by calculating the expectation value of the dipole operator to determine the material polarization, which then acts like a source for the detected radiation. The quantum-field method calculates the signal in a different fashion. Below we describe two quantum-field approaches for calculating the optical signals. These methods differ in their perspective: the first considers the signal from the field degrees point of view whereas the second method considers the signal from the perspective of the material degrees.

#### 1.7.1.1 Field Perspective

In a first quantum-field approach, the signal is calculated by looking at the field. This method equates the optical signal directly to the change in the number of photons. The photon number in mode $s$ is given by the expectation value of the photon occupation number operator, which is given by:

$$\hat{N}_s = \hat{a}_s^\dagger \hat{a}_s \tag{1.78}$$

The eigenvalues of $\hat{N}_s$ correspond to the number of photons in mode $s$:

$$\hat{N}_s = \hat{a}_s^\dagger \hat{a}_s \left|\psi_F^s(n)\right\rangle = n_s \left|\psi_F^s(n)\right\rangle \tag{1.79}$$

The signal detected in this mode is then defined by:

$$S_s = \frac{d}{dt}\langle \hat{N}_s \rangle \tag{1.80}$$

We see that the signal in Equation 1.80 has a very intuitive form, as it simply represents the change in the number of photons of a certain frequency $\omega_s$. The expectation value of the photon occupation number operator can be calculated by solving the density matrix for the total system $\hat{\rho}_{tot}(t)$, which now includes both material and field degrees. The signal then becomes:

$$\begin{aligned} S_s &= \frac{d}{dt}\mathrm{tr}[\hat{N}_s \hat{\rho}_{tot}(t)] \\ &= -\frac{2}{\hbar}\mathrm{Im}\{\mathrm{tr}[\hat{E}_s(\mathbf{r},t)\hat{V}^\dagger]\} \end{aligned} \tag{1.81}$$

In the next sections, we will use this expression to calculate the spontaneous Raman signal and the coherent Raman signals.

#### 1.7.1.2 Material Perspective

In a second quantum-field approach, the material's perspective is chosen. Instead of focusing on changes in the photon number, the transitions between states in the material are explicitly considered. Because the field and material degrees are coupled, a transition between states implies a corresponding change in the field degrees, i.e., energy has been exchanged between the fields and the material. For this reason, this approach is only useful for the calculation of dissipative signals. The transition rate $R_{a\to n}$ between state $|a\rangle$ and state $|n\rangle$ is given by Fermi's golden rule as:

$$R_{a\to n} = \frac{2\pi}{\hbar^2}\sum_n \left|\langle n|\hat{H}_{int}|a\rangle\right|^2 \delta(\omega_n - \omega_a) \tag{1.82}$$

The term within brackets can be interpreted as the transition amplitude. This expression considers the transition between states as mediated by one-photon interactions only. Fermi's golden rule can be expanded in the field-matter interaction to include higher order photon processes. The rate of a $k$-photon process is given by a generalized Kramers–Heisenberg form [16]:

$$R_{a\to n} \propto \left|T_{na}^{(k)}(\omega_1,\ldots,\omega_k)\right|^2 \delta\left(\sum_{i=1}^{k}\omega_k - \omega_{na}\right) \tag{1.83}$$

where $T_{na}^{(k)}$ are the $k$th-order transition amplitudes. The first three orders are given by:

$$T_{na}^{(1)}(\omega_{na}) = -\int d\omega_1 \tilde{T}_{na}^{(1)}(\omega_1)\delta(\omega_{na} - \omega_1) \tag{1.84}$$

$$T_{na}^{(2)}(\omega_{na}) = \frac{2\epsilon_0}{\hbar}\int d\omega_1 d\omega_2 E(\omega_1)E(\omega_2) \times \tilde{T}_{na}^{(2)}(\omega_1,\omega_2)\delta(\omega_{na} - \omega_1 - \omega_2) \tag{1.85}$$

$$T_{na}^{(3)}(\omega_{na}) = -\frac{4\epsilon_0^2}{\hbar^2}\int d\omega_1 d\omega_2 d\omega_3 E(\omega_1)E(\omega_2)E(\omega_3) \times \tilde{T}_{na}^{(3)}(\omega_1,\omega_2,\omega_3)\delta(\omega_{na} - \omega_1 - \omega_2 - \omega_3) \tag{1.86}$$

where $E(\omega_i)$ is the expectation value of the optical field operator and

$$\tilde{T}_{na}^{(1)}(\omega_1) = \mu_{na} \tag{1.87}$$

$$\tilde{T}_{na}^{(2)}(\omega_1,\omega_2) = \sum_v \frac{\mu_{nv}\mu_{va}}{\omega_1 - \omega_{va} + i\gamma} \tag{1.88}$$

$$\tilde{T}_{na}^{(3)}(\omega_1,\omega_2,\omega_3) = \sum_{v,w} \frac{\mu_{nw}\mu_{wv}\mu_{va}}{(\omega_1 + \omega_2 - \omega_{wa} + i\gamma)(\omega_1 - \omega_{va} + i\gamma)} \tag{1.89}$$

Because the field is quantized, the material transitions are directly correlated with the change in the number of photons. For a given photon mode $s$, we can thus write for the $a \rightarrow n$ transition:

$$S_s \propto \pm R_{a \rightarrow n} \tag{1.90}$$

where the minus sign indicates the emission of a $\omega_s$ photon and the plus sign the absorption of a $\omega_s$ photon.

The above description examines directly the transitions in a molecule, and thus provides an intuitive picture of the underlying physical process during the Raman excitation. Although the changes in the material degrees of freedom are coupled to changes in the field degrees of freedom, this description does not necessarily specify which particular field mode is affected. Hence, for processes that involve multiple field modes, additional information is required to determine which field modes are affected by the transitions in the material. Note also that the transition amplitude approach is not suitable to describe parametric processes such as homodyne-detected CARS, because of the lack of an effective molecular transition.

## 1.7.2 Quantum Description of Spontaneous Raman Scattering

The spontaneous Raman scattering process involves a strong driving field $\omega_1$ and a spontaneously emitted field at $\omega_2$. The spontaneous emission process cannot be accounted for with a classical description of the field. Therefore, the expression obtained for the Raman signal in Equation 1.11 is inaccurate. The classical model is also unable to reveal the similarity between the field-matter interaction in the spontaneous Raman process and the interactions of a $\chi^{(3)}$ process. A quantum-field description of the scattering process provides new insights on both counts.

We will first calculate the Raman signal using Equation 1.81. Since the incoming optical field mode $\omega_1$ is strong and relatively unattenuated by the Raman scattering process, this field can be treated as classical. The emitted field mode $\omega_2$ is quantized, and its density matrix is initially in the vacuum state $(|\psi_F\rangle\langle\psi_F| = |0\rangle\langle 0|)$. The lowest order density matrix operator that contributes to the signal $S_2$ is $\hat{\rho}^{(3)}_{tot}$, which involves two interactions with $\omega_1$ and two interactions with $\omega_2$. The pathway of the density matrix that contributes to the incoherent Raman response is sketched in Figure 1.9. Similar to the Raman active $\chi^{(3)}$ processes, the system resides in an $ab$ coherence after two field interactions. The final emission is at $\omega_2 = -\omega_1 + \omega_2 + \omega_1$. In this process, the density matrix of the $\omega_2$ field has been raised from the vacuum state to the one photon state $|1\rangle\langle 1|$, which is the radiated photon. Based on this pathway, the incoherent spontaneous Raman emission rate is obtained from Equation 1.81 as:

$$S_2 = \frac{1}{\hbar^4} \sum_{k_2} \frac{\hbar\omega_2}{2\epsilon_0 V} 2\,\mathrm{Im}\{R(\omega_2, \omega_2 - \omega_1, -\omega_1)\} |E_1|^2 \tag{1.91}$$

where the sum is over all the $\omega_2$ modes with wave vector $k_2$ within volume $V$. This equation shows that although the Raman response involves four field interactions, the

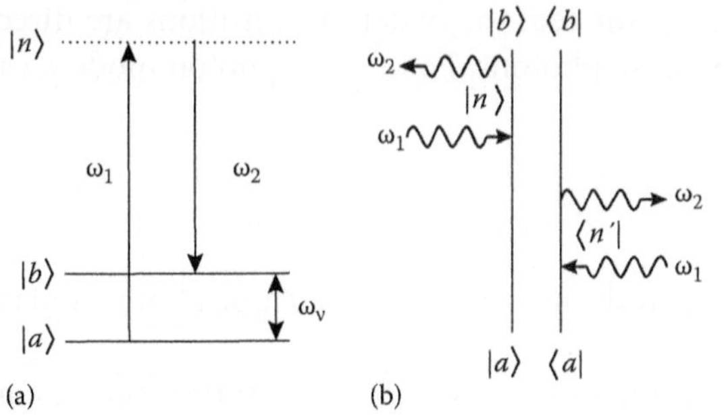


**FIGURE 1.9** Quantum mechanical picture of the spontaneous Raman process. (a) Energy level diagram. (b) Double-sided Feynman diagram of the spontaneous light scattering response with a Raman resonance.

detected signal scales linearly with the intensity of the incident field $I_0 = |E_1|^2$. Using Equations 1.91 and 1.12, the strength of the Raman signal can also be expressed in terms of the differential Raman cross section:

$$\sigma(\omega_2) = \frac{1}{9}\frac{\omega_1\omega_2^3}{\pi^2\epsilon_0^2\hbar^2c^4}\,\text{Im}\{R(\omega_2,\omega_2-\omega_1,-\omega_1)\} \tag{1.92}$$

The factor 1/9 results from averaging over molecular orientations. We see that the Raman signal scales linearly with $\omega_1$ and to the third power with $\omega_2$. The third-order power dependence includes the $\omega_2^2$ dependence of the density of field modes in the cavity with volume $V$. The cross section can be approximated as:

$$\sigma(\omega_2) \simeq C_1\,\text{Im}\{\chi^{(3)}(\Omega)\} \tag{1.93}$$

where $C_1$ is a proportionality constant. This expression is an approximation because $\chi^{(3)}(\Omega)$ contains more pathways than the one that contributes to the spontaneous Raman process. Discrepancies may arise, for instance, in case of additional electronic resonances that may lift certain terms in the $\chi^{(3)}(\Omega)$ signal that are not part of the spontaneous Raman response. However, far from electronic resonances, only ground state Raman resonances contribute and Equation 1.93 generally holds.

The Raman response can alternatively be derived from the Kramers–Heisenberg formalism. At least two transitions are required to establish the $a \rightarrow b$ transition. The lowest order transition amplitude that describes this process is $T_{ba}^{(2)}$. The transition rate can then be written as:

$$R_{a\rightarrow b} \propto \left|T_{ba}^{(2)}(\omega_{ba})\right|^2 \delta(\omega_1-\omega_2-\omega_{ba}) \tag{1.94}$$

In this transition process, the $\omega_1$ mode is changed from occupation number $n$ to $n-1$ (absorption), while the $\omega_2$ mode changes from the vacuum state $|0\rangle$ to field state $|1\rangle$ (emission). The transition rate can then be recast as:

$$R_{a\rightarrow b} \propto \frac{\omega_1\omega_2}{V^2}|E_1|^2\left|\sum_n \frac{\mu_{bn}\mu_{na}}{\omega_1-\omega_{na}+i\gamma}\right|^2 \delta(\omega_1-\omega_2-\omega_{ba}) \tag{1.95}$$

where $E_1$ is the expectation value of the field amplitude of the $\omega_1$ mode. Accounting for the field mode density of the scattered field, we can deduce the differential cross section at the Raman resonance as:

$$\sigma(\omega_2) \propto \frac{\omega_1\omega_2^3}{c^4}|\alpha|^2 \tag{1.96}$$

The transition amplitude $\alpha$ has the form of a (single pathway) Raman transition polarizability:

$$\alpha = \sum_n \frac{\mu_{bn}\mu_{na}}{\omega_1 - \omega_{na} + i\gamma} \tag{1.97}$$

In electronically resonant Raman experiments, the transition polarizability is large whenever the incident beam is close to an electronic transition of the molecule.

The virtue of this latter description is that it clearly shows what happens to the molecule. Before the interaction with the excitation field, the molecule is in its ground state $|a\rangle$. After the Raman scattering process, the molecule is in the vibrationally excited state $|b\rangle$. The molecule has thus gained energy in this process, during which one $\omega_1$ photon was annihilated and one $\omega_2$ photon was emitted. The energy gain of the molecule corresponds to the loss in the total light field, which amounts to $\hbar\omega_1 - \hbar\omega_2 = \hbar\omega_v$.

## 1.7.3 Quantum Description of Coherent Raman Signals: Interference of Pump-Probe Paths

From the previous section it is clear that the Raman process involves the transition from the ground state to the vibrational excited state in the molecule, while the total light field losses energy. This is a dissipative process, which under certain conditions is proportional to Im $\chi^{(3)}$, as shown in Equation 1.93. A similar analysis can be applied to interpret pump-probe type coherent Raman scattering signals (stimulated Raman scattering), which include the SRL and the SRG signals.

In SRL and SRG, we need to consider two field modes, $\omega_1$ and $\omega_2$, both of which are initially occupied by photons. During the pump-probe process, an $\omega_1$ mode is absorbed and an $\omega_2$ mode is emitted, while the molecule undergoes a transition from the ground state $a$ to the vibrationally excited state $b$. In SRL, the photon loss in the $\omega_1$ mode is detected, whereas in SRG the photon gain in the $\omega_2$ mode is detected. This process is identical to the Raman process sketched in Figure 1.9. The important difference between spontaneous Raman and stimulated Raman is that the $\omega_2$ mode in the stimulated process is occupied, whereas it is empty in the spontaneous Raman case. The stimulated Raman process is dissipative and the signal can be written in a Kramers–Heisenberg form [16]. Selecting terms that contain the $\omega_{ba} = \omega_1 - \omega_2$ resonance and ignoring contributions form electronic resonances, the transition rate can be written as:

$$R_{a\to b} \propto |E_1|^2 |E_2|^2 \left|\tilde{T}_{ba}^{(2)}\right|^2 \delta(\omega_1 - \omega_2 - \omega_{ba})$$

$$\propto \frac{\omega_1\omega_2}{V^2}|E_1|^2|E_2|^2 \left|\sum_n \frac{\mu_{bn}\mu_{na}}{\omega_1 - \omega_{na} + i\gamma}\right|^2 \delta(\omega_1 - \omega_2 - \omega_{ba}) \tag{1.98}$$

Comparing the transition rate of the stimulated Raman process with the rate of the spontaneous Raman process in Equation 1.95, we see that the stimulated Raman process exhibits a higher rate due to the factor $|E_2|^2$. This is a direct consequence of the fact that the $\omega_2$ mode is initially occupied, thereby enhancing the transition probability between the states $a$ and $b$. The enhanced transition rate corresponds to an enhanced rate of change in the field modes, i.e., a higher rate of photon loss in the $\omega_1$ detection channel and a higher rate of photon gain in the $\omega_2$ channel. Consequently, as long as the stimulated Raman photon flux is above the shot-noise limit, the SRL and SRG optical signals from a particular molecular transition can be many orders of magnitude higher than the corresponding optical signals measured in a spontaneous Raman experiment.

### 1.7.4 Quantum Description of Heterodyne Coherent Raman Signals

Because homodyne-detected CARS probes a non-dissipative process, the corresponding signal cannot be written in a generalized Kramers–Heisenberg form. Heterodyne-detected CARS, on the other hand, can probe dissipative processes. In the following, we will discuss the case of the heterodyne CARS signal as it can be conveniently described within the quantum field framework of this chapter. The quantum field description allows for an intuitive interpretation of the dissipative and the parametric contributions to the CARS signal.

In the pump-probe type coherent Raman processes, we considered two field modes. In the heterodyne CARS experiment, the number of field modes is higher, introducing multiple pathways. Here, we consider the general case of four input modes, $\omega_1$, $\omega_2$, $\omega_3$, and $\omega_4$. We will assume that $\omega_1 - \omega_2 + \omega_3 = \omega_4$, and that $\omega_1 - \omega_2 = \omega_{ba}$ and $\omega_4 - \omega_3 = \omega_{ba}$. This situation is sketched in Figure 1.10. We will also assume that all modes have high photon occupation numbers.

We will first focus on the dissipative contribution from the material point of view. We will be concerned with pathways that contribute to an $a \rightarrow b$ transition in the molecule. In Figure 1.10a, we identify two pathways that mediate a transition

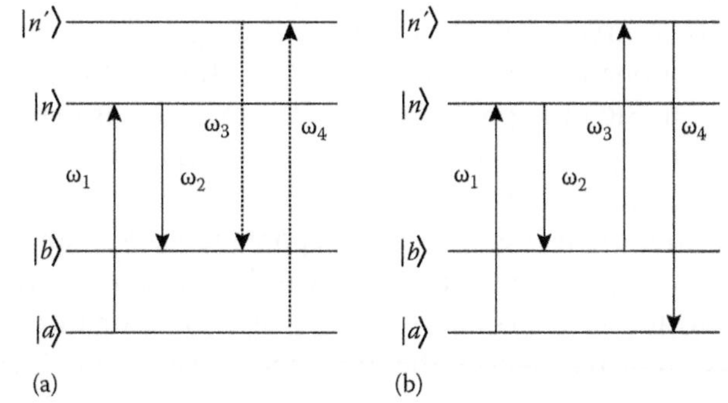


**FIGURE 1.10** Energy level diagram representation of dissipative and parametric contributions in a heterodyne CARS experiment. (a) Dissipative contribution. Two $a \rightarrow b$ pathways can be distinguished: the $(\omega_1, -\omega_2)$ and the $(\omega_4, -\omega_3)$ pump-probe pathways, which constructively interfere. Both $\omega_1$ and $\omega_4$ photons are absorbed. (b) Parametric contribution. A photon in the $\omega_1$ mode is absorbed and a photon in the $\omega_4$ mode is emitted. The initial and final states of the molecule are the same, i.e., no effective $a \rightarrow b$ transition is made. Arrows are not time-ordered.

in the molecule. The first pathway, indicated by the solid arrows, is a pump-probe process in which one $\omega_1$ photon is absorbed and one $\omega_2$ photon emitted, leaving the molecule in the $b$ vibrationally excited state. The second pathway, indicated by the dashed arrows, is a similar pump-probe process, in which one $\omega_4$ photon is absorbed and one $\omega_3$ photon emitted. The total transition *probability* in the heterodyne CARS experiment can be written as:

$$P_{a\to b} \propto |\mu_{bn}|^2 |\mu_{na}|^2 \left| \frac{E_1 E_2^*}{\omega_1 - \omega_{na} + i\gamma} \delta_\Delta(\omega_1 - \omega_2 - \omega_{ba}) + \frac{E_4 E_3^*}{\omega_4 - \omega_{na} + i\gamma} \delta_\Delta(\omega_4 - \omega_3 - \omega_{ba}) \right|^2 \tag{1.99}$$

where

$E_1$ are the expectation values of the field operators
$\delta_\Delta$ are slightly broadened delta functions

When calculating the square modulus, we find three processes that contribute to the dissipative signal:

$$P_{a\to b} = P_{a\to b}^{12} + P_{a\to b}^{34} + P_{a\to b}^{1234} \tag{1.100}$$

with

$$P_{a\to b}^{12} \propto |\mu_{bn}|^2 |\mu_{na}|^2 \left| \frac{E_1 E_2^*}{\omega_1 - \omega_{na} + i\gamma} \right|^2 \delta_\Delta^2(\omega_1 - \omega_2 - \omega_{ba}) \tag{1.101}$$

$$P_{a\to b}^{34} \propto |\mu_{bn}|^2 |\mu_{na}|^2 \left| \frac{E_4 E_3^*}{\omega_4 - \omega_{na} + i\gamma} \right|^2 \delta_\Delta^2(\omega_4 - \omega_3 - \omega_{ba}) \tag{1.102}$$

$$P_{a\to b}^{1234} \propto |\mu_{bn}|^2 |\mu_{na}|^2 \, 2\,\mathrm{Re}\left[ \frac{E_1 E_2^* E_3 E_4^*}{(\omega_1 - \omega_{na} + i\gamma)(\omega_4 - \omega_{na} + i\gamma)} \right] \times \delta_\Delta(\omega_1 - \omega_2 - \omega_{ba}) \delta_\Delta(\omega_4 - \omega_3 - \omega_{ba}) \tag{1.103}$$

We see that first two processes are pump-probe type. $P_{a\to b}^{12}$ can be probed by detecting the loss in the $\omega_1$ channel or the gain in the $\omega_2$ channel. Similarly, $P_{a\to b}^{34}$ can be probed by detecting the loss in the $\omega_4$ channel or the gain in the $\omega_3$ channel. The third term is an interference term that represents the mutual interference of the two pump-probe pathways. The transition probability is higher when the two pathways are in phase and lower when the pathways are out of phase. Hence, by controlling the phase of the two pathways, the dissipative part of the signal can be either enhanced or suppressed.

In the CARS experiment, the $\omega_4$ mode is detected, which we will denote as $S_4$. We will next write the CARS signal in terms of transition amplitudes and identify the differences between the dissipative and parametric parts. The total signal can be written as:

$$S_4 = S_4^{dis} + S_4^{par} \tag{1.104}$$

The dissipative part of the heterodyne CARS signal is closely related to the interference term $P_{a\to b}^{1234}$ of the transition probability, and can be rewritten in terms of transition amplitudes. Assuming electronically off-resonance conditions and no thermal population in the vibrationally excited state, the dissipative signal can be written as [16]:

$$S_4^{dis} \propto \delta(\omega_1 - \omega_2 + \omega_3 - \omega_4)\delta(\omega_1 - \omega_2 - \omega_{ba})$$
$$\times P(a)\left[ E_1 E_2^* E_3 E_4^* \tilde{T}_{ba}^{(2)}(-\omega_2, \omega_1)\tilde{T}_{ba}^{(2)*}(-\omega_4, \omega_3) + \text{c.c.} \right] \tag{1.105}$$

where $P(a)$ is the equilibrium probability that the system is state $a$. From this expression it is evident that the dissipative CARS signal depends on the interference between two second-order transition processes. Note that, similar to spontaneous Raman and stimulated Raman scattering, the dissipative part is described by a product of two second-order transition amplitudes.

The parametric part to the signal, $S_4^{par}$, involves the process in which the initial and final states of the molecule are identical. This situation is sketched in Figure 1.10b. In this process, the second pathway is reversed. A $\omega_3$ photon is absorbed and a $\omega_4$ photon is emitted, bringing the molecule back into the initial state. The lowest order transition amplitudes that contribute to this process are $\tilde{T}_{aa}^{(4)}$, a four-photon scattering process. For the CARS channel, the relevant transition amplitude is proportional to $\tilde{T}_{aa}^{(4)}(-\omega_4, \omega_3, -\omega_2, \omega_1)$. The parametric signal is written as:

$$S_4^{par} \propto \delta(\omega_1 - \omega_2 + \omega_3 - \omega_4)P(a)\text{Im}\left\{E_1 E_2^* E_3 E_4^*\right\}\text{Re}\left\{\tilde{T}_{aa}^{(4)}(-\omega_4, \omega_3, -\omega_2, \omega_1)\right\} \tag{1.106}$$

We see that both the $S_4^{dis}$ and $S_4^{par}$ contributions to the heterodyne CARS signal have the expected linear dependence on the field amplitude $E_4$, which corresponds to the classical local oscillator field. In addition, the quantum field description shows that the $S_4^{dis}$ contribution necessarily consists of second-order scattering processes in which the $a \to b$ transition is made. The $S_4^{par}$ contribution, on the other hand, is a fourth-order scattering process in which the molecule has not made an effective $a \to b$ transition. This latter information is not clearly expressed in the classical description, while the quantum field approach naturally shows the physical difference between the parametric and dissipative parts of the heterodyne CARS signal.

It is interesting to examine the parametric signal detected in other channels as well. In the $\omega_1$ channel, which detects $S_1^{par}$, the relevant transition amplitude is $\tilde{T}_{aa}^{(4)}(\omega_4, -\omega_3, \omega_2, -\omega_1)$, which is of identical amplitude. However, in the parametric process, for each emitted $\omega_4$ photon, there is an absorbed $\omega_1$ photon. This implies that the detected parametric signals in these channels are related as $S_1^{par} = -S_4^{par}$. Note that the same relation does not hold for dissipative signals. The dissipative contribution in

CARS results from the photon pairs ($\omega_1$, $-\omega_2$) and ($\omega_4$, $-\omega_3$), in which both $\omega_1$ and $\omega_4$ are absorbed. These are two Stokes processes that constructively interfere. It is assuming that the heterodyne CARS experiment can be understand as the interference of two Stokes processes rather than an anti-Stokes process as the acronym implies. Hence, $S_1^{dis}$ and $S_4^{dis}$ have the same sign. From this simple relation we see that:

$$S_1 + S_4 = S_1^{dis} + S_4^{dis} \tag{1.107}$$

where the parametric signal is cancelled out. Measuring the photon change in the sum of both channels is thus equal to measuring only dissipative contributions to the coherent Raman process. The same relation holds for the $S_2$ and $S_3$ channels. More generally, the total dissipative signal $D$ can be obtained by detecting all the modes simultaneously:

$$D = S_1 + S_2 + S_3 + S_4 \tag{1.108}$$

The interpretation of this latter result is straightforward: any energy loss in the combined field modes must correspond to an energy gain of the material, which constitutes the dissipative process.

## 1.8 Concluding Remarks

In this chapter, we have discussed the basics of the CRS process. We have seen that the classical, semi-classical, and quantum-field description each offer insight into several aspects of the CRS light-matter interaction. The classical model provides an intuitive picture in terms of oscillating electron clouds perturbed by harmonic nuclear modes. It offers a good framework for qualitatively interpreting the CRS signals measured in microscopy studies. The semi-classical model adds the actual quantum-mechanical mode structure of the molecule to the picture, which enables a direct connection between CRS experiments and quantum mechanical calculations of molecular vibrations. In addition, the semi-classical model offers a solid framework for dissecting ultrafast, time-resolved CRS experiments. Finally, the quantum-field approach takes into account the quantized energy exchange between light and matter. This latter description correctly predicts Raman cross sections and introduces additional insight into the origin of the parametric and dissipative contributions to the CRS signal.

## Acknowledgments

E.O.P. acknowledges support from the National Science Foundation (NSF), grants CHE-0802913 and CHE-0847097. S.M. gratefully acknowledges support from the NSF, grant CHE-1058791, and the National Institutes of Health, grant GM059230 and 287 GM091364, and from the Department of Energy (DOE), Division of Chemical Sciences, Geosciences, and Biosciences, Office of Basic Energy Sciences.

## References

1. R. W. Boyd, *Nonlinear Optics* (Academic Press, San Diego, CA, 2003).
2. D. Débarre and E. Beaurepaire, Quantitative characterization of biological liquids for third-harmonic generation microscopy, *Biophys. J.* **92**, 603–612 (2007).

3. G. R. Meredith, B. Buchalter, and C. Hanzlik, Third-order susceptibility determination by third harmonic generation. II, *J. Chem. Phys.* **78**, 1543–1551 (1983).
4. U. Gubler and C. Bosshard, Optical third-harmonic generation of fused silica in gas atmosphere: Absolute value of the third-order harmonic nonlinear optical susceptibility $\chi^{(3)}$, *Phys. Rev. B* **61**, 10702–10710 (2000).
5. T. Hashimoto, T. Yoko, and S. Sakka, Sol-gel preparation and third-order nonlinear optical properties of $TiO_2$ thin films, *Bull. Chem. Soc. Jpn.* **67**, 653–660 (1994).
6. Y. Wang, C. Y. Lin, A. Nikolaenko, V. Raghunathan, and E. O. Potma, Four-wave mixing microscopy of nanostructures, *Adv. Opt. Photon.* **3**, 1–52 (2011).
7. G. Placzek, Rayleigh-Streuung und Raman-Effekt, in *Handbuch der Radiologie*, E. Marx, Ed. (Akademische Verlagsgesellschaft, Leipzig, Germany, 1934).
8. E. Garmire, F. Pandarese, and C. T. Townes, Coherently driven molecular vibrations and light modulation, *Phys. Rev. Lett.* **11**, 160 (1963).
9. S. A. J. Druet and J. P. E. Taran, CARS spectroscopy, *Prog. Quantum Electron.* **7**, 1 (1981).
10. J. A. Armstrong, N. Bloembergen, J. Ducuing, and P. S. Pershan, Interactions between light waves in a nonlinear dielectric, *Phys. Rev.* **127**, 1918 (1962).
11. S. Mukamel, *Principles of Nonlinear Optical Spectroscopy* (Oxford University Press, New York, 1995).
12. L. A. Carreira, T. C. Maguire, and T. B. Malloy, Excitation profiles of the coherent anti-Stokes resonance Raman spectrum of β-carotene, *J. Chem. Phys.* **66**, 2621–2626 (1977).
13. W. Min, S. Lu, G. R. Holtom, and X. S. Xie, Triple-resonance coherent anti-Stokes Raman scattering microspectroscopy, *ChemPhysChem* **10**, 344–347 (2009).
14. R. Loudon, *The Quantum Theory of Light* (Oxford University Press, New York, 2000).
15. M. Scully and M. S. Zubairy, *Quantum Optics* (Cambridge University Press, Cambridge, U.K., 1997).
16. S. Mukamel and S. Rahav, Ultrafast nonlinear optical signals viewed from the molecule's perspective: Kramers–Heisenberg transition amplitudes versus susceptibilities, *Adv. Atom. Mol. Opt. Phys.* **59**, 233–263 (2010).

# 2. Coherent Raman Scattering under Tightly Focused Conditions

**Eric Olaf Potma, Xiaoliang Sunney Xie, Andreas Volkmer, and Ji-Xin Cheng**



*Coherent Raman Scattering Microscopy.* Edited by Ji-Xin Cheng and X. Sunney Xie © 2013 CRC Press/ Taylor & Francis Group, LLC. ISBN: 978-1-4398-6765-5.

## 2.1 Introduction

Coherent Raman scattering (CRS) microscopy makes use of many of the theoretical and experimental tools developed since the 1960s in the field of coherent Raman spectroscopy [1–5]. While the laser light sources, filters, detection hardware, and software have tremendously improved over the years, the basic ingredients of performing a CRS experiment share a great deal of similarity with the earliest experiments of the CRS pioneers. Perhaps one of the most essential differences is the size of the nonlinear interaction volume: in CRS microscopy the interaction volume is on the order of an optical wavelength, whereas in traditional coherent Raman experiments the probing volume is many orders of magnitude larger. At first sight, a smaller probing volume may seem a salient detail. However, the tight focus is more than a necessity for creating a measurable nonlinear signal, as it has emphasized a renewed appreciation of the physics of coherent signal generation.

In CRS microscopy, the probing volume is formed by the overlap of the focal fields of the incident beams. Using high numerical aperture objectives, such volumes approach sub-micrometer scales in both lateral and axial dimensions. This implies that a model that describes nonlinear signal generation with tightly focused fields needs to take into account its three-dimensional character. Hence, while conventional macroscopic coherent Raman scattering experiments are commonly described in terms of plane wave solutions in only one dimension and in a bulk homogeneous nonlinear medium, the situation in the optical microscope calls for more complex three-dimensional descriptions of both the optical fields and the sample material.

In this chapter, we will study the generation of coherent Raman scattering signal by using tightly focused collinearly propagating laser beams [6]. Although variants of collinear CARS microscopy have also been developed, including a BOXCARS geometry [7], wide-field CARS microscopy [8], and CARS near-field scanning optical microscopy [9,10], the collinear input beam geometry became the configuration of choice in CARS and SRS microscopy exhibiting superior image quality and added simplicity. To model the nonlinear signal generation under this condition, we will start with examining ways in which the spatial properties of the electromagnetic field can be described. Such a discussion necessarily starts with considering Maxwell's equations and using these for the purpose of modeling propagating light fields. From there, we will discuss the formation of the focal volume by a focusing lens, the generation of CRS fields within the focal volume and their subsequent detection in the far field.

Although many of the theoretical tools employed for describing the spatial aspects of nonlinear signal generation have been developed in the twentieth century, the much younger field of CRS microscopy has in many ways brought these tools together and managed to paint a more complete picture of nonlinear signal generation. An amount of theoretical work was published that takes the distinct features of signal generation in CARS microscopy into account. As such, the generation of the forward-detected CARS signal with tightly focused excitation beams was theoretically depicted in the thick-sample limit [11], the optical transfer function of CARS microscopy was derived [12], and a description of the total CARS field generated by an ensemble of coherently induced Hertzian dipoles while taking into account the wave vector mismatches for forward- and backward-detected CARS signals was provided [13]. The latter work was further generalized by use of the Green's function method to calculate the CARS signal generated with tightly focused Gaussian beams for samples of arbitrary shape and size [14]. From this improved picture, several new insights have emerged, including the nature of backward-propagating nonlinear radiation and the role of the Gouy phase shift. In this chapter, by using the angular spectrum representation of the focal fields, we will try to do justice to this generalized picture of CRS generation and highlight some of the unique features of nonlinear signal generation with tightly focused fields.

## 2.2 Classical Wave Propagation

In optical microscopy, the light field is focused to a tight focal spot. The signal radiation is generated in the vicinity of the focal spot and it subsequently propagates into the far field where it is detected with a photodetector. In this process, the light field, and its interaction with matter, is highly spatially dependent. Hence, to describe this process, the spatial dependence of the light field needs to be taken into account explicitly. A general discussion of the spatial properties of macroscopic electromagnetic fields starts with classical field theory in the form of Maxwell's equations. In the following sections, we will discuss the basic properties of electromagnetic fields and provide useful expressions for the propagating electric field, which will prove convenient for describing the formation of the focal volume.

### 2.2.1 Electric Field and Material Properties

In classical field theory, we are concerned with the behavior of the electric field **E** and the magnetic field **H** in the presence of charge and current densities. The description of these fields is performed at a macroscopic level, as the charges and currents are not considered to be point sources but rather averaged values expressed per unit volume (density). Hence, the contributions of individual molecules are not modeled as discrete point sources but rather as charge and current densities in a given volume, where the volume is much larger compared to the size of the molecules [15]. To model fields at the molecular level, a microscopic description is required, which is beyond the scope of this chapter. Nonetheless, in CRS microscopy applications, the radiation typically emanates from ensembles of molecules and is detected in the far field, a situation which is modeled well with macroscopic classical field theory.

The electric and magnetic fields in vacuum are influenced by the presence of charges and currents, which are defined in terms of the charge density ρ and the current density **J**. In addition, when the fields are present in a medium, the medium responds to the presence of the fields. The response of the medium, in turn, affects the electric and magnetic fields. It is then useful to define two new fields, called electric displacement **D** and magnetic induction **B**, which are written as

$$\mathbf{D} = \varepsilon_0 \mathbf{E} + \mathbf{P} \tag{2.1}$$

$$\mathbf{B} = \mu_0 \mathbf{H} + \mu_0 \mathbf{M} \tag{2.2}$$

where

$\varepsilon_0$ and $\mu_0$ are the electrical and magnetic permittivity of the vacuum
**P** and **M** are the polarization density and the magnetization density of the medium

The polarization and the magnetic densities depend on the strength of the field and are given to the first order as

$$\mathbf{P}^{(1)} = \varepsilon_0 \chi_e^{(1)} \mathbf{E} \tag{2.3}$$

$$\mathbf{M}^{(1)} = \chi_m^{(1)} \mathbf{H} \tag{2.4}$$

where
$\chi_e^{(1)}$ is the first order electric susceptibility
$\chi_m^{(1)}$ is the first order magnetic susceptibility

Both $\chi_e^{(1)}$ and $\chi_m^{(1)}$ are properties of the material. It also proves convenient to define the relative permittivities $\varepsilon$ and $\mu$ as

$$\mathbf{D} = \varepsilon_0 \varepsilon \mathbf{E} \tag{2.5}$$

$$\mathbf{B} = \mu_0 \mu \mathbf{H} \tag{2.6}$$

To first order accuracy, the relative permittivities and the susceptibilities of the medium are related to each other as

$$\varepsilon = 1 + \chi_e^{(1)} \tag{2.7}$$

$$\mu = 1 + \chi_m^{(1)} \tag{2.8}$$

and the linear refractive index is $n = n(\omega) = \sqrt{\varepsilon\mu} = \sqrt{(1+\chi_e^{(1)})(1+\chi_m^{(1)})}$.

For our discussion here, we are interested in electromagnetic light waves that interact with matter to produce a nonlinear optical response. In this optical regime, the magnetic response of the medium is generally negligible and we will be primarily concerned with only the electric field component of the electromagnetic radiation.

The response of the medium to the presence of an electric field is described by the polarization density **P**. The stronger the electric field, the stronger the polarization of the material. When the fields are strong, higher order polarizations of the medium may occur, as discussed in Chapter 1. If the medium is isotropic, centrosymmetric, and nonmagnetic, and we consider nonlinearities up to the third order, then the effect of the applied fields on the induced polarization density is written as

$$\begin{aligned} \mathbf{P} &= \mathbf{P}^{(1)} + \mathbf{P}^{(3)} \\ &= \varepsilon_0 \chi_e^{(1)} \mathbf{E} + \varepsilon_0 \chi_e^{(3)} \mathbf{E} \cdot \mathbf{E} \cdot \mathbf{E} \end{aligned} \tag{2.9}$$

Note that in the previous equation, the first term is the linear response of the system, that is, the resulting polarization density oscillates at the same frequency as the incident field. The second term is the nonlinear term, which is responsible for generating new frequencies of the polarization density. Through this term, electric fields at different frequencies can interact in the medium to generate a polarization density at a frequency, which is a linear combination of the incident frequencies.

From the previous equations, the relative electrical permittivity of the medium can be written as

$$\varepsilon = 1 + \chi_e^{(1)} + \chi_e^{(3)} \mathbf{E} \cdot \mathbf{E} \tag{2.10}$$

$$= \varepsilon_{lin} + \chi_e^{(3)} \mathbf{E} \cdot \mathbf{E} \tag{2.11}$$

where $\varepsilon_{lin}$ is same as the expression given in Equation 2.7. Now that we have defined the fields and the relevant macroscopic material properties, in the next section, we will examine how the spatial distribution of the electric field can be described.

### 2.2.2 Wave Equation

The spatial properties of the electric and magnetic field, in the presence of charge and current sources, are found by solving Maxwell's equations:

$$\nabla \times \mathbf{E} = -\frac{\partial}{\partial t} \mathbf{B} \tag{2.12}$$

$$\nabla \times \mathbf{H} = \mathbf{J} + \frac{\partial}{\partial t} \mathbf{D} \tag{2.13}$$

$$\nabla \cdot \mathbf{B} = 0 \tag{2.14}$$

$$\nabla \cdot \mathbf{D} = \rho \tag{2.15}$$

By substituting Equations 2.1 and 2.2 into Maxwell's equations, the inhomogeneous wave equation of the electric field is obtained:

$$\nabla \times \nabla \times \mathbf{E} + \frac{1}{c^2} \frac{\partial^2}{\partial t^2} \mathbf{E} = -\mu_0 \frac{\partial}{\partial t} \left( \mathbf{J} + \frac{\partial}{\partial t} \mathbf{P} + \nabla \times \mathbf{M} \right) \tag{2.16}$$

where $c = 1/\sqrt{\varepsilon_0 \mu_0}$ is the speed of light in vacuum. Assuming that the medium is non-magnetic ($\mathbf{M} = 0$; $\chi_m = 0$; $\mu = 1$) and that there are no sources of current ($\mathbf{J} = 0$) and no free charges ($\rho = 0$) in the medium, the wave equation for a given frequency and linear refractive index $n$ can be written as

$$\nabla \times \nabla \times \mathbf{E} + \frac{1}{c^2} \frac{\partial^2}{\partial t^2} \mathbf{E} = -\mu_0 \frac{\partial^2}{\partial t^2} \mathbf{P} \tag{2.17}$$

This equation makes it clear that the spatial and temporal propagation of the electric field in a medium depends on the polarization properties **P** of the material. Using the expressions for the material's nonlinear polarization density in Equation 2.9 and permittivity in Equation 2.7, the wave equation for the electric field can be recast as

$$\nabla^2\mathbf{E} - \frac{n^2}{c^2}\frac{\partial^2}{\partial t^2}\mathbf{E} = \frac{1}{\varepsilon_0 c^2}\frac{\partial^2}{\partial t^2}\mathbf{P}^{(3)} \tag{2.18}$$

This form of the wave equation is useful for describing the generation and propagation of waves that originate from a third-order nonlinear polarization in the material. This nonlinear wave equation forms the starting point for most models concerned with the spatial and temporal evolution of four-wave mixing signals, including the generation and propagation of CARS and SRS [2,16,17,18]. We will return to this form of the wave equation in the following sections.

Here, however, we are interested in the propagation of incident waves in the context of forming a tight focal volume. The nonlinear source term in Equation 2.18 is only relevant in the vicinity of the focal volume where the electric field strengths are high. Outside the focal volume, the nonlinear polarization is generally negligible. Therefore, to model the propagation of incident light toward the focal spot, we can set the nonlinear source term to zero.

The resulting wave equation is a homogeneous differential equation. To further simplify the wave equation, we assume that the fields are time-harmonic, that is, they oscillate harmonically with a given angular frequency $\omega$. This allows us to write the time-dependence of the electric field explicitly as

$$\mathbf{E}(\mathbf{r},t) = \mathrm{Re}\left\{\mathbf{E}(\mathbf{r})e^{i\omega t}\right\} = \frac{1}{2}\left[E(\mathbf{r})e^{i\omega t} + E^*(\mathbf{r})e^{-i\omega t}\right] \tag{2.19}$$

where $\mathbf{E}(\mathbf{r})$ is the (complex) spatial amplitude of the field. The wave equation can then be rewritten in terms of the spatial amplitude of the field:

$$\nabla^2\mathbf{E}(\mathbf{r}) + k^2\mathbf{E}(\mathbf{r}) = 0 \tag{2.20}$$

where $k = (\omega/c)n$ is the magnitude of the wave vector of the electric field. Equation 2.20 is the well-known Helmholtz equation. It describes the behavior of the spatial amplitude of the electric field in an isotropic, homogeneous, and source-free medium. The Helmholtz equation can be solved for a wide variety of boundary conditions. In the next section, we will be concerned with finding general solutions to the Helmholtz equation for the purpose of describing propagating light fields.

### 2.2.3 Angular Spectrum Representation of the Electric Field

We are interested in finding solutions to the Helmholtz wave equation that bear similarity to plane waves. A plane wave exhibits a main propagation axis, which we here define as the $z$-direction, in Cartesian coordinates. Here, we allow the plane normal to the propagation axis ($z$ = constant) to contain a spatial field distribution $\mathbf{E}(x, y)$, as sketched

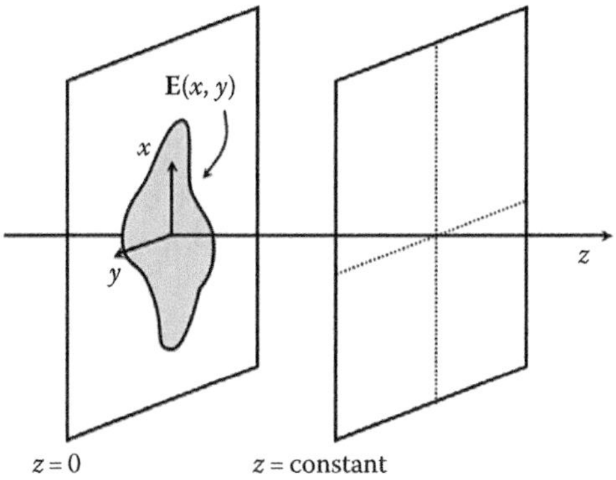


FIGURE 2.1 Schematic of the propagation of the electric field along the chosen $z$-axis. The transverse component of the electric field $\mathbf{E}(x, y)$ is defined in the plane perpendicular to the considered propagation direction.

in Figure 2.1. We will call $\mathbf{E}(x, y)$ the transverse field distribution of the wave. This formulation of the electric field can thus be written as

$$\mathbf{E}(x, y, z) = \mathbf{E}(x, y)e^{(\pm ik_z z)} \tag{2.21}$$

where the transverse field distribution is expressed as the two-dimensional Fourier transform of the spatial frequencies $k_x$ and $k_y$, taken at the reference plane at $z = 0$:

$$\mathbf{E}(x, y) = \iint_{-\infty}^{\infty} \hat{\mathbf{E}}(k_x, k_y; 0)e^{i(k_x x + k_y y)}dk_x dk_y \tag{2.22}$$

Hence, the real space coordinates $(x, y)$ and the spatial frequencies $(k_x, k_y)$ are Fourier conjugates. This expression of the field is known as the angular spectrum representation of the field. In this representation, the field is described from the perspective of the spatial frequency spectrum $\hat{\mathbf{E}}(k_x, k_y; 0)$ contained in the transverse plane and its corresponding evolution along the $z$-axis. For propagation in materials with a positive and real refractive index, the wavevector $k_z$ is real and the solution given in Equation 2.21 constitutes a free-space propagating wave in the $z$-direction. For plane wave propagation, the spatial frequencies $k_x$ and $k_y$ are restricted by the condition $k_x^2 + k_y^2 \leq k^2$. This plane wave expression of the electric field is a valid solution of the Helmholtz equation, and, by fulfilling the condition $\mathbf{k} \cdot \hat{\mathbf{E}} = 0$, is also compatible with Maxwell's equations [19,20].

There are good reasons for choosing this form of permissible solutions for the electric field. In particular, the angular spectrum representation forms a natural link with the results obtained from Fourier optics and geometrical optics. This link becomes clear when considering the far-field solution of the electric field in a homogeneous medium. In the angular spectrum representation, the electric field $\mathbf{E}_{far}$ at a point $\mathbf{r}$ in the far field $(\mathbf{r} \to \infty)$ is found as [19,21]

$$\mathbf{E}_{far} = -2i\pi k_z \hat{\mathbf{E}}(k_x, k_y; 0)\frac{e^{ikr}}{r} \tag{2.23}$$

with $k_z$ the wave vector along the $z$-propagation axis and $r$ the distance from the origin. We see from Equation 2.23 that the far field is defined by the spatial frequency spectrum, that is, the Fourier transform of the electric field distribution $\mathbf{E}(x, y)$ in the $z = 0$ reference plane. This is an important result, which states that the field and its far-field conjugate are related to one another through a Fourier transformation. We can now recast the expression of the electric field, as given in Equation 2.21, in terms of the far-field solution:

$$\mathbf{E}(x,y,z) = \frac{ire^{ikr}}{2\pi} \int \int_{k_x^2+k_y^2 \le k^2} \mathbf{E}_{far} e^{i(k_x x + k_y y \pm k_z z)} \frac{1}{k_z} dk_x dk_y \quad (2.24)$$

This final expression, known as the Debye–Wolf integral, describes the electric field at point $(x, y, z)$ in terms of a Fourier transform of the far-field electric field distribution. As we will see in Section 2.4, this latter result is very useful for describing the electric field in the focus of a high numerical aperture lens.

## 2.3 Coherent Raman Scattering in the Plane Wave Approximation

The CRS signal follows from the nonlinear wave equation given in Equation 2.18. The wave equation predicts the amplitude and phase of the nonlinear field $\mathbf{E}$, which has the nonlinear polarization $\mathbf{P}^{(3)}$ as a source term. The nonlinear polarization, as defined in Equation 2.9, is driven by three incoming fields. Because in CRS microscopy the incoming fields are focused, the spatial dependence of the nonlinear polarization implies that we need to seek three-dimensional solutions of the nonlinear wave equation, which will be derived in Section 2.5. However, much of the essential physics of the problem can be captured in a grossly simplified picture. In this section, we will discuss the plane wave solution of the wave equation to highlight some of the basic concepts in CRS signal generation.

We will assume that the incident fields are plane waves of constant amplitude and phase in the transverse $(x, y)$ plane that collinearly co-propagate in the forward direction along the $z$ axis, as sketched in Figure 2.2. For simplicity, we will also assume that the fields are linearly polarized in the $\hat{\mathbf{x}}$ direction. In four-wave mixing experiments, the third-order polarization is induced by three incoming fields $\mathbf{E}_l$ with $l = (1, 2, 3)$, which in turn form the source for the signal field $\mathbf{E}_4$. From Equations 2.19 and 2.24 we can write for the incoming fields

$$\mathbf{E}_j(z,t) = \frac{1}{2}\left(A_j(z)e^{i(k_j z - \omega_j t)} + A_j^*(z)e^{-i(k_j z - \omega_j t)}\right)\hat{\mathbf{x}} \quad j = 1,2,3 \quad (2.25)$$

where $\omega_j$, $k_j = \omega_j n_j / c$, and $A_j$ are the angular frequency, scalar wave vector, and amplitude of the incident wave, respectively. We will assume that the nonlinear medium

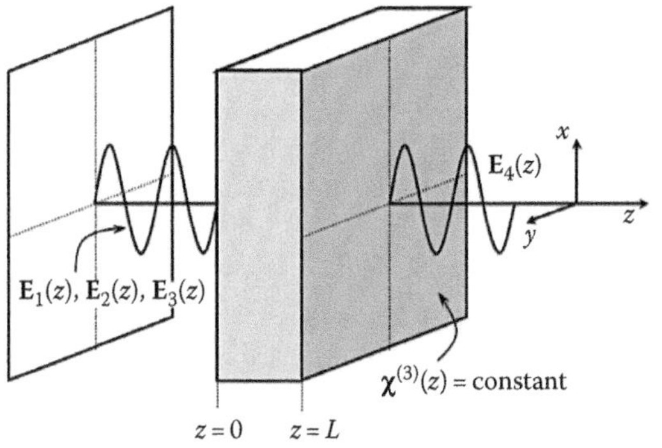


**FIGURE 2.2** Plane wave approximation for signal generation in a homogeneous $\chi^{(3)}$ medium. The three incoming fields $\mathbf{E}_1$, $\mathbf{E}_2$, and $\mathbf{E}_3$ have constant wavefronts and collinearly co-propagate in the direction of the positive $z$ axis. The fields are incident on a nonlinear material of thickness $L$ and constant $\chi^{(3)}$, which forms the source of the four-wave mixing field $\mathbf{E}_4$.

has a thickness $L$ with a homogeneous $\chi^{(3)}$ along the $z$ dimension but is otherwise invariant in the transverse plane. The nonlinear susceptibility $\chi_e^{(3)}(\Omega)$ is a function of the difference frequency $\Omega = \omega_1 - \omega_2$ and contains a two-photon accessible Raman resonance at $\Omega$:

$$\chi_e^{(3)}(\Omega) = \frac{G_v}{\omega_v - \Omega - i\gamma_v} \tag{2.26}$$

where

$G_v$ denotes the spectral amplitude

$\gamma_v$ is related to the dephasing time of the vibrational level at the resonance frequency $\omega_v$

We can now write a general expression of the induced third-order polarization in the nonlinear Raman-active material as

$$\mathbf{P}^{(3)}(z,t) = \frac{1}{8}\varepsilon_0 \chi_e^{(3)}(\Omega) A_1(z) A_2^*(z) A_3(z) e^{i[(k_1-k_2+k_3)z-\omega_4 t]}\hat{\mathbf{x}} + \text{c.c.} \tag{2.27}$$

From Equation 2.27 we see that the induced polarization acts as a source of coherent radiation that oscillates at the angular frequency $\omega_4 = \omega_1 - \omega_2 + \omega_3$ and emerges from the medium's boundaries in the form of forward- and backward-propagating waves with a wave vector amplitude $k_4$ along the $\hat{z}$ and $-z$ directions, respectively [22]. The spatial phase of the polarization then evolves as $\Phi_{pol}(z) = k_4 z$ in the forward direction and as $-k_4 z$ in the backward direction. Based on these characteristics for the nonlinear polarization, we may adopt tentative solutions to the wave equation that resemble forward- and backward-propagating plane waves with frequency $\omega_4$. The $x$-component of the forward-propagating plane wave with wave vector $k_4$ is now given by

$$\mathbf{E}_4(z,t) = \frac{1}{2} A_4(z) e^{i(k_4 z - \omega_4 t)} + \text{c.c.} \tag{2.28}$$

We can now insert this tentative solution into the wave equation. For the left hand side of the wave equation in Equation 2.18 we can write

$$\nabla^2 \mathbf{E}_4(z,t) - \frac{n_4^2}{c^2}\frac{\partial^2}{\partial t^2}\mathrm{E}_4(z,t) = \frac{1}{2}\left[\left(\frac{\partial^2}{\partial z^2} + 2ik_4\frac{\partial}{\partial z}\right)A_4(z)\right]e^{i(k_4 z - \omega_4 t)} + \text{c.c.}$$

$$\approx ik_4 \frac{\partial A_4(z)}{\partial z} e^{i(k_4 z - \omega_4 t)} + \text{c.c.} \tag{2.29}$$

The approximation made earlier is known as the slowly varying envelope approximation (SVEA), which assumes that the change of the amplitude $A_4$ of the wave is incrementally small over a distance on the order of an optical wavelength. This assumption is quite a stretch for the situation relevant to optical microscopy, where the nonlinear field is generated within a volume that is on the order of an optical wavelength. Despite the evident ambiguity of the SVEA, it has been shown that even in the case of tightly focused interaction volumes the approximation made in Equation 2.29 still renders solutions that deviate only minimally from a full solution of the wave equation [11].

With these approximations in mind, and using the expression for the nonlinear polarization in Equation 2.27, the wave equation for the field with amplitude $A_4$ can be recast as

$$\frac{\partial A_4(z)}{\partial z} = i\frac{\omega_4}{8n_4 c}\chi_e^{(3)}(\Omega)A_1(z)A_2^*(z)A_3(z)e^{i\Delta kz} \tag{2.30}$$

where $\Delta k = k_1 - k_2 + k_3 - k_4$ is the wave vector mismatch along the $z$ axis between the effective wave vector of the polarization field ($k_1 - k_2 + k_3$) and the wave vector of the radiating field ($k_4$). A solution to the resulting differential equation can be found if we assume that the amplitudes of the incident fields do not change significantly during the four-wave mixing process in the material. This is a very reasonable assumption in the case of optical microscopy, where the amount of energy exchanged among the participating fields is negligible relative to the total amount of energy carried by the incident fields. Under these conditions, the amplitudes $A_j$ ($j = 1, 2, 3$) can be considered as constants during the signal generation process. Integrating the differential equation in Equation 2.30 from $z = 0$ to $z = L$ yields [2]

$$A_4(L) = i\frac{\omega_4}{8n_4 c}\chi_e^{(3)}(\Omega)A_1 A_2^* A_3 L\,\text{sinc}\left(\frac{\Delta kL}{2}\right)e^{i\Delta kL/2} \tag{2.31}$$

where $\text{sinc}(x) = \sin(x)/x$. Inspection of Equation 2.31 reveals the significance of the extent of the wave vector mismatch for the efficient build-up of the coherent radiation field amplitude. It is maximized if the phase mismatch between the induced nonlinear polarization and the radiating field, $\Delta\Phi = \Delta kL$, is zero, and sinc (0) = 1. This condition implies that the nonlinear field can build up efficiently only if the phase of the nonlinear polarization throughout the medium remains in step with the phase of the propagating signal field. Because of dispersion in the medium, the wave vector mismatch is generally not zero. For nonzero $\Delta k$, the radiating field will run out of phase with the induced polarization. Effective coherence is only maintained over a length scale of $L_c = 2\pi/|\Delta k|$. Thus, the sinc function in Equation 2.31 acts as a filter function and defines an effective thickness of the $\chi^{(3)}$ medium along the $z$ axis, in which efficient nonlinear signal amplitude builds up.

In solving the wave equation, we had assumed a solution in the form of a plane wave that propagates in the direction of the positive $z$ axis. However, the backward-propagating wave with spatial phase $\Phi(z) = -k_4 z$ is also a solution of the wave equation. Under what conditions may we expect a backward-propagating wave? In addressing this question within our simple plane wave approximation, the wave vector mismatch plays an important role. While phase matching in the CRS microscope is automatically fulfilled in the forward direction ($\Delta k \simeq 0$), the wave vector mismatch for the backward-propagating wave amounts to $\Delta k = 2k_4$, and can thus be significant. Consequently, the effective coherence length is much smaller than the wavelength $\lambda_4$ of the radiating field, that is, $L_c = \lambda_4/2n_4$, and no appreciable amplitude accumulates in the backward-propagating direction in homogeneous bulk media. Physically, the radiating field is out of phase with the nonlinear polarization over the length scale of the interaction volume, and destructive interference prevents the build-up of the backward signal. However, for a very thin medium of thickness $L \leq L_c$, the destructive interference can be incomplete, and backward-propagating radiation may be observed. In CRS microscopy, where we typically deal with a heterogeneous sample composed of micron-sized objects embedded in bulk media, the distinct magnitudes of wave vector mismatches for the forward-propagating and backward-propagating waves will thus result in distinct contrast mechanisms for forward-collected and backward-collected (or epi-detected) CRS signals [13]. We will return to this issue of signal propagation direction under the conditions of CRS microscopy in Section 2.5.

### 2.3.1 CARS in the Plane Wave Approximation

In the following, we will first consider the plane wave solution for the CARS four-wave mixing process in the degenerate case of one pump beam ($\mathbf{E}_1 = \mathbf{E}_3$) at $\omega_1$ and one Stokes beam ($\mathbf{E}_2$) at $\omega_2$ ($\omega_2 < \omega_1$), collinearly co-propagating in the forward direction along the $z$ axis. In this dual-color CARS process, the pump beam participates twice in the four-wave mixing interaction. The radiating field ($\mathbf{E}_{as}$) oscillates at the anti-Stokes frequency $\omega_{as} = 2\omega_1 - \omega_2$ and has a scalar wave vector $k_{as} = 2k_1 - k_2$. Since the CARS signal is detected as the square amplitude of the field, by using Equation 2.31 we can write for its intensity

$$I(\omega_{as}) \propto |A_{as}(L)|^2 \propto \left|\chi_e^{(3)}(\Omega)\right|^2 I_1^2 I_2 L^2 \operatorname{sinc}^2\left(\frac{\Delta k L}{2}\right) \quad (2.32)$$

where the third-order nonlinear susceptibility is defined as $\chi_e^{(3)}(\Omega) = \chi^{(3)}(\omega_{as}; \omega_1, -\omega_2, \omega_1)$. Equation 2.32 summarizes several important features of the CARS signal. First, the CARS signal is proportional to the square amplitude of the nonlinear susceptibility. Because $\chi_e^{(3)}$ is proportional to the density of scattering molecules, the CARS signal depends quadratically on the number of molecules in the interaction volume. The quadratic dependence implies that the CARS signal is particularly strong for media that contain a high concentration of scattering molecules. Also, since the third-order susceptibility contains both vibrationally resonant and vibrationally nonresonant parts, $\chi_e^{(3)} = \chi_r^{(3)} + \chi_{nr}^{(3)}$, the square modulus operation produces interference terms between the resonant and nonresonant contributions. As detailed in Chapter 1, these interference terms give rise to the well known dispersive line shapes in CARS. Second, the CARS signal scales quadratically with the pump and linearly with the Stokes laser intensities. And finally, the signal depends on the propagation direction of the CARS field because of the distinct magnitudes of wave vector mismatch in the forward- and backward-propagating directions.

In the case of forward-detected (F-CARS) CARS signals, a small wave vector mismatch is caused by the dispersion of the medium. For example, for a pump wavelength of $\lambda_1 = 800\,\text{nm}$, a Stokes wavelength of $\lambda_2 = 1064\,\text{nm}$ and an anti-Stokes wavelength of $\lambda_{as} = 641\,\text{nm}$, $|\Delta k|$ in aqueous samples amounts to $0.00722\,\mu\text{m}^{-1}$, which corresponds to a coherence length of 0.87 mm. For microscopy applications, the nonlinear interaction length is generally on the order of a micrometer, which is much smaller than the coherence length. Consequently, while relevant for macroscopic CARS experiments, the dispersion induced phase mismatch introduces no significant effects in forward-detected CARS microscopy [11,13,14]. In the case of epi-detected (E-CARS) CARS signals, a significant wave vector mismatch of $|\Delta k| = 2|k_{as}|$ is introduced for the backward-propagating wave, which corresponds to an effective coherence length of $L_c = 0.24\,\mu\text{m}$ in water. Consequently, the phase-mismatch induced backward propagation is only relevant in E-CARS microscopy applications where a nanoscopic object is being studied whose axial dimension, which here determines the nonlinear interaction length, does not exceed the coherence length [13,14]. Like in the case of E-CARS, a large wave vector mismatch is also established in a geometry with a forward-propagating pump beam, a counter-propagating Stokes beam, and CARS detection in the forward direction with respect to the propagation direction of the incident pump beam. In this configuration, denoted counter-propagating CARS (C-CARS), the wave vector mismatch along the $z$ axis amounts to $|\Delta k| = 2|k_2|$ [14]. In the previous example of excitation wavelengths, this corresponds to an effective coherence length of $L_c = 0.40\,\mu\text{m}$ in water.

## 2.3.2 SRS in the Plane Wave Approximation

Within the plane wave approximation, we can find similar solutions to the wave equation for the different stimulated Raman scattering (SRS) processes in the degenerate case of one pump beam at $\omega_1$ and one Stokes beam at $\omega_2$ ($\omega_2 < \omega_1$), collinearly co-propagating in the forward direction along the $z$ axis. For instance, for the stimulated Raman gain (SRG) process, we may define the third-order nonlinear susceptibility as $\chi_e^{(3)}(-\omega_2;-\omega_1,\omega_2,\omega_1) = \{\chi_e^{(3)}(\Omega)\}^*$ where the $\omega_2$ field mode is detected [1,2]. Hence, the radiating field oscillates at a frequency similar to the frequency of the incident Stokes field. The wave vector mismatch for the forward-propagating field is $\Delta k = -k_1 + k_2 + k_1 - k_2 = 0$, which implies that phase mismatch due to material dispersion is completely absent in forward-detected SRG. Using Equation 2.31, the plane wave solution for the amplitude of the forward-propagating stimulated Raman wave can then be written as

$$A_2^{stim}(L) = i\frac{\omega_2}{8n_2c}\left\{\chi_e^{(3)}(\Omega)\right\}^* A_1^* A_2 A_1 L \tag{2.33}$$

Intrinsic to coherent radiation, the stimulated Raman wave exhibits a coherent relationship with the incident waves. Given that the stimulated field has an angular frequency similar to the frequency of the Stokes field, the two contributions will interfere at the detector. Hence, the forward-detected SRG intensity of the radiation at frequency $\omega_2$ is

$$I(\omega_2) \propto \left|A_2(L) + A_2^{stim}(L)\right|^2$$

$$= I_2 + \left|\chi_e^{(3)}\right|^2 I_1^2 I_2 L^2 + 2\,\mathrm{Im}\left\{\chi_e^{(3)}(\Omega)\right\} I_1 I_2 L \tag{2.34}$$

The first term on the right hand side is the intensity of the Stokes beam, which we assumed remains virtually unattenuated during the nonlinear process. The second term is the intrinsic SRS signal, which shows a similar dependence on the incident beam intensities, the $\chi^{(3)}$ of the material, and the propagation length as the CARS signal. Indeed, the intrinsic SRG signal is of similar magnitude as the CARS signal. The last term is the interference between the incident Stokes field and the stimulated Raman field at the detector. Note that the induced $A_2^{stim}$ field exhibits an intrinsic $\pi/2$ phase shift relative to the driving field $A_2$ (the induced field is $\pi/2$ ahead of the driving field), which is inherent to the plane wave approximation. In case $\chi_e^{(3)}$ is purely real, that is, far from any vibrational or electronic resonances, the interference term disappears. However, close to resonances, when $\mathrm{Im}\left\{\chi_e^{(3)}\right\} \neq 0$, the interference term is non-vanishing. Using modulation techniques, this interference term can be electronically isolated from the remaining terms in Equation 2.34. Under the reasonable assumption that the amplitude of the incident Stokes laser field exceeds the weak stimulated Raman field amplitude by orders of magnitude ($A_2 \gg A_2^{stim}$), the quantity physically detected in the forward SRG experiment is then the change in intensity of the Stokes beam:

$$\Delta I(\omega_2) \propto 2\,\mathrm{Im}\left\{\chi_e^{(3)}(\Omega)\right\} I_1 I_2 L \tag{2.35}$$

From Equation 2.35 we see that the plane wave model correctly predicts the linear scaling of the SRG interference term with the length $L$ of the interaction volume, as well as the linear dependence on the average intensities of the pump and Stokes beams ($I_1$ and $I_2$). Importantly, the SRG interference term is directly proportional to the imaginary part of $\chi_e^{(3)}$. At the vibrational resonance, $\Omega = \omega_v$, we see from Equation 2.26 that the imaginary part of $\chi_e^{(3)}$ is positive, which translates into a gain of the intensity in the $\omega_2$ (Stokes) channel. The plane wave solution thus describes the energy gain in the Stokes channel as a constructive interference between the radiated SRG field and the incident Stokes field. The analogous equation describing the forward-detected stimulated Raman loss (SRL) signal is obtained by exchanging the subscripts 1 and 2 and by inverting the sign of the imaginary part of the third-order susceptibility. For the quantity physically detected in the forward SRL experiment we then find

$$\Delta I(\omega_1) \propto -2\,\mathrm{Im}\left\{\chi_e^{(3)}(\Omega)\right\} I_1 I_2 L \tag{2.36}$$

Since $\chi_e^{(3)}$ is a positive quantity at resonance, the SRL signal represents a loss in intensity in the $\omega_1$ channel. The negative interference term indicates that the driving field and the induced field destructively interfere. These results are similar to the expressions for SRS found in Chapter 1. However, in Chapter 1, we imposed values for the phase shift between the driving field and the induced field. In the analysis that includes wave propagation, we see that the correct form of the SRS signals can be obtained without imposing this phase shift. Nonetheless, in the plane wave propagation model, the relevant phase shift is implicit. The plane wave model does not offer a physically intuitive picture for the origin of the phase differences between the fields. In Section 2.5.4, we will see that the radiating dipole model provides a more direct interpretation of wave interference in SRS.

In the derivation of Equations 2.35 and 2.36, we have restricted our discussion to the forward-detected SRS (F-SRS) signal. While the forward-propagating stimulated Raman field is always perfectly phase-matched with its driving field, a significant wave vector mismatch is introduced for the epi-detected SRS (E-SRS) signal. From the discussion of Equation 2.31 follows that the wave vector mismatch $|\Delta k|$ results in $2|k_2|$ and $2|k_1|$ for SRG and SRL detection in the backward direction, respectively. Using our previous example, for a pump wavelength of $\lambda_1 = 800\,\text{nm}$ and a Stokes wavelength of $\lambda_2 = 1064\,\text{nm}$, the effective coherence length $L_c$ in the epi-detected SRG and SRL experiment amounts to 0.40 and 0.30 μm, respectively, in water. Consequently, as in the case of E-CARS, the backward-propagation-induced phase mismatch in E-SRS is only relevant in microscopy applications where a nanoscopic object is being studied whose axial dimension does not exceed the coherence length.

Although the plane wave model qualitatively describes several of the most defining properties of the CARS and SRS signal generation and propagation processes, it naturally glosses over the intrinsic three-dimensionality of the tightly focused focal interaction volume. The plane wave model also considers the molecules in the sample as a part of a continuum, captured by a macroscopic polarization, and does not do justice to the actual radiation characteristics of individual point sources. In the following sections, we will evaluate the signal generation by recognizing the three-dimensional character of the focal spot as well as recognizing the radiation profiles of individual dipole sources.

## 2.4 Formation of the Focal Volume

### 2.4.1 Transformation by a Lens

The electric field $\mathbf{E}(x, y, z)$ found in Equation 2.24 is a vectorial field expressed in Cartesian coordinates. It has polarization components along each of the Cartesian axes, that is, $\mathbf{E}(x, y, z) = (E_x, E_y, E_z)$. Here, $E_x(x, y, z)$ and $E_y(x, y, z)$ are the field distributions that are polarized in the transverse plane and $E_z(x, y, z)$ is the field distribution, that is polarized in the longitudinal direction.

In CRS microscopy, the high numerical aperture lens used in the microscope transforms each incoming collinear field, which is polarized in the transverse plane, into a focused beam that has its geometric focus at a distance $f$ from the lens. If the incoming beam has a flat phase front then the lens converts it into a beam with a spherical phase front. The effective action of the lens is to rotate the field vector $\mathbf{E}_{inc}$ over an angle $\theta$ at the surface of a reference sphere of radius $f$, as shown in Figure 2.3. The maximum angle of the resulting light cone is called $\theta_{max}$, which is determined by the numerical aperture of

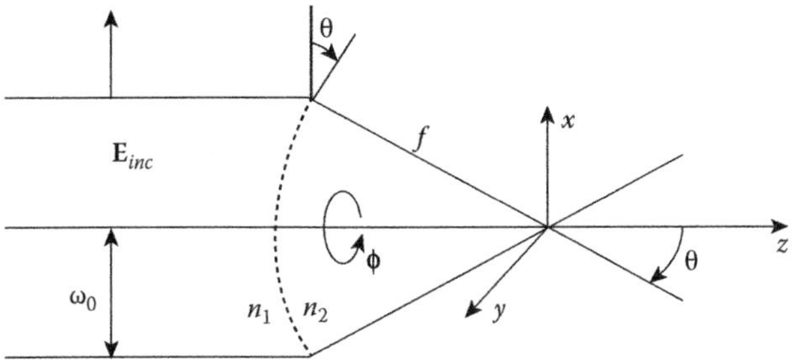


FIGURE 2.3 Definition of parameters for the transformation of the incoming field by a lens. Shown here is the refraction of the $x$-component of the $\mathbf{E}_{inc}$ field by the reference sphere of radius $f$. $\omega_0$ represents the waist of the incident Gaussian beam. $\theta$ and $\phi$ are the spherical coordinates.

the lens, $N.A. = n \sin \theta_{max}$, where $n$ is the index of refraction of the immersion medium. For instance, a water immersion objective lens with $N.A. = 1.2$ has a maximum angle of $\theta_{max} = 64.5°$, and an oil immersion lens with $N.A. = 1.4$ results in $\theta_{max} = 67.5°$.

To model the field in the vicinity of the focus of our lens, we will first write the angular spectrum representation of the electric field as given in Equation 2.24 as an integral over the spherical coordinates $\theta$ and $\phi$. The integral thus transforms the incoming far field $\mathbf{E}_{far}$ that is present to the right of the reference sphere of the lens into the field at point $(x, y, z)$ near the focus, $\mathbf{E}_f(x, y, z)$:

$$\mathbf{E}_f(x,y,z) = \frac{ikfe^{-ikf}}{2\pi} \int_{\phi=0}^{2\pi} \int_{\theta=0}^{\theta_{max}} \mathbf{E}_{far}(\theta,\phi) e^{ik(x\sin\theta\cos\phi + y\sin\theta\sin\phi + z\cos\theta)} \sin\theta \, d\theta \, d\phi \tag{2.37}$$

To find $\mathbf{E}_{far}(\theta, \phi)$, the refraction of the incoming field $\mathbf{E}_{inc}(\mathbf{r})$ at the reference sphere of the lens needs to be taken into account. The refractive action of the lens on each of the vectorial components of $\mathbf{E}_{inc}$ can be modeled by using transformation matrices [23,24]:

$$\mathbf{E}_{far}(\theta,\phi) = \left(\frac{n_1}{n_2}\right)^{1/2} \sqrt{\cos\theta} \mathbf{R}^{-1}\mathbf{L}\mathbf{R} \left|\mathbf{E}_{inc}(\mathbf{r})\right| \tag{2.38}$$

where the transformation matrices are defined as

$$\mathbf{R}(\phi) = \begin{pmatrix} \cos\phi & \sin\phi & 0 \\ -\sin\phi & \cos\phi & 0 \\ 0 & 0 & 1 \end{pmatrix} \tag{2.39}$$

$$\mathbf{L}(\theta) = \begin{pmatrix} \cos\theta & 0 & \sin\theta \\ 0 & 1 & 0 \\ -\sin\theta & 0 & \cos\theta \end{pmatrix} \tag{2.40}$$

To simplify our discussion, here we will consider only the $x$-polarized component of the electric field by setting the $y$ and $z$-polarized components to zero. This corresponds to an incoming electric field $\mathbf{E}_{inc}(\mathbf{r})$ that is linearly polarized along the $x$-axis. The refracted electric field at the reference sphere can then be written as

$$\mathbf{E}_{far}(\theta,\phi) = \left(\frac{n_1}{n_2}\right)^{1/2} \sqrt{\cos\theta} \begin{pmatrix} \cos^2\phi\cos\theta + \sin^2\phi \\ -(1-\cos\theta)\sin\phi\cos\phi \\ -\sin\theta\cos\phi \end{pmatrix} \left|\mathbf{E}_{inc}(\theta,\phi)\right| \quad (2.41)$$

where

$n_1$ is the refractive index of the propagation medium of $\mathbf{E}_{inc}$ (typically air)
$n_2$ is the refractive index of the immersion medium

The components of the vector in this expression correspond to the $E_x$, $E_y$, and $E_z$ polarization components of the electric field. Note that even though we have assumed that the incoming light is linearly polarized along the $x$-axis, we can see from Equation 2.41 that the refracted field contains field components polarized along the $y$ and $z$ directions as well. This is a direct consequence of the $\phi$-dependent $\theta$ rotation of the incoming rays carried out by the lens, which generates field projections along the $y$ and $z$ directions. It also proves convenient to express the coordinates in the vicinity of the focal volume in terms of polar coordinates. Using $x = \rho\cos\varphi$, $y = \rho\sin\varphi$, and $\rho = (x^2 + y^2)^{1/2}$, the focal field is then found as [19]

$$\mathbf{E}_f(\rho,\varphi,z) = \frac{ikfe^{-ikf}}{2\pi} \int_{\phi=0}^{2\pi} \int_{\theta=0}^{\theta_{max}} \mathbf{E}_{far}(\theta,\phi) e^{ikz\cos\theta} e^{ik\rho\sin\theta\cos(\phi-\varphi)} \sin\theta \, d\theta \, d\phi \quad (2.42)$$

From Equation 2.42, we see that the focal field is an effective Fourier transform of the refracted electric field at the reference sphere of our lens.

## 2.4.2 Tightly Focused Field

Using Equations 2.41 and 2.42, we can now write solutions for the focal field. The resulting integral formulation for the components of the electric field follows the derivation by Richards and Wolf [25]. Assuming that the incoming beam has a zeroth order Hermite–Gaussian amplitude and phase profile, the focal field can be written as [19,26]

$$\mathrm{E}_f(\rho,\varphi,z) = \mathrm{E}_0 e^{-ikf} \begin{pmatrix} I_{00} + I_{02}\cos 2\varphi \\ I_{02}\sin 2\varphi \\ -2iI_{01}\cos\phi \end{pmatrix} \quad (2.43)$$

In the previous equations, the quantities $I_{mn}$, ($m = 0$, 1 and $n = 0, \ldots, 4$) are the one-dimensional diffraction integrals with respect to the polar angle and are given by

$$I_{mn}(\rho,z) = \int_0^{\theta_{max}} f_\omega(\theta)\sqrt{\cos\theta}\, g_{mn}(\theta) J_i(k\rho\sin\theta) e^{ikz\cos\theta} \sin\theta \, d\theta \quad (2.44)$$

where

$l = n$ if $n \leq m$, and $l = n - m$ if $n > m$

$g_{0n} = 1 + \cos\theta, \sin\theta, 1 - \cos\theta$ for $n = 0, 1, 2$

$g_{1n} = \sin^2\theta, \sin\theta(1 + 3\cos\theta), \sin\theta(1 - \cos\theta), \sin^2\theta, \sin\theta(1 - \cos\theta)$ for $n = 0, 1, 2, 3, 4$, respectively

The functions $J_l$ are $l$th order Bessel functions. The cone angle of the high numerical aperture lens is indicated by $\theta_{max}$. We assumed that the incident field has a fundamental Gaussian profile with a beam waist of $\omega_0$ before the lens (c.p. Figure 2.3), which is accounted for by the apodization function $f_\omega(\theta)$ that is defined as [19]

$$f_\omega(\theta) = \exp\left[-f^2 \frac{\sin^2\theta}{\omega_0^2}\right] \tag{2.45}$$

A cross-sectional view ($xy$) of the focal field amplitude of a focused 650 nm laser beam is depicted in Figure 2.4A. The corresponding phase distribution is shown in Figure 2.4B.

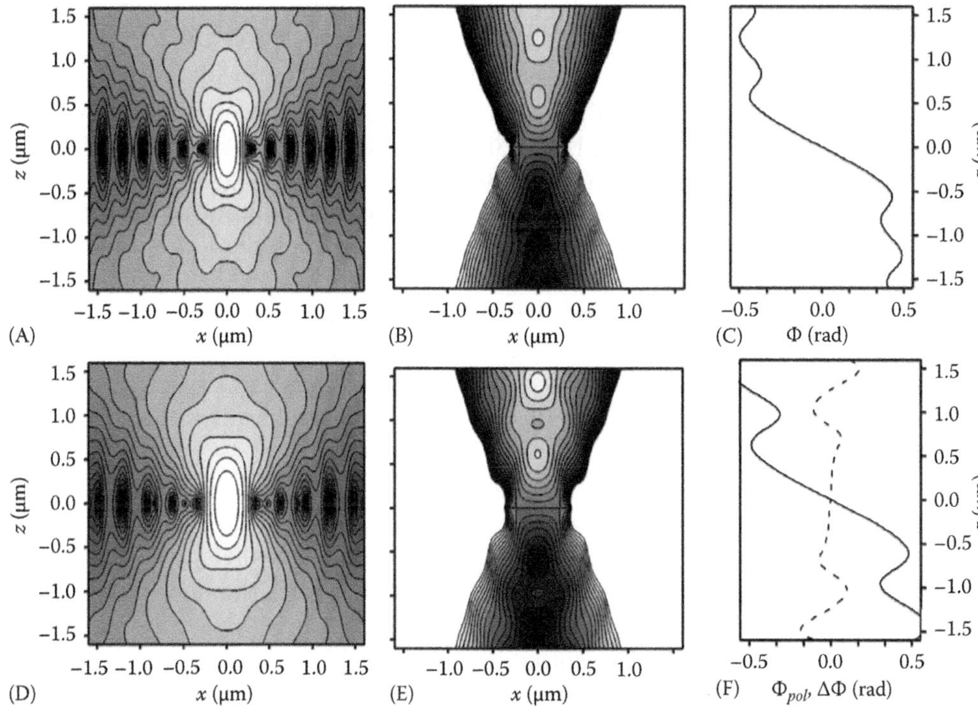


**FIGURE 2.4** Calculated amplitude and phase distributions of focal fields formed by a high-numerical-aperture lens (*N.A.* = 1.1, water immersion). (A) Focal field amplitude $E_f$ and (B) phase distribution $\Phi$ of a focused 650 nm laser beam. The propagation phase has been subtracted for clarity. Intensity is plotted on a logarithmic scale. (C) Phase profile along the optical axis in (B), showing the Gouy phase shift. Gray scale runs from $-\frac{1}{2}\pi$ (white) to $\frac{1}{2}\pi$ (black). (D) Amplitude of a focused CARS excitation field $\mathbf{E}_1^2\mathbf{E}_2^*$ with $\lambda_1$ = 807 nm and $\lambda_2$ = 1064 nm (anti-Stokes wavelength is 650 nm). (E) Phase distribution $\Phi_{pol}$ of the excitation field given in (D). (F) Phase profile (solid line) along the optical axis in (E). The dashed line indicates the phase difference $\Delta\Phi = \Phi_{pol} - \Phi$ between the induced polarization and the propagating field at the anti-Stokes wavelength. Note that $\Delta\Phi$ is very small in the region of relevant CARS excitation amplitude.

The phase profile along the optical axis is given in Figure 2.4C, where the linear propagation phase along the $z$-axis ($e^{ikz}$) has been subtracted for clarity. It is evident that the phase before the focal plane and the phase after the focal plane ($z = 0$) are not the same. Along the optical axis, the phase undergoes a full $\pi$-phase swing. This phase swing is known as the Gouy phase-shift [27,28].

The Gouy phase shift is a necessary consequence of the spatial confinement of the field in the lateral dimension in real space, while it ensures that all the **k** components of the wavefront remain in phase upon traversing the focal region [28]. The Gouy phase shift is a phase advance, and can be interpreted as the reduced path length of the diffracted field relative to the paths of the rays of the geometric focus [29]. This phenomenon is not unique to microscopic focal volumes, as each focused beam undergoes a Gouy phase shift, independent of the numerical aperture of the focusing lens. Unlike in incoherent optical microscopies, the Gouy phase shift plays an important role in coherent optical microscopies, because it can affect the amplitude and direction of the nonlinear signal generated in the focus.

In CRS microscopy, the molecules in focus respond to three incoming fields. As explained in more detail in the next section, the incident fields set up a nonlinear polarization in the material whose spatial profile is dictated by the product of the amplitudes of the excitation fields. For a typical dual-color CARS interaction, the effective driving field is $\mathbf{E}_1^2\mathbf{E}_2^*$, where the subscript 1 is used for the pump beam with frequency $\omega_1$ and 2 for the Stokes beam with frequency $\omega_2$. In Figure 2.4D, the CARS excitation field amplitude is plotted for a pump field of 807 nm and a Stokes field of 1064 nm, which corresponds to an anti-Stokes wavelength of 650 nm. For this particular excitation profile, the lateral full width half maximum of the excitation amplitude is 0.45 μm and the longitudinal width is 1.4 μm. The longitudinal extent of the excitation volume is on the order of an optical wavelength, which implies that phase mismatch between the incident waves and the emitted wave is minimal within the interaction volume. As we will see in the next section, phase mismatching effects due to material dispersion, which can be prominent in interaction volumes associated with low numerical aperture lenses, are not a major factor in the signal generation process when high numerical aperture lenses are used [11,13,14,30].

The phase distribution $\Phi_{pol}$ of the CARS excitation field in the focal volume is shown in Figure 2.4E, with the axial cross section shown in Figure 2.4F. Similar to the phase distribution $\Phi$ of the focused 650 nm laser beam, the spatial phase of the excitation field undergoes a $\pi$ phase shift. This implies that the nonlinear polarization within the focal volume exhibits a focal phase shift roughly similar to that of a forward-propagating beam. The dotted line in Figure 2.4F represents the difference in the phase profiles $\Delta\Phi = \Phi_{pol} - \Phi$ of the excitation field and the propagating field at 650 nm (from Figure 2.4C). It is clear that the phase difference is very small in the region of the interaction volume ($-0.5\,\mu m \le z \le 0.5\,\mu m$). Consequently, the CARS radiation is not particularly affected by a Gouy phase shift. A Gouy phase mismatch is, however, rather prominent for the other four-wave mixing processes, such as the third-harmonic generation, which has a profound effect on the spatial distribution of the emitted third-order signal [31,32].

## 2.5 Coherent Raman Scattering in Tightly Focused Volumes

### 2.5.1 Spatially Dependent Polarization

The focused pump and Stokes beams induce a spatially dependent polarization in the material. This induced third-order polarization can be written as

$$\mathbf{P}^{(3)}(\mathbf{r}) = \varepsilon_0 \chi^{(3)}(\mathbf{r})\mathbf{E}_1(\mathbf{r})\mathbf{E}_2(\mathbf{r})\mathbf{E}_3(\mathbf{r}) \quad (2.46)$$

For regular CRS microscopy, the spatial dependence of the focused driving fields is given by Equation 2.43. The nonlinear polarization of the material depends on both the spatial profile of the focal fields as well as the nonlinear susceptibility of the medium. The effective excitation field, as given in Figure 2.4D for CARS, confines the spatial distribution of the material polarization to the focal region. In addition, within the focal excitation volume, the magnitude of the polarization at any point $\mathbf{r}$ is determined by the nonlinear susceptibility of the material $\chi^{(3)}(\mathbf{r})$.

The driving fields are focused laser fields. The tight optical focus is a direct result of the high degree of spatial coherence in the laser fields. Hence, when these incident fields induce a nonlinear polarization in the focal volume, the resulting polarization exhibits spatial coherence as well. This implies that the polarization at a given point $\mathbf{r}_1$ in focus has a well-defined phase relation with the polarization at a different point $\mathbf{r}_2$. On a molecular level, we can think of this coherence as molecules that, although located at different positions $\mathbf{r}$ in the focal excitation volume, oscillate in unison.

This spatial coherence of the polarization has important implications for the imaging properties of the CRS microscope. Because the spatially coherent third-order polarization is the source of CRS radiation, the resulting radiation is coherent as well. As we will see in the next section, the detected CRS signal depends on the coherent addition of the waves emanating from each point $\mathbf{r}$ in the focal volume. The magnitude, direction and transverse profile of the propagating CRS field is defined by the mutual interference of all these waves emitted by the point oscillators in focus.

### 2.5.2 Far-Field Dipole Radiation

The third-order polarization serves as a source for the nonlinear radiation. The emitted nonlinear field is described by the nonlinear wave equation of Equation 2.18. The nonlinear field can thus be found by entering the nonlinear polarization of Equation 2.46 as a source in the nonlinear wave equation. In principle, the amplitude and phase of the nonlinear field at any point in space can be determined using the wave equation. However, to solve the resulting wave equation for each observation point in space proves highly impractical. Fortunately, not every point is space is relevant in our optical problem at hand. In CRS microscopy, the photodetector is commonly placed in the far field. Hence, we are primarily interested in finding far-field solutions of the nonlinear wave equation.

To describe the CRS signal in the far field, it is helpful to consider each point **r** in the excitation volume as an oscillating dipole source. As stated in the introductory paragraphs of this chapter, the macroscopic Maxwell equations are primarily concerned with polarization densities rather than with point sources. Indeed, classical field theory provides unsatisfactory solutions for the electric field at the location of a point dipole source, which yields a singularity. Nonetheless, the electric field of a dipole source can be found at locations in space away from the origin. In the far field, the dipole field is found to be a radiating field, which is a satisfactory solution to our problem. It is desirable to describe the focal volume as a collection of such radiating dipoles, as it provides an intuitive connection with an ensemble of molecular scatterers in focus. The problem is now reduced to finding the far-field solution of the radiation resulting from a dipole source located at point **r**. The electric field $E_d$ at a far-field location **R** resulting from a single radiating dipole situated at position **r** in the focal volume is written as [15]

$$E_d(\mathbf{R};\mathbf{r}) = \frac{e^{ik|\mathbf{R}-\mathbf{r}|}}{4\pi|\mathbf{R}-\mathbf{r}|^3}[(\mathbf{R}-\mathbf{r})\times\mathbf{P}^{(3)}(\mathbf{r})]\times(\mathbf{R}-\mathbf{r}) \tag{2.47}$$

where $k$ is the wave vector amplitude of the emitted field. Here we have considered only the radiation that results from the third-order polarization of the dipole oscillator. Note that, in the far field, the electric field of a single dipole emitter resembles a spherical wave that propagates with a phase factor $e^{ikr}$. The total field detected at point **R** is, however, dependent on the fields emitted by all the dipole oscillators in focus. To find **E(R)** at the detector, we have to coherently add the dipole fields emanating from all points **r**. Expressed in integral form, we can write

$$E(\mathbf{R}) = \int_V E_d(\mathbf{R};\mathbf{r})d^3\mathbf{r} \tag{2.48}$$

where the integration is performed over the volume $V$ that encloses the region of the focal excitation volume.

Figure 2.5 represents a basic schematic of the problem. The incident fields induce a nonlinear polarization in the vicinity of the focal volume. A dipole oscillator located

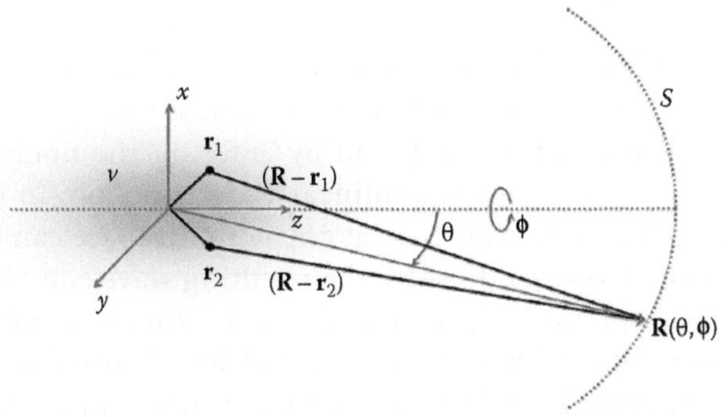


**FIGURE 2.5** Radiation in the far field. The radiation emanating from two dipoles sources, located at positions $\mathbf{r}_1$ and $\mathbf{r}_2$ within the focal volume $V$, is measured at position $\mathbf{R}(\theta, \phi)$. The point **R** lies on the far-field hemispherical surface $S$.

at a given point $\mathbf{r}_1$ contributes to the observed field in point $\mathbf{R}$ in the far field, located on the hemispherical surface $S$. The total detected field is a coherent sum of the dipole fields originating from $\mathbf{r}_1$, $\mathbf{r}_2$, etc. It is clear that the total field is highly dependent on the amplitudes and relative phases of the dipole fields, which are, in turn, dictated by the spatial and material characteristics of the sample. In the next section, we will see that the strength and spatial properties of the CRS radiation are strongly influenced by the nature of the object in focus.

The amplitude and phase of the nonlinear field at the far-field location $\mathbf{R}$ is given by Equation 2.48. When the signal is detected in a homodyne manner, as is the case in regular CARS microscopy, the signal intensity measured within the surface area $dS$ is proportional to $|\mathbf{E}(\mathbf{R})|^2 dS$. The total signal registered by the detector is then written as [13,14]

$$\mathbf{I} \propto \int_{\theta_1}^{\theta_2}\int_{\phi_1}^{\phi_2} |\mathbf{E}(\mathbf{R})|^2 R^2 \sin\theta \, d\theta \, d\phi \tag{2.49}$$

where $R = |\mathbf{R}|$.

In the case of a large aperture photodetector, the integration is performed over the range $\phi = \{0, 2\pi\}$ and $\theta = \{0, \alpha_{max}\}$, where $\alpha_{max}$ is determined by the maximum cone angle of the collimating lens. When the detector is placed in the epi-direction, the integration over $\theta$ is within the range $\{\pi - \theta_{max}, \pi\}$. Alternatively, if the signal is detected in a heterodyne manner, as is the case for SRS or heterodyne CARS microscopy, the signal intensity measured at surface element $dS$ is given by $|\mathbf{E}(\mathbf{R}) + \mathbf{E}_{LO}(\mathbf{R})|^2 dS$, where $\mathbf{E}_{LO}$ is the local oscillator field.

### 2.5.3 CARS Far-Field Radiation

We will focus first on the homodyne CARS signal, and study its dependence on the size and shape of the object in focus [13,14]. In Figure 2.6A the angular resolved CARS radiation profile of a $d$ = 50 nm nanocube is shown. The coherent emission pattern resembles the radiation profile of a Hertzian dipole, and the amount of signal emitted in the forward and backward direction is roughly similar. Backward-propagating radiation is

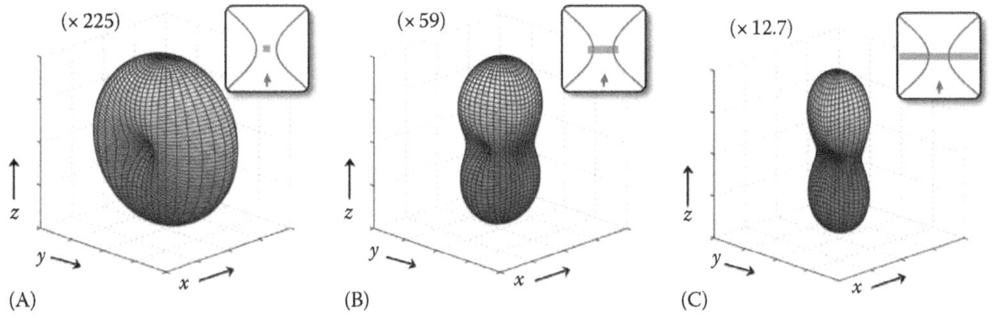


**FIGURE 2.6** Far-field radiation patterns of the CARS amplitude for different objects placed at the center of the focus: (A) 50 × 50 × 50 nm$^3$ nanocube; (B) nanowire with 50 × 50 nm$^2$ ($x$, $z$) cross section and infinite length along the $y$-axis; and (C) 50 nm thick nanolayer. Incident fields $\mathbf{E}_1$ and $\mathbf{E}_2$ are linearly polarized along $x$. A 1.1 $N.A.$ water immersion lens was used in the calculations.

expected because the object size $d$ is smaller than the coherence length for epi-directed light ($d \ll L_c$). In Figure 2.6B, the radiation profile of a nanowire with a 50 nm × 50 nm cross section is depicted. The wire extends along one dimension ($y$-axis) in the focal plane. Relative to the nanoparticle, radiating dipoles over a larger segment of the focal volume contribute to the overall radiation pattern. The mutual far-field interference of the waves emitted by the dipoles increases the directionality of the signal and leads to a tightening of the radiation pattern in the $y$-direction. Further tightening of the radiation profile is observed if a 50 nm thick two-dimensional nanolayer is considered, as illustrated in Figure 2.6C. Since all the structures considered here have an axial dimension much smaller than the coherence length, radiation is observed in both the forward and backward directions.

The forward-to-backward ratio increases significantly when the size of the object is expanded in the axial direction. In Figure 2.7A, the CARS radiation pattern of a semi-infinite material is displayed. The radiation of dipoles aligned in the axial direction constructively interferes, producing strong forward-directed CARS radiation. In the backward direction, the radiation is much weaker because the axial extent $d$ of the material that overlaps with the excitation volume is slightly larger than the coherence length $L_c$ for epi-directed radiation. In other words, backward-propagating radiation emanating from dipoles at different axial positions is interfering destructively in the far field, yielding weak epi-detected signals. When more dipole radiators are added in the axial dimension, as in Figure 2.7B where the material occupies the entire focal volume, the destructive interference between the backward-propagating waves is such that the epi-signals are almost completely suppressed. In contrast, in the forward direction, the dipole radiation from all locations within the focal volume arrives in-phase in the far field, producing a strong and highly directional coherent signal. Note that the strong directionality of the signal facilitates the collection of the signal in the forward direction.

The earlier discussion emphasizes that the strength of the epi-detected signal strongly depends on the axial size of the object in focus [13]. In Figure 2.8 the relative forward- and backward-detected CARS signals are plotted for a spherical object with increasing diameter $d/\lambda_1$. The forward-propagating signal is fully phase-matched and grows up to

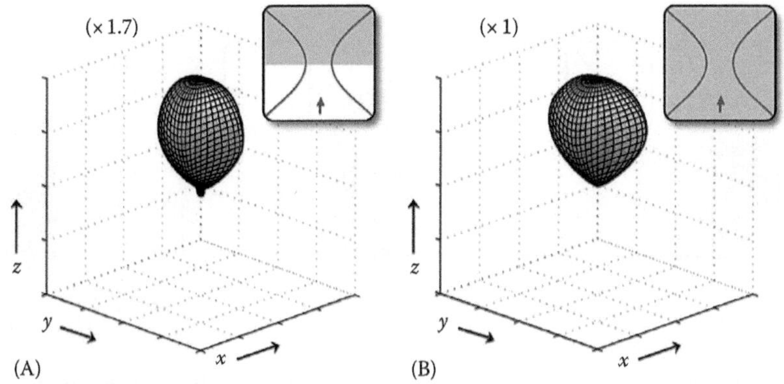


**FIGURE 2.7** Far-field radiation patterns of the CARS amplitude for bulk objects in focus: (A) semi-infinite material that occupies one half of the focal volume and (B) bulk material that occupies the entire focal volume. Simulation parameters are the same as in Figure 2.6.

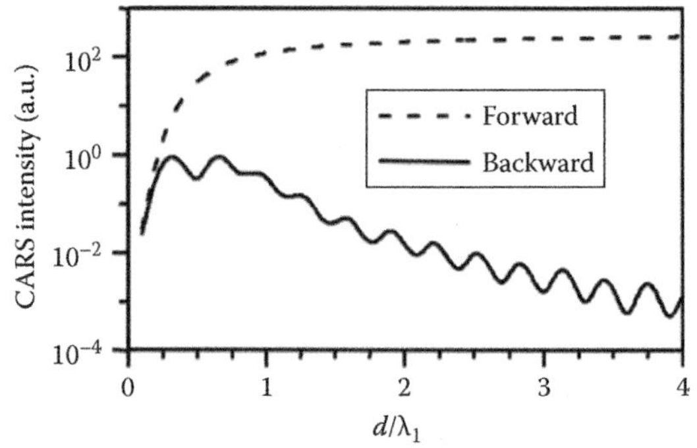


FIGURE 2.8 Relative CARS intensity of a sphere of variable diameter $d/\lambda_1$ detected in the forward and backward direction. A 1.4 *N.A.* oil immersion lens was used in these simulations. (Modified from Volkmer, A. et al., *Phys. Rev. Lett.*, 87, 023901, 2001.)

the point where the size of the sphere and the axial extent of the focal volume become comparable. The epi-detected signal initially grows with the number of dipole radiators but levels off due to phase mismatching. A maximum is seen, followed by a minimum. This is a direct result of the phase mismatch between the induced polarization and the backward-propagating wave. Because of the phase mismatch, the signal oscillates with a periodicity of $\sim\lambda_{as}/2n$, where $n$ is the refractive index of the material. Likewise, a large phase mismatch is established in a geometry with collinearly counter-propagating pump and Stokes beams, denoted counter-propagating CARS (C-CARS), with a signal dependence on the interaction length similar to that of E-CARS [14].

From Figure 2.8 we see that for smaller objects the destructive interference in the epi-direction is incomplete and a weak but detectable backward-propagating signal is observed. The results in the figure were simulated by considering an object with a $\chi_{obj}^{(3)}$ that is higher than the $\chi_{surr}^{(3)}$ of its surrounding. However, incomplete interference also occurs when a similar sized object is embedded in a matrix with a higher $\chi_{surr}^{(3)}$ relative to the object [14]. Incomplete destructive interference occurs whenever there is a difference in the nonlinear susceptibility, that is, $\left|\chi_{obj}^{(3)} - \chi_{surr}^{(3)}\right|$, that occurs on a length scale of $\sim\lambda/2n$. An example of such a $\chi^{(3)}$ discontinuity is a simple interface between two bulk materials of different $\chi^{(3)}$, as shown in Figure 2.7A. More generally, $\chi^{(3)}$ discontinuities are commonly encountered in biological imaging, where many structures with variable $\chi_{obj}^{(3)}$ manifest themselves on the sub-wavelength length scale. However, this incomplete interference model is not the only mechanism by which radiation in the epi-direction can be observed. In Section 2.6, we will see that the effect of linear scattering can also profoundly affect the light registered in the backward-propagating direction.

### 2.5.4 SRS Far-Field Radiation

From the plane wave analysis in Section 2.3, we have seen that the total SRS signal can be interpreted as a sum of (1) a weak intrinsic signal of similar magnitude and properties as the CARS signal and (2) a signal-dominating contribution from the constructive (SRG) or destructive (SRL) interference of the induced nonlinear field **E** with the driving field. Although the former signal contribution is generally negligible, it is instructive to inspect its far-field radiation patterns for different objects in the focus. As a homodyne,

CARS-like process, the far-field radiation pattern for an object smaller than the coherence length for the intrinsic, epi-directed SRS field resembles the radiation pattern of a Herztian dipole (cf. Figure 2.6A), whereas for a bulk material occupying the entire focal volume the radiation pattern is characterized by a suppressed epi-signal and a strong, highly directional forward signal (cf. Figure 2.7B) [33]. This discussion emphasizes that the strength of the intrinsic SRS radiation in the epi- and forward direction depends on the axial size of the object in focus.

However, it is important to realize that the quantity physically detected in an SRS experiment is strongly dominated by the signal contribution resulting from the interference between **E** and the driving field, which acts as a local oscillator $\mathbf{E}_{LO}$. This signal measured at a point **R** in the far field, can be written as

$$\mathbf{I}_{SRS}(\mathbf{R}) \propto \mathbf{E}(\mathbf{R})\cdot\mathbf{E}_{LO}^{*}(\mathbf{R})e^{i\phi(\mathbf{R})} + \mathbf{E}^{*}(\mathbf{R})\cdot\mathbf{E}_{LO}(\mathbf{R})e^{-i\phi(\mathbf{R})} \tag{2.50}$$

where $\phi(\mathbf{R})$ is the total phase shift between the induced field and the driving field at **R**. In the plane wave approximation, we found that in SRG the induced field is in-phase with the driving field, implying that $\phi = 0$. On the other hand, in SRL we found that the induced field is out-of-phase with the driving field, indicating that $\phi = \pm\pi$. What is the physical origin of $\phi$? The dipole radiation picture provides an intuitive explanation for the existence of this phase shift.

The total phase shift is determined by the spectral response of the material as well as by geometric factors. We can thus write $\phi = \phi(\omega) + \phi_{gm}$, where $\phi(\omega)$ is the spectral phase shift and $\phi_{gm}$ is a geometric phase shift. We first consider the spectral phase shift for the case of SRL. For the SRL process, the $x$-component of the induced third-order polarization $p^{(3)}(t)$ of a single dipole oscillator can be written as

$$p^{(3)}(t) = p^{(3)}(\omega_1)e^{-i\omega_1 t} + \text{c.c.} \tag{2.51}$$

The frequency dependent polarization $p^{(3)}(\omega_1)$ is proportional to $\chi_e^{(3)}(\Omega)$, which is purely imaginary on resonance, that is, $p^{(3)} \propto i\,\text{Im}\{\chi_e^{(3)}\}$. Because $\text{Im}\{\chi_e^{(3)}\}$ is a positive quantity, the phase angle corresponds to $+\pi/2$. The polarization can thus be written as follows:

$$p^{(3)}(t) \propto \text{Im}\{\chi_e^{(3)}(\Omega)\}e^{i\pi/2}e^{-i\omega_1 t} + \text{c.c.} = 2\,\text{Im}\{\chi_e^{(3)}(\Omega)\}\cos(\omega_1 t - \pi/2) \tag{2.52}$$

Given that the driving field $\text{E}_1 \propto A_1 \cos\omega_1 t$ (see Equation 2.25), we notice from Equation 2.52 that the dipole oscillation is $-\pi/2$ shifted with respect to $\text{E}_1$. This is analogous to a classical harmonic oscillator, which exhibits a $\pi/2$ phase lag relative to the driving field when driven at resonance.

We next consider the geometric phase. The geometric phase depends on the spatial configuration of the sample and the position of the detector. We will first consider the case of plane wave excitation of a medium that is invariant in the transverse dimension.

We are interested in determining the total field radiated by the ensemble of dipoles in an observation point **R** in the far field. The far-field emission of each dipole is given by Equation 2.47. To simplify the discussion, here we consider only the $x$-component of the field with the dipole axis aligned in the transverse ($xy$) plane. We can then write the far-field contribution from a single dipole located at the origin as

$$E_d(\mathbf{R};0) \propto \frac{\omega^2}{4\pi\varepsilon_0 c^2} p^{(3)} \frac{e^{ik|\mathbf{R}|}}{|\mathbf{R}|} \tag{2.53}$$

In Figure 2.9A, a plane wave excitation field and the radiation from a single dipole is sketched. The plane wave excitation of a homogeneous $\chi_e^{(3)}$ medium at a given plane $z$ is modeled by a collection of dipole radiators, aligned in an infinite sheet $S$ in the $xy$ plane. In the plane wave excitation limit, all dipoles in the sheet are driven with the same phase. The total field in **R** is obtained by summing over the contribution of each dipole:

$$E(\mathbf{R}) = \int_S E_d(\mathbf{R};x,y)dxdy \tag{2.54}$$

The field in **R** is found by solving the integral in Equation 2.54, which yields [34]

$$E(\mathbf{R}) \propto -ip^{(3)}e^{ik|\mathbf{R}|} \tag{2.55}$$

Comparing Equations 2.53 and 2.55, we see that there is a factor of $-i$ (= $-\pi/2$ phase shift) difference in the far-field between the field of an individual oscillating dipole and

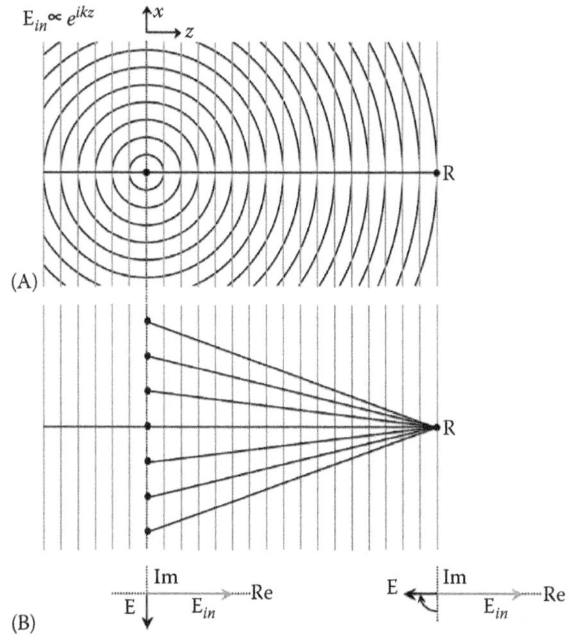


**FIGURE 2.9** Far-field radiation of dipole oscillators. (A) A single dipole excited by a plane wave. In this sketch, the induced polarization is assumed real. (B) A sheet of oscillating dipoles mimics the situation of plane wave excitation of a homogeneous medium. The total dipole field results from the summation over all dipole contributions in **R**. The dipole axis is aligned along the $y$-axis in the transverse plane. The insets in (B) show the orientation of the incoming field and the induced field in the complex plane at the origin and at point **R** for the case of SRL.

the field of a collection of dipoles in a planar sheet. This geometric phase shift is the result of the mutual far-field interference of radiating dipoles distributed on the planar surface S. This value of $\phi_{gm}$ is related to the phase shift seen in Equation 2.23, and the phase shift found in the Huygens–Fresnel expression for the propagating electric field. Richard Feynman, who discussed the same problem in his Lectures on Physics, mentioned that this extra phase lag "is a little strange, ... but that is the way it comes out ..." [34]. In our problem, at point **R**, the geometric phase shift puts the total dipole field $-\pi/2$ behind the plane wave excitation field. The combination of the spectral and the geometric phase shift thus yields $\phi = -\pi$, which implies that the induced field measured at **R** is out-of-phase with the driving field $E_1$.

Next, we consider the case of a single dipole positioned in the focal plane of a focused excitation field. This situation is sketched in Figure 2.10. Because there is only one oscillating dipole, the dipole radiation field E(**R**) is given by Equation 2.53. In this case, there is no extra $-\pi/2$ phase shift in the far-field. However, in the focused field geometry, the excitation field undergoes a $\pi/2$ phase *advance* when propagating from the focal plane to the far-field location **R** because of the Gouy phase shift. Hence, as before, an effective geometric phase shift dictates that the phase of the dipole field lags $\pi/2$ behind the excitation field at point **R**. Again, we find for the total phase $\phi = -\pi$ for SRL, and thus, destructive interference is observed at **R**. Table 2.1 summarizes the phase shifts contributions relevant to the classical wave description of SRS of both the plane wave geometry as well as the single dipole located in the geometric focal plane. Note, however, that in the case of a single dipole in a focused field geometry, the total phase shift $\phi$ depends

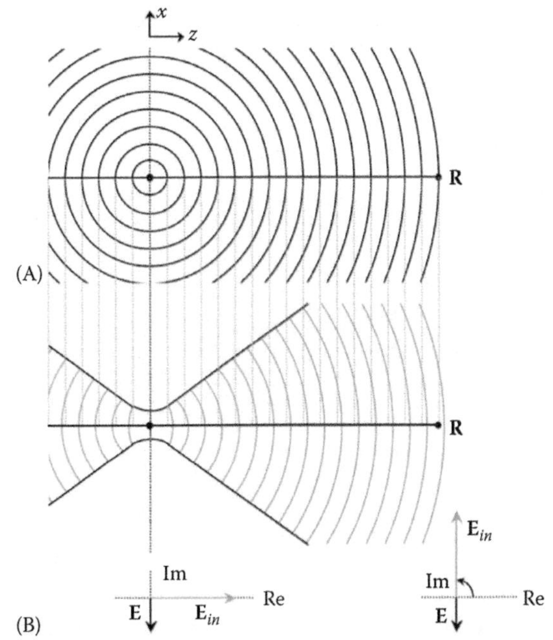


**FIGURE 2.10** A dipole excited by a focused excitation field. (A) Dipole radiation. (B) Focused excitation field. Note that in point **R**, the excitation field is $\pi/2$ ahead of the dipole field. The dipole axis is aligned along the *y*-axis in the transverse plane. The insets in (B) show the orientation of the incoming field and the induced field in the complex plane at the origin and at point **R** for the case of SRL.

Table 2.1 Phase Shifts of the Induced Field Relative to the Driving Field for a Single Dipole on Resonance, Positioned in the Focal Plane, as Relevant to SRS Experiments

| Method | Det. Mode | $\phi_{gm}$ | $\phi(\omega)$ | Total Phase | Signal |
|---|---|---|---|---|---|
| SRL | $\omega_1$ | $-\pi/2$ | $-\pi/2$ | $-\pi$ | $\propto -2\,\mathrm{Im}\{\chi^{(3)}\}$ |
| SRG | $\omega_2$ | $-\pi/2$ | $\pi/2$ | 0 | $\propto +2\,\mathrm{Im}\{\chi^{(3)}\}$ |

on the actual location of the dipole oscillator in focus. This also implies that the SRS signature of a single dipole radiator is dependent on its position along the axial dimension of the focal volume. As discussed in Section 2.6.3, in case the (single) dipole is positioned above or below the focal plane, the SRS response is correspondingly affected.

## 2.6 Imaging of Objects

### 2.6.1 Imaging with CRS Signals

In laser scanning optical microscopy, an image is formed by moving the laser spot over the sample. In CRS, the contrast is provided by the spatial variation of the nonlinear susceptibility $\chi_e^{(3)}(\mathbf{r})$. What is the response of the microscope as the laser spot is scanned across the sample? To model this, we will consider a microscopic sample object positioned at location $\mathbf{r}'$ in the vicinity of the focal volume. We assume that the electric field is detected in the far field and that $|\mathbf{R}| \gg |\mathbf{r}|$, which allows us to write $|\mathbf{R} - \mathbf{r}| \approx |\mathbf{R}| - \hat{\mathbf{n}}_R \cdot \mathbf{r}$, where $\hat{\mathbf{n}}_R$ is the unit vector along $\mathbf{R}$. We will also consider only the $x$-polarized component of the radiated field. The electric field detected at the far-field location $\mathbf{R}$ can then be written as

$$E(\mathbf{R};\mathbf{r}') \propto \frac{e^{ik|\mathbf{R}|}}{|\mathbf{R}|} M(\theta,\phi) \int_V e^{-ik\hat{\mathbf{n}}_R\cdot\mathbf{r}} h_{ex}(\mathbf{r}) \chi_e^{(3)}(\mathbf{r}-\mathbf{r}') d^3\mathbf{r}$$

$$= \frac{e^{ik|\mathbf{R}|}}{|\mathbf{R}|} M(\theta,\phi) \left\{ e^{-ik\hat{\mathbf{n}}_R\cdot\mathbf{r}'} h_{ex}(\mathbf{r}') \otimes \chi_e^{(3)}(\mathbf{r}') \right\} \tag{2.56}$$

where

$\otimes$ denotes the convolution operation

$h_{ex}(\mathbf{r})$ is the effective focal excitation function

For CARS, $h_{ex}(\mathbf{r})$ can be written as $\mathbf{E}_1^2(\mathbf{r})\mathbf{E}_2^*(\mathbf{r})$. The function $M(\theta, \phi)$ projects the radiative field onto the far-field hemispherical surface $S$ [14]. From Equation 2.49, we can find the total CARS intensity for the situation where the object is positioned in point $\mathbf{r}'$ as

$$I(\omega_{as})(\mathbf{r}') \propto \int_S M^2(\theta,\phi) \left| e^{-ik_{as}\hat{\mathbf{n}}_R\cdot\mathbf{r}'} h_{ex}(\mathbf{r}') \otimes \chi_e^{(3)}(\mathbf{r}') \right|^2 \sin\theta\, d\theta\, d\phi \tag{2.57}$$

Equation 2.57 illustrates that the detected signal depends on a convolution of the object function $\chi_e^{(3)}(\mathbf{r})$ and the effective excitation function $h_{ex}$. When the object function is a point, that is, $\chi_e^{(3)}(\mathbf{r}) = \delta(\mathbf{r})$, the CARS intensity scales as $|h_{ex}|^2$. The function $h_{ex}$ can thus be defined as an amplitude point spread function. However, this point spread function does not play the same role as the point spread functions commonly used in fluorescence microscopy. Several differences between CARS and incoherent imaging methods can be identified. First, the CARS intensity is dependent on a position sensitive phase function ($e^{-ik\hat{\mathbf{n}}\mathbf{R}\cdot\mathbf{r}}$) that does not have a counterpart in incoherent imaging. When the signal is detected in the forward-propagating, phase-matched direction, this phase function does not significantly affect the total signal. Second, the CARS signal scales as $|\chi_e^{(3)}|^2$, which means it depends nonlinearly on the number of scatterers. Because of interferences with the nonresonant background, the dependence of the CARS signal on the number of scatterers can vary from linear to quadratic. The nonlinear relation between the object function and the detected intensity implies that linear deconvolution techniques cannot be used to improve the imaging contrast of the CARS microscope [11,14].

In SRS, the demodulated signal scales as $\text{Im}\{\chi_e^{(3)}\}$, and a linear relationship between the number of scatterers and the detected intensity is maintained. As in CARS, when detected in the forward-propagating, phase-matched direction, the demodulated SRS signal can be approximated as $I(\mathbf{r}') \propto H_{ex}(\mathbf{r}') \otimes \text{Im}\{\chi_e^{(3)}(\mathbf{r}')\}$ with $H_{ex} = I_1 I_2$ the intensity point spread function. Under such conditions, the imaging properties of the SRS microscope are compatible with the use of linear deconvolution techniques. Hence, although both CARS and SRS depend on the third-order susceptibility, the different dependence on the number of Raman scatterers dresses the techniques with very different imaging properties.

### 2.6.2 Linear Scattering Effects

The primary source of contrast in CRS microscopy is the spatial variation of $\chi_e^{(3)}$ in the sample. As we have seen, most of the $\chi_e^{(3)}$ induced radiation propagates in the forward direction, while smaller amounts of intrinsic backward radiation can be observed from $\chi_e^{(3)}$ variations on the $\sim\lambda/2n$ length scale. When the sample is thick relative to the axial dimensions of the focal volume, the forward-propagating radiation generally continues on a path through the sample. While traversing the sample, the incident fields and the coherent Raman radiation are subject to linear scattering due to possible refractive index variations in the specimen. Consequently, the generation and subsequent propagation of the CRS field is impeded, and the forward-detected signal experiences a loss due to linear scattering, which gives rise to weaker CRS signals at the detector. This situation is particularly relevant when imaging tissues, which contain significant refractive index changes on the micrometer to sub-micrometer scale throughout the tissue matrix. When the scattering in the sample is strongly dependent on the position of the focus, the transmission of the forward-propagating radiation is spatially variant. The forward-detected image is correspondingly dressed with a spatially varying transmission function.

On the other hand, linear scattering in tissues generally increases the signal collected in the epi-direction. Light scattering at tissue structures steers part of the nonlinear radiation in directions different than the initial forward propagation direction, among

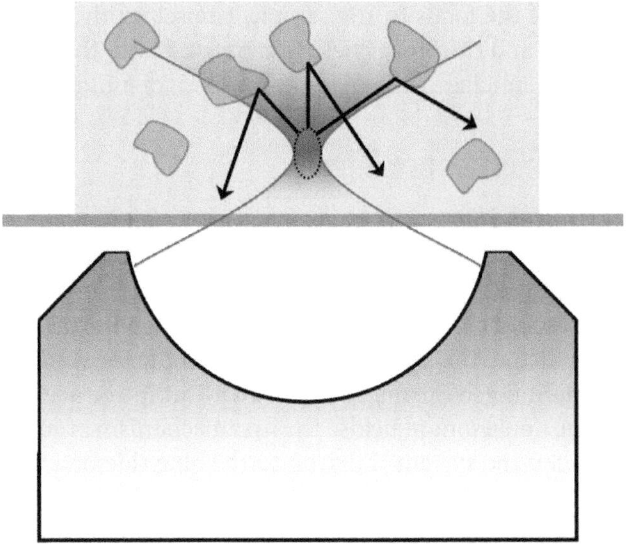

FIGURE 2.11 Illustration of back-scattered coherent Raman radiation in tissues. A portion of the forward-propagating light is back-scattered into the epi-direction. The back-scattered light retains important coherent qualities.

which is the backward direction. This mechanism allows efficient collection of the signal in the epi-direction with the excitation objective lens, as shown in Figure 2.11.

Monte Carlo calculations based on typical tissue scattering parameters relevant to skin indicate that 40%–45% of the forward-propagating CRS radiation is backscattered in thick tissue samples [35]. Some of this back-scattered radiation results from single-scattering events while another portion of the light has undergone multiple scattering events. When a high numerical aperture lens is used for both excitation and collection, a fraction of the backscattered light (~10%) can be captured, while the multiple-scattered, diffusive light generally lies outside the cone angle of the lens. This fraction of back-scattered light is sufficient to record lipid-based CARS images in the epi-channel at video rate [36]. Larger portions of the back-scattered light can be intercepted when large area detectors are used close to the tissue surface. The latter method has been used to achieve video rate SRS imaging of live tissues [35].

Note that the spatial resolution of the microscope is not affected by the fact that the light is back-scattered. The resolution results primarily from the size of the focal volume. Upon focusing in the tissue, scattering of the incident light can occur, which may slightly broaden the focal volume. Nonetheless, after formation of the focal interaction volume, the resolution is set. Any scattering of the CRS radiation beyond the nonlinear interaction volume is not related to the size of the focal spot and thus has a limited effect on the spatial resolution on the imaging experiment. In fact, a larger amount of back-scattered light results in stronger signals in the epi-channel, which generally improves the signal-to-noise ratio. Note also that the back-scattered light still retains the coherent properties of the signal field. The back-scattered field displays a spatial interference pattern called speckle, which is a direct manifestation of coherence. Although speckle patterns are generally complex, they are the result of a deterministic scattering process. Hence, in the case of SRS, the back-scattering for both the intrinsic SRS field and the Stokes field is identical for a given position of the focus in the sample. Consequently, the far-field interference between the SRS field and the Stokes field is conserved, and the stimulated Raman gain can be detected in the epi-channel even in the presence of multiple scattering [37].

### 2.6.3 Nonlinear Interference Effects

Besides linear scattering effects, the image contrast in CRS imaging is affected by nonlinear interference effects. These interference effects are especially prevalent in CARS microscopy. The origin of these effects lies in the different phases of the vibrationally resonant and nonresonant CARS contributions. To simplify the discussion in the following, we assume that all the signals are detected in the forward direction.

There are two main mechanisms in CARS that can introduce a phase shift between the resonant and the nonresonant fields. The first mechanism is due to the spectral properties of $\chi^{(3)}(\Omega)$. When the system is driven to the blue side of a vibrational resonance, the resonant field may acquire a phase shift of up to $\pi$ relative to the nonresonant field. Destructive interference may occur when the focal volume simultaneously contains resonant and nonresonant materials, producing a signal that is lower than what may be expected from each individual material. When an object is scanned axially through focus, this spectral phase effect produces a depleted signal at the top and bottom interface of the resonant object surrounded by a nonresonant medium. Because the nonresonant background is absent in SRS techniques, similar effects cannot be observed in SRL and SRG imaging.

A second mechanism is related to the Gouy phase shift. Whenever an object is small enough such that its radiation profile can be understood as a single dipole emitter, a geometric phase shift between the fields radiated by the object and the (bulk) medium can be observed. The phase of dipole field, $\phi_d$, is dictated by Equation 2.47. It is dependent on the phase of the driving fields. Consequently, $\phi_d$ is a function of the axial position of the object in focus, because of the Gouy phase shift carried by the excitation fields. On the other hand, the phase of the field from the medium, $\phi_m$, is fixed. $\phi_m$ is defined by the geometry of the focus and has, similar to the Gouy phase of the focused excitation fields, accrued a $\pi$-shift when detected in the far field. Hence, the effective geometric phase shift, $\phi = \phi_m - \phi_d$, is dependent on the position of the object in focus. If the object is located in the upper half of the focal volume, $\phi$ is small, and constructive interference will ensue. If the object is positioned in the lower half of the focal volume, $|\phi|$ may be up to $\pi$, and destructive interference will occur. This effect is sketched in Figure 2.12. The geometric phase shift can dress the signal from a particle, which is axially scanned through focus, with a dark shadow or dip. Other than the depleted signal due to the spectral phase shift, which produces a symmetric signal profile along the axial dimension, the geometric phase shift introduces an asymmetric profile [38].

This effect is also active in SRS, albeit in a less dramatic form. As discussed in Section 2.5.4, in SRS we need to consider the phase between the excitation field and the induced polarization. The phase shifts relevant for a small particle positioned in the focal plane are listed in Table 2.1. Relative to the situation in the focal plane, the total phase shift may change by as much as $\pm\pi/2$ when the particle is positioned below or above the focal plane. In SRL, for instance, the position-dependent phase shift renders the total phase shift between the induced field and the driving field as $\phi = -\pi \pm \pi/2$. Although the SRL signal

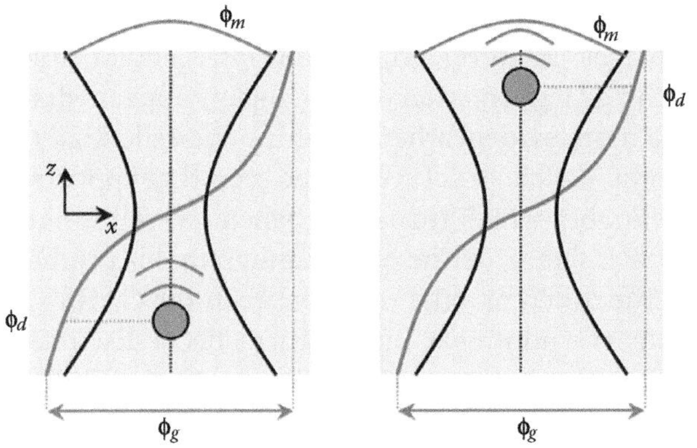


FIGURE 2.12 Interference between the field from a small object and the (nonresonant) field from the medium in forward-detected CARS microscopy. In the left panel, the object is situated in the lower half of the focal volume. Due to the Gouy phase shift $\phi_g$, the phase of the dipole field ($\phi_d$) is different from the phase of the nonresonant field ($\phi_m$), which can lead to destructive interference. In the right panel, the particle is located in the upper half of the focal volume, and $\phi_d$ is close to $\phi_m$, leading to constructive interference.

is affected by this extra phase shift, it puts the fields in quadrature at most, and thus it does not lead to extra destructive or constructive interference. Consequently, the strong destructive interference effects that affect CARS images are not similarly present in SRS images.

In Figure 2.13, a comparison is made between the CARS and SRL images of a cultured Chinese hamster ovary (CHO) cell. The CRS signals are tuned into resonance

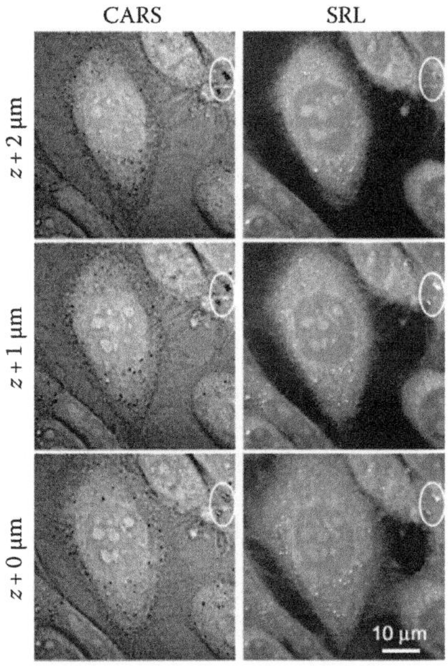


FIGURE 2.13 Different imaging properties between CARS and SRL due to nonlinear interference effects. In the left column, a Chinese hamster ovary cell is visualized with CARS contrast at several depths ($z$). In the right column, the same cell is visualized with SRL contrast. In these measurements, the pump was set to 820 nm and the probe to 665 nm, amounting to a Raman shift of 2842 cm$^{-1}$. Note the markedly different response in the circled area. (Images were generated by Junjie Li, Purdue University, West Lafayette, IN.)

with the $CH_2$ stretching mode. Several lipid-rich structures are highlighted in both CARS and SRL images. However, whereas all intracellular lipid droplets in the SRL image appear bright, a large fraction of the droplets appears dark in the CARS image. This effect is made more evident when scanning the cell axially through focus. In the encircled region, a lipid-rich structure can be seen that appears bright in CARS close to the glass coverslip, but which turns dark when the cell is axially detuned by a few micrometers. The SRL signal, on the other hand, remains bright during the full extent of the axial scan. The different axial trend between the CARS and SRL images can be largely explained by the nonlinear interference effects discussed earlier. In addition, the effects of linear scattering are more easily seen in the CARS image as depressions in the ubiquitous nonresonant background, whereas the nonresonant background in SRS is absent.

## 2.7 Conclusion

In this chapter, we have discussed the mechanism of CRS signal generation and signal propagation. These mechanisms are described well with classical field theory. Using a vectorial expression for the electric field, the qualities of the focal fields can be studied in great detail. The generation of CRS signals is intimately related with the spatial coherence of the induced polarization in the material, which produces coherent radiation that is spatially directed in the phase-matched directions. The small size of the focal volume, which is on the order of an optical wavelength, highlights some unique features of the coherent signal generation process. In particular, coherent radiation from small objects, on the order of $\sim\lambda/2n$, can be observed in the backward-propagating direction, a feature not commonly encountered in bulk spectroscopy measurements. Such observations can be fully understood by considering the far-field interference of the fields emanating from a collection of coherent dipole radiators in focus. Such models for signal generation and propagation help us appreciate the rich optical physics that underlies the CRS microscope.

## Acknowledgments

E.O.P. acknowledges support from the National Science Foundation (NSF), grants CHE-0802913 and CHE-0847097. A.V. is grateful to financial support from the Deutsche Forschungsgemeinschaft (DFG: VO 825/1-4), the European Commissions 7th Framework program (HEALTH-F5-2008-200820 CARS EXPLORER), and the German Federal Ministry of Education and Research (BMBF 13N10776 MEDICARS). X.S.X. thanks the NIH TR01 program and the Gates Foundation for support.

## References

1. G. L. Eesley, *Coherent Raman Spectroscopy* (Pergamon Press, Oxford, U.K., 1981).
2. S. A. J. Druet and J. P. E. Taran, CARS spectroscopy, *Prog. Quantum Electron.* 7, 1–72 (1981).
3. Y. R. Shen, *The Principles of Nonlinear Optics* (John Wiley & Sons, New York, 1984).
4. M. D. Levenson and S. S. Kano, *Introduction to Nonlinear Laser Spectroscopy* (Academic Press, San Diego, CA, 1988).

5. J. S. Gomez, Coherent Raman spectroscopy, in *Modern Techniques in Raman Spectroscopy*, ed. J. J. Laserna (John Wiley & Sons, New York, 1996), p. 305.
6. A. Zumbusch, G. R. Holtom, and X. S. Xie, Three-dimensional vibrational imaging by coherent anti-Stokes Raman scattering, *Phys. Rev. Lett.* **82**, 4142 (1999).
7. M. Müller, J. Squier, C. A. D. Lange, and G. J. Brakenhoff, CARS microscopy with folded BoxCARS phasematching, *J. Microsc.* **197**, 150–158 (2000).
8. C. Heinrich, S. Bernet, and M. Ritsch-Marte, Wide-field coherent anti-Stokes Raman scattering microscopy, *Appl. Phys. Lett.* **84**, 816–818 (2004).
9. R. D. Schaller, J. C. Johnson, K. R. Wilson, L. F. Lee, L. H. Haber, and R. J. Saykally, Chemically selective imaging of subcellular structure in human hepatocytes with coherent anti-Stokes Raman scattering (CARS) near-field scanning optical microscopy (NSOM), *J. Chem. Phys. B* **106**, 8489-8492 (2002).
10. T. Ichimura, N. Hayazawa, M. Hashimoto, Y. Inouye, and S. Kawata, Tip-enhanced coherent anti-Stokes Raman scattering for vibrational nanoimaging, *Phys. Rev. Lett.* **92**, 220801 (2004).
11. E. O. Potma, W. P. de Boeij, and D. A. Wiersma, Nonlinear coherent four-wave mixing in optical microscopy, *J. Opt. Soc. Am. B* **17**, 1678–1684 (2000).
12. M. Hashimoto and T. Araki, Three-dimensional transfer functions of coherent anti-Stokes Raman scattering microscopy, *J. Opt. Soc. Am. A* **18**, 771–776 (2001).
13. A. Volkmer, J. X. Cheng, and X. S. Xie, Vibrational imaging with high sensitivity via epidetected coherent anti-Stokes Raman scattering microscopy, *Phys. Rev. Lett.* **87**, 023901 (2001).
14. J.-X. Cheng, A. Volkmer, and X. S. Xie, Theoretical and experimental characterization of coherent anti-Stokes Raman scattering microscopy, *J. Opt. Soc. Am. B* **19**, 1363–1375 (2002).
15. J. D. Jackson, *Classical Electrodynamics* (John Wiley & Sons, New York, 1975).
16. H. Wang, Y. Fu, and J. X. Cheng, Experimental observation and theoretical analysis of Raman resonance-enhanced photodamage in coherent anti-Stokes Raman scattering microscopy, *J. Opt. Soc. Am. B* **24**, 544–552 (2007).
17. A. Nikolaenko, V. V. Krishnamachari, and E. O. Potma, Interferometric switching of coherent anti-Stokes Raman scattering signals in microscopy, *Phys. Rev. A* **79**, 013823 (2009).
18. P. Nandakumar, A. Kovalev, A.Volkmer, Vibrational imaging based on stimulated raman scattering microscopy, *New J. Phys.*, **11**, 033026 (2009).
19. L. Novotny, and B. Hecht, *Principles of Nano-Optics* (Cambridge University Press, New York, 2006).
20. C. F. Bohren and D. R. Huffman, *Absorption and Scattering of Light by Small Particles* (Wiley-VCH Verlag GmbH, Weinheim, Germany, 2004).
21. L. Mandel and E. Wolf, *Optical Coherence and Quantum Optics* (Cambridge University Press, New York, 1995).
22. N. Bloembergen, *Nonlinear Optics* (W. A. Benjamin, New York, 1965).
23. P. Török, P. D. Higdon, and T. Wilson, On the general properties of polarised light conventional and confocal microscopes, *Opt. Commun.* **148**, 300–315 (1998).
24. P. D. Higdon, P. Török, and T. Wilson, Imaging properties of high aperture multiphoton fluorescence scanning optical microscopes, *J. Microsc.* **193**, 127–141 (1999).
25. B. Richards and E. Wolf, Electromagnetic diffraction in optical systems. II. Structure of the image field in an aplanatic system, *Proc. R. Soc. London Ser. A* **253**, 358–379 (1959).
26. A. Boivin and E. Wolf, Electromagnetic field in the neighborhood of the focus of a coherent beam, *Phys. Rev. B* **138**, 1561–1565 (1965).
27. L. G. Gouy, Sur une propriété nouvelle des ondes lumineuses, *C. R. Acad. Sci. Paris* **110**, 1251–1253 (1890).
28. S. Feng and H. G. Winful, Physical origin of the Gouy phase shift, *Opt. Lett.* **26**, 485–487 (2001).
29. R. W. Boyd, Intuitive explanation of the phase anomaly of focused light beams, *J. Opt. Soc. Am* **70**, 877–880 (2001).
30. G. C. Bjorklund, Effects of focusing on the third-order nonlinear process in isotropic media, *IEEE J. Quantum Electron.* **QE-11**, 287–296 (1975).
31. J.-X. Cheng and X. S. Xie, Green's function formulation for third-harmonic generation microscopy, *J. Opt. Soc. Am. B* **19**, 1604–1610 (2002).
32. D. Débarre, N. Olivier, and E. Beaurepaire, Signal epi-detection in third-harmonic generation microscopy of turbid media, *Opt. Express* **15**, 8913–8924 (2007).
33. M. Marrocco, Vectorial descriptions of nonlinear Raman microscopy, *J. Raman Spectrosc.* **41**, 882–889 (2010).

34. R. Feynman, *The Feynman Lectures on Physics*, Volume I, Sec 30-7 (Addison-Wesley, Reading, MA, 1964).
35. B. Saar, C. W. Freudiger, J. Reichman, C. M. Stanley, G. Holtom, and X. S. Xie, Video-rate molecular imaging in vivo with stimulated Raman scattering, *Science* **330**, 1368–1370 (2010).
36. E. L. Evans, E. O. Potma, M. Puoris'haag, D. Côté, C. Lin, and X. S. Xie, Chemical imaging of tissue in vivo with video-rate coherent anti-Stokes Raman scattering (CARS) microscopy, *Proc. Natl. Acad. Sci. USA* **102**, 16807–16812 (2005).
37. P. Wang, M. N. Slipchenko, B. Zhou, R. Pinal, and J.-X. Cheng, Mechanisms of epi-detected stimulated Raman scattering microscopy, *IEEE Journal of Selected Topics in Quantum Electronics* **18**, 384–388 (2011).
38. K. I. Popov, A. F. Pegoraro, A. Stolow, and L. Ramunno, Image formation in CARS microscopy: Effect of the Gouy phase shift, *Opt. Express* **19**, 5902–5911 (2011).

# II Platforms

# 3. Construction of a Coherent Raman Microscope

**Brian G. Saar and Xiaoliang Sunney Xie**



## 3.1 Introduction

This chapter is devoted to describing the process of building a coherent Raman microscope. A great variety of approaches have been reported in the literature with different systems tailored for different applications. This chapter focuses primarily on the specific narrow-band, picosecond-laser-pumped, beam-scanning configuration that we have found to be optimal for high speed imaging in biological samples. Variations on this, including miniaturized scanning systems (Chapter 6), multiplex detection systems (Chapters 13 through 16) and systems utilizing one or more femtosecond lasers (Chapters 5, 10, 13 through 16, and 22) are reported elsewhere in this volume. Although there has been some debate about the most optimal configuration for coherent Raman scattering, it is clear and uncontroversial that this approach offers the best sensitivity and highest image acquisition speed,

*Coherent Raman Scattering Microscopy*. Edited by Ji-Xin Cheng and X. Sunney Xie © 2013 CRC Press/Taylor & Francis Group, LLC. ISBN: 978-1-4398-6765-5.

which are two key figures of merit for coherent Raman imaging. However, the major trade-off is that only a single Raman-active vibrational mode is probed in a given image, and scanning the laser wavelength to obtain a full Raman spectrum can be labor and time intensive.

## 3.2 Laser System

The bulk of the cost and complexity of a coherent Raman microscopy system is in the laser excitation. Two laser wavelengths must be utilized with some control over the absolute wavelengths and precise control over the difference frequency to within the width of an individual Raman line (10 $cm^{-1}$). Historically, the first coherent Raman microscope made use of pulsed dye lasers in the visible region of the spectrum [1]. More recent systems have made use of solid state lasers systems, either based on electronically synchronized titanium:sapphire lasers [2] or mode-locked lasers that synchronously pump optical parametric oscillators (OPOs) [3,4]. A new generation of fiber-based systems, either based on nonlinear frequency conversion in a photonic crystal fiber [5] or active fiber lasers [6] promise increased ease-of-use and lower cost, but currently trade-offs in performance are required to make use of these systems. The following sections describe the optimal parameters of a laser source for imaging, the previous and current generations of laser systems and finally touch on the future developments that are poised to significantly widen the availability of coherent Raman imaging systems.

### 3.2.1 Optimal Source Parameters

Excitation of coherent Raman microscopy requires (at least) two laser wavelengths, one of which must be tunable in order to match the difference frequency to the molecular vibrational frequency. In addition, excitation of both CARS and SRS with laser pulse widths of a few picoseconds has been previously shown to ideally balance the need for high peak power to efficiently generate nonlinear signals with the requirement for relatively narrow spectral bandwidth (<1 nm) in order to match the intrinsic linewidth of molecular vibrations [7]. For high speed imaging, repetition rates of at least 10 MHz are required, and should ideally be higher. This is because in video rate imaging, data is acquired at 10 megapixels per second, and at least one laser shot per pixel is required for CARS (and at least two for SRS with modulation transfer detection, as described later). In addition, laser excitation in the near-infrared region of the spectrum has been shown to minimize generation of the nonresonant background in CARS [8], to offer reduced photodamage compared to visible excitation, and also to offer good penetration into tissue for nonlinear microscopy [9]. Finally, because the optical excitation path in a CARS or SRS microscope typically has relatively low transmission (10%–20% is commonly observed from laser output to the sample), watt-level average power is required.

A number of groups have published papers that report using 50–200 fs pulse widths rather than the 2–6 ps pulses [5,10–12]. Although CRS processes may be excited by femtosecond pulses, this comes at the cost of decreased signal levels, limited tunability, loss of spectral selectivity, and an increased non-resonant background in CARS. There are two main reasons for this. First, the typical width of Raman spectral features is about 15 $cm^{-1}$. Near 800 nm, this corresponds to a bandwidth of about 1 nm. For any laser system, the inverse relationship between laser pulse width and laser spectral bandwidth means that for a given spectral bandwidth, there is a fundamental limit to the shortest pulse that can be achieved. In the case of a 1 nm bandwidth, this pulse width is about 0.95 ps, assuming a Gaussian pulse shape. Shortening the pulse below this value will result in an increase in the nonresonant background/resonant signal ratio in CARS, degrading the contrast and reducing image quality.

In SRS, the result will simply be that no additional signal will be generated even though the peak power (and therefore the nonlinear photodamage) will increase because the frequency components that are not on resonance with the Raman-active transition will not contribute signal. In addition, if two nearby resonances occur, the broader bandwidth will mean that the spectral resolution will be lower and the obtained images will be contaminated by signal from both resonances. For the typical 8 nm bandwidth obtained from the commercial mode-locked femtosecond ti:sapphire lasers used in multiphoton microscopy, this means that only about 1/8 of the laser energy applied to the sample is productively used by the CRS process. In contrast, for pulses of a few picoseconds, all the laser intensity is concentrated into narrower frequency bands that are completely matched to the Raman resonance, which can be well resolved. Although spectrally resolved detection with broadband femtosecond lasers can recover the CARS or SRS spectrum at high resolution, it typically requires a multi-element detector such as a CCD camera that has very long (>10 ms) readout time at each pixel, severely limiting the imaging speed [13–15].

A second feature of slightly longer pulses with higher average power but reduced peak power is that the nonlinear photodamage is reduced. This actually has the practical benefit of allowing for more total SRS signal by excitation with 6 ps pulses than with 150 fs pulses, even in the case of a broad resonance. The reason is that in many samples, the nonlinear photodamage increases faster than the signal of interest with decreasing laser pulse width (e.g., in work on imaging calcium activity, it was determined that the damage scaled with $1/\tau^{1.5}$ [16]). Since the SRS signal scales with $1/\tau$, the photodamage is clearly expected to rise more rapidly than the signal level of SRS as shorter pulses are employed. Of course, the actual scaling and damage threshold is highly sample-dependent, so it is difficult to make absolute statements about safe power levels. However, to test this experimentally, we imaged a collection of bacterial cells that had incorporated deuterium from deuterated carbon sources in their growth media. The carbon–deuterium stretching resonance is exceptionally broad: it has a width of over 100 $cm^{-1}$, and thus, even femtosecond pulses may be used to excite it with good efficiency. We then compared the maximum average power that could be applied to bacterial cells without obvious morphological changes in the sample due to the optical power. This is an extreme test of the sample damage limit, but gives some insight when applied using similar laser parameters with only the pulse width varying. At 180 fs, an average power of 25 mW was found to be usable. At 1 ps the power level was 80 mW, and at 6 ps the power level was 275 mW without obviously burning the sample. In all cases, the laser repetition rate was 76 MHz, the Stokes wavelength was between 1040 and 1064 nm, and the pump wavelength was between 800 and 816 nm. The same beam parameters and microscope optics were used in all cases. At the maximum power at 6 ps, the absolute SRS signal is higher than the signal at the maximum power at 150 fs by a factor of 4, even for a broad resonance. In the case of a narrow resonance, the difference is expected to be a factor of 30–40 because of the combination of the damage limits and the spectral widths. Because the major advantage of coherent Raman techniques is the imaging speed and sensitivity, this large factor in signal level obtainable is important and the proper choice of laser pulses in the 1–10 ps range (as opposed to 100–200 fs) is critical to obtaining the best results. Within this range, the exact choice makes only a small difference in the obtainable signal (less than a factor of 2) while maintaining good spectral resolution.

### 3.2.2 Historical Development of Coherent Raman Laser Systems

The first report of CARS microscopy in 1982 by Duncan and co-workers at the Naval Research Laboratory utilized pulsed dye lasers in the visible, because broadly tunable ultra-short-pulsed solid-state lasers were not widely available [1]. However, these systems generated high nonresonant background signal levels in CARS because they operated in the visible portion of the spectrum, and were inconvenient to operate. In addition, near-IR excitation increases penetration depth into biological samples and causes less photodamage, as electronic resonances are avoided. In contrast to spontaneous Raman scattering, longer wavelength excitation does not sacrifice signal.

Later generations have been based on solid-state and fiber laser systems. After the first generation amplified laser system utilized by Zumbusch et al. in their 1999 report [8], which does not satisfy the requirements of the previous section, the next system used involved synchronized, mode-locked titanium:sapphire lasers. These lasers were capable of producing mode-locked pulse trains with 80 MHz repetition rates and pulse widths of 2–5 ps. Watt-level average power is available and tunability over essentially the entire range of Raman shifts can be achieved. The major drawback of these systems is that two separate mode-locked laser systems are required, and these systems cannot be passively synchronized. This necessitates the use of electronic synchronization [2]. The synchronization process involves continuous adjustment of the cavity length of one of the lasers to match the cavity length of the other by means of continuous feedback on the laser repetition rate. This system is highly sensitive to mechanical vibrations and instabilities in the environment, and also to the stability of mode-locking. In addition, because it requires an error signal in order to correct the cavity length, some temporal jitter is inevitable, and this translates into amplitude fluctuations in the resulting images.

The next generation laser system solved this issue by using a passively mode-locked $Nd{:}YVO_4$ laser with ideal temporal properties and >10 W level average power to synchronously pump an OPO. An OPO is a resonant nonlinear frequency conversion device that makes use of a $\chi^{(2)}$ process known as optical parametric generation (OPG) in a nonlinear crystal material. This nonlinear process involves splitting a pump photon into two photons, known as the signal and idler. The signal and idler wavelength do not have to be equal (by definition, the signal is shorter), but the process must satisfy both energy conservation ($\hbar\omega_{signal} + \hbar\omega_{idler} = \hbar\omega_{pump}$) and phase-matching ($\vec{k}_{signal} + \vec{k}_{idler} = \vec{k}_{pump}$) to have reasonable efficiency (Figure 3.1). In order to maximize the conversion, the nonlinear crystal is placed into an optical cavity whose cavity length is matched to the pump laser cavity round trip time. This means that pump laser pulses that are injected into the OPO cavity circulate around the cavity and meet subsequent

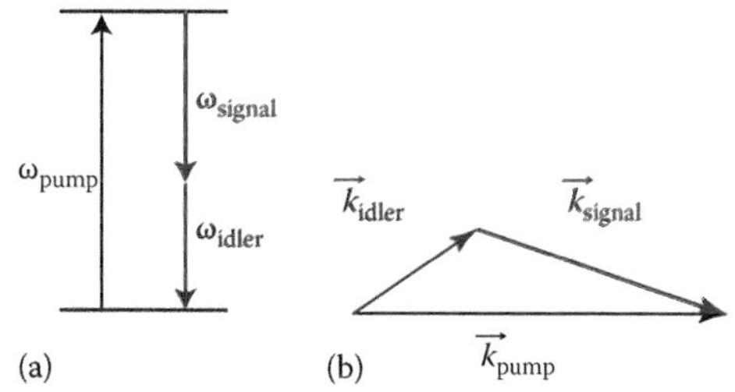


**FIGURE 3.1** Energy conservation (a) and the phase-matching condition (b) must be satisfied for efficient OPG.

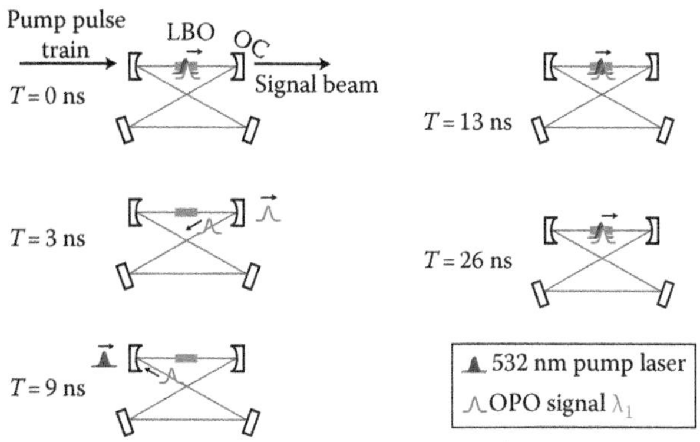


**FIGURE 3.2** In a synchronously pumped OPO, the round trip time of the OPO cavity (based on a lithium triborate, LBO, nonlinear optical crystal) is matched to the period of the pump laser repetition rate. In this configuration, the light pulse circulated in the cavity meets with each subsequent pump laser pulse in the gain medium, ensuring high peak intensity and therefore efficient conversion efficiency.

pulses back in the conversion medium (Figure 3.2). This high instantaneous intensity leads to efficient nonlinear frequency conversion. In this configuration, synchronization is automatic because the pump laser input pulses exactly drive the OPO output. Thus temporal overlap is achieved by means of a passive delay stage on the optical table, and once optimized, does not require adjustment for days or weeks in a stable instrument.

The original implementation of a synchronously pumped OPO for CARS microscopy involved a 1064 nm pump laser and a periodically poled potassium titanyl phosphate (KTP) crystal conversion medium. This generated longer wavelength signal waves, typically around 1.6 μm, which were intracavity frequency-doubled to reach the desired 816 nm pump wavelength for CARS imaging of the $CH_2$ stretching vibrations. The major challenge with this OPO was that the parameters of two separate crystals (one for frequency conversion through OPG, one for intracavity doubling) had to be optimized at each wavelength, making tuning extremely labor intensive. Nevertheless the implementation of passive synchronization significantly increased the stability of the laser system and made experiments easier.

A subsequent generation of OPO [3] made use of the frequency-doubled output of the 1064 nm pump laser at 532 nm for OPO pumping. This has the major advantage that the first frequency conversion step, doubling the 1064 nm laser frequency, always occurs at fixed wavelengths, pulse widths, power levels, etc. Thus it may be stably optimized once. Then, the OPO signal wave may be directly used as it appears in the range of 800–1000 nm without intra-cavity frequency doubling inside the OPO. This system is used for many recent publications in the coherent Raman field. A commercial version, known as the Levante Emerald, is sold by APE GmbH (Berlin, Germany). In its original implementation [3], the signal and idler waves of the OPO were used directly as the pump and Stokes beams for CARS. This has the advantage that spatial and temporal overlap occurs automatically, greatly simplifying the optical system. However, it has the disadvantage that the absolute wavelengths used are relatively long. For example, to probe the $CH_2$ stretching vibration at 2845 $cm^{-1}$, used throughout the coherent Raman imaging field, wavelengths of 924.2 and 1253.8 nm would be used in the signal and idler configuration. This same wavenumber shift could be probed by using the 1064 nm fundamental of the pump laser system in combination with 816.7 nm from the signal wave of the OPO. The advantages of using shorter wavelength excitation include higher spatial resolution and better quantum efficiency for the CARS photomultiplier detectors (discussed later in this chapter) which rapidly drop in efficiency above 800 nm and become essentially unusable. Secondarily, because SRS requires one of the two excitation beams to be modulated, as described in the following, the use of a system with automatically overlapped outputs is not advantageous because the beams must be separated and recombined for modulation in any event.

### 3.2.3 Future Developments: Toward Turn-Key Systems

Recently, the commercial supplier of the OPO systems has introduced a "turn-key" laser system which consists of the pump laser, frequency doubling unit, OPO and beam combination for CARS and SRS (including an integrated modulator). This system promises huge advantages in ease of use because wavelength control is completely automated and the system is designed to be permanently aligned with minimal user adjustment. Initial trials of this system in our laboratory have produced good quality images, and improvements to the software and mechanical stability of the system promise to make it a robust tool for researchers in the CRS community who are focused on developing applications and answering scientific questions rather than optimizing CRS techniques themselves.

While further development of these OPO-based systems promises maximum flexibility, a broad tuning range and high average power, the reliability and compactness of these systems is not ideal for clinical applications in which experts in ultra-fast laser systems may not be present during measurement. The development of highly reliable, computer-controlled laser systems is a major challenge to the introduction of coherent Raman scattering to the clinic. While we can envision the potential for clinical applications in dermatology and neurosurgery, these will require the development of a new generation of optical fiber-based laser systems, the topic of a number of recent papers [6,17–19]. Fiber lasers offer major advantages over solid state lasers because the gain medium is a doped optical fiber and, ideally, the entire laser system is contained within fiber-based components. This means that the system is automatically and permanently aligned by the waveguide action of the optical fiber. In addition, thermal management is easier because of the high surface-area-to-volume ratio of fiber, and robust components developed for high reliability at low cost by the telecommunications industry can be used. The development of a fiber pump laser to replace the

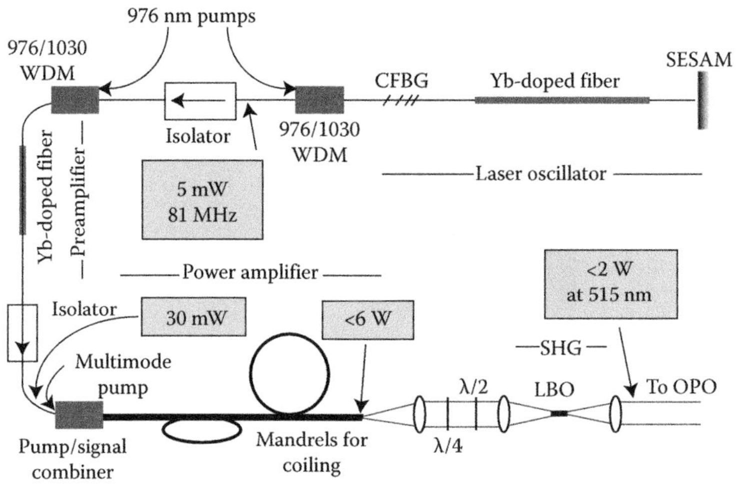


FIGURE 3.3 Schematic diagram of the picosecond laser system. Output power levels after each stage are indicated in gray boxes. Also shown is the LBO crystal used for SHG. WDM: wavelength division multiplexer; λ/2, λ/4: half- and quarter-wave plate, respectively.

solid state Nd:$YVO_4$ laser that we typically use (Figure 3.3) has been reported recently. The fiber laser system generates 3.5 ps pulses with 6 W average power at 1030 nm. Frequency doubling yields more than 2 W of green light, which can be used to pump an OPO to produce the pump and the Stokes beams for CRM. Detailed performance data on the laser and the various wavelength conversion steps are discussed in the literature [6].

However, efficient frequency conversion based on all-fiber devices has not yet been demonstrated. Several technology paths are available for generation of two colors, including synchronization based on a time-lens system [20], fiber-based OPOs [21], electronic synchronization [2] and all-optical synchronization. Ultimately, as the performance of fiber laser systems improves to the point where they can be utilized for coherent Raman imaging, the major barrier to entry for new researchers in the field, which is the cost and complexity of the laser system, will be removed.

## 3.3 Beam Combination

For both coherent Raman techniques, the two laser beams must be combined in time and space. Spatially overlapping the beams is relatively straightforward using a dichroic mirror (e.g., the 1064dcrb from Chroma Technology, which reflects 1064 nm and transmits up to 1000 nm) and several steering mirrors for fine adjustment. Typically, the overlap at two apertures spaced by about 1 m in the combined beam path may be used to verify the spatial alignment. Fine adjustment can be made based on the CARS or SRS signal strength.

Temporal overlap in the OPO-based systems is accomplished by means of a passive delay stage based on a retro-reflector which allows for path-length adjustment of one of the two beams while preserving the spatial alignment (Figure 3.4). Because the repetition rate of the laser systems that we use is typically $f \sim 80$ MHz, the temporal period between two pulses is $p = 1/f = 12.5$ ns. By multiplying this period with

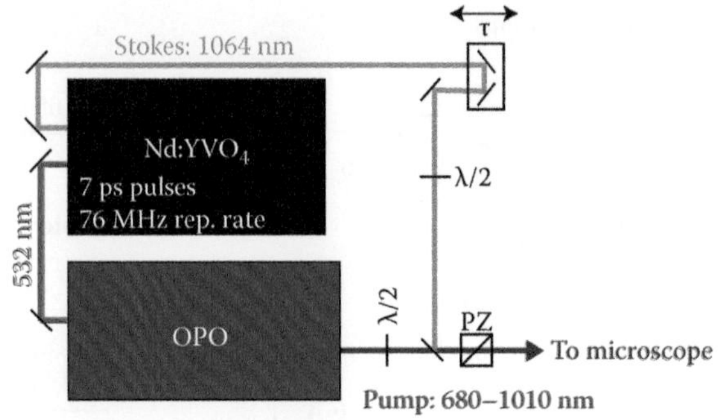


**FIGURE 3.4** Schematic of beam combination for CARS using a synchronously pumped OPO. A passive delay stage in the 1064 nm Stokes beam path is used to ensure temporal overlap, and a dichroic mirror is used to obtain spatial overlap. Each beam's linear polarization is adjustable by a half-waveplate ($\lambda/2$) which results in independent intensity adjustment after a polarizing beamsplitter cube (PZ) which ensure that both beams have parallel polarization state when entering the microscope.

the speed of light, we obtain $cp \approx 3.75$ m. Thus to find the temporal overlap, we must search pathlength differences that are up to ±1/2 of that distance. The spatial precision with which we must overlap the beams is determined by the spatial extent of the laser pulse, which has a duration of $\tau \sim 6$ ps. Multiplying, we obtain $c\tau \sim 1.8$ mm. To find the temporal overlap with this precision, we utilize a two-step procedure. First, the overlap is optimized as well as possible using a high bandwidth photodiode and an oscilloscope. Because the bandwidth of the fastest available real-time oscilloscopes is a few GHz, this allows us to find the temporal overlap to within ~500 ps, which translates to within 150 mm. The best measurements with the oscilloscope can be obtained by triggering from the laser pulse train using a stable internal detector within the pump laser. Then, individual pulse trains can be measured on the optical table and the centroid of the laser pulse measured using the cursors. By comparing the offset of the two laser pulse train centroids, a higher precision than the oscilloscope bandwidth can be obtained.

After the pulse trains are temporally overlapped as well as possible using an oscilloscope, an autocorrelator may be used for fine adjustment. Typically, the autocorrelator has a total range of ~50 ps, meaning that the pulse overlap must be found by trial-and-error in the area between the 500 ps precision of the oscilloscope and this 50 ps dynamic range. An autocorrelator is an optical instrument used to characterize very short laser pulses. It works by using the laser pulse itself as a measurement tool. In an autocorrelator, the input beams are split by a beamsplitter and sent into two arms of a (typically) Mach–Zender interferometer (Figure 3.5). One arm of the interferometer has a precision delay stage which is rapidly scanned. After the delay, the two beams are recombined and measured using a readout nonlinear process such as sum frequency generation that only provides signal when both beams are present. By recording the readout signal as a function of the delay in one arm of the interferometer and converting the delay distance to time using the known speed of light, the temporal delay between two beams may be inferred with high precision (easily <50 fs). Once the autocorrelator records the temporal overlap of the two beams, this will be sufficient to generate CARS or SRS signal.

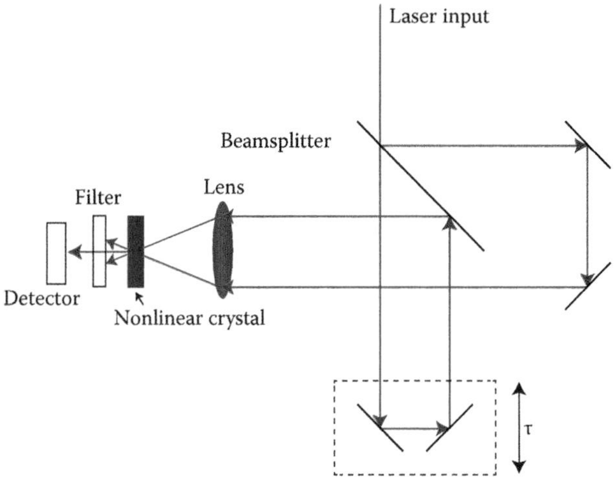


**FIGURE 3.5** Schematic of a Mach–Zender type autocorrelator. The incoming laser beam is split into two arms. One arm has a controllable delay stage (τ). The recombined beams are focused into a nonlinear crystal and the SFG signal (in green) is detected as a function of τ.

## 3.4 Image Formation via Raster Scanning

There are two major classes of image formation in coherent Raman microscopy. This chapter deals exclusively with raster scanning, that is, moving a tightly focused laser spot over the sample and recording the signal point-by-point for reconstruction with a computer. Recently, wide-field approaches to CARS imaging have been developed by Ritsch-Marte and co-workers. These techniques offer the intriguing possibility of very high frame rates but require complex phase-matching geometries and are typically performed with nanosecond lasers that have much different parameters that those described earlier for picosecond lasers. They are the subject of a later chapter in this work.

Because coherent Raman scattering techniques are nonlinear in the incident optical intensity, it is clear that compressing all the laser energy into a spot and moving that spot with high speed over the sample will maximize the CARS or SRS signal for a given total average power to be applied to the sample. Additionally, raster-scanning makes it easy to take advantage of the intrinsic optical sectioning capability of coherent Raman scattering because the signal is only generated at the tightest part of the focus. There are two common methods of raster scanning: sample scanning and beam scanning. Sample scanning offers a simpler apparatus but generally lower speeds and smaller fields of view, while beam scanning is more complex to implement and is more demanding on the optical system performance, but offers larger fields of view and higher imaging speed.

### 3.4.1 Sample Scanning

In sample scanning, the entire coherent Raman optical setup is fixed in place and the sample is translated relative to the focal spot. This means that the optical system can be aligned to a fixed laser beam, which is easier than aligning the system over a range of possible beam positions. In order to obtain high spatial resolution, a translation stage with high precision and repeatability is required. Typically, piezo-electrically actuated flexure stages are used. These stages offer step sizes and repeatability far in excess of what is required for optical microscopy (typically <5 nm) and maximum translation of a few hundred microns. There are two main disadvantages of such a system:

- The maximum field of view for an image is determined by the maximum travel of the stage, not by the optics. Thus switching to a lower magnification lens does not offer a large field of view. Typically fields of view of >1 mm are available in an optical microscope using a 10× magnification objective lens, but these are not accessible using a piezo stage.
- The mechanical resonance frequencies of these stages typically limit the maximum scan speeds to at least tens of milliseconds per line (or higher), meaning that they are at least an order of magnitude slower than beam scanning systems.

Despite these limitations, the simplicity of sample scanning makes it a viable option in many situations. The optical throughput of sample scanning systems is also very high, as the only optic required is the objective lens. Sample scanning can also be of advantage when the emission beam is further analyzed, for example, by a spectrometer, where beam movement can cause artifacts.

### 3.4.2 Beam Scanning

Images may also be formed by scanning the combined laser focus over the sample and recording the CARS or SRS signal as a function of position. Laser scanning is accomplished using a pair of galvanometer mirrors which are angularly deflected by electric current flowing through a coil. Ordinary nonresonant galvanometers are scanned with line rates up to ~1 kHz, while a resonant galvanometer has a line rate up to ~8 kHz. The line rate divided by the number of lines per image (typically 512) determines the frame rate of the imaging system. The period of the line (the inverse of the line rate) divided by the number of pixels per line (again, typically 512) determines the pixel dwell time in the scanning system. For an ordinary nonresonant galvanometer, imaging at ~2 frames per second with a pixel dwell time of ~2 μs is possible.

The scanning system consists of the scan mirrors, a pair of relay lenses, and the objective lens. The purpose of the relay lenses, known as the scan lens and the tube lens (in order of beam propagation), is to produce a magnified image of the scan mirrors on the back aperture of the microscope objective. This is necessary because scanning of the focal spot in the focal plane is accomplished by scanning the incoming angle of the laser beam on the back aperture with minimal movement of the laser spot at that position. Motion of the laser spot on the objective back aperture would yield uneven illumination intensity because of beam clipping (known as "vignetting"). In addition, the magnification of these two lenses in combination is used to increase the incoming beam diameter, which must be <4 mm to avoid clipping on the galvo mirrors but should be large enough to fill the objective back aperture (~5–10 mm) for tight focusing. Figure 3.6 shows the optical path inside a laser scanning microscope for two different scan mirror positions, shown as different colors. The imaging role of the tube lens and scan lens system is clear because both beams from both scan mirror angles strike the same position, and the

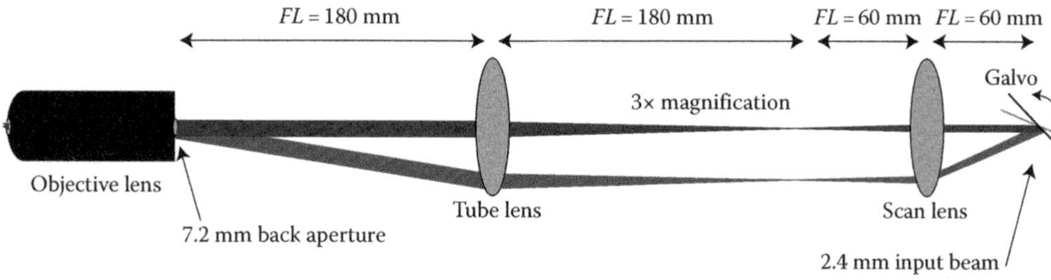


FIGURE 3.6 Schematic of the optical path of a laser scanning microscope. Approximate focal lengths (*FL*) for an Olympus system are shown. The beam path through the microscope is shown for two positions of the galvo mirror. The black mirror corresponds to the red path through the center of the optical system. The gray mirror corresponds to the orange path through the edge of the optical system. The scan lens/tube lens system serves a dual purpose: magnifying the input laser beam to match the objective back aperture and imaging the scan mirror position onto the back aperture to avoid beam clipping. Although this diagram only shows one-dimensional scanning, a second galvo mirror positioned close to the first one can be used to scan the orthogonal axis, or a two-axis scanning mirror may be used in the configuration shown here. Although simple singlet lenses are shown here, a real scanning optical system consists of multi-element lenses to obtain chromatic correction and a flat scan field.

magnification role is shown as the change in beam size. Scanning is typically accomplished under computer control, and the scan controller computer also records the pixel intensity versus time in order to draw the image onscreen.

## 3.5 Detection of the Coherent Raman Signal

### 3.5.1 CARS Microscopy Detection

CARS detection is similar to detection of two-photon fluorescence from the perspective of instrumentation. The two spatially and temporally overlapped beams are scanned through the sample as described earlier. The intensity at the newly generated anti-Stokes wavelength must be detected as a function of position in the sample. This is done by collecting the light emitted from the sample in either the forward or backward direction (Figure 3.7), separating the anti-Stokes wavelength from the pump and Stokes wavelengths using dichroic mirrors and optical filters (Figure 3.8) and detecting the light intensity using a photomultiplier tube. The PMT photocurrent as a function of position yields the CARS image.

Because of the relatively high intensity of the input pump and Stokes beams, very high optical density throughout the near infrared region is required. In addition, the detected anti-Stokes wavelengths range from 600 to 800 nm (or even higher). Thus the use of multi-alkali or GaAs-based photomultiplier tubes is preferred. These typically have quantum efficiencies of <25% and above 800 nm are typically <10%. Thus detection of CARS at low wavenumber shifts using a fixed 1064 nm Stokes beam as described earlier is problematic. As described in the following, SRS does not suffer from this limitation. Because of the fixed anti-Stokes wavelength, a combination of two optical filters can detect essentially the entire range of anti-Stokes wavelengths covered by the detector quantum efficiency. However, in the case of strongly two-photon-autofluorescent samples, additional narrowband filters may be required to reduce the contribution of autofluorescence to the detected CARS signal, even when the pump laser is completely suppressed. Since the spectral width of the CARS emission in a picosecond system is

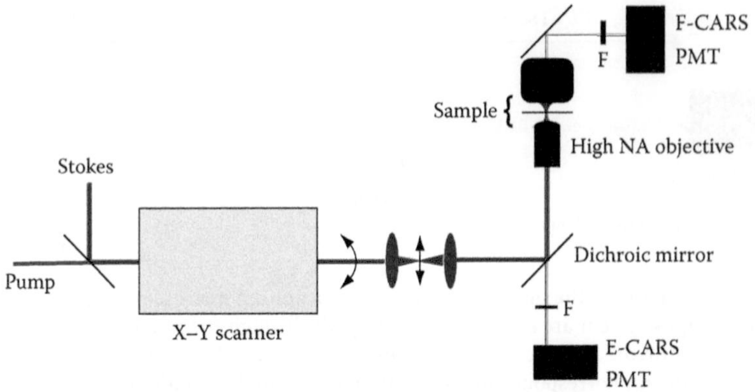


**FIGURE 3.7** Schematic detection of CARS in the forward- or epi-direction. In the forward direction, the forward-traveling CARS signal is collected by a condenser lens, the incident lasers are blocked by an optical filter (F, Figure 3.8) and a photomultiplier tube (PMT) detects the signal. In the epi-direction, light is collected with the excitation objective and a longpass dichroic mirror transmits the anti-Stokes light onto a similar detection system consisting of a filter and PMT. Typically the dichroic is positioned before the detection light has passed through the scan and tube lens and the scanning mirrors. While such non-descanned detection causes residual beam movement on the PMT, collection efficiencies are increased a factor of 5–10, typically.

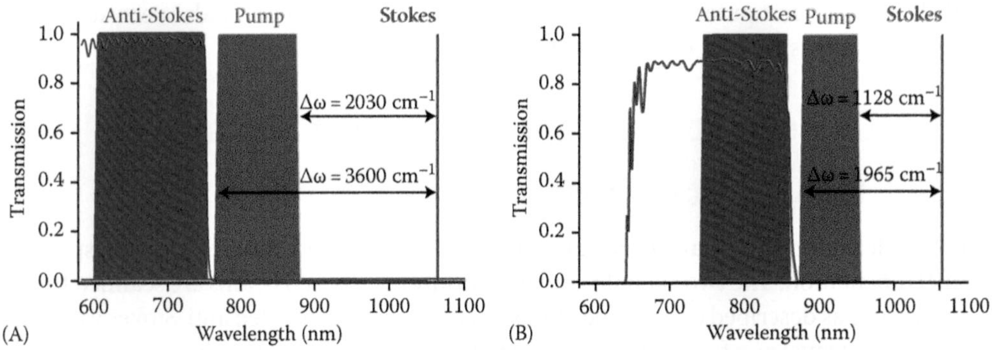


**FIGURE 3.8** Sample optical filters for detection of the CARS signal. (A) ET750sp-2p8 from Chroma Technology is useful for the higher wavenumber region. (B) HQ750-210m from Chroma Technology is useful for lower wavenumber detection. In both plots, the filter transmission is plotted as a red curve and the pump and anti-Stokes wavelengths that can be detected are shown in yellow and blue, respectively.

approximately the same as the incident laser spectral width (<1 nm), while autofluorescence emission widths are typically fairly broad (>10 nm), careful choice of this narrowband filter makes CARS imaging possible even in highly autofluorescent samples.

### 3.5.2 Modifications for SRS Microscopy

For SRS microscopy, the same temporally and spatially overlapped laser beams are applied to the sample and the same Raman difference frequency is probed. However, rather than detecting a small amount of light generated at a new wavelength, the SRS signal appears as an intensity gain on the Stokes beam and a simultaneous intensity loss on the pump beam. This accompanies vibrational excitation in the sample. Because the intensity changes, $\Delta I$, are small compared to the absolute intensity of the laser beam, $I$ ($\Delta I/I < 10^{-3}$), they are buried in the laser noise, which is typically a few percent.

To circumvent this problem, we use a modulation transfer detection scheme for SRS [22]. In this approach, one beam is amplitude-modulated at a high reference frequency. The two beams are then combined and interact with the sample at the focal spot. After the sample, the modulated beam is blocked by an optical filter and the originally unmodulated beam is measured by a detector with high dynamic range. An electrical circuit, most often a lock-in amplifier, is used to sensitively detect a small modulation at the reference frequency. The origin of this modulation in SRS is that when both beams are present, intensity is transferred from one beam to the other, while this cannot occur when only one beam is present. Thus if the modulation frequency is chosen where the originally unmodulated laser is quiet, then any detected amplitude modulation is attributable to SRS. Figure 3.9 shows a schematic of modulation transfer detection of SRS.

The key to this approach is the careful choice of modulation frequency. The laser repetition rate places a practical upper bound on the modulation frequency, since at least two pump laser pulses are required for modulation (in that case, one associated with a Stokes pulse train and one without). Thus ~40 MHz is the highest modulation frequency possible with an 80 MHz pump laser. Practically, detection of 40 MHz is challenging for the large area detectors that we use, so modulation in the 10–20 MHz range is commonly used. For video rate scanning, 20 MHz is required because the pixels are scanned at higher than 10 megapixels per second. For normal imaging with pixel dwell times >1 μs, 10 MHz modulation is acceptable. High frequency modulation is important not only to modulate faster than the pixel dwell time, but also because laser noise exists primarily at low frequencies, due to thermal and mechanical fluctuations which are limited to <100 kHz. In addition, scanning the laser beam through a turbid sample also causes intensity modulation of the laser beam. If the modulation frequency is close to the pixel dwell time of the scanner, then scanning across a turbid sample can actually cause leakage of spurious modulation signal that is detected by the lock-in amplifier *even* when no Stokes beam is present at all. All of these considerations point to modulation frequencies >10 MHz. Secondarily, the availability of extremely high quality electrical bandpass filters, particularly at 10.7 MHz

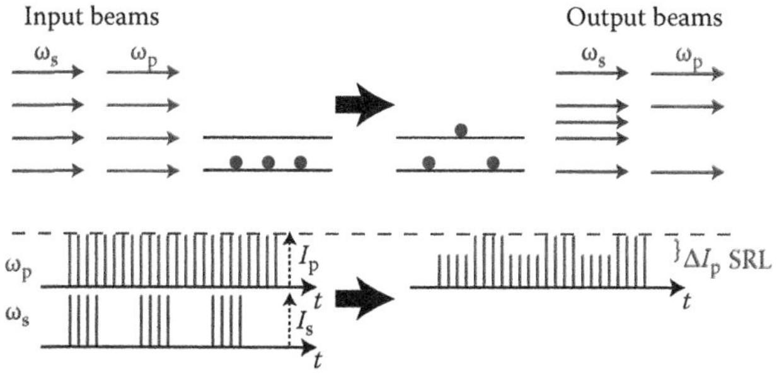


**FIGURE 3.9** The SRS process involves detection of an intensity transfer from the pump laser beam to the Stokes laser beam and the associated excitation of a molecular vibrational transition. This process is shown schematically here as the annihilation of a pump photon and creation of a Stokes photon together with the promotion of a (blue) molecule from the ground vibrational state to the first vibrational excited state. The modulation transfer process is depicted with two laser pulse trains. In this case, the Stokes beam is amplitude-modulated with a period of eight laser pulses (10 MHz modulation frequency for an 80 MHz laser repetition rate). Because SRS occurs when both pulse trains are applied to the sample and leads to a drop in pump beam intensity, while SRS does not occur when only the pump laser is present, amplitude modulation is transferred to the pump beam. This is detected with a lock-in amplifier.

and its second harmonic, 21.7 MHz (because these are commonly used as intermediate frequencies in heterodyne receivers) makes imaging easier because the use of a passive filter reduces saturation at the lock-in amplifier input.

Given these considerations, we need to modulate the laser at 10–20 MHz and detect this amplitude modulation. These topics are treated in the following two sections.

#### 3.5.2.1 Modulation

Amplitude modulation of a high peak and average power laser beam at high frequency is challenging. The peak powers used with our multi-watt level 1064 nm Stokes laser beam can damage acousto-optic modulators (AOMs) when tightly focused. Unfortunately, for high speed modulation, a tight focus is required by the AOM. This is because the sound wave that actuates the modulation must cross the laser beam waist in a perpendicular fashion. Given the speed of sound in commonly used acousto-optic materials, 10 MHz modulation would require a focal spot of <100 μm, and the resulting peak intensity is too high.

The use of broadband electro-optic modulators can also be problematic. This is because broadband electro-optic modulators make use of high power radiofrequency amplifiers with long cables to the relatively bulky modulator. These cables can emit electromagnetic interference which overwhelms lock-in amplifiers. Thus careful placement of the cables and amplifiers and good shielding is required. "Spooky" effects can be observed in which the noise level of the system depends on where individuals stand in the room are also observed, because human bodies can reflect electromagnetic radiation.

For this reason, the preferred modulator is a resonant Pockel cell. In this case, the capacitance of a small nonlinear crystal, in combination with a carefully chosen inductor, form a resonant "tank" circuit whose preferred frequency is the modulation frequency of interest. The resonance frequency of an inductor/capacitor tank circuit may be calculated according to the formula

$$f = \frac{1}{2\pi\sqrt{LC}} \tag{3.1}$$

where $L$ and $C$ represent the chosen inductance and crystal self-capacitance, respectively. At the resonance frequency, the impedance of the circuit becomes almost infinite, meaning that high voltages can be obtained across the capacitor (the nonlinear crystal) with relatively modest input power. This is highly desirable because it means that small radiofrequency amplifier (with output powers of <1 W) can be used to obtain high voltage modulation, and the entire system can be sealed in a metal enclosure to shield electromagnetic interference. If the nonlinear crystal is chosen appropriately, the achievable alternating current (AC) drive voltage will reach the ±1/4 wave voltage of the crystal for the input wavelength. Then, the modulator can be combined with a static $\lambda/4$ plate. In this configuration, when the modulator is at +1/4 wave, this retardance adds with the static waveplate to produce $\lambda/2$ retardance in total, effecting a 90° polarization rotation. When the modulator is at −1/4 wave, then the two retardances cancel and the net polarization rotation is zero. This whole optical system can be combined with a polarization analyzer so that the unrotated light is transmitted and the rotated light is suppressed. Thus, in total, with modest input power and a compact instrument, modulation at the chosen reference frequency is achieved with essentially 100% modulation depth. Given that the SRS signal scales linearly with modulation depth, maximizing that depth is

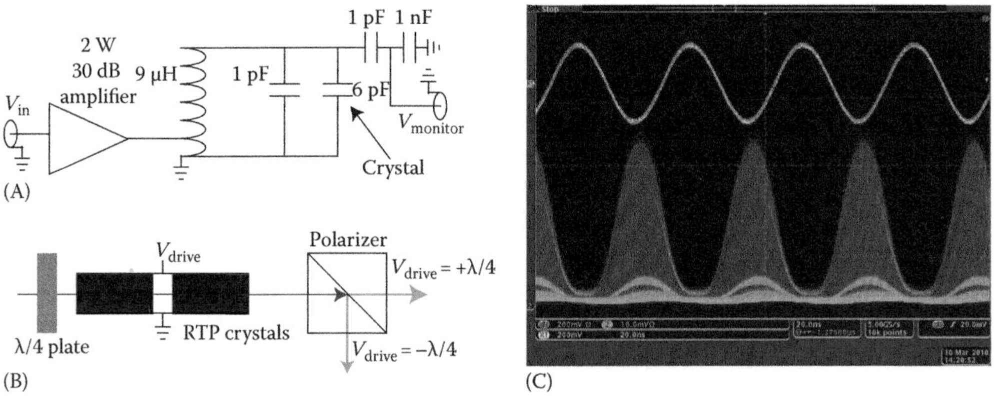


**FIGURE 3.10** Schematic of the resonant Pockel cell for high frequency modulation. (A) The *LC* tank circuit as described earlier is tuned to give a resonance at the desired frequency. The self-capacitance of the crystal, together with stray capacitance in the circuit (represented by a 1 pF lumped element) forms the *C*. The inductor is chosen to match the resonance and driven on a tap to obtain a voltage step-up. A low power RF amplifier drives the circuit, which is compact. A 1000:1 pickoff ratio formed by two additional capacitors is used to monitor the high voltage. (B) Schematic of the polarization optics that comprise the modulator, as described in the text. The polarizer transmission is determined by the voltage applied to the rubidium titanyl phosphate (RTP) nonlinear crystals. (C) Oscilloscope trace of the monitor voltage (yellow) and the laser pulse train shows 20 MHz modulation with high modulation depth.

important to obtain the highest signal levels for a given average power on the sample. Figure 3.10 shows a schematic modulator drive circuit with sample component values and an oscilloscope trace of the modulated waveform.

#### 3.5.2.2 Detection and Demodulation

Details of the modulation transfer detection scheme and its variations are found throughout this book and in a recent publication [22]. Here we will emphasize several items of general importance.

First, collection of light for transmission detection of SRS is different from that for CARS because high NA collection is essential. The reason is that if the complete cone angle emitted from the sample is not collected, then clipping of the beam can lead to a spurious background due to cross-phase modulation [22]. This arises from Kerr-lensing in or near the focal volume, because the presence of the modulated beam can influence the divergence of the unmodulated beam through the nonresonant Kerr effect. In that case, the emitted, originally unmodulated beam will have an amplitude modulation due to SRS (in the case of a vibrationally resonant sample) and a divergence modulation due to this cross-phase modulation effect (whether or not the sample is vibrationally resonant). If the beam then clips on any aperture, the divergence modulation will become amplitude modulation and will appear as a spurious background signal in the SRS detector.

Collection of light for epi-detection is challenging in SRS. The efficient collection requires specialized non-imaging optics and a large-area detector and is detailed in a recent publication [23].

After collection, the light is relayed onto a large active area detector. An optical filter, such as the Chroma Technology CARS890/220 must be used to very efficiently block the Stokes beam while transmitting the pump beam with high efficiency. Because the Stokes beam has 100% amplitude modulation, even a small fraction (<0.1%) will overwhelm the lock-in amplifier and prevent imaging. Again, to avoid cross-phase modulation, the beam must not clip on the edges of the detector. Practically, this makes the use of a large area (>5 × 5 mm) advantageous because then the optical system does not have to be perfect and a small amount of beam motion at the detector will not cause clipping and therefore cross-phase modulation. Because of the relatively high light levels (>50 mW is detected), highly sensitive detectors that are capable of photon counting such as PMTs or avalanche photodiodes are not needed (and would saturate, so they cannot be used). Instead, ordinary silicon PIN photodiodes are used. These diodes have large dynamic ranges, are inexpensive and can be obtained with large active areas (easily up to 10 × 10 mm). They have high quantum efficiency throughout the near-infrared. Because of the difference in quantum efficiency of silicon photodiodes compared to the multi-alkali PMTs used for CARS, SRS with a fixed 1064 nm Stokes wavelength can cover a far wider range of vibrational frequencies despite the fact that CARS always detects the shorter anti-Stokes wavelength compared to the longer pump wavelength in SRS for the same Raman shift. Figure 3.11 compares the quantum efficiency of the SRS photodiode with photomultiplier tubes used for CARS on a wavenumber basis (considering the pump wavelength detection for SRS and the anti-Stokes for CARS) to demonstrate this difference.

Two key parameters of PIN photodiodes are not ideal under zero bias voltage, however. First, they saturate easily, producing nonlinear response, and second, their intrinsic capacitance limits the temporal response of the circuit. Both of these problems can be solved by back-biasing the diode to 10–100 V, to achieve “photoconductive” operation. The back bias of the diode increases the thickness of the depletion layer, which both reduces the capacitance and increases the saturation threshold because a large volume of electron–hole pairs can be produced. The photodiode is back-biased through an electrical low pass filter whose corner frequency is much lower than the modulation frequency. This is critical for low noise operation because electrical noise can be picked up by the bias cables and then is detected as spurious amplitude modulation by the lock-in amplifier. The photodiode output is demodulated by a lock-in amplifier. Figure 3.12 shows the

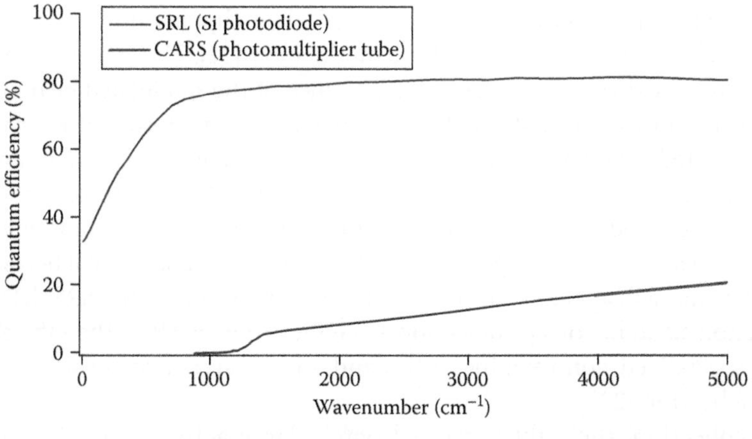


**FIGURE 3.11** Comparison of detector quantum efficiency for CARS and SRS. Because SRS makes use of a silicon PIN photodiode (FDS1010, Thorlabs) which has higher quantum efficiency throughout the near infrared, it can be extended to probe lower wavenumber shifts than CARS with a fixed 1064 nm Stokes wavelength, which uses a multi-alkali photomultiplier tube (R3896, Hamamatsu).

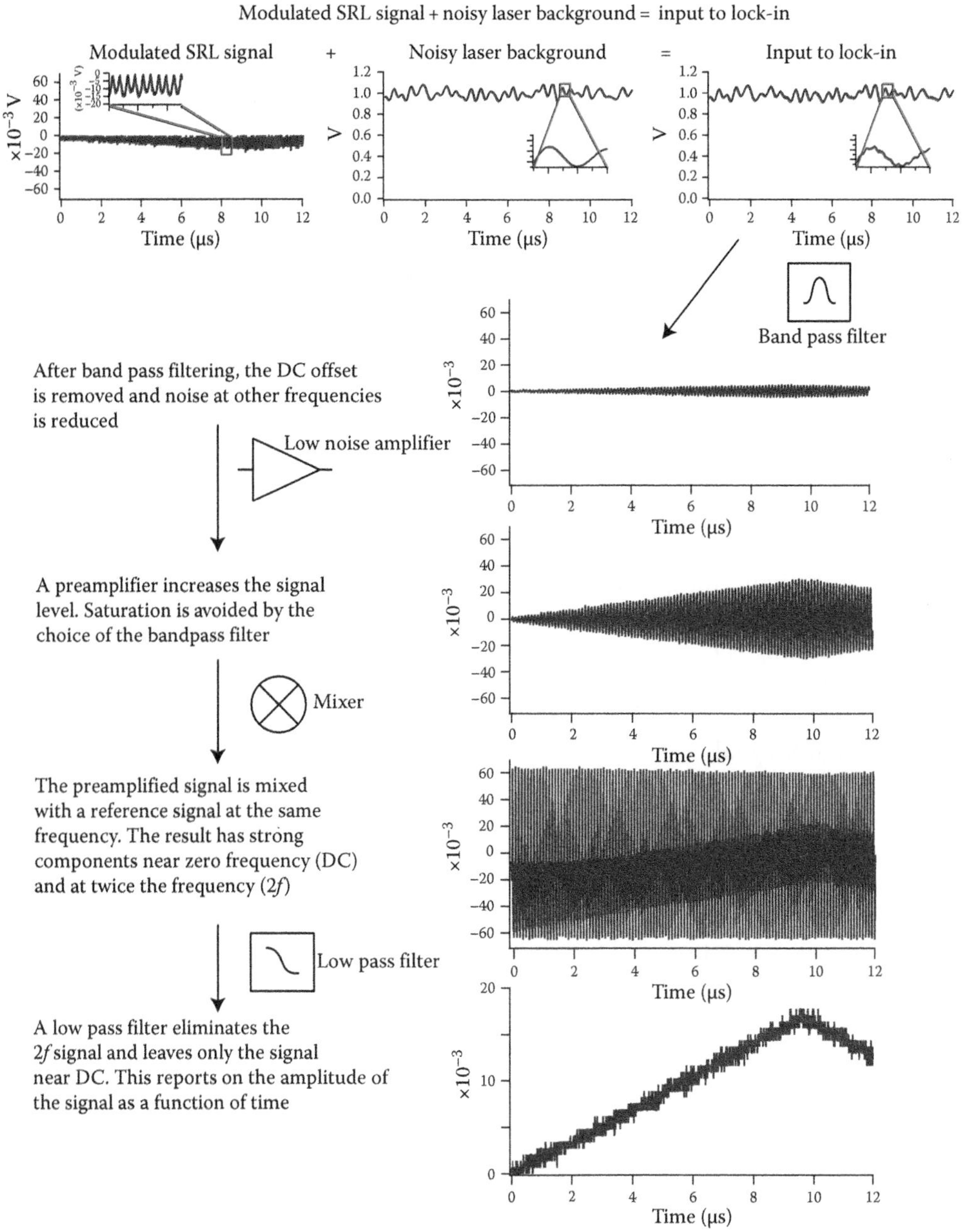


**FIGURE 3.12** The signal processing chain of a lock-in amplifier. The signal of interest is the SRS signal which is a triangular wave. The original modulated signal, together with random laser noise, and added together at the top to produce the lock-in amplifier input, where the modulation can barely be discerned in the inset. The resulting signal is band-pass filtered, pre-amplified, mixed with the reference wave to produce the sum and difference frequencies, and then low-pass filtered to produce the DC-coupled signal which reports on the envelope of the amplitude modulation in the original data on the top left. This schematically depicts the key steps of signal processing.

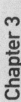

signal-processing chain of a generic lock-in amplifier, which takes a modulated AC waveform in and reports the amplitude and phase of the modulation signal while suppressing noise at other frequencies. In this configuration, stable operation at high gain (>80 dB) is possible because of the relatively narrow bandwidth.

## 3.6 Verifying the Presence of CARS and SRS Signals

When first building a CARS or SRS microscope, it can be difficult to determine the origin of an observed signal on either the PMT or the lock-in amplifier detector. However, a brief checklist can be utilized to verify the identity of a signal. Typically, a strongly resonant sample (e.g., a thin layer of dodecane between two cover slips) should be utilized and the maximum available power (at least 100 mW in each beam for a 6 ps laser system at 80 MHz repetition rate) should be applied to the sample.

For a CARS signal,

- Does the signal scale quadratically with pump beam power and linearly with Stokes beam power?
- Does the signal appear only at the anti-Stokes frequency?
- Does the signal disappear completely when either beam is blocked or the time delay is offset by an amount corresponding to the laser pulse duration?
- By scanning the incident laser frequency difference, do characteristic peaks and dips appear that compare with CARS spectra reported in the literature?

For an SRS signal,

- Does the signal scale linearly with both pump and Stokes power? Does it disappear when blocking either beam or detuning the time delay?
- Does the signal disappear completely when shutting off the modulator drive waveform?
- Does the signal disappear completely when the photodiode is unbiased?
- Does the signal disappear completely when tuning the lasers off resonance with the vibrational mode of interest?
- Does the excitation spectrum obtained when tuning the laser difference frequency across the Raman band of interest match the literature Raman spectrum?

## 3.7 Summary

The goal of this chapter is to familiarize the reader with the practical essentials for constructing a narrowband CRS imaging system capable of performing rapid, high sensitivity chemical imaging. The historical and technical comments on the laser system serve to offer perspective on the development of the field, but an important point is that laser systems for CRS are already commercially available, and the rapid developments in fiber laser technology and nonlinear frequency conversion promise that CRS laser sources will continue to improve in ease-of-use while dropping in price for years to come. Integration of these laser sources with laser scanning microscope is also becoming increasingly straightforward, with several major manufacturers offering either CRS-specific products, or confocal/multiphoton microscopes that can be readily adapted by making simple modifications to the optical filters and detection systems. By following some of the suggestions given throughout this chapter, the reader can benefit from significant improvements in the performance and ease-of-use of CRS imaging systems that have been realized over the past decade. Overall, the barriers to entry into the CRS field are rapidly crumbling, making this powerful label-free chemical imaging technology available to an increasingly wide array of scientists, engineers and even clinicians.

## References


1. M.D. Duncan, J. Reintjes, and T.J. Manuccia. Scanning coherent anti-Stokes Raman microscope. *Optics Letters*, 7(8):350–352, 1982.
2. D.J. Jones, E.O. Potma, J. Cheng, B. Burfeindt, Y. Pang, J. Ye, and X.S. Xie. Synchronization of two passively mode-locked, picosecond lasers within 20 fs for coherent anti-Stokes Raman scattering microscopy. *Review of Scientific Instruments*, 73:2843, 2002.
3. F. Ganikhanov, S. Carrasco, X.S. Xie, M. Katz, W. Seitz, and D. Kopf. Broadly tunable dual-wavelength light source for coherent anti-Stokes Raman scattering microscopy. *Optics Letters*, 31(9):1292–1294, 2006.
4. C.L. Evans, E.O. Potma, M. Puoris' haag, D. Côté, C.P. Lin, and X.S. Xie. Chemical imaging of tissue in vivo with video-rate coherent anti-Stokes Raman scattering microscopy. *Proceedings of the National Academy of Sciences of the United States of America*, 102(46):16807, 2005.
5. S. Murugkar, C. Brideau, A. Ridsdale, M. Naji, P.K. Stys, and H. Anis. Coherent anti-Stokes Raman scattering microscopy using photonic crystal fiber with two closely lying zero dispersion wavelengths. *Optics Express*, 15(21):14028–14037, 2007.
6. K. Kieu, B.G. Saar, G.R. Holtom, X.S. Xie, and F.W. Wise. High-power picosecond fiber source for coherent Raman microscopy. *Optics Letters*, 34(13):2051–2053, 2009.
7. C.L. Evans and X.S. Xie. Coherent anti-Stokes Raman scattering microscopy: Chemical imaging for biology and medicine. *Annual Review of Analytical Chemistry*, 1:883–909, 2008.
8. A. Zumbusch, G.R. Holtom, and X.S. Xie. Three-dimensional vibrational imaging by coherent anti-Stokes Raman scattering. *Physical Review Letters*, 82(20):4142–4145, 1999.
9. W. Denk, J.H. Strickler, and W.W. Webb. Two-photon laser scanning fluorescence microscopy. *Science*, 248(4951):73, 1990.
10. A.F. Pegoraro, A. Ridsdale, D.J. Moffatt, Y. Jia, J.P. Pezacki, and A. Stolow. Optimally chirped multimodal CARS microscopy based on a single Ti:sapphire oscillator. *Optics Express*, 17:2984–2996, 2009.
11. H. Chen, H. Wang, M.N. Slipchenko, Y.K. Jung, Y. Shi, J. Zhu, K.K. Buhman, and J.X. Cheng. A multimodal platform for nonlinear optical microscopy and microspectroscopy. *Optics Express*, 17(3):1282–1290, 2009.
12. Y. Ozeki, F. Dake, S. Kajiyama, K. Fukui, and K. Itoh. Analysis and experimental assessment of the sensitivity of stimulated Raman scattering microscopy. *Optics Express*, 17(5):3651–3658, 2009. ISSN: 1094-4087.
13. J. Cheng, A. Volkmer, L.D. Book, and X.S. Xie. Multiplex coherent anti-Stokes Raman scattering microspectroscopy and study of lipid vesicles. *Journal of Physical Chemistry B*, 106(34):8493–8498, 2002.
14. G.W.H. Wurpel, J.M. Schins, and M. Müller. Chemical specificity in three-dimensional imaging with multiplex coherent anti-Stokes Raman scattering microscopy. *Optics Letters*, 27(13):1093–1095, 2002.
15. S.H. Lim, A.G. Caster, and S.R. Leone. Fourier transform spectral interferometric coherent anti-Stokes Raman scattering (FTSI-CARS) spectroscopy. *Optics Letters*, 32(10):1332–1334, 2007.
16. A. Hopt and E. Neher. Highly nonlinear photodamage in two-photon fluorescence microscopy. *Biophysical Journal*, 80(4):2029–2036, 2001.
17. E.R. Andresen, C.K. Nielsen, J. Thøgersen, and S.R. Keiding. Fiber laser-based light source for coherent anti-Stokes Raman scattering microspectroscopy. *Optics Express*, 15:4848–4856, 2007.

18. G. Krauss, T. Hanke, A. Sell, D. Träutlein, A. Leitenstorfer, R. Selm, M. Winterhalder, and A. Zumbusch. Compact coherent anti-Stokes Raman scattering microscope based on a picosecond two-color Er: Fiber laser system. *Optics Letters*, 34(18):2847–2849, 2009.
19. A.F. Pegoraro, A. Ridsdale, D.J. Moffatt, J.P. Pezacki, B.K. Thomas, L. Fu, L. Dong, M.E. Fermann, and A. Stolow. All-fiber CARS microscopy of live cells. *Optics Express*, 17:20700–20706, 2009.
20. Y. Dai and C. Xu. Generation of high repetition rate femtosecond pulses from a CW laser by a time-lens loop. *Optics Express*, 17(8):6584–6590, 2009.
21. J.E. Sharping, M. Fiorentino, P. Kumar, and R.S. Windeler. Optical parametric oscillator based on four-wave mixing in microstructure fiber. *Optics Letters*, 27(19):1675–1677, 2002.
22. C.W. Freudiger, W. Min, B.G. Saar, S. Lu, G.R. Holtom, C. He, J.C. Tsai, J.X. Kang, and X.S. Xie. Label-free biomedical imaging with high sensitivity by stimulated Raman scattering microscopy. *Science*, 322(5909):1857, 2008.
23. B.G. Saar, C.W. Freudiger, J. Reichman, C.M. Stanley, G.R. Holtom, and X.S. Xie. Video-rate molecular imaging in vivo with stimulated Raman scattering. *Science*, 330:1368, 2010.

# 4. Stimulated Raman Scattering Microscopy

**Christian Freudiger and Xiaoliang Sunney Xie**



## 4.1 Label-Free Microscopy Based on Coherent Raman Scattering

While vibrational spectroscopy is an ideal candidate for optical contrast with chemical specificity and offers a label-free alternative to fluorescent labeling and dye staining (Turrell et al., 1996; Gremlich and Yan, 2001), microscopies based on infrared (IR) absorption and spontaneous Raman scattering are limited by low spatial resolution and intrinsically feeble signals, respectively. As discussed in previous chapters, coherent anti-Stokes Raman scattering (CARS) can overcome these limitations by providing coherent signal amplification of the weak spontaneous Raman scattering signals by orders of magnitude.

CARS was first discovered by Maker and Terhune (1965) and has become a widely used spectroscopy technique (Levenson and Kano, 1988). It was first utilized as a contrast mechanism for microscopy by Duncan et al. (1982). However the requirement for synchronized lasers prevented further development at the time. In 1999, CARS microscopy was implemented with near-IR laser beams and in a more straightforward collinear geometry by Zumbusch et al. (1999). Since then, optimized laser systems have made CARS microscopy more straightforward (Potma et al., 2002; Ganikhanov et al., 2006a) and have enabled the

*Coherent Raman Scattering Microscopy*. Edited by Ji-Xin Cheng and X. Sunney Xie © 2013 CRC Press/ Taylor & Francis Group, LLC. ISBN: 978-1-4398-6765-5.

first biomedical applications (Cheng and Xie, 2004; Evans and Xie, 2008) with imaging speeds up to video-rate (Evans et al., 2005).

Despite the major advantages in imaging speed of CARS compared to spontaneous Raman scattering microscopy, it has not yet been widely accepted in biomedical research. Compared to other microscopy techniques, CARS does not allow for straightforward image interpretation due the difficulties associated with the non-resonant background signal caused by the electronic response of the sample and the coherent signal addition. Specific limitations in CARS microscopy are

- Image artifacts due to spatial interference (Cheng et al., 2002)
- Spectral distortions due to spectral interference (Rinia et al., 2006)
- Nonlinear dependence of the signal on the concentration of the target species (Li et al., 2005)
- Limited sensitivity (weak resonant signals are buried in the laser-noise associated with the non-resonant background [Ganikhanov et al., 2006b])

These limitations have motivated a series of technical improvements to the original implementation of CARS microscopy (Zumbusch et al., 1999) over the last 10 years, with the aim of overcoming the non-resonant background. These include epi-detected CARS (E-CARS) (Cheng et al., 2001a), polarization-sensitive CARS (P-CARS) (Cheng et al., 2001b), time-resolved CARS (T-CARS), interferometric CARS (I-CARS) (Evans et al., 2004), and frequency-modulated CARS (FM-CARS) microscopy (Ganikhanov et al., 2006b; Saar et al., 2009; Chen et al., 2010). However, while adding significant technical complexity to the measurement, these new methods have not fully overcome the limitation of CARS, because the technique either does not fully suppress the background (E-CARS), sacrifices signal (P-CARS and T-CARS), cannot be applied to the imaging of turbid biological samples (I-CARS), or is extremely challenging to implement in a robust fashion (FM-CARS). Thus, despite many years of technical improvements, CARS microscopy has remained limited by the non-resonant background and is commonly criticized as only being good for the imaging of lipids.

This chapter summarizes the development of stimulated Raman scattering (SRS) (Woodbury and Ng, 1962; Owyoung, 1977; Levine et al., 1979; Levenson and Kano, 1988; Kukura et al. 2007) as a contrast for label-free microscopy (Ploetz et al., 2007; Freudiger et al., 2008; Ozeki et al., 2009; Nandakumar et al., 2009). SRS microscopy has overcome all limitations of CARS microscopy mentioned earlier, superseding CARS and enabling broader applications of CRS microscopy.

## 4.2 Principle of Stimulated Raman Scattering

SRS is another coherent Raman scattering (CRS) process, which is excited under the same conditions as resonant CARS (Levenson and Kano, 1988). In contrast to spontaneous Raman scattering, in which the sample is illuminated with one excitation field, in SRS two excitation fields at the pump frequency, $\omega_p$, and Stokes frequency, $\omega_S$, coincide on the sample. If the difference frequency of the excitation beams, $\Delta\omega = \omega_p - \omega_S$, matches a vibrational frequency, $\Omega$, of a molecule in the focus, the molecular transition

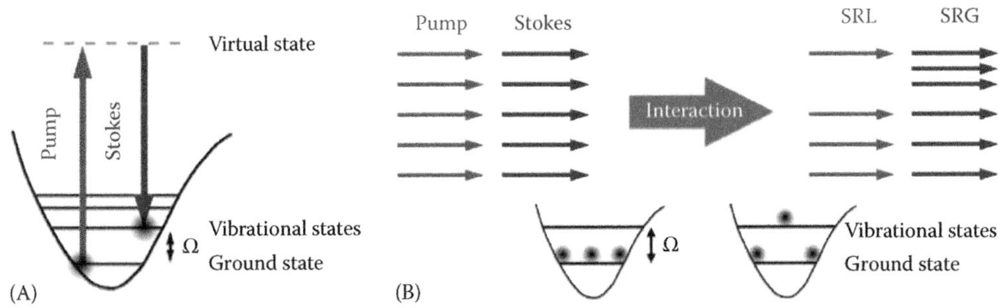


FIGURE 4.1 Principle of stimulated Raman scattering. (A) Energy diagram of SRS. The combined action of pump and Stokes beams effectively transfers the molecules in the sample from the ground state into the first vibrationally excited state by passing through a virtual state. The vibrational state to be excited can be selected by tuning the difference frequency between the pump and Stokes beams. (B) SRS as an energy transfer process. As a consequence of the excitation of a molecular vibration, a pump photon is absorbed and a Stokes photon is generated, which results in SRL and SRG of the transmitted pump and Stokes beams, respectively.

rate is enhanced due to the stimulated excitation of molecular transitions. The molecular population is transferred from the ground state through a virtual state to the vibrationally excited state of the molecule (Figure 4.1A). This is in contrast with spontaneous Raman scattering, in which the transition from the virtual state to the vibrationally excited state is spontaneous, resulting in much weaker signal.

As a result of the coherent excitation of a molecular vibration (Figure 4.1B), a pump photon is absorbed by the sample and a Stokes photon is generated. This results in loss (stimulated Raman loss, SRL) and gain (stimulated Raman gain, SRG), $\Delta I_p$ and $\Delta I_S$, of the intensity of the transmitted pump and Stokes beams, $I_p$ and $I_S$, respectively:

$$\Delta I_S \propto N \cdot \sigma_{Raman} \cdot I_p \cdot I_S \tag{4.1}$$

$$\Delta I_p \propto -N \cdot \sigma_{Raman} \cdot I_p \cdot I_S \tag{4.2}$$

where

$N$ is the number of molecules in the probe volume

$\sigma_{Raman}$ is the molecular Raman scattering cross-section (Levenson and Kano, 1988; Boyd, 2003)

In SRS microscopy we measure either $\Delta I_p$ (SRL) or $\Delta I_S$ (SRG) as a function of the position in the sample. Because $\Delta I \propto N$, that is, the signal is proportional to the concentration, $c$, of the target species, it is now possible to generate a quantitative chemical map of the sample. Different chemical species can be targeted based on different vibrational frequencies as documented in the spontaneous Raman literature as $\Delta I \propto \sigma_{Raman}$.

Because of the nonlinear dependence of the signal on the excitation intensities (overall quadratic), SRS allows for intrinsic optical sectioning similar to two-photon microscopy (Denk et al., 1990), eliminating the need for a confocal pinhole. This is especially useful for the imaging of thick tissue samples (Helmchen and Denk, 2005).

CARS is generated under the same conditions as SRS, but it differs in the method of detection. In SRS the gain and loss of intensity in the excitation beams is detected, while in

**Table 4.1** Comparison of Spontaneous Raman Scattering and Coherent Raman Scattering

| Spontaneous Raman Scattering | SRS |
|---|---|
| One-photon process | Multi-photon process |
| Very slow imaging (>20 min/frame) | Fast imaging up to video-rate speed (30 fps) |
| No intrinsic *z*-resolution | Optical sectioning |
| Excitation with visible/UV beams for enhanced scattering | Excitation with near-IR beam for enhanced imaging depth |
| Vulnerable to background fluorescence | Immune to background fluorescence |
| Full spectra | Selected spectral information |

**Table 4.2** Comparison of CARS and SRS

| CARS | SRS |
|---|---|
| Parametric process | Energy transfer process |
| Signal at new optical frequency ($\omega_{aS}$) | Intensity gain and loss of transmitted excitation beams |
| Unspecific non-resonant background | No non-resonant background |
| Distorted spectra | Spectra identical to spontaneous Raman |
| Coherent image artifacts | Signal is the convolution of the object with a point-spread function |
| Nonlinear concentration dependence | Linear concentration dependence |

CARS, new radiation at the anti-Stokes frequency, $\omega_{aS} = 2\omega_p - \omega_S$, is measured (Cheng and Xie, 2004; Evans and Xie, 2008). CARS is generated due to the optical parametric process known as four wave mixing, in which energy is exchanged between the optical fields. This is in contrast to SRS, which is an energy transfer process between the optical fields and the sample. This explains why SRS cannot occur if $\Delta\omega$ does not match a vibrational frequency of the sample and therefore does not suffer from the non-resonant background, because the sample has no eigenstate to absorb the quantum of vibrational energy.

In summary, SRS has all the advantages of coherent Raman scattering techniques when compared with spontaneous Raman scattering (see Table 4.1), but overcomes the limitations of CARS microscopy (see Table 4.2).

## 4.3 SRS Microscopy under Biocompatible Excitation Conditions

The phenomenon of SRS was first observed in 1962 by Woodbury and Ng (Woodbury and Ng, 1962), even before CARS was discovered. It has been used extensively in spectroscopy applications which require high sensitivity (Owyoung, 1977; Levine et al., 1979; Kukura et al., 2007). Although the advantages of SRS over CARS were well understood in the 1970s and 1980s (Levenson and Kano, 1988), SRS as a contrast mechanism for microscopy was overlooked. This was most likely because the intensity changes of the transmitted beams due to SRS are very small compared to changes in the sample transmission due to linear scattering and absorption in the heterogeneous samples that are typically examined under a microscope.

The original implementation of SRS microscopy by Ploetz et al. (2007) overcomes this problem by following the route of SRS spectroscopy (Kukura et al., 2007) and using an amplified laser system to boost the nonlinear signal. However such an approach is not suitable for biomedical imaging because the large peak power causes sample damage (Hopt and Neher, 2001; Fu et al., 2006) and the low laser repetition rate limits the image acquisition speed.

## 4.3.1 High-Frequency Modulation Transfer

We (Freudiger et al. 2008) and independently Ozeki et al. (2009) and Nandakumar et al. (2009) chose a different approach. Instead of an amplified laser system, we use the same high repetition rate (e.g., 76 MHz) lasers as for previous CARS imaging, which have orders of magnitude lower peak power than the amplified systems mentioned earlier. While bio-compatible, these excitation conditions create an SRS signal of $\Delta I_p/I_p$ and $\Delta I_S/I_S < 10^{-4}$. To detect such small signals with high sensitivity, we implement a high-frequency phase-sensitive detection scheme. The basic idea is to modulate the minute SRS signal at a known frequency $f$ and extract it from the much larger laser intensity (with its characteristic fluctuations) with a lock-in amplifier (Owyoung, 1977; Levine et al., 1979; Levenson and Kano, 1988). For SRL we modulate the intensity of the Stokes beam and detect the resulting modulation transfer to the originally unmodulated pump beam at the same frequency (Figure 4.2A). Similarly, SRG can be measured by modulating the pump and detecting the Stokes beam (Figure 4.2B).

Realizing that laser noise occurs primarily at low frequencies due to the $1/f$ noise and low frequency thermal and mechanical fluctuations (Figure 4.2C), we chose a modulation frequency higher than 1 MHz. In simple terms, we determine the difference between the intensity of one of the transmitted beams with "SRS on" and "SRS off" and the sensitivity is the highest when we make this measurement so rapidly that the laser intensity has not yet varied due to characteristic low frequency noise.

With this approach, $\Delta I_p/I_p = 2 \times 10^{-8}$ can be achieved with a 1 s time constant. For imaging, we typically use pixel dwell times <100 μs, thus the typical sensitivity is $6 \times 10^{-7}$ (see Figure 4.2D). This is within a factor two of the theoretical shot noise limit for bio-compatible average power levels.

## 4.3.2 Instrumentation

Since the original implementation of narrowband SRS microscopy (Freudiger et al., 2008), we have optimized various components of the experimental design, but the main instrument as shown in Figure 4.3A has remained unchanged. Typically, we detect SRL instead of SRG, because the responsivity of silicon photo-diodes is much higher for the pump than for the Stokes beam (InGaAs photo-diodes can overcome this problem but are less robust). As such, the SRS microscope in Figure 4.3A is based on the modulation scheme for SRL as shown in Figure 4.2A.

In transparent or semi-transparent samples, the transmitted pump or Stokes beams are typically detected in the forward direction. In turbid samples, backward (epi)

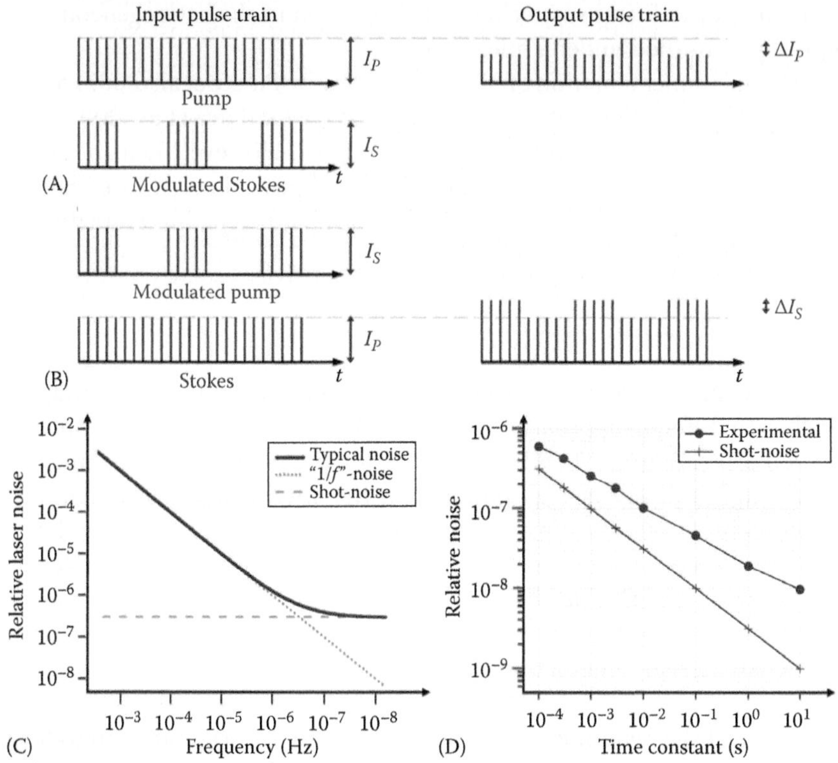


**FIGURE 4.2** High-frequency detection scheme for SRS to extract the relatively small SRS signal from the intensity fluctuations of the lasers and varying sample transmission. (A) For the detection of SRL, the Stokes beam is modulated at high frequency (>1 MHz), at which the resulting amplitude modulation of the pump beam due to SRL can be detected. (B) For the detection of SRG, the pump beam is modulated and the modulation transfer to the Stokes-beam due to SRG is measured. (C) Schematic of the frequency dependence of laser noise. The main sources of laser noise are the "1/*f*"-noise, which arises from cavity fluctuations and decreases at high frequency, since every laser pulse has several round-trips in the cavity, and the shot-noise, which arises from the quantization of the electromagnetic field into photons and is white noise. This shows that in order to measure the SRS signal with high sensitivity, it has to be detected at high frequency. The schematic does not indicate laser intensity fluctuations due to varying sample transmission while scanning the beams through the turbid sample and the 76 MHz peak due to the laser repetition rate. (D) Detection sensitivity as a function of the time constant of the lock-in amplifier (i.e., averaging time) for 1.7 MHz modulation rate. The blue curve shows the measured relative noise floor as recorded with 0.8 V at the photo-diode (corresponds to 40 mW of 800 nm light in focus) and the red curve the theoretical estimate of the shot-noise limit. At the short time constants relevant for imaging, SRS microscopy with high-frequency phase-sensitive detection is close to shot-noise limited.

detection is possible because multiple scattering events redirect a significant portion of the forward-propagating pump and Stokes beams to the backward direction. Such epi-detection will be discussed in a later section.

Typically, the Stokes beam is provided from a fixed wavelength Nd:$YVO_4$ laser (picoTRAIN, High-Q) at 1064 nm with 7 ps pulsewidth and 76 MHz repetition rate. To provide a synchronized and continuously tunable pump beam, a portion of this 1064 nm beam is frequency-doubled to synchronously pump an optical parametric oscillator (OPO) (Levante Emerald, APE-Berlin, Germany). The OPO uses a temperature-tuned

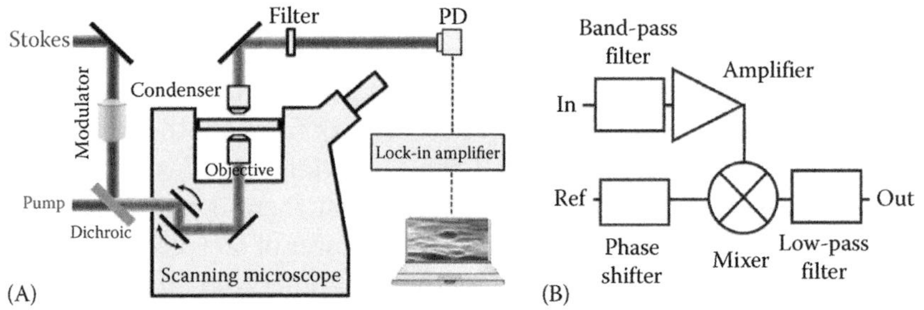


FIGURE 4.3 Experimental realization of narrowband SRS microscopy. (A) Basic setup of an SRL microscope. Synchronized pump and Stokes beams are spatially overlapped with a dichroic mirror, collinearly aligned into a beam scanning microscope, and focused into a common focal spot within the sample with an objective lens. The Stokes beam is modulated by an electro-optic or acousto-optic modulator. The transmitted beams are detected with a high-NA condenser, filtered to block the Stokes beam and detected by a large-area photo-diode. The SRL signal is extracted from the laser intensity with a lock-in amplifier to provide the intensity of a pixel. 3D images are obtained by raster-scanning the laser focus across the sample and micro-spectroscopy can be performed by automated tuning of the OPO wavelength. (B) Basic design of a lock-in amplifier. The output from the photo-diode is fed into the input (In) of the lock-in amplifier, coarsely filtered with a band-pass filter around the modulation frequency, and mixed with a reference from the modulator (Ref). The phase of the reference can be shifted to maximize signal. As a result of the mixing, the modulation frequency is shifted to DC, such that it can be selectively detected with a low-pass filter, amplified, and sent to the A/D converter of the microscope (Out). Typical lock-in detectors have multiple gain and filter stages.

non-critically phase-matched LBO crystal to allow coarse tuning of the difference frequency with the Stokes beam from 500 to 3400 $cm^{-1}$ (Ganikhanov et al., 2006a). The fine tuning with <1 $cm^{-1}$ resolution within the phase-matching bandwidth for a particular temperature is achieved with a stacked Lyot filter. The idler beam of the OPO is blocked with an interferometric filter (CARS 890/220M, Chroma Technology). The pump and Stokes beams, that is, the signal beam from the OPO and the 1064 nm fundamental, are spatially overlapped with a dichroic beam-combiner (1064DCRB, Chroma Technology) and the temporal overlap is ensured with a delay-stage that is positioned with the help of an optical autocorrelator (PulseCheck, APE Berlin, Germany) (Evans and Xie, 2008; Freudiger et al., 2008).

The Stokes beam is modulated with an electro-optic modulator (EOM) (EO-AM-NR-C2, Thorlabs or homebuilt version) or an acousto-optic modulator (AOM) (AOMO 3080-197, Crystal Technology). Acousto-optic modulators have the advantage that the drive voltage is relatively low, so RF shielding to minimize coherent pickup in the photodiode, cabling, and the lock-in detector is less critical. However we have realized that AOMs have long term damage issues with Watt level input power. EOMs have the additional advantage, that 100% modulation depth can be achieved at modulation frequencies of 20 MHz and higher. Thus the combination of a Pockel cell with a polarization analyzer is the preferred choice for SRS microscopy. Care has to be taken to use a resonant Pockels cell (such as EO-AM-NR-C2) rather than a broadband Pockels cell (such as 360-80, ConOptics) to avoid RF pickup, because in a resonant cell the total drive power is comparatively modest and the high currents may be confined inside the electrically shielded box around the crystal.

The modulation frequency (>1 MHz) is chosen somewhat arbitrarily, in many cases determined by the availability of RF filters after the photo-diode from vendors such as Minicircuits. Original experiments were performed at >1.7 MHz, but we have achieved slightly better signal-to-noise at ≈10 MHz. For video-rate imaging, a modulation rate of about ≈20 MHz is required in order to provide enough modulation cycles per pixel (see the following). Recent work from another group (Ozeki et al., 2010) suggests that modulation at half the repetition rate (Saar et al., 2009) of the lasers is most advantageous. However we find that there is no photodiode available that can handle both this speed and the average power of our picosecond laser system. We also found that no synchronization of the laser repetition rate and the modulation rate is required. Typically the modulator can be driven directly by the synchronization output of the lock-in amplifier or by a two-channel function generator (such as the AFG3252, Tektronix, Beaverton, OR).

Collinear pump- and Stokes-beams are coupled into a multi-photon laser scanning microscope, optimized for near-IR throughput. We find that multi-photon microscopes from the commercial vendors work well. The initial work (Freudiger et al., 2008) was performed on an upright Olympus microscope (BX61WI/FV300). For video-rate imaging we relied on the resonant scanner in a Leica microscope (TCS SP5) (Saar et al., 2010a). Initial testing of an inexpensive scanner from Thorlabs (VCMS-F) also yielded promising results. The choice of objective is more critical, as chromatic aberrations can cause the two excitation fields to focus at different depths in the sample, resulting in weakened SRS signal. We have found the Olympus UPlanApo/IR 60X 1.2 NA and Leica HCX IRAPO 25X 0.95 NA water immersion objectives to give large signals and large fields of view. The beam sizes should be matched to fill the back-aperture of the chosen objective.

Because the SRS signal only depends on the difference frequency of the pump and Stokes beams, the absolute wavelengths can be chosen independently. Typically, we use excitation within the "optical window" of tissue from 800 to 1100 nm, where both absorption and scattering are minimized and optical penetration depth is therefore maximized in tissue. Thus the combination of the OPO signal beam and the 1064 nm fundamental of the pump lasers as described earlier is ideal. Previously, signal and idler from the OPO have also been used for CARS, providing improved penetration (Ganikhanov et al., 2006a; Evans et al., 2007). Spurious background signal from two-color two-photon absorption (TPA), which results in modulation transfer similar to SRS but is independent of the Raman shift, can arise in biological samples. TPA has previously been used as label-free contrast in microscopy (Fu et al., 2007a,b) to image hemoglobin and melanin. To minimize its contribution to SRS, we use long-wavelength excitation. For a strongly absorbing sample, the combination of 1064 nm and idler would be possible and might provide a reduction in TPA signal.

Light is collected in transmission with a high-NA condenser (Nikon, 1.4 NA), which is aligned with white-light transmission from a lamp. It is critical to use a condenser with an NA that is higher than that of the excitation objective in order to minimize spurious background due to cross-phase-modulation (Ekvall et al., 2000). Cross-phase modulation originates from the non-resonant Kerr effect, which causes an index of refraction change as a response to strong light fields. Because the Stokes beam is focused into the sample, it has a gradient in intensity and thus causes a gradient in the index of refraction, that is, it causes a micro-lens, which changes the divergence of the pump beam, whenever the Stokes beam is switched on. Thus a modulation of the Stokes beam results in a modulation of the divergence of the pump beam after passing through the focus. Any type of aperture, including a low-NA condenser, can transform the divergence modulation into an amplitude modulation that is detected by the lock-in amplifier. For this reason, care has to be taken to collect all the light from the sample in order to avoid the spurious background signal (Freudiger et al., 2008).

A telescope is used to image the scanning mirrors onto the photodiode to avoid beam-movement due to laser scanning. A high OD band-pass filter (CARS 890/220, Chroma Technology) is used to block the Stokes-beam (at 1064 nm) and transmit the pump-beam. This filter works for imaging of all Raman-shifts from 500 to 3400 $cm^{-1}$. No leak-through of the modulated Stokes beam could be measured.

For detection of the pump beam we use a 1 cm × 1 cm large-area photo-diode (FDS1010, Thorlabs) to minimize effects of residual beam movement and the saturation of the detector. With reverse bias as high as ≈120 V, fast time response up to 20 MHz can be achieved (Freudiger et al., 2008). For video-rate imaging, we use a slightly smaller photo-diode (PDB-C613-2-ND, DigiKey) (Saar et al., 2010a). Proper thermal management is important in the latter case. For Raman gain measurements we use InGaAs photo-diodes (New England Photoconductors, I5-3-5) with a reverse bias of <12 V (Freudiger et al., 2011). In all cases the bias is applied through a passive low pass filter (BLP-1.9+, Minicircuits) to reduce noise pick-up in the system.

The photo-diode is terminated with 50 Ω and the output is band-pass filtered around the modulation frequency with a passive band-pass filter (such as BBP-10.7, Minicircuits) to avoid overloading the front-end of the lock-in amplifier with strong signals at the laser repetition frequency and low frequency fluctuations from scanning the beams through the sample with varying sample transmission.

To demodulate the signal, we use a high-frequency lock-in amplifier. The basic principle of a lock-in amplifier is shown in Figure 4.3B. The idea is that the time-dependent input signal is mixed with a reference sine-wave at the modulation frequency, $f$. Thus the signal contribution due to modulation transfer is mixed down to DC (and also mixed up to the first harmonic), according to trigonometric relations. By low-pass filtering the mixed signal, it is thus possible to extract the target signal from laser intensity fluctuation at frequencies other than $f$, which are not shifted to DC. The bandwidth of the low-pass filter is inversely proportional to the output time constant and can typically be adjusted. By shifting the phase of the reference wave, phase-sensitive detection can be achieved, reducing the noise from out-off-phase contributions at $f$. Variable amplification of the signal is provided at various stages before and after the mixer. The analog output of the lock-in amplifier is fed into the input of the microscope's data acquisition system in place of the analog signal that would normally be obtained from, for example, one of the photomultiplier tubes that are typically used for fluorescence detection.

In the original implementation of SRS microscopy we used a commercially available high-frequency lock-in amplifier (SR844RF, Stanford Research Instruments), which limited SRS imaging to a speed of ≈30 s per frame with 512 × 512 sampling (Freudiger et al., 2008). While this speed is sufficient for certain samples, it is much too slow for the imaging of living animals and humans, which inevitably move on the microscopic scale, resulting in image blur. To achieve high-speed imaging, we designed an all-analog lock-in amplifier with a response time of ≈100 ns. This allowed for imaging at video rate speed by beam scanning with a resonant galvanometer mirror with a line rate of 8 kHz (100 ns per pixel at 512 × 512 pixels with up to 30 frames/s) (Saar et al., 2010a).

This is about three orders of magnitude faster than our first implementation of SRS microscopy (pixel dwell time ~ 100 μs) (Freudiger et al., 2008), more than five orders of magnitude faster than spontaneous Raman scattering microscopy (pixel dwell time > 10 ms) and seven orders of magnitude faster than the first implementation of SRS microscopy (pixel dwell time 1 s) (Ploetz et al., 2007), demonstrating the unique capabilities of narrowband SRS microscopy.

## 4.3.3 Characterization

In this section, important properties of narrowband SRS as a contrast for microscopy will be evaluated. In particular, the validity of Equation 4.2 will be confirmed (Freudiger et al., 2008).

Figure 4.4A shows the comparison of SRS (blue circles) and CARS excitation spectra (green triangles) with the spontaneous Raman spectrum (black stars) for an isolated fingerprint peak of the molecule *trans*-retinol. SRS and spontaneous Raman spectra are nearly identical because they probe the same dependence $\sigma = \sigma(\Omega) = \sigma(\Delta\omega)$. The use of excitation lasers with a bandwidth of 2–3 $cm^{-1}$ further leads to minimal spectral broadening. The CARS spectrum suffers from a non-resonant background that is independent of the Raman shift and shows spectral distortion due to interference between the resonant and non-resonant signals.

Figure 4.4B shows good agreement between the SRL (blue circles), SRG (red squares) and spontaneous Raman spectra (black stars) with multiple peaks (Figure 4.4B). This highlights the fact that SRS spectra are identical to spontaneous Raman spectra even for more complex spectra with multiple peaks, allowing for straightforward identification based on Raman literature.

The lack of the non-resonant background in SRS microscopy can be translated from spectroscopy to imaging. Figure 4.5 compares on- and off-vibrational-resonance SRS and CARS images of lung cancer cells (Figure 4.5A) and *stratum corneum* in mouse skin (Figure 4.5B). When $\Delta\omega$ is tuned from on-resonance (2845 $cm^{-1}$) to off-resonance (2780 $cm^{-1}$) with the $CH_2$ stretching mode, the SRS signal vanishes completely whereas the non-resonant CARS background still exhibits undesirable contrast. This "false" signal complicates image interpretation and limits sensitivity.

We verified that SRS is indeed linear in both the pump and Stokes intensities (Figure 4.4C and D). Such overall quadratic intensity dependence is somewhat surprising for a third-order nonlinear process, but is a direct effect of heterodyne detection in SRS. Similar to multi-photon fluorescence (Denk et al., 1990), it allows for intrinsic three-dimensional sectioning in microscopy.

In comparison, CARS has a cubic intensity dependence. As such SRS has a slightly lower spatial resolution than CARS, but is less susceptible to power fluctuations caused by OPO tuning for micro-spectroscopy (see Figure 4.4A), suffers less from "shadowing" effects due to light scattering between the sample surface and focal plane in turbid samples, and has more even intensity across the field of view.

The linear dependence of the SRS signal on the concentration of the target molecules is shown in Figure 4.4E and F for the two molecules *trans*-retinol and methanol, which

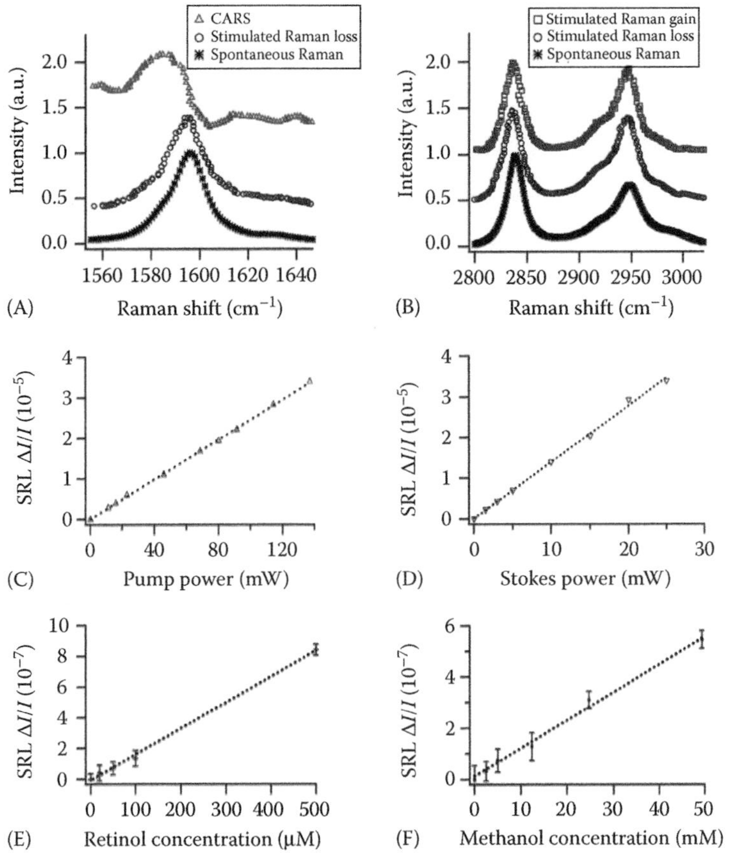


**FIGURE 4.4** Characterization of the narrowband SRS signal. (A) SRL spectrum (blue circles), spontaneous Raman spectrum (black stars), and CARS spectrum (green triangles) of an isolated Raman peak (1595 cm$^{-1}$) of 10 mM retinol in ethanol solution. (B) Both SRG (red squares) and SRL (blue circles) spectra coincide with the spontaneous Raman spectrum (black stars) of pure methanol in the CH-stretching region. Dependence of SRL signal on the pump beam (C) and Stokes beam (D) powers, respectively. The SRL signal is taken from 10% methanol/water solution tuned to the $CH_3$ stretching mode at 2840 cm$^{-1}$. The dependence of the SRL signal on concentrations of methanol/water (E) and retinol/ethanol (F). The linear dependence is valid up to much higher concentrations (data not shown here). The difference frequency is tuned to 2840 cm$^{-1}$ for methanol (E) and 1595 cm$^{-1}$ for retinol (F), respectively. The pump and Stokes beams have 40 mW power at the focus. For a 1s time constant at the lock-in amplifier, a noise level of $\approx 10^{-8}$ can be achieved. Error bars show the standard deviation of the signals for a 1 min recording. The detection limits were determined to be 5 mM for methanol and 50 μM for retinol with a signal-to-noise ratio of >1.6. (From Freudiger, C.W. et al., *Sci. Mag.*, 322(5909), 1857, 2008.)

are chosen because *trans*-retinol is an extremely strong Raman scatterer and because methanol represents a molecule with a single $CH_3$ oscillator. This is in contrast to a nonlinear concentration dependence of CARS, which makes interpreting CARS intensities challenging (Li et al., 2005). As such, SRS provides a better contrast for quantitative, chemical microscopy (see, e.g., Saar et al., 2010b).

Based on the error bars in Figure 4.4E and F it is possible to determine the sensitivity of SRS microscopy. The detection limit is 50 μM for retinol solutions and 5 mM for methanol, with <40 mW average laser power in each of the beams and a 1s time constant on the lock-in amplifier. This corresponds to about 3,000 retinol and 300,000 methanol

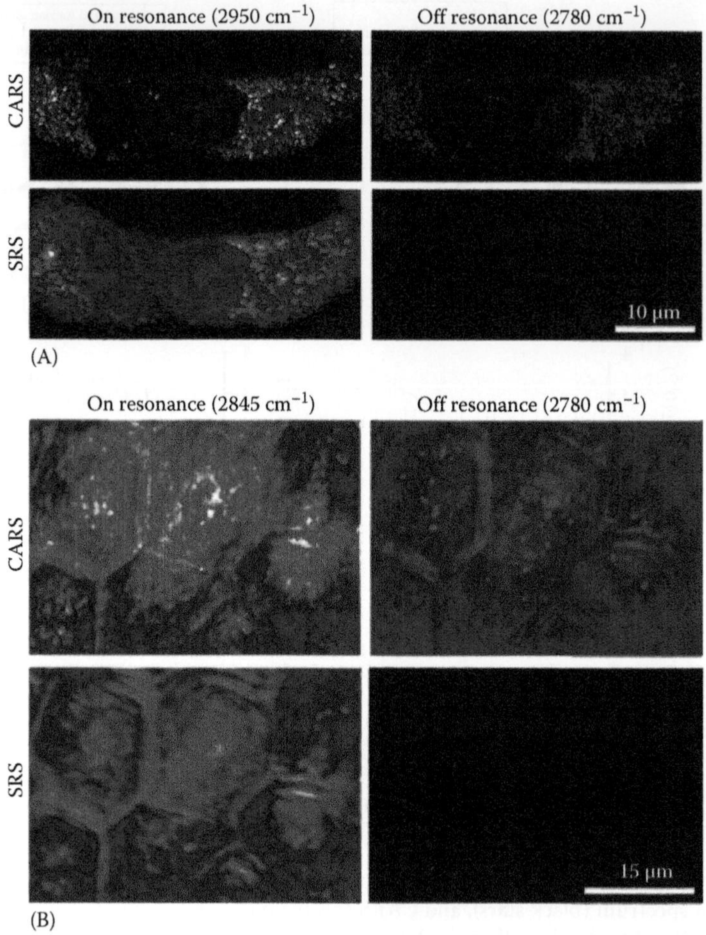


**FIGURE 4.5** SRS does not suffer from a non-resonant background. Vibrationally on- and off-resonant SRS and CARS images of (A) lung cancer cells and (B) *stratum corneum* of a mouse ear. Unlike the CARS signal, the SRS signal vanishes completely, when $\Delta\omega$ is tuned from on resonance ($CH_3$ stretching vibration at 2950 cm$^{-1}$ in (A) or $CH_2$ stretching vibration at 2845 cm$^{-1}$ in (B)) to off resonance (2780 cm$^{-1}$). While SRS microscopy is background free, the nonresonant background often leads to imaging artifacts for CARS microscopy. (From Freudiger, C.W. et al., *Sci. Mag.*, 322(5909), 1857, 2008.)

molecules in focus (with an excitation volume of about 0.1 fL), respectively. The error-bars were determined from the standard deviation of the SRS signals recorded for 1 min and the detection limit is defined at a signal-to-noise-ratio of 1.6 (Freudiger et al., 2008).

Detection limits for shorter time constants, which are more relevant for imaging, can be estimated based on Figure 4.2D. For example, with a 100 μs time constant, the noise increases by 30×, resulting in a detection limit for imaging of 1.5 mM for the strong Raman scatterer *trans*-retinol and 150 mM for a single vibrational oscillator in methanol. For more complex biomolecules with *n* oscillators per molecule, the order of magnitude of the detection limit can be estimated by 150 mM/*n* (this does not account for resonant effects such as in *trans*-retinol). For example, for cholesterol with 12 $CH_2$-bonds, the detection limit is ≈10 mM.

While this sensitivity limit of SRS microscopy is still much less than fluorescence microscopy (because Raman cross sections are much smaller than absorption cross sections), it has surpassed the detection limit reported for regular CARS microscopy (Ganikhanov et al., 2006b). It is also important to realize that SRS is sensitive to broad classes of molecular vibrations and the effective concentration of modes can thus be high. For example, somewhat surprisingly, the SRS signal from a single red blood cells due to the SRS of $CH_3$-vibrations is more than 10 times stronger (Saar et al., 2010a) than the two-photon absorption signal due to the hemoglobin absorption (Fu et al., 2007a,b), simply because per heme-group there is one absorption center for 141–146 amino acid residues. For the same reason, autofluorescence signals originating from flavins or NAD(P)H, which have been promoted for medical diagnostics (Konig et al., 2007), can be weaker than the SRS signal from tissue probing all proteins and lipids at the same excitation intensities and laser pulse durations.

Sensitivity limits in SRS are also largely independent of the size of the focal spot, that is, independent of the NA of the objective. In a simple estimation, the focal volume $V_{foc} \propto A_{foc} * Z_{foc} \propto R_{foc}^2 * Z_{foc}$, where $Z_{foc}$ is the focal extension and $A_{foc}$ is the focal area with radius $R_{foc}$. With $R_{foc} \propto 1/\text{NA}$ and the focal extension $Z_{foc} \propto 1/\text{NA}^2$, $V_{foc} \propto 1/\text{NA}^4$. On the other hand the power density of the excitation fields $I_{ex} \propto 1/A_{foc} \propto \text{NA}^2$. Thus the SRS signal is proportional to $I_{ex}^2 \cdot V_{foc} \propto \frac{\text{NA}^2 \cdot \text{NA}^2}{\text{NA}^4} = \text{constant}$ and is thus independent of the excitation NA. This estimation holds as long as the focus is smaller than the structure that contains the target vibrational mode and as long as the focal extension is shorter than the scattering mean free path of the sample.

In SRS microscopy, the spatial resolution is limited by diffraction to about 500 nm laterally and 1.5 μm axially, similar to two-photon fluorescence microscopy (Denk et al., 1990). Because SRS is measured at the same frequencies as the excitation fields, phase-matching is automatically fulfilled (Boyd, 2003). Thus SRS microscopy does not suffer from image artifacts due to phase-matching (Figure 4.6). Therefore image interpretation

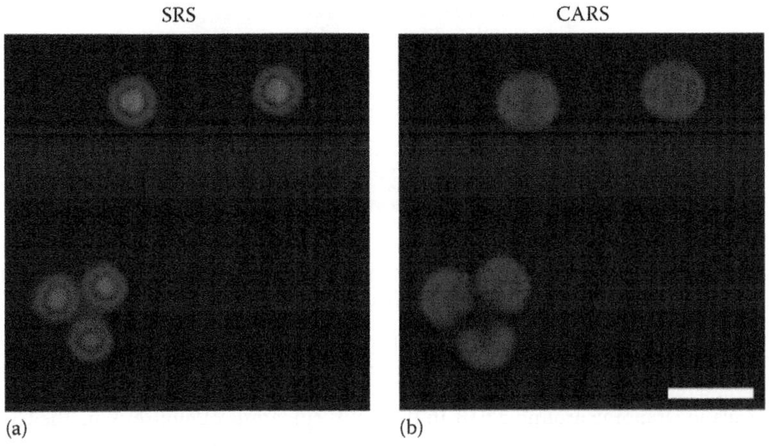


**FIGURE 4.6** SRS microscopy does not suffer from coherent image artifacts. Figure shows the images of 1 μm size polystyrene beads as acquired with SRS (a) and epi-CARS microscopy (b). While SRS shows the beads correctly, the CARS image has an intensity dip in the center of bead. This artifact is due to destructive interference of the CARS signal due to the large phase mismatch in the center of the bead, where the physical size of the bead is larger than the interaction length.

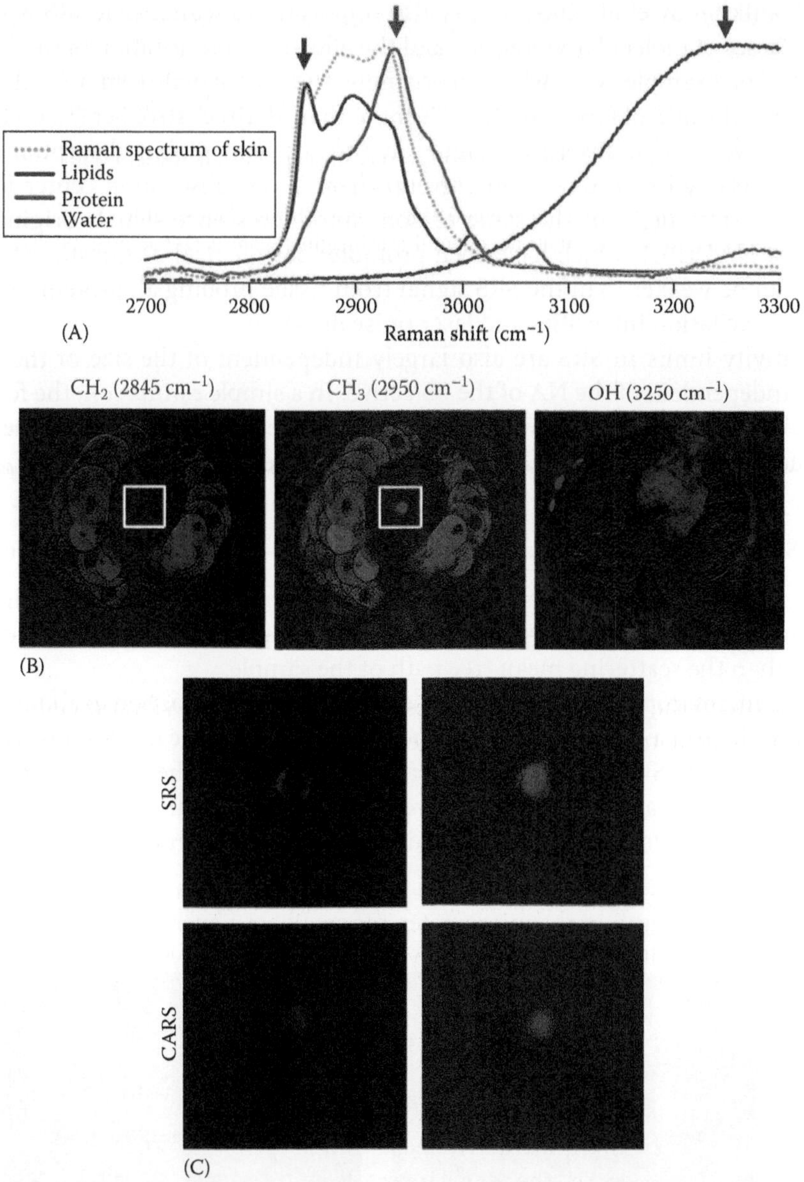


**FIGURE 4.7** SRS tissue imaging. (A) Typical Raman spectrum of mouse skin with contributions from lids (red), protein (green), and water (blue). Raman shifts at which SRS imaging is performed are indicated by the colored arrows. The $CH_2$-stretching vibration (2845 cm$^{-1}$) and the OH-stretching vibration (3250 cm$^{-1}$) are specific for lipids and water. The $CH_3$-stretching vibration (2950 cm$^{-1}$) has contributions due to lipids and proteins and ratiometric imaging is required to isolate the two contributions. (B) Label-free images of a sebaceous gland wrapping around a hair in the viable epidermis of mouse skin. The $CH_2$-image shows the lipid-rich gland cells with sub-cellular resolution. Nuclei appear as dark circles due to the lack of lipids. The $CH_3$-image shows residual lipid signal but also new protein-rich structures such as a hair in the center of the image, collagen fibers surrounding the gland, and red blood cells in the top left. The OH-image due to water shows inverse image contrast from the sebaceous gland. (C) Comparison of SRS and CARS imaging based on the hair in (B).

is straightforward in SRS microscopy and should allow deconvolution with a point spread function similar to fluorescence microscopy, which is not possible with CARS microscopy (Potma et al., 2000).

### 4.3.4 Example of Forward-Detected SRS Imaging

As a typical example of SRS microscopy we show imaging of all the structural components (lipids, protein and water) in mouse skin based on the $CH_2$-, $CH_3$-, and OH-stretching vibrations (Figure 4.7A). In particular, the lipid-rich sebaceous gland is shown in Figure 4.7B, with positive contrast in the lipid channel and inverse contrast in the water channel. Based on $CH_3$–$CH_2$-ratio-imaging, contributions that are only due to protein, such as the hair or collagen fibers, can be isolated. Doing so relies on the fact that neighboring vibrational modes are not spectrally distorted in an SRS microscope. This contrast can also be used for medical diagnostics, mimicking the most widely used pathology stains. The comparisons of CARS and SRS images (Figure 4.7C) of the hair in the center of Figure 4.7B further shows that even in strongly scattering samples, CARS images show artifacts due to the non-resonant background. While the SRS $CH_2$-image shows the sebaceous lipids surrounding the solid, protein-rich hair ($CH_3$-image) correctly, CARS cannot reproduce these results.

## 4.4 Epi-SRS Microscopy

Probably the biggest limitation of the original implementations of SRS microscopy (Ploetz et al., 2007; Freudiger et al., 2008; Ozeki et al., 2009; Nandakumar et al., 2009) was that SRS is typically performed in transmission. In CARS or fluorescence, signal collection in the backward (epi) direction is straightforward, because the techniques offer intrinsic emission at a new optical frequency in the backward direction, which can be collected by the objective lens and separated from the incident light using a dichroic mirror. In contrast, SRS microscopy involves measurement of the intensity loss of the transmitted pump beam, which travels in the forward direction. Thus, despite the advantages of SRS over CARS as a contrast mechanism for microscopy, it could not be used for the imaging of thick, non-transparent samples (e.g., a human arm).

### 4.4.1 Efficient Collection of Back-Scattered Signal

Epi-detection of SRS in thick, turbid samples is nonetheless possible, because in many biological samples scattering dominates over absorption (Cheong et al., 1990) and multiple scattering events redirect a significant portion of the forward-propagating pump and Stokes beams to the backward direction (Evans et al., 2005). While the light is back-reflected by the tissue after passing through the focal spot, the SRS signal still originates from the focal plane. The experimental challenge is to detect this signal with high efficiency.

We have explored different geometries for the detection of this back-reflected signal. In the original implementation of epi-SRS microscopy (Freudiger et al., 2008) we used collection with the excitation objective. To separate the emission from the excitation beams, which have the same frequency, we used a polarizing beam-splitter and an achromatic quarter-waveplate. After double passing the quarter-wave plate, the back-scattered light has a perpendicular polarization to the excitation light (assuming the scattering does not affect the light polarization) and is reflected to the photodiode positioned in the epi-direction.

With this setup, epi-SRS imaging from a ~1 mm thick slice of mouse brain was demonstrated (Freudiger et al., 2008) and a similar configuration was used by Slipchenko et al. (2010) for the imaging of drug tablets. While this approach works very well in the case of samples with extremely high scattering, such as a white tablet, in biological samples, the epi-directed SRS signal is about 1% of the forward-directed signal. Thus even with 1 min integration time the signal-to-noise of images obtained from biological tissue was poor and high-speed imaging of living specimens was impossible with this approach.

To quantitatively understand the poor light collection in this configuration of epi-SRS of biologically relevant samples, we performed non-sequential ray tracing simulations (Figure 4.8A and B) (Saar et al., 2010a). We found while 40%–45% of the light is

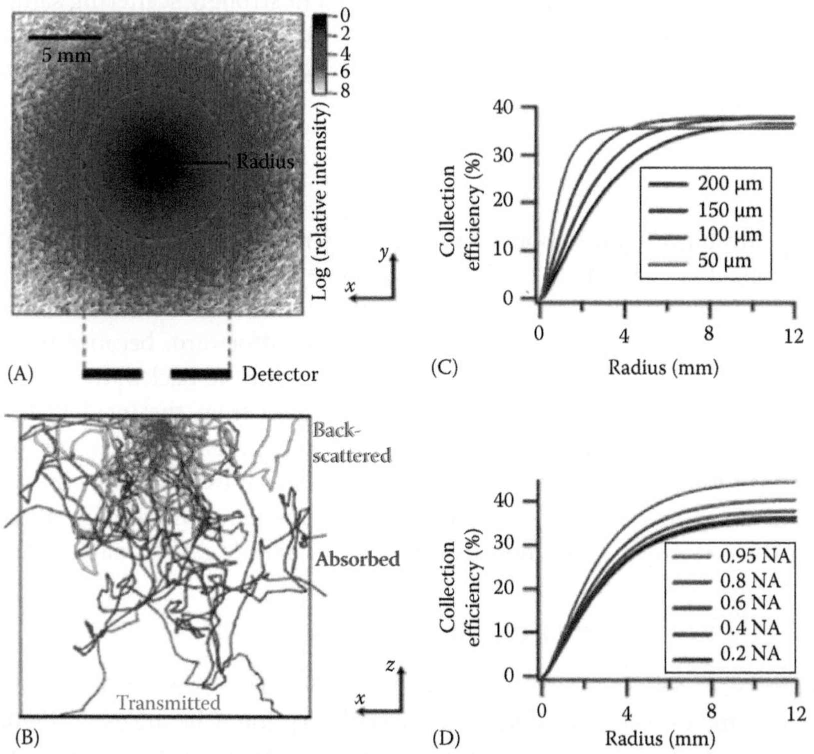


**FIGURE 4.8** Non-sequential ray-tracing simulations of collection efficiency with an annular detector. (A) Logarithmic plot of the distribution of the relative intensity of the backscattered light at the tissue surface from a focus at a depth of 100 μm into the tissue (scattering mean free path = 200 μm, anisotropy = 0.9) emitting in the forward direction with a 0.4 numerical aperture. (B) Depth profile of tissue showing sample ray trajectories colored according to the final outcome. Green traces are back-scattered to the tissue surface, red traces are scattered within the tissue until they are absorbed, and blue traces are transmitted through the 2 cm thickness of the sample. (C,D) Simulations of collection efficiency of back-scattered forward-traveling light versus detector radius for (C) different scattering mean free paths of the sample and (D) different numerical apertures (NA) of the excitation objective. (From Saar, B.C. et al., *Sci. Mag.*, 330(6009), 1368, 2010.)

backscattered in a thick tissue sample, the diffuse cloud of back-scattered light can be as large as 5 mm at the front aperture of the objective lens (Figure 4.8C), depending on the scattering mean free path of the tissue. These results are largely independent of the excitation numerical aperture (Figure 4.8D). A typical microscope objective has a front aperture radius of 1–2 mm, so more than 90% of the backscattered light does not even enter the objective.

## 4.4.2 Instrumentation

We solved this problem by placing the photo-detector directly in front of the objective lens, and exciting through a hole in the center of the detector (Figure 4.9). A similar geometry has been proposed for multi-photon microscopy of thick samples (Vucinic et al., 2006; Combs et al., 2007; McMullen and Zipfel, 2010), but requires further optics to collect the diffuse light onto a remote photo-multiplier tubes, because the PMT is bulky and cannot be put into close proximity with the sample. The laser system and the excitation path did not have to be altered from the standard video-rate imaging setup. However, one challenge was to design a filter to block the modulated Stokes beam while transmitting the pump beam, given that the back-scattered light is diffuse (Saar et al., 2010a).

In this geometry, we experimentally found that in mouse skin that we were able to collect 28% of the laser light impinging onto the sample (Saar et al., 2010a).

## 4.4.3 *In Vivo* SRS Imaging

Using this system, we imaged skin *in vivo* in mice. Figure 4.10 shows single SRS video rate frames obtained using the $CH_2$ stretching (primarily lipids, Figure 4.10A and D), OH stretching (primarily water, Figure 4.10E), and $CH_3$ stretching (primarily protein,

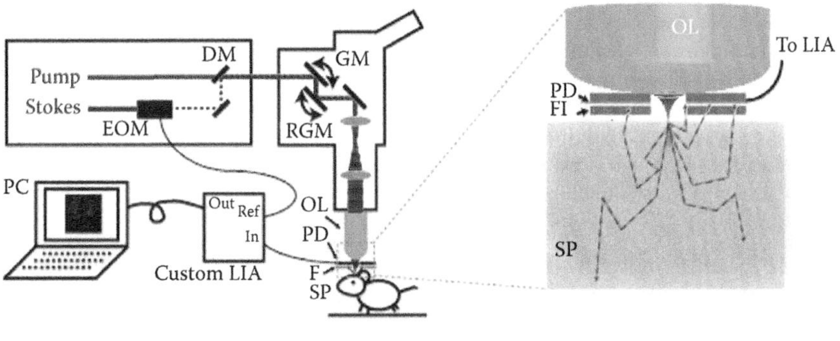


**FIGURE 4.9** Annular detector design for epi-SRS microscopy. The Stokes beam is modulated with an electro-optic modulator (EOM), spatially overlapped with the pump beam with a dichroic mirror (DM) and aligned into a laser scanning microscope. The beams are focused by the objective lens (OL) and the common focal spot is scanned through the specimen (SP) by a galvo mirror (GM) and a resonant galvo mirror (RGM). The detected intensity of the backscattered pump beam is demodulated with a custom high-speed lock-in amplifier (LIA) to provide the SRS signal to the computer (PC). The inset depicts the epi-detector assembly. The sample is excited by focusing light through a small hole in the center of the large-area epi-detector (PD). Scattering redirects a significant portion of the forward-traveling light to illuminate the detector active area. The modulated Stokes beam is blocked by an optical filter (FI), and the transmitted pump beam is detected. (From Saar, B.C. et al., *Sci. Mag.*, 330(6009), 1368, 2010.)

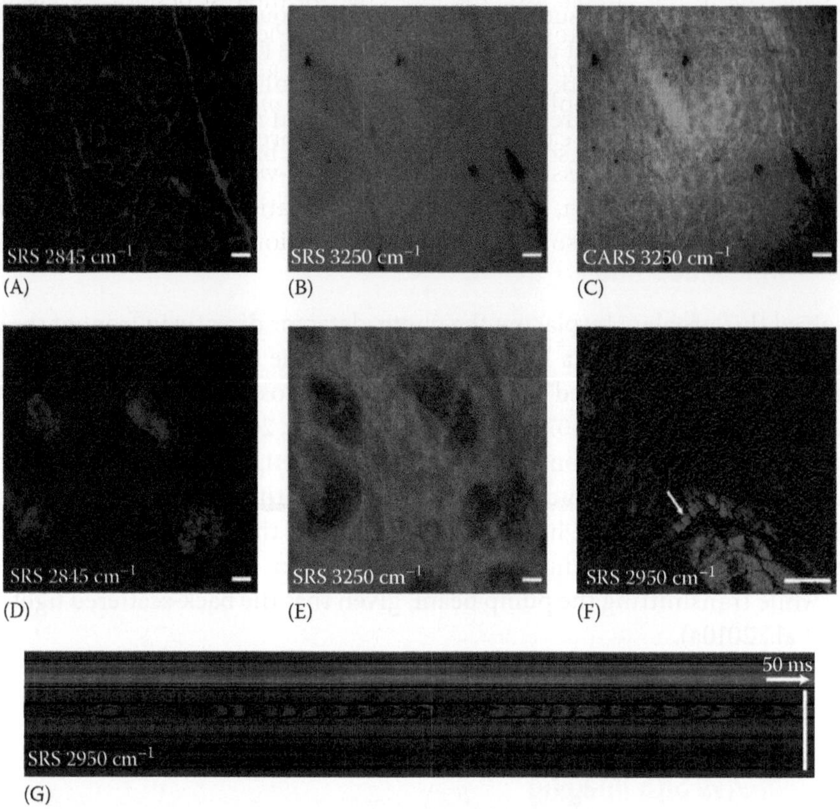


**FIGURE 4.10** SRS skin imaging in living mice. (A) SRS image of lipids of the *stratum corneum* shows inter cellular spaces between hexagonal corneocytes and (B) SRS water image (3250 cm$^{-1}$) of the same region shows a homogeneous distribution of water. (C) A CARS water image acquired simultaneously with (B) shows artifacts from the non-resonant background of lipids. (D) SRS lipid and (E) water images of the viable epidermis show sebaceous glands with positive and negative contrast, respectively. (F) SRS images of the viable epidermis at the $CH_3$ stretching vibration (2950 cm$^{-1}$) mainly highlight proteins as well as residual lipid-rich structures. A capillary with individual red blood cells (arrow) is visible. The cells are imaged without motion blur due to video rate acquisition speed. (G) SRS *in vivo* flow cytometry. An $x$–$t$ plot acquired by line-scanning across a capillary at the position of the arrow in (F). Individual red blood cells are captured on the fly. (A–E) are acquired in epi-direction, while (F and G) are acquired in transmission, all with 37 ms/frame acquisition speed and 512 × 512 pixel sampling. Scale: 25 μm. (From Saar, B.C. et al., *Sci. Mag.*, 330(6009), 1368, 2010.)

Figure 4.10F and G) vibrations. The lipid distributions are as expected from previous work (Freudiger et al., 2008), but water can only be measured reliably *in vivo* because the skin hydration changes in excised tissue. Imaging water is of particular interest in studying the transport properties of water-soluble drugs and their effect on the hydration of the skin barrier (Rawlings and Harding, 2004).

Figure 4.10C highlights the fact that CARS imaging of water is distorted by the non-resonant background, which introduces an image artifact: it shows positive contrast for the lipid-rich areas of the *stratum corneum* layer, which do not contain water. Thus the contrast in CARS is inverted compared to the real water distribution. This effect is not observed in SRS because it is free from this background (Freudiger et al., 2008).

The protein image (Figure 4.10F and G) shows red blood cells moving in a capillary of the viable epidermis. By performing a line-scan over time we reconstructed a plot of the cells as they pass through the scan line (Figure 4.10G), allowing for *in vivo* flow cytometry based on intrinsic chemical contrast.

## 4.5 Conclusions

In conclusion, SRS has superseded CARS as a contrast mechanism for microscopy. SRS microscopy allows label-free chemical imaging based on vibrational spectroscopy at video-rate speed in both the forward direction and epi-direction. Compared to CARS,

- SRS is free from non-resonant background
- SRS has much improved sensitivity
- SRS does not suffer from spectral distortions
- SRS signal is linear in the concentration of the target species
- SRS does not have imaging artifacts due to phase-matching

Applications in biology, material science and medicine, capitalizing on the advantageous properties of SRS microscopy, are presented in Chapter 24.

## References

Boyd, R.W. *Nonlinear Optics*. Academic Press, New York, 2003.

Chen, B.C., J. Sung, and S.H. Lim. Chemical imaging with frequency modulation coherent anti-Stokes Raman scattering microscopy at the vibrational fingerprint region. *The Journal of Physical Chemistry B*, 114(50): 16871–16880, 2010. ISSN 1520-6106.

Cheng, J.X. and X.S. Xie. Green's function formulation for third-harmonic generation microscopy. *Journal of the Optical Society of America B*, 19(7): 1604–1610, 2002.

Cheng, J.X. and X.S. Xie. Coherent anti-Stokes Raman scattering microscopy: Instrumentation, theory, and applications. *Journal of Physical Chemistry B*, 108(3): 827–840, 2004.

Cheng, J., A. Volkmer, L.D. Book, and X.S. Xie. An epi-detected coherent anti-Stokes Raman scattering (E-CARS) microscope with high spectral resolution and high sensitivity. *Journal of Physical Chemistry B*, 105(7): 1277–1280, 2001a.

Cheng, J.X., L.D. Book, and X.S. Xie. Polarization coherent anti-Stokes Raman scattering microscopy. *Optics Letters*, 26(17): 1341–1343, 2001b.

Cheng, J.X., A. Volkmer, and X.S. Xie. Theoretical and experimental characterization of coherent anti-Stokes Raman scattering microscopy. *Journal of Optical Society of America B*, 19(6): 1363–1375, 2002. ISSN 1520-8540.

Cheong, W.F., S.A. Prahl, and A.J. Welch. A review of the optical properties of biological tissues. *IEEE Journal of Quantum Electronics*, 26(12): 2166–2185, 1990.

Combs, C.A., A.V. Smirnov, J.D. Riley, A.H. Gandjbakhche, J.A.Y.R. Knutson, and R.S. Balaban. Optimization of multiphoton excitation microscopy by total emission detection using a parabolic light reflector. *Journal of Microscopy*, 228(3): 330–337, 2007.

Denk, W., J.H. Strickler, and W.W. Webb. Two-photon laser scanning fluorescence microscopy. *Science*, 248(4951): 73, 1990.

Duncan, M.D., J. Reintjes, and T.J. Manuccia. Scanning coherent anti-Stokes Raman microscope. *Optics Letters*, 7(8): 350–352, 1982.

Ekvall, K., P. van der Meulen, C. Dhollande, L.E. Berg, S. Pommeret, R. Naskrecki, and J.C. Mialocq. Cross phase modulation artifact in liquid phase transient absorption spectroscopy. *Journal of Applied Physics*, 87: 2340, 2000.

Evans, C.L. and X.S. Xie. Coherent anti-Stokes Raman scattering microscopy: Chemical imaging for biology and medicine. *Annual Review of Chemistry*, 2008. http://www.annualreviews.org/doi/abs/10.1146/annurev.anchem.1.031207.112754?journalCode=anchem

Evans, C.L., E.O. Potma, and X.S. Xie. Coherent anti-Stokes Raman scattering spectral interferometry: Determination of the real and imaginary components of nonlinear susceptibility $\chi(3)$ for vibrational microscopy. *Optics Letters*, 29(24): 2923–2925, 2004.

Evans, C.L., E.O. Potma, M. Puoris' haag, D. Cote, C.P. Lin, and X.S. Xie. Chemical imaging of tissue in vivo with video-rate coherent anti-Stokes Raman scattering microscopy. *Proceedings of the National Academy of Sciences of the United States of America*, 102(46): 16807, 2005.

Evans, C.L., X. Xu, S. Kesari, X.S. Xie, S.T.C. Wong, and G.S. Young. Chemically-selective imaging of brain structures with CARS microscopy. *Optics Express*, 15(19):12076–12087, 2007.

Freudiger, C.W., W. Min, B.G. Saar, S. Lu, G.R. Holtom, C. He, J.C. Tsai, J.X. Kang, and X.S. Xie. Label-free biomedical imaging with high sensitivity by stimulated Raman scattering microscopy. *Science*, 322(5909): 1857, 2008.

Freudiger, C.W., W. Min, G.R. Holtom, B. Xu, M. Dantus, and X.S. Xie. Spectral imaging by stimulated Raman scattering. *Nature Photonics*, 2011.

Fu, Y., H. Wang, R. Shi, and J.X. Cheng. Characterization of photodamage in coherent anti-Stokes Raman scattering microscopy. *Optics Express*, 14: 3942–3951, 2006.

Fu, D., T. Ye, T.E. Matthews, B.J. Chen, G. Yurtserver, and W.S. Warren. High-resolution in vivo imaging of blood vessels without labeling. *Optics Letters*, 32(18): 2641–2643, 2007a.

Fu, D., T. Ye, T.E. Matthews, G. Yurtsever, and W.S. Warren. Two-color, two-photon, and excited-state absorption microscopy. *Journal of Biomedical Optics*, 12: 054004, 2007b.

Ganikhanov, F., S. Carrasco, X. Sunney Xie, M. Katz, W. Seitz, and D. Kopf. Broadly tunable dual-wavelength light source for coherent anti-Stokes Raman scattering microscopy. *Optics Letters*, 31(9): 1292–1294, 2006a.

Ganikhanov, F., C.L. Evans, B.G. Saar, and X.S. Xie. High-sensitivity vibrational imaging with frequency modulation coherent anti-Stokes Raman scattering (FM CARS) microscopy. *Optics Letters*, 31(12): 1872–1874, 2006b.

Ganikhanov, F., C.L. Evans, B.G. Saar, and X.S. Xie. High-sensitivity vibrational imaging with frequency modulation coherent anti-Stokes Raman scattering (FM CARS) microscopy. *Optics Letters*, 31(12): 1872, 2006c.

Gremlich, H.U. and B. Yan. *Infrared and Raman Spectroscopy of Biological Materials*. Marcel Dekker Inc, New York, 2001.

Helmchen, F. and W. Denk. Deep tissue two-photon microscopy. *Nature Methods*, 2(12): 932–940, 2005.

Hopt, A. and E. Neher. Highly nonlinear photodamage in two-photon fluorescence microscopy. *Biophysical Journal*, 80(4): 2029–2036, 2001.

Konig, K., A. Ehlers, I. Riemann, S. Schenkl, R. Buckle, and M. Kaatz. Clinical two-photon microendoscopy. *Microscopy Research and Technique*, 70(5): 398–402, 2007. ISSN: 1097-0029.

Kukura, P., D.W. McCamant, and R.A. Mathies. Femtosecond stimulated Raman spectroscopy. *Annual Review of Chemistry*, 2007.

Levenson, M.D. and S.S. Kano. *Introduction to Nonlinear Laser Spectroscopy*. Academic Press, New York, 1988.

Levine, B., C. Shank, and J. Heritage. Surface vibrational spectroscopy using stimulated Raman scattering. *IEEE Journal of Quantum Electronics*, 15(12): 1418–1432, 1979.

Li, L., H. Wang, and J.X. Cheng. Quantitative coherent anti-Stokes Raman scattering imaging of lipid distribution in coexisting domains. *Biophysical Journal*, 89(5): 3480–3490, 2005. ISSN 0006-3495.

Maker, P.D. and R.W. Terhune. Study of optical effects due to an induced polarization third order in the electric field strength. *Physical Review*, 137: A801–A818, 1965.

McMullen, J.D. and W.R. Zipfel. A multiphoton objective design with incorporated beam splitter for enhanced fluorescence collection. *Optics Express*, 18(6): 5390–5398, 2010. ISSN: 1094-4087.

Nandakumar, P., A. Kovalev, and A. Volkmer. Vibrational imaging based on stimulated Raman scattering microscopy. *New Journal of Physics*, 11: 033026, 2009.

Owyoung, A. Sensitivity limitations for CW stimulated Raman spectroscopy. *Optics Communications*, 22(3): 323–328, 1977.

Ozeki, Y., F. Dake, S. Kajiyama, K. Fukui, and K. Itoh. Analysis and experimental assessment of the sensitivity of stimulated Raman scattering microscopy. *Optics Express*, 17(5): 3651–3658, 2009.

Ozeki, Y., Y. Kitagawa, K. Sumimura, N. Nishizawa, W. Umemura, S. Kajiyama, K. Fukui, and K. Itoh. Stimulated Raman scattering microscope with shot noise limited sensitivity using subharmonically synchronized laser pulses. *Optics Express*, 18(13): 13708–13719, 2010. ISSN: 1094-4087.

Ploetz, E., S. Laimgruber, S. Berner, W. Zinth, and P. Gilch. Femtosecond stimulated Raman microscopy. *Applied Physics B: Lasers and Optics*, 87(3): 389–393, 2007.

Potma, E.O., W.P. de Boeij, and D.A. Wiersma. Nonlinear coherent four-wave mixing in optical microscopy. *Journal of the Optical Society of America B*, 17(10): 1678–1684, 2000.

Potma, E.O., D.J. Jones, J.X. Cheng, X.S. Xie, and J. Ye. High-sensitivity coherent anti-Stokes Raman scattering microscopy with two tightly synchronized picosecond lasers. *Optics Letters*, 27(13): 1168–1170, 2002.

Rawlings, A.V. and C.R. Harding. Moisturization and skin barrier function. *Dermatologic Therapy*, 17: 43–48, 2004.

Rinia, H.A., M. Bonn, and M. Muller. Quantitative multiplex CARS spectroscopy in congested spectral regions. *Journal of Physical Chemistry B*, 110(9): 4472–4479, 2006.

Saar, B.G., G.R. Holtom, C.W. Freudiger, C. Ackermann, W. Hill, and X.S. Xie. Intracavity wavelength modulation of an optical parametric oscillator for coherent Raman microscopy. *Optics Express*, 17(15): 12532–12539, 2009. ISSN: 1094-4087.

Saar, B.G., C.W. Freudiger, C.M. Stanely, G.R. Holtom, and X.S. Xie. Video-rate molecular imaging in vivo with stimulated Raman scattering. *Science*, accepted, 2010a. http://www.sciencemag.org/content/330/6009/1368.short

Saar, B.G., Y. Zeng, C.W. Freudiger, Y.S. Liu, M.E. Himmel, X.S. Xie, and S.Y. Ding. Label-free, real-time monitoring of biomass processing with stimulated Raman scattering microscopy. *Angewandte Chemie International Edition*, 2010b.

*Model SR844 RF Lock-In Amplifier—Manual*. Stanford Research Systems, Sunnyvale, CA, 1997.

Slipchenko, M.N., H. Chen, D.R. Ely, Y. Jung, M.T. Carvajal, and J.X. Cheng. Vibrational imaging of tablets by epi-detected stimulated Raman scattering microscopy. *The Analyst*, 2010.

Turrell, G., J. Corset, and H.G.M. Edwards. *Raman Microscopy: Developments and Applications*. Academic Press, London, U.K., 1996.

Vucinic, D., T.M. Bartol, Jr., and T.J. Sejnowski. Hybrid reflecting objectives for functional multiphoton microscopy in turbid media. *Optics Letters*, 31(16): 2447, 2006.

Woodbury, E.J. and W.K. Ng. Ruby laser operation in the near IR. *Proceedings of the IRE*, 50(11): 2367, 1962.

Zumbusch, A., G.R. Holtom, and X.S. Xie. Three-dimensional vibrational imaging by coherent anti-Stokes Raman scattering. *Physical Review Letters*, 82(20): 4142–4145, 1999.

# 5. Femtosecond versus Picosecond Pulses for Coherent Raman Microscopy

**Mikhail N. Slipchenko, Delong Zhang, and Ji-Xin Cheng**



Chapter 5

*Coherent Raman Scattering Microscopy.* Edited by Ji-Xin Cheng and X. Sunney Xie © 2013 CRC Press/ Taylor & Francis Group, LLC. ISBN: 978-1-4398-6765-5.

# 5.1 Introduction

## 5.1.1 Pros and Cons of Femtosecond and Picosecond Laser Excitation

Coupling of nonlinear optical (NLO) signal generation and scanning microscope has generated a panel of imaging tools for biology and materials research. These tools include the one-beam modality such as two-photon excited fluorescence (TPEF), second harmonic generation (SHG), and third harmonic generation (THG) microscopy. The two-beam modality includes coherent anti-Stokes Raman scattering (CARS), four-wave mixing (FWM), stimulated Raman scattering (SRS), photothermal, and pump-probe microscopy (for a review, see Yue et al. [2011]). Due to nonlinear nature of interaction high peak power laser sources are required to generate strong signals. Although femtosecond (fs) laser sources provide highest peak power, the picosecond (ps) pulse excitation provides high spectral resolution for the Raman-based NLO modalities such as CARS and SRS. The advantages and disadvantages of both femtosecond and picosecond excitations are summarized in Table 5.1.

Historically laser pulses of a few picoseconds in duration were promoted for CARS imaging in order to increase the ratio of resonant signal to nonresonant background (Cheng et al., 2001). Such requirement is eliminated in SRS microscopy because the SRS signal is free of the nonresonant background. Previously, femtosecond pulse excitation has been used for SRS spectroscopy (McCamant et al., 2003, Kukura et al., 2007). In the following, we compare theoretically CARS and SRS modality and show experimentally that compared to picosecond excitation, femtosecond pulse excitation results in the increase of the SRS signal-to-noise ratio (SNR) by one order of magnitude. Such signal enhancement allows high-speed, bond-selective imaging of isolated Raman bands, including the C–H and C–D stretch vibrations.

## 5.1.2 Description of SRS and CARS in Frequency Domain

Both CARS and SRS processes occur simultaneously in a sample and differ only by the detection scheme. To understand the fundamental differences between these two processes let us consider the nonlinear interaction of two electrical waves, namely pump, $E(\omega_p)$, and Stokes, $E(\omega_s)$ in the sample having molecular vibration at frequency $\Omega$,

**Table 5.1** Femtosecond versus Picosecond Laser Source for Multimodal NLO Microscopy

| | Pros | Cons |
|---|---|---|
| *fs pulse* | • Optimal for TPEF, SHG, THG, FWM<br>• High signal level for CARS and SRS<br>• Available for spectral shaping | • Without processing, low spectral resolution for vibrational imaging |
| *ps pulse* | • High signal to background ratio for CARS<br>• High spectral resolution for vibrational imaging and Raman analysis<br>• Simplicity for single frequency imaging | • Low signal level for TPEF, SHG, THG, FWM<br>• Low signal level for CARS and SRS<br>• Lack of capacity for spectral shaping |

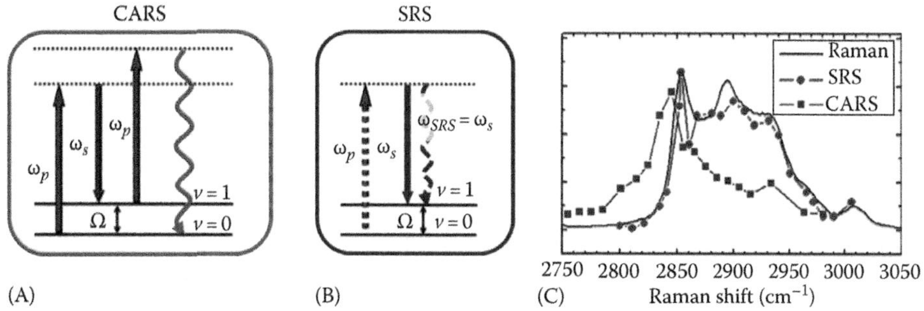


**FIGURE 5.1** Energy diagram of CARS and SRS and spectral relevance to spontaneous Raman. (A) and (B) Energy diagram of CARS and SRS. Solid lines represent electronic and vibrational states of molecules, dashed lines are virtual states. The straight arrows are excitation beams, the wavy arrows are output signal beams. $\omega_p$ and $\omega_s$ are pump and Stokes beams from the laser sources, respectively. $\Omega$ is a frequency of vibrational transition between vibrational ground state $v = 0$ and vibrationally excited state $v = 1$. (C). Raman (black solid line), SRS (filled red circles), and CARS (filled blue squares) spectra of lipids in the CH stretch region.

so that $\omega_p - \omega_s = \Omega$. Figure 5.1 shows the energy diagrams of CARS and SRS. Here we only consider stimulated Raman gain (SRG) process detected at $\omega_s$. The stimulated Raman loss (SRL) signal intensity equals to that of SRG but detected at $\omega_p$. Detailed description of SRG and SRL can be found in (Shen, 1984, Levenson and Kano, 1988, Mukamel, 1995).

The two electrical fields induce in the sample third-order polarizations $P^{(3)}_{CARS}$ and $P^{(3)}_{SRG}$, which in the frequency domain differ by frequency:

$$P^{(3)}_{CARS}(\omega_{aS} = 2\omega_p - \omega_S) \propto N\chi^{(3)}(-\omega_{aS}, \omega_p, \omega_p, -\omega_S)\left|E(\omega_p)\right|^2 E^*(\omega_S) \tag{5.1}$$

$$P^{(3)}_{SRG}(\omega_S) \propto N\chi^{(3)}(-\omega_S, \omega_p, -\omega_p, \omega_S)\left|E(\omega_p)\right|^2 E(\omega_S) \tag{5.2}$$

where

$N$ is a number density

$\chi^{(3)}$ is a third-order susceptibility of the medium

$$\chi^{(3)} = \chi^{(3)}_{NR} + \chi^{(3)}_R(\omega_p, \omega_S) = \chi^{(3)}_{NR} + \frac{A}{(\omega_p - \omega_S) - \Omega + i\Gamma} \tag{5.3}$$

The third-order susceptibility consists of two parts, the first is the nonresonant part $\chi^{(3)}_{NR}$, which arises from electronic motion and does not depend on vibrational modes of the molecules, and the second is the resonant part, which is sensitive to detuning from Raman transition. Here $A$ is a coefficient dependent on Raman cross section and $\Gamma$ is the half width of a Raman transition. The induced third-order polarization drives the electric field:

$$E_{CARS}(\omega_{aS}) \propto iP^{(3)}_{CARS}(\omega_{aS} = 2\omega_p - \omega_S) \tag{5.4}$$

$$E_{SRG}(\omega_S) \propto iP^{(3)}_{SRG}(\omega_S) \tag{5.5}$$

The CARS signal is homodyne in nature and we can combine Equations 5.1 and 5.4 to write the CARS intensity as

$$I_{CARS}(\omega_{aS}) = |E_{CARS}|^2 \propto N^2 |\chi^{(3)}|^2 |E(\omega_p)|^4 |E(\omega_S)|^2 \tag{5.6}$$

The SRG electric field $E_{SRG}(\omega_S)$, however, has the same frequency and propagation direction as an electric field of the Stokes beam $E(\omega_S)$, wherefore the Stokes electric field acts as a local oscillator $E_{LO}(\omega_S) = E(\omega_S)$ and interferes with SRG electric field at the detector:

$$I_{SRG}(\omega_S) \propto |E_{LO}(\omega_S) + E_{SRG}(\omega_S)|^2 = |E_{LO}(\omega_S)|^2 + |E_{SRG}(\omega_S)|^2 + 2\,\mathrm{Re}(E_{SRG}E_{LO}^{*}) \tag{5.7}$$

The third term in Equation 5.7 represents the heterodyne signal which is detected via modulation of the pump beam and phase sensitive detection of the small changes $(\Delta I(\omega_S)/I(\omega_S) < 10^{-4})$ in the Stokes beam. Combining Equations 5.2, 5.5, and 5.7 gives the SRG signal:

$$I_{SRG}(\omega_S) = \Delta I(\omega_S) \propto N\,\mathrm{Im}(\chi^{(3)}) |E(\omega_p)|^2 |E(\omega_S)|^2 \tag{5.8}$$

Now we can compare Equations 5.6 and 5.8 to analyze the differences between origins of CARS and SRG signals. (1) CARS signal has quadratic dependence on pump beam intensity while SRG signal has linear dependence on each beam; (2) CARS signal is quadratically dependent on concentration compared while SRG signal has linear dependence on concentration; (3) CARS signal contains contribution from both resonant and nonresonant terms of third-order susceptibility, which results in dispersive spectral line shapes different from spontaneous Raman (see Figure 5.1 lower panel). In contrast, SRG signal depends only on the imaginary component of third-order susceptibility and therefore has no contribution from the nonresonant part, which results in spectral line shapes equal to that of spontaneous Raman (see Figure 5.1 right panel).

## 5.2 Signal Dependence on Spectral Width of Laser Pulse

### 5.2.1 Broad Band versus Narrow Band CARS Signal and Discussion of Nonresonant Contribution

Laser pulses of a few picoseconds in duration were promoted for CARS imaging in order to increase the ratio of resonant signal to nonresonant background (Cheng et al., 2001). Figure 5.2A shows the dependence of resonant and nonresonant CARS signals on excitation pulses bandwidths calculated for $A/\Gamma = 3\chi_{NR}^{(3)}$ (see Equation 5.3). The resonant CARS signal begins to saturate once the bandwidth reaches the Raman transition spectral width. In contrast, the nonresonant CARS signal continues to grow with increase of bandwidth. The maximum contrast, which is determined as ratio of resonant to

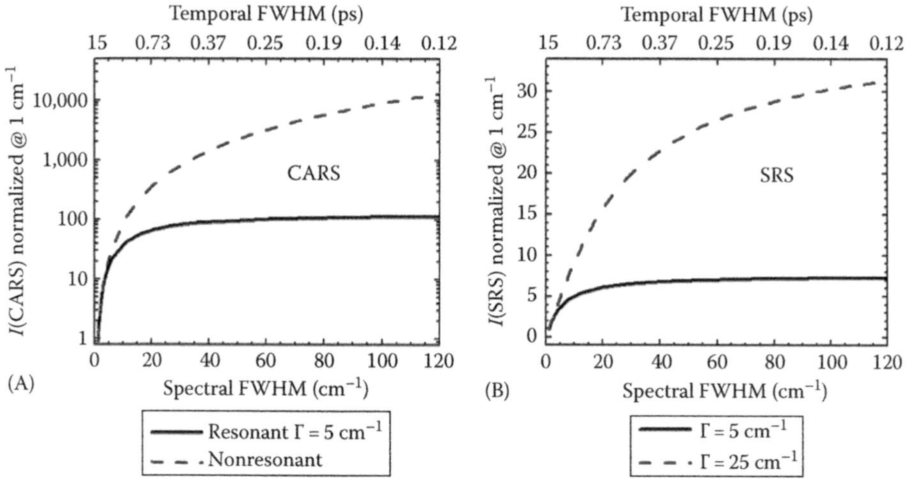


**FIGURE 5.2** The resonant and nonresonant CARS components intensity and heterodyne detected SRS intensity as a function of pulse spectral width. (A) Pure resonant and nonresonant CARS intensities. (Adapted from Cheng, J.X. et al., *J. Phys. Chem. B*, 105, 1277, 2001.) (B) Calculated SRS intensities for narrow $\Gamma = 5$ cm$^{-1}$ and broad $\Gamma = 25$ cm$^{-1}$ Raman transitions. The intensities are normalized by the values at the pulse spectral width of 1.0 cm$^{-1}$. $\Gamma$ is the Raman line half-width.

nonresonant signals has a maximum value at width of the excitation pulses equal to the spectral width of the Raman mode (Cheng et al., 2001, Pegoraro et al., 2009). Shorter broadband pulses produce stronger CARS signal but also stronger nonresonant background, while longer pulses generate lower signal but have better spectral resolution and specificity. Although picosecond pulse excitation gives better CARS contrast, the femtosecond laser sources have higher peak power that is required to generate strong NLO signals for one-beam modalities such as TPEF and THG. Recently, coupling of CARS modality with widely used multiphoton modalities was demonstrated with femtosecond laser pulses (Chen et al., 2009, Pegoraro et al., 2010).

## 5.2.2 Dependence of SRS Signal on Pulse Width and Raman Line Width

Because shot noise limit has been reached with mode-locked picosecond lasers (Freudiger et al., 2008), an effective way to enhance the detection sensitivity is to increase the SRS signal level. In SRS microscopy the contrast dependence on excitation pulse bandwidth is eliminated because the SRS signal is free of the nonresonant background. To evaluate the relationship between the coherent Raman scattering intensity and pulse width we have adopted a theoretical model from (Cheng et al., 2001). The induced third-order polarization for SRG can be integrated over excitation pulses spectral profiles and written as

$$P^{(3)}(\omega_{sig}) = \int_{-\infty}^{+\infty} d\omega_p \int_{-\infty}^{+\infty} d\omega_p' \int_{-\infty}^{+\infty} d\omega_s \chi^{(3)} E_p(\omega_p) E_p(\omega_p') E_s(\omega_s) \delta(\omega_p - \omega_p' + \omega_s - \omega_{sig}) \quad (5.9)$$

Chapter 5

where $E_s(\omega_s)$ and $E_p(\omega_p)$ denote the Stokes and pump fields, respectively. Here the delta function represents the conservation of energy. We assume that the pump and Stokes fields are two temporally overlapped pulses with Gaussian spectral profiles,

$$E_p(\omega_p) = \frac{e_p}{\Delta_p^{1/2}} \exp\left( \frac{-2\left(\omega_p^0 - \omega_p\right)^2 \ln 2}{\Delta_p^2} \right)$$
$$E_s(\omega_s) = \frac{e_s}{\Delta_s^{1/2}} \exp\left( \frac{-2\left(\omega_s^0 - \omega_s\right)^2 \ln 2}{\Delta_s^2} \right) \tag{5.10}$$

where

- $\omega_p^0$ and $\omega_s^0$ are central wavelengths of the pump and Stokes fields
- $e_p$ and $e_s$ are constants related to peak intensities
- $\Delta_p$ and $\Delta_s$ are the spectral full widths at half maximum (FWHM) of the pump and Stokes fields, respectively

To simplify the calculation, we assume $\Delta_p = \Delta_s$. The prefactors in Equation 5.10 ensure that the pulse energy is independent of the pulse spectral width. Based on Equations 5.5, 5.8, and 5.9 the heterodyne-detected SRG intensity is written as

$$\Delta I_s(\omega_{sig}) = 2\int_{-\infty}^{+\infty} d\omega_p \int_{-\infty}^{+\infty} d\omega_p' \int_{-\infty}^{+\infty} d\omega_s \int_{-\infty}^{+\infty} d\omega_s' \,\mathrm{Im}(\chi^{(3)}) E_p(\omega_p) E_p(\omega_p') E_s(\omega_s) E_s(\omega_s')$$
$$\times\, \delta(\omega_p - \omega_p' + \omega_s - \omega_{sig})\delta(\omega_p - \omega_p' + \omega_s - \omega_s') \tag{5.11}$$

where $\Delta I_s(\omega_{sig})$ is a modulated component of the Stokes beam, which is detected via phase sensitive detection. The second delta function represents the requirement for mixing between the local oscillator and the SRS field. Based on the aforementioned theoretical model we have computed the relationship between the SRS intensity and the pulse spectral width. Our calculation (Figure 5.2B) indicates that $\Delta I_s(\omega_{sig} = \omega_s^0)$ strongly depends on the Raman line half-width and the pulse spectral width. For $\Gamma = 5$ cm$^{-1}$, $\Delta I_s$ is amplified by six times when the spectral width increases from 1 to 20 cm$^{-1}$ and becomes saturated thereafter. However, for $\Gamma = 25$ cm$^{-1}$, $\Delta I_s$ is amplified by 30 times when the pulse spectral width increases from 1 cm$^{-1}$ (~14 ps) to 120 cm$^{-1}$ (~120 fs). Therefore, by use of shorter pulses, we expect to increase the SRS signal level by one order of magnitude at the same excitation energy for spectrally broad Raman lines.

### 5.2.3 Dependence of Photodamage on Excitation Wavelength for Femtosecond Pulse Excitation

A concern of using femtosecond excitation is the photodamage caused by the high peak power of the pulse train. Because the photodamage threshold is wavelength dependent, CARS and SRS techniques are different in terms of photodamage. To understand the differences between CARS and SRS techniques we need to consider the optimal ratio between pump and Stokes beams intensities. We define such an optimal ratio as the ration $I_p/I_S$ at which the total power at the sample $I_p + I_S$ is minimal for a given CARS or SRS signal. Due to quadratic dependence of CARS signal on pump beam intensity the optimal ratio $I_p/I_S = 2$, while for SRS $I_p/I_S = 1$ due to linear dependence of SRS signal on pump and Stokes. In experiment the SNR is a more important parameter to consider. In case of negligible nonresonant and fluorescence background the noise in CARS arises from the electronics and does not depend on excitation powers, therefore the optimal ratio remains the same, $I_p/I_S = 2$. In contrast, SRS noise arises from shot noise of a local oscillator. As a result the optimal ratio is $I_p/I_S = 2$ for SRG and $I_p/I_S = 1/2$ for SRL. Therefore, compared to CARS and SRG, in which more excitation power is carried by the pump beam, SRL uses more power from the longer wavelength Stokes beam and less power from the pump beam for excitation. Such power configuration minimizes the photodamage in femtosecond SRL microscopy. To explore the phototoxicity, we determined the cell damage threshold at different wavelengths (Figure 5.3). We used plasma membrane blebbing (Fu et al., 2006) as an indicator of cell damage. At the wavelengths of 680, 830, 880, and 1000 nm, the power density at the sample to cause CHO cell membrane blebbing was found to be 1.6, 8.9, 10.5, and 15.8 MW/cm², respectively. At 1100 nm, we did not observe photodamage with the maximum power (9.5 MW/cm² at sample)

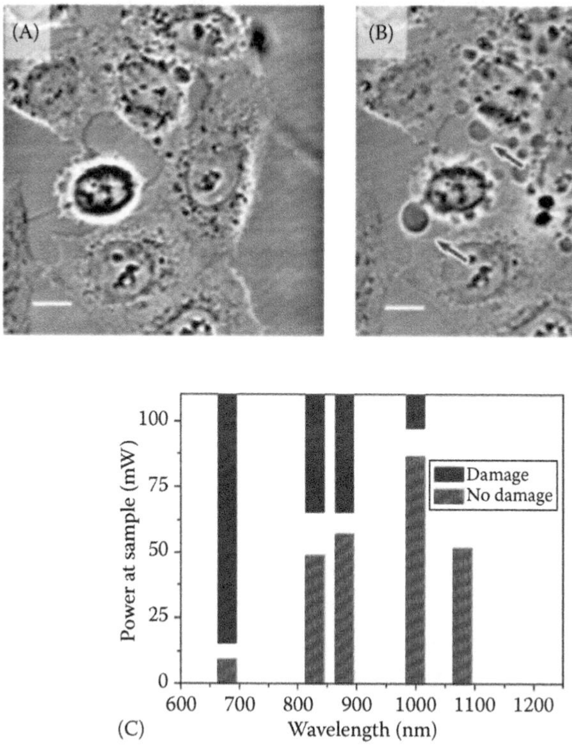


**FIGURE 5.3** Characterization of photodamage as a function of femtosecond laser wavelength. (A) Non-damaged cells. (B) Damaged cells indicated by blebbings of the cell membranes (arrows). (C) Damage threshold at different wavelengths. Bar: 20 μm. In all experiments, CHO cells were continuously scanned for 2 min.

Chapter 5

provided by the optical parametric oscillator (APE OPO, Coherent). In our typical SRL imaging experiments, the maximum power density we use at the sample is below 3 MW/cm$^2$ for 830 nm, 6 MW/cm$^2$ for 1000 nm, and 8 MW/cm$^2$ for 1100 nm, which is close to the optimal ratio and far below the damage threshold.

### 5.2.4 Applications Associated with Pico- versus Femto-Pulse Excitation

The picosecond excitation provides high spectral resolution and therefore high chemical selectivity for SRS and CARS microscopy. The narrow band SRS microscopy has been applied to vibrational imaging of tablets and biomass based on fingerprint Raman bands (Saar et al., 2010b, Slipchenko et al., 2010), ranging from 500 to 2000 cm$^{-1}$. Meanwhile, direct femtosecond pulse excitation is suitable for broad Raman bands such as the CH stretching vibrations around 2900 cm$^{-1}$ (FWHM ~ 100 cm$^{-1}$), CD stretching vibrations around 2100 cm$^{-1}$ (FWHM ~ 25 cm$^{-1}$),OH stretching vibrations around 3300 cm$^{-1}$ (FWHM ~ 400 cm$^{-1}$), and OD stretching vibrations around 2450 cm$^{-1}$ (FWHM ~ 300 cm$^{-1}$). Notably, the CH and OH modes and their isotope substituted CD and OD modes are the most abundant molecular vibrations in biological tissues and correspond to lipids, proteins and water molecular signatures. The study of metabolism pathways using isotopic substitution should be target applications for femtosecond SRS microscopy as well as imaging of low concentration lipid-based structures. Due to quadratic dependence of CARS signal on concentration (see Equation 5.6), the signal intensities of CARS are higher than SRS for highly concentrated lipid compartments, whereas at low concentrations the nonresonant background limits the CARS sensitivity.

## 5.3 Construction of a Multimodal Femtosecond SRL Microscope

A key advantage of femtosecond SRL is that it can be implemented on a multiphoton microscope platform. The schematic of a typical SRL microscope is shown in Figure 5.4. A Ti:Sapphire laser (Chameleon Vision, Coherent) with up to 4W (~140 fs pulse width) is

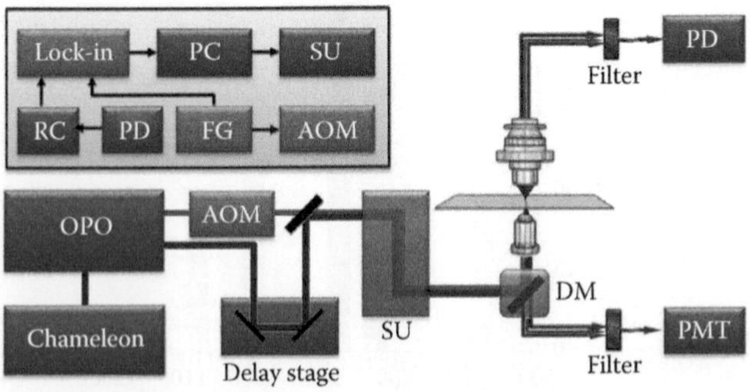


**FIGURE 5.4** Block diagram of a multimodal SRS microscope. OPO is an optical parametric oscillator. AOM is an acousto-optical modulator. SU is a scanning mirror unit. DM is an exchangeable dichroic mirror. PD is a photodiode for forward SRS detection. PMT is a photomultiplier tube for backward detection of multiphoton modalities. The block diagram shows the links between different components of the setup. Lock-in is a lock-in amplifier. RC is a resonant circuit. PC is a computer. FG is a function generator.

used to pump an optical parametric oscillator (Chameleon Compact OPO, Angewandte Physik & Elektronik GmbH), providing the pump and Stokes beams (Stokes beam is tunable from 1.0 to 1.6 μm). The pump and Stokes pulses are collinearly overlapped and directed into an inverted (IX71, Olympus) or upright (BX51, Olympus) laser-scanning microscope. The Stokes beam intensity is modulated by an acousto-optic modulator (AOM, 15180-1.06-LTD-GAP Gooch & Housego) at 5.4 MHz. The SRL signals are detected by a photodiode (818-BB-40, Newport) and then sent to a lock-in amplifier (HF2LI, Zurich Instrument). The laser scan and data acquisition are controlled by a PC. A special 60X/IR objective lens (1-U2B8931R, Olympus Inc) is used to transmit NIR wavelengths up to 1.5 μm. An air condenser (NA = 0.55) or 60x objective (NA = 1.2) is used to collect the forward signal in inverted microscope, an oil condenser (NA = 1.40) is used to collect forward signal in upright microscope.

### 5.3.1 Choice of Laser System

In recent years the advances in solid state and fiber laser technologies led to a wide range of commercially available turnkey pulsed picosecond and femtosecond laser systems with the repetition rates of tens of MHz. The ideal laser system for CRS microscope, which provides pump and Stokes beams, satisfies the following requirements: (1) inherently time synchronized (e.g., pump + optical parametric oscillator (OPO) laser system; (2) femtosecond pump is in near infrared (NIR) for optimal multiphoton microscopy; (3) easy tunable Stokes beam, which allows to access Raman modes in the range from 500 to 4000 cm$^{-1}$; (4) both pump and Stokes beams are high power (>100 mW). At present, the solid state systems or combinations of fiber pump + solid state OPO systems are the best choice to satisfy aforementioned requirements. This, however, comes at a high price in a multi-hundred thousand dollar range. The complete fiber-based systems are considerably less expensive but still lack enough power to compete with solid state designs.

### 5.3.2 Choice of Modulator: AOM versus EOM

In order to perform phase sensitive detection of SRS signal one of the beam has to be modulated. To avoid low-frequency laser noise the modulation of several MHz is desirable. There are two types of devices capable of high-frequency modulation, namely electro-optical modulator (EOM) and acousto-optical modulator (AOM). Due to differences in basic principles the implementation and performance of these two types of devises are different. The EOM has the advantage of a large aperture and does not require beam focusing. In addition EOM has large modulation depth of up to 100% and more than 90% throughput. The EOM has a disadvantage for the femtosecond lasers due to large dispersion, which, however, can be compensated. The major disadvantage of EOM originates from high driving voltage, which results in high cost at high modulation frequency. In addition the higher the wavelength the higher voltage is required. The AOM advantages are lower cost and lower dispersion for femtosecond pulses compared to EOM as well as operation at longer wavelengths without special modifications. The disadvantage of AOM is that 100% modulation can be achieved only in deflected beam, but the deflection angle is wavelength dependent. The operation in zero order or using modulation of undeflected beam results in modulation depth below 85%. However, due to low dispersion, high flexibility in wavelength and modulation frequency range as well as low price we choose AOM (15180-1.06-LTD-GAP, Gooch & Housego) as a modulation device.

### 5.3.3 Choice of Oil Condenser for Maximum Signal Collection

Although the SRS signal is free from the nonresonant background, there are a number of optical processes, including cross phase modulation and the thermal lensing effect (Uchiyama et al., 2000), which alter the refractive index of media due to high peak power pump beam and change the propagation of Stokes beam. In addition scattering tissue distorts the forward propagating radiation due to diffusion. To avoid background due to change in refractive index and to efficiently collect the diffused light the condenser with large numerical aperture and large field of view is needed. Often the microscope objective with characteristics similar to excitation objective is used to collect forward photons (Freudiger et al., 2008; Slipchenko et al., 2009). However, the small field of view makes the collection of diffused light inefficient (Saar et al., 2010a). To overcome small field of view limitation of high NA objectives we used oil condenser (U-AAC; achromatic/aplanatic condenser NA = 1.40 APERTIRIS,10X-100X, Olympus).

### 5.3.4 Choice of Photodiode and Optimal Focusing of the Beam after Condenser

Both electrical and geometrical parameters of photodiode are important for optimal detection of SRS signal. Due to non-descanned nature of detected beam the large area photodiodes with area from 20 to 100 mm$^2$ are necessary for large field of view detection. For optimal detection of the signal the photodiode surface should be uniformly illuminated. In addition the movement of the beam due to laser scanning should be minimized. Figure 5.5 shows the example of relative position of condenser and photodiode for uniform illumination and minimal beam motion, which is achieved by imaging the back aperture of objective on photodiode surface (see ray tracing in Figure 5.4). For efficient detection of the beam with large dynamic range and excellent linearity the

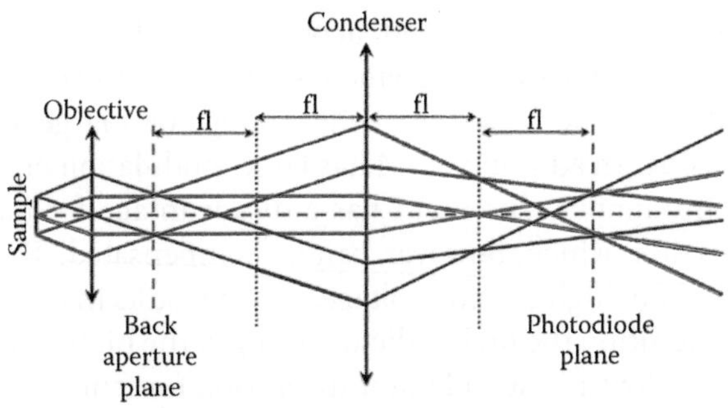


**FIGURE 5.5** Ray tracing of optimal condenser position. fl is a effective focal length of the condenser lens. The dashed vertical lines show the relayed image planes. The short dash vertical lines show the focal planes of the condenser.

biased photodiodes with high reverse voltage are used. For high-frequency operation PIN photodiodes are the best choice due to minimized junction capacitance compared to PN type of photodiodes.

# 5.4 Performance of Femtosecond SRL Microscopy

## 5.4.1 Shot-Noise-Limited Detection

The detected SRS intensity scales linear with each beam power (see Equation 5.8). However, the noise in the signal is only contributed by beam being detected. In order to achieve ultimate SNR, the noise in the detected beam should be limited only by photons shot noise, which scales as square root of number of detected photons. Due to micro-environmental changes and mechanical vibrations most of the laser sources are far from shot noise limit at below 100 kHz frequencies and their noise scales linear with laser power. At higher frequencies the environment become silent and shot-noise-limited stability can be achieved. Figure 5.6 illustrates the advantage of high-frequency modulation. The noise detected at 1.7 MHz has close to linear dependence and indicate the large contribution of laser noise to total detected noise. If detection at 1.7 MHz employed the increase in detected beam power would result in only marginal improvement of overall S/N of SRS signal. On the other hand at 5 MHz the noise scales as a square root of laser power, which is an indication of shot-noise-limited laser stability. Therefore, the modulation frequency of 5.5 MHz was used for all experiments discussed later.

## 5.4.2 Signal Enhancement due to Femtosecond Excitation

To confirm the computational results which predict large enhancement of SRS signal for femtosecond versus picosecond excitation (see Figure 5.2B), we compared the SRL intensity from olive oil generated by 5 ps and 200 fs laser pulses, respectively. The 5 ps laser system for SRS imaging was described in our previous paper (Slipchenko et al., 2010). We used ~6 mW for pump and ~6 mW for Stokes at the sample. The intensity profiles below the images (Figure 5.7) show that the signal level increased for more than 12 times when the lasers are changed from 5 ps to 200 fs.

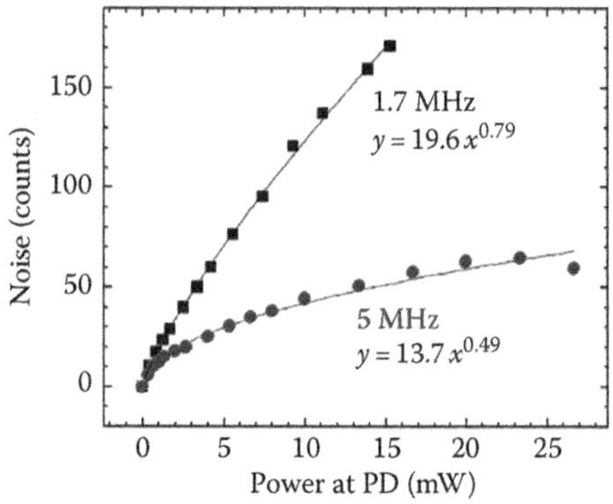


**FIGURE 5.6** The dependence of laser noise on power of femtosecond Ti:Sapphire oscillator for two modulation frequencies. The contribution of constant electronic noise is removed.

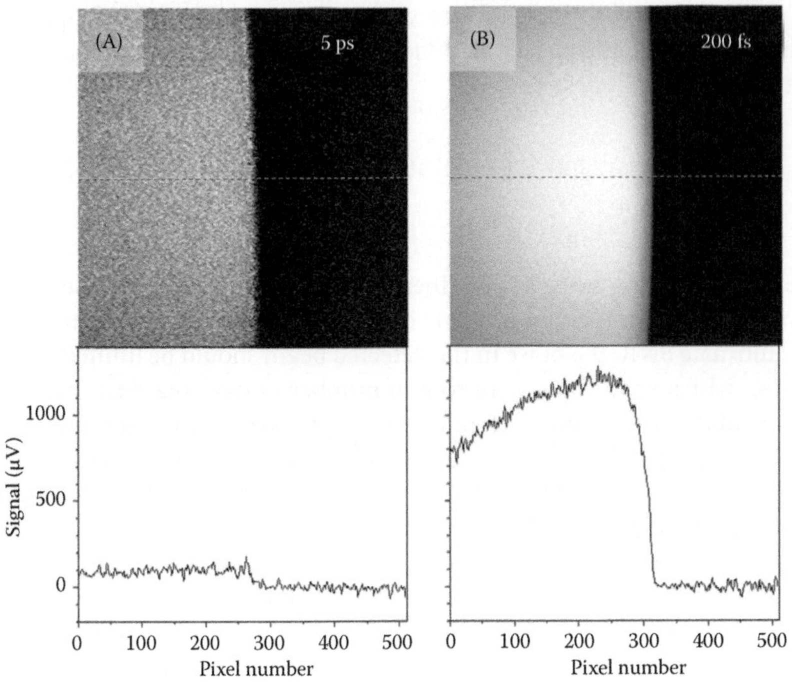


**FIGURE 5.7** Comparison of SRL signal levels between picosecond and femtosecond excitation at the same power. (A) SRL imaging at the interface of olive oil and air, with 5 ps excitation. (B) SRL imaging with 200 fs with the same sample and the same laser power (pump and Stokes, respectively). Shown below each image are the intensity profiles along the dashed lines in the images.

### 5.4.3 Multimodal Imaging of *C. elegans*

A challenge for multimodal imaging is simultaneous acquisition of various NLO signals in a convenient way. Due to heterodyne nature of detected signal SRS is insensitive to incoherent background such as fluorescence because it cannot be amplified by the local oscillator field. As a result SRS modality is fully compatible with two photon excited fluorescence (TPEF), which can be effectively detected in backward direction with appropriate dichroic mirror (see Figure 5.4). The SHG and THG modalities are generated in UV and can be easily separated from SRS beam in both backward and forward directions. Note that in the current setup of fSRL the Stokes beam generates SHG signal in visible range and careful selection of bandpass filters is required to prevent mixing of TPEF and SHG channels. We demonstrate simultaneous fSRL and TPEF imaging of live *Caenorhabditis elegans* (*C. elegans*). As a label-free imaging modality, CARS microscopy has been employed to selectively visualize LDs in *C. elegans* (Le et al., 2010; Yen et al., 2010). More recently, SRS microscopy has been used to map lipids in the worm but without 3D sectioning (Wang et al., 2011). By using an objective lens of 1.2 NA, we obtained 3D fSRL images of a wild type *C. elegans* using the signal from C–H bond (Figure 5.8A). Such 3D sectioning is critical to distinguish the fat stored in LDs from the membrane lipids. As an additional advantage, our system with femtosecond pulse excitation is highly compatible with other modalities such as TPEF. This advantage allows simultaneous forward SRL imaging of lipid and

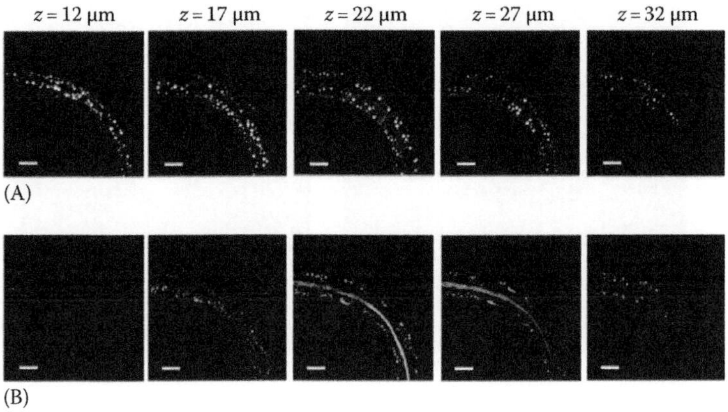


**FIGURE 5.8** Multimodal imaging of *C. elegans*. Depth-resolved fSRL images of a live *C. elegans* based on the signal from C–H stretching band (A) and simultaneous TPEF images of autofluorescence (B). A total of 50 frames were taken at the axial step of 1.0 μm. The images of 512 × 512 pixels were acquired at the speed of 2 μs per pixel. The pump and Stokes power at the sample was 10 and 40 mW, respectively. The pump wavelength was 830 nm. The Stokes wavelength was 1090 nm. Bar: 20 μm.

backward TPEF imaging of autofluorescence in *C. elegans*. As shown in Figure 5.8B, the autofluorescence arose from the small intestine and some particles that were not overlapped with the LDs.

### 5.4.4 Imaging of Fatty Acid Uptake by Cells

With the fSRL microscope, we explored the uptake of palmitic acid and its intracellular fate in CHO cells. Palmitic acids have been shown to induce lipo-toxicity in mammalian cells (Listenberger et al., 2003). However, direct visualization of the fatty acids was not accessible with existing imaging tools. To selectively detect the molecule by SRL, we used deuterated palmitic acid-$d_{31}$ (Aldrich) which gives an isolated Raman band at 2110 $cm^{-1}$ based on the C–D bond stretch vibration. For the cellular uptake study, deuterated palmitic acid was dissolved in DMSO and added to the cell culture medium, with a final concentration of 100 μM. After 6 h incubation, the cells were imaged with the SRL microscope. The deuterated compound is clearly visualized in lipid bodies and also in the cellular membranes (Figure 5.9A). We also imaged the same cells based on the signal from C–H vibration (Figure 5.9B). It was found that the C–D rich droplets were overlapped with the C–H abundant droplets. These results implied that a portion of the palmitic acid has been converted into triacylglyceride and stored in the lipid droplets. Without addition of deuterated palmitic acid, we did not detect SRL signals at the C–D vibration frequency (Figure 5.9C), which confirms the bond-selective imaging capability of our setup.

### 5.4.5 *Ex Vivo* Imaging of Lipid Storage in Small Intestine of Mouse

As another application, we demonstrate 3D sectioning of lipid storage in villi of small intestine of mouse. Previously CARS imaging has been used to map lipid storage in small intestine of mouse and image processing was used to remove the intrinsic nonresonant background (Zhu et al., 2009; Uchida et al., 2011). By using

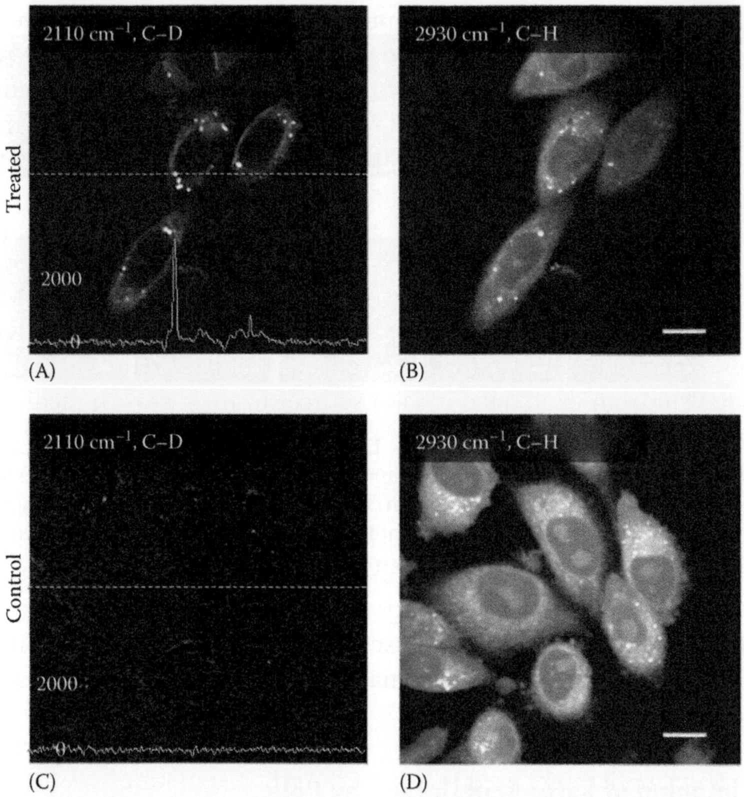


**FIGURE 5.9** Femtosecond SRL imaging of deuterated palmitic fatty acid in live CHO cells. (A and B) SRL imaging of C–D and C–H stretch vibrations in cells treated with deuterated palmitic fatty acid. (B–D) SRL imaging of C–D and C–H stretch vibrations in untreated cells. The insets are intensity profiles along the dash lines. Scale bar: 10 μm. The image speed is 2 μs/pixel.

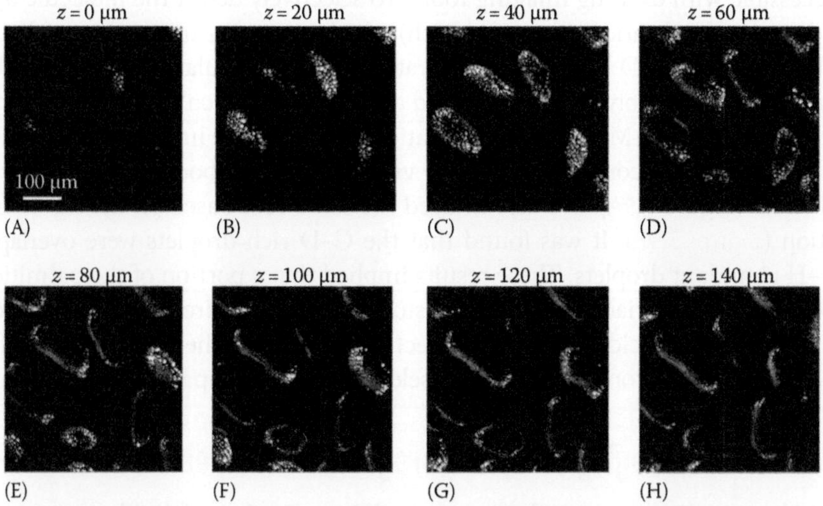


**FIGURE 5.10** Femtosecond SRL imaging of lipid in small intestine of wild-type mouse (*ex vivo*). (A–H) 3D SRL sectioning (every 20 μm from the coverslip surface) of SRL imaging on villi in small intestine at C–H vibration band Scale bar is indicated. The image speed is 2 μs/pixel. Pump and Stokes powers at the sample are 25 and 30 mW, respectively. Images are obtained with 10× NA = 0.4 objective.

the background-free SRL modality from C–H vibration at 2850 $cm^{-1}$, we performed 3-D sections of villi covering inner surface of small intestine. The results are shown in Figure 5.10A through H. The lipid droplets show a bright signal over the weak background from other structures.

## 5.5 Conclusions and Discussion

We presented a theoretical comparison between CARS and SRS regarding the signal dependence on excitation bandwidth. Theoretical calculations indicate that femtosecond excitation increases the SRS signal intensity by more than one order of magnitude compared to picosecond laser excitation. We have experimentally compared a femtosecond SRL setup with a SRS system pumped with 5 ps laser at same power and found a 12-fold increase of signal level for imaging C–H bonds. We employed the broad tunable OPO (from 1.0 to 1.6 μm) which permits bond-selective SRL imaging of C–D, O–D, S–H, C–H, N–H, and O–H bonds based on their stretching vibrations. We also demonstrated that the use of MHz modulation frequency and a very low pump beam power (a few mW at sample) enables SRL imaging at the laser shot noise limit. Because the majority of the excitation power is carried by the Stokes beam at wavelength above 1.0 μm photodamage to the sample is eliminated. The femtosecond SRL modality is highly compatible with other NLO modalities including TPEF and SHG which facilitate multimodal imaging of tissue samples. The drawback of the femtosecond excitation is a lack of the spectral resolution required for selective imaging of narrow Raman modes. However, this disadvantage can be overcome by pulse shaping of the femtosecond pulses. Simple pulse shaping can be done in the time domain by chirping the pulses, which allows spectral focusing for narrowband excitation (Hellerer et al., 2004; Pegoraro et al., 2009). To summarize, femtosecond excitation allows larger NLO signal to be generated compared to picosecond excitation, and spectral selectivity can be achieved with pulse shaping techniques (Silberberg, 2009; Freudiger et al., 2011).

## References

Chen, H., Wang, H., Slipchenko, M. N. et al. (2009) A multimodal platform for nonlinear optical microscopy and microspectroscopy. *Optics Express*, 17, 1282–1290.

Cheng, J. X., Volkmer, A., Book, L. D., and Xie, X. S. (2001) An epi-detected coherent anti-Stokes Raman scattering (E-CARS) microscope with high spectral resolution and high sensitivity. *Journal of Physical Chemistry B*, 105, 1277–1280.

Freudiger, C. W., Min, W., Holtom, G. R. et al. (2011) Highly specific label-free molecular imaging with spectrally tailored excitation-stimulated Raman scattering (STE-SRS) microscopy. *Nature Photonics*, 5, 103–109.

Freudiger, C. W., Min, W., Saar, B. G. et al. (2008) Label-free biomedical imaging with high sensitivity by stimulated Raman scattering microscopy. *Science*, 322, 1857–1861.

Fu, Y., Wang, H., Shi, R., and Cheng, J. X. (2006) Characterization of photodamage in coherent anti-Stokes Raman scattering microscopy. *Optics Express*, 14, 3942–3951.

Hellerer, T., Enejder, A. M. K., and Zumbusch, A. (2004) Spectral focusing: High spectral resolution spectroscopy with broad-bandwidth laser pulses. *Applied Physics Letters*, 85, 25–27.

Kukura, P., Mccamant, D. W., and Mathies, R. A. (2007) Femtosecond stimulated Raman spectroscopy. *Annual Review of Physical Chemistry*, 58, 461–488.

Le, T. T., Duren, H. M., Slipchenko, M. N., Hu, C. D., and Cheng, J. X. (2010) Label-free quantitative analysis of lipid metabolism in living *Caenorhabditis elegans*. *Journal of Lipid Research*, 51, 672.

Levenson, M. D. and Kano, S. S. (1988) *Introduction to Nonlinear Laser Spectroscopy*. San Diego, CA: Academic Press.

Listenberger, L. L., Han, X., Lewis, S. E. et al. (2003) Triglyceride accumulation protects against fatty acid-induced lipotoxicity. *Proceedings of the National Academy of Sciences of the United States of America*, 100, 3077.

Mccamant, D. W., Kukura, P., and Mathies, R. A. (2003) Femtosecond broadband stimulated Raman: A new approach for high-performance vibrational spectroscopy. *Applied Spectroscopy*, 57, 1317–1323.

Mukamel, S. (1995) *Principles of Nonlinear Optical Spectroscopy*. New York: Oxford University Press.

Pegoraro, A. F., Ridsdale, A., Moffatt, D. J. et al. (2009) Optimally chirped multimodal CARS microscopy based on a single Ti:sapphire oscillator. *Optics Express*, 17, 2984–2996.

Pegoraro, A. F., Slepkov, A. D., Ridsdale, A., Pezacki, J. P., and Stolow, A. (2010) Single laser source for multimodal coherent anti-Stokes Raman scattering microscopy. *Applied Optics*, 49, F10–F17.

Saar, B. G., Freudiger, C. W., Reichman, J. et al. (2010a) Video-rate molecular imaging in vivo with stimulated Raman scattering. *Science*, 330, 1368.

Saar, B. G., Zeng, Y. N., Freudiger, C. W. et al. (2010b) Label-free, real-time monitoring of biomass processing with stimulated Raman scattering microscopy. *Angewandte Chemie International Edition*, 49, 5476–5479.

Shen, Y. R. (1984) *The Principles of Nonlinear Optics*. New York: John Wiley & Sons Inc.

Silberberg, Y. (2009) Quantum coherent control for nonlinear spectroscopy and microscopy. *Annual Review of Physical Chemistry*, 60, 277–292.

Slipchenko, M. N., Chen, H., Ely, D. R. et al. (2010) Vibrational imaging of tablets by epi-detected stimulated Raman scattering microscopy. *Analyst*, 135, 2613–2619.

Slipchenko, M. N., Le, T. T., Chen, H., and Cheng, J. X. (2009) High-speed vibrational imaging and spectral analysis of lipid bodies by compound Raman microscopy. *Journal of Physical Chemistry B*, 113, 7681–7686.

Uchida, A., Slipchenko, M. N., Cheng, J.-X., and Buhman, K. K. (2011) Fenofibrate, a peroxisome proliferator-activated receptor α agonist, alters triglyceride metabolism in enterocytes of mice. *Biochimica et Biophysica Acta (BBA)—Molecular and Cell Biology of Lipids*, 1811, 170–176.

Uchiyama, K., Hibara, A., Kimura, H., Sawada, T., and Kitamori, T. (2000) Thermal lens microscope. *Japanese Journal of Applied Physics* 1, 39, 5316–5322.

Wang, M. C., Min, W., Freudiger, C. W., Ruvkun, G., and Xie, X. S. (2011) RNAi screening for fat regulatory genes with SRS microscopy. *Nature Methods*, 8, 135–138.

Yen, K., Le, T. T., Bansal, A. et al. (2010) A comparative study of fat storage quantitation in nematode *Caenorhabditis elegans* using label and label-free methods. *PloS One*, 5, 387–398.

Yue, S., Slipchenko, M. N., and Cheng, J. X. (2011) Multimodal nonlinear optical microscopy. *Laser and Photonics Reviews*, 5, 496–512.

Zhu, J., Lee, B., Buhman, K. K., and Cheng, J. X. (2009) A dynamic, cytoplasmic triacylglycerol pool in enterocytes revealed by ex vivo and in vivo coherent anti-Stokes Raman scattering imaging. *Journal of Lipid Research*, 50, 1080–1089.

# 6. Miniature Coherent Raman Probes for *In Vivo* Biomedical Imaging

**Gangjun Liu, Mihaela Balu, Zhongping Chen, and Eric Olaf Potma**



## 6.1 Introduction

Coherent Raman microscopy uniquely combines selective molecular contrast with fast image acquisition times. The rapid image acquisition time is arguably one of the most important merits of the coherent Raman technique. Whereas the vibrational contrast is no better than what can be obtained with a spontaneous Raman microscope, it is the fast scanning capability of CRS methods that has given these techniques their unique place among optical imaging tools. In no area of application are these combined traits more deployed as in the area of biomedical imaging of tissues *in vivo*. The powerful merger of molecular contrast and rapid imaging has permanently opened the door toward real-time vibrational imaging of living animals and human subjects [1,2]. This transition from

*Coherent Raman Scattering Microscopy.* Edited by Ji-Xin Cheng and X. Sunney Xie © 2013 CRC Press/ Taylor & Francis Group, LLC. ISBN: 978-1-4398-6765-5.

*ex vivo* to *in vivo* studies is important, as it brings the discipline of vibrational imaging into the realm of live animal research and clinical studies.

CRS microscopy techniques have already demonstrated great potential for clinical applications. For instance, the exquisite capability of CARS and SRS to detect lipids at high speed renders these techniques the primary tool for the examination and diagnosis of lipid abnormalities in live superficial tissues [3]. Studies of organ tissues accessible through minimally invasive surgery, such as the small intestine and spinal cord in mouse models, have shown that CARS is capable of detecting sub-cellular variations in lipid organization, composition, and quantity in live animals. These studies have provided hitherto inaccessible microscopic information on intestinal lipid absorption [4,5] and myelin-related neurodegenerate diseases [6–8]. Beyond the analysis of lipids, coherent Raman techniques have shown clinical potential for the detection of topically applied agents to skin and hair [1,2,9,10]. In particular, video-rate SRS has been successfully used to monitor the diffusion of small molecular agents through live human skin [2]. No competing imaging techniques are currently capable of selectively visualizing the microscopic distribution of topically applied molecular compounds in a noninvasive manner. The CRS technique thus fills an important void in the existing arsenal of biomedical imaging tools. These examples illustrate that, in terms of molecular contrast and imaging speed, coherent Raman microscopy constitutes an imaging approach suitable for clinical applications, including the diagnosis of lipid related diseases and the diffusion of unlabeled pharmaceutical drugs in superficial tissues.

Despite the clear clinical potential of CRS imaging techniques, the actual implementation of these methods in a clinical setting is currently limited by technical difficulties. Whereas the CRS techniques have matured in the optics research lab, these state-of-the-art technologies are not automatically compatible with the experimental requirements relevant to clinical studies. When applied to human subjects, the imaging technology needs to comply with additional criteria beyond the conditions of high imaging contrast and speed. Among these criteria are patient comfort, ease of use, and robustness. In this regard, the powerful ultrafast laser technologies that have raised the level of eminence of CRS methods in optics laboratories are generally ill-matched for use in clinical settings. The relatively large footprints of conventional ultrafast lasers and their sensitivity to temperature fluctuations, acoustic vibrations, and mechanical disturbances make such light sources unfit for reliable performance in a clinical examination room. The development of smaller and more robust laser light sources is an important direction of activity that will facilitate the translation of CRS technologies into the clinic. Some of the recent developments in ultrafast fiber lasers are particularly encouraging and provide glimpses of tomorrow's compact light sources that exhibit reliable performance at the patient's bedside.

Similar to the shortcomings of current laser light sources, the optical layouts of typical CRS microscopes are not optimized for examination of human subjects. A conventional upright or inverted optical microscope, designed for interrogation of cell tissue cultures, does not allow for comfortable and flexible imaging of live superficial tissues. Rapid examination of large tissue areas necessitates a less rigid approach toward imaging where the image forming microscope lens can be moved flexibly to selected areas of interest in and on the body. This strategy departs from the fixed space optical configuration of the traditional microscope and calls for a design that allows for movable parts of the imaging system.

What are the required optical ingredients for a CRS imaging system? First, a focusing element is needed to produce a tightly focused spot that warrants the efficient generation of nonlinear Raman radiation. The formation of a good quality focal spot is key to the success of the imaging device, which puts stringent constraints on the design of the focusing element. Although CRS wide-field illumination schemes have been developed as well [11,12], the current illumination geometry used in such approaches is less suitable for live tissue imaging. Besides the generation of a tight focal spot, the focusing element may also be used to collect the signal from the tissue in an epi-detection geometry. Second, the preferred point illumination scheme requires a beam-scanning mechanism for image formation. The beam-scanning unit imparts an angular deviation to the light beam, which is translated by the focusing element to a lateral shift of the focused spot in the sample. The necessity for beam-scanning significantly complicates the design of a CRS imaging probe, as it involves mechanically moveable parts that are controllable with a high degree of precision. Third, the imaging system incorporates relay optics to shape the excitation radiation and direct it to and between the beam-scanning module and the focusing element. These relay optics may include the optical elements that guide the light from the source to the scanner and microscope lens, the imaging optics between the scanning module and the focusing element, and any dichroic beamsplitters and optical filters. Fourth, the system includes a photodetector, which exhibits sufficient sensitivity for converting the presences of signal photons into an electrical signal. Taken together, these elements constitute the basic functional components of an optical imaging system. The challenge for designing an optical imaging system compatible for clinical use is to fulfill the criteria of robustness, flexibility and patient's comfort without significantly compromising the quality of the images.

There is no single solution toward optimizing the imaging system for clinical use, as such criteria are highly dependent on the specific clinical application. Nonetheless, it is instructive to consider the common features of the optical designs of several existing scanning laser beam-based imaging technologies that are used for clinical examination. For our discussion here, two general categories can be identified. The first category is formed by optical imaging systems that are based on free-space optics and that bear a large degree of similarity to the layout of conventional microscopes. In such systems, the major design effort is in optimizing the relay optics such that the focusing element, a microscope objective, can be conveniently guided to the area of interest on the tissue. Examples include commercial solutions such as a confocal reflectance microscope [13,14] and a nonlinear optical microscope [15–17], which both feature relay optics that enable flexible arms to move the objective lens to the desired location.

The second category constitutes optical imaging systems that are miniaturized to comply with the needs for remote sensing and intravital probing. The design of such systems can depart significantly from the layout of conventional imaging microscopes. For instance, miniaturization of several key elements, most notably the focusing element and associated relay optics, enables the construction of small probes that are suitable for endoscopic examinations. In addition, delivery of excitation light through optical fibers and the miniaturization of the beam-scanning module constitute design challenges that are notably different from free-space-based optical microscopes. In this chapter, we are concerned with several aspects of the miniaturization of a CRS imaging system. We will focus on the optical elements of a miniature CRS imaging system rather than the design criteria for a clinical ultrafast laser source.

In the following sections, we will briefly discuss recent developments in the field of miniature probe design in nonlinear optical microscopy. We will then lay out several specific design criteria for CRS endoscopic imaging, followed by discussions on fiber delivery and detection, miniaturization of the focusing optics and design solutions for scanners compatible with the dimensions required for endoscopic probing. We will zoom in on several prospective applications and conclude this chapter by providing an outlook of the future of CRS endoscopy.

## 6.2 Miniaturization of Multiphoton Microscopes

Over the past decade, tremendous progress has been made in the development of miniaturized probes suitable for imaging with nonlinear optical contrast. One of the first examples of a miniaturized nonlinear imaging system was developed for the purpose of probing two-photon excited fluorescence (TPEF) from fluorophores injected in the brain tissue of a living mouse [18]. The design consisted of a fiber-coupled miniature microscope system that included a piezoelectric fiber tip scanner, assembled millimeter-sized lenses, a dichroic and a mounted photomultiplier tube. The microscope assembly was mounted directly onto the head of a mouse. Although this system produced impressive TPEF images at the time, subsequent developments and alternative designs over the years have significantly improved the performance of individual components, reduced their size and produced fiber-based miniature probes of significantly less weight while boosting the image quality [19,20]. These advances have paved the way for nonlinear endoscopic probing systems based on TPEF [21] or second harmonic generation (SHG) contrast [22].

In terms of probe design, nonlinear optical techniques share several important features. First, nonlinear optical methods generally employ pulsed excitation. The use of pulse excitation poses certain challenges when such radiation is propagated through optical fibers. Effects such as temporal and spectral broadening effects of the pulses need to be considered when choosing a particular fiber for pulse delivery. Some of these considerations will be discussed in Section 6.3.1. Second, the point illumination mechanism used in nonlinear microscopy implies that proper beam scanners have to be incorporated in the probe design. In many cases, this involves the miniaturization of beam scanning optics or the design of specialized relay optics. Finally, the tight focusing required for generating nonlinear optical signals calls for a focusing element with a high numerical aperture but with minimum optical aberrations.

In all of these areas, that is, the optimization of pulse propagation through fibers, the design of miniature scanners, and the fabrication of micro-lenses, the precedent developments in multiphoton fluorescence and SHG microscopy contain useful lessons for designing a miniature CRS probe [19,20]. In the following sections, we will discuss individual optical components and design strategies used for assembling miniature nonlinear probes. We will put special emphasis on the additional design criteria relevant to coherent Raman scattering and discuss various successful embodiments of CRS probes.

# 6.3 Fiber Delivery and Detection in CRS Microscopy

## 6.3.1 Pulse Propagation in Optical Fibers

Optical fibers enable flexible delivery of excitation radiation from a laser light source to the distal tip of a miniature probe. The mechanical flexibility offered by optical fibers has accelerated their integration in imaging and sensing systems that align well with the clinical requirements of ruggedness and patient's comfort. Standard optical fibers are composed of a higher refractive index solid silica core and a lower refractive index cladding surrounding the core. Light is guided in the core through the mechanism of total internal reflection, which puts restrictions on the angular variability of the entering rays. Radiation that falls within the angular acceptance cone of the fiber, which is usually called the numerical aperture of the fiber, is supported for propagation. Propagating light is confined to transverse modes, the number of which is strongly dependent on the size of the core and the refractive index difference between the core and the cladding. Fibers that support only one transverse mode are called single-mode fibers (SMFs), whereas fibers that support multiple transverse modes are referred to as multimode fibers. SMFs operating in the visible and near-infrared range typically have core diameters on the order of a few micrometers (2–10 μm), with a cladding of up to a few hundred micrometers. The superior mode quality of SMFs make these fibers good candidates for microscopy applications where the tightness of the focal spot crucially depends on the quality of the transverse beam mode.

When ultrafast laser radiation is used, however, the choice of optical fiber is subject to several constraints. Both linear and nonlinear effects in the fiber can affect the quality of the pulsed excitation light when propagated through a SMF. In the linear regime, the optical dispersion of the fiber material introduces temporal broadening of the pulses. The different frequencies within an optical pulse travel with different group velocities, a phenomenon known as group velocity dispersion (GVD). GVD increases the duration of a Gaussian transform-limited pulse with original temporal width $T_0$ to a new temporal width $T_1$ according to [23]

$$T_1(L) = T_0\left[1+\left(\frac{L}{L_D}\right)^2\right]^{1/2} \tag{6.1}$$

where

$L$ is the fiber length

$L_D = T_0^2/|\beta_2|$ is the dispersion length

$\beta_2$ is the GVD parameter

The temporal broadening lowers the peak power of the pulses, which reduces the magnitude of the induced nonlinear signal upon focusing into the sample. The effects of temporal broadening become more significant for shorter pulse. For instance, after propagation in a 1 m long fused silica fiber, GVD stretches a transform-limited pulse of 100 fs at 800 nm to 375 fs whereas a 10 fs pulse increases in duration to 3.6 ps. The negative effects of linear temporal broadening can generally be mitigated by pre-conditioning the pulses with prism and/or grating-based compensators or adaptive pulse shapers [24].

Besides linear broadening effects, the temporal and spectral profile of the pulses are affected by nonlinear optical effects within the fiber. A prominent third-order nonlinear process in the fiber is self-phase modulation (SPM). SPM is a result of the time-varying intensity dependence of the medium's refractive index. A high intensity ultrashort pulse propagating in a medium can induce a nonlinear response through a time-varying refractive index:

$$n = n_0 + n_2 I(t) \tag{6.2}$$

where

$n_0$ is the linear refractive index
$n_2 = 2.6 \times 10^{-16}$ cm$^2$/W is the second order nonlinear refractive index of silica [25]
$I(t)$ is the time dependent pulse intensity

The time-varying refractive index will result in steep changes to the temporal profile of the optical phase of the pulse. In the frequency domain, the temporal compression and stretching, due to the nonlinear optical phase changes, correspond to the generation of new frequency components. Hence, the frequency spectrum of the pulse generally widens as a result of SPM as it propagates through the fiber. The time dependence of the changes in instantaneous optical frequency across the pulse is known as frequency chirping. The further the pulse travels, the more frequency chirping is induced through SPM and the more frequencies components are generated as the pulse propagates in the fiber. A broad spectrum generated this way is called a supercontinuum. While the generation of a supercontinuum is very useful for broadband CRS applications, it is generally an undesired effect when optical fibers are used for pulse delivery in miniature probes. The combination of SPM and GVD generally lowers the nonlinear excitation efficiency in the sample and reduces the effective penetration depth [26,27]. Minimization of SPM effects is, therefore, an important criterion when choosing an optical fiber for the design of a miniature optical probe.

A measure for the impact of SPM on the pulse spectrum is found in considering the magnitude of the nonlinear phase shift. Spectral broadening becomes significant when the nonlinear phase shift approaches $2\pi$. The pulse energy at which a $2\pi$-phase shift is induced can be estimated as

$$E_{2\pi} \approx \frac{\pi \lambda t_p D_{eff}^2}{4 n_2 L} \tag{6.3}$$

where

$\lambda$ is the center wavelength of the pulse
$t_p$ is the temporal pulse width
$D_{eff}$ is the effective mode diameter

For instance, for a 800 nm, 7 ps pulse in a 1 m long SMF with a 5 μm mode field diameter, SPM-induced broadening can be expected at pulse energies of ~4 nJ per pulse.

In addition to SPM, other nonlinear effects may affect the pulse characteristics. In particular, stimulated Raman scattering within the fiber constitutes another mechanism for spectral broadening. SRS in the fiber is caused by the Raman resonance of the stretching and bending modes of Si–O bonds, which form a band of overlapping modes that peaks around 420 $cm^{-1}$ and is broad due to the noncrystalline nature of fused silica glasses [28]. Consequently, the SRS mechanism generates new frequency components that are red-shifted by ~420 $cm^{-1}$ relative to the input frequencies contained in the pulse. Cascaded SRS can produce an array of new frequencies that can, in combination with SPM and assisted by GVD, effectively broaden the pulse spectrum.

As is exemplified by Equation 6.3, the impact of nonlinear broadening mechanisms grows with the peak energy of the incident pulses. Therefore, keeping the pulse power at a minimum is a straightforward approach to suppress unwanted nonlinear pulse broadening effects. When using near-infrared picosecond pulses in CRS imaging, with pulse energies below ~4 nJ within a 1 m fiber, broadening in SMFs is generally negligible. These pulse energies correspond to an average power of 300 mW at 76 MHz, which is generally more than sufficient for highly efficient CRS imaging of tissues. Légaré et al. found that above such pulse energies, SPM is significant and spectral broadening gains prominence [29]. When femtosecond pulses are used for CRS excitation at similar average powers, however, SMFs generally introduce significant broadening of the pulses. Since further lowering of the incident pulse energy is not always an option, alternative fiber designs are required to manage temporal and spectral broadening of the excitation light.

### 6.3.2 Photonic Crystal Fibers

One approach to minimize the temporal and spectral broadening of the excitation pulses in delivery fibers is to select fibers with larger core diameters. A larger core diameter implies a lower peak intensity per unit area in the fiber, which reduces the impact of optical nonlinearities during propagation. For instance, the efficiency of SPM is lowered for larger core areas. Since the light is less confined in larger core diameters, single-mode propagation cannot always be maintained as such fibers can support multiple transverse modes. It was shown that multimode fibers (MMFs) with core diameters of ~9 μm enable propagation of picosecond pulses through several meters of fiber without significant spectral broadening [30]. Allowing more modes to propagate through the fiber is a feasible solution as long as the higher-order modes are present at relatively small amplitudes. Nonetheless, with increasing diameter, higher-order modes may gain prominence and affect, therefore, the quality of the focal spot in the sample.

An alternative solution is found in the use of photonic crystal fibers (PCFs). A photonic crystal structure, formed from air holes in a silica matrix, constitutes a mechanism for confining and guiding the light. The major advantage of using such a guiding mechanism is that the optical properties of the fiber can be tailored by changing the pattern of the photonic crystal structure. This flexibility in the design has produced a wide variety of PCFs with optical properties that are beyond what can be attained in SMFs.

In high-index core PCFs, the photonic crystal exhibits a lower effective refractive index than the silica core. Hence, the photonic crystal acts like a cladding that confines the light to the region of the solid core. The photonic crystal can be designed such that single-mode propagation is achieved in the core, even for larger core diameters [31]. The so-called large-mode-area (LMA) PCF features single-mode propagation at core diameters of up to 35 μm, which significantly reduces optical nonlinear effects in the fiber. Besides the larger core diameter, the dispersive properties of the PCF can be

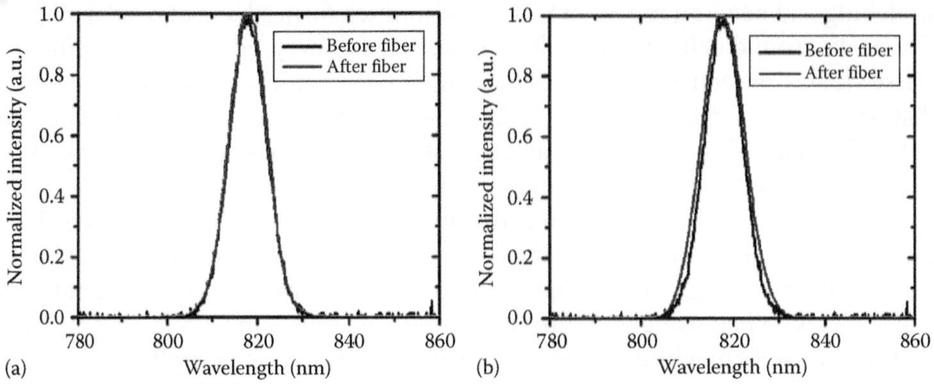


**FIGURE 6.1** Intensity spectra of a 817 nm, 280 fs pulse train measured before and after (a) a 80 cm long double-clad fiber (DCPCF16, crystal fiber) with FWHM = 9.7 nm before the fiber and FWHM = 9.7 nm after the fiber; (b) a 50 cm long large-mode area PCF (LMA-20, crystal fiber) with FWHM = 10 nm before the fiber and FWHM = 11 nm after the fiber.

carefully designed. In particular, the balancing between anomalous and normal dispersion of the PCF permits the design of fibers that exhibit zero dispersion for certain wavelengths. When pulses with a wavelength close to a zero-dispersion wavelength are launched in the fiber, GVD broadening is almost completely suppressed. Hence, PCFs allow the control of pulse distortion due to both GVD and SPM [32]. LMA fibers have been shown to successfully deliver near-infrared picosecond pump and Stokes pulses to a CARS microscope with limited distortion at average powers of less than 100 mW [33]. At similar powers, undistorted propagation a 280 fs pump pulses (817 nm) along with 7.5 ps Stokes pulses (1064 nm) was achieved in a 50 cm long, 20 μm core LMA, allowing pulse delivery for efficient CARS generation [34]. Figure 6.1 shows the spectra of 280 fs near-infrared pulses before and after propagation through photonic crystal fibers, demonstrating that SPM-induced broadening effects can be efficiently managed.

Another type of PCF is the photonic bandgap fiber (PBF), in which light is confined to a hollow core surrounded by a photonic bandgap material. Since the radiation is confined to the air core, optical nonlinear effects in the fiber are virtually absent. In addition, the GVD properties of the fiber can be tuned through the design of photonic bandgap structure, allowing dispersion free propagation in the near-infrared window. PBF are well suited for distortion-free delivery of femtosecond pulses of several nanoJoules [35,36], and have been successfully used in TPEF microscopy applications [37]. The ability to deliver pulses with high peak power makes the PBF a good candidate for CRS implementations as well.

### 6.3.3 Additional Third-Order Optical Effects

Whereas in TPEF and SHG microscopy the selection of a fiber is based on its guiding properties of a single excitation beam, in picosecond CRS microscopy additional selection criteria apply because of the delivery of both the pump and Stokes beams. The preferred approach is to propagate both beams simultaneously through one fiber, which implies that the fiber needs to have good guiding properties over a wide wavelength range. Several fibers, including LMA PCFs and MMF fibers, were found to support the propagation of both pump and Stokes pulses, which were shifted by 2845 cm$^{-1}$, at acceptable peak powers with minimum SPM and GVD distortion [29,30,33,34,38].

However, in addition to the propagation effects of the individual beams, new nonlinear optical effects come into play when the pump and Stokes beams temporally overlap in the fiber. Cross-phase modulation (XPM) is a third-order Kerr-type optical process in which the effective refractive index experienced by one pulse depends on the intensity envelope of the co-propagating pulse [39]. XPM can introduce further spectral broadening and temporal distortions of the pulses. Similar to SPM, XPM can be reduced by minimizing the peak energy per square area in the fiber, which can be achieved with large mode area fibers. Another third-order effect that is enhanced due to the presence of two excitation beams within the fiber is SRS. Whenever the frequency difference between the pump and the Stokes beams is in the vicinity of the 420 cm$^{-1}$ Si–O Raman resonance, energy will be transferred from the pump to the Stokes. Because SRS is phase-matched in the fiber, this effect can be quite substantial. Nonetheless, outside the Si–O vibrational window, the two-beam SRS effect is negligible.

Although XPM and two-beam SRS are effects with minimum impact for most CRS experiments, nonlinear generation of radiation at the anti-Stokes frequency ($2\omega_1 - \omega_2$) through four-wave mixing in the fiber can substantially interfere with CARS imaging experiments [34,38]. This anti-Stokes FWM process results from the nonresonant electronic $\chi^{(3)}$ response of silica and is present whenever the pump and Stokes beams temporally overlap in the fiber. When the frequency difference between the pump and Stokes is substantial (as is the case for CRS imaging of the 2845 cm$^{-1}$ $CH_2$ mode), this form of FWM is generally not phase-matched due to dispersion in the fiber, unless specialized fibers are used [40]. Nonetheless, anti-Stokes frequency generation can occur at the glass–air interfaces at the fiber tips, where symmetry breaking allows for incomplete destructive interference of the non-phase-matched FWM process. This mechanism is supported by the observation that the FWM signal is suppressed when the fiber tip is immersed in a refractive index matching fluid [30]. Figure 6.2 shows the spectrally resolved FWM signal from the fiber when using an 817 nm pump and a 1064 nm Stokes beam. The clean spectral profile indicates that, relative to other fiber nonlinear effects, FWM is the dominant nonlinear mechanism in the fiber under typical CRS excitation conditions.

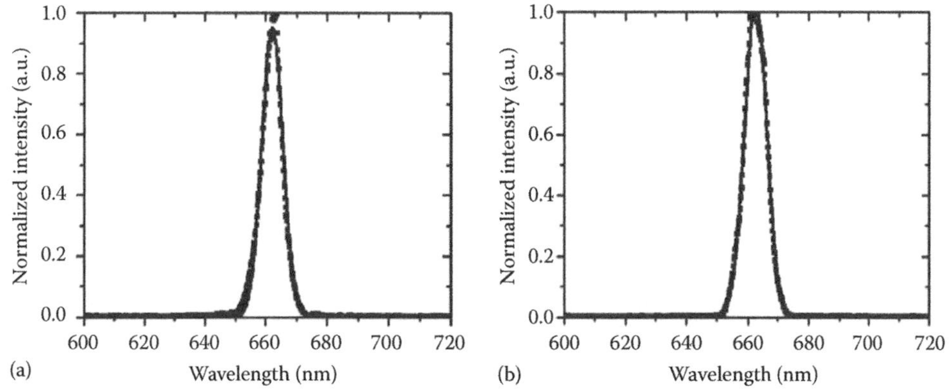


**FIGURE 6.2** Spectrally resolved four-wave-mixing signal resulting from a 817 nm pump and a 1064 nm Stokes beam measured at the output of (a) a 50 cm long LMA PCF fiber and (b) a 40 cm long silica SMF fiber. A strong FWM signal is found for all solid core silica fibers.

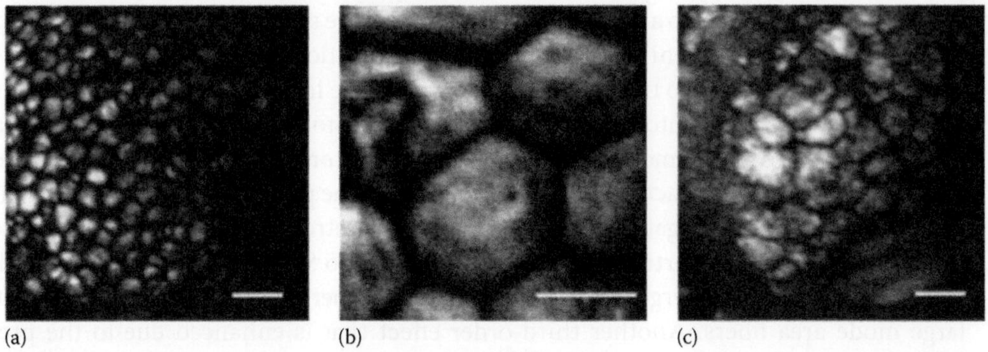


**FIGURE 6.3** CARS images of tissue samples *ex vivo* at 2845 cm$^{-1}$, obtained with a fiber-coupled probe. (a) Small adipocytes in mouse skin. (b) Adipocytes in the subcutaneous layer of rabbit skin. (c) Meibomian gland in a mouse eyelid. Images were acquired in 2 s. Scale bar is 50 μm.

The fiber-generated FWM radiation can back-scatter in the sample and overwhelm and interfere with weak CARS signatures at the detector. This unwanted background radiation can be suppressed by spectral filtering at the distal end of the fiber. For instance, Balu et al. used dichroic optics in an optical probe to reject the FWM radiation out of the main beam path, which enabled fiber-coupled CARS imaging of tissues as shown in Figure 6.3 [34]. Several schemes have recently been pointed out to reduce the FWM fiber contribution [41]. Note, however, the FWM generation in the fiber is only detrimental for CARS microscopy and does not impact SRS imaging.

### 6.3.4 Light Collection with Optical Fibers

Besides delivering the light to the miniature probe, optical fibers are also used to collect and guide the signal generated at the distal site to a detector located at the proximal site. In this regard, the core size and numerical aperture of the fiber are important parameters that define the efficiency of signal collection. Because the CRS radiation generated in the sample is backscattered in the tissue, the light captured by the focusing lens exhibits a wide angular and spatial distributions of photons [2]. Naturally, a small fiber core size will act as a pinhole and reject a large portion of the back-scattered signal. In CARS imaging, this pinhole-effect not only reduces signal but also affects the imaging contrast [29].

Similarly, a small NA limits the collection of a light present in a wide angular cone. The optimization of the collection parameters is not easily achieved with standard SMFs.

Relative to SMFs, the larger cores of MMF and LMA PCFs can improve the collection of back-scattered signal. However, the low NA of the most PCFs put a limit on the collection efficiency of the LMA PCF fibers. Double-clad PCFs provide an attractive solution to this problem. Double-clad PCFs consist of a LMA core surrounded by a micro-structured inner cladding and a silica outer-cladding. Between the inner and outer cladding are air-gaps and microscopic silica bridges, which transform the inner-cladding into a multimode waveguide. The NA of the inner-cladding is high, up to 0.7, which ensures efficient collection of light over a large area. Double-clad PCFs thus enable undistorted delivery of pulsed light through the LMA core as well as excellent signal collection by the inner cladding [21,42,43].

While the simplest configuration for a fiber coupled probe uses a single fiber for both pulse delivery and signal collection, alternative schemes exist that mitigate some of the problems associated with single fiber designs. In particular, the use of separate optical fibers for delivery and collection allows for an independent optimization of pulse delivery and signal detection. For instance, a large core diameter (400 μm), high NA (0.39) multimode fiber has been used for efficient collection of the CARS signal alongside a single-mode LMA PCF fiber for undistorted delivery the pump and Stokes pulses. In addition, the use of two fibers and dichroic splitters enables a straightforward rejection of the FWM radiation generated in the delivery fiber, which is essential for CARS imaging [34]. Such suppression of the FWM radiation is more difficult to accomplish in single fiber configurations.

While detection of the CRS signal at the proximal site is conveniently achieved with sensitive photomultiplier tubes (CARS) or fast photodiodes (SRS), the fiber collection of back-scattered light inadvertently leads to loss of signal photons. This is because the coupling of light into an optical fiber is always associated with losses. Even when large core diameter optical fibers are used, complete coupling of the diffusely back-scattered light into the fiber will remain out of reach. An alternative scheme that avoids such intrinsic losses is to position miniature detectors directly at the distal site. Millimeter-sized photodiode detectors located at the output face of the probe ensure direct and efficient collection of scattered light in SRS imaging [44]. Similarly, integrated CMOS detectors in the probe head have recently shown excellent detection sensitivity and may pave the way for optimized distal detection of CARS signal [45].

## 6.4 Miniaturization of Focusing Optics

Commercially available objective lenses exhibit focusing properties that are very close to the theoretical limits as dictated by diffraction theory. A careful design based on a variety of lenses can effectively nullify the negative effects of other prominent aberrations such as chromatic and spherical aberrations. The optical elements in such microscope objective designs are often a few millimeters in size, and can, in principle, be re-assembled and packaged in a miniature probe [46,47]. However, further miniaturization of the optical elements is nontrivial and requires the use of new designs and optical materials. One particular problem is found in the challenge to achieve the high numerical apertures required for tight focusing when using miniaturized optics. An additional problem in CRS is that the two excitation beams are of different colors, introducing significant chromatic aberrations when unaccounted for.

Several commercial microscope objectives are available with significantly smaller diameters relative to conventional microscope lenses. The Olympus MicroProbe series features objective lenses with probing tips as small as 1.3 mm. These miniature microscope objectives have been successfully used for *in vivo* TPEF [48] and CARS [49] imaging in living mice and rat models. The small diameter of the probing tip enables insertion of the imaging optics into the tissue, bringing into view deeper lying structures that cannot be reached by focusing light from the surface into the tissue. An example of the use of such microscope mounted miniature probes is shown in Figure 6.4, demonstrating the ability to generate CARS images of the myelin in the spinal cord of a living rat.

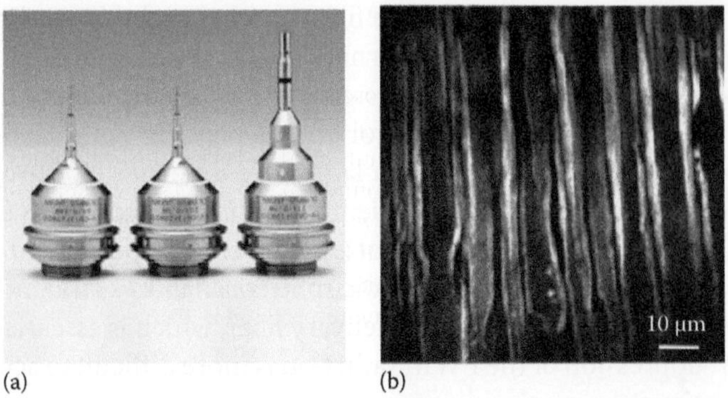


**FIGURE 6.4** Miniature microscope objective probes for enhanced penetration in tissues. (a) Microscope mountable microprobe objectives. (Image courtesy of Olympus, America Inc., Tempe, AZ.) (b) CARS image of axonal myelin in a rat spinal cord acquired with a microprobe objective, visualized at 2845 cm$^{-1}$. (Reprinted from Wang, H. et al., Increasing the imaging depth of coherent anti-Stokes Raman scattering microscopy with a miniature microscope objective, *Opt. Lett.*, 32, 2212–2214, 2007. With permission of Optical Society of America.)

Currently, no commercial miniature probes are available that are suitable for direct application to freely moving animals. For such applications, the probe head needs to operate independently from the microscope frame and the mass of the probe head needs to be significantly reduced [19,50]. In Figure 6.5, a lightweight miniature probe is shown that makes use of ~1.8 mm diameter optics [51]. Different glass types (SF4 and FK51)

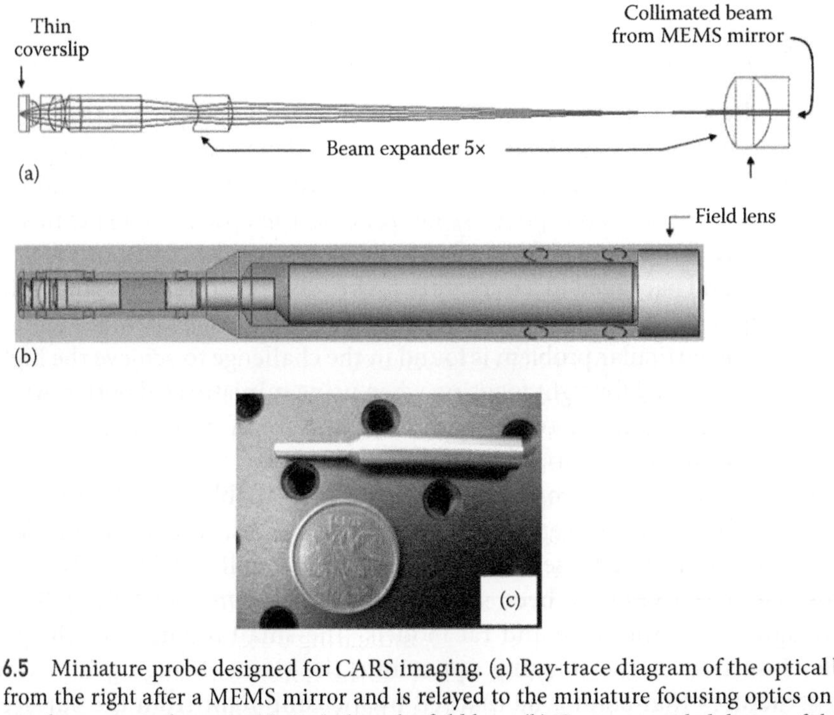


**FIGURE 6.5** Miniature probe designed for CARS imaging. (a) Ray-trace diagram of the optical beam as it enters from the right after a MEMS mirror and is relayed to the miniature focusing optics on the left, through a 5× beam expander assembly including the field lens. (b) Computer-aided design of the barrel. (c) Photograph of the fully packaged barrel. A penny is included for size comparison. (Reprinted from Murugkar, S. et al., Miniaturized multimodal CARS microscope based on MEMS scanning and single laser source, *Opt. Express*, 18, 23796–23804, 2010. With permission of Optical Society of America.)

were used to compensate for the chromatic aberration at the pump and Stokes wavelengths. This design features a relatively high NA of 0.6, which ensures sufficiently tight focusing for the generation of nonlinear CRS signals. This miniature probe has been used for CARS imaging of tissues and offers opportunities for mounting to freely moving small animals.

Further miniaturization of the probe head can be challenging as the compensation for spherical and chromatic aberrations typically increases the number of single lens elements that need to be incorporated in the design. In this regard, incorporation of aspheric lenses into miniature probes has been an important strategy to minimize the number of optical elements. Aspheric lenses are less prone to spherical aberration and can replace multiple spherical lenses in the probe design, resulting in a lighter and more compact imaging system. Despite their miniature size, aspheric lenses are available with relatively high numerical apertures (NA > 0.5), which makes these elements an attractive source for incorporation into a miniature nonlinear optical probe. It has been shown that chromatic aberrations in aspherical compound lenses are relatively small, which has been shown to improve the collection efficiency of TPEF and SHG signals [52]. Such reduced levels of chromatic aberrations are even more crucial for CRS imaging, where the signal strength is directly dependent on the overlap of the pump and Stokes foci. In Figure 6.6, the results of a ray tracing calculation are shown for a standard miniature aspheric lens when focusing both the pump and Stokes beams. As is evident from the figure, the pump and Stokes beams are focused at different position along the optical axis: a direct manifestation of chromatic aberration. Hence, compared to TPEF and SHG probes, additional optical elements may be needed to further suppress chromatic effects for efficient CRS generation in the sample.

Optical elements with a gradient refractive index (GRIN) form another class of microlenses that are popular candidates for integration into a miniature CRS probe. A GRIN lens focuses the light through a gradual refractive index change along the radius of the cylindrical optical element. Because the focusing properties of the lens do not result from refractive effects at curved surfaces, which are increasingly more difficult to manufacture for smaller sized optics, the GRIN technology offers a flexible route for

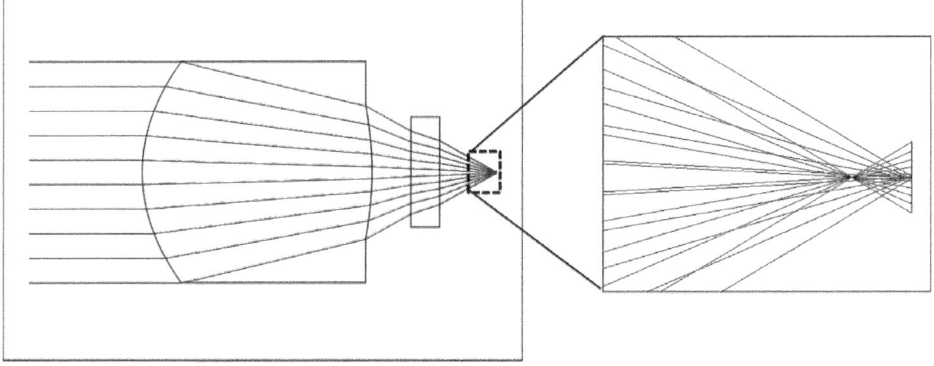

FIGURE 6.6 Ray tracing calculation for a miniature aspheric lens (Geltech 352150) with an NA of 0.45, a diameter of 2 mm and a focal length of 2 mm. A borosilicate coverslip of 0.25 mm has been placed after the mirror. The zoomed-in area shows the axial offset between the pump (800 nm, blue line) and the Stokes (1064 nm, red line). Calculations were performed using the Zemax package.

designing compact miniature lenses with a wide range of focusing properties and NA values of up to 0.6. GRIN lenses with diameters as small as 0.35 mm have been successfully used in miniature nonlinear optical probes [43,50,53–56]. It was recently shown that in combination with a miniature plano-convex lens, the NA of the focusing system could be further raised to 0.85, which significantly improved the TPEF imaging resolution [57]. For CRS applications, a combination of different optical elements may be required to counter the chromatic aberration of GRIN elements.

## 6.5 Miniaturization of Beam Scanners

Depending on the location of the scanning device relative to the two ends of the delivery media, we can classify the scanning mechanism as either proximal scanning or distal scanning mechanisms [19]. In a proximal scanning setup, the scanning device is located at the proximal end of the beam delivery medium. The delivery medium can consists of, for example, a fiber bundle or a gradient lens as sketched in Figure 6.7. In a distal scanning setup, the scanning device is located at the distal end of the beam delivery media. Proximal scanning offers several advantages. In particular, scanning devices located at the proximal end of beam delivery media do not have to be miniaturized, permitting the use of conventional scanners and detectors. On the other hand, a scanning device at the distal end is generally small and packaged into a miniature probe head. In the following, we briefly review the main features of proximal and distal scanning mechanisms.

### 6.5.1 Proximal Scanning Mechanism

For proximal endoscopic scanning, the choice of beam-scanner is flexible and virtually any bench-top scanning module can be used [58]. The most common device used is a scanning unit that employs galvanometer-based scanning mirrors for controlling the angular deviation of the laser beam. This kind of scanner can provide large deflection angles, fast scanning speeds, and flexible scanning patterns. The scanning pattern can be controlled by the voltage waveforms applied to each of the two axes. The most

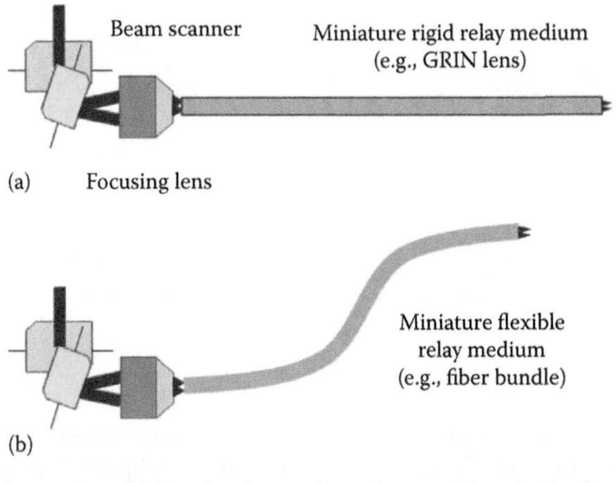


**FIGURE 6.7** Proximal scanning. (a) Proximal scanning scheme with a rigid relay medium. (b) Proximal scanning scheme with a flexible relay medium.

common pattern is the raster scan pattern, which can be achieved by applying sawtooth waveforms to both axes. For fast scanning speeds, the galvo mirrors must be driven by a sinusoidal waveform with frequency equal to the galvo mirror's resonant frequency. Because a sinusoidal and not a linear waveform is used for resonant scanning, correction is needed to reconstruct the image. Commercially available resonant scanning mirrors can achieve a resonant frequency of up to 12 kHz and deflection angles greater than 40°. Acousto-optic deflectors are widely used for ultrafast scanning in confocal and two-photon microscopes. However, for CRS applications acousto-optic deflectors are less attractive because the deflection angle is wavelength dependent.

In proximal scanning endoscopic systems, a beam delivery medium is needed that relays the scanning pattern made at the proximal site to the distal site of the imaging system. The two most common types of relay optics are the fiber bundle and the GRIN lens relay. Fiber bundles consist of tens of thousands of fibers with diameters on the order of a few micrometers. The diameter of the total bundle may vary from several hundred micrometers to a few millimeters, which allows for flexible delivery of excitation light from a proximal location to the distal end. Fiber bundles relay the scanning pattern projected onto the proximal end to the distal end of the probe. Hence, the image quality depends on the number of fibers in the bundle and the suppression of crosstalk between the fibers. This form of proximal scanning has been successfully used in TPEF microscopy [59], and may offer a viable strategy for CRS-based endoscopy as well. Similarly, a compact GRIN lens relay can be used to transfer the scanning pattern and scanning spot from the focusing objective side to the sample side. Unlike the fiber bundle, the GRIN lens relay is suitable for use in rigid millimeter-sized probes, which can be inserted directly into the tissue or mounted onto a freely moving animal. This kind of scheme has been demonstrated in two-photon fluorescence endoscopic applications [57,60,61]. Longer GRIN lens relay systems (>10 cm) are good solutions for the fabrication of rigid probes suitable for the examination of hollow tracts of the human body, including the oral cavity and the esophagus. In this regard, endoscopes based on long GRIN relays have been successfully used in optical coherence tomography (OCT) applications [62], setting an inspiring precedent for CRS implementations as well.

### 6.5.2 Distal-End Scanning Mechanism

In the distal-end scanning scheme, a single fiber is typically used to deliver the excitation beam to the probe head. In this configuration, both the scanning device and the focusing components are an integral part of the probe head. Naturally, the components should be small enough to minimize the total mass of the probe and to facilitate its insertion into tissues and body cavities. There is an arsenal of scanning mechanisms that is compatible with the strict requirements for size in endoscopy. In the following, we will discuss several of these mechanisms that hold great promise for integration into a CRS miniature probe.

A scanning mechanism very suitable for miniaturized probes is the fiber vibration actuator. In this scheme, the distal side of the delivery fiber is attached or glued to a miniature actuator in such a way that a fiber cantilever is formed. By oscillating the actuator, the fiber tip is physically shaken into motion in a controlled manner. The vibrating fiber tip is placed in the image plane of the miniature focusing optics. The position of

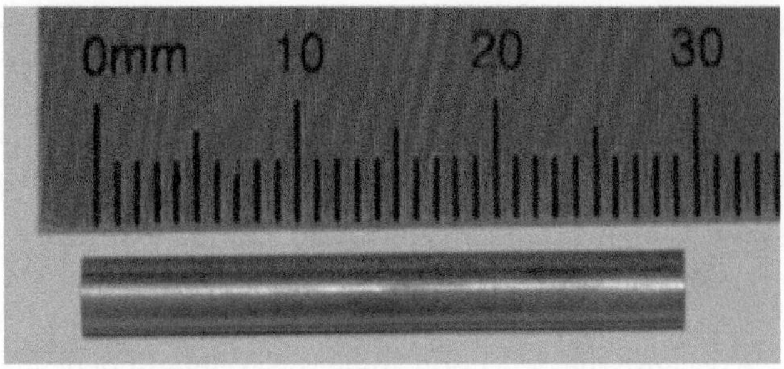


**FIGURE 6.8** A PZT tube of 3 mm diameter and with quadrant electrodes on the outer surface.

the fiber tip is mapped onto the conjugate sample plane, enabling rapid scanning of the focal spot through the oscillatory motion of the fiber cantilever. Both piezoelectric and electromagnetically driven actuators can be used. A piezoelectric material can produce a mechanical deformation when an electric field is applied. Lead zirconate titanate (PZT) is a ferroelectric ceramic that shows a strong piezoelectric effect. PZT-based tube actuators are widely used in scanning tunneling microscopy, scanning probe microscopy, and atomic force microscopy [63]. A PZT tube with quadrant electrodes is shown in Figure 6.8. The diameter of this tube is 3 mm, but similar actuators can be manufactured with much smaller diameters. The use of PZT tubes for driving the fiber cantilever has been successfully demonstrated in the field of TPEF microscopy, OCT [18,21,52,56,64,65], and recently also in CRS [44].

When used for fiber optical endoscopy, the PZT tubes usually exhibit an outside diameter of a few hundred micrometers to a few millimeters. The outer metal coating of the PZT tube is sectioned into four quadrants and two pairs of electrodes are formed. Figure 6.5 shows a photo of a 3 mm diameter PZT tube. When voltage is applied to the electrodes, the tube will bend. The final deflection of the PZT tube is governed by the applied voltage, the piezoelectric coefficient, the outer diameter of the tube, the wall thickness and the length of the tube [66]. In addition to the tube parameters mentioned before, the scanning range of the fiber tip is also affected by the frequency of the applied voltage. When the driving voltage frequency matches the mechanical resonant frequency of the fiber cantilever, the scanning range reaches its maximum. The resonant frequency of a fiber cantilever can be described by [65]:

$$f_R = \frac{\beta}{4\pi}\frac{R}{L^2}\sqrt{\left(\frac{E}{\rho}\right)}$$

where

$L$ and $R$ are the length and radius of the fiber cantilever, respectively
$E$ is Young's modulus
$\rho$ is the mass density
$\beta$ is a constant

When both axes of the PZT tube are driven with constant amplitude and resonant frequency voltages, the 2D scanning pattern is a Lissajous pattern and algorithms or look-up tables can used to reconstruct the image. A spiral scanning pattern is also possible by modulating the resonant frequency sinusoidal driven waveform with a triangular envelope [65]. Similarly, electromagnetically driven actuators have been used for fiber vibration scanning in endoscopic applications [67,68]. In this latter configuration, the fiber is coated or attached to a magnetic material. The fiber is forced to vibrate at specific resonance frequencies by electromagnetic forces. Most fiber vibration-based endoscopes provide a field of view of a few hundred micrometers by a few hundred micrometers.

An alternative scheme for distal-end beam scanning is found in the use of miniature scanning mirrors. Recent advances in microelectromechanical systems (MEMS) have made miniature scanning mirrors an emerging laser beam scanning mechanism for endoscopic applications. MEMS technology extends integrated circuit technology by adding optical or (and) mechanical elements. The MEMS mirror technology has benefited from the development of portable digital displays, optical communications, adaptive optics and bar code recognition, which have brought forward miniature mechanical scanners that are low cost, highly reliable and have low power consumption. Today, MEMS mirrors with diameters ranging from several hundred micrometers to a few millimeters are commercially available.

There are various MEMS mirror actuation mechanisms such as electrothermal, electrostatic, magnetic, and pneumatic actuation. Among these mechanisms, electrostatic actuation enables the fabrication of a low mass device with low power consumption, making it a suitable candidate for integration into a miniature probe head. An attractive feature of MEMS mirrors for endoscopic applications is that 2D scanning can be achieved with a single micro-mirror. Figure 6.9 shows a photo of a packaged two-axis MEMS mirror. A single MEMS mirror can achieve 2D raster scanning with a high degree of accuracy and reproducibility, which avoids the need for image reconstruction. When driven at frequencies that correspond to mechanical resonances, MEMS mirrors can achieve video rate imaging speeds. At such high rates, the scanning pattern is typically a Lissajous pattern and image correction is thus needed. The resonant frequency of the MEMS mirror may be as high as 40 kHz and the maximum optical deflection angle can be more than 8° [55]. MEMS mirror–based scanners have been successfully implemented in confocal microscopy, OCT and TPEF imaging modalities [42,43,69–72].

The MEMS rotational motor is another scanning technology that is highly promising for intravascular endoscopic applications. Unlike the scanning mechanisms discussed earlier, the rotational motor provides an extended circular field of view. In this scheme,

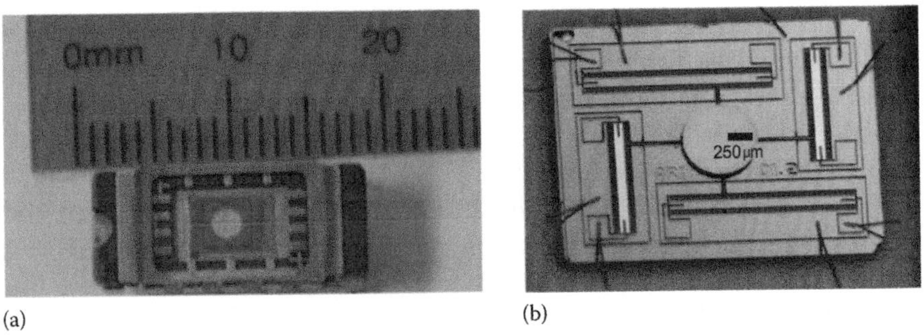


(a) (b)

FIGURE 6.9 (a) A packaged two-axis MEMS mirror with a 2 mm diameter. (b) A zoomed-in view of a MEMS mirror with a 0.7 mm diameter.

Chapter 6

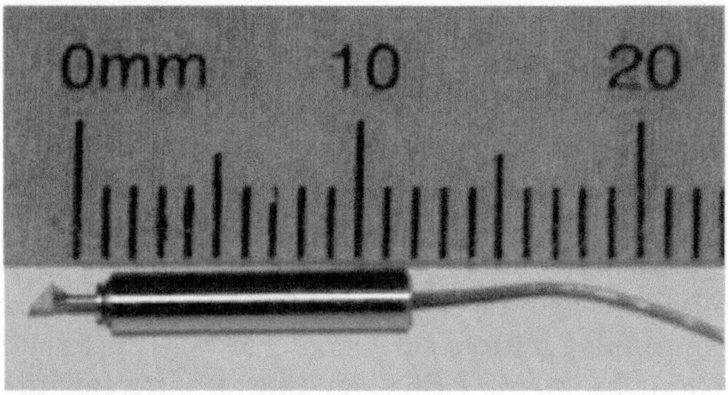


**FIGURE 6.10** A MEMS motor with a prism glued to the shaft.

a mirror or reflection prism is attached to the shaft of the rotational motor. Figure 6.10 displays a photo of such a MEMS motor with a diameter of 2.2 mm. Upon rotating the reflector, a 360° circular scanning pattern is produced. In combination with a linear motion of the probe head assembly in the direction of the fiber axis, a cylindrical field of view of the tissue can be achieved. The rotational speed of such MEMS motors can be as fast as several thousand revolutions per second without applying a load on the shaft, with somewhat reduced speeds when the shaft is loaded. With diameters as small as 1 mm, the rotational MEMS scanning mechanism represents a viable solution for intravascular probing of arterial disease. This scheme has already been integrated in OCT and TPEF fiber coupled probes [73–75], and can be readily implemented in CRS-based miniature probes.

## 6.6 Case in Point: Intravascular Probing of Atherosclerosis

Atherosclerosis is a progressive disease that is characterized by the accumulation of lipids, cholesterol, fibrous constituents, monocytes, and various other inflammatory cells in the arterial wall. These deposits form vascular lesions known as atheromatous plaques, which may contain necrotic cores and are separated from the arterial intima by a fibrous cap made up of collagen and smooth muscle cells [76,77]. Upon plaque maturation, the fibrous caps become thin and increasingly susceptible to tearing, increasing the vulnerability to rupture. Rupture of these vulnerable plaques releases the inflammatory elements of the necrotic core into the artery, causing thrombosis. This leakage may lead to obstruction of arterial blood flow and angina and/or myocardial infarction, which can be lethal [78]. Atherosclerosis is the single most important contributor to cardiovascular disease-related deaths.

Early detection of plaque lesions is a first and necessary step in preventing the lethal consequences of atherosclerosis. Unfortunately, atherosclerosis exhibits an asymptomatic nature, as vulnerable plaques grow without causing any detrimental side effects until rupture [76]. Due to this complication, the information provided by current clinical arterial imaging techniques is often insufficient to diagnose vulnerable plaque formation at an early stage. For instance, angiography and angioscopy report only on later stages of the disease (e.g., stenosis and thrombus formation, respectively).

Intravascular ultrasound (IVUS) and magnetic resonance imaging (MRI) are capable of revealing information on intima-media thickness and the presence of lipid-rich tissue, but the diagnostic capabilities of these techniques are compromised by insufficient spatial resolution and chemical contrast [79–81].

Optical techniques are prime candidates for breaking new ground in atherosclerosis research. The optical detection approach constitutes several fundamental advantages. First, images can be acquired at high (sub-) cellular resolution. The ability to visualize the plaque morphology and structure with cellular detail significantly increases the diagnostic capabilities in terms of structure compared to non-optical approaches. For example, intravascular OCT, which is sensitive to structural density variations in the arterial wall, was clinically proven a sensitive method for determining the thickness of the fibrous cap [82,83]. Second, several optical modalities provide chemical selectivity. Of particular interest are optical probing techniques that are label-free. In particular, endoscopic near-infrared reflectance spectroscopy (NIRS) has been used to characterize the intra-plaque lipid content, and is currently under investigation in large-scale clinical studies [84,85]. Nonlinear optical techniques, including SHG imaging of collagen, TPEF imaging of elastin, and CRS imaging of lipids have not yet reached the stage of clinical studies but have shown great potential for atherosclerotic research [86–91].

CARS microscopy, in particular, has been very successful at identifying key components of atherosclerotic lesions in excised arterial tissue. Based on a swine animal model, Wang et al. were able to identify plaque regions in excised aortas using label-free CARS imaging [89,92]. Moreover, the lipid content per plaque could be determined, which enabled a reliable assessment of maturity of the lesion. Similarly, Lim et al. used CARS microscopy to perform a quantitative analysis of early atheromatous plaques in an ApoE deficient mouse model, and demonstrated the technique's sensitivity to Type II and Type III plaques (Figure 6.11, [93]). The spectral content of the CARS signal was furthermore shown to be able to differentiate between different intracellular and extracellular lipophilic structures in the plaque, including crystalline-like depositions [94,95]. By combining the contrast from SHG and TPEF with CARS, Mostaço-Guidolin et al. developed an index that provides a quantitative measure of plaque burden in a rabbit model system [96]. This index was shown to exhibit a strong correlation with the age of

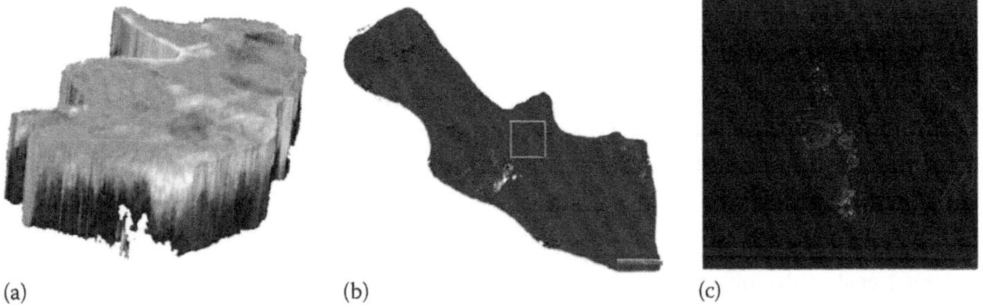


**FIGURE 6.11** CARS imaging of atherosclerosis in an ApoE-/- model system. (a) Complete 3D ($x$, $y$, $z$) data stack of a ~0.5 cm long excised aorta (en face). Green is lipid (CARS), pink is elastin (TPEF) and blue is collagen (SHG) contrast. This 3D rendering is a mosaic composed of more than 500 individual frames. (b) One ($x$, $y$) slice showing the CARS contrast. Bright regions correspond to regions of high lipid content. (c) Zoom of the boxed area in (b).

the animal and the severity of the atherosclerotic plaques. Such indices could play a very important role in assessing disease progression when used in diagnostic *in vivo* probing. A more detailed discussion of the biomedical implications of CARS imaging in atherosclerosis research can be found in Chapter 25.

Although the results obtained in excised tissue suggest a very important role for CRS imaging methods in atherosclerosis studies, the impact of this role strongly depends on whether CRS measurements can be performed in an intravascular manner. Precedent technologies, including ultrasound, OCT and NIRS, have shown the feasibility of generating detailed morphological images from the intima toward the adventitia layers of the arterial wall *in vivo*. The much higher resolution and chemical selectivity of CRS technology are very likely to improve the diagnostic capabilities of current intravascular probes. In this regard, the miniaturization of fiber-coupled CRS probes is a crucial step in translating coherent Raman technologies into reliable instruments that benefit human health.

## 6.7 Outlook and Conclusions

In this chapter we have considered several design strategies for the miniaturization of the CRS microscope. Compared to existing nonlinear miniature probes, the implementation of CRS techniques into probes suitable for endoscopic studies brings additional design criteria to the drawing table. Among these extra criteria are the undistorted propagation of two pulse trains through optical fibers, the suppression of chromatic aberrations in the miniature optics, and, in the case of CARS imaging, the rejection of a FWM background generated in relay optics. Nonetheless, despite these additional challenges, tremendous progress has been made in designing miniature probes suitable for CRS imaging, and some of the current designs were discussed in this chapter. There is no doubt that the probe designs will continue to improve, yielding smaller and lighter assemblies that are compatible with the strict requirements for clinical examination. The further miniaturization of the CRS probe head and its merger with fiber optics and micro-detectors will set the stage for viable endoscopic systems for intravascular inspection. With such optimized systems in hand, it is likely that the CRS techniques will assert themselves as indispensable clinical tools.

## References


1. Evans, C.L. et al., Chemical imaging of tissue in vivo with video-rate coherent anti-Stokes Raman scattering (CARS) microscopy. *Proc. Natl. Acad. Sci. USA*, 2005. **102**: 16807–16812.
2. Saar, B.G. et al., Video-rate molecular imaging in vivo with stimulated Raman scattering. *Science*, 2010. **330**: 1368–1370.
3. Le, T.T., S. Yue, and J.X. Cheng, Shedding new light on lipid biology with coherent anti-Stokes Raman scattering microscopy. *J. Lipid Res.*, 2010. **51**: 3091–3102.
4. Zhu, J. et al., A dynamic cytoplasmic triacylglycerol pool in enterocytes revealed by ex vivo and in vivo coherent anti-Stokes Raman scattering imaging. *J. Lipid Res.*, 2009. **50**: 1080–1089.
5. Lee, B. et al., Intestine specific expression of acyl CoA:diacylglycerol acyltransferase 1 (DGAT1) reverses resistance to diet-induced hepatic steatosis and obesity in Dgat1-/- mice. *J. Lipid Res.*, 2010. **51**: 1770–1780.
6. Henry, F. et al., Real-time in vivo assessment of the nerve microenvironment with coherent anti-Stokes Raman scattering microscopy. *Plastic Reconstr. Surg.*, 2009. **123**: 123S–130S.

7. Huff, T.B. and J.X. Cheng, In vivo coherent anti-Stokes Raman scattering imaging of sciatic nerve tissues. *J. Microsc.*, 2007. **225**: 175–182.
8. Huff, T.B. et al., Multimodel nonlinear optical microscopy and applications to central nervous system. *IEEE J. Sel. Topics Quantum Electron.*, 2008. **14**: 4–9.
9. Zimmerley, M. et al., Quantitative detection of chemical compounds in human hair with coherent anti-Stokes Raman scattering microscopy. *J. Biomed. Opt.*, 2009. **14**: 044019.
10. Zimmerley, M. et al., Following dimethyl sulfoxide skin optical clearing with quantitative nonlinear multimodal microscopy. *Appl. Opt.*, 2009. **48**: D79–D87.
11. Heinrich, C., S. Bernet, and M. Ritsch-Marte, Wide-field coherent anti-Stokes Raman scattering microscopy. *Appl. Phys. Lett.*, 2004. **84**: 816–818.
12. Toytman, I. et al., Wide-field coherent anti-Stokes Raman scattering microscopy with non-phase-matching illumination. *Opt. Lett.*, 2007. **32**: 1941–1943.
13. Lucid, *VivaScope*: Rochester, New York.
14. Rajadhyaksha, M., R.R. Anderson, and R.H. Webb, Video-rate confocal scanning laser microscope for imaging human tissues in vivo. *Appl. Opt.*, 1999. **38**: 2105–2115.
15. JenLab, *MPTflex*: Jena, Germany.
16. König, K. and I. Riemann, High-resolution multiphoton tomography of human skin with subcellular spatial resolution and picosecond time resolution. *J. Biomed. Opt.*, 2003. **8**: 432–439.
17. Koehler, M.J. et al., In vivo assessment of human skin aging by multiphoton laser scanning tomography. *Opt. Lett.*, 2006. **31**: 2879–2881.
18. Helmchen, F. et al., A miniature head-mounted two-photon microscope: High-resolution brain imaging in freely moving animals. *Neuron*, 2001. **31**: 903–912.
19. Flusberg, B.A. et al., Fiber-optic fluorescence imaging. *Nat. Methods*, 2005. **2**: 941–950.
20. Fu, L. and M. Gu, Fibre-optic nonlinear optical microscopy and endoscopy. *J. Microsc.*, 2007. **226**: 195–206.
21. Myaing, M.T., D.J. MacDonald, and X. Li, Fiber-optic scanning two-photon fluorescence endoscope. *Opt. Lett.*, 2006. **31**: 1076–1078.
22. Bao, H. et al., Second harmonic generation imaging via nonlinear endomicroscopy. *Opt. Express*, 2010. **18**: 1255–1260.
23. Agrawal, G.P., *Nonlinear Fiber Optics*. 2007, Burlington, MA: Academic Press.
24. Xi, P. et al., Two-photon imaging using adaptive phase compensated ultrashort laser pulses. *J. Biomed. Opt.*, 2009. **14**: 014002.
25. Boyd, R.W., *Nonlinear Optics*. 2003, San Diego, CA: Academic Press.
26. Wolleschensky, R. et al., Characterization and optimization of a laser-scanning microscope in the femtosecond regime. *Appl. Phys. B*, 1998. **67**: 87–94.
27. Bird, D. and M. Gu, Fibre-optics two-photon scanning fluorescence microscopy. *J. Microsc.*, 2002. **208**: 35–48.
28. Shuker, R. and R.W. Gammon, Raman-scattering selection-rule breaking and the density of states in amorphous materials. *Phys. Rev. Lett.*, 1970. **25**: 222–225.
29. Légaré, F. et al., Towards CARS endoscopy. *Opt. Express*, 2006. **14**: 4427–4432.
30. Wang, Z. et al., Delivery of picosecond lasers in multimode fibers for coherent anti-Stokes Raman scattering imaging. *Opt. Express*, 2010. **18**: 13017–13028.
31. Bjarklev, A., J. Broeng, and A.S. Bjarklev, *Photonic Crystal Fibers*. 2003, Norwell, MA: Kluwer Academic Publishers.
32. Ouzounov, D.G. et al., Delivery of nanojoule femtosecond pulses through large-core microstructured fibers. *Opt. Lett.*, 2002. **27**: 1513–1515.
33. Wang, H., T.B. Huff, and J.X. Cheng, Coherent anti-Stokes Raman scattering imaging with a laser source delivered by a photonic crystal fiber. *Opt. Lett.*, 2006. **31**: 1417–1419.
34. Balu, M. et al., Fiber delivered prove for efficient CARS imaging of tissues. *Opt. Express*, 2010. **18**: 2380–2388.
35. Göbel, W., A. Nimmerjahn, and F. Helmchen, Distortion-free delivery of nanojoule femtosecond pulses from a Ti:sapphire laser through a hollow-core photonic crystal fiber. *Opt. Lett.*, 2004. **29**: 1285–1287.
36. Humbert, G. et al., Hollow core photonic crystal fibers for beam delivery. *Opt. Express*, 2004. **12**: 1477–1484.
37. Tai, S.P. et al., Two-photon fluorescence microscope with a hollow-core photonic crystal fiber. *Opt. Express*, 2004. **12**: 6122–6128.
38. Jun, C.S. et al., Investigation of a four-wave mixing signal generated in fiber-delivered CARS microscopy. *Appl. Opt.*, 2010. **49**: 3916–3921.

39. Akhmanov, S.A., R.V. Khokhlov, and A.P. Sukhorukov. Self-focusing, self-defocusing and self-modulation of laser beams. *Laser Handbook*, eds. T. Arecchi and E.O. Schulz-Dubois, Vol. 2, 1972. North-Holland: Amsterdam, the Netherlands.
40. Andresen, E.R., S.R. Keiding, and E.O. Potma, Picosecond anti-Stokes generation in a photonic-crystal fiber for interferometric CARS microscopy. *Opt. Express*, 2006. **14**: 7246–7251.
41. Wang, Z. et al., Coherent anti-Stokes Raman scattering microscopy imaging with suppression of four-wave mixing in optical fibers, *Opt. Express*, 2011. **19**: 7960–7970.
42. Fu, L. et al., Nonlinear optical endoscopy based on a double-clad photonic crystal fiber and a MEMS mirror. *Opt. Express*, 2006. **14**: 1027–1032.
43. Jung, W. et al., Miniaturized probe based on microelectromechanical system mirror for multiphoton microscopy. *Opt. Lett.*, 2008. **33**: 1324–1326.
44. Saar, B.G. et al., Coherent Raman scattering fiber endoscopy, *Opt. Lett.*, 2011. **36**: 2396–2398.
45. Faramarzpour, N. et al., CMOS photodetector systems for low-level light applications. *J. Mater. Sci.*, 2009. **20**: S87–S93.
46. Liang, C. et al., Design of a high-numerical aperture miniature microscope objective for an endoscopic fiber confocal reflectance microscope. *Appl. Opt.*, 2002. **41**: 4603–4610.
47. Rouse, A.R. et al., Design and demonstration of a miniature catheter for a confocal microendoscope. *Appl. Opt.*, 2004. **43**: 5763–5771.
48. Williams, R.M. et al., Strategies for high-resolution imaging of epithelial ovarian cancer by laparoscopic nonlinear microscopy. *Transl. Oncol.*, 2010. **3**: 181–194.
49. Wang, H. et al., Increasing the imaging depth of coherent anti-Stokes Raman scattering microscopy with a miniature microscope objective. *Opt. Lett.*, 2007. **32**: 2212–2214.
50. Flusberg, B.A. et al., High-speed, miniaturized fluorescence microscopy in freely moving mice. *Nat. Methods*, 2008. **5**: 935–938.
51. Murugkar, S. et al., Miniaturized multimodal CARS microscope based on MEMS scanning and single laser source. *Opt. Express*, 2010. **18**: 23796–23804.
52. Wu, Y. et al., Scanning fiber-optic nonlinear endomicroscopy with miniature aspherical compound lens and multimode fiber collector. *Opt. Lett.*, 2009. **34**: 953–955.
53. Bird, D. and M. Gu, Two-photon fluorescence endoscopy with a micro-optic scanning head. *Opt. Lett.*, 2003. **28**: 1552–1554.
54. Fu, L. and M. Gu, Double-clad photonic crystal fiber coupler for compact nonlinear optical microscopy imaging. *Opt. Lett.*, 2006. **31**: 1471–1473.
55. Piyawattanametha, W. et al., Fast-scanning two-photon fluorescence imaging system based on a microelectromechanical systems two-dimensional scanning mirror. *Opt. Lett.*, 2006. **31**: 2018–2020.
56. Engelbrecht, C.J. et al., Ultra-compact fiber-optic two-photon microscope for functional fluorescence imaging in vivo. *Opt. Express*, 2008. **16**: 5556–5564.
57. Barretto, R.P.J., B. Messerschmidt, and M.J. Schnitzer, In vivo fluorescence imaging with high-resolution microlenses. *Nat. Methods*, 2009. **6**: 511–512.
58. Pawley, J.B., *Handbook of Biological Confocal Microscopy*, 3rd edn., 2006, Springer: New York.
59. Göbel, W. et al., Miniaturized two-photon microscope based on a flexible coherent fiber bundle and gradient-index lens objective. *Opt. Lett.*, 2004. **29**: 2521–2523.
60. Jung, J.C. and M.J. Schnitzer, Multiphoton endoscopy. *Opt. Lett.*, 2003. **28**: 902–904.
61. Jung, J.C. et al., In vivo mammalian brain imaging using one- and two-photon fluorescence microendoscopy. *J. Neurophysiol.*, 2004. **92**: 3121–3133.
62. Xie, T. et al., GRIN lens rod based probe for endoscopic spectral domain optical coherence tomography with fast dynamic focus tracking. *Opt. Express*, 2006. **14**: 3238–3246.
63. Binnig, G. and D.P.E. Smith, Single-tube three-dimensional scanner for scanning tunneling microscopy. *Rev. Sci. Instrum.*, 1986. **57**: 1688–1689.
64. Flusberg, B.A. et al., In vivo brain imaging using a portable 3.9 gram two-photon fluorescence microendoscope. *Opt. Lett.*, 2005. **30**: 2272–2274.
65. Liu, X. et al., Rapid-scanning forward-imaging miniature endoscope for real-time optical coherence tomography. *Opt. Lett.*, 2004. **29**: 1763–1765.
66. Chen, C.J., Electromechanical deflections of piezoelectric tubes with quartered electrodes. *Appl. Phys. Lett.*, 1992. **60**: 132.
67. Dhaubanjar, N. et al., A compact optical fiber scanner for medical imaging. *Proc. SPIE*, 2006. **6414**: 64141Z.
68. Min, E.J. et al., Single-body lensed-fiber scanning probe actuated by magnetic force for optical imaging. *Opt. Lett.*, 2009. **34**: 1897–1899.

69. Piyawattanametha, W. et al., In vivo brain imaging using a portable 2.9 g two-photon microscope based on a microelectromechanical system scanning mirror. *Opt. Lett.*, 2008. **34**: 2309–2311.
70. Pan, Y., H. Xie, and G.K. Fedder, Endoscopic optical coherence tomography based on a microelectromechanical mirror. *Opt. Lett.*, 2001. **26**: 1966–1968.
71. Zara, J.M. et al., Electrostatic micromachined scanning mirror for optical coherence tomography. *Opt. Lett.*, 2003. **28**: 628–670.
72. Jung, W. et al., Three-dimensional endoscopic optical coherence tomography by use of a two-axis microelectromechanical scanning mirror. *Appl. Phys. Lett.*, 2006. **88**: 163910.
73. Tran, P.H. et al., In vivo endoscopic optical coherence tomography by use of a rotational microelectromechanical system probe. *Opt. Lett.*, 2004. **29**: 1236–1238.
74. Su, J. et al., In vivo three-dimensional microelectromechanical endoscopic swept source optical coherence tomography. *Opt. Express*, 2007. **15**: 10390–10396.
75. Liu, G. et al., Rotational multiphoton endoscopy with a 1 μm fiber laser system. *Opt. Lett.*, 2009. **34**: 2249–2251.
76. Narula, J. and H.W. Strauss, Imaging of unstable atherosclerotic lesions. *Eur. J. Nucl. Med. Mol. Imaging*, 2005. **32**(1): 1–5.
77. Virmani, R. et al., Atherosclerotic plaque progression and vulnerability to rupture: Angiogenesis as a source of intraplaque hemorrhage. *Arterioscler. Thromb. Vasc. Biol.*, 2005. **25**(10): 2054–2061.
78. Marcu, L. et al., In vivo detection of macrophages in a rabbit atherosclerotic model by time-resolved laser-induced fluorescence spectroscopy. *Atherosclerosis*, 2005. **181**(2): 295–303.
79. Kaufmann, B.A. and J.R. Lindner, Molecular imaging with targeted contrast ultrasound. *Curr. Opin. Biotechnol.*, 2007. **18**(1): 11–16.
80. Amirbekian, V. et al., Detecting and assessing macrophages in vivo to evaluate atherosclerosis noninvasively using molecular MRI. *Proc. Natl. Acad. Sci. USA*, 2007. **104**(3): 961–966.
81. Waxman, S., F. Ishibashi, and J.E. Muller, Detection and treatment of vulnerable plaques and vulnerable patients. *Circulation*, 2006. **114**: 2390–2411.
82. Cilingiroglu, M. et al., Detection of vulnerable plaque in a murine model of atherosclerosis with optical coherence tomography. *Catheter. Cardiovasc. Interv.*, 2006. **67**(6): 915–923.
83. Tearney, G.J., I.K. Jang, and B.E. Bouma, Optical coherence tomography for imaging vulnerable plaque. *J. Biomed. Opt.*, 2006. **11**(2): 021002.
84. Moreno, P.R. et al., Detection of lipid pool, thin fibrous cap, and inflammatory cells in human aortic atherosclerotic plaques by near-infrared spectroscopy. *Circulation*, 2002. **105**(8): 923–927.
85. Wang, J. et al., Near-infrared spectroscopic characterization of human advanced atherosclerotic plaques. *J. Am. Coll. Cardiol.*, 2002. **39**: 1305–1313.
86. Boulesteix, T. et al., Micrometer scale ex vivo multiphoton imaging of unstained arterial wall structure. *Cytometry*, 2005. **69A**: 20–26.
87. Campagnola, P.J. et al., Three-dimensional high-resolution second-harmonic generation imaging of endogenous structural proteins in biological tissue. *Biophys. J.*, 2002. **81**: 493–508.
88. Lilledahl, M.B. et al., Characterization of vulnerable plaques by multiphoton microscopy. *J. Biomed. Opt.*, 2007. **12**(4): 044005.
89. Wang, H.W. et al., Imaging and quantitative analysis of atherosclerotic lesions by CARS-based multimodal nonlinear optical microscopy. *Artherioscler. Thromb. Vasc. Biol.*, 2009. **29**: 1342–1348.
90. Wang, H.W., T.T. Le, and J.X. Cheng, Label-free imaging of arterial cells and extracellular matrix using a multimodal CARS microscope. *Opt. Commun.*, 2008. **281**: 1813–1822.
91. Zoumi, A. et al., Imaging coronary artery microstructure using second-harmonic and two-photon fluorescence microscopy. *Biophys. J.*, 2004. **87**: 2778–2786.
92. Le, T.T. et al., Label-free molecular imaging of atherosclerotic lesions using multimodal nonlinear optical microscopy. *J. Biomed. Opt.*, 2007. **12**: 054007.
93. Lim, R. et al., Multimodal CARS microscopy determination of the impact of diet on macrophage infiltration and lipid accumulation on plaque formation in ApoE-deficient mice. *J. Lipid Res.*, 2010. **51**(7): 1729–1737.
94. Kim, S.H. et al., Multiplex coherent anti-Stokes Raman spectroscopy images intact atheromatous lesions and concomitantly identifies distinct chemical profiles of atherosclerotic lipids. *Circ. Res.*, 2010. **106**: 1332–1341.
95. Lim, R.S. et al., Identification of cholesterol crystals in plaques of atherosclerotic mice using hyperspectral CARS imaging, *J. Lipid. Res.*, 2011. **52**: 2177–2186.
96. Mostaço-Guidolin, L.B. et al., Differentiating atherosclerotic plaque burden in arterial tissues using femtosecond CARS-based multimodal nonlinear imaging. *Biomed. Opt. Express*, 2010. **1**: 59–73.

# 7. Wide-Field CARS-Microscopy

Alexander Jesacher, Gregor Thalhammer, Stefan Bernet, and Monika Ritsch-Marte



*Coherent Raman Scattering Microscopy.* Edited by Ji-Xin Cheng and X. Sunney Xie © 2013 CRC Press/Taylor & Francis Group, LLC. ISBN: 978-1-4398-6765-5.

# 7.1 Introduction: A Different Approach

CARS microscopy (CARSM) is typically carried out by point-scanning the sample with tightly focused beams (Zumbusch et al., 1999; Müller and Zumbusch, 2007), but this is not an intrinsic necessity. However, if one wants to implement a wide-field approach, i.e., collect a coherent CARS signal at once from a volume of dimension larger than the wavelength ($V \gg \lambda^3$), one has to keep in mind that the phase-matching condition now plays a more important role. In particular, it is not sufficient to simply excite the entire field of view with high laser intensity: An efficient CARS signal build-up from an extended sample volume at a tolerable illumination intensity requires that the phase-matching condition is satisfied.

One way of dealing with this is to tailor the excitation geometry such that the propagation directions of the incoming beams fulfill the phase-matching condition of a chosen resonance (Heinrich et al., 2004a, 2006b). Another one relies on structures in the sample to redistribute the light such that phase-matching is at least partially satisfied (Toytman et al., 2007, 2009). Note that for both methods of wide-field CARS-microscopy (WF-CARSM) phase-matching is actually required for efficient signal generation, the difference between the approaches being that in phase-matched WF-CARSM phase-matching is already achieved in the absence of any sample structures, whereas in non-phase-matched WF-CARSM it is not.

In the following we will give a concise review on WF-CARSM, explaining the differences compared to scanning CARSM, describing the experimental realizations that have been used so far, exemplifying the performance of WF-CARSM (with respect to e.g., spatial and chemical resolution, sensitivity, and image acquisition time), and discussing the range of possible applications for which WF-CARSM seems predestined.

# 7.2 The CARS Process under Wide-Field Excitation

## 7.2.1 Resonance and Phase-Matching Conditions

The energy level diagram for WF-CARS is the same as for scanning CARS with narrowband frequency-degenerate excitation fields (cf. Figure 7.1a), where solid and dashed lines depict real and virtual energy levels, respectively. The CARS process, a coherent four-photon interaction, is resonantly enhanced whenever the difference in photon

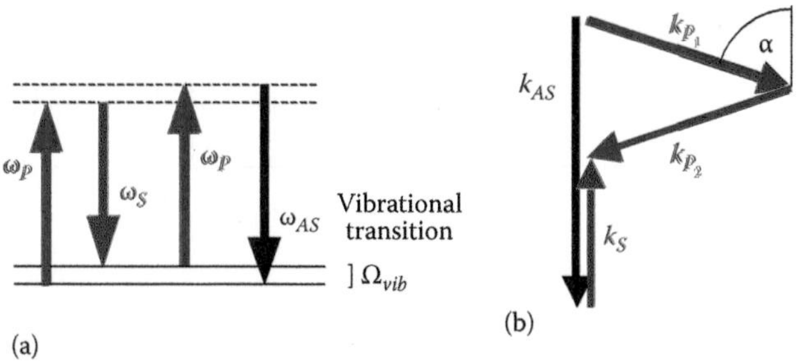


**FIGURE 7.1** Photon energy and photon momentum conservation in WF-CARSM: (a) level scheme, (b) phase-matching diagram (extremely folded BoxCARS).

energy between the pump and the Stokes beams matches the vibrational frequency $\Omega_{vib}$ of a Raman-active transition of molecules in the sample, i.e.,

$$\Omega_{vib} = \omega_P - \omega_S. \tag{7.1}$$

The excitation wavelengths typically lie in the visible or near infrared (NIR), and the vibrational wavelength in the micrometer-range. Photon energy conservation allows us to calculate the emitted CARS frequency to be

$$\omega_{AS} = 2\omega_P - \omega_S. \tag{7.2}$$

The CARS signal is blue shifted, away from the (auto)fluorescent background, which represents one of the many attractive features of CARS-microscopy.

A major difference in the experimental requirements for WF-CARS and for (picosecond) scanning CARS lies in the stringency of phase-matching or conservation of photon momentum (cf. Figure 7.1b): Phase-matching means that the sum of the photon momenta of the two absorbed pump photons $\vec{k}_{P_1} + \vec{k}_{P_2}$ has to equal the sum of the two emitted photons $\vec{k}_S + \vec{k}_{AS}$, i.e.,

$$\Delta\vec{k} = \vec{k}_{AS} + \vec{k}_S - \vec{k}_{P_1} - \vec{k}_{P_2} = 0. \tag{7.3}$$

Phase-matching is not a necessity for a CARS process to take place in an individual molecule, since momentum conservation can usually be accommodated by the molecule in many ways. But it is required for the build-up of a *coherent* CARS amplitude proportional to $N$, with $N$ being the density of Raman-active molecules in the volume, that is emitted into a particular direction, e.g., along the $z$-axis.

The CARS intensity behind a homogeneous sample of interaction length $L$ is given by the expression (Zheltikov, 2000; Volkmer et al., 2001; Boyd, 2008):

$$I(\omega_{AS}; z = L) \propto L^2 \left| \sum_m N_m \chi_m^{(3)}(\omega_P, \omega_S, \omega_{AS}) \right|^2 I_P^2 I_S \operatorname{sinc}^2\left(\frac{|\Delta\vec{k}|L}{2}\right), \tag{7.4}$$

as briefly outlined in the Appendix. Here $N_m$ denotes the density of Raman-active molecules of type $m$, and the sum over $m$ is included, since different types of targeted molecules may be in the sample, e.g., the molecules of interest besides solvent molecules with vibrational levels not too far from resonance. Please note that contrary to other notations, we have singled out the number density $N_m$ as a factor preceding $\chi^{(3)}$, which in our notation denotes the nonlinear susceptibility *per molecule.*

Using tightly focused beams implies a large spread of $k$-vectors, which means that there is a wealth of possible combinations of $k$-vectors that are compatible with conservation of momentum, but wide-field illumination demands specific excitation geometries. The details of some specific configurations will be given in Sections 7.3 and 7.4.

### 7.2.2 Excitation Light

Another peculiarity of non-scanning CARS compared to scanning CARS relates to the laser systems that can be used. Nonlinear processes generally require high intensities and thus excitation by pulsed light sources. To date, depending on the experimental method, nanosecond (ns), picosecond (ps), and femtosecond (fs) laser sources are employed in CARS microscopy. It has been shown (Cheng and Xie, 2004) that the best signal-to-background ratio for scanning CARS with co-propagating beams is achieved with excitation pulse durations in the ps rather than in the fs-pulse regime, despite the higher peak powers of the latter.

This relates to the fact that the spectral width of pulses with duration of a few ps better matches the typical width of about 10 $cm^{-1}$ of vibrational resonances in biological samples (Müller and Zumbusch, 2007); thus, the excitation power of ps-pulses is more efficiently used for the nonlinear conversion, whereas—if no special "tricks" as e.g., spectral focusing are employed—the higher peak intensity of the spectrally broader fs-pulses is to a large extent wasted on non-resonant background generation, for example for exciting the broad O–H stretching vibration of water around 3300 $cm^{-1}$ (Potma et al., 2001).

Comparing scanning and non-scanning CARSM, however, the experimental requirements on peak intensity and pulse energy are not entirely the same: expanding the beam to illuminate an extended field of view with a single pulse (ultimately with the intention to record snap-shot images, which are obtained by a single set of excitation pump and Stokes pulses) substantially lowers the *intensity*—and thus the efficiency of the CARS signal generation, which has to be compensated by larger pulse energies than typically utilized by picosecond scanning system. Such pulse energies are for instance delivered by Nd:YAG lasers in the ns-regime, which were utilized in early demonstrations of WF-CARSM. In fact, it turns out that for images of the same size the *fluence* that is deposited to record the entire image is on the same order of magnitude, about 1 $J/cm^2$, for scanning and non-scanning CARS-microscopes (Evans et al., 2005; Heinrich et al., 2006a; Toytman et al., 2009).

To date WF-CARSM has been successfully performed with either ns-lasers (Heinrich et al., 2004a, 2006a) or with bandwidth limited ps-lasers (Toytman et al., 2007, 2009), both in combination with an optical parametric oscillator (OPO). ns-pulses have the advantage that they need not be bandwidth-limited to match the typical width of a Raman transitions of about 10 $cm^{-1}$ in the region around 2900 $cm^{-1}$ in a liquid sample at room temperature.

In a recent paper the special requirements for lasers for WF-CARSM were discussed in general terms (Simanovskii et al., 2009), but it is difficult to argue what the perfect light source for WF-CARSM is, as it largely depends on details of the setup and the sample. Longer pulse duration on one hand greatly facilitates the spatio-temporal overlap of the beams and high spectral resolution is more readily achieved. However, this comes at the disadvantage that ns-pulses are more disruptive than ps-pulses in samples containing water because of stronger shock wave emission and cavitation bubble expansion following optical breakdown, despite higher threshold for plasma formation (Vogel et al., 1994). Thus, whenever the sample requires high power excitation, "hot spots" of high intensity (e.g., due to backreflection and focusing in the sample) have to be avoided, for example, by scrambling the wavefront of the excitation beams with a diffuser or a multimode fiber in the optical path.

On the other hand it is possible to improve the spectral resolution of fs-lasers for CARSM, for instance to about 3 cm$^{-1}$ using chirped pulses for spectral focusing (Hellerer et al., 2004a,b).

# 7.3 Implementation Concepts

## 7.3.1 Phase-Matched WF-CARS Microscopy

In phase-matched WF-CARS phase-matching is globally maintained in the interaction volume by a particular choice of incidence angles of the exciting beams. In the absence of refracting and scattering structures, e.g., for monitoring chemical reactions in a flow cell, globally enforced phase-matching in the solvent is a necessity.

Note that—in contrast to dilute gases—in dispersive media such as liquids, phase-matching cannot be achieved for collimated beams in *collinear* geometry, since it is typically not possible to accommodate both conditions $\omega_{AS} = 2\omega_P - \omega_S$ and $\Delta\vec{k} = 0$ simultaneously, due to the dependence of the solvent refractive index on the different wavelengths that are involved. Thus it is necessary for the excitation beams to propagate with a certain relative angle to achieve $\Delta\vec{k} = 0$ (cf. Figure 7.1b).

If one illuminates the sample from above with two opposing pump beams and from below through the microscope objective with the Stokes beam, as indicated in Figure 7.1b, this leads to a counter-propagating anti-Stokes signal, which is optimally separated from the other beams and thus easily filtered out. For a large angle $\alpha$ (approaching 90°) one can achieve a kind of "sheet of light" illumination (irrespective of whether the light forms a cone all around or whether it consists of only two beams coming from opposite directions as, for example, generated by a Ronchi phase grating). In reference to Müller et al. (2000) the configuration in Figure 7.1b is sometimes called extremely folded BoxCARS configuration.

For a targeted vibrational transition $\Omega_{vib}$, a specific angle of incidence for the pump beams, given by

$$\cos(\alpha) = \frac{|\vec{k}_{AS}| - |\vec{k}_S|}{2|\vec{k}_P|} \approx \frac{\Omega_{vib}}{\omega_P}, \tag{7.5}$$

satisfies the phase-matching condition such that the emitted anti-Stokes (signal) beam counter-propagates with respect to the Stokes beam into the direction of the objective, which is then used for imaging. Here the dispersion of the refractive index of the surrounding medium can be almost neglected, since in this excitation geometry it has a much smaller effect on the phase-matching condition than in the case of the quasi co-propagating excitation geometry.

In water typical angles cover a range of 64°–90° after refraction at the glass-water boundary. Equation 7.5 allows us to calculate the range of vibrational resonances that can be accommodated. Using a pump field at 532 nm this range is 0–8200 cm$^{-1}$, and for $\lambda_P$ = 810 nm it is 0–5400 cm$^{-1}$, which includes all biologically relevant regions, e.g., the so-called fingerprint region 500–2000 cm$^{-1}$, which is more

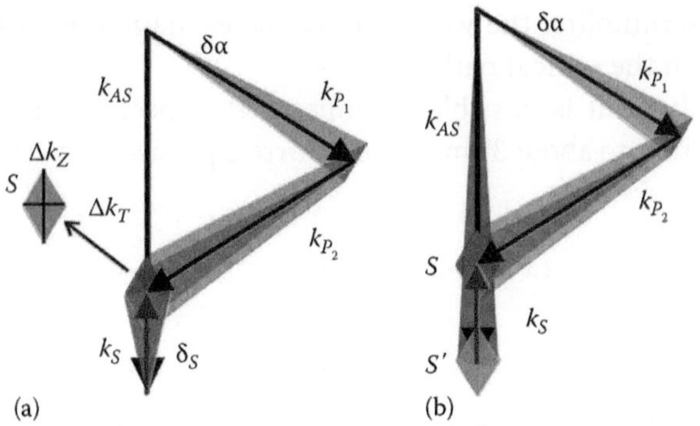


**FIGURE 7.2** Tolerance of the phase-matching condition: (a) spread in pump and Stokes beams, (b) sharp Stokes beam with spread in anti-Stokes signal.

specific to an individual compound, but also more "crowded" than other regions in Raman spectra of organic molecules.

Let us briefly assess the tolerance of the phase-matching condition in the folded BoxCARS configuration with fixed wavelength, if the incident Stokes and pump beams have an angular distributions instead of being plane waves. An angular spread in the pump beams, i.e., $\vec{k}_{P_1}$ and $\vec{k}_{P_2}$ have a range of incidence angles in the intervals $\alpha_1 = \alpha \pm \delta\alpha$ and $\alpha_2 = -(\alpha \pm \delta\alpha)$, respectively, results in a maximal longitudinal (i.e., along the Stokes beam) phase mismatch of $\Delta k_Z = 2k_P\delta\alpha \sin\alpha$, whereas the maximal transverse phase mismatch (the deviation from the vertical) is $\Delta k_T = 2k_P\delta\alpha \cos\alpha$.

As can be seen from the schematic diagram Figure 7.2a the transverse phase mismatch can be compensated by a corresponding angular spread $\delta_S$ of the Stokes beam which should satisfy $k_S\delta_S = \Delta k_T$. In this case the phase-matching condition can still be satisfied by combinations of photon triples (i.e., two pump and one Stokes photon) from almost all incident photons, thus resulting in an efficient signal along the *z*-axis. The area *S* shown in the picture confines all points that can be reached by combinations of pump beam vectors. The diagonals of *S* are $\Delta k_T$ and $\Delta k_z$, respectively.

Figure 7.2a also demonstrates that the longitudinal phase mismatch cannot be compensated entirely: The upper part of *S* is not accessible by tilting $\vec{k}_S$ (which can only *decrease* not *increase* the *z*-component), whereas the lower part can be compensated by increasing the opening angle in the Stokes beam directions up to a maximal value determined by $k_S\delta_S^2 = \Delta k_Z$. If the Stokes bandwidth is sufficiently broad for this, phase matching is possible for incidence angles which are smaller than the optimal phase matching angle α. This also implies that even for a Stokes beam with a large angular distribution, a continuous symmetric increase of the pump beam incidence angle α results in an abrupt vanishing of the signal as soon as the incidence angle exceeds its optimal value. Figure 7.2b illustrates that for a *sharp* Stokes *k*-vector a spread in the pump beams will give rise to a CARS signal with a spread of $\Delta k_T$ (with $S'$ being the area *S* translated by $-\vec{k}_S$). If the pump fields from both sides are arbitrary wave-packets in *k*-space with a certain bandwidth in Δ*k*, the efficiency of the CARS signal generation depends on the overlap integral of the wave-packets, as is shown in the Appendix.

It is quite intuitive to visualize phase-matched WF-CARSM in the context of Bragg scattering (cf. Figure 7.3): WF-CARSM may be interpreted as inelastic (blue-shifted) Bragg scattering of a pump photon from a moving grating formed by the beating

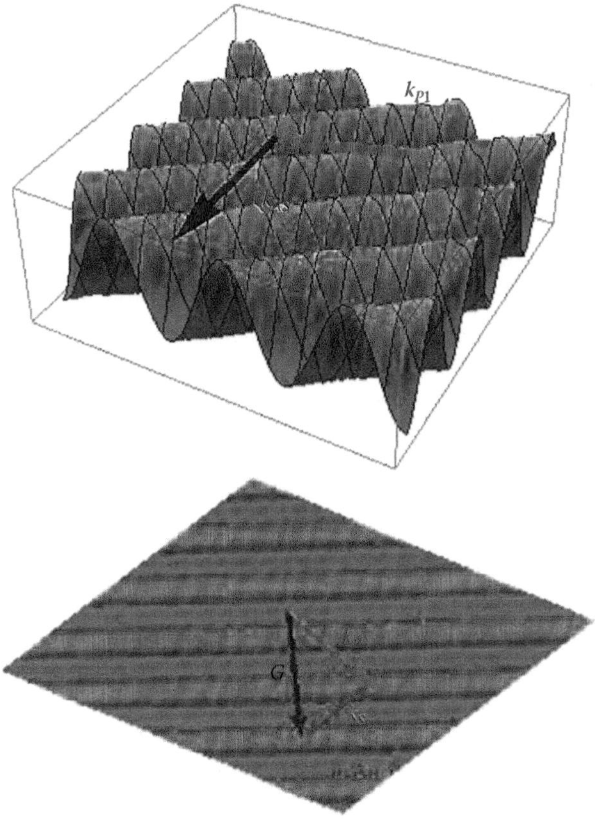


**FIGURE 7.3** Visualization of phase-matched WF-CARS as Bragg scattering.

of the pump and the Stokes beam. In "simplistic" terms of plane waves, the fields $E_{P_1} = \exp(ik_P \sin\alpha x - ik_P \cos\alpha z + i\omega_P t)$ and $E_S = \exp(ik_S z + i\omega_S t)$ at any time form an intensity grating which scatters photons from $E_{P_2} = \exp(-ik_P \sin\alpha x - ik_P \cos\alpha z + i\omega_P t)$ into $E_{AS} = \exp(-ik_{AS} z + i\omega_{AS} t)$.

The energy for the inelastic scattering is provided by the resonantly excited vibrational transition. $\vec{k}_{vib} = \vec{k}_{P_1} - \vec{k}_S$ (with $k_{vib} = \Omega_{vib}/c$) plays the role of the "grating vector" $\vec{G}$. Using the phase-matching condition (Equation 7.3) the grating vector $\vec{G} = \vec{k}_{vib}$ can also be written as $\vec{k}_{vib} = \vec{k}_{AS} - \vec{k}_{P_2}$. At the bottom of Figure 7.3 the "Bragg planes" of the moving intensity grating formed by the beating of $E_{P_1}$ and $E_S$ are shown for some specific time. The $k$-vectors $\vec{k}_{P_1}$ and $\vec{k}_S$ forming this intensity grating, as well as the grating vector $\vec{G}$, are graphically indicated as arrows (of arbitrary scaling). The Bragg planes are seen to be inclined at the angle $\alpha$ given by Equation 7.5. The upper part of the figure schematically depicts the inelastic scattering from $E_{P_2}$ into $E_{AS}$ by arrows indicating the directions of $\vec{k}_{P_2}$ and $\vec{k}_{AS}$ with respect to the intensity pattern.

In this picture phase-matching takes on the form of the Bragg resonance condition, which is dictated by momentum conservation. Please note, however, that CARS is a coherent four-photon process, which does *not* comprise of two *separate* two-photon steps following each other, and it is equally valid to form "Bragg gratings" from other pairs of beams. The depicted visualization of Figure 7.3 only seems particularly intuitive.

*Remark*: Note that comparing this situation to holographic writing in photorefractive materials (Goodman, 1968), the CARSM situation neither corresponds to writing in transmission (i.e., to the so-called Laue geometry, with the grating planes being orthogonal to the crystal surface and the signal and reference beam travelling in the same direction) nor to writing in reflection (the so-called Bragg geometry, with the grating planes parallel to the crystal surface and the signal and reference beam travelling in opposite directions). Here we have a "mixed" case, with pump and Stokes beams acting as "reference" and "signal" beams, respectively, in "writing" the hologram, and the pump beam from the opposite side and the anti-Stokes beams acting as "reference" and "signal" in the "read-out."

## 7.3.2 Non-Phase-Matched WF-CARS

In the presence of strongly scattering or refracting objects it is actually not necessary to implement phase-matching via the direction of the incident beams—one can also rely on the sample structures to redirect the light such that the condition is satisfied by at least a part of the incident photons. This has the advantage that the solvent will not be phase-matched and thus will not contribute significantly with a non-desired background signal. In particular, for *small* objects such as nanoparticles, phase-matching is less important and a CARS signal can be generated without controlling the directions of the excitation beams. Figure 7.4 shows a set-up which was recently used to experimentally demonstrate non-phase-matched WF-CARSM (Toytman et al., 2009) involving a normally incident pump and a converging coaxial annular Stokes beam: the pump beam comes from the side and illuminates the sample from above after reflection from a dichroic mirror, while the Stokes beam is focused on the sample by a Cassegrain objective.

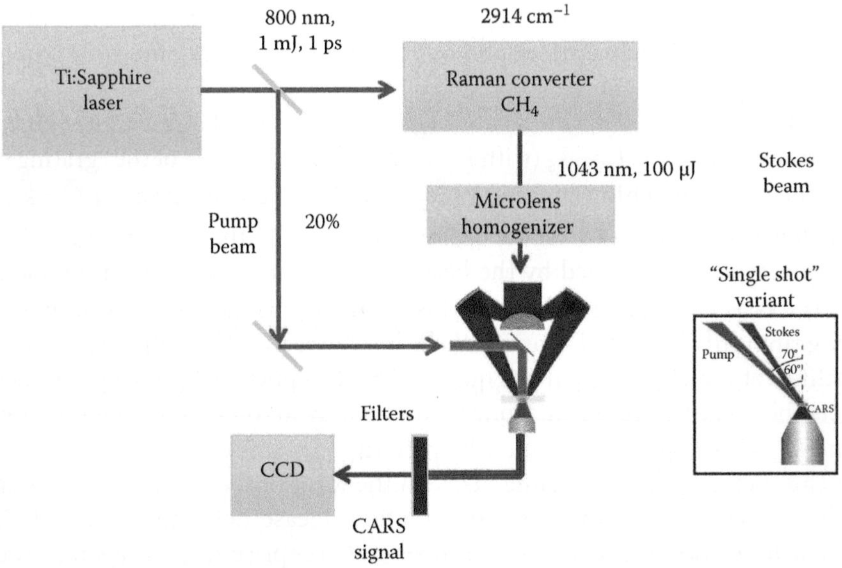


**FIGURE 7.4** Set-up for non-phase-matched WF-CARSM. (Adapted from Toytman, I. et al., *Opt. Express*, 17(9), 7339, 2009. With permission.)

In this configuration pump and Stokes beam both come from above, in earlier implementations (Toytman et al., 2007) they came from opposite sides. The 800 nm ps-pulses from a Ti:Sapphire amplifier were attenuated and collimated with a long focal length lens, producing a 500 μm diameter Gaussian spot of which only the central part was taken in order to have homogeneous illumination. The Stokes beam was created from a part of the pump beam by a Raman converter cell filled with methane ($CH_4$) and focused on the sample with a Cassegrain objective (NA = 0.28), providing a range of angles of 9°–16° with respect to the pump beam. The CARS signal around 650 nm was propagating downwards and imaged onto the camera by the objective lens (100×, NA = 0.78) of an inverted microscope. The pump and Stokes pulses were reported to be tens of μJ and the average fluences over the 200 μm wide illuminated area were 100 and 18 mJ/cm$^2$ per pulse for pump and Stokes beams, respectively.

In order to get rid of speckle and other coherence artifacts in the 200 μm diameter spot at the focal plane, an irregular divergence was introduced by means of a microlens array that was translated from shot to shot by about 100 μm (this corresponds to the shaking multimode fibers or rotating diffuser disk utilized in phase-matched WF-CARSM). About 100 shots were required to average interference effects out. This speckle averaging of course limits the image acquisition speed. Thus the set-up was modified (Toytman et al., 2009) to allow for single pulse image acquisition. Now pump and Stokes beams were collimated by long focal length lenses and directed onto the sample in a common plane of incidence at 70° and 60° with respect to the microscope objective axis (cf. Figure 7.4). Shutters were introduced into the paths of both beams to ensure that only one pair of pulses at a time was delivered to the sample. Since the excitation beams were outside the collection angle of the microscope objective, higher intensities could be applied without endangering the optics. In this "single-shot" approach pump and Stokes pulses with fluences of 750 and 100 mJ/cm$^2$, respectively, were applied, which was comparable to the total fluence used for a single frame acquisition in a modern video-rate scanning CARS system (Evans et al., 2005).

Very recently Zumbusch and coworkers have presented a compact WF-CARSM system with *collinear* non-phase-matched illumination that is capable to deliver images at video-rate (Lei et al., 2011). It utilizes NIR pulses of a few ps at 76 MHz repetition rate. The power of the pump and Stokes beam are 150 and 100 mW, respectively, which corresponds to a fluence *per pulse* in the range of tens of μJ/cm$^2$. The corresponding intensities are about 2000 times lower than those used in scanning CARSM. The EMCCD camera allowed imaging speeds as high as 33 frames/s at a resolution of 512 × 512 pixels. The approach was applied to image—and chemically differentiate—a mixture of μm-sized polystyrene and PMMA beads. PMMA is only visible in the image taken at a Raman shift of 2940 cm$^{-1}$ (aliphatic $CH_3$-stretch), but not at 3050 cm$^{-1}$ (aromatic CH-stretch). Good quality images of *C. elegans* embryos at the aliphatic $CH_2$ resonance at 2850 cm$^{-1}$ were recorded in half a second.

Zumbusch et al. summarize that while their setup has shortcomings such as the lack of 3D resolution (the *z*–resolution is worse than 10 μm) and the need for heterogeneous samples (as phase-matching relies on refractive index changes and scattering in the sample), it has advantages over conventional WF-CARSM systems, i.e., simple implementation, very low excitation powers and high imaging speed.

## 7.4 Experimental Realizations of Phase-Matched WF-CARS

Possible realizations of the general idea of phase-matched WF CARS include a darkfield condenser, a grating, or a spatial light modulator (SLM) to impose phase-matching. These possibilities will now be discussed briefly.

### 7.4.1 Phase-Matched WF-CARS with an Ultra-Darkfield Condenser

The earliest experimental realization of WF-CARSM, first with negative contrast, i.e., chemically targeting the solvent (Heinrich et al., 2004), and later imaging oil droplets with positive contrast (Heinrich et al., 2006a), utilized a so-called ultra-darkfield condenser, an oil immersion darkfield condenser of high numerical aperture (NA), to satisfy phase-matching in the region of interest. The internal mirrors of the condenser shape the pump light into a cone of light in the object plane with a large opening angle (see Figure 7.5). For structures immersed in water and covered by a (0.17 mm) thin cover slip, the numerical aperture range of the darkfield condenser (NA = 1.2–1.4) provides a distribution of pump beam angles between 64° and 90° after refraction at the glass water interface. Assuming a plane Stokes beam from below, this can accommodate a very broad phase-matching bandwidth, as discussed in Section 7.3.1.

The excitation of the CARS process was carried out with a Nd:YAG laser system and an optical parametric oscillator (OPO). The Stokes beam at 1064 nm was generated by a flash-light amplified Nd:YAG laser (Coherent, Infinity 100) emitting pulses of 3 ns duration, at a spectral bandwidth of 0.03 $cm^{-1}$ and at a repetition rate up to 100 Hz. The Stokes beam was guided through a multimode fiber of 0.7 mm core diameter before entering the epi-fluorescence port of the microscope, and in order to obtain a homogeneously illuminated region of interest in the sample, the fiber aperture was imaged by the microscope objective sharply into the object plane in a Köhler-like illumination scheme.

Altogether this gives rise to a "sheet-of-light"-like overlap region of pump and Stokes fields, as shown in the numerical simulation of Figure 7.6. The 3D graphics visualize

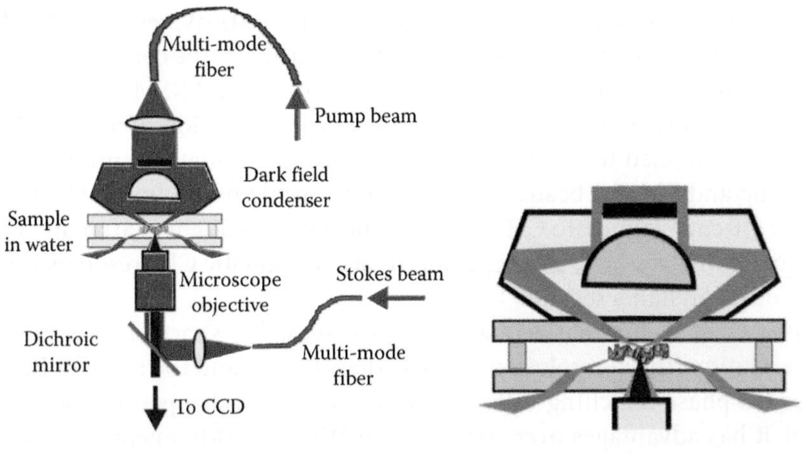


**FIGURE 7.5** Experimental setup for phase-matched WF-CARS with an ultra-darkfield condenser.

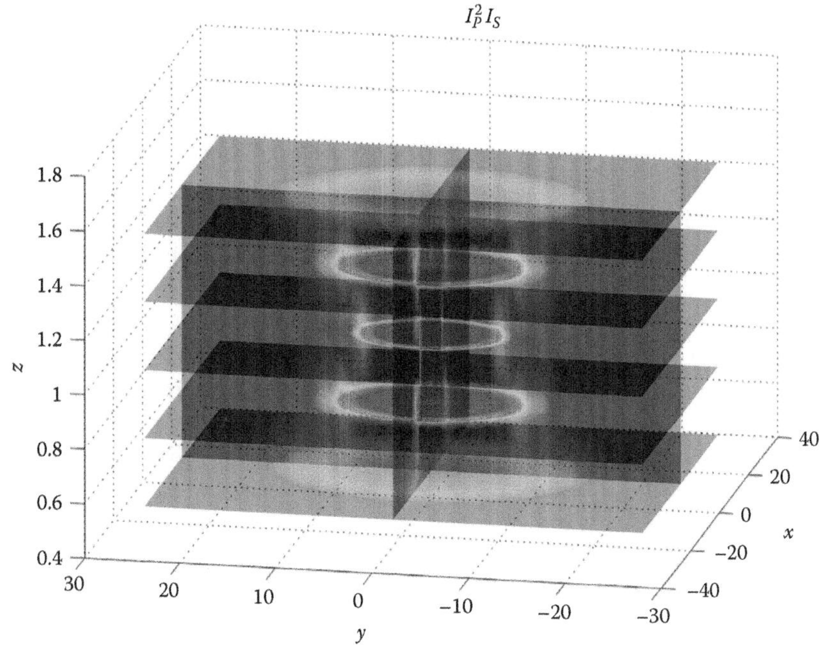


**FIGURE 7.6** Numerical simulation of the "sheet-of-light" overlap region of the pump and Stokes light (axes labeled in μm).

the overlap region for the nonlinear interaction which is created by sharply imaging the circular fiber aperture of the pump field through the darkfield condenser from above and the Stokes field fiber aperture through the microscope objective from below. Cross-sections of $I_P^2 I_S$ are depicted as a function of axial distance from the focal plane in the middle. The cross-sections of the cone-of light of the pump-field emerging from the darkfield condenser are seen to be shrinking rings that contract into a circular image of the pump-fiber aperture, which is brought to overlap with the circular image of the Stokes fiber aperture. By measuring the CARS signals of randomly distributed polystyrene beads fixed in agarose gel, we could roughly determine the axial thickness of the effective nonlinear interaction region to be on the order of <4 μm, which matches the prediction of numerical simulations. Experimentally the illumination is never perfectly homogeneous, but this can be compensated by normalizing the images with previously recorded calibration images, if necessary.

Note that the nonlinear four-wave-mixing process takes place in a well-confined interaction region where pump and Stokes fields overlap and that there are no considerable signal intensities originating from above or below the focal plane.

In a more recently used configuration, the pump beam was created by pumping a mid-band optical parametric oscillator (VISIR, GWU) by the frequency tripled (355 nm) pulses of a Nd:YAG laser. The OPO has a broad tuning range of 410–2500 nm, but typically a 40 nm broad wavelength range around 800 nm was used in order to excite the strong vibrational resonances around 3000 cm$^{-1}$ in polystyrene. The pulses emitted by the OPO were 3 ns long with a pulse energy of a few mJ and a spectral bandwidth of less than 5 cm$^{-1}$. For the Stokes beaam the fundamental wavelength 1064 nm of the Nd:YAG laser was employed. The energies of the pump and the Stokes pulses in the sample plane were less than 1 mJ.

The coherent CARS signal (typically around 650 nm) is emitted in the direction determined by phase-matching, i.e., it counter-propagates with respect to the Stokes beam through the microscope objective and is efficiently separated by a dichroic mirror and a band pass optical filter (645 nm center wavelength). The images are taken with an electron multiplying CCD camera (Andor iXon X3 888), which has a resolution of 1024 × 1024 pixels.

As already mentioned, in many situations, in order to avoid coherence artifacts such as speckle, it is necessary to decrease the spatial coherence of the excitation beams. This can be done by sending the beams through multimode fibers or a diffuser plate. By averaging over several images it is possible to obtain smoother images with averaged-out speckle noise by dynamically modulating the excitation fields, e.g., by vibrating the multimode fiber or by rotating the diffuser plate. However, speckle may also be used to increase the axial optical sectioning (Heinrich et al., 2008a) in a technique called Dynamic Speckle Illumination Microscopy (Ventalon et al., 2007).

### 7.4.2 Phase-Matched WF-CARS with a Grating or a Spatial Light Modulator

It is actually not necessary to have a full 360° cone of light impinging on the sample to satisfy phase-matching over an extended region. In order to create a coherent anti-Stokes signal that counter-propagates with respect to the Stokes beam towards the camera, one only needs two beams crossing the sample from opposite sides with the correct angles. This can also be implemented by a phase grating or by a spatial light modulator (SLM), as shown in Figure 7.7.

A Ronchi phase grating is a binary phase grating with phase steps of $\pi$ between the grooves and the baseline. Ronchi phase gratings have the advantage that the zeroth-order diffraction is almost totally suppressed and the majority of the light is sent into only two directions, the positive and the negative first diffraction orders. This makes this type of grating ideally suited to the given task. In order to have the two diffraction orders still overlapping in the sample volume, one may use a transmission grating which is directly immersed in the sample volume, with a thin glass cover-slip on the grating side to avoid pollution. When using a rotating diffuser plate in the optical path of the pump beam, the interference pattern of the two first order diffracted beams

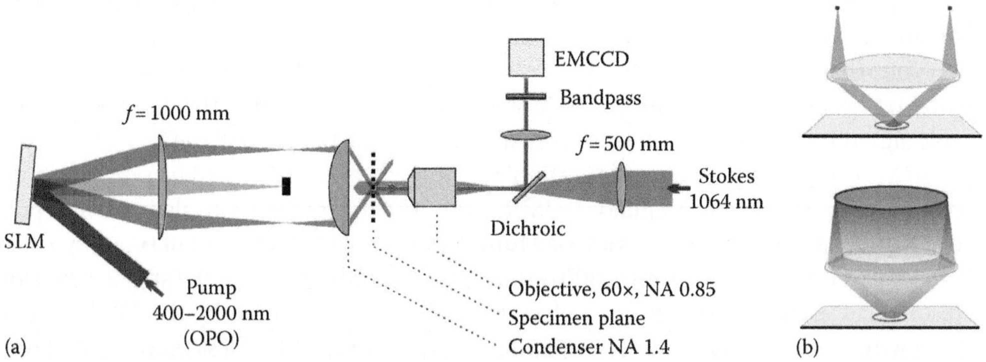


**FIGURE 7.7** WF-CARSM with a spatial light modulator (SLM). (a) depicts the setup, (b) shows two possible illumination configurations in close-up (see text for details).

behind the grating washes out when averaged over a number of laser shots. However, the important thing is that the beam directions needed for satisfying the phase-matching condition are present in the light field.

The grating constant to be used depends on the particular CARS resonance, and thus on the pump and Stokes wavelengths. For a grating constant $d$ the diffraction angle of the two first order beams (in a solvent with refractive index $n$) is given by $\sin\alpha = 2\pi/(nk_P d)$. According to the phase-matching condition Equation 7.3, the required angles of incidence of the two pump beams with respect to the axial direction have to satisfy $\cos\alpha = \Omega_{vib}/\omega_P = k_{vib}/k_P$ with $k_P = 2\pi/\lambda_P$ and $k_{vib} = 2\pi/\lambda_{vib}$. Using the identity $\sin^2\alpha + \cos^2\alpha = 1$ we thus arrive at

$$k_G^2 = n^2\left(k_P^2 - k_{vib}^2\right) \tag{7.6}$$

with $k_G = 2\pi/d$ denoting the magnitude of the grating vector. Alternatively, one may substitute the pump field for the Stokes field by means of $k_P = k_S + k_{vib}$ and write

$$k_G^2 = n^2\left(k_S^2 + 2k_S k_{vib}\right). \tag{7.7}$$

Assuming quasi-plane waves in water for the pump and Stokes beams, one finds a required grating frequency $1/d$ in lines per mm which is $1450\,\text{mm}^{-1}$ for a CARS resonance in the fingerprint region around $1500\,\text{cm}^{-1}$, or of $1600\,\text{mm}^{-1}$ for the strong and popular CARS resonances around $3000\,\text{cm}^{-1}$. However, since there typically exists an intrinsic angular spread in the incident pump beams, which converts itself into a corresponding spread of the incidence angles, as has been discussed in Section 7.3.1, and therefore allows CARS excitation of a larger vibrational frequency range, it practically suffices to utilize a grating with an average spatial frequency around 1500 lines/mm to accommodate both regions.

Spatial light modulators (SLMs) are devices one can use to shape optical fields by setting the phase shift of individual pixels to sculpture the wave-front. Typically, any phase shifts in an interval between 0 and $2\pi$ can be performed at micron-resolution, encoded in a gray-value pattern sent to the SLM. These SLMs are widely applied, e.g., to create holographic optical tweezers (Liesener et al., 2000), in adaptive optics (Love, 1997) and as flexible Fourier filters for contrast-enhanced microscopy methods (Maurer et al., 2011). Naturally a SLM can also be employed to generate appropriate excitation geometries for CARS. One may replace the Ronchi phase grating on top of the sample by a SLM some distance away and optical elements to image the SLM grating into the specimen plane. Due to their relatively large pixel size (approximately $10\,\mu\text{m}$ side length), SLM-gratings necessarily have a much larger period than required in the sample plane. Consequently, the imaging optics have to be strongly demagnifying. A demagnification of about 100 was found to be practicable and can for instance be realized by a 4-f telescope consisting of a lens with 1 m focal length and a high NA condenser with a focal length of about 10 mm.

The SLM has the huge advantage that—simply by exchanging the gray-level value on its screen—one can realize a variety of different pump fields or tune the phase-matching (Jesacher et al., 2011). One can tailor and compare different excitation configurations, or "design" phase-matching configurations. Two examples are shown in Figure 7.7b: The upper image shows the case of displaying a Ronchi grating on the SLM, creating two dominant diffraction orders (higher orders are not shown in the figure and blocked in the experiment). The lower image illustrates the "cone-type" illumination geometry as discussed in Section 7.4.1. This illumination scheme requires the use of a rotating diffuser in the back aperture of the condenser in order to avoid the formation of a Bessel beam in its focal plane and to suppress coherence artifacts.

## 7.5 Performance

In the following we will exemplify the performance of WF-microscopes, in terms of signal-to-noise ratio, resolution, image acquisition time, and sensitivity, and—whenever meaningfully possible—compare it with standard CARS-microscopes.

### 7.5.1 Image Acquisition Time

A major difference between scanning and non-scanning CARSM lies, of course, in the—theoretically achievable—image acquisition time. Scanning has become considerably faster during recent years: Using a rotating polygon mirror for one lateral axis and a galvanometric mirror for the other, real-time imaging is possible (Evans et al., 2005), but it typically takes many minutes to record the entire image of a 800 μm × 600 μm region. Naturally there exists a trade-off between scanning speed and image quality and/or field of view.

WF-CARS is potentially orders of magnitude faster, since it allows one to take snapshots, that is, entire images recorded with just a single pair of excitation pulses. Single-shot CARS images have successfully been recorded within 3 ns (Heinrich et al. 2006a) and later even within 1 ps (Toytman et al., 2009). An example is shown in Figure 7.8. This can also be used to record movies of dynamically changing samples at video rate by synchronizing the (adjustable) laser repetition rate with the camera frame rate (typically 30 Hz).

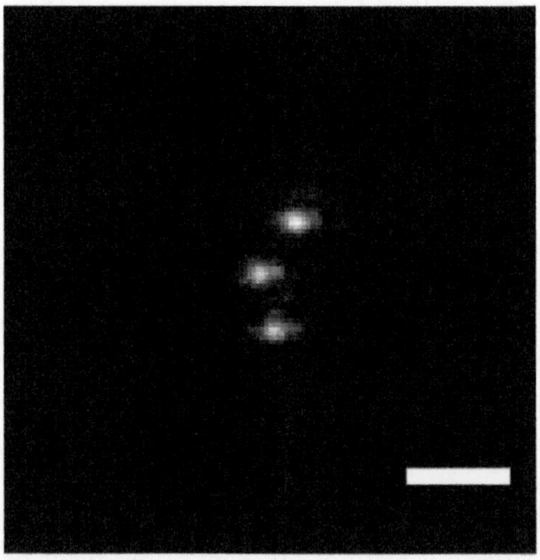

**FIGURE 7.8** Snapshot image with WF-CARSM of 3 μm polystyrene beads, (scale bar = 10 μm).

However, in practice, for weak resonances, single-shot images may have an insufficient image quality (low signal-to-noise ratio) and therefore one averages over several pulses (typically 100 pulses).

When several pulses have to be averaged, the acquisition time is determined more by the repetition rate than by the pulse duration. Typical repetition rates that were utilized to date have been on the order of kHz for non-phase-matched WF-CARSM (with picosecond pulses) and on the order of 10–40 Hz (limited by the damage threshold of the OPO crystal in this energy regime) for phase-matched WF-CARSM with ns-pulses, a system with a larger repetition rate would be favorable in this respect. The achievable field of view can be quite large, e.g., 200 μm × 200 μm (Toytman et al., 2009).

## 7.5.2 Non-Resonant CARS-Background and Signal-to-Noise-Ratio

Even without background subtraction or further processing, a good signal-to-noise ratio is achievable with WF-CARSM, as can be seen in Figure 7.9, originally published in Heinrich et al. (2007). Polystyrene beads with a diameter of 3 μm immersed in water were imaged exciting the 3050 cm$^{-1}$ aromatic C-H stretch resonance ($\omega_P$ = 803.3 nm, $\omega_S$ = 1064 nm). The images were recorded using 100 pulse pairs, no background subtraction or image processing was carried out, only the camera dark current frame was subtracted.

In Figure 7.9 the contrast between regions 1 and 2 in image A is 24:1, and comparing with image B, where the pump beam was detuned to $\lambda_P$ = 800.4 nm, regions 1 and 3 have a contrast of 40:1. Image D is a CARS-image of a living alveolar type II cell with 300–1000 nm wide vesicles containing pulmonary surfactant inside, the contours of which are clearly outlined in the darkfield image C. For this image 100 pulses with $\lambda_P$ = 816.4 nm and $\lambda_P$ = 1064 nm were utilized to excite the vibrational resonance at 2850 cm$^{-1}$. Here the contrast between the areas 4 and 5 is 5:1, the contrast between resonant and off-resonant excitation ($\lambda_P$ = 804 nm) is 7:1.

The non-resonant background (cf. Appendix) is usually a severe problem in CARSM. A multitude of techniques have been developed to avoid—or, on the contrary, utilize(!)—the non-resonant background (Müller and Zumbusch, 2007). P-CARS, for instance,

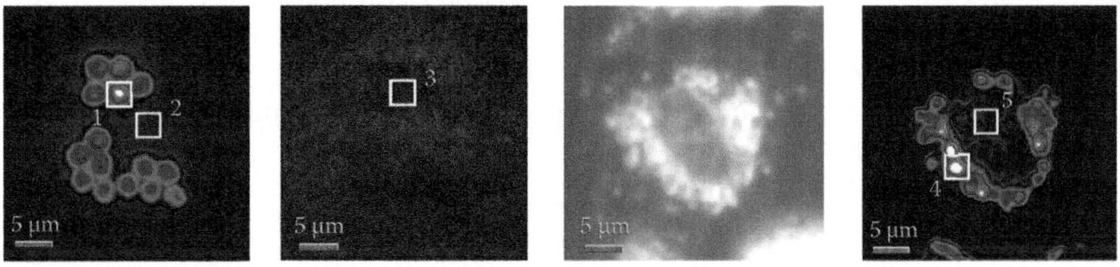


(A) (B) (C) (D)

**FIGURE 7.9** Signal to noise in WF-CARSM images: beads and cells. (From Heinrich, C. et al., *Opt. Lett.*, 32(23), 2468, 2007. With permission.) The images show polystyrene beads imaged with WF-CARSM on resonance (A) and off-resonance (B) and a live alveolar cell containing pulmonary surfactant vesicles imaged by darkfield microscopy (C) and resonant WF-CARSM (D). See text for a quantitative comparison of contrast in the specified regions.

using polarization to reduce the background (Cheng et al., 2001a) efficiently suppresses the background (but also sacrifices some of the signal), but is rather complicated to implement. The resonant $\chi_R$ and the non-resonant $\chi_{NR}$ contributions are typically in quadrature, i.e., if the non-resonant part $\chi_{NR}$ of the complex $\chi^{(3)}$ is real, the resonant part $\chi_R$ contains a real and an imaginary part, which makes heterodyning techniques possible (Jurna et al., 2007, 2008).

Moreover, a strong non-resonant signal may in fact be beneficial in some cases, since it "pushes" the term $2\chi_{NR}\mathrm{Re}\chi_R$. In multiplex-CARS (Wurpel et al., 2002), for instance, where one addresses a significant spectral range of vibrational resonances simultaneously with a combination of broad-band and narrow-band pulses, this fact can be utilized to get quantitative spectra (after a normalization to a reference non-resonant background recorded away from the sample [Wurpel et al., 2004]). Lock-in modulation techniques have also been employed and lately Stimulated Raman Scattering (SRS) has emerged as an alternative with less non-resonant background (Nandakumar et al., 2009; Ozeki et al., 2009; Ozeki and Itoh, 2010; Saar et al., 2010). Especially for submicron particles epi-detection (Cheng et al., 2001b; Volkmer, 2005) is another possibility: for particles smaller than the wavelengths phase-matching is less important and a largely back-ground free signal can be detected in back-reflection.

For phase-matched WF-CARSM it is of great importance that the pump light impinges at the correct excitation angles, which are almost perpendicular to the optical axis. This means that there is no pump light above or below the focal plane, and the excited volume is therefore restricted to a narrow sheet of light only a few microns thick. This effect strongly reduces non-desired CARS signal from the solvent above and below the imaged plane. In fact, in a more collinear version of WF-CARSM that was recently employed by H. Rigneault et al. (private communication), the non-resonant background was found to be considerably larger, but apparently still considerably smaller than in scanning CARSM, as was shown very recently (Lei et al., 2011).

Concerning scanning CARSM it should be remarked that image formation in a confocal CARS-microscope is different from e.g., confocal imaging of the incoherent light emitted by fluorescent molecules, since the CARS signal—including the non-resonant background—is a coherent field. Therefore it cannot be reasoned that the weak non-resonant light (or the off-resonant CARS-signal from the solvent) generated by the excitation beams propagating a considerable distance collinearly through the sample above and below the imaging plane does not appear in the image due to the good optical sectioning capability of the confocal microscope. This contribution is weak because of the strong dependence on the intensities, which decay according to $1/z^2$, away from the focus, but in contrast to fluorescence imaging it is directional and it is integrated over a considerable length.

### 7.5.3 Spatial and Chemical Resolution

The spatial resolution of the WF-CARS microscopes that have been presented here is diffraction limited like that of ordinary light microscopes, depending on the numerical aperture of the microscope objective. They cannot match the spatial resolution that is available from scanning CARS-microscopes based on confocal microscope systems, which is about 30% better than the resolution achievable by the Rayleigh criterion $\Delta x = 0.6\lambda/\mathrm{NA}$

of conventional light microscopes (Müller, 2006). The lateral resolution of WF-CARSM depends on the numerical aperture of the objective, and on the wavelength of the anti-Stokes field, which typically restricts it to around 350 nm. The axial resolution is that of a standard wide-field microscope, and in our case typically is about 3–4 μm.

Concerning the chemical resolution, it has already been argued that the efficiency of the conversion process dictates that the spectral width of the excitation beams should not be larger than the spectral width of the vibrational resonance. The near infrared (NIR) pulses coming from the OPO utilized in the WF-CARS implementations explained in Sections 7.3 and 7.4 have a spectral width <5 $cm^{-1}$, and thus match this criterion, even without being transform limited. Please note that in our case pump and Stokes beam do not have about the same spectral width, which would give rise to an effective factor of 2 in the bandwidth. Since the Nd:YAG laser emitting the Stokes beam at 1064 nm is spectrally very narrow (0.03 $cm^{-1}$), in our case the pump beam bandwidth alone sets the limit for achievable chemical resolution.

## 7.5.4 Sensitivity

To estimate the sensitivity is a difficult issue in CARSM, since its actual limit depends on a large number of parameters, including the strength of the Raman-active resonance or the size (and—for a certain particle-size range—because of refraction effects also the shape) of the particles in the sample. C–H vibrations usually give the strongest CARS signals, especially if they are highly concentrated, therefore lipid-enriched structures such as e.g., liposomes or myelin fibers are normally the easiest structures to target by CARSM. Just to give an idea what can typically still be imaged by WF-CARSM, one can state that it has been possible to record images of individual polystyrene particles with a diameter of 200 nm at 3052 $cm^{-1}$. Aggregates of smaller particles are also easily addressable.

**Table 7.1** Performance of WF-CARS Microscope in the Darkfield Condenser Configuration for a Test Sample Consisting of Polystyrene Beads of Various Size Embedded in Agarose

| | |
|---|---|
| Lateral spatial resolution | 600 nm |
| Axial resolution | 3 μm |
| Spatial contrast: | |
| $I_{polystyrene}/I_{water}$ at 3052 $cm^{-1}$ | >18 for 3 μm sized beads |
| Spectral contrast (polystyrene): | |
| $I_{3052\,cm^{-1}}/I_{3100\,cm^{-1}}$ | >35 for 3 μm sized beads |
| Detection limit (bead size) | >200 nm diameter |
| Field of view | 70 μm × 70 μm |
| Spectral resolution | 5 $cm^{-1}$ (limited by OPO) |
| Wavelength scanning speed | 100 $cm^{-1}$/s; accuracy: 1 $cm^{-1}$ |
| Imaging speed | Typically 10 s for a high quality image, video rate (30 Hz) or 3 ns single-shot images possible |

Table 7.1 summarizes the performance of a WF-CARS microscope in the dark-field condenser configuration as determined by a quantitative analysis of the system parameters with a test sample consisting of polystyrene beads of various size (e.g., 2–5 μm) embedded in agarose. The vibrational resonance at 3052 $cm^{-1}$, corresponding to the antisymmetric C–H stretching vibration in polystyrene was excited with a Stokes beam at 1064 nm and a pump beam at 803.2 nm.

## 7.6 Applications of WF-CARS Microscopy

### 7.6.1 Sample Requirements

One of the attractive features of CARSM is the fact that the anti-Stokes signal is blue-shifted, away from fluorescent background. For the excitation light one can utilize NIR (Cheng and Xie, 2004), which has the advantage of large penetration depth in biological tissue.

Generally speaking, the same class of sample structures that are suitable for chemical imaging with CARS in terms of chemical constituents or particle size, may also be imaged in WF-CARSM. Structures containing densely packed lipids usually give the strongest CARS signals. An example for this has already been given in Section 7.5.2, which depicts the CARS image of living alveolar cells containing internal vesicles filled with pulmonary surfactant, a lipoprotein complex with dipalmitoylphosphatidylcholine (DPPC) as the main lipid component. Even small (sub-micron) vesicles can be discerned without a problem when averaging over about 100 pulses.

Figure 7.10 shows another biological example for chemically selective WF-CARS imaging of lipid-rich structures: the lipid-rich myelin sheath in thin unstained cross-sections of rat-spinal cord show up with bright contrast (Toytman et al., 2009). In the figure, (a) shows a slice stained with a myelin-specific dye (FluoroMyelin green), (b) the histological preparation with toluidine blue, (c) the CARS-image at the strong CH-stretch vibration, and (d) the white light image. All scalebars represent 10 μm, and images (b–c) depict the same structure. The CH-vibration at 2884 $cm^{-1}$ was excited by a pump wavelength of $\lambda_P = 800$ nm and a Stokes wavelength of $\lambda_P = 1040$ nm, which gives rise to an anti-Stokes signal emerging at $\lambda_{AS} = 650$ nm.

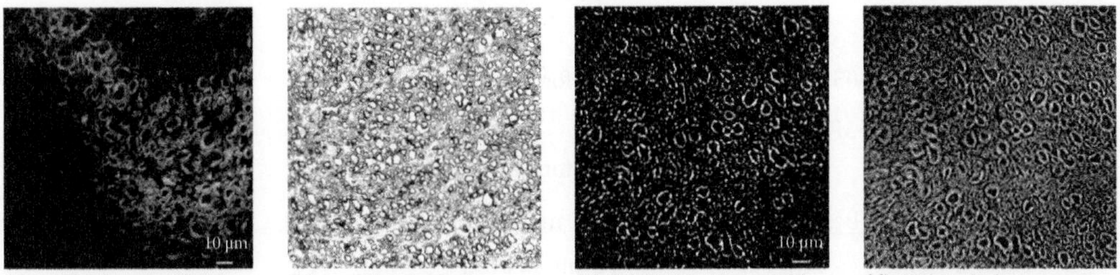


(a) Myelin-specific fluorescent stain (b) Histological preparation (c) CARS (C–H stretch) (d) White light image

**FIGURE 7.10** WF-CARS images of cross-sections of rat spinal cord. (a) Shows a slice stained with a myelin-specific dye (FluoroMyelin green), (b) the histological preparation with toluidine blue, (c) the CARS-image at the strong CH-stretch vibration, and (d) the white light image. (Adapted from Toytman, I. et al., *Opt. Express*, 17(9), 7339, 2009. With permission.)

However, there are some specific sample requirements intrinsic to the wide-field approach. For example, the necessity to satisfy phase-matching over the region of interest makes the approach necessarily a microscopy method that works only in transmission. If one wants to record the anti-Stokes (and not the red-shifted Stokes signal), then it follows immediately from the simultaneous energy and momentum conservation that is required for the coherent signal build-up (cf. Section 7.2) that there exists no configuration with all fields confined to the same side of the sample. This, of course, is not true anymore, if the size of the particles to be imaged is considerably smaller than the wavelengths, where backscattering becomes important and epi-detection is possible (Volkmer et al., 2001). For larger sample structures such as cell cultures or tissue slices, however, we have the sometimes restrictive requirement that WF-CARSM needs thin samples to be imaged in transmission. On the other hand, WF-CARSM offers the huge advantage of delivering snapshot images of fast processes with (in CARSM unparalleled) time-resolution. Chemical images with single ns- or ps-pulses have successfully been recorded (cf. Section 7.5.1).

### 7.6.2 Example: Visual Differentiation of Saturated and Unsaturated Lipids

The low non-linear background achievable in WF-CARSM with spectrally narrow "sheet-of-light" excitation also has consequences for the CARS spectra and the achievable chemical resolution: As has been mentioned, a small $\chi_{NR}$ leads to smaller interference effects between resonant and non-resonant contributions which normally give rise to the typical asymmetric (dispersive-like) lineshape of CARS spectra, and therefore the WF-CARS-spectra resemble the Raman spectra more closely. Vibrational modes of solvent molecules are normally quite far from the addressed resonances (if not so, one may shift them by using e.g., a deuterated solvent), and they can only contribute to the "non-resonant background" when present in the light paths of both the pump and Stokes beams. Therefore off-resonant contributions from the solvent above and below the focal plane do not significantly contribute to the coherent sum in Equation 7.4 for a quasi "sheet-of-light" excitation from the side. Of course also the bandwidth of the excitation fields plays a vital role for the non-resonant background, making narrow-band excitation more favorable in this context.

Compared to illumination with collinear beams, for "sheet of light" illumination from the side less pump light reaches the solvent outside the targeted objects in the focal plane, and thus the interference terms which are weighted by the density of scatterers in the solvent $N_{solvent}$ that are illuminated by both, pump and Stokes beams are small. For micro-beads of several microns (e.g., 2–3 μm radius) this effect can be observed directly, as the depletion of the solvent from the bead volume is clearly visible in Figure 7.11c.

Figure 7.11 demonstrates this for the example of mixed micro-emulsions containing various lipids. Figure 7.11a shows the measured CARS spectra for various vegetable oils in the spectral region of Raman shifts of 2750–3050 cm$^{-1}$. Each data point represents the total CARS signal of a micrometer sized droplet in an oil-in-water emulsion, normalized by the cross sectional area of the droplet. Each spectrum was acquired in 4 min and consists of 50 data points recorded at spectral distances of 5 cm$^{-1}$. For comparison Figure 7.11b shows the Raman spectrum for linseed oil. As indicated, the line profiles in the region 2750–2950 cm$^{-1}$ stem from various vibrational states of symmetric and

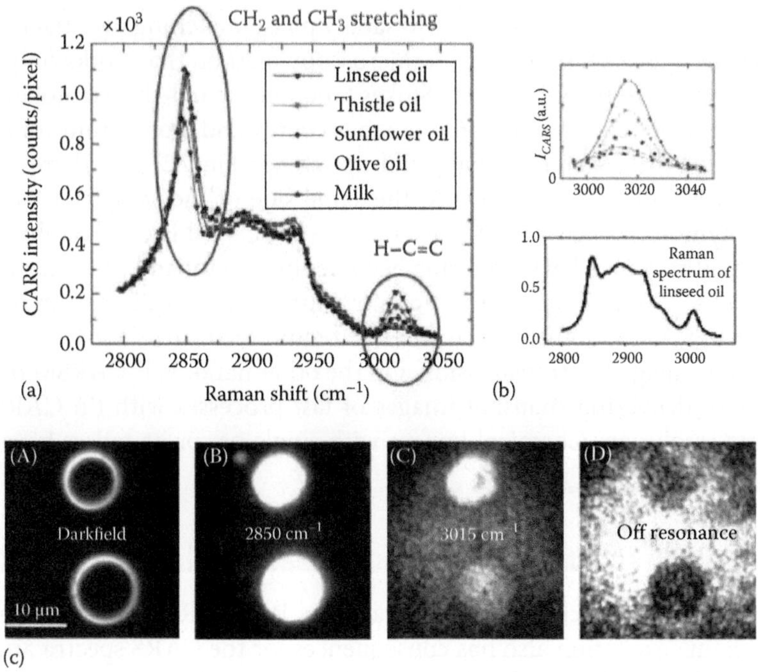


FIGURE 7.11 CARS spectra to determine the degree of saturation of image structures. (a) Measured CARS spectra for various vegetable oils in the spectral region of Raman shifts, (b) Raman spectrum for linseed oil for comparison, (c) images of a mixed olive oil and thistle oil emulsion, left to right: darkfield image (A), WF-CARSM images taken at the strong CH-stretch resonance (B), the weaker double-bond related resonance (C) and off-resonance (D).

asymmetric $CH_2$ and $CH_3$ stretching vibrations with the maximum at 2850 cm$^{-1}$ and are thus expected to be very similar for different vegetable oils. Hence a differentiation of vegetable oils in this frequency region is obviously not possible. On the contrary, the peak in the Raman shift region 2990–3050 cm$^{-1}$, which corresponds to the C–H stretching frequency of a hydrogen attached to a C=C group, is seen to vary strongly for different vegetable oils.

The peak at 3015 cm$^{-1}$, which is a signature of the C=C double bond, is clearly resolved in the WF-CARS spectra (Heinrich et al., 2008b). It would be challenging to record such a spectrum in standard CARSM without any background-reduction procedure, but stimulated Raman imaging, however, can deliver this kind of spectrum as has recently been demonstrated (Freudiger et al., 2008). WF-CARS allows one to directly visualize (un)saturation in a complex sample by the recording of two CARS images, one at the strong, unspecific $CH_2/CH_3$ stretching resonance, and one at the weak, but C=C double-bond dependent resonance, and then analyzing the ratio of these images.

Figure 7.11c demonstrated this for a mixed olive oil and thistle oil emulsion. Both kinds of droplets give a similarly large CARS signal at 2850 cm$^{-1}$, but since olive oil contains less (poly)unsaturated fatty acids than thistle oil, it clearly produces a fainter image at 3015 cm$^{-1}$. In the off-resonant image on the right side, one can also clearly see a negative contrast (darker area at the position of the droplet) originating from the displacement of the solvent by the oil droplet, which nicely shows that there are practically no off-resonant contributions from the solvent from outside the sheet-of-light region which is only a few microns thick, and which in this case is narrower than the bead diameter.

Thus WF-CARSM provides us with a tool for direct visualization of the saturation of lipids in heterogeneous biological samples. This can, for instance, be used to monitor lipid-metabolism on the single cell level. This approach has recently been applied (Heinrich et al., 2008b) to monitor the lipid-uptake of living (cultivated) mouse pre-adipocytes cells fed on different diets. Using the ratio between the strong CARS signal at 2850 cm$^{-1}$ and the resonance around 3015 cm$^{-1}$ for distinction, it was possible to quantify differences of a few percent of the total lipid cell content in the uptake of C=C bonds, in order to visually differentiate cell populations fed on linoleic acid as compared to arachidonic acid. High performance liquid chromatography (HPLC) was employed to check the assessment of the quantitative CARS visualization which was determined by analyzing a population of cells that showed visual signs (such as liposome size) of having incorporated the lipid diet. Good agreement between CARSM and HPLC was found, once a systematic error had been corrected, which was a baseline shift in the HPLC results which had to be carried out, because HPLC analyzed all cells in the sample, including those 20% which were not fully differentiated and thus did not take up any lipids at all. Figure 7.12 gives an example for the CARSM visualization of the internalized fatty acid diets.

In the future the intrinsically narrow-bandwidth light and the low non-resonant background of WF-CARSM systems might be valuable assets in trying to selectively detect disease-related chemical changes in cell cultures or tissue slices in the fingerprint region. Such a challenge could be to selectively detect protein aggregates in brain

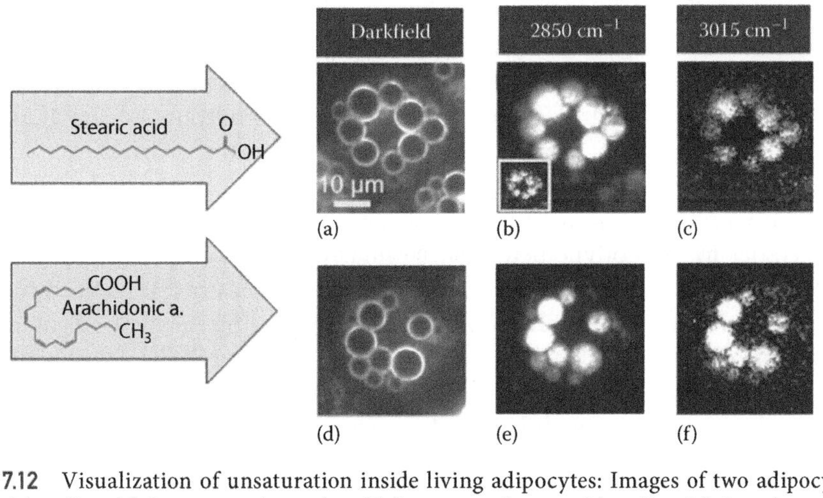


**FIGURE 7.12** Visualization of unsaturation inside living adipocytes: Images of two adipocyte cells, one fed with a diet rich in saturated stearic acid (images a–c), one with a diet rich in polyunsaturated arachidonic acid (images d–f). Comparison of darkfield images (a and d), images at a strong resonance which is independent of the saturation in the fatty acid (b and e) and at a resonance which carries a signature of the C=C double bonds (c and f) Taking the ratio of the images at the two resonances allowed one to quantify the difference in C=C uptake in the two cell populations. The insert in (b) shows a single-shot image.

Chapter 7

tissue slices in the Raman shift region around 1650 cm$^{-1}$ and to reliably differentiate them e.g., from the spectrally nearby signals from lipids.

As already mentioned, WF-CARS operates at a lower intensity level compared to scanning CARSM. This is beneficial in keeping possible damaging effects to biological samples at a low level, but to date not much is known about actual damage threshold levels in the wide-field approach as no systematic study has been carried out. Especially high-power ns-lasers have a bad reputation for their destructive power (Vogel et al., 1994), and—in fact—as already mentioned in the phase-matched realizations of WF-CARSM with ns-lasers one has to take care to reduce the spatial coherence in order to avoid "hot spots" that can destroy the imaging optics or the sample. Palanker and coworkers state that in order to avoid damage to the sample and to the microscope objective, the pump beam was attenuated so that its fluence in the sample plane did not exceed 100 mJ/cm$^2$ (Simanovskii et al., 2009). In particular for the set-up geometry where the coherent Stokes beam is weakly focused in the back aperture of the objective lens, damaging the lens can be a risk for pulse energies above approximately 100 μJ.

## 7.7 Discussion and Outlook

A wealth of possible applications is still waiting for WF-CARSM, especially in the finger-print region, where the resonances are considerably weaker, but more specific to particular biomolecules of interest. Moreover, the coherent nature of the CARS signal could be utilized in many clever ways. One application which has been demonstrated first in scanning CARS is homodyne/heterodyne detection of the CARS signal for obtaining a strong signal contrast enhancement (Jurna et al., 2008). This may be achieved by coherently mixing the CARS signal with a reference beam of the same wavelength and a fixed relative phase, which can be obtained either by generating a second CARS signal in a reference sample, or by using the working principle of the optical parametric oscillators (OPOs) which are typically used to generate one of the CARS excitation beams. These OPOs produce two outputs—a signal and an idler beam—where only one of them is typically used as one of the CARS excitation beams. However, in some configurations the second output automatically has the same frequency and a fixed phase with respect to the CARS signal, and can be directly used as a reference for homodyne detection. Similarly it has been previously demonstrated in WF-CARSM that such a coherent CARS reference beam can be used to produce a hologram of the sample, which can then be analyzed using the methods of holographic microscopy (Shi et al., 2010; Xu et al., 2010). Similarly it might be possible in the future to implement more of the successful methods of coherent microscopy in CARS imaging, like phase contrast- or differential interference contrast microscopy or to utilize focus-engineering (Krishnamachari and Potma, 2007).

## Acknowledgment

The authors thank D. Simanovskii and D. Palanker from Stanford University for the material for Figures 7.4 and 7.10.

## 7.A Appendix

Here we briefly outline the derivation of Equation 7.4 for the CARS intensity behind a homogeneous sample, following Boyd's 2008 book, *Nonlinear Optics*. Starting from the wave-equation with a nonlinear polarization term, one can derive the so-called coupled-wave equations by inserting the slowly-varying amplitudes $A(z)$ defined by

$$E_i(z;t) = A_i(z)e^{ik_i z - i\omega_i t} + \text{c.c.}, \tag{7.8}$$

for monochromatic fields and neglecting terms $|d^2A_i/d^2z|$ in comparison to terms of the order $\left|k_i \frac{dA_i}{dz}\right|$ (cf. Chapter 2 in Boyd [2008]). In these expression the transverse coordinates $x$ and $y$ have been dropped, because we have assumed a homogeneous sample (which is normally not the case in CARSM, which thus actually requires a more involved treatment including the $x$ and $y$ coordinates). One usually also assumes that the exciting light fields are not significantly depleted by the generation of the CARS signal, which is weak in comparison.

Under all of these assumptions the CARS amplitude $A_{AS}(z)$ that is emitted into the direction $\vec{k}_{AS}$ satisfies

$$\frac{dA_{AS}(z)}{dz} = \frac{1}{2}\frac{3i\omega_{AS}}{n(\omega_{AS})c}\sum_m N_m \chi_m^{(3)}(\omega_{AS})A_P^2 A_S^* e^{i\Delta kz}, \tag{7.9}$$

with $n(\omega_{AS})$ being the refractive index of the anti-Stokes wave in the medium and $N_m$ the number density of molecules contributing to the Raman-active transition $m$. Here it was assumed that all dipole moments are optimally aligned with the linear polarization of the laser beams, deviations from this reduce the CARS signal accordingly. Let us also remark that in Equation 7.9 a term related to linear gain from stimulated Raman scattering has been neglected.

Note that the sum over the four-wave-mixing susceptibilities $\chi_m^{(3)}$ may also contain non-resonant vibrational modes belonging to *virtual* energy levels (shown above the upper vibrational level in Figure 7.1a). These non-resonant contributions, which form the non-resonant background, are not specific to a particular vibrational transition and therefore contain no chemical information (Lotem et al., 1976; Mukamel, 1995; Zheltikov, 2000; Müller and Zumbusch, 2007). One usually writes

$$\chi^{(3)}(\omega_P, \omega_S, \omega_{AS}) = \chi_R^{(3)}(\omega_{AS} = \omega_P + \omega_P - \omega_S) + \chi_{NR}^{(3)}. \tag{7.10}$$

Apart from this, if the *solvent* (e.g., water or agarose) around the sample structures possesses vibrational transitions that are in the vicinity of the targeted transition, this may also lead to a background signal around the structures.

The CARS intensity $I_{AS}$ behind the interaction region (i.e., behind the sample of thickness $L$) is the time-averaged Poynting vector. We find

$$I_{AS}(L) = 2n(\omega_{AS})\epsilon_0 c\left|A_{AS}(L)\right|^2, \tag{7.11}$$

where $A_{AS}(L)$ is derived by integrating Equation 7.9 from 0 to the interaction length $L$. Using the identity

$$\left|\frac{e^{i\Delta kL}-1}{\Delta k}\right|^2 = L^2\frac{\sin^2(\Delta kL/2)}{(\Delta kL/2)^2} =: L^2\text{sinc}^2\left(\frac{\Delta kL}{2}\right). \tag{7.12}$$

and neglecting dispersion (i.e., assuming $n(\omega_{AS}) = n(\omega_P) = n(\omega_S)$, all fields have the same conversion factor from electric field amplitudes (in V/m) into intensities (in W/m²), $I_i = p|A_i|^2$ with $p = 2n\epsilon_0 c$), we get the expression (Boyd, 2008)

$$I(\omega_{AS}) = \left(\frac{3}{2np}\frac{L\omega_{AS}}{c}\right)^2 \left|\sum_m N_m\chi_m^{(3)}(\omega_{AS})\right|^2 I_P^2 I_S \text{sinc}^2\left(\frac{\Delta kL}{2}\right) \tag{7.13}$$

for the CARS intensity as a function of pump and Stokes intensities, interaction length $L$ and phase mismatch $\Delta k$ with

$$\Delta k = \left|\Delta\vec{k}\right| = \left|\vec{k}_{AS} + \vec{k}_S - \vec{k}_{P_1} - \vec{k}_{P_2}\right|. \tag{7.14}$$

The argument of the sinc-function, $L/L_C$ with $L_C = 2/\Delta k$, is responsible for the efficiency of the CARS process, therefore $L_C$ is called the "coherent build-up length."

When comparing different spatial patterns for the excitation beams, e.g., created by a SLM, it is the overlap in $k$-space what matters for the efficiency of the CARS signal generation: Let us for instance assume the extremely folded BoxCARS configuration with wave-packets around the mean vectors $\vec{k}_{P_1} = k_P(\sin\alpha, 0, -\cos\alpha)$ and $\vec{k}_{P_2} = k_P(-\sin\alpha, 0, -\cos\alpha)$ in $k$-space for the pump fields coming from left and right in Figure 7.1b. For the sake of simplicity let us for the moment consider a spreading of the $k$-vectors in $k_x$ only, i.e., we introduce variables $k_1$ and $k_2$ which represent the variation of the $x$-component of $\vec{k}_{P_1}$ and $\vec{k}_{P_2}$ around a mean value of $k_P\sin\alpha$ and $-k_P\sin\alpha$, respectively.

Note that we can have a "bandwidth" of wave-vectors for each pump field without any change in $|\vec{k}_P|$, and thus without any change in the targeted vibrational resonance, since variations in $k_x$ can be compensated for example by variations in $k_y$, e.g., when moving the tip of the $k$-vector along a circle in the $(k_x k_y)$-plane as in the emission cone of the ultra-darkfield condenser, or by a spread in the $k$-vector of the Stokes beam, as illustrated in Figure 7.2. A bandwidth in $k_x$ alone, without a corresponding bandwidth in $k_y$, however, is not consistent with an unchanged $|\vec{k}_P|$. Nevertheless, in order to keep it simple, we only include the $k_x$ direction and ignore the $k_y$-direction, and assume that the wave-packets are sufficiently narrow to be compatible with the narrow spectral bandwidth of the quasi-monochromatic fields that we assumed to start with.

With the wave-packet ansatz

$$A_{P_1}(x) = \int_{-\infty}^{+\infty} dk_1 F(k_1) e^{i\varphi(k_1)} e^{i(k_1 + k_P\sin\alpha)x} \tag{7.15}$$

$$A_{P_2}(x) = \int_{-\infty}^{+\infty} dk_2 G(k_2) e^{i\tilde{\varphi}(k_2)} e^{-i(k_2 + k_P \sin\alpha)x}, \tag{7.16}$$

where *both* functions $F$ and $G$ are taken to be centered around some positive mean value (e.g., two Gaussians peaked near $k_P \sin\alpha$), the CARS-amplitude is proportional to

$$A_{CARS} \propto A_{P_1}(x) A_{P_2}(x) = \int_{-\infty}^{+\infty}\int_{-\infty}^{+\infty} dk_1 dk_2 F(k_1) e^{i\varphi(k_1)} e^{i(k_1 - k_2)x} G(k_2) e^{i\tilde{\varphi}(k_2)}. \tag{7.17}$$

With the variable transform $k = (k_1 + k_2)/2$ and $K = k_1 - k_2$ we can rewrite this in the form

$$A_{CARS} \propto \int_{-\infty}^{+\infty}\int_{-\infty}^{+\infty} dk dK\, F(k + K/2) e^{i\varphi(k+K/2)} G(k - K/2) e^{i\tilde{\varphi}(k-K/2)} e^{iKx}$$

$$\approx \int_{-\infty}^{+\infty} dk\, F(k) G(k) e^{i\varphi(k) + i\tilde{\varphi}(k)}. \tag{7.18}$$

From this we see that it is the overlap-integral in $k$-space of the two opposing pump fields which is important for the CARS-signal generation. This simply means that the global shape of the wave-packet distribution is not important for the efficiency of the coherent four-photon process. The efficiency is proportional to the overlap in the support of the two wave-packets, i.e., it increases with the number of photons coming from one side that have a partner photon with opposite $k$-vector from the other side. Note that in the last step a "stationary phase" argument was used, which assumes a slowly varying phase profile of the prefactors of $\exp(iKx)$, which itself then acts as an approximate delta-function $\delta(K)$. This also shows that phase variations in this integral affect the CARS signal, if they matter on the scale of the supports of the wave-packet envelopes $F(k_1)$ and $G(k_2)$.

Additionally, if there was phase noise depending on the spatial coordinates $x$ and $y$ (which have been dropped, since a homogeneous situation was assumed so far), the CARS signal would be sent into different directions corresponding to different "local" phase-matching conditions, which can represent a problem for imaging, if the numerical aperture of the objective is not sufficient to collect all of the CARS signal.

## References


Boyd, R. W. (2008). *Nonlinear Optics*. San Diego, CA: Academic Press.

Cheng, J. X., Book, L. D., and Xie, X. S. (2001a). Polarization coherent anti-Stokes Raman scattering microscopy, *Optics Letters 26*(17), 1341–1343.

Cheng, J. X., Volkmer, A., Book, L. D., and Xie, X. S. (2001b). An epi-detected coherent anti-Stokes Raman scattering (e-cars) microscope with high spectral resolution and high sensitivity, *Journal of Physical Chemistry B 105*(7), 1277–1280.

Cheng, J. X. and Xie, X. S. (2004). Coherent anti-Stokes Raman scattering microscopy: Instrumentation, theory, and applications, *Journal of Physical Chemistry B 108*(3), 827–840.

Evans, C. L., Potma, E. O., Puoris'haag, M., Cote, D., Lin, C. P., and Xie, X. S. (2005). Chemical imaging of tissue in vivo with video-rate coherent anti-Stokes Raman scattering microscopy, *Proceedings of the National Academy of Sciences of the United States of America 102*(46), 16807–16812.

Freudiger, C. W., Min, W., Saar, B. G., Lu, S., Holtom, G. R., He, C. W., Tsai, J. C., Kang, J. X., and Xie, X. S. (2008). Label-free biomedical imaging with high sensitivity by stimulated Raman scattering microscopy, *Science 322*(5909), 1857–1861.

Goodman, J. W. (1968). *Introduction to Fourier Optics*. New York: McGraw Hill.

Heinrich, C., Bernet, S., and Ritsch-Marte, M. (2004). Wide-field coherent anti-Stokes Raman scattering microscopy, *Applied Physics Letters 84*(5), 816–818.

Heinrich, C., Bernet, S., and Ritsch-Marte, M. (2006a). Nanosecond microscopy with spectroscopic resolution, *New Journal of Physics 8*, 36.

Heinrich, C., Bernet, S., and Ritsch-Marte, M. (2007). Comment on "Wide field CARS microscopy with non phase matching illumination, *Optics Letters 32*(23), 3468.

Heinrich, C., Hofer, A., Bernet, S., and Ritsch-Marte, M. (2008a). Coherent anti-Stokes Raman scattering microscopy with dynamic speckle illumination, *New Journal of Physics 10*, 023029.

Heinrich, C., Hofer, A., Ritsch, A., Ciardi, C., Bernet, S., and Ritsch-Marte, M. (2008b). Selective imaging of saturated and unsaturated lipids by wide-field cars-microscopy, *Optics Express 16*(4), 2699–2708.

Heinrich, C., Meusburger, C., Bernet, S., and Ritsch-Marte, M. (2006b). Cars microscopy in a wide-field geometry with nanosecond pulses, *Journal of Raman Spectroscopy 37*(6), 675–679.

Hellerer, T., Enejder, A. M. K., and Zumbusch, A. (2004a). Spectral focusing: High spectral resolution spectroscopy with broad-bandwidth laser pulses, *Applied Physics Letters* 85, 25–27.

Hellerer, T., Enejder, A. M. K., Burkacky, O., and Zumbusch, A. (2004b). Highly efficient coherent anti-Stokes Raman scattering (cars)-microscopy, *Proceedings of SPIE* 5323, 223.

Jesacher, A., Roider, C., Khan, S., Thalhammer, G., Bernet, S., and Ritsch-Marte, M. (2011). Contrast enhancement in wide-field CARS microscopy by tailored phase-matching using a spatial light modulator, *Optics Letters 36*(12), 2245.

Jurna, M., Korterik, J. P., Otto, C., Herek, J. L., and Offerhaus, H. L. (2008). Background free cars imaging by phase sensitive heterodyne cars, *Optics Express 16*(20), 15863–15869.

Jurna, M., Korterik, J. P., Otto, C., and Offerhaus, H. L. (2007). Shot noise limited heterodyne detection of cars signals, *Optics Express 15*(23), 15207–15213.

Krishnamachari, V. V. and Potma, E. O. (2007). Imaging chemical interfaces perpendicular to the optical axis with focus-engineered coherent anti-Stokes Raman scattering microscopy, *Chemical Physics 341*(1–3), 81–88.

Lei, M., Winterhalder, M., Selm, R., and Zumbusch, A. (2011). Video-rate wide-field coherent anti-Stokes Raman scattering microscopy with collinear nonphase-matching illumination, *Journal of Biomedical Optics 16*(02), 021102.

Liesener, J., Reicherter, M., Haist, T., and Tiziani, H. J. (2000). Multi-functional optical tweezers using computer-generated holograms, *Optics Communications 185*(1–3), 77–82.

Lotem, H., Lynch, R. T., and Bloembergen, N. (1976). Interference between Raman resonances in 4-wave difference mixing, *Physical Review A 14*(5), 1748–1755.

Love, G. D. (1997). Wave-front correction and production of zernike modes with a liquid-crystal spatial light modulator, *Applied Optics 36*(7), 1517–1524.

Maurer, C., Jesacher, A., Bernet, S., and Ritsch-Marte, M. (2011). What spatial light modulators can do for optical microscopy, *Laser and Photonics Reviews 5*(1), 81–101.

Mukamel, S. (1995). *Principles of Nonlinear Optical Spectroscopy*. New York: Oxford University Press.

Müller, M. (2006). *Introduction to Confocal Fluorescence Microscopy*. Washington, DC: SPIE Press.

Müller, M., Squier, J., De Lange, C. A., and Brakenhoff, G. J. (2000). Cars microscopy with folded BoxCars phasematching, *Journal of Microscopy-Oxford 197*, 150–158.

Müller, M. and Zumbusch, A. (2007). Coherent anti-Stokes Raman scattering microscopy, *Chemphyschem 8*(15), 2156–2170.

Nandakumar, P., Kovalev, A., and Volkmer, A. (2009). Vibrational imaging based on stimulated Raman scattering microscopy, *New Journal of Physics 11*, 033026.

Ozeki, Y., Dake, F., Kajiyama, S., Fukui, K., and Itoh, K. (2009). Analysis and experimental assessment of the sensitivity of stimulated Raman scattering microscopy, *Optics Express 17*(5), 3651–3658.

Ozeki, Y. and Itoh, K. (2010). Stimulated Raman scattering microscopy for live-cell imaging with high contrast and high sensitivity, *Laser Physics 20*(5), 1114–1118.

Potma, E. O., de Boeij, W. P., van Haastert, P. J. M., and Wiersma, D. A. (2001). Real-time visualization of intracellular hydrodynamics in single living cells, *Proceedings of the National Academy of Sciences of the United States of America 98*(4), 1577–1582.

Saar, B. G., Freudiger, C. W., Reichman, J., Stanley, C. M., Holtom, G. R., and Xie, X. S. (2010). Video-rate molecular imaging in vivo with stimulated Raman scattering, *Science 330*(6009), 1368–1370.

Shi, K. B., Li, H. F., Xu, Q., Psaltis, D., and Liu, Z. W. (2010). Coherent anti-Stokes Raman holography for chemically selective single-shot nonscanning 3D imaging, *Physical Review Letters 104*(9), 093902.

Simanovskii, D., Toytman, I., and Palanker, D. (2009). Solid state lasers for wide-field cars microscopy, *Proceedings of SPIE* 7193, 719328.

Toytman, I., Cohn, K., Smith, T., Simanovskii, D., and Palanker, D. (2007). Wide-field coherent anti-Stokes Raman scattering microscopy with non-phase-matching illumination, *Optics Letters 32*(13), 1941–1943.

Toytman, I., Simanovskii, D., and Palanker, D. (2009). On illumination schemes for wide-field cars microscopy, *Optics Express 17*(9), 7339–7347.

Ventalon, C., Heintzmann, R., and Mertz, J. (2007). Dynamic speckle illumination microscopy with wavelet prefiltering, *Optics Letters 32*(11), 1417–1419.

Vogel, A., Busch, S., Jungnickel, K., and Birngruber, R. (1994). Mechanisms of intraocular photodisruption with picosecond and nanosecond laser-pulses, *Lasers in Surgery and Medicine 15*(1), 32–43.

Volkmer, A. (2005). Vibrational imaging and microspectroscopies based on coherent anti-Stokes Raman scattering microscopy, *Journal of Physics D-Applied Physics 38*(5), R59–R81.

Volkmer, A., Cheng, J. X., and Xie, X. S. (2001). Vibrational imaging with high sensitivity via epidetected coherent anti-Stokes Raman scattering microscopy, *Physical Review Letters 87*(2), 023901.

Wurpel, G. W. H., Schins, J. M., and Müller, M. (2002). Chemical specificity in three-dimensional imaging with multiplex coherent anti-Stokes Raman scattering microscopy, *Optics Letters 27*(13), 1093–1095.

Wurpel, G. W. H., Schins, J. M., and Müller, M. (2004). Direct measurement of chain order in single phospholipid mono- and bilayers with multiplex cars, *Journal of Physical Chemistry B 108*(11), 3400–3403.

Xu, Q., Shi, K. B., Li, H. F., Choi, K., Horisaki, R., Brady, D., Psaltis, D., and Liu, Z. W. (2010). Inline holographic coherent anti-Stokes Raman microscopy, *Optics Express 18*(8), 8213–8219.

Zheltikov, A. M. (2000). Coherent anti-Stokes Raman scattering: From proof-of-the-principle experiments to femtosecond cars and higher order wave-mixing generalizations, *Journal of Raman Spectroscopy 31*(8–9), 653–667.

Zumbusch, A., Holtom, G. R., and Xie, X. S. (1999). Three-dimensional vibrational imaging by coherent anti-Stokes Raman scattering, *Physical Review Letters 82*(20), 4142–4145.

# 8. Vibrational Spectromicroscopy by Coupling Coherent Raman Imaging with Spontaneous Raman Spectral Analysis

**Mikhail N. Slipchenko and Ji-Xin Cheng**





*Coherent Raman Scattering Microscopy*. Edited by Ji-Xin Cheng and X. Sunney Xie © 2013 CRC Press/Taylor & Francis Group, LLC. ISBN: 978-1-4398-6765-5.

## 8.1 Introduction

Recent advances in vibrational imaging based on coherent Raman scattering (CRS) are opening up exciting opportunities for dynamic, non-invasive, and compositional imaging. In CRS microscopy strong signals are generated by focusing picosecond (ps) or femtosecond (fs) excitation laser power on a single Raman mode, which allows imaging at video rate speed (Evans et al., 2005; Saar et al., 2010). In addition same excitation lasers can be used for multiphoton imaging on the same platform (for a review see Yue et al. [2011b]).

A drawback of single-frequency CARS and SRS microscopy is its lack of detailed spectral information. To overcome this shortcoming, multiplex CARS (M-CARS) microscopy has been devised (Cheng et al., 2002; Muller and Schins, 2002). In this technique the combination of narrowband pump and broadband Stokes beams simultaneously excite multiple Raman modes and spectrally resolved M-CARS signal is detected. Compared to single-frequency CARS, the M-CARS signal is, however, significantly weaker because the excitation energy is spread over a broad spectral window. Consequently, M-CARS has reduced sensitivity and requires minutes to obtain an image of 100 × 100 pixels (Kee and Cicerone, 2004; Rinia et al., 2008). Recently, hyperspectral CARS imaging was demonstrated by rapid acquisition of a single-frequency CARS images at large number of spectral points (Lim et al., 2011; Lin et al., 2011). This approach requires less total time per pixel in a moderate spectral range of approximately 200 $cm^{-1}$ (Lin et al., 2011). In contrast, spontaneous Raman microscopy gets full spectral information at each pixel but, due to extremely low Raman signal levels, the long acquisition time on the order of seconds per pixel restricts the use of Raman

**Table 8.1** Summary of Advantages and Disadvantages of Raman-Based Imaging Modalities

| | Pros | Cons |
|---|---|---|
| Single-frequency CARS | • High speed (μs/pixel)<br>• Simple detection at new frequency<br>• Efficient backward detection | • No spectral information<br>• Nonresonant background<br>• Background from fluorescence |
| Single-frequency SRS | • High speed, μs/pixel<br>• No nonresonant background<br>• Linear to molecular concentration<br>• Phase matched<br>• Not sensitive to incoherent background | • No spectral information<br>• Complicated detection<br>• Accompanied by other pump-probe contrasts |
| Multiplex CARS or SRS | • Spectrally resolved detection<br>• Background removed in post-processing | • Integration time, tens of ms/pixel |
| Hyperspectral CARS or SRS | • Spectrally resolved detection<br>• Fast acquisition, ms/pixel | • Narrow spectral window, 200 $cm^{-1}$ |
| Spontaneous Raman | • Shot noise limited<br>• Cost-effective (cw laser)<br>• Whole spectrum analysis | • Long integration time, s/pixel<br>• Very sensitive to incoherent background |

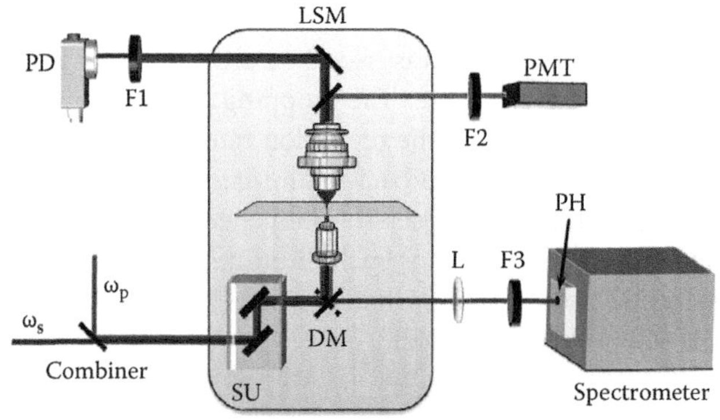


**FIGURE 8.1** Optical layout of the CRS microscope coupled with the Raman spectrometer. LSM is a commercial laser scanning microscope, SU is a laser scanning unit, DM is an exchangeable dichroic mirror, $\omega_p$ and $\omega_s$ are pump and Stokes laser beams, respectively, L is an achromatic lens of 100 mm focal length, and PH is a 100 μm pinhole. PD is large area photodiode for SRS detection and PMT is photomultiplier tube for CARS detection. F1, F2, and F3 are bandpass filters for SRS, CARS, and spontaneous Raman detection, respectively.

microscopy to relatively static samples. Recently, line illumination was employed to reduce the time per image from hours to minutes, allowing live cell imaging (Hamada et al., 2008).

Table 8.1 summarizes the advantages and disadvantages of coherent and spontaneous Raman modalities. From Table 8.1 it is clear that single-frequency CRS modalities provide high speed imaging capability while spontaneous Raman provides full spectral information. Therefore, it is natural that single-frequency coherent Raman and spontaneous Raman techniques are inherently complementary to each other: coherent Raman scattering permits high-speed vibrational imaging by focusing the excitation energy on a single Raman band (Cheng et al., 2001); Spontaneous Raman scattering covers the entire window of molecular vibrations and allows fast spectral analysis at a specified location. In this chapter we describe the coupling (see Figure 8.1) of high-speed coherent Raman imaging of a biological sample and confocal Raman spectral analysis at points of interest using a ps laser source (Slipchenko et al., 2009) and various applications enabled by this method.

# 8.2 Experimental Design

## 8.2.1 Choice of Spectrometer and CCD

The bandwidth of Raman modes in condense phase vary from less than 1.0 to more than 100 wavenumbers, however the majority of the modes in the fingerprint spectral region are around 10 cm$^{-1}$ (full width at half maximum). Therefore, the resolution of the spectrometer and the bandwidth of the laser should not extend beyond 10 cm$^{-1}$ (at 800 nm it corresponds to 0.8 nm). In our setup we use the 303 mm focal length spectrometer (Shamrock SR-303i, Andor) equipped with a 300 g/mm 500 nm blaze angle grating, which provide maximum resolution of 0.4 nm and a wide spectral range from 260 to 4000 cm$^{-1}$, which covers both the fingerprint and the CH-stretch and OH-stretch vibration regions. Due to extremely low Raman signal levels the CCD with low noise is needed in order to acquire spectrum with high signal to noise ratio. Initially we used back-illuminated

EMCCD (Newton DU970N-BV, Andor), which was TE cooled to −70°C in order to minimize dark current noise. The CCD was used in a crop mode to collect signal only from 20 rows to further decrease the noise. The cropping mode also decreases the CCD signal processing time and maximizes the repetition rate. Due to NIR nature of the Raman signal, the spectrum obtained from a back-illuminated CCD shows strong etalon effect, which requires complicated post-processing (Slipchenko et al. 2009). This effect can be avoided by using a front illuminated or back illuminated, deep depletion CCD detectors. In the current setup we use Back Illuminated, Deep Depletion CCD with fringe suppression (Newton DU-929N-BRDD, Andor).

### 8.2.2 Choice of Laser System

The laser source should have sufficient spectral resolution on the order of 10 $cm^{-1}$, and therefore, ps laser source is optimal. In case of fs laser system, which usually has a linewidth of >100 $cm^{-1}$, the spectral bandwidth can be reduced by narrowband filter placed in the laser beam path. Such filter can be a simple narrowband bandpass filter, a high-precision narrow-spaced etalon, or a 4-f pulse shaper with a narrow slit in the Fourier plane. The optimal wavelength for the Raman excitation depends on the sample but due to autofluorescence present in biological samples the near infrared excitation is preferable. However, due to the low efficiency of CCD beyond 1 μm, it is advantageous to use shorter than 800 nm excitation wavelength.

### 8.2.3 Addition of Spectrometer to Laser Scanning Microscope

Raman spectromicroscopy on the CRS microscope can be realized by adding a spectrometer to one of the optical ports of the microscope (see Figure 8.1) and using appropriate filters to reject the scattering of the excitation laser (Slipchenko et al., 2009). To acquire spectral data at a point of interest: first pump and Stokes lasers are tuned to excite single Raman mode and CRS image is obtained. After acquisition of a CRS image the Stokes beam is blocked and the long-pass dichroic mirror under objective is switched to a short-pass/long reflect dichroic mirror which directs the Raman signal toward the spectrometer. The utilization of the same laser for both CRS imaging and confocal Raman spectrometry eliminates the need for any spatial calibration. The spectrometer is externally triggered by a point-scan signal from the laser scanner. To maximize the efficiency of confocal detection, we have used a micro-positioning translational stage to position the point of interest to the center of the field of view for Raman spectral analysis.

### 8.2.4 Advantage of Pinhole over the Slit in front of Spectrometer

In commercial confocal Raman spectrometers, the collected Raman signal first focused on the confocal pinhole and then directed to the slit of the spectrometer. Such design allows separate control over depth of field and spectral resolution. To simplify the setup and light collection efficiency, we have placed the spectrometer close to a non-descanned port of the microscope (see Figure 8.1). In that case if we focus collected Raman signal directly to the slit of the spectrometer, the slit rejects background only in one dimension,

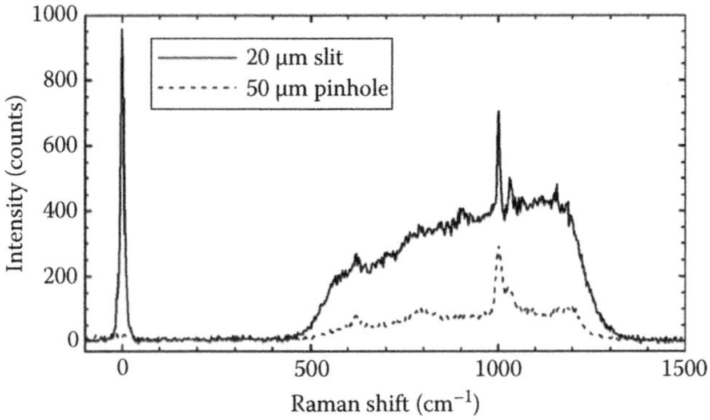


**FIGURE 8.2** Performance comparison of slit and pinhole for Raman spectromicroscopy of 2.2 μm polystyrene bead. A 5 ps laser at 707 nm was used for excitation. Power at the sample is 6 mW.

which results in the high background signal. To reject background signal and achieve 3D spatial resolution, the spectrometer slit assembly have been replaced with a pinhole. Figure 8.2 shows the spectra of the same 2.2 μm diameter polystyrene bead measured with 20 μm width slit and 50 μm diameter pinhole placed in front of the spectrometer. The pinhole setup demonstrates several times better rejection of the fluorescence background from the substrate.

## 8.2.5 Combination of cw and Pulsed Laser Systems

Because spontaneous Raman is a linear process, a simple and inexpensive cw laser source with sufficiently narrow linewidth could be coupled with CRS microscope for Raman excitation. Such combination can be an alternative to line narrowing, which otherwise required for femtosecond lasers. To demonstrate such combination we coupled cw 785 nm laser to a CRS microscope, using a dichroic mirror to combine pulse and cw laser beams. The cw laser was carefully aligned (collinear with pulsed beams) to ensure the excitation of the same focal volume. The alignment was confirmed by taking transmission images with both cw and pulsed lasers. To detect Raman signal we installed a 800 nm short pass dichroic in the microscope turret and an appropriate bandpass filter in front of the spectrometer. Figure 8.3 compares the performance of 785 nm cw and 707 nm ps

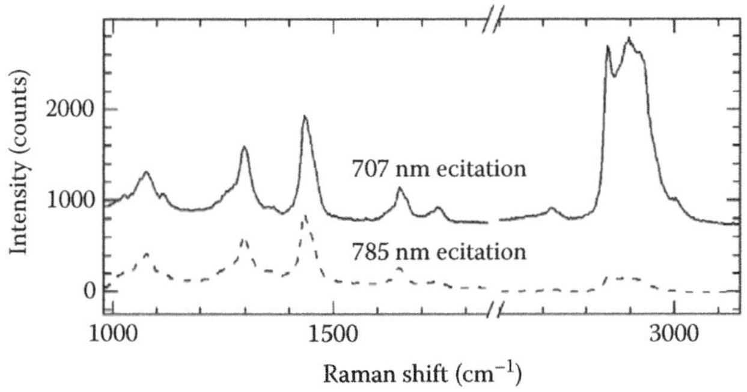


**FIGURE 8.3** Raman spectromicroscopy of olive oil with ps (solid curve) and cw (dashed curve) laser excitations. The power of 707 nm ps and 785 nm cw laser beams at the sample were 5 and 10 mW, respectively. Integration time per each Raman spectrum is 10 s.

laser excitation. The Raman spectra of olive oil in fingerprint region are almost identical with both lasers, whereas CH stretch intensity around 2900 $cm^{-1}$ is considerably decreased when excited with cw laser. This decrease is due to the drop of CCD quantum efficiency at longer wavelength. Note that for 707 nm excitation the Raman signal from CH stretch is detected at around 900 nm, while for 785 nm excitation the same Raman mode is detected at around 1000 nm. Therefore, a cw laser with shorter wavelength is preferable for Raman spectromicroscopy of CH and OH stretching modes.

## 8.3 Performance

### 8.3.1 Spatial Resolution

We first evaluated the performance of our setup using polystyrene (PS) beads of known diameters. Single PS beads of 2.2 μm and 110 nm in diameter spread on a coverslip were first visualized by epi-detected CARS (Figure 8.4a). The CARS intensity profiles cross the beads are displayed in Figure 8.4b. The lateral full-width-at-half-maximum (FWHM) resolution is 230 nm for the 110 nm PS bead, which is smaller than the 300 nm (FWHM) Airy disk calculated from the wavelength of the pump beam and assuming NA = 1.2 for the objective. We evaluated the depth resolution of the confocal Raman spectrometer by

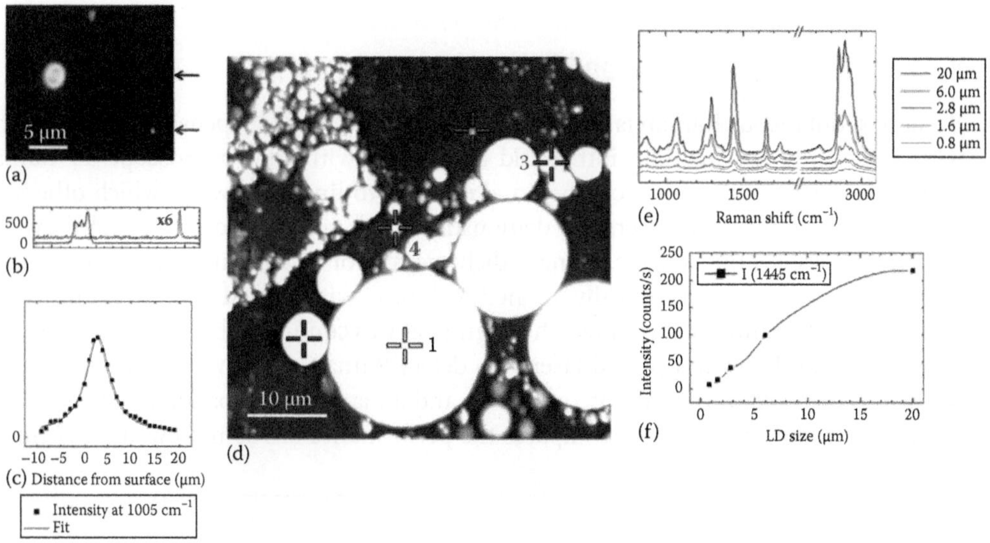


**FIGURE 8.4** Spatial resolution of Raman microanalysis. (a) Epi-detected CARS image of a mixture of PS beads of 2.2 μm and 110 nm diameter taken with a 60× W/IR objective (Olympus) and average pump and Stokes powers of 10 and 15 mW at the sample position. The total integration time per pixel is 150 μs. (b) Intensity profiles along the lines indicated by arrows in panel (a). (c) Intensity profile showing the depth resolution of confocal Raman measurement. (d) The CARS image of 3T3-L1 cells after 14 days of differentiation. Crosses indicate the LDs from which Raman spectra were obtained. (e) Raman spectra of LDs of different sizes as indicated in panel (d). (f) The dependence of intensity of 1445 $cm^{-1}$ peak on LD size. The image and Raman spectra were taken using a 60× water-immersed IR objective (Olympus). The integration time per Raman spectrum is 40 s. A lens of 100 mm focal length in combination with 100 μm pinhole was used to collect Raman signal. The interferometric intensity modulation caused by the back-illuminated EMCCD and the fluorescence background were removed by data processing. (From Slipchenko, M.N. et al., *J. Phys. Chem. B*, 113, 7681, 2009.)

obtaining spectra of a 2.2 μm PS bead at different depths. The intensities of the 1005 $cm^{-1}$ peak were fitted with a Lorentzian function (De Grauw et al., 1997), which yielded an axial FWHM of 6.4 μm (see Figure 8.4c). Figure 8.4d shows Raman spectra obtained from lipid droplets (LDs) of different sizes in 3T3-L1 cells. The same spectrum profiles are observed from different LDs (see Figure 8.4e). Due to an axial resolution of 6.4 μm, the intensity of Raman spectra is saturating for LDs with diameter larger than 6 μm (see Figure 8.4f).

## 8.3.2 Sensitivity of Raman Spectromicroscopy

We evaluated the sensitivity of the setup in two experiments. First the confocal Raman spectra of the 2.2 μm and 110 nm PS beads were acquired in 4 s. For the ring breathing band at 1005 $cm^{-1}$, the signal-to-noise ratio for 2.2 μm and 110 nm PS beads is 100 and 6, respectively (Figure 8.5a). Secondly, we measured the spectrum of 10 mM deuterated palmitic fatty acid dissolved in DMSO. Figure 8.5b *top panel* shows Raman spectra of pure DMSO and deuterated fatty acid. Note that the CD stretching region has no contribution from DMSO Raman modes. The *bottom panel* shows spectra (Y axis is expanded 100 times) of pure DMSO and DMSO with 10 mM deuterated fatty acid. The difference spectrum clearly indicates the presence of CD vibration with signal to noise ratio of about 4.

## 8.3.3 Chemical Selectivity

The ability of spontaneous Raman spectroscopy to obtain the whole spectrum results in high chemical selectivity compared to single frequency or even multiplexed coherent Raman techniques. To demonstrate such selectivity, we applied CARS microscopy coupled with Raman spectromicroscopy to monitor the uptake and storage of oleic acid by CHO cells incubated with 500 μM oleic acid for 6 h. Using CARS imaging at the speed of 10 μs/pixel, we visualized numerous cytoplasmic lipid droplets (LDs) with varying diameters up to 1.5 μm (Figure 8.6a). Three LDs (Figure 8.6a) within a CHO cell were analyzed and their Raman spectra are displayed in Figure 8.6b. For the $CH_2$ deformation peak at 1445 $cm^{-1}$, the Raman spectra of lipid droplets acquired in 4 s have a signal-to-noise ratio of 30. The Raman

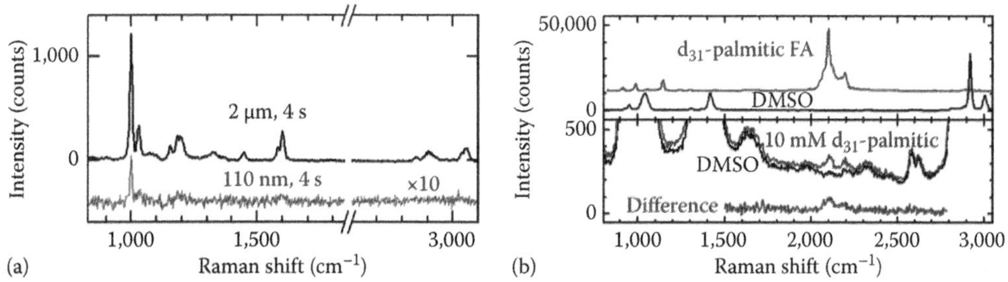


**FIGURE 8.5** Sensitivity of Raman spectromicroscope. (a) Raman spectrum obtained from the two PS beads indicated by arrows in Figure 8.4a at 3 mW of 707 nm ps pump laser at the sample position. For clarity the spectrum from 110 nm bead is offset and multiplied by factor 10. The PS beads were dried on a fused silica coverslip to reduce the fluorescence background. (b) *Top panel* shows Raman spectra of pure DMSO and $d_{31}$-palmitic fatty acid. (b) *Bottom panel* shows expanded view of spectra of 10 mM $d_{31}$-palmitic fatty acid dissolved in DMSO (red) and pure DMSO (black). The difference spectrum (green) is also shown. Raman spectra were taken using a 60x water-immersed IR objective (Olympus). The excitation power at the sample is 6 mW. The integration time per Raman spectrum is 40 s.

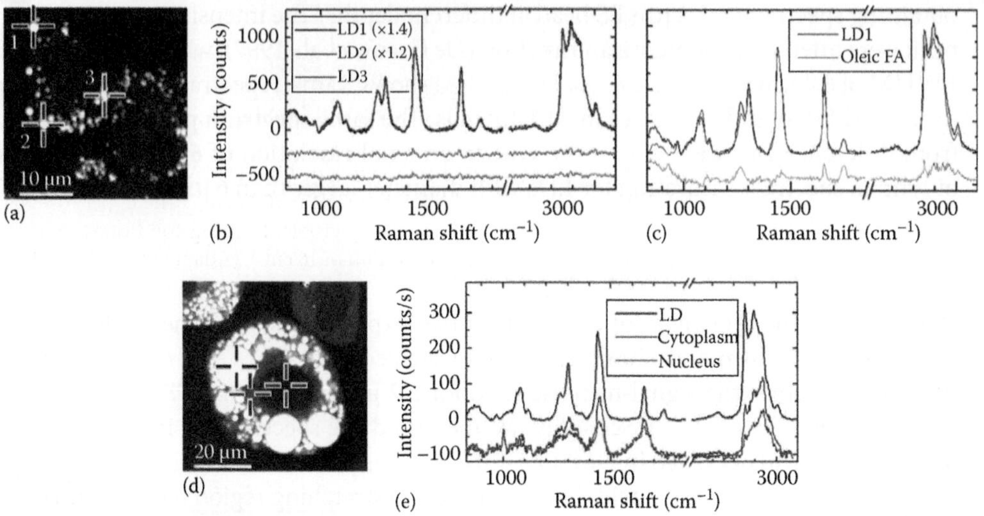


**FIGURE 8.6** Coherent Raman imaging and spontaneous Raman analysis of live cells. (a) CARS image of CHO cells incubated for 6 h in medium contained 500 μM of oleic FA. Each image is a stack of four 512 × 512 pixels images with 1.0 μm depth separation acquired at the speed of 10 μs/pixel. The CARS images were obtained using a 60× IR objective (Olympus), 10 mW of pump and 15 mW of Stokes at the sample. The crosses numbered 1–3 indicate the position where Raman point scans were performed. (b) The obtained Raman spectra from three LDs marked in panel (a) together with differences shown in green. The difference spectra are offset for clarity. (c) Confocal Raman spectrum of LD #1 from panel (a) together with the spectrum of pure oleic FA. Both spectra are normalized on the $CH_2$ deformation band around 1445 cm$^{-1}$. The acquisition time for each Raman spectrum in panels (b) and (c) is 4 s. (d) SRS images of 3T3-L1 cell at 2850 cm$^{-1}$. (e) Raman spectra of LD, cytoplasm and nucleus indicated by crosses in panel (d). The spectra of cytoplasm and nucleus are multiplied by 5 and off set for clarity. Cells were pre-differentiated for 5 days. The acquisition time for each Raman spectrum in panel (e) is 20 s. The acquisition time for the SRS image is 53 s.

spectra of cytoplasmic LDs are identical to each other (Figure 8.6b) and resemble the spectrum of pure oleic FA in solution (Figure 8.6c red curve). However, a few distinctive spectral features between the LDs and oleic FA are observed. First, the LD spectra exhibit a higher intensity of the =CH deformation band at 1265 cm$^{-1}$ and of the C=C stretching band at 1654 cm$^{-1}$, indicating a higher level of unsaturation per chain as compared to the oleic FA. This result agrees with the fact that exogenous fatty acids may be desaturated and elongated in cells before being stored in cytoplasmic LDs. Second, only the LD spectra exhibit peaks at 1742 cm$^{-1}$, which is assigned to the C=O carbonyl stretching vibration. Such peak suggests the conversion of free FA into the esterified form, possibly into triglyceride which in the LDs. To demonstrate the selectivity of full spectrum analysis we also recorded the whole vibrational spectra of individual LD, cytoplasm, and nucleus in a 3T3-L1 cell (Figure 8.6d and e). Our data showed that the intensity ratio of the $CH_2$ symmetric stretch at 2850 cm$^{-1}$ to the $CH_3$ symmetric stretch at 2935 cm$^{-1}$ in the LD is higher than that in the cytoplasm and nucleus. A possible explanation is that the lipids stored in the LDs have a higher density of $CH_2$ groups than proteins which are abundant in the cytoplasm and nucleus. Another interesting spectral feature is the amide I band of proteins at around 1650 cm$^{-1}$. However, it coincides with the stronger C=C vibration of FAs at 1654 cm$^{-1}$. Together, these data demonstrate high chemical selectivity of spontaneous Raman spectrum, which permits compositional analysis of objects visualized by CARS or SRS microscopy.

## 8.4 Applications

### 8.4.1 Chemical Imaging and Spectral Analysis of Lipid Species in *C. elegans*

By simultaneous CARS and TPEF imaging of living *C. elegans*, we can visualize neutral lipid droplets and autofluorescent particles (Figure 8.7a and b) (Le et al., 2010). Visualization of neutral lipid species with CARS has been previously reported by Hellerer et al. (2007). Further confocal Raman spectral analyses of the neutral lipid droplets reveal strong

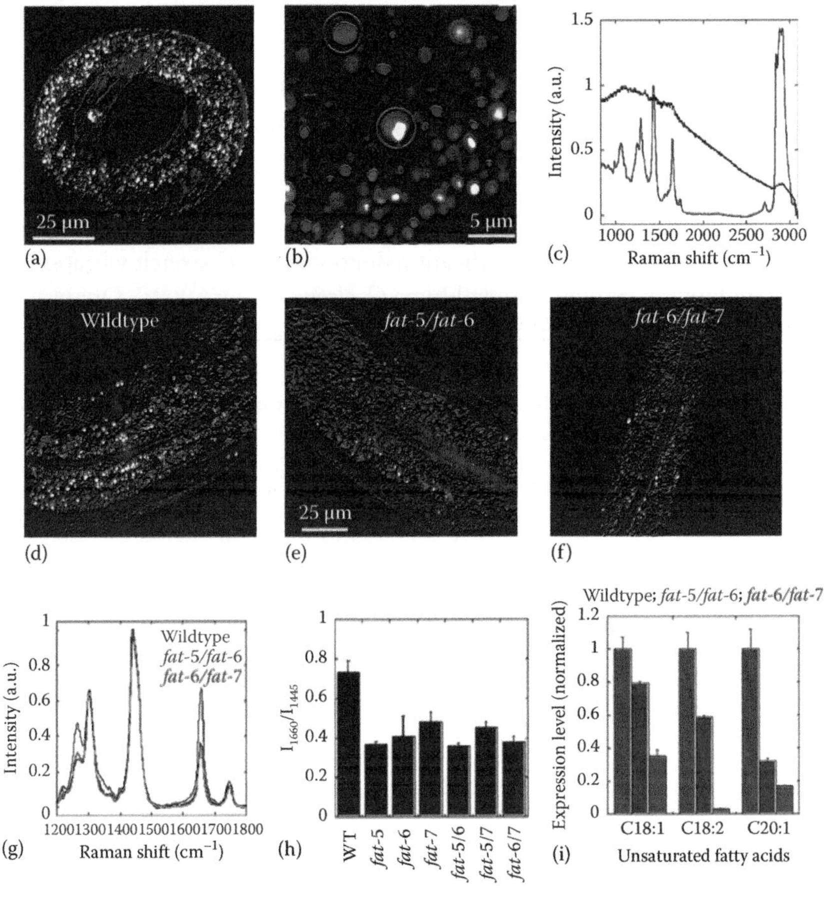


**FIGURE 8.7** Chemical imaging and spectral analysis of lipid species in *C. elegans*. (a) Overlaid CARS (red) and TPEF (blue) imaging of a larval L2 *C. elegans*. Images presented as a 3-D projection of 25 frames taken along vertical axis at 1 μm interval. (b) An enlarged image of lipid species in *C. elegans*. (c) Spontaneous Raman spectral analysis of neutral lipid droplets (red) and autofluorescent particles (blue). (d–f) CARS imaging of lipid (red) and TPEF imaging of autofluorescent lipid species (blue) of adult N2 wild-type and mutant *C. elegans*. Images are presented as 3-D projections of 36 frames taken along the vertical axis at 1 μm interval. (g) Representative Raman spectra of neutral lipid droplets in wildtype and mutant *C. elegans*. (h) Raman quantitative analyses of lipid-chain unsaturation of lipid droplets in wildtype and Δ9 desaturase mutants. Error bars represent distribution across 18 lipid droplets measured in 6 adult wildtype or mutant *C. elegans*. (i) GC-MS analysis of fatty acids composition of wildtype and Δ9 desaturase double mutants. The ratio of unsaturated fatty acids over C18:0 is normalized to 1 for wild-type and comparatively for mutants. Error bars represent distribution across three repeated experiments.

chemical signatures typical of triacylglycerides (Figure 8.7c). In contrast, the fluorescence from the autofluorescent lipid droplets dominates the Raman spectra (Figure 8.7c). Several previous studies have also identified such autofluorescent particles and associated them with lipids, oxidative stress, and lifespan of *C. elegans* (Clokey and Jacobson, 1986; Hosokawa et al., 1994). To explore the potential of using the neutral lipid species and autofluorescent particles as a readout of lipid metabolism, we evaluated their expression levels in wildtype and mutant *C. elegans*. For example, deletion of Δ9 desaturases (palmitoyl-CoA desaturase *fat-5* and stearoyl-CoA desaturases *fat-6* and *fat-7*) (Brock et al., 2007) represses neutral lipid-droplet formation and promotes autofluorescent particle formation (see Figure 8.7d through f). By CARS imaging and spontaneous Raman analysis of single lipid droplets, we further evaluated the degree of lipid-chain unsaturation in wide-type and mutant *C. elegans*. The degree of lipid-chain unsaturation can be measured using three Raman-active bands including 1280, 1660, and 3015 $cm^{-1}$ (Freudiger et al., 2008; Rinia et al., 2008; Le et al., 2010). Because the signal-to-noise ratio for C=C stretch is highest at 1660 $cm^{-1}$ band, we select this band to evaluate lipid-chain unsaturation. We observe that $I_{1660}/I_{1445}$ is linearly correlated with lipid-chain unsaturation (Le et al., 2010). Using $I_{1660}/I_{1445}$ as a reliable measure of Δ9 desaturase enzymatic activity, we systematically evaluated lipid-chain unsaturation of neutral lipid droplets. We observed significant reduction in C=C stretch vibration signal in Δ9 desaturase mutants as compared to wildtype *C. elegans* (Figure 8.7g). Quantitative analysis of lipid droplet $I_{1660}/I_{1445}$ in six desaturase mutants reveals up to twofold reduction in lipid-chain unsaturation in single and double Δ9 desaturase mutants (Figure 8.7h). Our Raman spectral analyses are further supported by gas chromatography coupled to mass spectrometry (GC-MS) measurements of lipid-chain unsaturation of total lipid extracts. We find a dramatic decrease in the ratios of unsaturated oleic, linoleic, and eicosenoic fatty acids over saturated stearic acid in *fat-5/fat-6* and *fat-6/fat-7* mutants as compared to wildtype (Figure 8.7i). Complete analyses of lipid composition of Δ9 desaturase mutants using GC-MS have been described previously (Brock et al., 2007). However, unlike GC-MS, the combination of CARS imaging and confocal Raman spectral analysis enabled us to measure lipid-chain desaturation non-invasively with single lipid-droplet sensitivity. This capability should allow real-time dynamic studies of the activity of desaturases and other lipid metabolism enzymes in living *C. elegans*.

## 8.4.2 Analysis of Breast Tissue Polarity

An important area of investigation is the alteration of tissue architecture that precedes tumor development. Models that mimic phenotypically normal breast glandular differentiation exist and have been used to unravel pathways involved in the initiation of tumors. Cells are cultured in the presence of Engelbreth-Holm-Swarm (EHS)-derived ECM, which resembles a basement membrane (BM), and functionally reproduce mammary glandular structures (acini) with, for certain cell lines like the HMT-3522 S1, complete tissue polarity axis (Nelson and Bissell, 2005; Lee et al., 2007; Lelievre, 2010). This axis largely determines the functional integrity of epithelia and includes both apical polarity and basal polarity (Rodriguez-Boulan and Nelson, 1989; Fish and Molitoris, 1994; Adissu et al., 2007). Tight junctions are formed by transmembrane proteins connected to cytoplasmic proteins at the most apical side of cell–cell junctions (Mitic and Anderson, 1998; Furuse, 2010). The redistribution of tight junction proteins away from apical sites is an indicator of the disruption of apical polarity (Cereijido et al., 1998; Matter et al., 2005; Matter and Balda, 2007). Therefore the status of tissue polarity constitutes a critical readout to assess epithelial integrity and homeostasis. Here we show how lipid ordering of ApM and BaM could be used to characterize apical polarity in live mammary acini (Yue et al., 2012). To measure polarity in live acini we acquired information on lipid organization in the plasma membranes of acinar cells by simultaneous mapping of lipid distribution by CARS and analysis of lipid phases by Raman spectromicroscopy. CARS permits selective imaging of lipid-rich structures, therefore it was used to visualize the ApM and BaM in the equatorial plane of live mammary acini. Substantial CARS signals arose from the lipid-rich ApM, BaM, and cytoplasm of the mammary acinus, whereas weak CARS signals came from the lipid-poor cell nucleus (Figure 8.8a through c). This strong contrast easily distinguished, in a label-free manner, the different compartments of individual cells that form the single-layered epithelium. An important characteristic of the morphogenesis of acinar structures is the narrower width of the apical pole compared to the basal pole (Plachot et al., 2009). This characteristic was clearly outlined by CARS (Figure 8.8c). Confocal Raman spectromicroscopy was further used to analyze lipid phases of ApM and BaM. Although the fingerprint regions on Raman spectra of ApM and BaM revealed no clear difference (Yue et al., 2012), the C–H stretching regions were very distinct (Figure 8.8d), indicating unique lipid compositions of these cellular compartments. Moreover, based on the Lorentzian curve fitting, the degrees of lipid ordering represented by the ratio between the areas under Raman bands at 2885 $cm^{-1}$ ($A_{2885}$) and at 2850 $cm^{-1}$ ($A_{2850}$), $A_{2885}/A_{2850}$ (R), were largely different among ApM, BaM, cytoplasm and nucleus (Figure 8.8e through h). In particular, R was much higher in ApM relative to BaM in the majority of *P* acini (n = 40) (Figure 8.8i). The polarized status was confirmed by the apical location of ZO-1 (Figure 8.8j). Notably, the derived ratio of the degree of lipid ordering appears superior to immunofluorescence-base analysis of fixed samples to determine the presence or disruption of apical polarity (Yue et al., 2012). The combined coherent Raman microscopy and spontaneous Raman spectromicroscopy provide an unprecedented label-free screening platform for rapid identification of risk factors that initiate the very early stage of epithelial neoplasia. Moreover, the use of the membrane lipid phase as readout to assess polarity should be applicable to other types of epithelia where over 90% of cancers originate. Indeed polarity is a common feature of the homeostasis of such tissues and all epithelial neoplasms studied so far for early changes report polarity loss. This method offers new opportunities for the study of environmental factors that influence epithelial homeostasis.

### 8.4.3 Vibrational Imaging of Tablets

Proper chemical imaging tools are critical to the pharmaceutical industry due to growing regulatory demand for intermediate and end-product content uniformity testing. Coherent Raman microscopy and spontaneous Raman spectromicroscopy was combined for chemical mapping of active pharmaceutical ingredient (API) and excipients within tablets (Slipchenko et al., 2010). Amlodipine besylate (AB) tablets from Pfizer (Norvasc) contain API and four excipients: microcrystalline cellulose (MCC), dibasic calcium phosphate anhydrous (DCPA), sodium starch glycolate (SSG), and magnesium

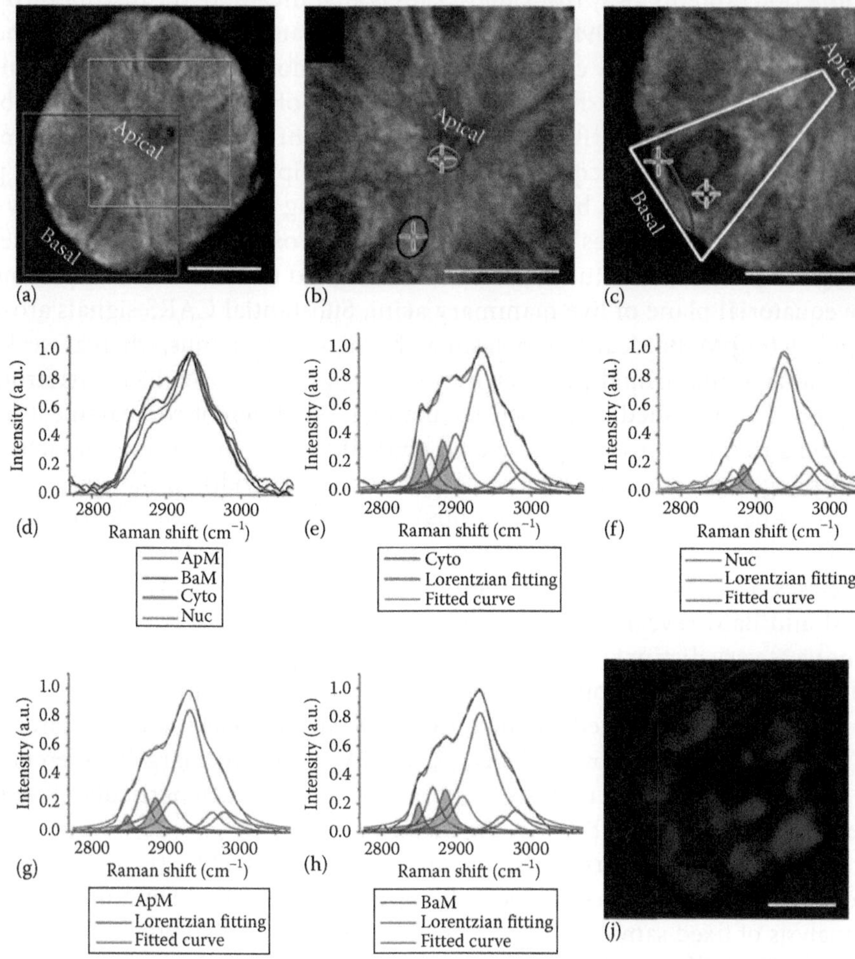


| | ApM | | BaM |
|---|---|---|---|
| $R = A_{2885}/A_{2850}$ | 3.32 | > | 2.31 |

(i)

**FIGURE 8.8** Label-free analysis of lipids in polarized mammary acini. (a) CARS image of a mammary acinus with apical and basal poles. (b) CARS image of the orange square in (a) shown at higher magnification. Crosses within purple and black circles indicate apical membrane (ApM) and cytoplasmic regions of the cell, respectively, where Raman spectra were recorded. (c) CARS image of the blue square in (a) shown at higher magnification. Crosses within blue and green circles indicate basal membrane (BaM) and nucleus regions of the cell, respectively, where Raman spectra were recorded. (d) Normalized Raman spectra of regions within ApM (purple line), BaM (blue line), cytoplasm (Cyto, black line), and nucleus (Nuc, green line). (e–h) Lorentzian curve fitting for Raman spectra of ApM, BaM, Cyto, and Nuc. Original spectra are shown in the same color as in (d). The Lorentzian fitting curves are shown in green. The areas under Raman bands around 2850 and 2885 $cm^{-1}$ are highlighted with green stripes. The cumulative fitted curves are shown in orange. (i) Ratio $A_{2885}/A_{2850}$ (R) for ApM and BaM (R for Cyto equals 1.56, and R for Nuc equals 4.11). (j) Fluorescence image of the mammary acinus in A labeled for apical polarity marker ZO-1(red) and DNA (DAPI, blue; nine nuclei are seen in this focal plane). Size bar, 10 μm. For CARS imaging, pump and Stokes lasers were tuned to 14,140 and 11,300 $cm^{-1}$, respectively, to be in resonance with the $CH_2$ symmetric stretch vibration at 2840 $cm^{-1}$. The combined Stokes and pump laser power at the specimen was 40 mW. Each Raman spectrum was acquired in 10 s, and pump laser power at the specimen was maintained at 15 mW.

stearate (MS). The tablets were directly placed on coverslip for epi-detected SRS and confocal Raman imaging. Based on the known spectra of pure components, the CC stretching band of AB (Szab et al., 2009) at 1650 cm$^{-1}$ and the PO stretching band of DCPA (Xu et al., 1999) at 985 cm$^{-1}$ can be used directly for SRS imaging since they do not overlap with bands of other excipients (see Figure 8.9a2 and a3). The CH stretching region around 2900 cm$^{-1}$ has overlapping bands from MCC, SSG and MS. In order to obtain SRS images of these three components we imaged the tablet at 2850 cm$^{-1}$ and at 2900 cm$^{-1}$ (see Figure 8.9a1 and a4). Based on the known Raman intensities of excipients

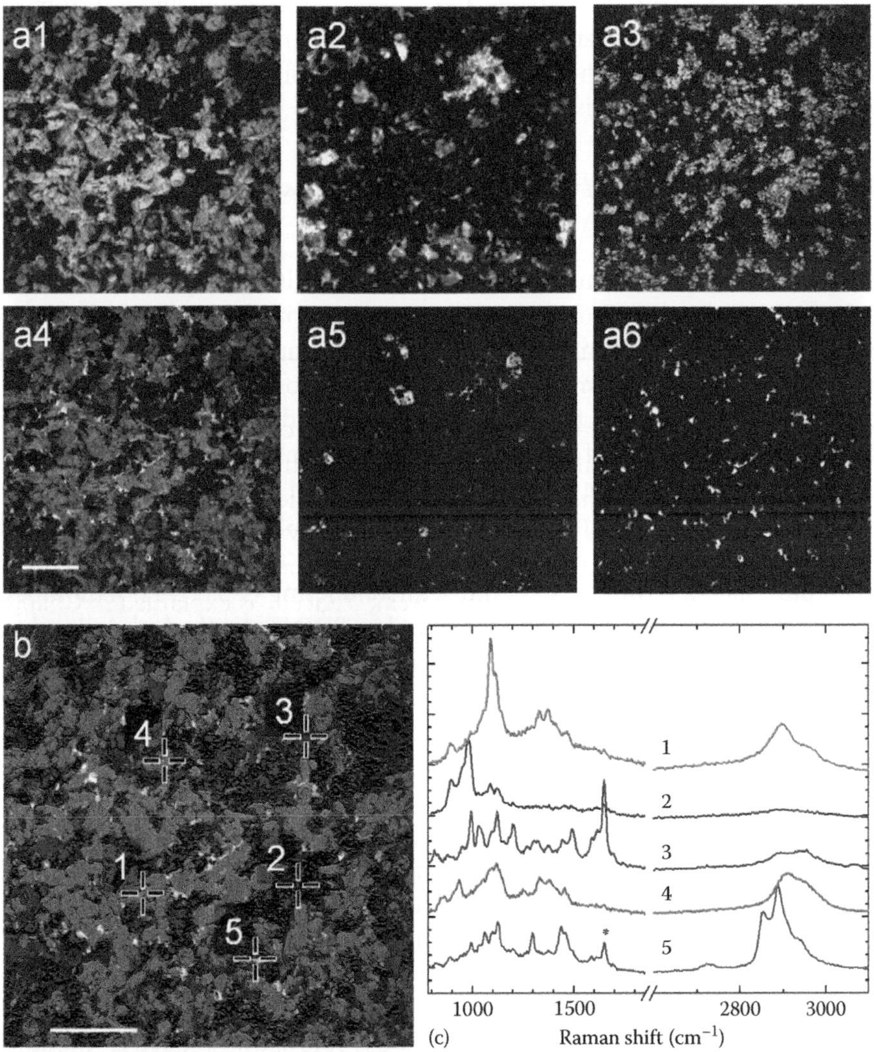


FIGURE 8.9 SRS imaging and Raman spectral analysis of tablet. (a1–a4) SRS images obtained at 2900, 1650, 985, and 2850 cm$^{-1}$. The images (a5) and (a6) were obtained from (a1) and (a4) by deconvolution based on the known ratios of Raman intensities at 2850 cm$^{-1}$ and at 2900 cm$^{-1}$ for MCC, DCPA, and MS. The power of the pump and Stokes beams at the sample was 20 and 15 mW, respectively. (b) Overlaid images (a1–a6) showing distribution of MCC (green), DCPA (blue), AB (red), SSG (yellow/orange), and MS (magenta). (c) Raman spectra obtained at positions indicated in (b). Raman excitation wavelength is 707 nm, 6 mW power at the sample. 20× air objective (Olympus, NA = 0.75) was used. Scale bars are 100 μm.

at these Raman shifts we obtained images of MS and SSG (see Figure 8.9a5 and a6). The SRS imaging of SSG can be also done at 840 $cm^{-1}$ since its Raman spectrum has no overlap from other components at that Raman shift. The overlaid images of all components give image shown in Figure 8.9b. To confirm that the observed signals are from the drug molecules and excipients, confocal Raman spectromicroscopy was performed at the points of interest (see Figure 8.9c). Raman spectra obtained at positions which correspond to particles of MCC, DCPA, AB, SSG, and MS are similar to those of pure components (Slipchenko et al., 2010). Due to the large depth of field the confocal Raman spectra show some contribution of Raman peaks from different components. For example, the spectrum 5 from MS particle in Figure 8.9c has a contribution from spectrum 3 from AB particle as indicated by asterix. To summarize, confocal Raman spectromicroscopy complements the high-speed imaging capability of CRS microscopy through full spectral analysis at pixels of interest.

### 8.4.4 Integration of Raman Spectroscopy with Transient Absorption Microscopy

Raman analysis is virtually useful for any NLO modalities. Jung et al. demonstrated the coupling of spontaneous Raman and transient absorption microscopy to discriminate metallic and semiconducting single-walled carbon nanotubes (SWNTs) (Jung et al., 2010). Transient absorption microscopy is a novel platform, in which a pump laser is used to excite molecules to an electronically excited state and a probe beam is used to monitor a change in the population of ground or excited states following the pump pulse (Ye et al., 2009). In the setup shown in Figure 8.9a, the transient absorption signal is obtained by intensity modulation of the pump field and phase-sensitive detection of the probe field. To determine whether the transient absorption signal can be used to map metallicity, we studied pure metallic (m-SWNTs) and semiconducting (s-SWNTs) nanotubes. Figure 8.10b and c show the images of s-SWNTs and m-SWNTs by the lock-in in-phase channel signals, respectively. We observed a positive contrast in the in-phase-channel image of the s-SWNTs (Figure 8.10b). Therefore, the signals from the s-SWNTs have the phase of 0°, which corresponds to an in-phase modulation of the probe (or reduced transient absorption). Different from the s-SWNTs, the m-SWNTs show a negative contrast in the in-phase-channel image (Figure 8.10c). These signals correspond to the phase of 180°, which indicates an anti-phase modulation of the probe field (or enhanced transient absorption). Collectively, we observed opposite contrast in the in-phase-channel between s- and m-SWNTs. The different phases in the transient absorption signals arise from the different electronic energy structures of the s-SWNTs and the m-SWNTs used in our experiments (see Figure 8.10d). To confirm the assignments, we recorded the spontaneous Raman spectra of the s- and m-SWNTs indicated with a cross mark in Figure 8.10b and c. The Raman spectrum of s-SWNTs shows a sharp $G^+$ mode peaked at 1587.6 $cm^{-1}$ (Figure 8.10e). The Raman spectrum of m-SWNTs shows a broad $G^-$ mode peaked at 1559.4 $cm^{-1}$ and a $G^+$ mode peaked at 1585.1 $cm^{-1}$ (Figure 8.10f). These results are consistent with previous studies (Dresselhaus et al., 2002; Jorio et al., 2003; Krupke et al., 2003) and validate our transient absorption data. Our mapping method demonstrated here provides great potential to investigate metallicity, chirality, and carrier dynamics within single nanotubes and other nanostructures.

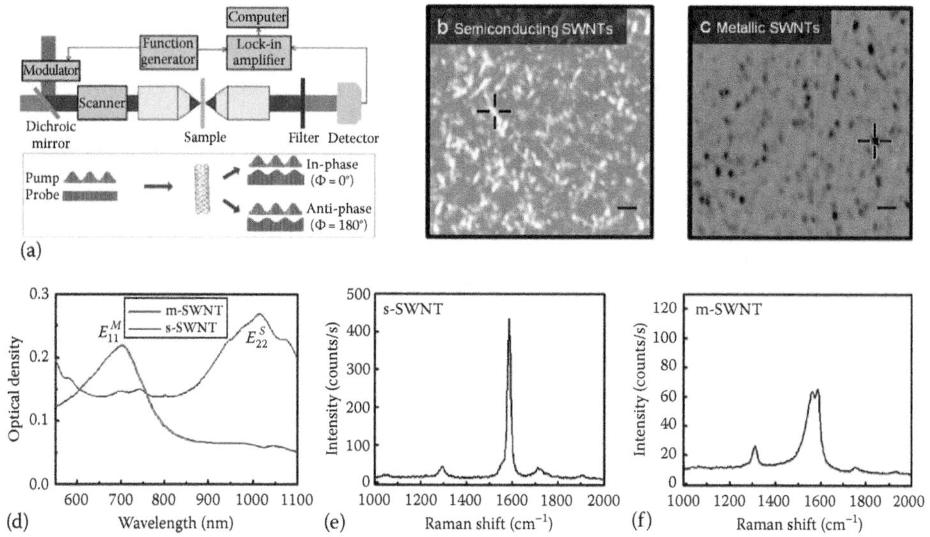


**FIGURE 8.10** Coupling of spontaneous Raman with transient absorption microscopy. (a) Schematic drawing of a transient absorption microscope. The rectangular box shows that the in-phase and anti-phase modulations of the probe field, relative to the modulation of the pump field, corresponds to a negative and a positive transient absorption, respectively. (b and c) In-phase images of semiconducting and metallic SWNTs, respectively. (d) Extinction spectrum of the metallic and semiconducting SWNTs in aqueous solution. (e and f) Raman spectrum of the G band in semiconducting and metallic SWNTs, respectively. The data acquisition positions are indicated by cross in (b) and (c). The images in (b) and (c) were acquired with a speed of 50 μs per pixel, resulting in 13 s per frame. Scale bar = 2 μm for all images.

## 8.5 Conclusions and Outlook

In this chapter, we described the integration of spontaneous Raman capability into coherent Raman microscope for both high-speed chemical imaging and quantitative spectral analysis of samples. The ps laser excitation can be directly used for Raman excitation, while the broadband fs laser has to be narrowed down to attain high spectral resolution. As an alternative approach the cw laser source can be combined with pulsed lasers for Raman excitation; however, such a combination requires careful alignment. We have evaluated the performance of the integrated Raman spectrometer and demonstrated micrometer-range spatial resolution, millimolar-range sensitivity, and high chemical selectivity of the setup. The integrated coherent Raman imaging and spontaneous Raman spectral analysis was applied to lipid bodies of cultured cells, live animals, and pharmaceutical samples. Important information including lipid body abundance, size, degree of carbon chain unsaturation, and lipid-packing density could be obtained within a few seconds. In addition, we showed that spontaneous Raman spectromicroscopy can be integrated with other NLO modality such as transient absorption microscopy. There are two major roles of spontaneous Raman spectromicroscopy integrated into NLO microscope: (1) to analyze chemical composition of points of interest in complicated biological samples and (2) to validate single-frequency NLO imaging of defined samples such as pharmaceutical dosage forms. The high-speed data acquisition capability of integrated spontaneous and coherent Raman techniques offers exciting possibilities for *in vivo* studies of lipid metabolism as well as chemical imaging of samples in pharmaceutical, environmental, and other research fields.

## References


Adissu, H. A., Asem, E. K., and Lelièvre, S. A. (2007) Three-dimensional cell culture to model epithelia in the female reproductive system. *Reproductive Science*, 14, 11–19.

Brock, T. J., Browse, J., and Watts, J. L. (2007) Fatty acid desaturation and the regulation of adiposity in *Caenorhabditis elegans*. *Genetics*, 176, 865–875.

Cereijido, M., Valdes, J., Shoshani, L., and Contreras, R. G. (1998) Role of tight junctions in establishing and maintaining cell polarity. *Annual Review of Physiology*, 60, 161–177.

Cheng, J.-X., Volkmer, A., Book, L. D., and Xie, X. S. (2001) An epi-detected coherent anti-Stokes Raman scattering (E-CARS) microscope with high spectral resolution and high sensitivity. *Journal of Physical Chemistry B*, 105, 1277–1280.

Cheng, J. X., Volkmer, A., Book, L. D., and Xie, X. S. (2002) Multiplex coherent anti-Stokes Raman scattering microspectroscopy and study of lipid vesicles. *Journal of Physical Chemistry B*, 106, 8493–8498.

Clokey, G. V. and Jacobson, L. A. (1986) The autofluorescent lipofuscin granules in the intestinal-cells of *Caenorhabditis-elegans* are secondary lysosomes. *Mechanisms of Ageing and Development*, 35, 79–94.

De Grauw, C. J., Sijtsema, N. M., Otto, C., and Greve, J. (1997) Axial resolution of confocal Raman microscopes: Gaussian beam theory and practice. *Journal of Microscopy-Oxford*, 188, 273–279.

Dresselhaus, M. S., Dresselhaus, G., Jorio, A., Souza Filho, A. G., and Saito, R. (2002) Raman spectroscopy on isolated single wall carbon nanotubes. *Carbon*, 40, 2043–2061.

Evans, C. L., Potma, E. O., Mehron, P. H. et al. (2005) Chemical imaging of tissue in vivo with video-rate coherent anti-Stokes Raman scattering microscopy. *Proceedings of the National Academy of Sciences of the United States of America*, 102, 16807–16812.

Fish, E. M. and Molitoris, B. A. (1994) Alterations in epithelial polarity and the pathogenesis of disease states. *New England Journal of Medicine*, 330, 1580–1588.

Freudiger, C. W., Min, W., Saar, B. G. et al. (2008) Label-free biomedical imaging with high sensitivity by stimulated Raman scattering microscopy. *Science*, 322, 1857–1861.

Furuse, M. (2010) Molecular basis of the core structure of tight junctions. *Cold Spring Harbor Perspectives in Biology*, 2, a002907.

Hamada, K., Fujita, K., Smith, N. I. et al. (2008) Raman microscopy for dynamic molecular imaging of living cells. *Journal of Biomedical Optics*, 13, 4.

Hellerer, T., Axang, C., Brackmann, C. et al. (2007) Monitoring of lipid storage in *Caenorhabditis elegans* using coherent anti-Stokes Raman scattering (CARS) microscopy. *Proceedings of the National Academy of Sciences of the United States of America*, 104, 14658–14663.

Hosokawa, H., Ishii, N., Ishida, H. et al. (1994) Rapid accumulation of fluorescent material with aging in an oxygen-sensitive mutant Mev-1 of *Caenorhabditis-elegans*. *Mechanisms of Ageing and Development*, 74, 161–170.

Jorio, A., Pimenta, M., Souza Filho, A. et al. (2003) Characterizing carbon nanotube samples with resonance Raman scattering. *New Journal of Physics*, 5, 139.

Jung, Y., Slipchenko, M. N., Liu, C. H. et al. (2010) Fast detection of the metallic state of individual single-walled carbon nanotubes using a transient-absorption optical microscope. *Physical Review Letters*, 105, 4.

Kee, T. W. and Cicerone, M. T. (2004) Simple approach to one-laser, broadband coherent anti-Stokes Raman scattering microscopy. *Optics Letters*, 29, 2701–2703.

Krupke, R., Hennrich, F., Lohneysen, H. V., and Kappes, M. M. (2003) Separation of metallic from semiconducting single-walled carbon nanotubes. *Science*, 301, 344–347.

Le, T. T., Duren, H. M., Slipchenko, M. N., Hu, C. D., and Cheng, J. X. (2010) Label-free quantitative analysis of lipid metabolism in living *Caenorhabditis elegans*. *Journal of Lipid Research*, 51, 672–677.

Lee, G. Y., Kenny, P. A., Lee, E. H., and Bissell, M. J. (2007) Three-dimensional culture models of normal and malignant breast epithelial cells. *Nature Methods*, 4, 359–365.

Lelievre, S. A. (2010) Tissue polarity-dependent control of mammary epithelial homeostasis and cancer development: An epigenetic perspective. *Journal of Mammary Gland Biology and Neoplasia*, 15, 49–63.

Lim, R. S., Suhalim, J. L., Miyazaki-Anzai, S. et al. (2011) Identification of cholesterol crystals in plaques of atherosclerotic mice using hyperspectral CARS imaging. *Journal of Lipid Research*, 52, 2177–2186.

Lin, C. Y., Suhalim, J. L., Nien, C. L. et al. (2011) Picosecond spectral coherent anti-Stokes Raman scattering imaging with principal component analysis of meibomian glands. *Journal of Biomedical Optics*, 16, 9.

Matter, K., Aijaz, S., Tsapara, A., and Balda, M. S. (2005) Mammalian tight junctions in the regulation of epithelial differentiation and proliferation. *Current Opinion in Cell Biology*, 17, 453–458.

Matter, K. and Balda, M. S. (2007) Epithelial tight junctions, gene expression and nucleo-junctional interplay. *Journal of Cell Science*, 120, 1505–1511.
Mitic, L. L. and Anderson, J. M. (1998) Molecular architecture of tight junctions. *Annual Review of Physiology*, 60, 121–142.
Muller, M. and Schins, J. M. (2002) Imaging the thermodynamic state of lipid membranes with multiplex CARS microscopy. *Journal of Physical Chemistry B*, 106, 3715–3723.
Nelson, C. M. and Bissell, M. J. (2005) Modeling dynamic reciprocity: Engineering three-dimensional culture models of breast architecture, function, and neoplastic transformation. *Seminars in Cancer Biology*, 15, 342–352.
Plachot, C., Chaboub, L., Adissu, H. et al. (2009) Factors necessary to produce basoapical polarity in human glandular epithelium formed in conventional and high-throughput three-dimensional culture: Example of the breast epithelium. *BMC Biology*, 7, 77.
Rinia, H. A., Burger, K. N. J., Bonn, M., and Muller, M. (2008) Quantitative label-free imaging of lipid composition and packing of individual cellular lipid droplets using multiplex CARS microscopy. *Biophysical Journal*, 95, 4908–4914.
Rodriguez-Boulan, E. and Nelson, W. J. (1989) Morphogenesis of the polarized epithelial cell phenotype. *Science*, 245, 718–725.
Saar, B. G., Freudiger, C. W., Reichman, J. et al. (2010) Video-rate molecular imaging in vivo with stimulated Raman scattering. *Science*, 330, 1368–1370.
Slipchenko, M. N., Chen, H., Ely, D. R. et al. (2010) Vibrational imaging of tablets by epi-detected stimulated Raman scattering microscopy. *Analyst*, 135, 2613–2619.
Slipchenko, M. N., Le, T. T., Chen, H., and Cheng, J. X. (2009) High-speed vibrational imaging and spectral analysis of lipid bodies by compound Raman microscopy. *Journal of Physical Chemistry B*, 113, 7681–7686.
Szab, L., Chis, V., Pîrnău, A. et al. (2009) Spectroscopic and theoretical study of amlodipine besylate. *Journal of Molecular Structure*, 924–926, 385–392.
Xu, J. W., Butler, I. S., and Gilson, D. F. R. (1999) FT-Raman and high-pressure infrared spectroscopic studies of dicalcium phosphate dihydrate (CaHPO4 center dot 2H(2)O) and anhydrous dicalcium phosphate (CaHPO4). *Spectrochimica Acta Part A-Molecular and Biomolecular Spectroscopy*, 55, 2801–2809.
Ye, T., Fu, D., and Warren, W. S. (2009) Nonlinear absorption microscopy. *Photochemistry and Photobiology*, 85, 631–645.
Yue, S., Cárdenas-Mora, J. M., Chaboub, L. S., Lelièvre, S. A., and Cheng, J. X. (2012) Label-free analysis of breast tissue polarity by Raman imaging of lipid phase. *Biophysical Journal*, 102, 1215–1223.
Yue, S., Slipchenko, M. N., and Cheng, J. X. (2011b) Multimodal nonlinear optical microscopy. *Laser and Photonics Reviews*, 5, 496–512.

# 9. Coherent Control in CARS

**Jonathan M. Levitt, Ori Katz, and Yaron Silberberg**



## 9.1 Introduction

In CARS spectroscopy pump and Stokes photons at frequencies $\omega_p$ and $\omega_s$, respectively, coherently excite molecular vibration at the frequency $\Omega_{vib} = \omega_p - \omega_s$, and the vibration is subsequently probed by interaction with probe photons at frequency $\omega_{pr}$. The vibrational spectrum is resolved by measuring the blue shift of the generated anti-Stokes photons frequency from the probe frequency $\omega_{AS} = \omega_{pr} + \Omega_{vib}$. CARS spectroscopy is typically performed as a multi-beam, multi-source technique [1–3] (Figure 9.1a and b), which is experimentally challenging to implement due to the strict requirement of spatial and temporal overlap of the excitation beams. This paradigm has changed with the utilization of quantum coherent

*Coherent Raman Scattering Microscopy*. Edited by Ji-Xin Cheng and X. Sunney Xie © 2013 CRC Press/ Taylor & Francis Group, LLC. ISBN: 978-1-4398-6765-5.

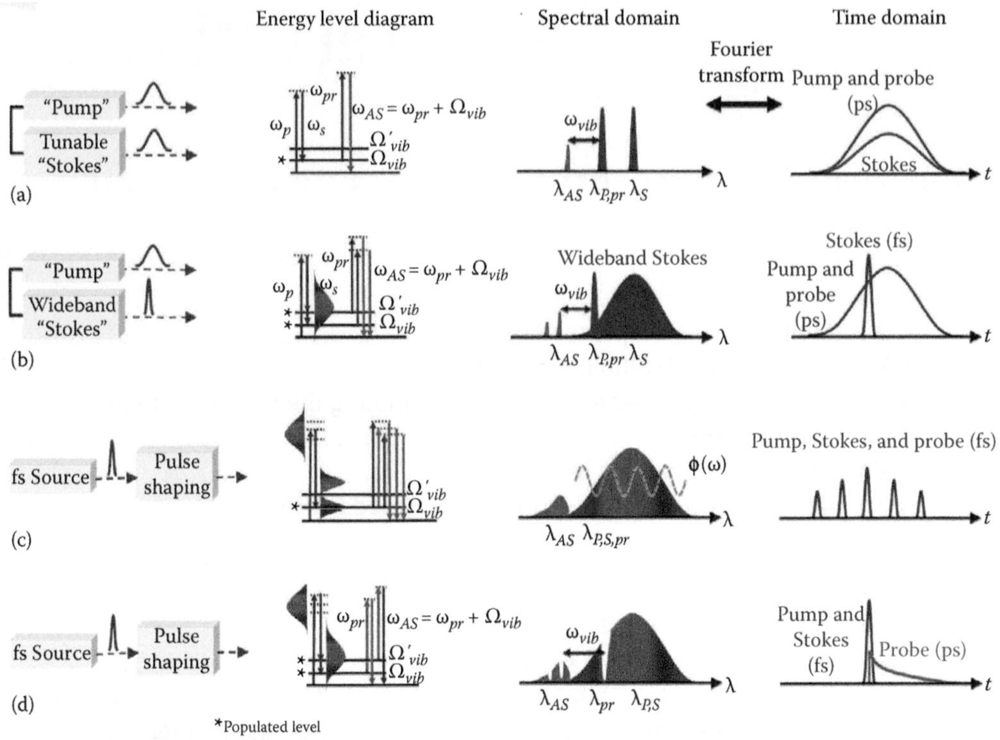


**FIGURE 9.1** Pulse shaping for femtosecond single-pulse CARS, side-by-side with the conventional multi-beam analogs. The corresponding energy-level diagrams (left), and the spectral- and time-domain pictures (center and right). (a) Conventional multi-beam CARS where a single vibrational level is excited and probed by narrowband pump and Stokes beams from synchronized sources; (b) conventional multi-beam multiplex CARS utilizing a wideband ultrashort Stokes pulse and a narrowband probe beam, simultaneously exciting and probing several vibrational levels; (c) single-pulse CARS by selective vibrational excitation: the wideband pulse is shaped such that vibrational excitation is optimized for a single vibrational line. In the depicted example, the spectral phase of the pulse is modulated to produce a train of impulses in the time domain, maximizing the excitation for the vibrational frequency $\Omega_{vib}$ (marked by *) with the period equal to the pulse separation. [7]. Different pulse shapes are used to excite different vibrational levels, in the same manner that different pump and Stokes frequencies are used in conventional multi-beam CARS as depicted in (a); (d) single-pulse CARS by spectrally shaped narrowband probe: in this technique a narrow probe is defined in the pulse bandwidth by spectral shaping. (Depicted here a notch amplitude shaping [18].) The wideband pulse coherently excites a band of vibrational levels, and the shaped narrowband probe effectively produces a time-delayed temporally extended probe, yielding a spectral resolution comparable to the probe spectral width. The vibrational spectrum is resolved by measuring the blue shift of the induced interference features in the CARS spectrum from the shaped probe frequency. Multiple probes can be shaped to tailor an enhanced anti-Stokes feature from contributions of several vibrational levels of a target chemical species [20].

control and pulse-shaping techniques in CARS spectroscopy. In brief, the goal of quantum control is to drive a quantum system from an initial state to a desired final state by exploiting constructive quantum-mechanical interferences that build up that state amplitude, while avoiding undesirable final states through destructive interferences [4]. This approach can be directly applied in CARS spectroscopy to selectively excite a single vibrational level of a specific chemical species, while minimizing the excitation of undesirable levels or species (Figure 9.1c). Today, the most common experimental approach to quantum control uses shaped femtosecond pulses. To effectively exploit quantum interferences, one must preserve the quantum-mechanical phase and hence avoid decoherence. It is therefore advantageous to use ultrafast interactions that are faster than the lifetimes of the relevant levels or, in equivalent terms, to use laser pulses that are spectrally broader than the width of those levels. This is in contrast with classical laser spectroscopy, where the desired laser line width is traditionally much narrower than the width (and spacing) of the level(s) being assessed (e.g., Figure 9.1a). The most common route to coherent control utilizes the technique of femtosecond pulse shaping. For control, one can apply a sequence of femtosecond pulses to drive the system impulsively, but the most versatile approach uses an optical setup known as a pulse shaper [5] in which the short pulses are modified to generate more complex, specifically tailored pulses by controlling the individual frequencies of the short pulse. The most common experimental configuration for pulse shaping is based on spatial dispersion and is often referred to as a 4-f pulse shaper, which closely resembles a back-to-back optical grating spectrometer (Figure 9.2). The first grating disperses the spectral components of the pulse in space, and the second grating packs them back together, while a pixelated spatial light modulator (SLM) applies a specified transfer function, thereby modifying the amplitudes, phases, or polarization states of the various spectral components. In essence, the shaper is an optical synthesizer that can generate complex pulses with a temporal resolution limited by the inverse of the spectral content of the pulse and a maximal controlled duration limited by the spectral resolution of the setup and the SLM pixel size. The configuration shown in Figure 9.2 is just one approach to pulse shaping, and a few other techniques have been employed, most notably an acousto-optical device known as the dazzler [6].

Utilizing pulse shaping, CARS spectroscopy can be converted from a multi-source, multi-beam technique to a single-source, single-beam technique, which employs a single femtosecond pulse for CARS spectroscopy and microscopy [7–10]. In such “single-pulse” schemes, a single wideband femtosecond pulse simultaneously provides the necessary pump, Stokes, and probe photons, and spectral selectivity is attained by pulse shaping (Figure 9.1c and d). Femtosecond single-pulse CARS techniques are attractive as only a single laser source is required and the spatiotemporal overlap of the pump, Stokes, and probe photons is inherently maintained. Moreover, because of their high peak intensity, femtosecond pulses are favorable over picosecond pulses for a variety of other nonlinear

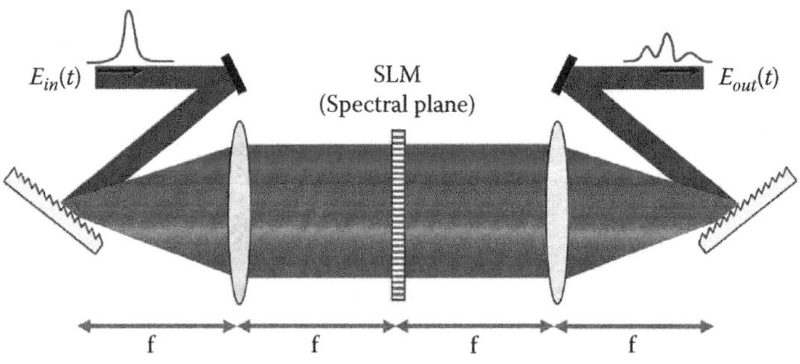


FIGURE 9.2 Schematic drawing of a Fourier-domain 4-f femtosecond pulse shaper. A transform-limited input pulse is dispersed by the input grating (left), and the different frequency components are modified by the mask function at the spectral plane using an SLM. The different spectral components are then recombined by the output grating (right) to form the shaped output pulse.

processes such as second- and third-harmonic generation (SHG, THG) and multiphoton fluorescence, an advantage for label-free multimodal microscopy [11]. Since first demonstrated [7], single-pulse CARS schemes have been utilized for vibrational imaging [12,13], time-resolved chemical microanalysis [14], and remote detection of hazardous materials [15,16]. The two main difficulties when implementing CARS using femtosecond pulses are the loss of spectral resolution and an increase of non-resonant contribution. First, the loss of spectral resolution is the result of the pulse bandwidth being much broader than the width and spacing of vibrational lines [7,17]. This can be intuitively portrayed as the entire pulse bandwidth serving as a probe. Second, the non-resonant contribution is significantly enhanced due to higher peak powers, compared to longer pulses of equivalent energy, and the inherent temporal overlap of the pump, Stokes, and probe photons. While both of these complications severely inhibit the use of unshaped femtosecond pulses in CARS, the use of pulse-shaping techniques can easily address them. Specifically, through careful pulse shaping, a spectroscopic resolution orders of magnitude better than the pulse bandwidth can be retained, and the non-resonant background can be reduced or strategically used as a local oscillator for signal amplification.

In this chapter, we review the different pulse-shaping schemes for femtosecond and single-pulse CARS. In particular we evaluate the effect of specific pulse shapes in terms of driving and probing molecular vibrations. Figure 9.1 briefly summarizes the main approaches for femtosecond single-pulse CARS, together with the corresponding conventional multi-beam analogs. The various different pulse-shaping schemes can be divided into two main approaches. In one approach, pulse shaping is used to gain control over the vibrational Raman *excitation* process (Figure 9.1c). Tailoring the pulse allows exciting *selectively* a single vibrational level instead of the multiple levels that would have been excited by an unshaped wideband pulse [7]. Different pulse shapes are then used to excite different vibrational levels, in the same manner that different pump and Stokes frequency pairs are used in conventional multi-beam CARS (Figure 9.1a). This is achieved by altering the excitation probability amplitude of a particular Raman mode $A(\Omega_{vib})$ (Equation 9.1):

$$A(\Omega) = \int_0^\infty d\omega E^*(\omega)E(\omega-\Omega) = \int_0^\infty d\omega \left|E(\omega)\right|\left|E(\omega-\Omega)\right| e^{i[\varphi(\omega)-\varphi(\omega-\Omega)]} \tag{9.1}$$

$A(\Omega)$ is the probability amplitude to populate a vibrational level with energy $\Omega$ via Raman excitation, function of the incident pulse electric field spectral amplitude $E(\omega)$ [10]. The population amplitude $A(\Omega)$ is the sum of all possible two-photon pairs leading to the same intermediate state and can be controlled by shaping the spectral phase of the incident electric field, $\varphi(\omega)$ (Figure 9.1c). The term $E(\omega)$ describes the pump field and the term $E(\omega - \Omega)$ is the Stokes field, both originating from the same wideband ultrashort pulse. In an alternative approach, pulse shaping is used to control the probing process (Figure 9.1d). By spectral phase, amplitude, or polarization shaping of a narrow frequency band, a narrowband probe can be defined in the wide pulse bandwidth, allowing for *multiplex* single-shot measurement of the vibrational spectrum [8,9,18,19] (Figure 9.1d). This is analogous to conventional multiplex CARS [1,8,9], where several vibrational levels are coherently excited by broadband pump and Stokes beams, and is subsequently probed by a narrowband probe beam, producing blue-shifted spectral peaks in the CARS spectrum (Figure 9.1b). In multiplex single-pulse CARS, the shaped narrowband probe produces narrow spectral interference features in the CARS spectrum, which are also blue-shifted from the shaped probe frequency by the molecular vibrational frequencies (Figure 9.1d). The interference features are due to the dependence of the resonant signal on spectral phase compared to the non-resonant signal (Equations 9.2 and 9.3). The vibrational spectrum is easily resolved by measuring the energy shift of these features from the probe frequency. An advantageous coherent-control feature of this approach is accessible by shaping multiple probes. A large coherent interference signal from contributions of several vibrational levels can be generated by shaping several probes spaced by the vibrational lines of a known substance. This produces an anti-Stokes signal that is tailored to a specific chemical species and is significantly larger than the linear sum of each individual vibrational contribution [20]:

$$P_r^{(3)}(\omega) = C_r \int_0^{\infty} d\Omega \frac{1}{\Omega - \Omega_R - i\Gamma_R} E(\omega - \Omega) A(\Omega) \tag{9.2}$$

$P_r(\omega)$ is the nonlinearly induced vibrationally resonant polarization, in a CARS process, from a single vibrational level, using a broadband pulse [10]. This polarization generates the measured electric fields at the anti-Stokes frequencies. $\Omega_R$ and $\Gamma_R$ are the Raman level frequency and line width, respectively, and $A(\Omega)$ is the probability amplitude to populate a vibrational level with energy $\Omega$ via Raman excitation (Equation 9.1). $C$ is a constant which includes the summation over the dipole moments. $E(\omega - \Omega)$ in equation is the probing field:

$$P_{nr}^{(3)}(\omega) = C_{nr} \int_0^{\infty} d\Omega \frac{1}{\Omega} E(\omega - \Omega) A(\Omega) \tag{9.3}$$

$P_{nr}(\omega)$ is the nonlinearly induced non-resonant polarization in a CARS four-wave-mixing process with a broadband pulse. The term $1/\Omega$ is a correction term accounting for the not fully instantaneous nature of the non-resonant process [10].

Although intuitively depicted in the frequency domain, the different single-pulse CARS approaches can be also simply portrayed in the time domain (Figure 9.1c and d, right column): In the selective excitation schemes, the tailored pulse shapes are manifested in the time domain by a train of two or more impulses, which maximize the excitation for the vibrational frequency with the period equal to the pulse separation, in a similar manner to pushing a resonant swing (Figure 9.1c). In the narrow spectral probing schemes, the narrow spectral phase-, amplitude-, or polarization-shaped probe is manifested in the time domain as a temporally extended probing pulse at the shaped wavelength. The overall outcome of such shaping in the time domain is an impulsive femtosecond excitation of coherent molecular motion by a femtosecond pulse, followed by narrowband probing with a much longer pulse (typically hundreds of femtoseconds to picoseconds). The resulting vibrational features in the CARS spectrum are the coherent interference of the CARS field induced by the time-extended probe and the non-resonant-dominated four-wave mixing field induced by the unshaped part of the excitation pulse. For a single shaped probe, the optimal probe pulse duration and delay are set by shaping a "matched-filter" probing with spectral width equal to the vibrational line width [21,22].

It is important to note that pulse shaping for femtosecond CARS does not necessarily require a programmable dynamic pulse shaper, which is typically a relatively complex and costly experimental setup, which requires careful calibration [5]. Selective excitation (Figure 9.1c) can be attained by spectral focusing of two chirped femtosecond pulses [22–24] or double-pulse excitation [25,26], albeit these schemes rely on a multi-beam experimental apparatus in contrast to the single beam used in the pulse-shaper-based techniques. Narrow probing (Figure 9.1d) can be attained by shaping a narrowband probe using simple notch filters [18], Fabry–Perot interferometers and even a thin wire in the spectral plane of a pulse compressor [19], a common apparatus in amplified femtosecond sources.

This chapter is organized as follows. The two major approaches, to obtain multiplexed measurements via probe control or through selective excitation, are discussed separately. In both approaches, the progression from conventional multi-beam experiments is briefly discussed with the progression to single-pulse experiments that employ coherent control. The major discussion points of the multiplexed techniques include methods for removing the non-resonant contribution from the CARS signal, alternative probe shaping techniques that are not based on the standard 4-f shaper design, and applications of single-pulse, multiplexed CARS. The major discussion points of the selective excitation techniques include alternative shaping functions for greater mode selectivity, alternative excitation sources for increasing the detectable Raman bandwidth, heterodyne amplification for enhanced sensitivity, two-color configurations with independent probe shaping for multi-octave frequency range, and fast pulse shaping techniques for rapid detection of selectively excited Raman levels.

## 9.2 Multiplexed Measurements with Probe Control

### 9.2.1 Multi-Beam Multiplexed CARS

Using a *combination* of both narrowband and broadband pulses, one can achieve multiplexed CARS measurements [25,27,28]. This is conventionally implemented with a broadband Stokes, typically through supercontinuum generation, and a narrowband pump, which also serves as a probe. In this configuration (Figure 9.1b), the pump and Stokes beams simultaneously drive molecular vibrations of multiple levels ($\Omega_R$). Following vibrational excitation, the levels are interrogated by a narrow probe, which inherently dictates the spectral resolution. The resulting CARS spectrum, accordingly, contains features from all the excited Raman modes at spectral locations corresponding to energy shifts from the probe frequency.

Multiplexed CARS measurements are also possible using broadband, femtosecond pulses *exclusively* by employing pulse-shaping techniques. In a multi-beam experiment, Oron et al. used transform-limited pump and Stokes beams to drive molecular vibrations and probed with a shaped beam to obtain spectral resolution one order

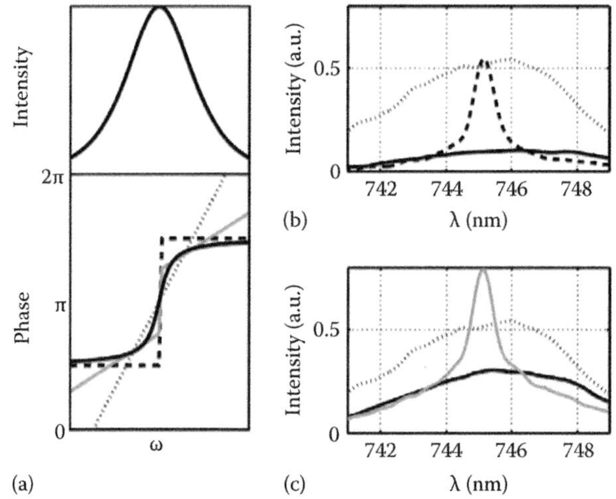


**FIGURE 9.3** (a) Schematic drawing of the spectral intensity and various phase functions; optimal phase function (solid line), probe delay (dotted line), π-phase step (dashed line), and π/2 phase step accompanied by some delay (solid gray line). (b) CARS spectrum from a $Ba(NO_3)_2$ sample at zero probe delay for a transform-limited probe (solid line), probe pulse with a π-phase step at 808 nm (dashed line), and transform-limited probe at 300 fs delay (dotted line). (c) CARS spectrum from a $Ba(NO_3)_2$ sample at 100 fs probe delay for a transform-limited probe (solid line), probe pulse with a π/2 phase step at 808 nm (solid gray line), and transform-limited probe at 300 fs delay (dotted line). (From Oron, D. et al., *Phys. Rev. Lett.*, 88, 063004, 2002.)

of magnitude better than the probe bandwidth [29]. Specifically, it was shown that a probe shaped with a phase step can be used to coherently sum off-resonant quantum paths driving molecular vibrations above and below the resonant frequency (Figure 9.3). While still applying multiple laser sources, this work was the first to use pulse shaping and phase control as a tool for spectroscopy, and in particular as a mean to enhance spectral resolution.

The concept of shaping the probe beam with a phase step is based on the inherent phase inversion at the resonance of driven harmonic systems. As seen in Equation 9.2, while the numerator of the resonant polarization is real and positive, the denominator term $(\omega_R - \Omega) + i\Gamma$ inverts the phase over a range of Γ about the resonant frequency ($\omega_R$). Such systems are driven in phase with the driving force below the resonant frequency, in quadrature at the resonance, and inversely above it. Consequently, when probed with a transform-limited probe, destructive interference is induced between pathways leading to driving fields above and below the resonant frequency. By applying probe phase function of $\Phi = \arctan[(\omega - \omega_R)/\Gamma]$ (Figure 9.3), one can achieve an enhancement in the CARS signal at a given frequency ω in the presence of a Raman level due to constructive interference of off-resonant terms [29]. While the applied phase inverts the relative sign of the frequencies much above and below $\omega - \omega_R$, nearly linear phase is added to the frequencies in the range of Γ about $\omega - \omega_R$. The linear phase component introduces a small delay of the probe pulse with respect to the pump and Stokes, which has been shown useful to reduce non-resonant contribution [21,29–32]. More recently, the Offerhaus group has developed a similar multi-beam approach using both negative and positive phase steps to more accurately retrieve spectral line widths. They report a maximum resonant signal when the phase step is on the order of the width of the vibrational resonance [33,34].

### 9.2.2 Single-Pulse Multiplexed CARS

In a seminal experiment by Oron et al., it was demonstrated that a *single* pulse can serve to provide the necessary pump, Stokes, and probe photons to obtain multiplexed CARS measurements [8]. In this configuration, a narrow spectral band, encoded with a π-phase shift (π-gate), serves as an effective probe for multiplexed CARS (Figure 9.4). This technique takes advantage of the coherent relationship between the resonant and non-resonant signal and utilizes the strong resonant background as a local oscillator in a homodyne amplification scheme. Since the measured CARS signal is the coherent sum of the resonant and non-resonant signal, amplification is maximized when the phase difference is 0 or π. Encoding a narrow π-gate in the excitation pulse results in a ±π/2 phase shift between the resonant signal and the non-resonant background, which are initially in quadrature. However, similar to the previously discussed multi-beam experiment that shaped a phase step, there is an enhancement of the resonant signal due to the off-resonant contributions as a result of the inherent phase shift of the driven oscillator. Specifically, enhancement occurs at a given frequency $\omega$ in the presence of a resonant signal from a level ($\Omega_R$), generated by a narrowband probe centered at $\omega - \Omega_R$. Since the π-gate essentially affects the resonant signal as two π-steps with inverse signs, the resulting resonant spectrum has two features with opposite phase (Figure 9.5). The resulting measured CARS spectrum, that is, interference of the resonant and non-resonant signals, has a peak-dip feature due to the edges of the π-gate. Since the π-gate

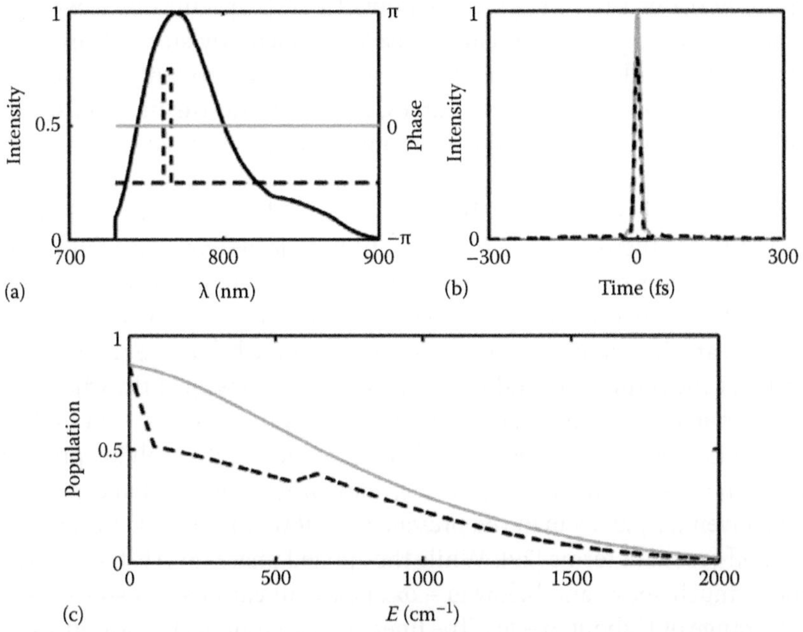


**FIGURE 9.4** Schematic drawing of single-pulse multiplexed CARS; (a) the excitation pulse spectral intensity (solid black line) and spectral phase of a transform-limited pulse (gray line) and a 10-pixel π-phase gate pulse (dashed line). (b) Temporal intensity of a transform-limited pulse (gray line) and a π-phase-gate-shaped pulse (dashed line). (c) Calculated population amplitude $A(\Omega)$ for the transform-limited pulse (gray line) and the shaped pulse (dashed line). (From Dudovich, N. et al., *J. Chem. Phys.*, 118, 9208, 2003. With permission.)

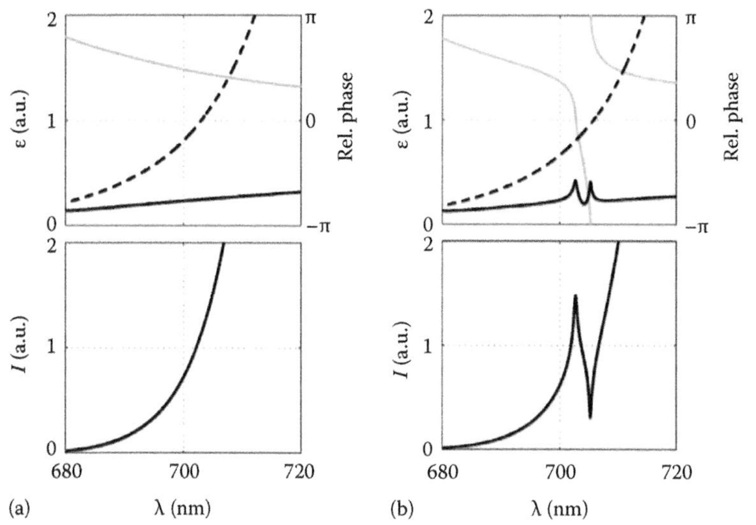


**FIGURE 9.5** Comparison of (a) transform-limited and (b) phase-gate-shaped pulses in a single-pulse CARS scheme. The top images show the calculated CARS electric field as a function of frequency for the resonant signal (solid line) and the non-resonant background (dashed line) for $Ba(NO_3)_2$. Also shown is the relative phase between the two (gray line). The bottom images show the calculated resulting CARS spectrum. (From Dudovich, N. et al., *J. Chem. Phys.*, 118, 9208, 2003. With permission.)

simultaneously probes multiple excited vibrational levels, the resulting spectrum is multiplex with a peak-dip feature corresponding to the various Raman modes.

Although this technique is complimentary to the multi-beam approach (Figure 9.3) where the probe beam is spectrally encoded with a *phase step*, the single-pulse configuration (Figure 9.4) alternatively employs the use of a narrow *phase gate* (π-gate). The benefit of using a π-gate is the limited effect on the excitation probability (Figure 9.4c). Since the same pulse provides the pump, Stokes, and probe photons, it is essential to limit the pulse deviance from the transform limit in order to preserve maximal excitation probability $A(\Omega)$. This method has been shown to yield spectral resolution two orders of magnitude better than the pulse bandwidth [8]. As in the case of the multi-beam phase-step configuration, the resolution of this technique is dependent on the slope of the π-gate rather than the spectral width. Furthermore, this technique is particularly attractive because it relaxes the need to suppress the non-resonant background. By strategically creating resonant features with similar phase to the large non-resonant background, the non-resonant signal can consequently be used to amplify the resonant CARS signal.

Additional methods, based on narrow spectral probing, have been developed to simplify measurements and improve sensitivity. Due to the coherent nature of CARS signals, it is possible to use more advanced probing methods that interact with photons generated from multiple levels. This technique alleviates the need for full spectral acquisition to obtain multiplexed spectra by coherently summing the contribution from multiple levels to a single measured frequency. This method, coined as "all optical processing," has been demonstrated using multiple probes in a configuration where $\omega_{AS} = \omega_{p1} + \Omega_1 = \omega_{p2} + \Omega_2 = \cdots \omega_{pN} + \Omega_N$ where the probes are centered around $\omega_{pn}$ for each given level $\Omega_n$ of a target chemical species [20]. Under this condition, the enhanced signal at frequency $\omega_{AS}$ is due to the coherent contribution from multiple levels (Figure 9.6a and b).

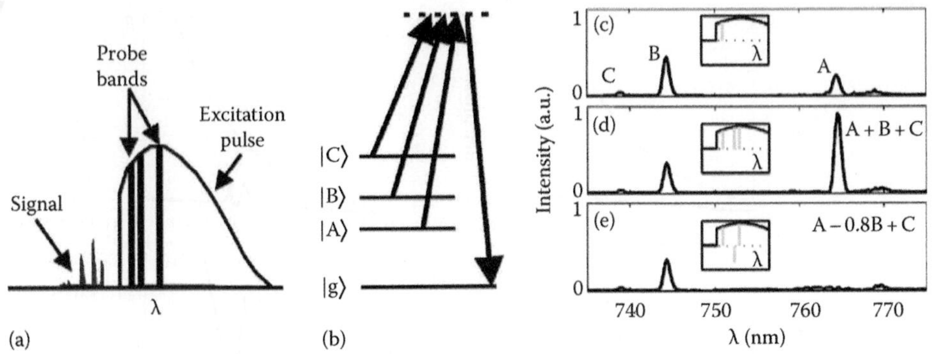


**FIGURE 9.6** Graphic description of a multiple-probe CARS process. (a) Schematic description of the excitation pulse, blocked at its high-frequency (short-wavelength) end, the polarization-shifted spectral bands serving as probe in the CARS process, and the resulting signal. (b) Schematic description of the interference in the CARS spectrum when several Raman bands are excited and probed by correspondingly chosen narrow spectrum probe bands. (c–e) CARS spectra from 1,2-dichloroethane. The insets denote the spectral masks used, where the excitation spectrum is shown in black and the polarization-shifted probe bands are shown in gray. (c) CARS spectrum using a single probe band (at 782 nm), showing three peaks corresponding to the Raman levels at 298 cm$^{-1}$ (A), 652 cm$^{-1}$ (B), and 750 cm$^{-1}$ (C). (d) CARS spectrum using a pulse with three probe bands (at 782, 804, and 811 nm) designed to induce constructive interference between the contributions of the three levels to the peak at 764 nm. Note that the peaks at 739 and 744 nm contain only a contribution from the probe band at 782 nm and are therefore almost unchanged relative to (c). (e) CARS spectrum using a pulse with three probe bands designed to induce destructive interference at 764 nm between the contributions of the three levels. (Adapted from Oron, D. et al., *Phys. Rev. A*, 70, 023415, 2004. With permission.)

The signal is measured with enhanced sensitivity, compared to a single probe, as the total signal is greater than the linear sum of the individual contributions. This technique is particularly useful when interrogating compounds with overlapping lines. As seen in Figure 9.6c through e, a configuration can be implemented where the CARS signal from a particular level is diminished by destructively interfering with the contribution from different level. This essentially negates the contribution of a specific Raman mode for a particular compound, while allowing for the detection of another with similar level structure. This method can also prove useful in samples with unknown spectra where a cost function can be satisfied to maximize the signal at a given spectral range while applying numerous probes at various spectral locations [20].

### 9.2.3 Techniques for Separating and Eliminating the Non-Resonant CARS Signal

Although the necessity of removing the non-resonant background is greatly relaxed in homodyne amplification schemes, it is still beneficial for faithful reproduction of Raman spectra. The elimination can be achieved through either experimental technique or computational post-processing. In the former scenario, it has been shown that the non-resonant contribution can be almost entirely eliminated using both phase- and polarization-shaping techniques [9]. By shaping a narrow probe in the pulse bandwidth, at an orthogonal polarization ($P_y$), the ultrafast pulse is essentially broken into two distinct pulses separable by polarization. The resulting resonant spectrum, probed by

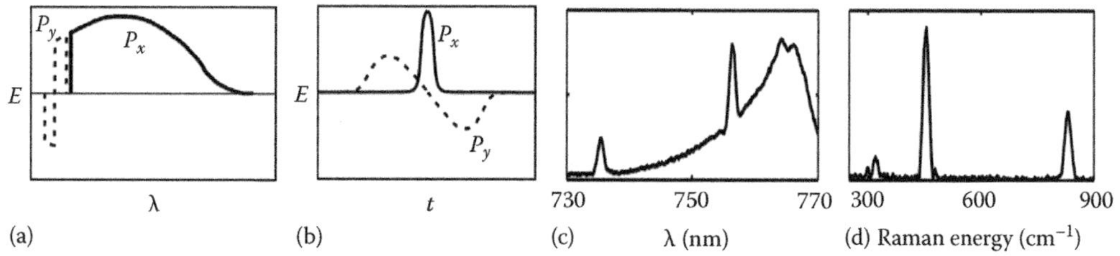


FIGURE 9.7 Schematic drawings of a single-pulse CARS scheme, with phase and polarization shaping; (a) spectral and (b) temporal electric field amplitudes in both the $x$ polarization (solid line) and the $y$ polarization (dashed line) for phase- and polarization-shaped excitation pulses. A shift of the spectral phase by $\pi$ (in (a)) is equivalent to a sign inversion of the electric field amplitude in a band within the $y$-polarized component. For convenience, the temporal field amplitude (b) has been normalized. In practice, the peak field amplitude of the $x$-polarized component is an order of magnitude higher than that of the $y$ component. (c) Normalized measured CARS spectrum and (d) the extracted Raman spectrum after background subtraction from p-xylene (resonant at 313, 459, and 830 cm$^{-1}$). (Adapted from Oron, D. et al., *Phys. Rev. Lett.*, 90, 213902, 2003. With permission.)

the orthogonal polarization, can then be separated from the non-resonant signal at $P_y$. Further suppression of the non-resonant signal can be achieved with the addition of a $\pi$-phase step at the center of the probe spectrum. The phase step splits the probe into two spectrally longer pulses with opposite phase (Figure 9.7a and b), crossing zero intensity at zero delay, where the excitation intensity peaks. Due to the broad spectral response, the non-resonant background from the two probe pulses destructively interfere with a reduction of multiple orders of magnitude, allowing for straightforward detection of the Raman spectrum (Figure 9.7c and d).

Using similar experimental approach, Lim et al. developed a computational technique to extract pure Raman spectra by separating the resonant signal into the corresponding real and imaginary components ($I_{Raman} \alpha \, \mathrm{Im}[I_{CARS}]$). This is achieved by measuring the CARS signal at polarizations $\pm\pi/2$ from the pump/Stokes beams (also orthogonal to the probe) [35]. By controlling the phase of the probe over few realizations, the real and imaginary parts of the resonant signal can be numerically calculated, and hence, a more accurately represented Raman spectrum can be resolved. This technique has been demonstrated in various samples, and the computed spectra were found to be virtually background free [36]. In an alternative approach, the non-resonant background can be eliminated using a technique referred to as "Fourier transform spectral interferometry" (FTSI-CARS). Unlike the previously described method, FTSI-CARS can be implemented with phase only shaping and is applicable to single-pulse, multiplexed measurements. This technique is based on the causality of CARS process, that is, there is no signal before the laser excitation. In brief, the imaginary and real parts of the resonant signal are separated by zeroing the values at $t < 0$, attained by the Fourier transform of a normalized CARS spectrum [37]. Transforming this causal timetrace back to the frequency domain yields the imaginary part of the spectrum. Contingent on this calculation, it was found that the sum of the amplitude ($|I_{res}|$) and the imaginary part ($\mathrm{Im}[R_{esonant}]$) of the resonant signal accurately portrays the Raman vibrational spectrum [12]. It has also been demonstrated that the signal sensitivity can be further improved by 2 orders of magnitude by incorporating amplitude shaping after reduction of the laser repetition rate and optimization of the power, bandwidth, and spectral phase [12].

### 9.2.4 Alternative Shaping Techniques

One of the drawbacks of commonly employed single-pulse, multiplexed CARS techniques is the need for a pulse-shaping apparatus. Pulse shaping is typically achieved with the use of an SLM in a 4-f configuration [5]. The SLM is costly and requires precise alignment and calibration of wavelength-dependent induced phase. Recently, there have been advances in system configurations that alleviate the need for a pulse shaper entirely. In one such technique, the Milner group demonstrated how multiplexed spectra can be acquired by taking the autocorrelation of the detected CARS signal [38]. As in the previously described configurations, the pulse is shaped, however, randomly in phase and/or amplitude. Since the shaped pulse also acts as a probe, with numerous features, the CARS signal is effectively shaped according to similar features at spectral regions corresponding to the multiple Raman energy shifts. By taking an autocorrelation of the CARS signal, the Raman spectrum, which is convolved with the randomly shaped probe, is revealed with a resolution on the order of the noise correlation length. The only requirement is that the applied phase/amplitude function applied to the pulse is constant over the time required for an autocorrelation. The shaping can be achieved by the use of a random diffuser in the spectral plane or simply passing the pulse through a multimode fiber. In another study, Katz et al. also alleviated the use for a pulse shaper by demonstrating multiplexed measurements employing a narrow notch filter as a spectral amplitude and phase shaping device as shown in Figure 9.8 [18,39]. The spectral shaping

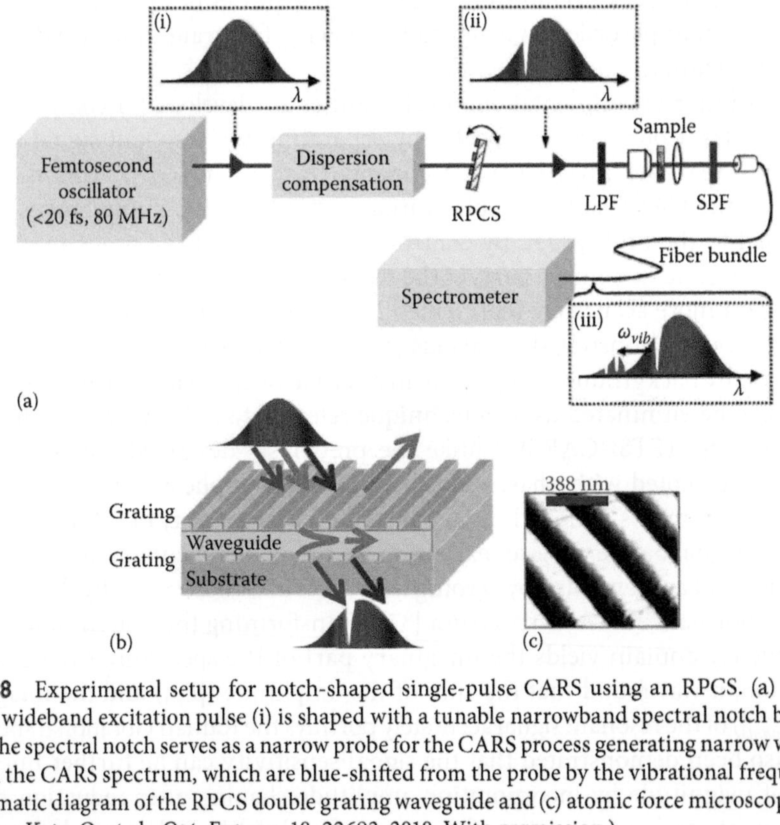


**FIGURE 9.8** Experimental setup for notch-shaped single-pulse CARS using an RPCS. (a) The optical setup: The wideband excitation pulse (i) is shaped with a tunable narrowband spectral notch by the RPCS filter (ii). The spectral notch serves as a narrow probe for the CARS process generating narrow well-defined features in the CARS spectrum, which are blue-shifted from the probe by the vibrational frequencies (iii). (b) A schematic diagram of the RPCS double grating waveguide and (c) atomic force microscopy measurement. (From Katz, O. et al., *Opt. Express*, 18, 22693, 2010. With permission.)

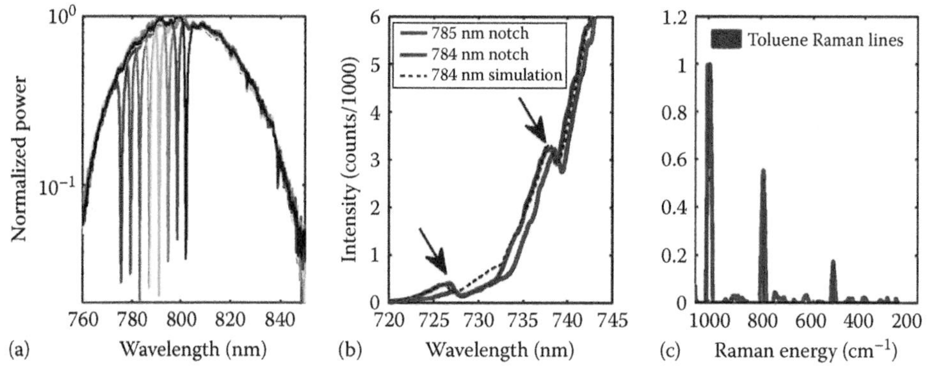


**FIGURE 9.9** Experimental results: (a) several RPCS notch-shaped excitation spectra. The notch location can be continuously tuned by the RPCS angle relative to the excitation beam. (b) Single-shot measured CARS spectra from toluene at two slightly shifted notch locations with peak-and-dip interference features corresponding to the 787 and 1005 $cm^{-1}$ vibrational lines (marked by arrows), which are in good agreement with numerical simulations. (c) Resolved vibrational spectrum of toluene retrieved from (b). (From Katz, O. et al., *Opt. Express*, 18, 22693, 2010. With permission.)

is achieved using a resonant photonic crystal slab (RPCS) filter, which is a grating waveguide structure comprised of a thin waveguide layer with an etched sub-wavelength grating. While most of the spectrum passes through the filter, a narrow band is coupled into the waveguide and destructively interferes with the pulse. This results in a narrow (<1.5 nm) notch at a central frequency, which is angularly dependent on the incident angle of the excitation beam to the filter (Figure 9.9a). Complimentary measurements can be taken corresponding to slightly different notch locations (Figure 9.9b), and the CARS spectrum can be easily calculated taking into account the non-resonant background and excitation probability (Figure 9.9c). This technique is particularly attractive because any conventional multi-photon microscope setup can be used for CARS measurements provided the pulse bandwidth is wide enough to excite the vibrational levels.

## 9.2.5 Applications of Single-Pulse Multiplexed CARS

Single-pulse, multiplexed CARS measurements have proven to be useful in a host of applications. Notably, the technique of shaping both phase and polarization has been to measure Raman transitions in gases. It has been shown that spectrally broad (sub-10 fs), single pulses can be shaped to probe lines greater than 2000 $cm^{-1}$. Particularly, the pressure dependence and percentage of $N_2$ in mixtures has been characterized [40]. In a similar study, the $CO_2$ was interrogated with visible Raman peaks at 1285 and 1388 $cm^{-1}$, corresponding to Fermi dyads [41]. A new realization of standoff chemical detection has also been born from multiplexed CARS using single pulses. With a phase-only shaped spectrum, experiments have been conducted to collect CARS signals at standoff distances greater than 10 m. These experiments demonstrated the ability to collect spectra at a distance from solids, liquids, and gases [15,16,42]. In a standoff, experiment, Katz et al. demonstrated that closely separated Raman lines could be detected at a distance by using a matched frequency filter corresponding to the shape caused by the interference of the phase gate (Figure 9.10) [15]. In a similar study, Natan et al. incorporated the concepts from standoff detection and shaper-free

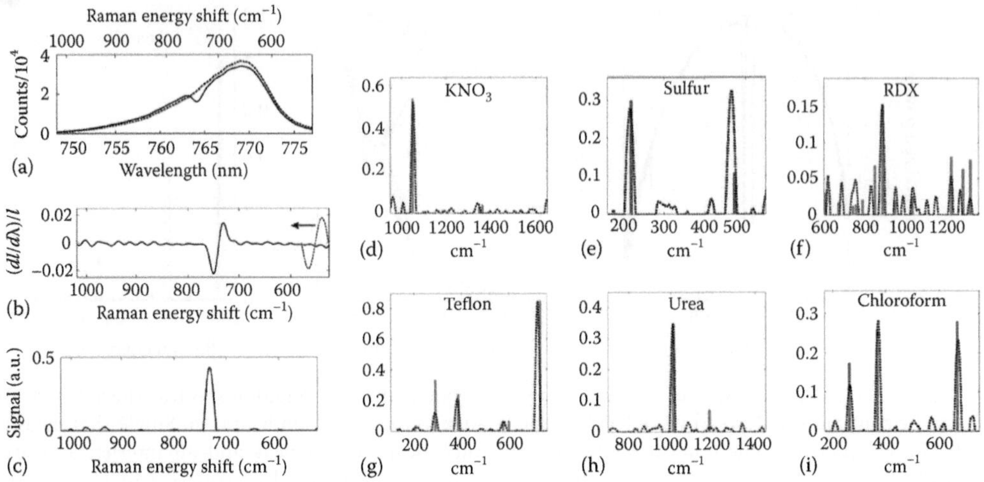


**FIGURE 9.10** Extracted vibrational spectra from standoff CARS measurements: (a) measured CARS signals from bulk polytetrafluoroethylene (Teflon) at a standoff distance of 5 m, from a transform-limited pulse (dashed black) and from a phase-gate-shaped pulse (solid blue). The distinctive peak-dip interference feature of a resonant vibrational level is apparent at 764 nm. (b) Normalized intensity derivative difference of the measured CARS spectra (blue). A convolution of this signal with a predetermined matched-filter function (dashed red) extracts the (c) vibrational levels spectrum. Resolved femtosecond CARS vibrational spectra of several scattering samples (dashed blue) at standoff distances of (d–f) 12 m and (g–i) 5 m with corresponding Raman lines in gray. (From Katz, O. et al., *Appl. Phys. Lett.*, 92, 171116, 2008. With permission.)

configurations; they demonstrated detection of explosive materials from a distance of 50 m using an RPCS filter as a shaping device [43].

Single-pulse multiplexed techniques have also been applied to microscopy. Specifically, Raman spectra were obtained across a 2-D plane, and images were then generated based on the contrast of a specific Raman levels. This has been shown in generating contrast between solids, including PDMS/amide [36] and polystyrene/polyethylene terephthalate [12]. More recently, this technique has been demonstrated using low average powers, less than 30 mW, in liquids and thinly sliced potato sections, where the strong skeletal mode of starch granules was exploited (Figure 9.11) [18]. The ability to successfully obtain images from samples using low intensities is a first step in moving toward single-pulse CARS techniques for biologically relevant applications.

## 9.3 Time-Domain and Selective Excitation CARS

### 9.3.1 Multi-Beam Selective Excitation

The earliest applications of sub-100 fs pulses for time-domain measurements of CARS date back to the late 1980s with Leonhardt's novel experiment measuring molecular quantum beats in liquids [44]. This technique exploits the fact that when molecules are excited to a given Raman level, there is molecular vibration at the frequency corresponding to that particular energy level. The main principle of time-domain approaches is that if one were able to identify at which frequency(s) a molecule is vibrating at, then the complete Raman spectrum can be identified via Fourier transform. One such

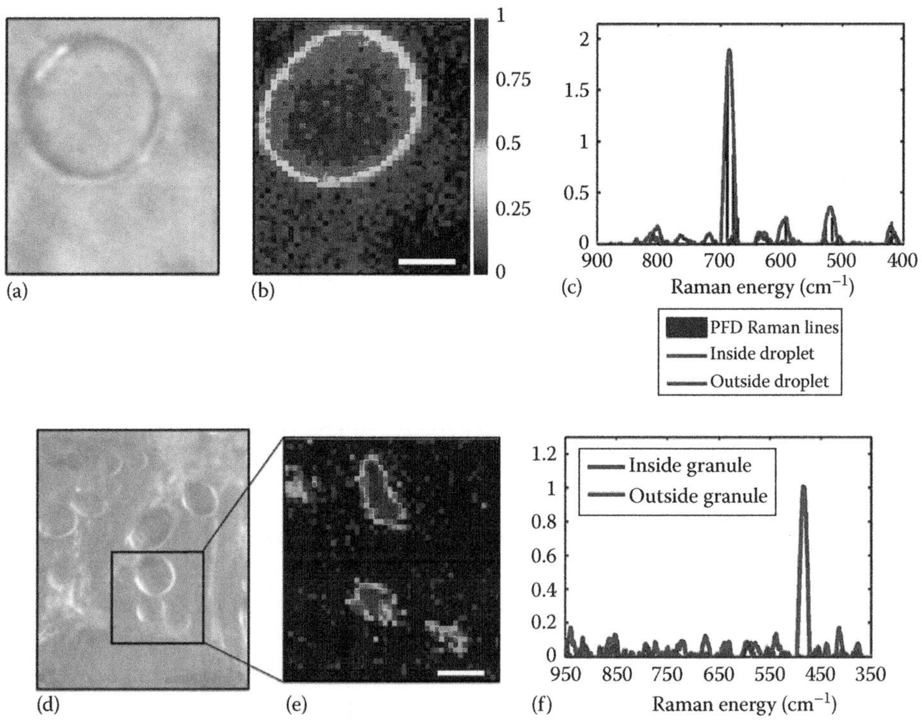


**FIGURE 9.11** Single-beam vibrational imaging using RPCS notch-shaped single-pulse CARS: In (a–c), the sample is a mixture of water and perfluorodecalin (671, 679, 689 cm$^{-1}$), and in (d–f), the sample is a potato slice containing starch granules (474 cm$^{-1}$): (a,d) transmission images, (b,e) vibrational contrast image (scale bar 10 μm), and (c,f) spatially resolved vibrational spectra inside the drop and starch granule. (From Katz, O. et al., *Opt. Express*, 18, 22693, 2010. With permission.)

technique for identifying the vibrational frequency components (Raman levels) is by using a series of pulses that provide pump/Stokes photons to excite molecular vibration, and a time-delayed probe pulse. Since a molecule vibrates in time, the probe experiences a time-dependent energy shift, which is due to the oscillating molecular polarization. By taking a series of measurements corresponding to different probe delays from the excitation, the vibrational beat frequencies can be resolved, which are a Fourier-related pair to the time-domain measurement. This technique, coined as "resonant femtosecond CARS," was originally carried out in a simplistic three-beam configuration and successfully observed vibrational modes of gases [45–47]. However, as the technique evolved, resonant femtosecond CARS benefited from the addition of pulse-shaping techniques and moved toward applications measuring solid materials. Zeidler et al. designed an experiment where a shaped Stokes pulse was used in combination with a delayed, transform-limited probe [48]. Using an iterative feedback algorithm, specific phase- and amplitude-shaping functions were found to selectively enhance the population of specific Raman modes. Continuing with the goal of selective excitation of Raman modes, Oron et al. showed that phase-only encoding of the pump and Stokes beams can lead to selectivity in the excited Raman modes and eliminate the need to scan the probe delay and spectrally resolve the measured signal [30]. This technique was first demonstrated in a three-beam setup using 100 fs second pulses corresponding to a bandwidth

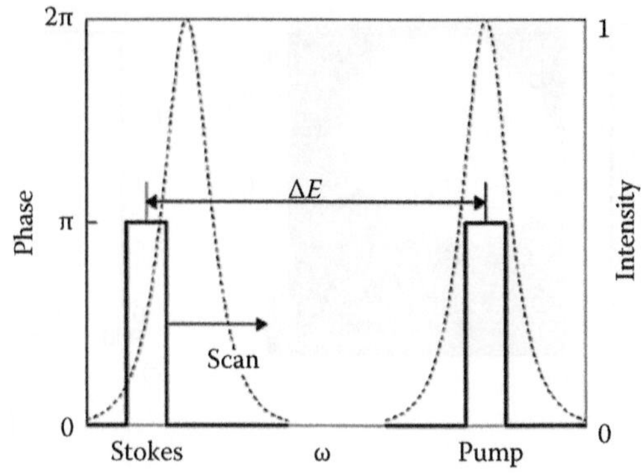


**FIGURE 9.12** Schematic of the pump and Stokes beams for a multi-beam selective excitation scheme. The Stokes and pump beams are encoded with a π-step separated by an energy $\Delta E$, which leads to an enhancement of the CARS signal when corresponding to a Raman energy level. (From Oron, D. et al., *Phys. Rev. A*, 65, 043408, 2002.)

of 120 $cm^{-1}$. Particularly, this work showed the advantage of encoding π-phase gates in the pump and Stokes beams with an energy difference ($\Delta E$) (Figure 9.12). In the case when $\Delta E$ corresponds to a vibrational level, an enhancement is observed in the CARS signal. This enhancement is observed because the non-resonant signal is diminished with the application of the π-gates and the vibrational mode is maximally excited when $\Delta E$ corresponds to the energy level, that is, the excitation photon pairs are in phase (see Equation 9.1). This can be simply thought of as carefully altering the pump and Stokes beams to produce a signal enhancement by constructively interfering with two narrow bands of frequencies with energy differences satisfying the population requirement of a specific energy level. Consequently, by scanning the spectral location of one of the gates, and hence $\Delta E$, one can interrogate a sample and identify the Raman modes. Since this technique selectively excites a specific Raman mode, a single detector can be used rather than a spectrometer. Although there is a trade-off between spectral resolution (narrowness of the gate) and total enhanced signal (width of the gate), this method can be used to gain spectral resolution two orders of magnitude better than the pulse bandwidth [30]. In this work, it was also shown that there was a benefit to applying a fixed probe delay to further reducing the non-resonant background.

## 9.3.2 Single-Pulse Selective Excitation

The concept of tailoring a *single* femtosecond pulse to selectively populate and then probe a specific Raman level has been studied intensively. In a pioneering experiment, Dudovich et al. first demonstrated how coherent control techniques can be applied to selectively excite specific vibrational levels and obtain full spectral information [7]. The underlying concept of selective excitation is the application of periodic spectral phase functions to the pulse, a concept that was demonstrated already a decade earlier on molecular crystals [49]. This method can be intuitively reasoned in both the frequency and time domain (Figure 9.13a and b). First, in the frequency domain, a specific vibrational level, $\Omega$, is populated by a combination of photons with frequencies such that the energy difference $h\omega_1 - h\omega_2 = \Omega$ with a relative phase $\phi(\omega) - \phi(\omega - \Omega)$. Therefore, to maximize the population amplitude of a particular vibrational level at a frequency $\Omega$: $A(\Omega)$,

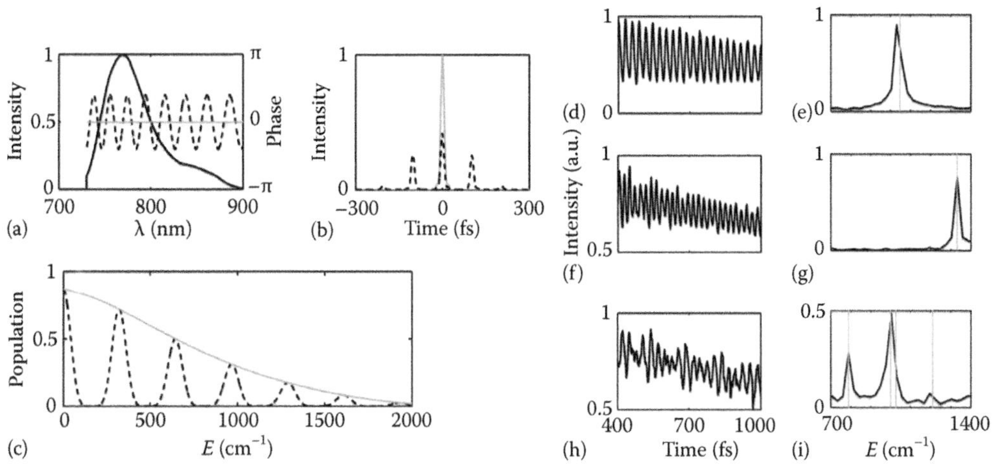


**FIGURE 9.13** Schematic drawing of a single-pulse selective excitation scheme showing the pulse in the (a) frequency domain (spectral intensity solid black) and (b) time domain with the corresponding (c) excitation probability $A(\Omega)$ for a transform-limited pulse (gray line) and a modulated phase-shaped pulse (dashed line). (d–i) Selective excitation of vibrational levels in various materials using a modulated spectral phase function. (d,f,h) The CARS signal intensity as a function of the time separation between the different pulses in the pulse train and (e,g,i) the extracted Raman spectra, derived by Fourier transformation of the intensity curves given for (d,e) $Ba(NO_3)_2$ (resonant at 1048 cm$^{-1}$), (f,g) diamond (resonant at 1333 cm$^{-1}$), and (h,i) toluene (resonant at 788, 1001, 1028, and 1210 cm$^{-1}$). (Adapted from Dudovich, N. et al., *J. Chem. Phys.*, 118, 9208, 2003. With permission.)

the electric fields at frequency pairs separated by the vibrational energy should be in phase to achieve highest constructive interference, that is, $\phi(\omega) = \phi(\omega - \Omega)$ as seen in Equation 9.1. While this condition is satisfied for all levels given a transform-limited pulse, there is selective population of $\Omega = N\Omega_m$ where $N$ is an integer and $\Omega_m$ is the spectral phase modulation period (Figure 9.13c). In simplicity, the periodic phase creates constructive interference between all the various spectral components, which drive a particular molecular vibration while destructively interfering with paths driving other frequencies. This is similar to the previously discussed multi-beam approach where narrow π-phase gates were encoded into the pump and Stokes beams, separated by the energy corresponding to the desired vibrational level [30].

In the time domain, one can intuitively reason the dynamics of selective excitation by thinking of the induced molecular vibration. As dictated by the Fourier transform, the corresponding effect of a periodic phase is temporal pulse splitting leading to a train of pulses equally separated by $\tau_m = 2\pi/\Omega_m$ (Figure 9.13b). When the period of the modulated phase pattern corresponds to an integer number of the vibrational energy, the temporal delay between the split pulses matches the period (or integer harmonic) of the molecule vibration. When this condition is satisfied, the pulse train coherently drives oscillations at resonant frequencies corresponding to vibrational periods $T = \tau_m/N$ (Figure 9.13c). This can be thought of as coherently driving a mass on a spring with the proper phase at the resonant frequency or periodically pushing a child on a swing such that the push is given at the same point in the motion. While the effect coherently adds energy to the resonant mode, the non-resonant signal has been shown to be reduced by 2 orders of magnitude due to the decrease in peak power as an effect of the temporal pulse splitting [7,10]. It has been shown that the addition of harmonic

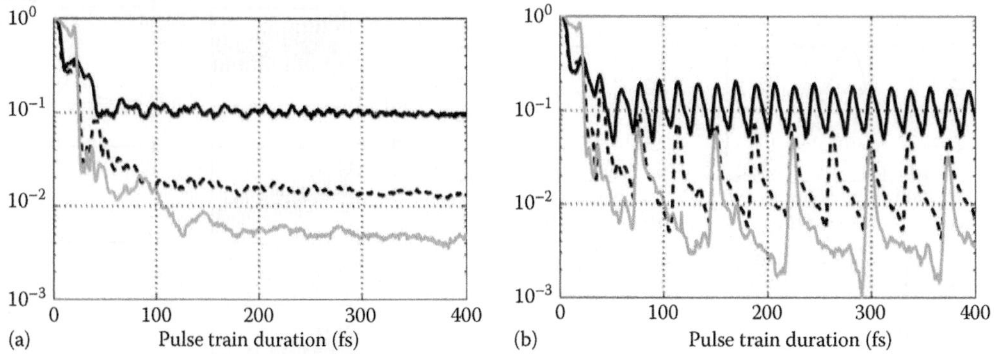


**FIGURE 9.14** Background suppression using periodic spectral phase functions with additional harmonic orders. The CARS intensity is shown as a function of the total pulse train duration for a sinusoidal phase function (solid line), a phase function containing two harmonics (dashed line), and four harmonics (gray line). The measured results are for (a) glass (non-resonant) and (b) $Ba(NO_3)_2$ (resonant at 1048 $cm^{-1}$). (From Dudovich, N. et al., *J. Chem. Phys.*, 118, 9208, 2003. With permission.)

content in the spectral domain creates a series satellite of pulses in the time domain, which can further reduce the non-resonant background while leaving the resonant term virtually unaffected (Figure 9.14a). When using a just single periodic frequency, one measures a predominantly resonant signal, which is greater than the non-resonant background by a factor of 4 [7]. Using a function with two or four harmonics results in a further attenuation of the non-resonant signal by a factor of ~80 and ~250, respectively (Figure 9.14b) [10].

As in the multi-beam approach, the technique of selectively exciting Raman levels using a single-pulse is particularly simple in terms of detection as a single detector can be used, and there is no need to spectrally resolve the CARS spectrum. Moreover, one can identify the Raman spectrum from a given sample by taking measurements of the total CARS signal from realizations corresponding to different pulse separations (Figure 9.13d,f, and h). From the time-domain CARS signal, the Raman spectrum can be simply calculated via Fourier transform (Figure 9.13e,g, and i), similar to conventional interferometric Fourier transform CARS (FT-CARS) techniques [26,50]. Traditionally, the Fourier transform is computed from realizations taken at pulse separations greater than the pulse width to limit the non-resonant signal contribution.

## 9.3.3 Advanced Selective Excitation Techniques

### 9.3.3.1 Alternative Shaping Functions for Enhanced Level Selectivity

Taking principles from FT-CARS and single-pulse experiments with periodic phase content, one can conduct full spectral analysis while suppressing or enhancing specific Raman lines. This concept was shown in a three-beam, femtosecond-pulse approach where the time delay between the first two (excitation) beams was scanned and was then subsequently probed [51]. In this configuration, it was shown that further mode selectivity can be achieved by adding periodic phase content in the excitation pulses to create satellite pulses. Specifically, this was done such that the periodic modulation matched the vibrational period. An enhancement was found for a particular Raman mode when the harmonic content and the delay between the two sets of pulses were

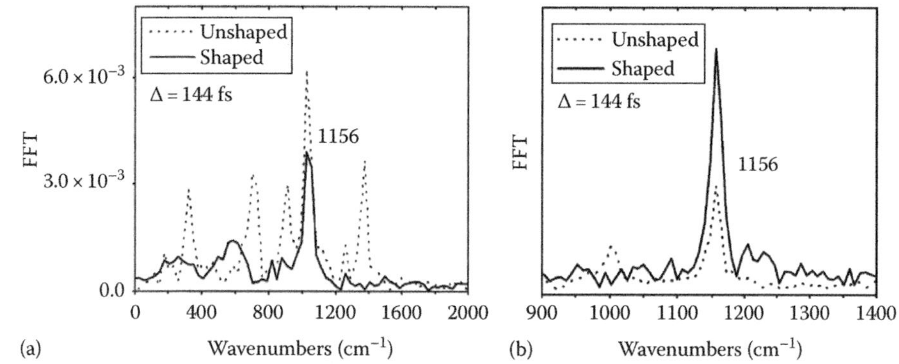


**FIGURE 9.15** FFT-spectrum comparing Raman intensities corresponding to Stokes and pump pulses separated by 144 fs (fifth Harmonic) to maximize the 1156 $cm^{-1}$ mode when shaped (dotted) and unshaped (line) to create satellite pulses in the (a) absence and (b) presence of an electronic level. (a) In the absence of an electronic level, a shaped pulse is more selective but less in total energy compared to an unshaped pulse with equal energy. (b) In the presence of an electronic level, an unshaped pulse is highly selective; however, there is an increase in the total energy when using a shaped pulse. (From Hauer, J. et al., *Chem. Phys. Lett.*, 421, 523, 2006. With permission.)

consistent with the vibrational period. Mode selectivity was attained throughout the scans of the delay between the two excitation pulses simply because even when the pulse separation matched the period of a specific level, the satellite harmonic pulses did not coherently contribute to the population. This technique has been shown to enhance the contribution from desired Raman levels while reducing the contribution from others as seen in the spectra computed by Fourier analysis (Figure 9.15a). The striking finding from this work is that the shaped pulses can lead to an increase in the energy and selectivity of a specific mode (compared to unshaped pulses) in the presence of a resonant electronic transition (Figure 9.15b).

A number of single-pulse, selective-excitation approaches have been developed based on the concept of applying cosine-based phase functions to alter the pathways leading to vibrational population. More recently, it has been demonstrated that selective excitation can be achieved using a variety of alternative periodic phase functions. One such variant to cosine modulation functions is the use of Pseudorandom Galois functions, which are strongly uncorrelated except at a specific frequency. A greater level of anti-correlation compared to cosine phase modulation among the off-resonant frequencies leads to higher selectivity of a particular desired frequency. This technique has been applied to various types of nonlinear optical applications, including two-photon excitation and stimulated Raman scattering [52]. It has also been shown that the use of periodic replica of random binary phases can be used to create constructive interference for a specific Raman mode [53]. Examples of binary Pseudorandom Galois functions applied to the excitation spectral phase are shown in Figure 9.16 where phase configurations can be tailored to populate specific levels.

#### 9.3.3.2 Alternative Excitation Sources for Increased Detectable Raman Bandwidth

From Figure 9.16, it is clear how the pulse bandwidth hampers the ability to excite higher-energy Raman modes (energy different between pump and Stokes photons). When using single-pulse techniques, the upper and lower limits of CARS detection are

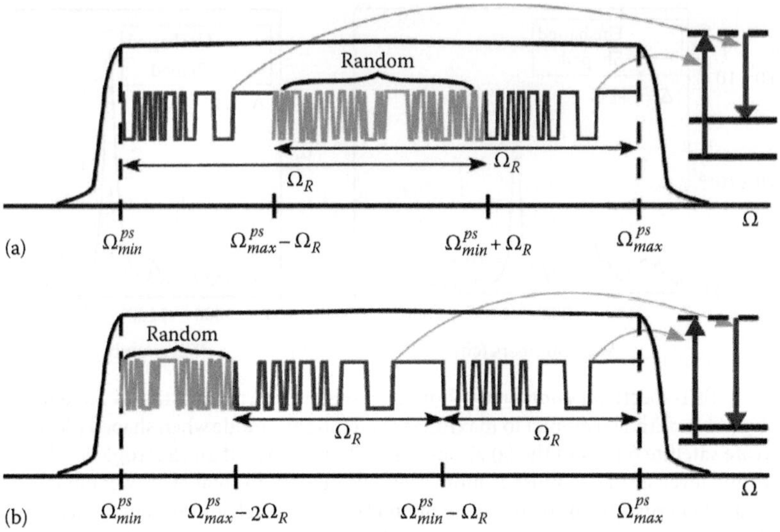


**FIGURE 9.16** Selective excitation of Raman-active vibrational modes via binary phase shaping: (a) binary phase mask design for $\Delta\Omega^{ps} < 2\Omega_R$; (b) binary phase mask design for $\Delta\Omega^{ps} \geq 2\Omega_R$. ($\Delta\Omega^{ps}$ is the bandwidth of the pump Stokes part of the laser spectrum and $\Omega_R$ is the transition Raman shift.) (Enhanced figure kindly provided by Marcos Dantus and colleagues, Wrzesinski, P.J. et al., *J. Raman Spectrosc.*, 42(3), 393, 2010. With permission.)

restricted by the pulse bandwidth and proximity to the excitation pulse that CARS signal can be detected, respectively. Although techniques based on selective excitation can be implemented using a single detector, it has been shown that the use of a spectrometer can further aid in spectral filtering to effectively collect CARS signal closer to the excitation pulse bandwidth and hence resolve lower Raman lines. This has been demonstrated by von Vacano et al. where the excitation spectrum was cut near the central frequency to allow for higher energy at the lower end of the spectrum and thus enhance the signal of the lower lines (i.e., lowest detectable frequencies are probed by the bluest part of the spectrum) [54]. Using this technique, Raman modes below 200 cm$^{-1}$ were readily resolved. Conversely, the upper limit of vibrational detection has been increased with the advent of ultra-broadband femtosecond sources. These octave spanning lasers have been employed in CARS measurements, allowing for detection of Raman lines higher than 4500 cm$^{-1}$, as demonstrated with enhanced sensitivity afforded by spectral focusing [13]. It should be noted that the mentioned techniques for resolving a greater Raman bandwidth are also applicable to multiplexed techniques.

#### 9.3.3.3 Heterodyne Amplification for Increased Sensitivity for Detection of Low Signals

One of the complications in conducting selective excitation experiments is that the CARS signals are typically very low. The low signal further increases measurement integration times and therefore limits the ability to use these techniques for imaging biological and other dynamic systems. One approach to enhance the inherently weak signals is to coherently use the excitation field as a local oscillator. Similar to multiplexed, phase-gated techniques, where CARS signals are amplified by the non-resonant background, it has been shown that part of the excitation field can effectively serve in a heterodyne amplification scheme. The Motzkus group has shown that allowing a narrow, attenuated portion of the excitation spectrum to interact with the CARS signal results in a functional local oscillator that can amplify the CARS signal over three orders of magnitude [55]. The resulting heterodyne CARS intensity ($S(\omega)$) is linearly dependent on concentration and the phase function of the local oscillator (Equation 9.4). By choosing an appropriate phase function for the local oscillator, the heterodyne CARS signal can be optimized, thus increasing the sensitivity limit. Using such optimized local oscillator fields, Müller et al. successfully demonstrated sensitivity in the attomole regime corresponding to $5 \times 10^6$ molecules in the focal volume [56]

$$S(\omega) = I_{CARS}(\omega) + 2\sqrt{I_{CARS}(\omega)}\sqrt{I_{LO}(\omega)} \times \cos[\Delta\varphi(\omega)] + I_{LO}(\omega) \quad (9.4)$$

$$S^{(HET)}(\omega) = 2\sqrt{I_{CARS}(\omega)}\sqrt{I_{LO}(\omega)} \times \cos[\Delta\varphi(\omega)]$$

$I_{CARS}(\omega) \, \alpha \, |E_{CARS}(\omega)|^2$ scales quadratically with the number $N$ of scattering molecules. Mixing $E_{CARS}(\omega)$ with the local oscillator field $E_{LO}(\omega)$ in a square-law detector yields the signal $S(\omega)$ where $I_{CARS}$ and $I_{LO}$ are the homodyne intensities of the CARS signal and the local oscillator, respectively. From this equation, $S^{(HET)}$ (the heterodyne interference term) scales linearly with $N$ and thus the concentration of Raman scatterers. Maximum amplification is consequently achieved when $\varphi_{LO}(\omega) = \varphi_{CARS}(\omega) + \Delta\varphi(\omega)$ [55].

#### 9.3.3.4 Two Color Single-Beam CARS with Independent Probe Shaping for Multi-Octave Frequency Resolution

A caveat in typical implementation of single-pulse CARS, via phase-only periodic modulations, is the limit of detection to one frequency octave. For the sake of simplicity, this complication arises due to the presence of multiple pulses in excitation train, which can excite a number of vibrational levels, namely, integer harmonics of all the possible frequencies generated by the multiple pulses, thus leading to ambiguity in which pulse serves at the probe. One solution to alleviate the single-octave restriction is to *spectrally* break a very broad pulse into distinct time-delayed pulses so that the probe is clearly separable from the pump and Stokes pulses in both the time and spectral frequency (Figure 9.17a). This can be done by applying a linear phase to a spectral band where the slope is directly related to the time delay as dictated by the Fourier relations. The result is two broadband pulses that are separated by a controlled time delay (Figure 9.17b). In this configuration, a transform-limited pulse provides the pump and Stokes photons to excite the molecular vibration while the second pulse acts as a probe. By scanning the delay between the pulses, the probe interacts with the molecule at various points during the oscillating polarization period and hence spectrally shifts the probe accordingly, that is, detected CARS signal. In contrast to conventional single-pulse CARS, the signal originating from the spectrally defined probe is exclusive detected. Specifically, the signal originating from probe photons in the first pulse is spectrally separated from the signal originating from probe photons in the second (spectrally different) pulse. The vibrational lines can then be resolved through analysis of the periodic, spectral oscillations of the detected CARS photons [57]. This approach can be further expanded upon in a technique that also incorporates selective excitation principles. For example, one can enhance the contribution from specific Raman line by applying a periodic

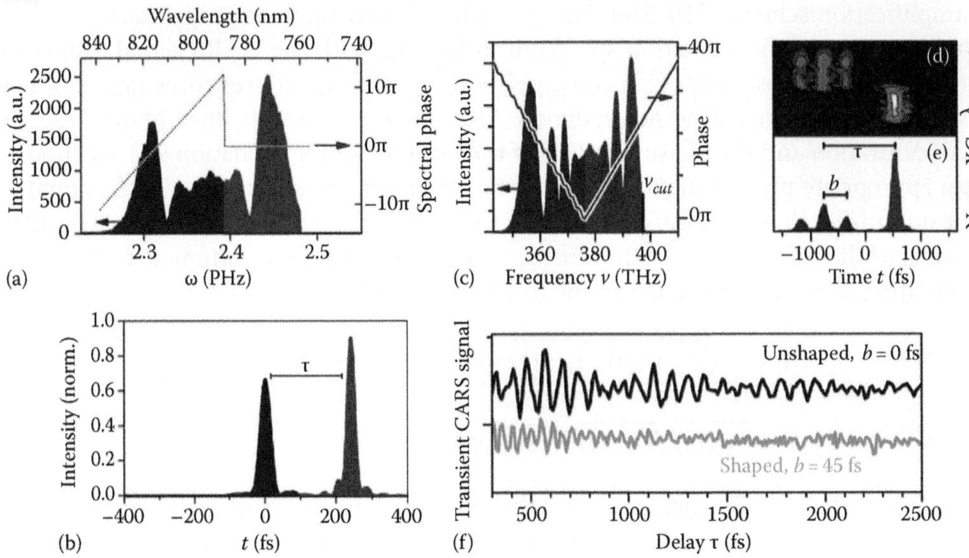


**FIGURE 9.17** Generation of two-color double pulses from compressed supercontinuum. (a) Experimental spectrum with spectral phase φ(ω) (dotted line). (b) Simulated double pulse with corresponding delay τ = 250 fs. (Adapted from von Vacano, B. and Motzkus, M., *Opt. Commun.*, 264, 488, 2006. With permission.) (c) Spectrum and spectral phase of the shaped excitation pulses, with the red part at lower frequencies acting as pump and Stokes, and the blue part as time-delayed probe. The spectral phase shown creates a temporal separation of the two spectra of 1300 fs, and an additional multi-pulse modulation on the pump/Stokes pulse with a temporal spacing of 420 fs. (d) A frequency-resolved cross correlation and (e) the integrated cross correlation measured *in situ* confirm the successful delivery of the desired pulse shape. (f) Measured CARS transients for the sample mixture ($CHCl_3$ and $CBrCl_3$) with unshaped black curve and shaped gray curve pulses show direct control over the molecular oscillations. (Adapted from von Vacano, B. and Motzkus, M., *J. Chem. Phys.*, 127, 144514, 2007. With permission.)

modulation to the spectral region of the initial pulse, which provides the pump and Stokes photons (Figure 9.17c through f). As in the previously described selective excitation method, a specific mode is enhanced when the vibrational period matches that of the desired vibrational mode. This technique gains the benefits of selective excitation while eliminating the restriction of single-octave detection since the CARS signal originates from an exclusively defined probe, which is particularly beneficial when observing low wave numbers [58]. The theoretical and experimental considerations of the two-color approach have been studied extensively, particularly the effects of choosing the optimal spectral region to break the pulse and the benefits of also incorporating polarization-based detection for enhanced sensitivity due to removal of the non-resonant contribution [14].

#### 9.3.3.5 Fast Pulse Shaping Techniques for Rapid Tunability of Excited Raman Levels

One of the ultimate goals of label-free techniques such as CARS is the application to biologically relevant problems. While a number of CARS studies have been performed on live specimens [2,59–62] and observed molecular dynamics [21], there is still a need for the development of tools to allow for faster scanning while using selective excitation techniques. It can be easily understood that rapid imaging can be achieved when interrogating a single specific Raman level; however, it is often desired to simultaneously probe multiple Raman levels or interrogate an unknown sample and collect a full Raman spectrum. Particularly, a major challenge in both conventional multi-beam and single-beam selective-excitation techniques is the ability to rapidly tune the specific Raman mode under interrogation. In conventional multi-beam approaches, the speed of a spectral scan is therefore limited by the desired resolution, that is, number of spectral acquisitions, and dictated by the time required to detune the pump and Stokes beams (seconds). In single-beam selective excitation approaches, based on pulse shaping, the speed is also limited by the resolution, however dictated by the ability to change the spectral mask, which can be on the order of milliseconds when using a conventional SLM or as low as nanoseconds with more advanced devices such as acousto-optical modulators.

The ability to rapidly shape pulses has been previously studied using SLM techniques [63–67] and applied in CARS schemes [63,68]. One of the simplest adaptations to a traditional SLM-based, 4-f shaper is the ability to change phase shapes by scanning the spectrally spread beam in an additional dimension on a 2-D SLM, transverse to the spectral axis (Figure 9.18 inset). Using a 2-D SLM, Frumker et al. have shown that rapid shape selection is possible by imaging a galvanometric mirror on the grating of a 4-f shaper, prior to the Fourier lens, which thus enables one to scan the location of the beam on the SLM and addresses a series of phase patterns (Figure 9.18). This technique allows for independent phase patterns to be addressed once fixed on the SLM and changed at rates limited by the galvanometer. Using this technique, Frumker demonstrated pulse shaping up to 100 kHz [66]. Such use of rapid phase scanning has demonstrated in a CARS and two-photon absorption schemes where the phase was alternated between shapes that maximize the constructive and destructive interference paths leading to

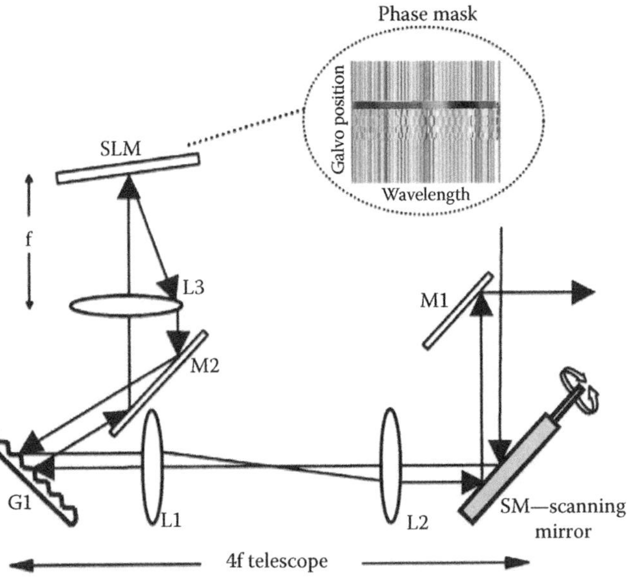


FIGURE 9.18 Schematic drawing of a 4-f pulse shaper with a 2-D SLM and corresponding phase pattern (inset) scanned by a galvanometric mirror (SM). The spectral content is dispersed in the horizontal dimension and scanned along the vertical dimension. (Adapted from Frumker, E. et al., *Opt. Lett.*, 29, 890, 2004. With permission.)

population of a vibrational [63,68] or electronic level [67], respectively. This technique readily allows for rapid acquisition with suppressed background using standard lock-in detection. Although this method has been used to rapidly observe a single Raman mode, full spectral information could be obtained by encoding the 2-D SLM with phases corresponding to a full range of temporally split pulses, and spectral information could be retrieved via real-time Fourier analysis with a spectrum analyzer.

Using a similar experimental configuration to Frumker et al., Levitt et al. demonstrated 80 kHz phase-only shaping for rapid selective-excitation measurements with multiplexed detection [68]. In this technique, a series of phase shapes are carefully chosen such that when scanned, the CARS signals from different chemicals are modulated at different frequencies (Figure 9.19a). This is achieved by first taking full time-domain scans of various pure chemicals and maximizing a cost function that identifies the necessary shapes (shapes corresponding to A–D) leading to the highest contrast between them upon rapid scanning, that is, lowest frequency correlation (Figure 9.19b). The various shapes are then updated on a 2-D SLM and scanned with a galvanometric mirror, resulting in the CARS signals modulated at different integer harmonics of the scan frequency, corresponding to the different chemical species (Figure 9.19c). The entire CARS signal is chemically discriminated using a dual-harmonic, lock-in amplifier and has been demonstrated with pixel dwell times of 500 μs in 3-D scans [68]. While this technique has proven capable to acquire multiplex measurements of two chemicals, it can be expanded to many more limited by the number of shapes that can be addressed on the SLM as the minimum number of shapes is two to the power of detectable chemicals.

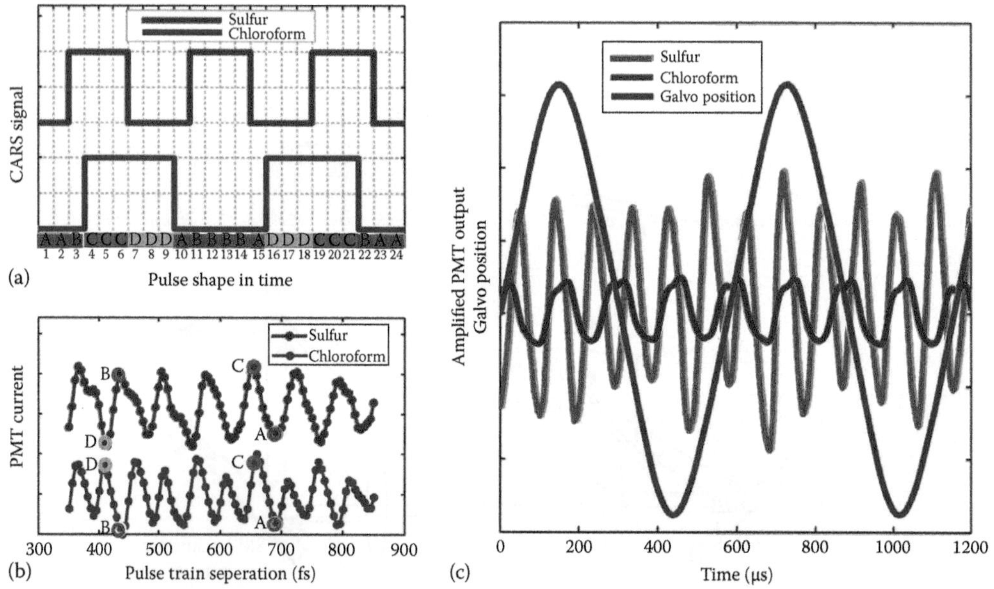


**FIGURE 9.19** Schematic diagram for rapid multiplexed CARS detection. (a) The desired CARS intensity for sulfur and chloroform, modulated by a series sequentially scanned alternated phase shapes comprised of (b) four independent shapes (A–D). (c) When rapidly scanned, the CARS signal from sulfur and chloroform is modulated at the specified harmonic frequencies of the galvanometric mirror and detected via lock-in detection.

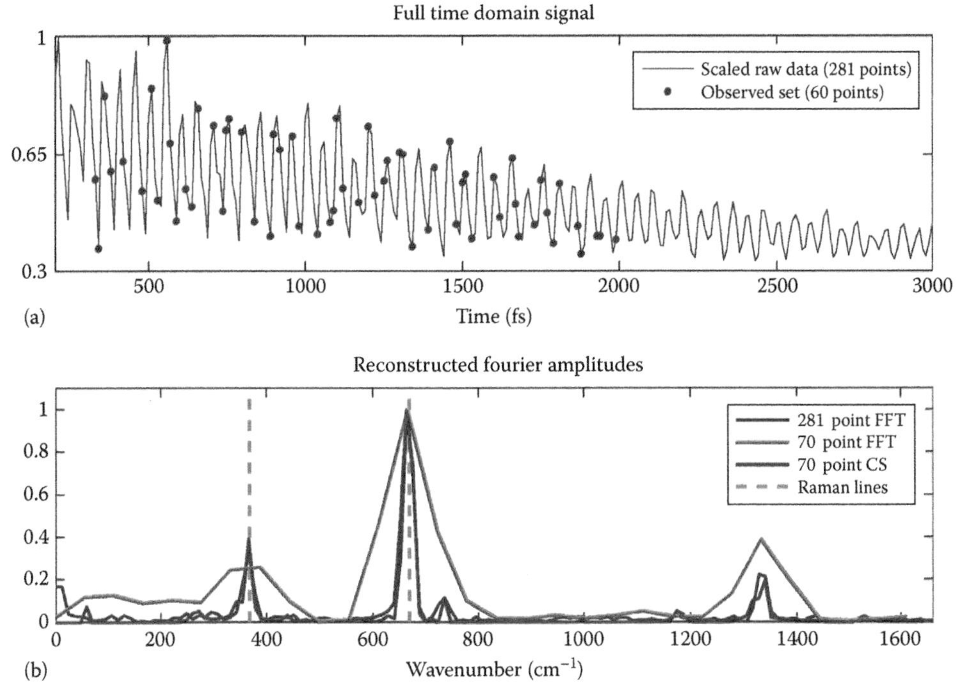


**FIGURE 9.20** Experimental results of compressive-sensing Fourier transform vibrational spectroscopy using under-sampled time-domain data. (a) Raw time-domain trace for chloroform, containing 281 points (blue) and the randomly sampled 60 points (red). (b) Resolved vibrational spectrum from full data (blue), 60-point data using standard FT (green) and 60-point data using CS reconstruction (red). (Adapted from Wrzesinski, P.J. et al., *J. Raman Spectrosc.*, 42(3), 393, 2010.)

There have also been advances in improving acquisition speeds using techniques that reduce the number of required independent acquisitions. One such technique uses the concepts developed in the field of *compressed sensing* (or compressive sensing). In brief, it has been shown that an $N$-point signal can be reconstructed from much less than $N$ Fourier measurements, under the assumption of sparse prevalence in the frequency domain [69,70]. This can be readily applied to the described selective-excitation CARS measurements as the number of vibrational spectra of simple molecules are usually sparse; therefore, the number of non-zero elements in the spectral domain is low and can be calculated from a similar number of temporal measurements. Using this technique, Katz et al. demonstrated a fourfold reduction in the number of temporal measurements necessary to accurately resolve sparse vibrational spectra (Figure 9.20) [71,72].

## 9.4 Future Outlook

While the concept of coherent control has been practiced for more than 20 years, the use of the technique for CARS has rapidly grown in the past decade. Throughout this time, the technical aspects of the implementations have been intensively studied and recently focused on applications. With further development of broadband sources, single-pulse CARS can be applicable to a greater range of applications afforded by the increased resolvable Raman bandwidth. Furthermore, development of techniques to preserve phase through scattering media can better allow shaped pulses to probe biologically interesting samples as done using conventional CARS methods. These advances in combination with fast shaping techniques can provide an attractive alternative to conventional multi-beam approaches. In the future, it would not be unlikely to see further development of single-beam techniques that employ pulse-shaping techniques due to the additional benefits afforded such as concurrent multi-modal detection of additional ultra-fast phenomena such as harmonic generation.

## References

1. Schrader, B. 1995. *Infrared and Raman Spectroscopy: Methods and Applications.* VCH, Weinheim, Germany.
2. Evans, C. L. and X. S. Xie. 2008. Coherent anti-Stokes Raman scattering microscopy: Chemical imaging for biology and medicine. *Annual Review of Analytical Chemistry* 1:883–909.
3. Volkmer, A. 2005. Vibrational imaging and microspectroscopies based on coherent anti-Stokes Raman scattering microscopy. *Journal of Physics D: Applied Physics* 38:R59.
4. Silberberg, Y. 2009. Quantum coherent control for nonlinear spectroscopy and microscopy. *Annual Review of Physical Chemistry* 60:277–292.
5. Weiner, A. M. 2000. Femtosecond pulse shaping using spatial light modulators. *Review of Scientific Instruments* 71:1929–1960.
6. Verluise, F., V. Laude, Z. Cheng, C. Spielmann, and P. Tournois. 2000. Amplitude and phase control of ultrashort pulses by use of an acousto-optic programmable dispersive filter: Pulse compression and shaping. *Optics Letters* 25:575–577.
7. Dudovich, N., D. Oron, and Y. Silberberg. 2002. Single-pulse coherently controlled nonlinear Raman spectroscopy and microscopy. *Nature* 418:512–514.
8. Oron, D., N. Dudovich, and Y. Silberberg. 2002. Single-pulse phase-contrast nonlinear Raman spectroscopy. *Physical Review Letters* 89:273001.
9. Oron, D., N. Dudovich, and Y. Silberberg. 2003. Femtosecond phase-and-polarization control for background-free coherent anti-Stokes Raman spectroscopy. *Physical Review Letters* 90:213902.
10. Dudovich, N., D. Oron, and Y. Silberberg. 2003. Single-pulse coherent anti-Stokes Raman spectroscopy in the fingerprint spectral region. *The Journal of Chemical Physics* 118:9208–9215.
11. Carriles, R., D. N. Schafer, K. E. Sheetz, J. J. Field, R. Cisek, V. Barzda, A. W. Sylvester, and J. A. Squier. 2009. Invited review article: Imaging techniques for harmonic and multiphoton absorption fluorescence microscopy. *Review of Scientific Instruments* 80:081101.
12. Chen, B. C. and S. H. Lim. 2008. Optimal laser pulse shaping for interferometric multiplex coherent anti-Stokes Raman scattering microscopy. *Journal of Physical Chemistry B* 112:3653–3661.
13. Isobe, K., A. Suda, M. Tanaka, H. Hashimoto, F. Kannari, H. Kawano, H. Mizuno, A. Miyawaki, and K. Midorikawa. 2009. Single-pulse coherent anti-Stokes Raman scattering microscopy employing an octave spanning pulse. *Optics Express* 17:11259–11266.
14. von Vacano, B. and M. Motzkus. 2008. Time-resolving molecular vibration for microanalytics: Single laser beam nonlinear Raman spectroscopy in simulation and experiment. *Physical Chemistry Chemical Physics* 10:681–691.
15. Katz, O., A. Natan, Y. Silberberg, and S. Rosenwaks. 2008. Standoff detection of trace amounts of solids by nonlinear Raman spectroscopy using shaped femtosecond pulses. *Applied Physics Letters* 92:171116–171113.
16. Li, H., D. A. Harris, B. Xu, P. J. Wrzesinski, V. V. Lozovoy, and M. Dantus. 2009. Standoff and arms-length detection of chemicals with single-beam coherent anti-Stokes Raman scattering. *Applied Optics* 48:B17–B22.
17. Scully, M. O., G. W. Kattawar, R. P. Lucht, T. Opatrný, H. Pilloff, A. Rebane, A. V. Sokolov, and M. S. Zubairy. 2002. FAST CARS: Engineering a laser spectroscopic technique for rapid identification of bacterial spores. *Proceedings of the National Academy of Sciences of the United States of America* 99:10994–11001.
18. Katz, O., J. M. Levitt, E. Grinvald, and Y. Silberberg. 2010. Single-beam coherent Raman spectroscopy and microscopy via spectral notch shaping. *Optics Express* 18:22693–22701.

19. Natan, A., O. Katz, S. Rosenwaks, and Y. Silberberg. 2009. Single-pulse standoff nonlinear Raman spectroscopy using shaped femtosecond pulses. In *Ultrafast Phenomena XVI*. P. Corkum, S. Silvestri, K. A. Nelson, E. Riedle, and R. W. Schoenlein, eds. Springer, Berlin, Germany, pp. 985–987.
20. Oron, D., N. Dudovich, and Y. Silberberg. 2004. All-optical processing in coherent nonlinear spectroscopy. *Physical Review A* 70:023415.
21. Pestov, D., R. K. Murawski, G. O. Ariunbold, X. Wang, M. Zhi, A. V. Sokolov, V. A. Sautenkov, Y. V. Rostovtsev, A. Dogariu, Y. Huang, and M. O. Scully. 2007. Optimizing the laser-pulse configuration for coherent Raman spectroscopy. *Science* 316:265–268.
22. Pegoraro, A. F., A. Ridsdale, D. J. Moffatt, Y. Jia, J. P. Pezacki, and A. Stolow. 2009. Optimally chirped multimodal CARS microscopy based on a single Ti:sapphire oscillator. *Optics Express* 17:2984–2996.
23. Langbein, W., I. Rocha-Mendoza, and P. Borri. 2009. Single source coherent anti-Stokes Raman microspectroscopy using spectral focusing. *Applied Physics Letters* 95:081109.
24. Hellerer, T., A. M. K. Enejder, and A. Zumbusch. 2004. Spectral focusing: High spectral resolution spectroscopy with broad-bandwidth laser pulses. *Applied Physics Letters* 85:25–27.
25. von Vacano, B., L. Meyer, and M. Motzkus. 2007. Rapid polymer blend imaging with quantitative broadband multiplex CARS microscopy. *Journal of Raman Spectroscopy* 38:916–926.
26. Ogilvie, J. P., E. Beaurepaire, A. Alexandrou, and M. Joffre. 2006. Fourier-transform coherent anti-Stokes Raman scattering microscopy. *Optics Letters* 31:480–482.
27. Müller, M. and J. M. Schins. 2002. Imaging the thermodynamic state of lipid membranes with multiplex CARS microscopy. *The Journal of Physical Chemistry B* 106:3715–3723.
28. Kee, T. W. and M. T. Cicerone. 2004. Simple approach to one-laser, broadband coherent anti-Stokes Raman scattering microscopy. *Optics Letters* 29:2701–2703.
29. Oron, D., N. Dudovich, D. Yelin, and Y. Silberberg. 2002. Narrow-band coherent anti-Stokes Raman signals from broad-band pulses. *Physical Review Letters* 88:063004.
30. Oron, D., N. Dudovich, D. Yelin, and Y. Silberberg. 2002. Quantum control of coherent anti-Stokes Raman processes. *Physical Review A* 65:043408.
31. Selm, R., M. Winterhalder, A. Zumbusch, G. Krauss, T. Hanke, A. Sell, and A. Leitenstorfer. 2010. Ultrabroadband background-free coherent anti-Stokes Raman scattering microscopy based on a compact Er:fiber laser system. *Optics Letters* 35:3282–3284.
32. Cui, M., B. R. Bachler, and J. P. Ogilvie. 2009. Comparing coherent and spontaneous Raman scattering under biological imaging conditions. *Optics Letters* 34:773–775.
33. Postma, S., A. C. W. van Rhijn, J. P. Korterik, P. Gross, J. L. Herek, and H. L. Offerhaus. 2008. Application of spectral phase shaping to high resolution CARS spectroscopy. *Optics Express* 16:7985–7996.
34. van Rhijn, A. C. W., S. Postma, J. P. Korterik, J. L. Herek, and H. L. Offerhaus. 2009. Chemically selective imaging by spectral phase shaping for broadband CARS around 3000 $cm^{-1}$. *Journal of the Optical Society of America B* 26:559–563.
35. Lim, S. H., A. G. Caster, and S. R. Leone. 2005. Single-pulse phase-control interferometric coherent anti-Stokes Raman scattering spectroscopy. *Physical Review A* 72:041803.
36. Lim, S. H., A. G. Caster, O. Nicolet, and S. R. Leone. 2006. Chemical imaging by single pulse interferometric coherent anti-Stokes Raman scattering microscopy. *Journal of Physical Chemistry B* 110:5196–5204.
37. Lim, S. H., A. G. Caster, and S. R. Leone. 2007. Fourier transform spectral interferometric coherent anti-Stokes Raman scattering (FTSI-CARS) spectroscopy. *Optics Letters* 32:1332–1334.
38. Xu, X. G., S. O. Konorov, J. W. Hepburn, and V. Milner. 2008. Noise autocorrelation spectroscopy with coherent Raman scattering. *Nature Physics* 4:125–129.
39. Levitt, J. M., O. Katz, E. Grinvald, and Y. Silberberg. 2009. Single-pulse CARS spectroscopy using a resonant photonic crystal slab filter (RPCS). In *Frontiers in Optics*, OSA Technical Digest (CD), Optical Society of America, paper FMK3.
40. Roy, S., P. Wrzesinski, D. Pestov, T. Gunaratne, M. Dantus, and J. R. Gord. 2009. Single-beam coherent anti-Stokes Raman scattering spectroscopy of $N_2$ using a shaped 7 fs laser pulse. *Applied Physics Letters* 95:074102–074103.
41. Roy, S., P. J. Wrzesinski, D. Pestov, M. Dantus, and J. R. Gord. 2010. Single-beam coherent anti-Stokes Raman scattering (CARS) spectroscopy of gas-phase $CO_2$ via phase and polarization shaping of a broadband continuum. *Journal of Raman Spectroscopy* 41:1194–1199.
42. Li, H., D. A. Harris, B. Xu, P. J. Wrzesinski, V. V. Lozovoy, and M. Dantus. 2008. Coherent mode-selective Raman excitation towards standoff detection. *Optics Express* 16:5499–5504.

43. Natan, A., J. M. Levitt, L. Graham et al. 2012. Standoff detection via single-beam spectral notch filtered pulses. *Applied Physics Letters* 100(5):051111.
44. Leonhardt, R., W. Holzapfel, W. Zinth, and W. Kaiser. 1987. Terahertz quantum beats in molecular liquids. *Chemical Physics Letters* 133:373–377.
45. Lang, T., K. L. Kompa, and M. Motzkus. 1999. Femtosecond CARS on $H_2$. *Chemical Physics Letters* 310:65–72.
46. Lang, T., M. Motzkus, H. M. Frey, and P. Beaud. 2001. High resolution femtosecond coherent anti-Stokes Raman scattering: Determination of rotational constants, molecular anharmonicity, collisional line shifts, and temperature. *The Journal of Chemical Physics* 115:5418–5426.
47. Schmitt, M., G. Knopp, A. Materny, and W. Kiefer. 1997. Femtosecond time-resolved coherent anti-Stokes Raman scattering for the simultaneous study of ultrafast ground and excited state dynamics: Iodine vapour. *Chemical Physics Letters* 270:9–15.
48. Zeidler, D., S. Frey, W. Wohlleben, M. Motzkus, F. Busch, T. Chen, W. Kiefer, and A. Materny. 2002. Optimal control of ground-state dynamics in polymers. *The Journal of Chemical Physics* 116:5231–5235.
49. Weiner, A. M., D. E. Leaird, G. P. Wiederrecht, and K. A. Nelson. 1990. Femtosecond pulse sequences used for optical manipulation of molecular motion. *Science* 247:1317–1319.
50. Cui, M., M. Joffre, J. Skodack, and J. P. Ogilvie. 2006. Interferometric Fourier transform coherent anti-Stokes Raman scattering. *Optics Express* 14:8448–8458.
51. Hauer, J., H. Skenderovic, K.-L. Kompa, and M. Motzkus. 2006. Enhancement of Raman modes by coherent control in [beta]-carotene. *Chemical Physics Letters* 421:523–528.
52. Lozovoy, V. V., B. Xu, J. C. Shane, and M. Dantus. 2006. Selective nonlinear optical excitation with pulses shaped by pseudorandom Galois fields. *Physical Review A* 74:041805.
53. Wrzesinski, P. J., D. Pestov, V. V. Lozovoy, B. Xu, S. Roy, J. R. Gord, and M. Dantus. 2010. Binary phase shaping for selective single-beam CARS spectroscopy and imaging of gas-phase molecules. *Journal of Raman Spectroscopy* 42(3):393.
54. von Vacano, B., W. Wohlleben, and M. Motzkus. 2006. Single-beam CARS spectroscopy applied to low-wavenumber vibrational modes. *Journal of Raman Spectroscopy* 37:404–410.
55. von Vacano, B., T. Buckup, and M. Motzkus. 2006. Highly sensitive single-beam heterodyne coherent anti-Stokes Raman scattering. *Optics Letters* 31:2495–2497.
56. Müller, C., T. Buckup, B. von Vacano, and M. Motzkus. 2009. Heterodyne single-beam CARS microscopy. *Journal of Raman Spectroscopy* 40:809–816.
57. von Vacano, B. and M. Motzkus. 2006. Time-resolved two color single-beam CARS employing supercontinuum and femtosecond pulse shaping. *Optics Communications* 264:488–493.
58. von Vacano, B. and M. Motzkus. 2007. Molecular discrimination of a mixture with single-beam Raman control. *The Journal of Chemical Physics* 127:144514.
59. Pegoraro, A. F., A. Ridsdale, D. J. Moffatt, J. P. Pezacki, B. K. Thomas, L. Fu, L. Dong, M. E. Fermann, and A. Stolow. 2009. All-fiber CARS microscopy of live cells. *Optics Express* 17:20700–20706.
60. Evans, C. L., E. O. Potma, M. Puoris'haag, D. Cote, C. P. Lin, and X. S. Xie. 2005. Chemical imaging of tissue in vivo with video-rate coherent anti-Stokes Raman scattering microscopy. *Proceedings of the National Academy of Sciences of the United States of America* 102:16807–16812.
61. Cheng, J. X., Y. K. Jia, G. F. Zheng, and X. S. Xie. 2002. Laser-scanning coherent anti-Stokes Raman scattering microscopy and applications to cell biology. *Biophysical Journal* 83:502–509.
62. Wang, H. F., Y. Fu, P. Zickmund, R. Y. Shi, and J. X. Cheng. 2005. Coherent anti-Stokes Raman scattering imaging of axonal myelin in live spinal tissues. *Biophysical Journal* 89:581–591.
63. Frumker, E., D. Oron, D. Mandelik, and Y. Silberberg. 2004. Femtosecond pulse-shape modulation at kilohertz rates. *Optics Letters* 29:890–892.
64. Frumker, E. and Y. Silberberg. 2007. Femtosecond pulse shaping using a two-dimensional liquid-crystal spatial light modulator. *Optics Letters* 32:1384–1386.
65. Frumker, E. and Y. Silberberg. 2007. Phase and amplitude pulse shaping with two-dimensional phase-only spatial light modulators. *Journal of the Optical Society of America B* 24:2940–2947.
66. Frumker, E. and Y. Silberberg. 2009. Two-dimensional phase-only spatial light modulators for dynamic phase and amplitude pulse shaping. *Journal of Modern Optics* 56:2049–2054.
67. Pillai, R. S., C. Boudoux, G. Labroille, N. Olivier, I. Veilleux, E. Farge, M. Joffre, and E. Beaurepaire. 2009. Multiplexed two-photon microscopy of dynamic biological samples with shaped broadband pulses. *Optics Express* 17:12741–12752.

68. Levitt, J. M., O. Katz, and Y. Silberberg. 2010. Shaperless single-pulse CARS for micro-spectroscopy and fast modulated temporal pulse splitting for selective excitation. In *Focus on Microscopy*, Shanghai, China. Manuscript in preparation.
69. Candes, E. J. and M. B. Wakin. 2008. An introduction to compressive sampling. *IEEE Signal Processing Magazine* 25:21–30.
70. Donoho, D. L. and J. Tanner. 2010. Precise undersampling theorems. *Proceedings of the IEEE* 98:913–924.
71. Katz, O., J. M. Levitt, and Y. Silberberg. 2010. Compressive Fourier transform spectroscopy. arXiv:1006.2553v1, submitted on June 13, 2010.
72. Katz, O., J. M. Levitt, and Y. Silberberg. 2010. Compressive Fourier transform spectroscopy. In *Frontiers in Optics*, OSA Technical Digest (CD), Optical Society of America, paper FTuE3.

# 10. Fourier Transform CARS Microscopy

**Jennifer P. Ogilvie**



## 10.1 Introduction

For imaging complex biological samples, high-resolution CARS spectra spanning the fingerprint region (~700–1800 $cm^{-1}$) offer high specificity and the exciting possibility of tracking multiple chemical species in space and time. The most commonly used mode of CARS microscopy, where a single vibrational mode is recorded to permit rapid imaging, can produce CARS spectra, but only by the time-consuming process of scanning the pump-Stokes frequency difference [1–3]. More practical "multiplex" CARS methods combine a narrowband pump and broadband probe to obtain CARS spectra directly in the frequency domain [4–7]. In this chapter we present a related time-domain approach for obtaining spectrally-resolved CARS images: Fourier transform coherent anti-Stokes Raman spectroscopy (FTCARS) microscopy [8–10]. FTCARS microscopy employs a single broadband laser to obtain spectrally-resolved images over the bandwidth of the laser source. It is a time-domain technique that offers straightforward removal of the non-resonant background, essentially arbitrary spectral resolution, and can be readily modified to permit interferometric (heterodyne) detection for enhanced sensitivity.

Following the initial work by the Silberberg group [11], there are now a number of single-laser methods for CARS microscopy with broadband femtosecond pulses [6,7,9,12–14],

*Coherent Raman Scattering Microscopy*. Edited by Ji-Xin Cheng and X. Sunney Xie © 2013 CRC Press/Taylor & Francis Group, LLC. ISBN: 978-1-4398-6765-5.

many of which are presented in this volume. While the use of broadband femtosecond pulses for CARS microscopy enhances the spectral information obtainable, it also necessitates more elaborate methods for achieving high spectral resolution and suppressing the nonresonant background, which increases with decreasing pulse duration [15]. Pulse-shaping of the spectral amplitude and/or phase of the broadband pulses has been shown to be an effective approach for suppressing the nonresonant background, enhancing spectral resolution and in some cases achieving mode selectivity. FTCARS falls into the broad category of methods that employ pulse-shaping of broadband femtosecond pulses to obtain spectrally-resolved CARS images. In the case of FTCARS, the pulse-shaper is a simple Michelson interferometer.

This chapter begins with a discussion of the theory behind FTCARS in Section 10.2. The addition of a local oscillator (LO) for interferometric FTCARS (IFTCARS) is described in Section 10.3. Section 10.4 addresses noise considerations for FTCARS imaging applications, while Section 10.6 compares the advantages and disadvantages of time- and frequency-domain detection.

## 10.2 Theoretical Description of FTCARS

FTCARS is based on the principle that a single Fourier transform-limited femtosecond pulse can simultaneously excite all Raman modes with frequencies within the laser bandwidth, as in Impulsive Stimulated Raman Scattering (ISRS), illustrated in Figure 10.1a [16,17]. Borrowing from ISRS spectroscopy, FTCARS is a time-domain approach in which an initial "pump" excitation pulse excites vibrational coherences via ISRS, while a second collinear "probe" pulse, depending on its relative time delay $\tau$, coherently enhances or suppresses the initial excitation, generating Stokes and anti-Stokes fields [9]. Recording the CARS emission as a function of the time delay between the two collinear excitation pulses then permits extraction of the FTCARS spectrum through a simple Fourier transform. In the frequency domain, the two-pulse sequence produces a sinusoidally modulated spectral amplitude, making FTCARS analogous to pulse-shaping CARS microscopy approaches. In FTCARS the nonresonant background can be readily removed by windowing out the contribution around zero time

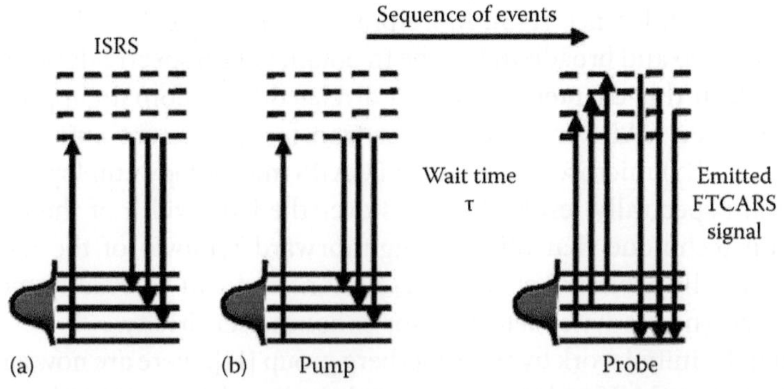


**FIGURE 10.1** (a) Energy diagram for impulsive stimulated Raman scattering (ISRS). (b) Energy diagram for FTCARS signal generation. An initial broadband pulse excites Raman coherences within the laser bandwidth (~1500 cm$^{-1}$). A time delayed probe pulse scatters from the coherence, producing the detected anti-Stokes signal. (Adapted from Cui, M. et al., *Appl. Opt.*, 47, 5790, 2008.)

delay, as demonstrated in time-resolved CARS microscopy at a fixed delay [18], and an interferometric CARS experiment that illustrated the difference in time evolution of resonant and nonresonant signals [19]. Figure 10.1b shows a schematic of the FTCARS signal generation. The anti-Stokes signal is selected via spectral filtering and recorded as a function of the time delay $\tau$. This permits retrieval of the CARS spectrum upon Fourier transform of the time domain data. Being a Fourier transform technique, the spectral resolution of FTCARS depends on the maximum time delay, providing the capability of achieving spectral resolution limited only by the sample itself. The nonresonant background generated by the combined pump and probe pulses is nonzero only during their temporal overlap, and can therefore be windowed out in the time domain trace.

To mathematically describe the FTCARS signal, we begin by considering the third order polarization created by the FTCARS pulse sequence [10]. This polarization is as a sum of vibrationally resonant and nonresonant contributions. Following the formalism outlined by Dudovich et al. [20], the third order vibrationally resonant polarization $P_R^{(3)}(\omega)$ created by a femtosecond laser pulse with electric field envelope $E(\omega)$ is given by

$$P_R^{(3)}(\omega) \propto \int_0^{\infty} d\Omega \frac{1}{\Omega - \Omega_R - i\Gamma_R} E(\omega - \Omega) A(\Omega) \tag{10.1}$$

where
$\Omega_R$ is the Raman frequency
$\Gamma_R$ is the linewidth

Here the term $A(\Omega)$ describes all possible two field interactions leading to the same intermediate state:

$$A(\Omega) = \int_0^{\infty} d\omega' E^*(\omega') E(\Omega + \omega') \tag{10.2}$$

The nonresonant polarization is given by

$$P_{NR}^{(3)}(\omega) \propto \int_0^{\infty} \frac{d\Omega}{\Omega} E(\omega - \Omega) A(\Omega) \tag{10.3}$$

where the $\Omega$ denominator accounts for the noninstantaneous nature of the nonresonant contribution.

We express the FTCARS pulse sequence as a total electric field given by

$$E(\tau, t) = E_{pump}(t) + E_{probe}(t - \tau) \tag{10.4}$$

where both pump and probe fields have been spectrally modified by a long-pass filter to facilitate separation of the blue-shifted FTCARS signal. The total detected FTCARS signal arising at time delay $\tau$ is given by

$$S_{FTCARS}(\tau) \propto \int_0^{\infty} \left|P_{FTCARS}(\tau,t)\right|^2 dt \tag{10.5}$$

where $P_{FTCARS}$ is the total polarization created by the laser pulse sequence, consisting of both resonant and nonresonant contributions:

$$P_{FTCARS}(\tau,t) = P_R^{(3)}(\tau,t) + P_{NR}^{(3)}(\tau,t) \tag{10.6}$$

Here $P_R^{(3)}(\tau,t)$ and $P_{NR}^{(3)}(\tau,t)$ are the Fourier transforms of $P_R^{(3)}(\omega)$ and $P_{NR}^{(3)}(\omega)$ respectively.

A description of FTCARS requires consideration of several different signals generated by the pump and probe pulses. Each pulse provides impulsive excitation, and thus each one produces both resonant and nonresonant signals. However, contributions solely from the pump or probe pulse alone produce a constant background signal: only those signals requiring both pump and probe fields are delay dependent and appear as oscillations in the time domain FTCARS data. At short delays, several combinations of pump and probe fields give rise to resonant and nonresonant contributions. When the pump and probe are well-separated, a resonant contribution is produced by two pump field interactions that create the Raman coherence, and a single probe field interaction that scatters from this coherence. The only other contributions that are time-coincident on the detector come solely from the probe pulse and thus have a third order dependence on the probe field. These contributions are independent of $\tau$, but interfere with the resonant signal produced by the combination of pump and probe pulses.

The polarization components that give rise to the oscillatory FTCARS signal can be expressed as:

$$P_{FTCARS}(\tau) \approx P_R^{both}(\tau) + P_{NR}^{both}(\tau) + P_R^{probe} + P_{NR}^{probe} \tag{10.7}$$

where $P_R^{both}(\tau)$ and $P_{NR}^{both}(\tau)$ are respectively the resonant and nonresonant third order polarizations created by the combination of pump and probe pulses. Similarly $P_R^{probe}$ and $P_{NR}^{probe}$ are respectively the resonant and nonresonant third order polarizations created by the probe pulse alone. We neglect the contributions arising from the pump pulse alone because they are not time-coincident with the other signal components. In addition, $P_{NR}^{both}(\tau)$ can be neglected for delays $\tau$ at which the pump and probe are well-separated. The field dependence of the remaining contributions is given by:

$$P_{FTCARS}(\tau) \approx \chi_R^{(3)} I_{pump} E_{probe}(\tau) + \chi_R^{(3)} I_{probe} E_{probe} + \chi_{NR}^{(3)} I_{probe} E_{probe} \tag{10.8}$$

For $I_{pump} \approx I_{probe}$ the dominant contribution to the polarization will be the time-independent nonresonant contribution due to the probe (the third term in Equation 10.8). This constant term will interfere with the $\tau$ dependent resonant signal (the first term in Equation 10.8) to produce the detected FTCARS signal that is linearly proportional to the intensity of the pump pulse and quadratically proportional to the probe pulse intensity:

$$S_{FTCARS} \approx \chi_R^{(3)} I_{pump} I_{probe}^2 \tag{10.9a}$$

The other contributions to the signal produce low frequency components upon Fourier transform.

We noted previously that FTCARS is similar to ISRS. The key difference is that in ISRS, the full probe field enters the detector, and the small pump-dependent change in probe transmission is typically detected by the use of a lock-in amplifier. In FTCARS, spectral filtering isolates the blue-shifted signal, removing the large background introduced by the probe field. This increases the signal contrast, producing signals that can be readily digitized by a high-speed data acquisition board, removing the need for lock-in detection.

### 10.2.1 Experimental Implementation of FTCARS

The experimental setup for FTCARS microscopy is shown in Figure 10.2i. A broadband Ti:sapphire oscillator (Femtolasers Synergy) produces a 12 fs, 75 MHz pulse train with a center wavelength of 790 nm. The pulse train is sent into a dispersion-balanced Michelson interferometer to create two replicas of the incident pulse. The time delay between the pump

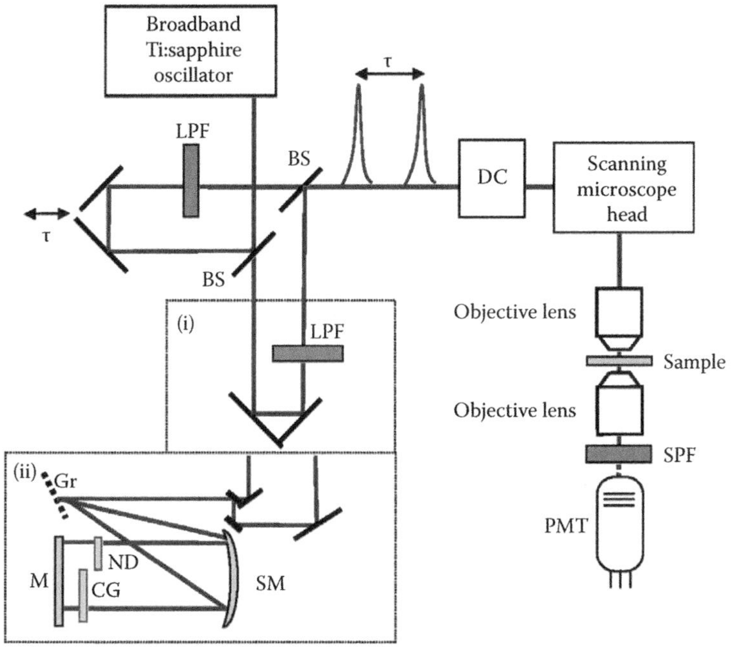


FIGURE 10.2 (i) Experimental implementation of standard FTCARS. BS: beamsplitter, DC: dispersion compensation (chirped mirrors), LPF: long-pass filter, SPF: short-pass filter, PMT: photomultiplier tube. (ii) Experimental setup for Heterodyne CARS. Gr: 600 lines/mm grating, ND: neutral density filter, CG: compensating glass, M: mirror, SM: spherical mirror ($f$ = 30 cm). Helium-neon laser used for fringe tracking not shown. (Adapted from Cui, M. et al., *Opt. Express*, 14, 8448, 2006.)

and probe pulses is accurately monitored using fringe-tracking of a helium-neon laser [21,22] during translation of a linear DC actuator in one arm of the interferometer. The resulting pump–probe sequence is sent into a laser scanning microscope (modified Prairie Technologies system) equipped with an NA 0.4 objective (Olympus; spot size ~1.5 μm; used for bulk samples). To reduce image acquisition time as well as sample damage due to long pixel dwell times, FTCARS images are acquired one line at a time by rapidly scanning one of the microscope's galvanometric mirror axes at ~200 Hz while scanning the time delay $\tau$ at ~1 Hz. A time series for each pixel is then constructed for each pixel in the line. The FTCARS signal is spectrally filtered and detected by a photomultiplier tube (PMT) (Hamamatsu R928). The spectral filtering is achieved by a combination of a long-pass filter (740 nm cutoff, Omega) placed before the sample and two short-pass filters (720 nm cutoff, Omega) placed before the PMT. The long-pass filter serves to introduce a sharp cutoff on the blue edge of the spectrum so that the FTCARS signal can be cleanly separated from the incident pump and probe pulses by the short-pass filter. Dispersion precompensation is performed with a pair of chirped mirrors (Femtolasers), and verified via interferometric autocorrelation using a two-photon GaAsP photodiode to produce approximately transform-limited pulses of 17 fs duration at the sample. The analog FTCARS signal from the PMT is digitized at 1 MHz (National Instruments PCI-6110E). Acquisition and analysis are performed via MATLAB®.

Before presenting imaging results we illustrate our theoretical description of FTCARS spectroscopy with calculations and experiments in solution, using pyridine, which has two dominant Raman modes at 990 and 1030 cm$^{-1}$ [23]. Figure 10.3 shows a comparison

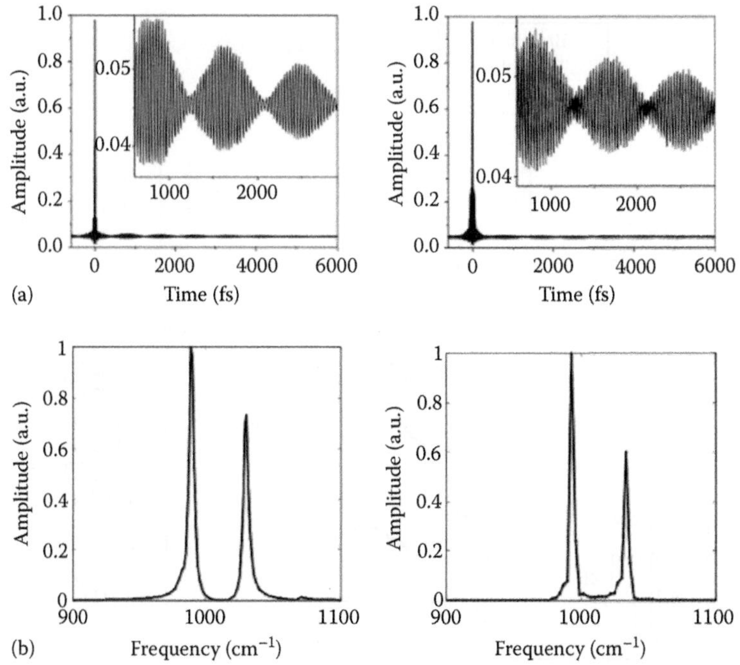


**FIGURE 10.3** FTCARS results for pyridine including the dominant Raman modes at 1000 and 1040 cm$^{-1}$, with smaller peaks at 983 and 1058 cm$^{-1}$. (a) Simulated (left) and experimental (right) time domain FTCARS signal for pyridine. The scan length was 12 ps, providing ~3 cm$^{-1}$ resolution. (b) FTCARS power spectrum obtained from the Fourier transform of the time domain data in (a) following windowing of the nonresonant background. (Adapted from Cui, M. et al., *Opt. Express*, 14, 8448, 2006.)

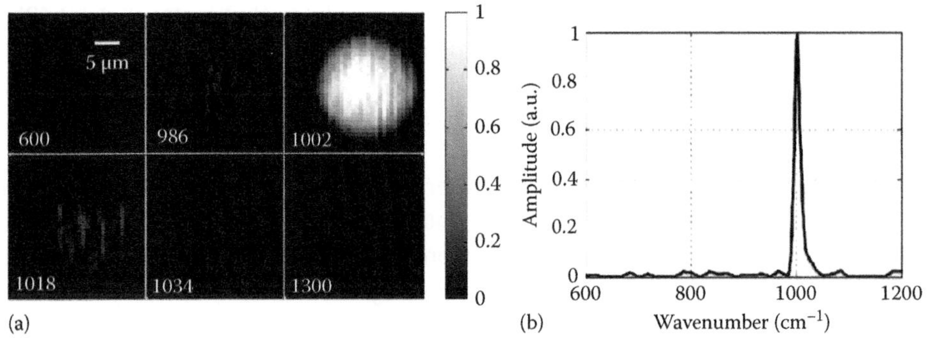


**FIGURE 10.4** (a) FTCARS image of a 15 μm polystyrene bead in water. The different resonant frequencies are indicated in each frame. (4 mW excitation in each beam, 36 × 36 pixels, 7 ms/pixel exposure, spectral resolution of 16 cm$^{-1}$) (b) FTCARS spectrum of a single pixel at the bead center, showing the expected dominant Raman mode near 1000 cm$^{-1}$. (Adapted from Cui, M. et al., *Appl. Opt.*, 47, 5790, 2008.)

between theoretical and experimental FTCARS data. The large signal at zero delay arises from the nonresonant background generated by the combined pump and probe pulses. This signal is windowed out (for delays of less than 300 fs) before obtaining the FTCARS spectrum via the Fourier transform. The spectral resolution of the measurement is determined by the maximum time delay and is ~3 cm$^{-1}$ in this case. The constant nonzero offset in the signal amplitude arises from the resonant and nonresonant signal contributions from the probe pulse. Here the combined pump and probe power used was 86 mW with a data acquisition time of 1.5 s, limited by the scanning speed of the stage which could be readily increased.

In Figure 10.4 we show several FTCARS images. Figure 10.4a is a spectrally-resolved FTCARS image of a 15 μm polystyrene bead immersed in water. This particular image was recorded using 4 mW of power in each pulse, with 7 ms/pixel exposure time and 16 cm$^{-1}$ spectral resolution. Higher spectral resolution is readily obtained by scanning longer $\tau$ delays. The corresponding FTCARS spectrum for a single pixel in the bead is shown in Figure 10.4b. Figure 10.5 shows FTCARS images at different frequencies for a

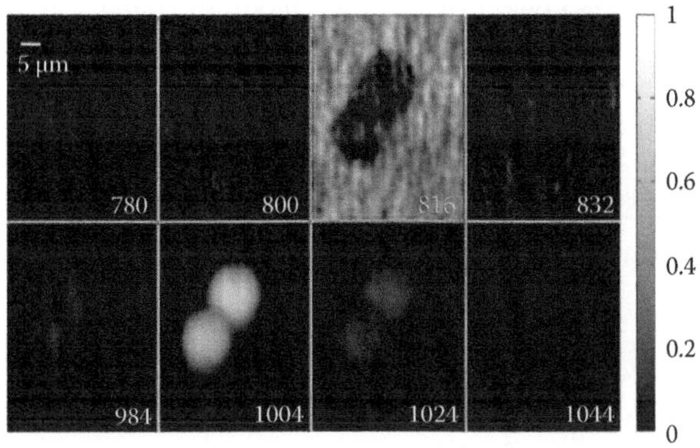


**FIGURE 10.5** FTCARS image of 15 μm polystyrene beads embedded in PMMA, showing the expected dominant contribution from PMMA at 820 cm$^{-1}$ and polystyrene near 1000 cm$^{-1}$. The upper row is multiplied by a factor of 4 to make the PMMA contribution readily visible. Imaging conditions were 4 mW excitation in each beam, 7 ms/pixel exposure, spectral resolution of 16 cm$^{-1}$. (Adapted from Cui, M. et al., *Appl. Opt.*, 47, 5790, 2008.)

sample of 15 μm polystyrene beads embedded in poly(methyl methacrylate) (PMMA) recorded under the same imaging conditions.

## 10.3 Interferometric FTCARS

FTCARS is an inherently interferometric (heterodyne)* method. As shown in Equation 10.8, the detected FTCARS signal arises from the interference between the delay-dependent resonant and nonresonant contribution arising from both pump and probe pulses, with the constant background signal generated by the probe-only contributions. We rewrite the detected signal as:

$$S_{FTCARS}(\tau) \approx \left|\chi_R^{(3)} I_{pump} E_{probe}(\tau) + \text{const.}\right|^2 \approx \left|E_{signal}(\tau) + \text{const.}\right|^2 \tag{10.9b}$$

In this form we see readily that the constant offset acts as a "local oscillator" (LO) that interferes with the signal of interest. If the amplitude of the LO and its phase relative to the signal is controllable, the frequently used method of interferometric signal amplification [24] can be used to improve detection sensitivity, and provide the ability to separate quadrature components of the signal field. A number of groups have demonstrated interferometric detection for CARS microscopy applications [25–31]. In the standard FTCARS implementation, control over the LO is not readily available. As well as requiring a controllable phase and amplitude, a LO for FTCARS must be time-coincident, and contain the appropriate frequencies to interfere with the blue-shifted resonant signal. The blue edge of the broadband spectrum, which is normally attenuated by a long-pass filter to enable effective separation of the FTCARS signal, can be used to provide a LO field for interferometric detection. In this case the overall signal for IFTCARS can be written as

$$\begin{aligned} S_{IFTCARS}(\tau) &\approx \left|E_{signal}(\tau) + \text{const.} + E_{LO}\right|^2 \\ &\approx \left|E_{LO}\right|^2 + 2c\sqrt{I_{signal} I_{LO}} \cos\left[\Delta\phi(\tau)\right] \end{aligned} \tag{10.10}$$

where $\Delta\phi(\tau)$ is the relative phase difference between the signal and LO fields. Here we have assumed that $|E_{LO}| \gg |\text{const.}|$ and $|E_{LO}| \gg |E_{signal}|$. $c \leq 1$ is a constant accounting for the fact that the signal and LO have different amplitude and phase profiles and therefore may not interfere perfectly. The cosine term reflects the need for a controlled and stable relative phase between the signal and the LO for effective signal amplification. As in FTCARS, the Fourier transform of $S_{IFTCARS}(\tau)$ yields the resonant CARS spectrum, provided the nonresonant contribution during pulse overlap is windowed out prior to the Fourier transform.

### 10.3.1 Experimental Implementation of IFTCARS

The experimental setup for IFTCARS is similar to that of FTCARS, with modifications to provide for the addition of a LO field. Figure 10.2ii shows that this can be readily accomplished by the addition of a modified 4f pulse-shaper in the probe arm in place

* We note that different communities often use the terms "heterodyne" and "homodyne" in the same context. We employ the term "interferometric" here as described earlier.

of the long-pass filter that sharply cuts off the blue end of the spectrum in standard FTCARS. In the 4f pulse-shaper a neutral density filter is placed at the Fourier plane to control the intensity of the blue edge of the spectrum that will act as the LO. A compensating plate in the remaining portion of the beam ensures temporal overlap between the probe and the LO.

Figure 10.6 illustrates the improved signal-to-noise ratio (SNR) that can be achieved with IFTCARS compared to standard FTCARS. Here we show a line scan taken from a 4 μm polystyrene bead, acquired with 13 pJ (1 mW) pulses with an acquisition time of 4 s. Figure 10.6a employs IFTCARS, while Figure 10.6b shows the corresponding image using standard FTCARS (where the ND filter is replaced by a razor blade to block the LO). The spectral resolution of both images is ~16 $cm^{-1}$, corresponding to a $\tau$ delay range of ~2100 fs. Under equivalent experimental conditions, IFTCARS produces a bright resonant image at the expected spectral peak position for polystyrene, while the FTCARS image is dominated by noise. Figure 10.6c compares the power spectrum for a single pixel in each image, demonstrating the considerable SNR improvement made with IFTCARS.

We note that in both FTCARS and IFTCARS, interferometer pathlength fluctuations are not a significant source of noise. We can image the frequency range of 400–1500 $cm^{-1}$, or equivalently, wavelengths of 25–6.7 μm. Since the pump–probe time delay is determined using a He–Ne fringe tracking system, optical pathlength fluctuations between the two arms of the interferometer are accounted for with an error of $\ll$6.3 nm; better than $\lambda/1000$ for even the highest frequency mode. Since both

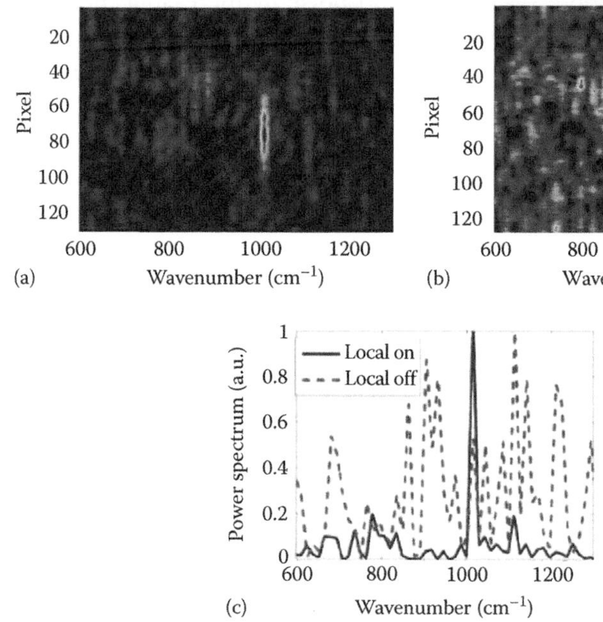

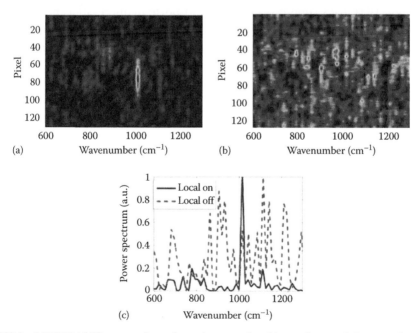


**FIGURE 10.6** (a) IFTCARS line scan through a polystyrene bead (4 μm diameter). Pump and probe powers were each 1 mW. The spectral resolution is 16 $cm^{-1}$ with a data acquisition time of 4 s. (b) Corresponding line scan image taken with standard FTCARS (where the LO is blocked). (c) Comparison of spectra from pixel 75 in the IFTCARS (local on) and FTCARS (local off) images. (Adapted from Cui, M. et al., *Appl. Opt.*, 47, 5790, 2008.)

FTCARS and IFTCARS are interferometric methods, a more serious issue is phase stability between the probe and the LO. In both methods the probe and LO traverse the same optical path (they do not travel different arms of the interferometer), thus they are inherently phase stable.

## 10.4 Noise Considerations

Three types of noise are present in FTCARS measurements: laser noise, detector noise, and electronic noise. Ti:sapphire oscillators are known to have low frequency amplitude fluctuations due to $1/f$ noise [32]. This type of noise can be suppressed by the use of balanced detectors, or by shifting the signal to higher frequencies via lock-in detection. For low light measurements and high data acquisition bandwidth, this noise source is often much smaller than detector noise, which is the case in our FTCARS measurements. Low light linear detectors such as PMTs and avalanche photodiodes typically have three noise components: dark current, shot noise, and Johnson noise. The latter component is much weaker than the first two and can often be neglected. The dark noise presents a constant background signal. The shot noise is given by $\sqrt{2qI_s\Delta f}$, where $q$ is the elementary charge, $I_s$ is the current, which is proportional to the incident intensity $I$ (when operating in the linear regime), and $\Delta f$ is the bandwidth. The total noise can therefore be written as $a+b\sqrt{I}$, where $a$ is the overall background noise (including electronic noise and detector dark current noise) and $b$ is a constant that describes the shot noise. In FTCARS, additional sources of noise include leakage from imperfect filters, and the nonresonant signal generated by the first pulse, all of which increase the background noise floor. IFTCARS suppresses background noise by increasing the LO field. As we can see from Equation 10.10, like the shot noise, the signal amplification is proportional to the square root of the LO intensity. We can therefore express the SNR as

$$\mathrm{SNR} = \frac{2c\sqrt{I_{local}I_{signal}}}{a+b\sqrt{I}} \tag{10.11}$$

where $I$ is the total incident intensity that can be approximated as $I_{local}$.

To characterize the SNR within the frequency range of our experiments, we perform noise measurements under the same conditions as those used for FTCARS imaging. Varying the incident intensity, we use a high speed data acquisition card (PCI6110) to digitize the PMT signal at a sampling rate of higher than twice the bandwidth of the detection system. The averaged amplitude of the Fourier transform around 100 Hz represents the noise at a given incident intensity, which is chosen to be close to typical signal levels (Note that our detection frequency is determined by the scanning speed of our time delay.) From the data we obtain, under typical FTCARS imaging conditions, values of $a = 3.1 \pm 0.2 \times 10^{-13}$ W and $b = 9.0 \pm 0.4 \times 10^{-7}$ $\mathrm{W}^{1/2}$ for the parameters in Equation 10.11. A plot of Equation 10.11 reveals two regimes: one in which background noise dominates (LO < 1 pW) and another where the detection is shot-noise-limited (LO > 100 pW) [8]. Equation 10.11 indicates that the upper limit of the SNR is $2c\sqrt{I_{signal}}/b$, which is independent of the background noise $a$. This limit is achieved when the shot-noise is much greater than background noise: in the shot-noise limit.

Above the threshold for shot-noise-limited detection, increasing the LO power does not increase the SNR due to the fact that the signal and shot noise both scale the same way with LO. We note that the Offerhaus group has discussed the capability of achieving shot-noise-limited detection of heterodyne CARS signals using different detectors [26]. For lower gain detectors, such as photodiodes, they find that heterodyne CARS shows a large degree of enhancement, while this enhancement is more modest for PMT detection. For FTCARS, our choice of detection with a PMT is motivated by the high gain and quantum efficiency of these detectors, providing shot-noise-limited operation at lower signal levels.

From the earlier discussion we can estimate the expected enhancement afforded by IFTCARS for the experimental conditions corresponding to Figure 10.6. In these experiments a total LO of 38 pW was used in IFTCARS, whereas the standard FTCARS produced a nonresonant background of only 4 pW to act as the LO. Leakage through the filters and contributions from the pump pulse produced a background signal power of 22 pW. Under these conditions Equation 10.11 predicts an SNR enhancement of ~4 in the signal intensity with IFTCARS, which is reasonably consistent with our measurements. In this study the magnitude of LO was chosen to operate as close to the shot-noise limit as possible, while taking into account the dynamic range of the 12 bit data acquisition electronics. Further increasing the LO without the need to consider the dynamic range of the data acquisition is expected to give an additional ~30% improvement in SNR.

To optimize the IFTCARS signal, another important parameter is the interference contrast constant $c$, which depends on the relative spectral profiles of the signal and LO. To investigate the effect of using the blue edge of the laser pulse (IFTCARS) and the nonresonant signal (standard FTCARS), we calculate the interference strength for these two cases and compare the result to interference using an ideal LO with the same spectral characteristics as the signal. We find that the interference contrast in both cases is within 5% of the optimum value for an 800 $cm^{-1}$ mode and within 25% for a 1400 $cm^{-1}$ mode. Thus, in agreement with our experimental observations, the difference between employing the nonresonant signal and the blue edge of the laser pulse as the LO is negligible. The benefit of IFTCARS comes from the ability to adjust the magnitude of the LO to ensure that imaging is performed within the shot-noise-limit.

## 10.5 Other Considerations

The FTCARS signal amplitude also depends on several other experimental parameters: the long-pass filter edge wavelength, the frequency of the Raman mode excited, and the power ratio between the two pulses. We explore these dependences by calculating the time-domain FTCARS signal as outlined in Section 10.2. The relative size of the resonant to nonresonant background is set to be 1:20 as measured from FTCARS data on 2-propanol [8]. The maximum time delay is 3 ps, yielding a spectral resolution of ~10 $cm^{-1}$. We assume transform-limited pump and probe pulses of 12 fs duration, centered at 790 nm, and perform spectral filtering with a separation of 5 nm between long and short-pass filters. Before taking the Fourier transform to yield the FTCARS spectrum, we window out the first 300 fs of the time-domain data. The results are summarized in Figure 10.7.

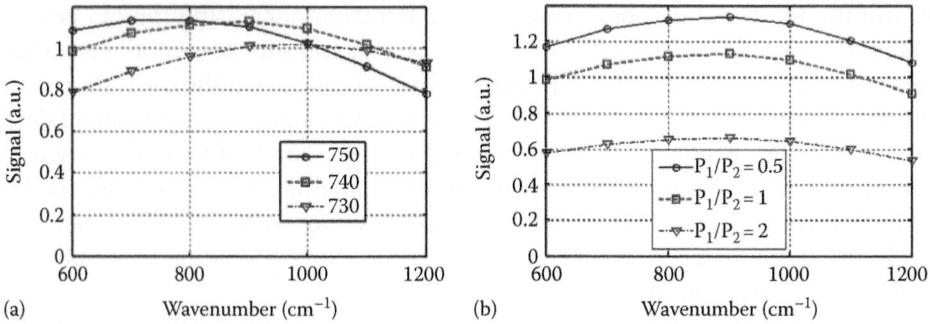


**FIGURE 10.7** (a) Dependence of the FTCARS signal amplitude on the frequency of the CARS mode. Curves for different choices of long-pass filter cutoff are shown. (b) Dependence of the FTCARS signal on the relative power of pump ($P_1$) and probe ($P_2$) pulses. (Adapted from Cui, M. et al., *Appl. Opt.*, 47, 5790, 2008.)

The calculations show that the resonant signal level is reduced for higher frequency modes. This is expected since fewer frequency combinations are available within the pulse bandwidth to excite the given mode. This is partly compensated by the fact that the higher frequency modes experience a larger blue shift and therefore pass the combination of long-pass and short-pass filters more efficiently than lower frequency modes. The result is a resonant signal level that is almost flat over the entire 600 cm$^{-1}$, demonstrating the broad effective bandwidth of FTCARS. The choice of long-pass filter wavelength also affects the relative efficiency with which different modes can be excited as illustrated by the different curves in Figure 10.7a. In standard FTCARS, the signal is linearly dependent on the pump pulse intensity and quadratically dependent on the probe pulse intensity. Thus as expected, the second pulse should be stronger than the first pulse for optimum excitation, as illustrated in Figure 10.7b. In IFTCARS, however, the second pulse need not be stronger than the first pulse to obtain a high SNR.

## 10.6 Time vs. Frequency Domain Detection

Fourier transform spectroscopy measurements have been reported to have a "multiplex advantage" over scanning monochromator measurements [33]. The origin of this advantage (also called "Fellgett's advantage") is the following: within a given measurement time $T$, during which time $N$ measurements are made, a Fourier transform approach measures $N/2$ frequencies, each of which has $N/2$ times longer data accumulation time than when using a scanning monochromator that measures $N/2$ frequencies one at a time. By employing CCD detectors, as in multiplex CARS, this advantage no longer exists: a detector with $N/2$ channels accumulates signal in each frequency channel over the full duration of the measurement, yielding equivalent total measurement time for both time and frequency domain methods. McGuire and Shen have compared the relative benefits of time domain vs. spectral domain techniques in sum-frequency generation spectroscopy [34]. They found a comparable SNR for the two methods in cases, where detector noise dominates. In the shot-noise-limited region, they found that the spectrally resolved methods have an $\sqrt{M}$ advantage, where $M$ is the number of channels with significant nonzero frequency content. These arguments are not directly applicable to CARS microscopy, where the nonresonant background guarantees nonzero frequency content across a broad spectrum, and the details of nonresonant background subtraction must be taken into account. The noise characteristics of the different detectors used in time and frequency domain measurements must also be considered. CCD detectors offer higher quantum efficiency over PMTs, which can provide some improvement (approximately a factor of 2) in SNR for shot-noise-limited CARS imaging. While FTCARS requires only a single detector, avoiding spectrometer losses, this benefit is somewhat offset by losses from the spectral filtering method (~45%).

Another important consideration for biological imaging is sample photodamage. Hell and co-workers have studied photodamage in cells using 75 fs to 3.2 ps pulse durations [35], finding that photodamage scales nonlinearly with incident intensity and providing motivation for using longer pulses. Further studies of the photodamage effects of short pulse excitation are needed. Another important consideration for minimizing photodamage may be the pixel dwell time. In FTCARS, images are acquired by rapidly scanning one line in the image while simultaneously scanning the time delay. Thus, in FTCARS, the excitation returns multiple times to make measurements at a given sample position, but has a short dwell time at each measurement. In multiplex CARS methods, spectral acquisition generally occurs point-by-point, where the sample is scanned and the speed of acquisition is dictated by the CCD readout time and required SNR. Depending on the dominant photodamage mechanism, shorter dwell times could make time domain measurements advantageous in some applications. We also note that when photodamage thresholds are low or the chemical species of interest is in low concentration, imaging via spontaneous Raman scattering may be competitive with spectrally-resolved CARS microscopy [36].

## 10.7 Conclusions

FTCARS microscopy offers spectrally-resolved imaging using a straightforward experimental approach that employs a Michelson interferometer and single broadband laser. FTCARS shares the advantages of other Fourier transform methods (no need for a spectrometer, arbitrary spectral resolution, and high throughput) and also permits easy removal of the nonresonant background. We have presented a theoretical description of FTCARS, showing that it is an inherently interferometric method. In its standard implementation, at low excitation levels (below the shot-noise-limit), IFTCARS offers improved sensitivity via the addition of an inherently phase-stable LO. We have demonstrated chemical imaging of polymer samples using FTCARS and IFTCARS and have discussed experimental considerations necessary for optimum imaging with these methods. Disadvantages of the FTCARS technique are the relative difficulty of managing dispersion for ultrashort femtosecond pulses compared to picosecond pulses, and the fact that FTCARS cannot readily image single CARS modes, which could be desirable in applications where detailed spectral information is not needed or rapid imaging is essential. In FTCARS applications requiring rapid image acquisition, a practical solution is to combine imaging modalities: faster methods such as single-mode CARS or fluorescence microscopy could be used for full sample characterization, while FTCARS could perform high-resolution microspectroscopy in selected sample regions.

## Acknowledgment

This material is based upon work supported by the National Science Foundation under Grant No. 0721370.

## References

1. M. Hashimoto, T. Araki, and S. Kawata, Molecular vibration imaging in the fingerprint region by use of coherent anti-Stokes Raman scattering microscopy with a collinear configuration, *Optics Letters* **25**, 1768–1770 (2000).
2. J. X. Cheng, A. Volkmer, L. D. Book, and X. S. Xie, An epi-detected coherent anti-Stokes Raman scattering (E-CARS) microscope with high spectral resolution and high sensitivity, *Journal of Physical Chemistry B* **105**, 1277–1280 (2001).
3. J. X. Cheng, L. D. Book, and X. S. Xie, Polarization coherent anti-Stokes Raman scattering microscopy, *Optics Letters* **26**, 1341–1343 (2001).
4. J. X. Cheng, A. Volkmer, L. D. Book, and X. S. Xie, Multiplex coherent anti-Stokes Raman scattering microspectroscopy and study of lipid vesicles, *Journal of Physical Chemistry B* **106**, 8493–8498 (2002).
5. M. Muller and J. M. Schins, Imaging the thermodynamic state of lipid membranes with multiplex CARS microscopy, *Journal of Physical Chemistry B* **106**, 3715–3723 (2002).
6. T. W. Kee and M. T. Cicerone, Simple approach to one-laser, broadband coherent anti-Stokes Raman scattering microscopy, *Optics Letters* **29**, 2701–2703 (2004).
7. H. Kano and H. Hamaguchi, In-vivo multi-nonlinear optical imaging of a living cell using a supercontinuum light source generated from a photonic crystal fiber, *Optics Express* **14**, 2798–2804 (2006).
8. M. Cui, J. Skodack, and J. P. Ogilvie, Chemical imaging with Fourier transform coherent anti-Stokes Raman scattering microscopy, *Applied Optics* **47**, 5790–5798 (2008).
9. J. P. Ogilvie, E. Beaurepaire, A. Alexandrou, and M. Joffre, Fourier-transform coherent anti-Stokes Raman scattering microscopy, *Optics Letters* **31**, 480–482 (2006).
10. M. Cui, M. Joffre, J. Skodack, and J. P. Ogilvie, Interferometric Fourier transform coherent anti-Stokes Raman scattering, *Optics Express* **14**, 8448–8458 (2006).
11. N. Dudovich, D. Oron, and Y. Silberberg, Single-pulse coherently controlled nonlinear Raman spectroscopy and microscopy, *Nature* **418**, 512–514 (2002).
12. S. H. Lim, A. G. Caster, and S. R. Leone, Single-pulse phase-control interferometric coherent anti-Stokes Raman scattering spectroscopy, *Physical Review A* **72**, 041803 (2005).
13. J. Sung, B. C. Chen, and S. H. Lim, Fast three-dimensional chemical imaging by interferometric multiplex coherent anti-Stokes Raman scattering microscopy, *Journal of Raman Spectroscopy* **42**, 130–136 (2010).
14. A. F. Pegoraro, A. Ridsdale, D. J. Moffatt, Y. W. Jia, J. P. Pezacki, and A. Stolow, Optimally chirped multimodal CARS microscopy based on a single Ti:sapphire oscillator, *Optics Express* **17**, 2984–2996 (2009).
15. J. X. Cheng and X. S. Xie, Coherent anti-Stokes Raman scattering microscopy: Instrumentation, theory, and applications, *Journal of Physical Chemistry B* **108**, 827–840 (2004).
16. Y. X. Yan and K. A. Nelson, Impulsive stimulated light-scattering. 2. Comparison to frequency-domain light-scattering spectroscopy, *Journal of Chemical Physics* **87**, 6257–6265 (1987).
17. R. Merlin, Generating coherent THz phonons with light pulses, *Solid State Communications* **102**, 207–220 (1997).
18. A. Volkmer, L. D. Book, and X. S. Xie, Time-resolved coherent anti-Stokes Raman scattering microscopy: Imaging based on Raman free induction decay, *Applied Physics Letters* **80**, 1505–1507 (2002).
19. D. L. Marks, C. Vinegoni, J. S. Bredfeldt, and S. A. Boppart, Interferometric differentiation between resonant and coherent anti-Stokes Raman scattering and nonresonant four-wave-mixing processes, *Applied Physics Letters* **85**, 5787–5789 (2004).
20. N. Dudovich, D. Oron, and Y. Silberberg, Single-pulse coherent anti-Stokes Raman spectroscopy in the fingerprint spectral region, *Journal of Chemical Physics* **118**, 9208–9215 (2003).
21. K. Naganuma, K. Mogi, and H. Yamada, General method for ultrashort light pulse chirp measurement, *IEEE Journal of Quantum Electronics* **25**, 1225 (1989).

22. J. P. Ogilvie, K. J. Kubarych, A. Alexandrou, and M. Joffre, Fourier transform measurement of two-photon excitation spectra: Applications to microscopy and optimal control, *Optics Letters* **30**, 911–913 (2005).
23. Sadtler, *The Sadtler Standard Raman Spectra* (Sadtler Research Laboratories, subsidiary of Block Engineering, Philadelphia, PA, 1973).
24. M. D. Levenson and G. L. Eesley, Polarization-selective optical heterodyne detection for dramatically improved sensitivity in laser spectroscopy, *Applied Physics* **19**, 1 (1979).
25. E. O. Potma, C. L. Evans, and X. S. Xie, Heterodyne coherent anti-Stokes Raman scattering (CARS) imaging, *Optics Letters* **31**, 241–243 (2006).
26. D. Oron, N. Dudovich, and Y. Silberberg, Single-pulse phase-contrast nonlinear Raman spectroscopy, *Physical Review Letters* **89**, 273001 (2002).
27. S.-H. Lim, A. G. Caster, and S. R. Leone, Single-pulse phase-control interferometric coherent anti-Stokes Raman scattering spectroscopy, *Physical Review A* **72**, 041803R (2005).
28. S. H. Lim, A. G. Caster, O. Nicolet, and S. R. Leone, Chemical imaging by single pulse interferometric coherent anti-Stokes Raman scattering microscopy, *Journal of Physical Chemistry B* **110**, 5196–5204 (2006).
29. G. W. Jones, D. L. Marks, C. Vinegoni, and S. A. Boppart, High-spectral-resolution coherent anti-Stokes Raman scattering with interferometrically detected broadband chirped pulses, *Optics Letters* **31**, 1543–1545 (2006).
30. C. L. Evans, E. O. Potma, and X. S. Xie, Coherent anti-Stokes Raman scattering spectral interferometry: Determination of the real and imaginary components of nonlinear susceptibility $\chi^{(3)}$ for vibrational microscopy, *Optics Letters* **29**, 2923–2925 (2004).
31. M. Greve, B. Bodermann, H. R. Telle, P. Baum, and E. Riedle, High-contrast chemical imaging with gated heterodyne coherent anti-Stokes Raman scattering microscopy, *Applied Physics B* **81**, 875–879 (2005).
32. W. Min, C. W. Freudiger, S. Lu, and X. S. Xie, Coherent nonlinear optical imaging: Beyond fluorescence microscopy, *The Annual Review of Physical Chemistry* **62**, 507–530 (2011).
33. L. Mertz, *Transformations in Optics* (John Wiley & Sons, Inc., New York, 1965).
34. J. A. McGuire and Y. R. Shen, Signal and noise in Fourier-transform sum-frequency surface vibrational spectroscopy with femtosecond lasers, *Journal of the Optical Society of America B-Optical Physics* **23**, 363–369 (2006).
35. H. J. Koester, D. Baur, R. Uhl, and S. W. Hell, Ca2+ fluorescence imaging with pico- and femtosecond two-photon excitation: Signal and photodamage, *Biophysical Journal* 77, 2226–2236 (1999).
36. M. Cui, B. R. Bachler, and J. P. Ogilvie, Comparing coherent and spontaneous Raman scattering under biological imaging conditions, *Optics Letters* **34**, 773–775 (2009).

# 11. CRS with Alternative Beam Profiles

Varun Raghunathan, Hyunmin Kim,
Stephan Stranick, and Eric Olaf Potma



## 11.1 Spatial Coherence in Coherent Raman Microscopy

One of the most defining features of the coherent Raman scattering (CRS) microscope is the mechanism by which the signal radiation emanates from the focal volume. True to its name, the coherent Raman signal is coherent. In a simple monochromatic picture, this coherence is best understood in terms of a spatial coherence. Spatial coherence implies that a molecule in the focal volume is oscillating with a phase that is not independent from the phase with which the other molecules in focus oscillate. Consequently, when the molecular

*Coherent Raman Scattering Microscopy.* Edited by Ji-Xin Cheng and X. Sunney Xie © 2013 CRC Press/ Taylor & Francis Group, LLC. ISBN: 978-1-4398-6765-5.

oscillations are converted into radiation, the radiative contributions from the molecules throughout the focal spot will interfere as a result of this spatial coherence.

The coherent nature of the coherent Raman technique is one of its most celebrated virtues. Without this coherence, the ensuing signals would not benefit from constructive interference of the radiating components: the signals would be incoherent, isotropically emitted and, most importantly, very weak. Indeed, as explained in Chapter 2, it is the coherent character of the radiation that produces signals many orders of magnitude stronger than those observed in incoherent Raman techniques.

While coherence is the key to success in coherent Raman microscopy, enabling for instance video-rate vibrational imaging, it is also one of the most problematic properties of the CRS microscope. This two-faced nature of coherence becomes evident when looking a little closer at the imaging properties of the CRS microscope. Spatial coherence is the reason why the coherent Raman signals are phase-matched in the forward propagating direction, resulting in strong forward-directed signals with limited signal directly radiated in the epi-direction.[1–3] The lack of strong epi-emitted CRS signals is a limitation for the efficient detection of coherent Raman signals from thick tissues, for instance, where the forward detection scheme is rendered impractical.[4] Another complication related to coherence is the difficulty of implementing resolution enhancement techniques in the CRS microscope. Many of the resolution enhancement methods developed for fluorescence microscopy, most notably the stimulated emission by depletion (STED) technique,[5–7] rely on the manipulation of a subset of emitters in focus in a spatially incoherent fashion, which does not affect their peers. Because of the spatial coherence in CRS, it is not straightforward to selectively address a subset of molecules without interfering with their neighbors in focus. Therefore, the coherent nature of CRS complicates a super-resolution implementation in the coherent Raman microscope that is directly modeled after successful fluorescence techniques.

Whereas coherence prevents a forthright carbon copy of fluorescence methods in the CRS microscope, it does offer additional opportunities that have no analog in fluorescence microscopy. An important consequence of coherence is that, when manipulated and controlled, contrast mechanisms can be attained that are different from the regular CRS operating mode. An excellent example of manipulating coherence in terms of the spectral content of the oscillatory motions in focus is the use of spectral phase masks. Numerous studies have shown the benefits of sculpting the spectral phase and amplitude in CARS microscopy, resulting in efficient nonresonant background rejection,[8–11] selective mode detection,[12–14] and maximized signal strength.[15,16] Similarly inspired applications in SRS are likely to follow. Beyond spectral amplitude and phase shaping, a somewhat less explored area of coherent manipulation is the direct phase shaping of spatial coherences. As we will see in this chapter, controlling the amplitude and phase with which the molecules in focus radiate offers several routes to phase-matching control, contrast enhancement and, in some cases, resolution enhancement.

Manipulation of spatial coherences in focus is achieved by shaping the spatial profile of the excitation beams. In early CARS spectroscopy work, alternative beam profiles were sometimes used to improve the phase-matching conditions in collinear excitation geometry. For instance, annular pump beams have been used in combination with regular Gaussian Stokes beams to enhance the spatial overlap between the excitation pulses over longer interaction lengths.[17,18] It was also recognized that the CARS process can be influenced when using combinations of higher-order beam profiles. De Boeij et al. used this principle to suppress nonresonant CARS contributions in waveguides.[19] The latter observation shows that by merely adopting alternative beam profiles, the balance between the vibrational and the nonresonant contributions to the CARS signal can be tweaked. This is a powerful principle, as a simple tweak of the beam profile may drastically change the CARS contrast, thus offering room for contrast improvement. Moreover, the advent of spatial light modulators has allowed a relatively easy manipulation of the beam profile, offering an unlimited array of transverse beam shapes. Therefore, spatial beam shaping is an attractive technology when combined with CRS imaging, as it is relatively easy to implement while adding another dimension of signal control to the existing CRS microscope.[20]

Although the field of spatial beam shaping is still in its infancy when applied to coherent microscopy, recent work has shown the potential of this approach in a variety of imaging applications. Inspired by the resolution improvement in linear optical microscopy,[21] radially polarized beams and annular detection pupils have been used to improve the imaging properties of the CARS microscope.[22,23] Higher-order Hermite-Gaussian beam modes have been used in CARS microscopy to suppress bulk contributions and increase the sensitivity to microscopic edges and interfaces,[24–27] whereas annular phase masks have been shown to improve the axial imaging properties[28] of the CARS microscope while providing a route toward tightening the lateral focusing properties.[29–31] Moreover, these recipes are not limited to CRS applications. For instance, similar principles have been applied to improve the imaging properties of third-harmonic generation (THG) microscopy, yielding images with higher resolution relative to the resolution observed when using conventional beam modes.[32]

In this chapter, we will discuss several opportunities that arise when adopting alternative beam profiles in CRS microscopy. Because the principle of spatial phase is inherent to this discussion, we will place an emphasis on the concept of spatial interference. We will discuss several exemplary beam profiles and connect the unique spatial interferences that ensue in focus when such beams are used and how they can be tailored to retrieve additional information from the sample. We will also include a discussion on the use of spatial beam shaping for unveiling nanoscopic information and for achieving resolution enhancement, in addition to its feasibility in the context of tissue scattering.

## 11.2 A Closer Look at the Amplitude and Phase of Tightly Focused Beams

As discussed in Chapter 2, focusing a Gaussian-shaped beam with a high numerical aperture lens produces a tightly focused spot. Similar to the confinement of the electric field amplitude to a sub-micrometer volume, the phase of the field is also spatially nonuniform. While the discussion in Chapter 2 was mostly focused on the existence of the Gouy phase shift, here we will explore the larger extent of the focal phase profile and how this phase profile can be controlled with beam shaping.

### 11.2.1 Gaussian Beams

In Figure 11.1, the amplitude and phase of the focal spot, formed by focusing 800 nm radiation with a 1.1 numerical aperture lens, is shown. Figure 11.1a depicts the amplitude of the electric field in the focal plane, whereas the corresponding phase is shown in Figure 11.1b. It is immediately clear that each dark ring in the amplitude pattern corresponds to an abrupt π-phase jump. In general, each node in the amplitude distribution corresponds to a π-shift in the phase distribution. Of interest to our discussion is that these phase jumps occur at definite locations and that they are very sharp. In fact, theoretically, these phase edges can be considered infinitely sharp or discontinuous. Despite these sharp phase edges, within each of the rings of the Airy disk, the phase is spatially flat.

Along the optical axis, the phase undergoes a relatively shallow π-phase shift, which is the Gouy phase shift. This is plotted in Figure 11.1d. Away from the optical axis, and away from the focal plane, we see that the phase in each subsequent lateral plane is not constant but a slowly varying function in space. Despite the rather complex phase pattern, the spatial phase is relatively constant within each lateral plane when we confine ourselves to the region of the highest field amplitude (Figure 11.1c).

### 11.2.2 Hermite-Gaussian Beams

Hermite-Gaussian (HG) beam profiles correspond to higher-order Gaussian beams. In particular, the focal volumes of the lower-order HG beams (HG01 and HG10) are characterized by discrete one-dimensional phase steps, introducing nodal lines in the

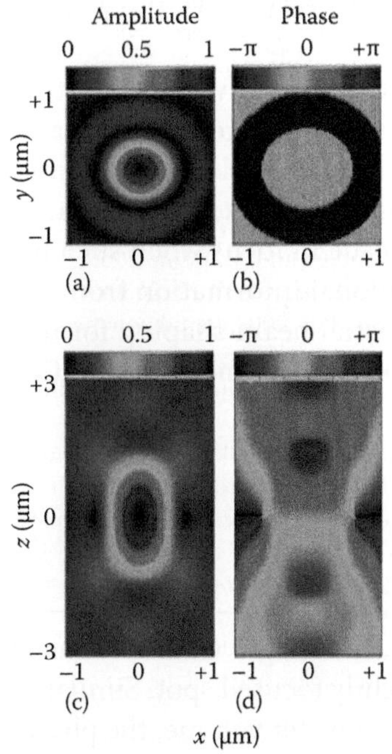


**FIGURE 11.1** Calculated amplitude and phase of a focused 800 nm laser beam by a 1.1 NA water immersion lens. (a) Amplitude of the field in the focal plane. (b) Phase of the field in the focal plane. (c) Amplitude in the *xz* plane. (d) Phase in the *xz* plane.

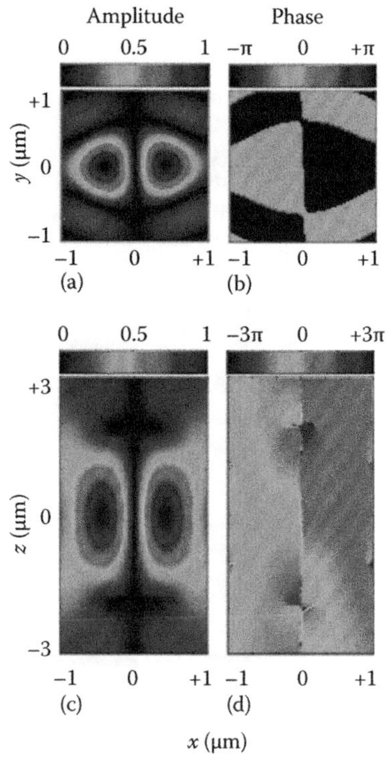


FIGURE 11.2 Amplitude and phase of an 800 nm focused laser beam with a HG01 phase profile. (a) Amplitude of the field in the focal plane. (b) Phase of the field in the focal plane. (c) Amplitude in the *xz* plane. (d) Phase in the *xz* plane.

amplitude distribution. Starting with a regular Gaussian transverse beam profile, a focal phase profile similar to a HG01 mode, for instance, can be easily realized by shaping the collimated beam with a phase mask that features a one-dimensional π-phase step. As can be gleaned from Figure 11.2b, the resulting phase step in focus is very sharp, dividing the total focal volume into two parts with equal amplitude but opposite phase.

Beam modes like HG01 are interesting because a phase pattern can be imposed onto the focal volume, producing a phase partitioned focal spot. Molecules can be driven at a particular phase in one of these partitions, while molecules in a different part are driven with an opposite phase. The ability to assign different phases to molecules according to their location in the focal volume adds a new degree of freedom for generating coherent signals. In subsequent sections we will discuss how a phase partitioning of the focal volume can be used to change the contrast in a CRS microscope.

### 11.2.3 Toraldo-Style Annular Beams

Applying an annular pattern of π-phase steps to a collimated beam generally produces a focal volume with a somewhat tighter central lobe at the expense of enhanced sidelobes. The use of such phase masks was first proposed by Toraldo,[33] who recognized that the tightening of the central lobe could be leveraged to improve the resolving power of the microscope. For our discussion here, we will focus primarily on the focal phase distribution produced by such annular phase masks.

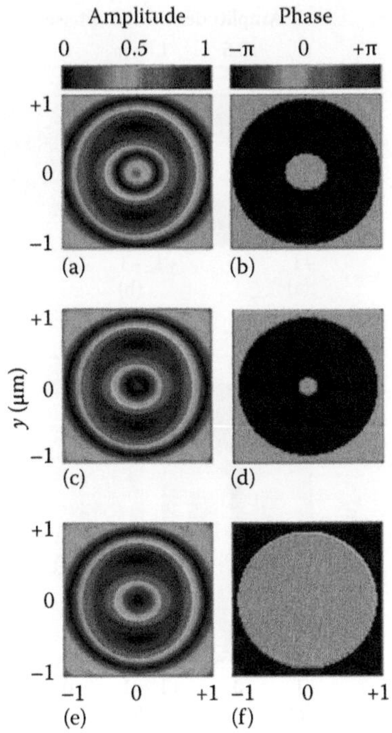


**FIGURE 11.3** Calculated amplitude (a, c, e) and phase (b, d, f) in the focal plane of a focused optical bottle beam for various settings of the normalized radius of the inner zone of the annular phase plate ($r$). For (a) and (b), $r = 0.275$, for (c) and (d), $r = 0.310$, and for (e) and (f), $r = 0.3175$.

The simplest annular phase profile consists of two concentric zones that are $\pi$ out-of-phase, yielding a corresponding phase distribution in the focal plane composed of concentric $0$–$\pi$ phase zones. Importantly, by changing the normalized radius $r$ of the inner zone of the phase mask, the size of the central phase zone can be tuned in the focal plane: note that $r$ is normalized to the radius of the back aperture of the objective. For the particular setting of $r = 0.3175$, the central phase zone becomes infinitely small, at which point the amplitude of the central lobe disappears, leaving a dark spot in the middle of the focal volume (Figure 11.3). The resulting profile is commonly referred to as an optical bottle beam (OBB)[34] and has found several applications in optical microscopy, most notably in STED microscopy. Figure 11.4 shows experimental pointspread functions of OBBs measured by probing the two-photon excited luminescence (TPEL) response of a gold nanoparticle.[29] In particular, Figure 11.4b clearly demonstrates that the width of the central lobe, along with the size of the central phase zone, changes by adjusting the $r$-setting of the phase mask.

The scaling of the central phase zone in the focus makes such beam profiles ideal candidates for improving the sensitivity in coherent microscopy. Although the overall volume of the focal spot is not considerably altered by applying the annular phase mask, the partitioning of the focal volume into smaller regions with a designated phase opens the door to retrieving information from the sample at length scales smaller than dictated by conventional diffraction-limited focal spots. In Section 11.7, we will discuss how control of the size of the central phase zone can be used to probe spatial information that is inaccessible with conventional beam profiles.

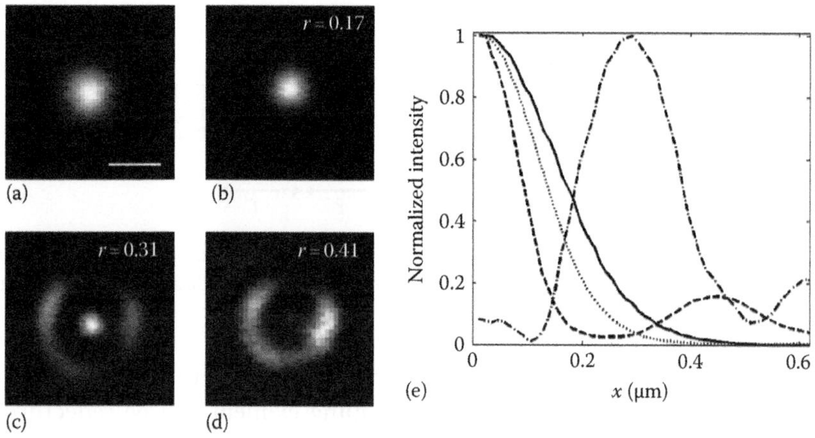


**FIGURE 11.4** Experimental pointspread functions (psf) of an optical bottle beam obtained by monitoring the two-photon excited luminescence (TPEL) from a 40 nm gold nanoparticle that is laterally scanned in the focal plane. (a) TPEL image of the nanoparticle when using conventional beams, (b) using OBB excitation with $r = 0.17$, (c) $r = 0.31$, and (d) $r = 0.41$. (e) Normalized one-dimensional cross sections through the pointspread functions with conventional (solid), $r = 0.17$ (dotted), $r$ 0.31 (dashed) and $r = 0.41$ (dot-dashed). Scale bar is 0.5 μm.

## 11.3 Creating Alternative CRS Excitation Volumes

In CRS microscopy, effective focal volume is shaped by the nonlinear interaction of the material with the pump and Stokes fields. The nonlinear excitation volume thus takes on the shape formed by the multiplicative effect of the pump and Stokes focal fields. By shaping either one of the incoming fields, or both, the effective CRS excitation volume can be sculpted beyond what can generally be achieved in linear microscopy. In the following we shall discuss several alternative CARS excitation volumes with similar strategies applicable to SRS excitation volumes as well.

### 11.3.1 Multiplicative Focal Volume Engineering

In CARS, the effective amplitude and phase of the excitation volume is given as $E_p^2 E_S^*$, where $E_p$ is the pump field and $E_S$ is the Stokes field. The corresponding spatial phase of the excitation field is given as $\phi_{ex}(r) = 2\phi_p(r) - \phi_S(r)$. This implies that the spatial phase at each position $r$ in the focal excitation volume is dictated by the phase of both the pump and the Stokes field. In the CARS excitation scheme, a phase pattern on the pump beam will contribute differently to the phase of the excitation field than the spatial phase of the Stokes beam. In particular, because the pump contributes twice, the phase information encoded in a 0–π phase pattern will translate in a flat (0–2π) phase pattern in focus. The linear interaction of the Stokes beam, on the other hand, ensures that the phase pattern on the Stokes beam is directly transferred to the excitation volume. Hence, by phase shaping the Stokes beam, and keeping the phase of the pump beam flat, the phase pattern of the excitation volume can be conveniently controlled.

A schematic of a CARS setup with spatial phase control is shown in Figure 11.5. A typical phase shaping setup incorporates a phase shaping element in a 4f arrangement. Such an arrangement ensures that the phase pattern imprinted onto the beam in the

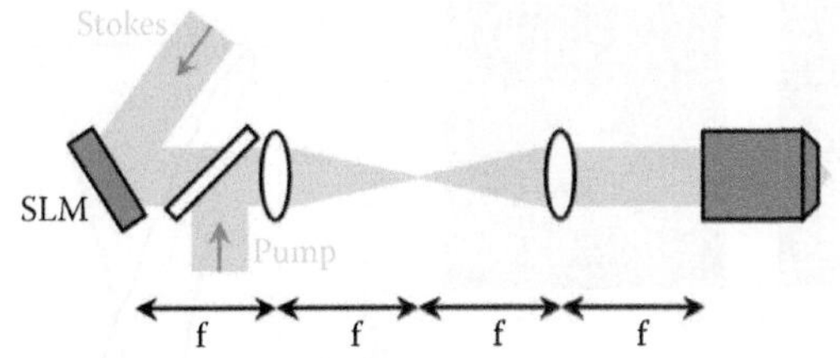


FIGURE 11.5 Schematic of a typical CARS setup with an SLM phase shaper for focus engineering. The 4f arrangement ensures a proper projection of the transverse phase profile at the SLM plane onto the back aperture of the objective lens.

plane of the phase shaping element is properly transferred to the back aperture of the high numerical aperture lens. The phase shaping element can be reflective or can be operating in transmission mode. Examples of transmission-based phase shapers include liquid crystal spatial light modulators. For 2D phase patterns, reflective-based elements are typically used, which can be either an actuator-controlled deformable mirror or a reflective liquid crystal display. With the advent of high-resolution reflective liquid crystal displays (1920 × 1080 pixels), the phase shaping element is no longer a limiting factor, and fine control of complex phase patterns can be achieved.

Figure 11.6 shows the calculated amplitude and phase of two different CARS excitation fields in the focal plane. In Figure 11.6a, the amplitude is shown of the CARS excitation field when the Stokes beam is dressed with the phase pattern reminiscent of a HG01 beam mode. As can be seen in Figure 11.6b, the sharp phase step carried by the Stokes field is imprinted onto the excitation volume, producing two lobes that are π-phase shifted relative to one another. Similarly, when the Stokes field is shaped with an annular phase mask, a concentric amplitude distribution of the CARS excitation field is obtained, as shown in Figure 11.6c. As can be gleaned from Figure 11.6d, the central

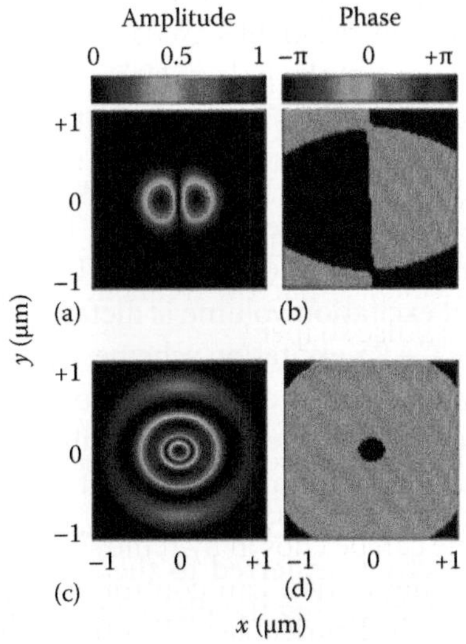


FIGURE 11.6 CARS excitation profiles in the focal plane using different phase masks for the Stokes beam. (a) Amplitude and (b) phase of the CARS excitation field when a HG01 phase profile is applied to the Stokes beam. (c) Amplitude and (d) of the CARS excitation field when an OBB (rho = 0.62) phase profile is applied to the Stokes beam.

lobe of the excitation field is π-phase shifted relative to the surrounding side lobe contribution. This implies that molecules in the central lobe are driven at a different phase than molecules situated in the other portions of the focal volume. Note that, because the central lobe can be scaled arbitrarily small with proper control of the annular phase mask, molecules can be driven with a designated phase in very small portions of the total diffraction-limited excitation volume. The amplitude and phase of CARS excitation volumes like the ones shown in Figure 11.11 have recently been realized and measured experimentally.[35]

It should also be noted that the segmentation of the focal volume into domains of alternating phases relies on the ability to construct high-quality focal spots in the sample. When sample scattering is significant, the quality of the focal volume (conventional or engineered) is compromised. In principle, the effects of sample scattering can be partially compensated for in CRS microscopy with the use of adaptive optics techniques.[36] Ultimately, the confinement of the focal spot is directly related to the sharpness of the focal phase profile. In typical scattering materials such as tissues, the amplitude in focus is strongly dependent on the focusing depth, yet the width of the focal volume is barely affected.[37–39] This preserved lateral confinement of the focal spot implies that, even in the presence of scattering, focal volumes with well-demarcated phase domains can be engineered. An example is found in STED imaging, which crucially depends on the quality of the phase-engineered focal profile of the depletion beam. Indeed, the dark focus of the depletion beam in STED microscopy is the direct result of the interference of focal contributions that are π-phase shifted in the center portion of the focal volume.[6] The successful implementation of STED in biological samples is a direct manifestation of the preservation of phase information in focus.

## 11.3.2 Subtractive Focal Volume Engineering

Whereas the multiplicative interaction of the pump and Stokes beams with the material allows for a direct shaping of the focal volume through manipulating the transverse phase profile of one (or both) of the beams, the resulting amplitude of the focal excitation volume is ultimately limited by the shapes allowed through the $E_p^2 E_S^*$ multiplication. Nonetheless, the nonlinear Raman interaction allows for alternative excitation schemes that go beyond shaping through the multiplicative field interaction.

One such scheme is shown in Figure 11.7a. In this scheme, the pump beam is split into two parts, $E_{p1}$ and $E_{p2}$. One arm is a simple relay while the other arm contains a phase shaping element. The pump beams are recombined on a beam splitter such that their relative phase delay is π/2. The two pump beams are then collinearly overlapped with the Stokes beam and focused to a diffraction limited spot by a high numerical aperture lens. The CARS polarization $P^{(3)}$ in focus can then be written as:

$$P_{as}^{(3)} \propto \chi^{(3)} \left\{ E_{p1}^2 E_S^* + E_{p2}^2 E_S^* e^{i\pi} + E_{p1} E_{p2} E_S^* e^{i\pi/2} \right\} \tag{11.1}$$

The polarization results from three different multiplicative field terms. The first two terms are π out-of-phase and combine to $(E_{p1}^2 - E_{p2}^2)E_S^*$. The resulting effect is that the effective excitation field is shaped by the subtraction of two fields $E_{p1}^2$ and $E_{p2}^2$. The subtractive

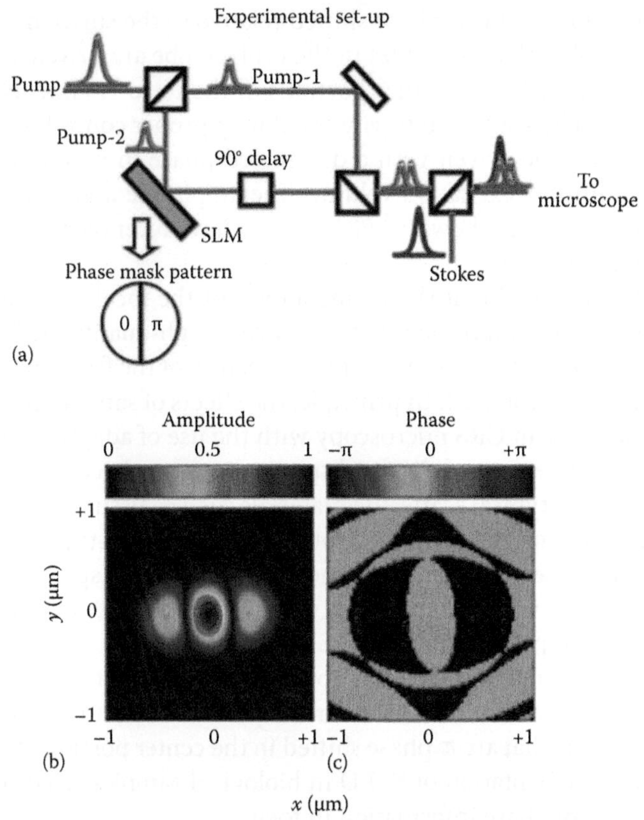


**FIGURE 11.7** (a) Schematic for the subtractive phase shaping method. (b) Amplitude of the in-phase component of the polarization, showing tightening of the central lobe in addition to the growth of side bands. (c) Corresponding focal phase profile.

operation enables field manipulations beyond what can be achieved with multiplicative interactions alone. The third term in Equation 11.1 is in quadrature with the first two terms and thus does not interfere with the former contributions. Within the approximation that the sample response is purely real, the latter term can be discriminated from the other components by using interferometric techniques.[20,40] Figure 11.7b and c show an example of an excitation field attained when using the subtractive phase shaping scheme.

## 11.4 Emission Profiles

Manipulation of the phase distribution in the focus has immediate consequences for the emission profile. Under standard CARS excitation conditions, the spatial phase distribution in the focal volume is such that the molecules radiate in unison in the direction of the optical axis direction, producing phase-matched, forward propagating radiation. In Figure 11.8a, the CARS emission profile from a bulk sample is shown when using regular Gaussian excitation fields with flat phase fronts (HG00 mode). The strongest signals are seen in the forward direction along the optical axis, $\theta = 0°$. At larger angles of $\theta$ away from the optical axis, the signal becomes increasingly more phase mismatched, effectively producing a highly directional forward propagating radiation field.

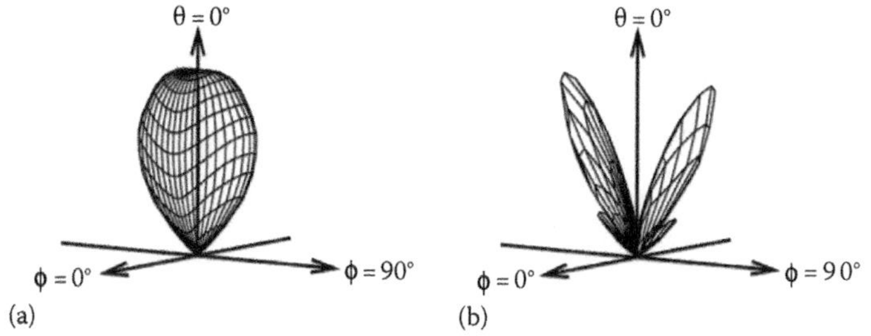


**FIGURE 11.8** Calculated CARS emission profiles using different phase profiles for the Stokes beam. (a) Regular (flat) Stokes phase profile. (b) HG01 Stokes phase profile. The calculations were done for a 1.1 NA water immersion lens with an 800 nm pump and a 1064 nm Stokes beam.

Whenever a portion of the molecules in the focus is driven at a phase different from the molecules in surrounding focal regions, the resulting radiation from the total focal volume is no longer necessarily in phase along the optical axis. In Figure 11.8b, for instance, the emission profile is shown from a CARS excitation volume shaped with a HG01 Stokes beam, similar to the excitation profile shown in Figure 11.6a and b. Because the molecules in one of the lobes of the excitation field give rise to optical waves that are out of phase with the waves emanating from the molecules in the other lobe, the CARS radiation along the optical axis is nullified by destructive interference. For larger angles $\theta$, however, the destructive interference along the optical axis can be compensated by the phase difference accumulated along the lateral dimension, yielding phase-matched emission lobes with a known tilt relative to the optical axis.

This simple example shows that the radiation pattern can be significantly altered by phase shaping the excitation beams, effectively controlling the direction of the emitted field. Consequently, phase shaping allows control of the phase matching conditions in focus. The concept of controlling the direction of propagation of the radiated field is not limited to small tweaks about the optical axis. By making use of more advanced beam profiles, the phase pattern in the direction of the optical axis can also be significantly altered, opening up opportunities, for instance, to tentatively generate strong phase-matched contributions emitted in the backward direction, $\theta = 180°$.

## 11.5 New Contrast Mechanisms

As we have seen, the alternative phase profiles of the excitation beams give rise to more complex focal volumes, which, in turn, yield unconventional radiation patterns. Such controllable focal volumes can be used to improve the imaging properties of the microscope. In this section we discuss a few brief examples of how spatial beam shaping can be used to alter the imaging contrast in the CARS microscope.

### 11.5.1 Interplay of Spatial and Spectral Phase

Thus far, we have discussed the deliberate application of sharp $\pi$-phase discontinuities in the spatial layout of the focal volume. The presence of such phase steps produces destructive interference of the CARS radiation in a controlled manner. However, the

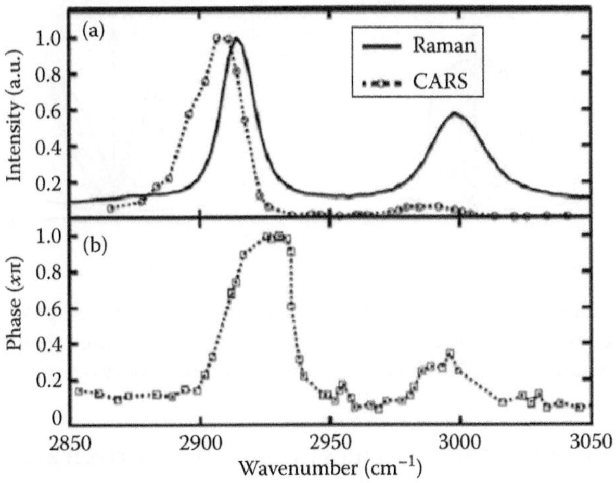


**FIGURE 11.9** Spectral amplitude and phase of a dimethyl sulfoxide (DMSO) solution. (a) Raman (solid curve) and CARS intensity (dashed curve with circles) spectra of a 1:5 DMSO:$H_2O$ mixture. (b) Spectral phase of the same sample measured using CARS interferometer.

spatial phase is not the only source of destructive interference between the radiative CARS contributions. As discussed in Chapter 1, the material response is complex, and thus generates CARS radiation with a frequency-dependent amplitude and phase. The phase of the nonlinear polarization changes in the vicinity of a vibrational resonance. An example is shown in Figure 11.9. In Figure 11.9a, the CARS amplitude is shown of the symmetric (2913 $cm^{-1}$) and asymmetric (3000 $cm^{-1}$) $CH_3$ stretching vibrations of aqueous dimethyl-sulfoxide. In Figure 11.9b, the corresponding phase change of the CARS radiation is plotted. It can be seen that for an isolated vibrational resonance such as the 2913 $cm^{-1}$ band, the phase of emitted field undergoes a $\pi$-phase step as a function of frequency. The phase on the red edge of the resonance is thus different from the phase at the blue side of the resonance. The relation between the amplitude and phase is inherent to any driven oscillator, including Raman-active oscillators, and is captured by the celebrated Kramers–Kronig transform. This intimate relationship between the spectral amplitude and phase is also the basis for vibrational phase imaging, which will be discussed further in Chapter 12.

The two sources of phase shifts in the CARS response, the spatial phase and the spectral phase, combine to control the degree of constructive and destructive interference of the ensuing radiation, respectively. The total CARS signal is thus the result of the interplay between these two phase-controlling mechanisms. In particular, the destructive interference introduced by a sharp spatial $\pi$-phase step in focus, as seen in Figure 11.8b, can be mitigated by a compensating spectral $\pi$-phase shift. This principle is schematically illustrated in Figure 11.10 for the case of a HG01 spatial phase profile. Whenever one half of the focus is occupied by an object that produces a vibrational $\pi$-phase shift, the phase difference between the two halves is lifted and constructive interference restored. In this configuration, the image contrast is dominated by vibrational resonances at interfaces, whereas the bulk signal is suppressed through destructive interference.

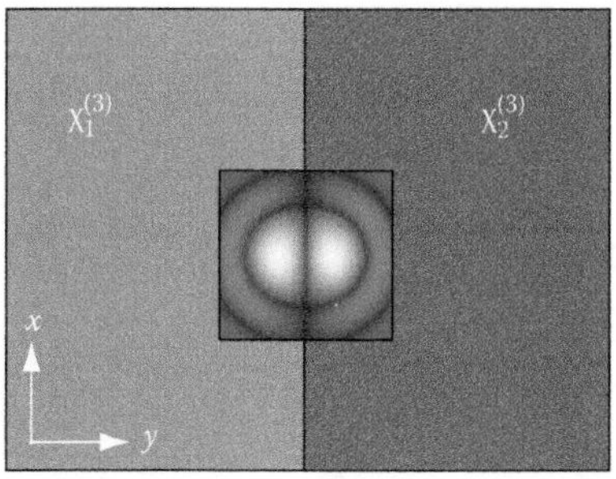


FIGURE 11.10 Principle of interface imaging using a HG01 phase-shaped CARS excitation volume. The interface is formed between two materials whose spectral phase of $\chi^{(3)}$ is $\pi$ out-of-phase with respect to the other material. A lateral projection of the CARS excitation volume is shown at the interface. Because the lobes of the excitation volume are also $\pi$ out-of-phase, the excitation phase can cancel the spatially variation in the spectral phase, yielding constructive interference of the radiation along the optical axis.

## 11.5.2 Detecting Chemical Interfaces

An interface constitutes a boundary between two materials. The optical properties of the materials, in the form of $\chi^{(3)}$, can be very different, each characterized by a particular amplitude and phase. For instance, one material may be vibrationally resonant while the other is not. Whenever the focal volume is positioned at such an interface, the fields generated in each of these materials will interfere according to the relative phase of the radiating contributions. The CARS emission pattern as well as the total detected intensity is drastically dependent on the mutual phase of the field contributions emanating from the different parts of the focus. An example is shown in Figure 11.11a, where the interface between two different liquids, deuterated DMSO and paraffin oil sandwiched between two glass coverslips, is visualized using regular CARS microscopy. At this particular frequency (2977 $cm^{-1}$), paraffin is driven to the blue side of the vibrational resonance where the phase of the CARS field is $\pi$-shifted relative to the nonresonant (real) contribution from the glass. On the other hand, the signal from the d-DMSO is far from a vibrational resonance, which corresponds to a CARS field that is in phase with the nonresonant glass contribution. The dark regions correspond to regions of two media that radiate with opposite phase, which destructively interfere and give rise to depleted CARS signals.[41]

The example above shows that at interfaces the coherent nature of the coherent Raman signal is evident in full glory. In leveraging the interference effects at properly configured interfaces, unwanted contributions can be destructively interfered while vibrational contributions can be enhanced.[42] Such effects can be even more pronounced when detecting selected portions of the CARS radiation pattern with the aid of detection masks.[43]

The effects at interfaces can be manipulated and controlled by using phase shaping of the excitation beams. As alluded to in the previous section, a phase partitioning of the focal volume can be counteracted by a spectral phase step across the interface. This contrast can be useful for imaging the contours, or edges, of vibrationally resonant objects surrounded by a nonresonant bulk. For instance, Figure 11.11b shows the contrast obtained from the d-DMSO:paraffin interface when the Stokes beam is dressed with a HG01 phase profile. Comparing this contrast with the one seen in Figure 11.11a, it is

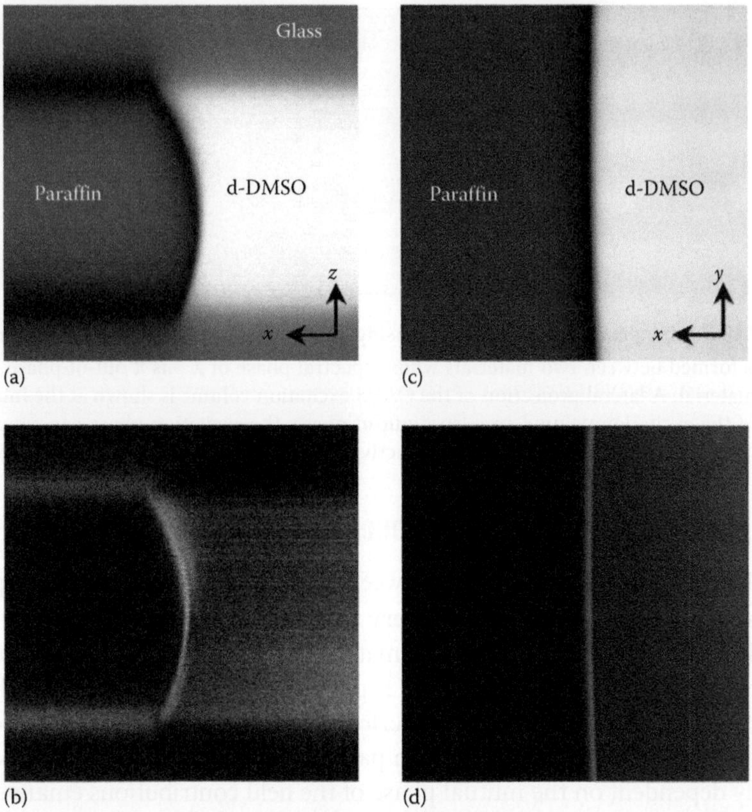


**FIGURE 11.11** Highlighting interfaces with focus-engineered CARS. An interface is formed between d-DMSO and paraffin oil, sandwiched between two glass coverslips. (a) CARS *xz*-cross-sectional image taken at 2977 $cm^{-1}$. (b) Same cross section with a HG01 Stokes beam. Note how the lateral and longitudinal interfaces are highlighted whereas the bulk signal is suppressed. (c) CARS *xy*-image of the d-DMSO:paraffin interface. (d) Same image now taken with a HG01 Stokes beam. Length of arrows is 2 μm.

evident that constructive interference between the two focal compartments is obtained right at the interface that is oriented either parallel or perpendicular to the optical axis.[25,26] Other phase masks, such as the annular phase patterns, have been shown to selectively highlight interfaces oriented perpendicular to the optical axis.[28]

It should be noted that these additional contrast mechanisms do not come at the expense of regular CARS contrast. A simple switch of the phase pattern on the computer-controlled phase shaping element enables one to quickly toggle between regular CARS contrast and enhanced interface contrast. The new contrast mechanisms enabled by phase shaping should be seen as complementary rather than as substitutes for conventional CARS imaging.

## 11.6 Interferometric Detection of CARS Radiation

Another interesting aspect of the coherent Raman radiation is that it can interfere with another phase-coherent light beam. The second light beam is typically called the local oscillator $E_{LO}$. There are two main reasons why interfering the coherent Raman radiation with a local oscillator beam is of interest. First, it offers a way to amplify the information carried by the signal. Second, it allows a direct detection of the emitted field rather than the intensity. Detecting the field implies that the signal is linearly proportional to the concentration of Raman scatterers. Besides detecting the amplitude of the field, interferometric detection also permits the characterization of the phase. This latter observation has far reaching consequences, as phase sensitive detection of the emitted field enables a full determination of the material response in terms of the real and imaginary parts of $\chi^{(3)}$. CARS interferometry was first implemented for spectroscopic applications in gases and liquids,[44–46] and was later shown to be useful for microscopic applications as well.[20,40,47–53] The details of how interferometric detection can be used to amplify the CARS signal, to suppress the nonresonant background and to selectively detect the amplitude and phase of the vibrational oscillators, is further discussed in Chapter 12. Here, we are interested in yet another aspect of CARS interferometry: the controlled energy exchange between radiation at the anti-Stokes frequency and radiation at the incident pump and Stokes frequencies.

Interferometric detection of a weak coherent Raman signal is a very general principle. Around the time when the first stimulated Raman gain (SRG)[54,55] and stimulated Raman loss (SRL)[56] were measured in the 1960s, it was realized that the strong coherent stimulated Raman effect was due to the coherent mixing of the radiation with a local oscillator beam.[57,58] Here, the local oscillator beams are the pump and Stokes beam themselves. In the coupled wave picture of the stimulated Raman scattering effect (SRG and SRL), the signal at $\omega_p$ interferes destructively with the pump beam, resulting in a loss of energy at the pump frequency (SRL), while the signal at $\omega_p$ interferes constructively with the Stokes radiation, producing an energy gain at the Stokes frequency (SRG). In stimulated Raman scattering, it is the imaginary part of material response that is in phase with the Stokes beam while the real part of the response is quadrature relative to the local oscillator beams. Consequently, the dissipative imaginary response of the material mediates an energy exchange from the pump beam to the Stokes beam. In a quantized description of the interacting fields, one pump photon is lost, one Stokes photon is gained, and the energy difference $\omega_p - \omega_S$ is absorbed by the vibrational eigenmode of the material.[59]

The interferometric interaction of the incident light and the material response in the SRS example discussed above illustrates that in the coherent Raman process several light beams can communicate coherently, i.e., exchange energy, in the material. In CARS, the emission occurs at a different frequency, and no interferometrically mediated energy exchange between the anti-Stokes field and the incident waves takes place. However, when a phase-coherent local oscillator with a frequency similar to the anti-Stokes radiation $\omega_{as}$ is applied in addition to the pump and Stokes beams, a similar energy exchange between the incident fields is observed. A simple example can illustrate this point. Consider the emitted anti-Stokes field as:

$$E_{as} \propto E_{in}\chi^{(3)}$$

where $E_{in} = E_p^2 E_S^*$ is the effective excitation field of the CARS process. When the anti-Stokes field is mixed with the local oscillator field, the signal at the detector is

$$I = |E_{LO}|^2 + |E_{as}|^2 + 2E_{LO}E_{in}\left[\mathrm{Re}\left\{\chi^{(3)}\right\}\cos\Theta + \mathrm{Im}\left\{\chi^{(3)}\right\}\sin\Theta\right]$$

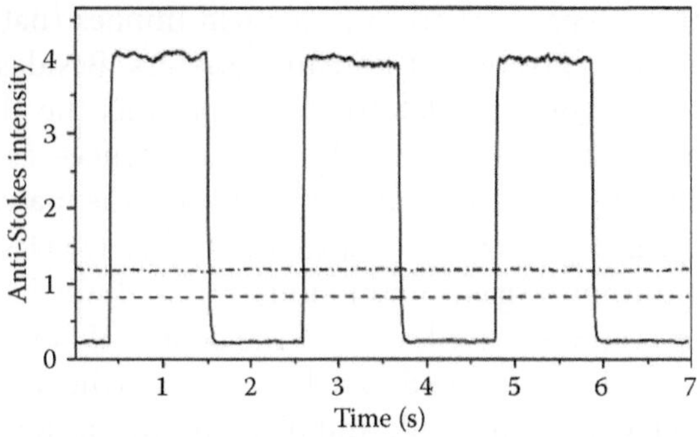


FIGURE 11.12 Dependence of the coherent anti-Stokes signal as a function of the local oscillator phase. Solid line is the total anti-Stokes signal. Dashed line is the CARS signal from DMSO at 2913 $cm^{-1}$ and dash-dotted line is the local oscillator intensity. Phase of the local oscillator was periodically toggled between 0 and $\pi$ by a computer-controlled interferometer. High plateau of the total anti-Stokes signal corresponds to $\Theta = 0$ and the low plateau corresponds to $\Theta = \pi$.

Here the phase between the local oscillator field and the anti-Stokes field is indicated as $\Theta$. In case the material's response is purely real, i.e., $\text{Im}\{\chi^{(3)}\} = 0$, and the phase is set to $\Theta = 0$, the total intensity at the detector is more than the sum of the intensities of the local oscillator and the CARS signal. Vice versa, when $\Theta = \pi$, the total intensity is less than the sum of the local oscillator and CARS intensities. An experimental example is shown in Figure 11.12. The phase-dependent loss of energy in the anti-Stokes radiation channel raises the question: where does the energy go? As can be shown by both a coupled wave equation analysis[60] and a quantized description of the fields,[61] the loss or gain of energy in the anti-Stokes channels is compensated by a gain or loss of energy in the incident frequency channels. In other words, controlling the phase of the local oscillator provides a handle for directing more or less energy into the anti-Stokes frequency channel.

The ability to control the energy flow and to resolve the amplitude and phase of the CARS field with a local oscillator beam adds a new dimension to focal volume control. In particular, CARS interferometry enables the possibility to selectively amplify the emission from certain portions of the focal volume (gain) while the emission from other portions of the focus is reduced (loss). In the following section, we will discuss some examples of how CARS interferometry can be used in combination with focal engineering to retrieve nanoscopic information that is inaccessible with regular coherent Raman microscopy.

## 11.7 Retrieving Nanoscopic Information with Phase-Shaped Beams

### 11.7.1 Imaging with Enhanced Resolution

The size of the focal volume is directly related to the resolution attainable in the CRS microscope. In the case of CARS, the effective excitation volume is defined through $E_{in} = E_p^2 E_S^*$, where spatial confinement of the focal fields $E_p$ and $E_S$ is dictated by diffraction theory. Engineering the focal volume of each of these incident fields by using phase-only shapers such as spatial light modulators or deformable mirrors by no means circumvents the diffraction limit. The spatial distribution of each focal field is still fully defined by the mechanism of diffraction.

Improving the spatial resolution of the CRS microscope is a formidable challenge. The success of fluorescence microscopy techniques that provide enhanced resolution suggests that similar mechanisms can be translated to CRS microscopy as well. However, as discussed before, the spatial coherence in CRS limits an independent manipulation of the radiation in some parts of the focal volume without affecting other portions of the focus. In addition, whereas the optical manipulation of populations of electronic states offers a very efficient mechanism for switching emitters on and off, an essential ingredient of resolution enhancement, the corresponding Raman-based manipulations of vibrational states are very inefficient. The power levels needed to populate or depopulate electronic states in typical molecular targets with high efficiency often exceed the photodamage levels when two-photon Raman transitions are used. An alternative scheme which uses direct one-photon (incoherent) excitation of the vibrational states using infrared radiation, followed by a Raman probing interaction, has been proposed to address this issue.[62] Other advanced schemes include the use of spatially dependent Rabi-oscillations to achieve an effective reduction of the focal volume.[63]

Here, we briefly discuss the kinds of information that can be obtained when using focus engineering of the pump and Stokes beams under regular excitation conditions and without the need for specialized energy states in the molecule. The attractive feature of this approach is that it can be relatively easily implemented in existing CRS microscopes without adding further imaging restrictions. Although the spatial distribution of the incoming fields remains determined by diffraction theory, it is important to realize that the CRS excitation volume is intrinsically nonlinear. This implies that by shaping the beams of pump and Stokes independently, several avenues are opened up for manipulating the size of the interaction volume beyond what can be achieved in regular CRS microscopy.[29]

As discussed in Section 11.3, the multiplication and subtraction procedures result in nonlinear excitation profiles that have a narrower central lobe, at the expense of an increasing contribution from the sidelobes. The smaller size of the central lobe opens up opportunities for improving the resolution of the CSR microscope. However, because the CRS emission is coherent, the presence of sidelobes introduces undesirable far-field interferences of these contributions with the radiation of the central lobe. Consequently, for imaging applications maximum reduction of the sidelobes is imperative in order to benefit from the reduced dimensions of the central lobe.

How much can the waist of the central lobe be tightened without introducing serious deteriorating imaging effects due to the sidelobes? Assuming a pump wavelength of 816 nm a Stokes wavelength of 1064 nm and a water immersion lens with a NA of 1.1, the intensity of the central lobe of the CARS excitation volume is 0.34 μm. Using an OBB profile for the Stokes beam with $r = 0.30$, the multiplicative CARS scheme offers a central lobe of 0.21 μm while keeping the intensity of the sidelobes at 4% relative to the central lobe. Hence, a 38% reduction of the center portion of the focal volume can be accomplished while the effect of the sidelobes is kept to a minimum.[30] Figure 11.13 shows experimental CARS data for an OBB-shaped excitation formed by shaping the pump

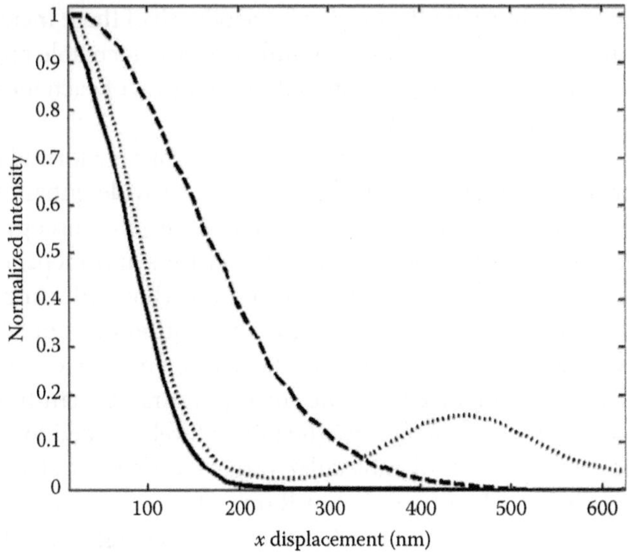


**FIGURE 11.13** Experimental measurement of CARS excitation volumes dressed with an OBB profile. Signals were measured as the electronic CARS response (718 nm) from a 40 nm gold nanoparticle (solid line). The pump beam (785 nm) is dressed with an OBB pattern of $r = 0.31$, and the resulting TPEL pointspread function is indicated by the dotted line. The TPEL pointspread function of the Stokes beam (852 nm) is shown as the dashed line.

beam with an annular phase mask of $r = 0.31$ while the Stokes beam exhibits a uniform phase. This figure shows the dramatic narrowing of the central lobe without introducing any significant sidelobes. Further reduction of the central lobe results in rapidly increasing contribution from the sidelobes. For instance, when $r = 0.60$, the central lobe shrinks to 0.15 μm, while the contribution from the sidelobes rockets to 65%.

For smaller Raman shifts, when the pump and Stokes wavelengths are closer in energy, the deteriorating effects of the sidelobes can be significantly suppressed while still benefiting from a reduction of the central lobe. For instance, it has been shown that the multiplicative beam-shaping scheme can lead to a central lobe below 0.13 μm for a pump wavelength of 754 nm and a Stokes wavelength of 785 nm, without significant sidelobes.[31] This approach offers high spatial resolution imaging in electronic FWM and CARS microscopy.

## 11.7.2 CARS Interferometric Focal Volume Partitioning

As discussed in Section 11.6, interferometric detection can translate phase differences between the nonlinear polarization in focus and the local oscillator into a loss or gain in the anti-Stokes radiation channel. This detection mechanism is very useful when combined with spatial phase shaping of the focal volumes. As we have seen in Section 11.3, the focal volumes can be partitioned into smaller segments of alternating phase. Although segmenting the focal volume does not lead to appreciable changes in the spatial extent of the integrated focal amplitude, the use of interferometric detection enables the recording of an otherwise inaccessible parameter: the spatial phase. By considering the focal volume not only in terms of its amplitude but also in terms of its phase, more spatial information can be retrieved relative to regular microscopy techniques.

In Figure 11.14, the CARS signal of a 50 nm nanoscopic particle is calculated when using an OBB Stokes beam in combination with interferometric detection. In Figure 11.14c, the anti-Stokes signal is plotted when the nanoparticle is laterally scanned through focus, with the central lobe shaped to a width of 0.20 μm. The phase of the local oscillator is tuned such that it is in phase with the center portion of the focal volume. The positive signal corresponds to a gain whereas the negative signal corresponds to a loss in the anti-Stokes channel. A loss signal occurs because the nonlinear polarization in the sidelobes is out-of-phase relative to the nonlinear polarization in the central lobe. As can be seen in the figure, the spatial locations where the transition from gain to loss occurs are separated by 0.20 μm, a distance smaller than the focal width of regular CARS excitation volumes. The gain or loss transition is very well defined when lock-in detection is used, enabling an effective "gain" imaging of the nanoparticle with an apparent resolution of 0.20 μm. By shrinking the central lobe further, the apparent resolution of the gain image can be reduced even more. In Figure 11.14f, the gain/loss anti-Stokes signal is given for a lateral scan of the nanoparticle through the focal volume with a central lobe of width 0.10 mm. The gain signal reveals spatial pattern of only 0.10 mm wide.

It should be emphasized that the apparent resolution improvement works well when a single isolated particle is scanned across the focal volume. However, this technique should not be regarded as a general imaging method for achieving CARS images with higher resolution. In particular, the presence of multiple particles or a large homogeneous sample in the focal volume will introduce interference of the anti-Stokes waves

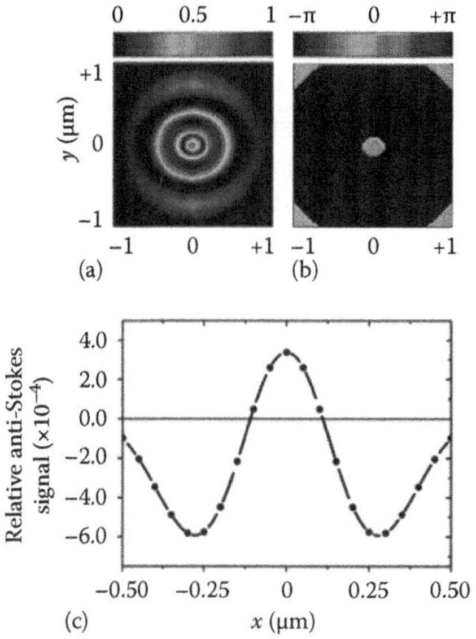

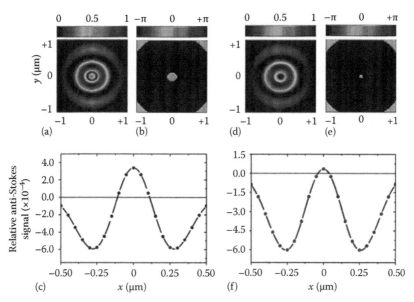


**FIGURE 11.14** Interferometric detection of the CARS signal leads to gain/loss in the anti-Stokes channel when a 50 nm particle is scanned through focus. In (a) and (b), the calculated amplitude and phase of an OBB CARS excitation field is shown, where *r* is set such that a central lobe of 200 nm is obtained. (c) Calculated gain and loss in the anti-Stokes channel when the particle is laterally scanned through focus. (d) and (e) show the amplitude and phase of an excitation volume with a central lobe of width 100 nm, and (f) depicts the calculated gain/loss in the anti-Stokes channel. Note that the gain/loss transitions occur at sub-diffraction limited distances, reflecting the width of the central lobe.

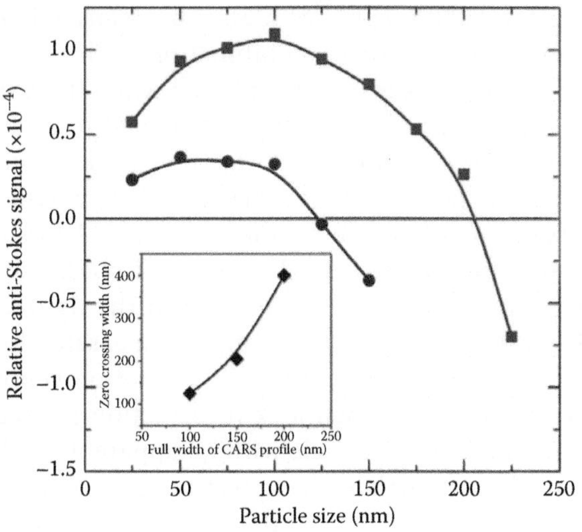


FIGURE 11.15 Calculated gain/loss in the anti-Stokes channel for particles of various sizes that are put in the center of the focal spot. The circles correspond to the anti-Stokes signal for an excitation profile with a central lobe of width 100 nm, whereas the squares correspond to a center lobe of 150 nm. Note that for a given width of the lobe, the point of zero gain in the anti-Stokes channel depends on the size of the particle. The inset shows the relation between the radius of the phase mask (width of lobe) and the particle size (point of zero gain).

from different focal segments, which complicates a straightforward decoding of the phase profile. Nonetheless, there are numerous problems in nanoscience in which the visualization of isolated nanostructures at their natural length scales can be useful. For instance, CARS excitation schemes have been used to detect the nonlinear response of metallic and semiconducting nanostructures.[64–69] The CARS focal volume partitioning combined with interferometric detection can provide spatial details of such nanostructures that are not easily observed with regular nonlinear imaging techniques.

Finally, the gain/loss transition when imaging individual nanoparticles can be used to determine the size of the particle. In Figure 11.15, the interferometric anti-Stokes signal is shown for particles of different size that are positioned in the center of the focal spot. For each particle size, the gain/loss transition occurs for a different setting of $r$. The inset of Figure 11.15 shows the relationship between the width of the central lobe (defined by $r$) and the particle size for which zero gain is achieved. Hence, by scanning $r$ on the SLM, the size of the particle can be determined by monitoring the gain/loss transitions. Using this recipe, the size of particles well below 100 nm can be determined. Consequently, this optical method enables size-determination of nanoparticles directly in the far-field optical microscope without the use of separate atomic force microscopy or electron microscopy measurements.

## 11.8 Conclusions

In this chapter, we have discussed several aspects of CRS imaging with alternative beam profiles. As the examples in this chapter show, focus engineering offers additional contrast mechanisms without fundamentally altering the design of the CRS microscope. Incorporation of an SLM into the setup is sufficient for realization of coherent Raman microscopy with more complex beam profiles, while a simple push of a button restores the regular imaging mode of the CRS microscope.

The detected signals in CRS are the result of the far-field interferences defined by the spatial coherences within the focal volume. Focus engineering in CRS takes full advantage of these coherent properties of the coherent Raman microscope. In fact, the contrast mechanisms in focus engineered CRS microscopy are a unique manifestation of the coherent oscillations in the focal volume and have no direct analog in fluorescence microscopy. Among the attractive features that focus engineering adds to the existing coherent Raman contrast are the highlighting of interfaces, a moderate improvement of spatial resolution, and the all-optical retrieval of sub-diffraction limited size information of isolated nanoscopic particles. Combined with the powerful capabilities of interferometric detection, it is expected that CRS microscopy with alternative beam profiles will find useful applications in the microscopic analysis of biological and engineered materials.

## References


1. G. C. Bjorklund, Effects of focusing on the third-order nonlinear process in isotropic media, *IEEE J. Quantum Electron.* **QE-11**, 287–296 (1975).
2. J.-X. Cheng, A. Volkmer, and X. S. Xie, Theoretical and experimental characterization of coherent anti-Stokes Raman scattering microscopy, *J. Opt. Soc. Am. B* **19**, 1363–1375 (2002).
3. E. O. Potma, W. P. d. Boeij, and D. A. Wiersma, Nonlinear coherent four-wave mixing in optical microscopy, *J. Opt. Soc. Am. B* **17**, 1678–1684 (2000).
4. C. L. Evans, E. O. Potma, M. Puoris'haag, D. Cote, C. Lin, and X. S. Xie, Chemical imaging of tissue in vivo with video-rate coherent anti-Stokes Raman scattering (CARS) microscopy, *Proc. Natl. Acad. Sci. USA* **102**, 16807–16812 (2005).
5. S. Hell and J. Wichmann, Breaking the diffraction limit by stimulated emission: Stimulated-emission depletion fluorescence microscopy, *Opt. Lett.* **19**, 780–782 (1994).
6. T. A. Klar, S. Jakobs, M. Dyba, A. Egner, and S. W. Hell, Fluorescence microscopy with diffraction resolution barrier broken by stimulated emission, *Proc. Natl. Acad. Sci. USA* **97**, 8206–8210 (2000).
7. K. I. Willig, R. R. Kellner, R. Medda, B. Hein, S. Jakobs, and S. W. Hell, Nanoscale resolution in GFP-based microscopy, *Nat. Methods* **3**, 721–723 (2006).
8. N. Dudovich, D. Oron, and Y. Silberberg, Single-pulse coherently controlled nonlinear Raman spectroscopy and microscopy, *Nature* **418**, 512–514 (2002).
9. S. H. Lim, A. Caster, and S. R. Leone, Single pulse phase-control interferometric coherent anti-Stokes Raman scattering (CARS) spectroscopy, *Phys. Rev. A* **72**, 041803 (2005).
10. H. Li, D. A. Harris, B. Xi, P. J. Wrzesinski, V. V. Lozovoy, and M. Dantus, Standoff and arms-length detection of chemicals with single-beam coherent anti-Stokes Raman scattering, *Appl. Opt.* **48**, B17–B22 (2009).
11. B. v. Vacano and M. Motzkus, Time-resolved two-color single-beam CARS employing supercontinuum and femtosecond pulse shaping, *Opt. Commun.* **264**, 488–493 (2006).
12. T. Hellerer, A. M. K. Enejder, and A. Zumbusch, Spectral focusing: High spectral resolution spectroscopy with broad-bandwidth laser pulses, *Appl. Phys. Lett.* **85**, 25–27 (2004).
13. A. F. Pegoraro, A. Ridsdale, D. J. Moffatt, Y. Jia, J. P. Pezacki, and A. Stolow, Optimally chirped multimodal CARS microscopy based on a single Ti:sapphire oscillator, *Opt. Express* **17**, 2984–2996 (2009).
14. W. Langbein, I. Rocha-Mendoza, and P. Borri, Coherent anti-Stokes Raman micro-spectroscopy using spectral focusing: Theory and experiment, *J. Raman Spectrosc.* **40**, 800–808 (2009).
15. A. C. W. v. Rhijn, S. Postma, J. P. Korterik, J. L. Herek, and H. L. Offerhaus, Chemically selective imaging by spectral phase shaping for broadband CARS around 3000 $cm^{-1}$, *J. Opt. Soc. Am. B* **26**, 559–563 (2009).
16. A. C. W. v. Rhijn, H. L. Offerhaus, P. v. d. Walle, J. L. Herek, and A. Jafarpour, Exploring, tailoring, and traversing the solution landscape of a phase-shaped CARS process, *Opt. Express* **18**, 2695–2709 (2010).

17. A. C. Eckbreth, G. M. Dobbs, J. H. Stufflebeam, and P. A. Tellex, CARS temperature and species measurements in augmented jet engine exhausts, *Appl. Opt.* **23**, 1328–1339 (1984).
18. D. R. Snelling, R. A. Sawchuk, and R. E. Mueller, Single pulse CARS noise: A comparison between single-mode and multimode pump lasers, *Appl. Opt.* **24**, 2771–2778 (1985).
19. W. P. d. Boeij, J. S. Kanger, G. W. Lucassen, C. Otto, and J. Greve, Waveguide CARS spectroscopy: A new method for background suppression, using dielectric layers as a model, *Appl. Spectrosc.* **47**, 723–730 (1993).
20. E. O. Potma and V. V. Krishnamachari, Imaging with phase-sensitive narrowband nonlinear microscopy, in *Biochemical Applications of Nonlinear Optical Spectroscopy*, V. V. Yakovlev, ed. (CRC Press, Boca Raton, FL, 2009).
21. C. J. R. Sheppard and A. Choudhury, Annular pupils, radial polarization, and superresolution, *Appl. Opt.* **43**, 4322–4327 (2004).
22. J. Lin, F. Lu, H. Wang, W. Zheng, C. J. R. Sheppard, and Z. Huang, Improved contrast radially polarized coherent anti-Stokes Raman scattering microscopy using annular aperture detection, *Appl. Phys. Lett.* **95**, 133703 (2009).
23. J. Lin, F. Lu, W. Zheng, and Z. Huang, Annular aperture-detected coherent anti-Stokes Raman scattering microscopy for high-contrast vibrational imaging, *Appl. Phys. Lett.* **97**, 083701 (2010).
24. V. V. Krishnamachari and E. O. Potma, Focus-engineered coherent anti-Stokes Raman scattering: A numerical investigation, *J. Opt. Soc. Am. A* **24**, 1138–1147 (2007).
25. V. V. Krishnamachari and E. O. Potma, Detecting lateral interfaces with focus-engineered coherent anti-Stokes Raman scattering microscopy, *J. Raman Spectrosc.* **39**, 593–598 (2008).
26. V. V. Krishnamachari and E. O. Potma, Multi-dimensional differential imaging with FE-CARS microscopy, *Vib. Spectrosc.* **50**, 10–14 (2009).
27. C. Liu and D. Y. Kim, Differential imaging in coherent anti-Stokes Raman scattering microscopy with Laguerre-Gaussian excitation beams, *Opt. Express* **15**, 10123–10134 (2007).
28. V. V. Krishnamachari and E. O. Potma, Imaging chemical interfaces perpendicular to the optical axis with phase-shaped coherent anti-Stokes Raman scattering microscopy, *Chem. Phys.* **341**, 81–88 (2007).
29. M. R. Beversluis and S. J. Stranick, Enhanced contrast coherent anti-Stokes Raman scattering microscopy using annular phase masks, *Appl. Phys. Lett.* **93**, 231115 (2008).
30. V. Raghunathan and E. O. Potma, Multiplicative and subtractive focal volume engineering in coherent Raman microscopy, *J. Opt. Soc. Am. A* **27**, 2365–2374 (2010).
31. H. Kim, G. W. Bryant, and S. J. Stranick, Superresolution four-wave mixing microscopy, *Opt. Express* **20**, 6042–6051 (2012).
32. O. Masihzadeh, P. Schlup, and R. A. Bartels, Enhanced spatial resolution in third-harmonic microscopy through polarization switching, *Opt. Lett.* **34**, 1240–1242 (2009).
33. G. T. d. Francia, Nuovo pupille superrisolvente, *Atti. Fond. Giorgio Ronchi.* **7**, 366–372 (1952).
34. J. Arlt and M. J. Padgett, Generation of a beam with a dark focus surrounded by regions of higher intensity: The optical bottle beam, *Opt. Lett.* **25**, 191–193 (2000).
35. V. Raghunathan, A. Nikolaenko, C.-Y. Chung, and E. O. Potma, Amplitude and phase of shaped nonlinear excitation fields in a four-wave mixing microscope, *Appl. Phys. Lett.* **99**, 171114 (2011).
36. A. J. Wright, S. P. Poland, J. M. Girkin, C. W. Freudiger, C. L. Evans, and X. S. Xie, Adaptive optics for enhanced signal in CARS microscopy, *Opt. Express* **15**, 18209–18219 (2007).
37. A. K. Dunn, V. P. Wallace, M. Coleno, M. W. Berns, and B. J. Tromberg, Influence of optical properties on two-photon fluorescence imaging in turbid samples, *Appl. Opt.* **39**, 1194–1201 (2000).
38. C. Y. Dong, K. Koenig, and P. So, Characterizing point spread functions of two-photon fluorescence microscopy in turbid medium, *J. Biomed. Opt.* **8**, 450–459 (2003).
39. C. Hayakawa, V. Venugopalan, V. V. Krishnamachari, and E. O. Potma, Amplitude and phase of tightly focused laser beams in turbid media, *Phys. Rev. Lett.* **103**, 043903 (2009).
40. E. O. Potma, C. L. Evans, and X. S. Xie, Heterodyne coherent anti-Stokes Raman scattering (CARS) imaging, *Opt. Lett.* **31**, 241–243 (2006).
41. D. Gachet, F. Billard, N. Sandeau, and H. Rigneault, Coherent anti-Stokes Raman scattering (CARS) microscopy imaging at interfaces: Evidence of interference effects, *Opt. Express* **15**, 10408–10420 (2007).
42. D. Gachet, F. Billard, and H. Rigneault, Focused field symmetries for background-free coherent anti-Stokes Raman spectroscopy, *Phys. Rev. A* **77**, 061802 (2008).
43. D. Gachet, S. Brustlein, and H. Rigneault, Revisiting the Young's double slit experiment for background-free nonlinear Raman spectroscopy and microscopy, *Phys. Rev. Lett.* **104**, 213905 (2010).
44. Y. Yacoby, R. Fitzgibbon, and B. Lax, Coherent cancellation of background in four-wave mixing spectroscopy, *J. Appl. Phys.* **51**, 3072–3077 (1980).

45. E. S. Lee and J. W. Hahn, Relative phase control between two successive coherent anti-Stokes Raman scattering signals for the recovery of spectral lines, *Appl. Opt.* **33**, 8302–8305 (1990).
46. G. Marowsky and G. Lüpke, CARS-background suppression by phase-controlled nonlinear interferometry, *Appl. Phys. B* **51**, 49–51 (1990).
47. C. L. Evans, E. O. Potma, and X. S. Xie, Coherent anti-Stokes Raman scattering interferometry: Determination of the real and imaginary components of the nonlinear susceptibility for vibrational microscopy, *Opt. Lett.* **29**, 2930–2932 (2004).
48. D. L. Marks and S. A. Boppart, Nonlinear interferometric vibrational imaging, *Phys. Rev. Lett.* **92**(12), 123905, 123901–123904 (2004).
49. C. Vinegoni, J. Bredfeldt, D. Marks, and S. Boppart, Nonlinear optical contrast enhancement for optical coherent tomography, *Opt. Express* **12**, 331–341 (2004).
50. M. Jurna, E. T. Garbacik, J. P. Korterik, J. L. Herek, C. Otto, and H. L. Offerhaus, Visualizing resonances in the complex plane with vibrational phase contrast coherent anti-Stokes Raman scattering, *Anal. Chem.* **82**, 7656–7659 (2009).
51. M. Jurna, J. P. Korterik, C. Otto, J. L. Herek, and H. L. Offerhaus, Vibrational phase contrast microscopy by use of coherent anti-Stokes Raman scattering, *Phys. Rev. Lett.* **103**, 043905 (2009).
52. E. S. Lee, J. Y. Lee, and Y. S. Yoo, Nonlinear optical interference of two successive coherent anti-Stokes Raman scattering signals for biological imaging applications, *J. Biomed. Opt.* **12**, 024010 (2007).
53. P. D. Chowdary, W. A. Benalcazar, Z. Jiang, D. M. marks, S. A. Boppart, and M. Gruebele, High speed nonlinear interferometric vibrational analysis of lipids by spectral decomposition, *Anal. Chem.* **82**, 3812–3818 (2010).
54. E. J. Woodbury and W. K. Ng, Ruby laser operation in the near IR, *Proc. Inst. Radio Eng.* **50**, 2347–2348 (1962).
55. G. Eckhardt, R. W. Hellwarth, F. J. McClung, S. E. Schwarz, D. Weiner, and E. J. Woodbury, Stimulated Raman scattering from organic liquids, *Phys. Rev. Lett.* **9**, 455–457 (1962).
56. W. J. Jones and B. P. Stoicheff, Inverse Raman spectra: Induced absorption at optical frequencies, *Phys. Rev. Lett.* **13**, 657–659 (1964).
57. J. A. Armstrong, N. Bloembergen, J. Ducuing, and P. S. Pershan, Interactions between light waves in a nonlinear dielectric, *Phys. Rev.* **127**, 1918–1939 (1962).
58. N. Bloembergen and Y. R. Shen, Coupling between vibrations and light waves in Raman laser media, *Phys. Rev. Lett.* **12**, 504–507 (1964).
59. R. W. Hellwarth, Theory of stimulated Raman scattering, *Phys. Rev.* **130**, 1850–1852 (1963).
60. A. Nikolaenko, V. V. Krishnamachari, and E. O. Potma, Interferometric switching of coherent anti-Stokes Raman scattering signals in microscopy, *Phys. Rev. A* **79**, 013823 (2009).
61. S. Rahav and S. Mukamel, Stimulated coherent anti-Stokes Raman spectroscopy (CARS) resonances originate from double-slit interference of two Stokes pathways, *Proc. Natl. Acad. Sci. USA* **107**, 4825–4829 (2010).
62. W. P. Beeker, P. Gross, C. J. Lee, C. Cleff, H. L. Offerhaus, C. Fallnich, J. L. Herek, and K. J. Boller, A route to sub-diffraction limited CARS microscopy, *Opt. Express* **17**, 22632–22638 (2009).
63. W. P. Beeker, C. J. Lee, K. J. Boller, P. Gross, C. Cleff, C. Fallnich, H. L. Offerhaus, and J. L. Herek, Spatially dependent Rabi oscillations: An approach to sub-diffraction limited coherent anti-Stokes Raman scattering microscopy, *Phys. Rev. B* **81**, 012507 (2010).
64. H. Kim, D. K. Taggart, C. Xiang, R. M. Penner, and E. O. Potma, Spatial control of coherent anti-Stokes emission with height-modulated gold zig-zag nanowires, *Nano Lett.* **8**, 2373–2377 (2008).
65. H. Kim, C. Xiang, A. G. Guell, R. M. Penner, and E. O. Potma, Tunable two-photon-excited luminescence in single gold nanowires fabricated by lithographically patterned nanowire electrodeposition, *J. Phys. Chem. C* **112**, 12721–12727 (2008).
66. Y. Jung, L. Tong, A. Tanaudommongkon, J. X. Cheng, and C. Yang, In vitro and in vivo nonlinear optical imaging of silicon nanowires, *Nano Lett.* **9**, 2440–2444 (2009).
67. Y. Jung, H. Chen, L. Tong, and J. X. Cheng, Imaging gold nanorods by plasmon-resonance-enhanced four-wave mixing, *J. Phys. Chem. C* **113**, 2657–2663 (2009).
68. Y. J. Lee, S. H. Parekh, J. A. Fagan, and M. T. Cicerone, Phonon dephasing and population decay dynamics of the G-band of semi-conducting single-wall carbon nanotubes, *Phys. Rev. B* **82**, 165342 (2010).
69. Y. Wang, C. Y. Lin, A. Nikolaenko, V. Raghunathan, and E. O. Potma, Four-wave mixing microscopy of nanostructures, *Adv. Opt. Photon.* **3**, 1–52 (2011).

# 12. Vibrational Phase Microscopy

**Martin Jurna, Cees Otto, and Herman L. Offerhaus**



Coherent processes like CARS, SRS, or variants thereof probe the molecular polarization and the phase of this polarization is imprinted on the generated signal. The phase carries information about the flow of energy; whether the driving field is below or above the molecular resonance or non-resonant. When only the intensity of the signal is detected, the information contained in the phase is lost and resonant signals cannot be separated from the non-resonant part. In mixtures of several compounds, the resonant CARS signal of less abundant constituents may be overwhelmed by the non-resonant background, preventing detection of resonant molecules.

In this chapter, we explore the phase as a source of contrast. We look at how the phase can be detected and exploited. The detection of the phase of the vibrational motion can be regarded as a vibrational extension of the linear (refractive index) phase contrast microscopy introduced by Zernike in 1933 [1].

## 12.1 Introduction

Due to the coherent nature of the CARS process, interference methods can be applied to improve the sensitivity and the selectivity of CARS. With a suitable local oscillator, heterodyne interferometric CARS detection can reach the sensitivity (shot noise) limit.

*Coherent Raman Scattering Microscopy*. Edited by Ji-Xin Cheng and X. Sunney Xie © 2013 CRC Press/Taylor & Francis Group, LLC. ISBN: 978-1-4398-6765-5.

The selectivity is enhanced because the phase separates the resonant, wavelength-dependent CARS signal from the non-resonant wavelength-independent signal. Similarly, mixtures of different materials with overlapping resonances can be separated based on the phase.

## 12.2 Heterodyne Interferometric Detection

Interferometric detection mixes a reference field, the so-called Local Oscillator (LO) field, with the generated CARS field at the anti-Stokes (AS) frequency. The total intensity on the detector can be expressed as

$$I_{detector} = |E_{LO}|^2 + |E_{AS}|^2 + 2E_{LO}E_{AS}. \tag{12.1}$$

The CARS field is proportional to the input fields and the resonant and non-resonant vibrational response $\chi^{(3)}$,

$$E_{AS} \propto E_{in}\left[\chi_R^{(3)} + \chi_{NR}^{(3)}\right], \tag{12.2}$$

where the input field is given by the combination of the Pump, Stokes, and Probe fields. For degenerate CARS the Probe equals the Pump so that:

$$E_{in} = E_{Pump}^2 E_{Stokes}^* \tag{12.3}$$

$\chi^{(3)}$ can be expressed as

$$\chi^{(3)} = \left\{\left[\chi_{NR}^{(3)} + \mathrm{Re}\left(\chi_R^{(3)}\right)\right]\cos(\varphi_\chi) + \left[\mathrm{Im}\left(\chi_R^{(3)}\right)\right]\sin(\varphi_\chi)\right\}, \tag{12.4}$$

where $\varphi_\chi$ is the vibrational phase of the molecules in the focal volume referenced to the phase of the non-resonant response. This is equal to the phase difference between the (complex) resonant part and the (purely real) non-resonant part.

The local oscillator must be phase- and wavelength-related to the generated CARS signal. It can be created in bulk media [2,3] adjacent to the CARS setup and combined with the pump and Stokes beams either before or after the sample. The non-resonant signal created in the CARS process can also be used as the local oscillator [4]; however, in that case the amount of amplification depends on the local amount of non-resonant signal created in the sample.

To avoid amplitude noise of the local oscillator and input fields, heterodyne detection can be used. For heterodyne detection the local oscillator field is shifted in frequency with respect to the CARS field, creating a beat signal on the detector. This beat can be separated by a lock-in amplifier. The lock-in amplifier provides the amplitude and phase of the interference signal, which corresponds to the amplitude of the generated CARS field and phase difference between that generated signal and the local driving field.

## 12.3 Cascaded Phase Preserving Chain

To generate a stable low-noise local oscillator, a cascaded phase preserving chain (PPC) can be used [5] as presented in Figure 12.2. The energy diagram shows the match between the wavelengths involved in creating the CARS signal and the wavelengths employed in the optical parametric oscillator (OPO). In Table 12.1 the wavelength and phase relations are listed. Note that the frequency doubling of the fundamental, 1064–532 nm, is phase coherent [6] and that the phases of the signal and idler are independent, but the sum of their phases is locked to the phase of the pump. The CARS wavelength is equal to the signal wavelength from the OPO. Furthermore, the phase of the resonant CARS signal is determined by the signal phase plus the phase of the vibrational response. The phase of the OPO signal is therefore locked to the phase of the CARS signal except for path-length variations between the point of creation and the point where they are combined. This phase preservation means that the signal interferes with the CARS signal in a predictable way and can thus be used for interferometric amplification and phase detection [7].

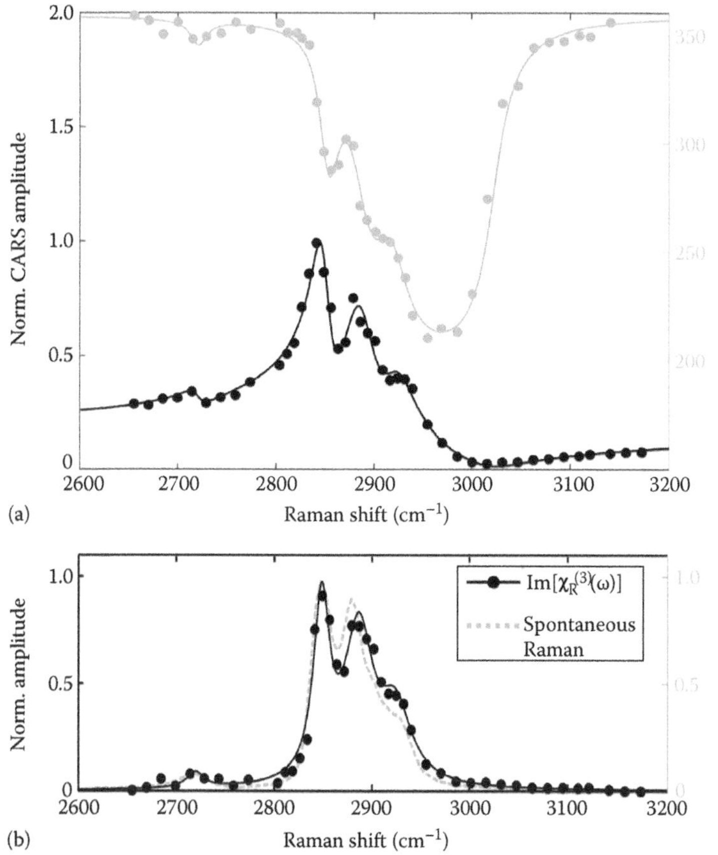


**FIGURE 12.1** (a) The CARS amplitude spectrum and associated phase of PE. The symbols are data and the solid curves are fits based on multiple Lorentzian bands. (b) The dotted line is the spontaneous Raman spectrum of PE and the solid curve is the imaginary part of the data in panel (a).

**Table 12.1** The Wavelength and Phase Relations of the Individual Stages in the Cascaded Phase Preserving Chain

| Process | Wavelength Relation | Phase Relation |
|---|---|---|
| 1. SHG | $2\omega_{1064} = \omega_{532}$ | $2\varphi_{1064} = \varphi_{532}$ |
| 2. OPO | $\Omega_{532} = \omega_{signal} + \omega_{idler}$ | $\varphi_{532} = \varphi_{signal} + \varphi_{idler}$ |
| 3. CARS | $\omega_{CARS} = 2\omega_{1064} - \omega_{idler}$ | $\varphi_{CARS} = 2\varphi_{1064} - \varphi_{idler} + \varphi_{\chi 3}$ |
| 4. PPC | $\omega_{CARS} = \omega_{signal}$ | $\varphi_{CARS} = \varphi_{signal} + \varphi_{\chi 3}$ |

## 12.4 Heterodyne Phase Detection

To verify the phase detection capabilities, a spectroscopic measurement of the phase over several overlapping vibrational resonances is presented. The sample is comprised of two adjacent areas, where one side consists of agarose gel, which contributes mainly to the non-resonant signal. The other side contains polyethylene (PE), which gives mostly resonant signal. Scanning the sample reveals a phase step between the non-resonant and resonant sides of the sample. The height of the phase step depends on the spectral position with respect to the vibrational resonance of PE. Due to the small scan area, there is no phase difference caused by the curvature of the field of view, as was verified by measurements on a sample containing only resonant material. While tuning the OPO through the high frequency vibrational spectrum of PE, images are recorded and averaged for the resonant region and the non-resonant region in the sample. Since there is no long-term stability to the absolute phase and the phase is certainly shifted after tuning the OPO, the phase on the non-resonant part of the image serves as the reference phase and the resonant phase can be measured with respect to the non-resonant part. The phase is determined as the difference in phase between the non-resonant and the resonant side of the image. The averages are corrected for intensity variations of the input wavelengths. The measured amplitude and phase are shown in Figure 12.1.

The phase response of the molecules is the difference between the phase of the driving field, at the difference frequency between pump and Stokes, and the phase of the motion of the dipoles in the sample. When the driving frequency is low (left) the response is dominated by the non-resonant response, which has a phase of zero. As the first resonance at ~2700 $cm^{-1}$ is approached, the phase of the response changes, but since this response is not much stronger than the non-resonant response, the phase does not rise to a full phase difference of $\pi$. Subsequent stronger resonances pull the phase further down. At frequencies above the resonances the non-resonant response starts to dominate again, returning the phase difference to zero.

To show that the phase data is in agreement with the amplitude data, the complex dataset, containing both the phase and amplitude information, is fitted to four vibrational levels of PE, modeled as complex Lorentzians bands, with a real offset for the non-resonant background. The fit results in the solid curves for the amplitude and phase as shown in Figure 12.1a, indicating that the measured phase and the spectral amplitude are related, as predicted. The phase has an additional offset of 8° due to the refractive index difference between PE and agarose gel. A direct comparison with the spontaneous Raman spectral intensity of PE can now be made. Figure 12.1b shows the comparison. The solid blue line shows the imaginary part of the complex CARS dataset and the purple dotted line shows the spontaneous Raman spectrum.

## 12.5 Detection of the Local Excitation Phase

The phase difference between the total CARS field and the excitation field is $\varphi_\chi$, see Equations 12.1 and 12.4. This is the phase difference that distinguishes the real (non-resonant) and the imaginary (resonant, Raman) parts. However, this is not the heterodyne phase measured in the focal volume; even for co-propagating fields this phase is disturbed by refractive index changes in the sample, curvature of the field-of-view, and interferometric instabilities. Detecting the local excitation phase and using this "base" phase to correct the phase of the CARS output gives the accurate phase of the oscillators in the focal volume, henceforth referred to as "vibrational phase". For the cascaded chain (Figure 12.2a), the phase difference ($\Delta_1$), obtained by the interference between the local oscillator (the OPO signal beam) and the CARS signal ($C_{HP}$) is given by:

$$\begin{aligned}\Delta_1 &= \varphi_S - \varphi_{CHP} \\ &= \varphi_S - (2\varphi_{F*} - \varphi_I + \varphi_\chi) \\ &= 2\varphi_F - 2\varphi_{F*} - \varphi_\chi,\end{aligned} \tag{12.5}$$

where F* is a frequency-shifted (by 50 kHz) version of F. Figure 12.2b shows the detection of the local excitation phase, where there are two separate routes to the same vibrational level (or continuum level) creating an interference that can be probed and detected as

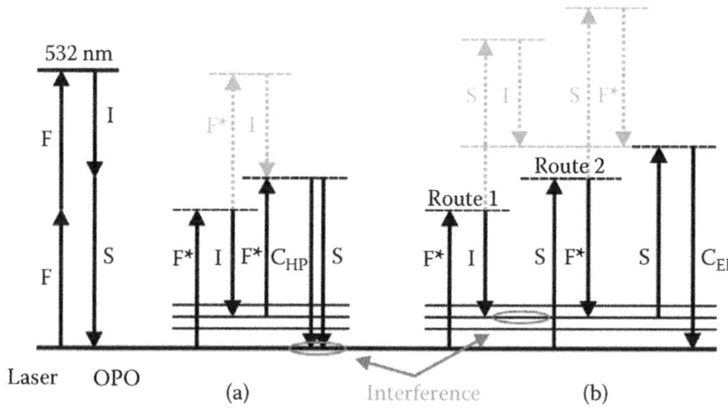


FIGURE 12.2 Schematic of the cascaded phase preserving chain between laser, OPO, and the CARS process. (a) The heterodyne phase detection scheme. (b) The local excitation phase detection scheme. F = laser fundamental (1064 nm), F* = frequency-shifted fundamental, I = OPO idler, S = OPO signal and C = CARS signal. Given in black and gray are the resonant and non-resonant CARS processes, respectively.

an intensity modulation at the $C_{EP}$ wavelength. The interference arises from the phase difference between the two pathways ($\Delta_2$):

$$\begin{aligned}\Delta_2 &= \Delta_{\text{Route 2}} - \Delta_{\text{Route 1}} \\ &= (\varphi_S - \varphi_{F^*}) - (\varphi_{F^*} - \varphi_I) \\ &= 2\varphi_F - 2\varphi_{F^*}.\end{aligned} \tag{12.6}$$

The difference between $\Delta_2$ and $\Delta_1$ is now purely due to the interaction with the vibrational phase response $\varphi_\chi$.

$$\begin{aligned}\textit{Vibrational phase} &= \Delta_2 - \Delta_1 \\ &= (2\varphi_F - 2\varphi_{F^*}) - (2\varphi_F - 2\varphi_{F^*} - \varphi_\chi) \\ &= \varphi_\chi.\end{aligned} \tag{12.7}$$

Phase disturbances will occur within the acquisition time of one frame and over the full field-of-view. Using phase detection on a point-by-point basis over the sample overcomes these interferometric instabilities [8].

## 12.6 Detection Setup

The key elements of the setup (see Figure 12.3) are the laser source emitting at 1064 nm, which is partially doubled to 532 nm, and a 532 nm synchronously pumped Optical Parametric Oscillator (OPO). The input powers of the fundamental and idler are several tens of mW, while the power of the local oscillator is only a few nW. Heterodyne interferometric detection is obtained by frequency shifting of the fundamental by an acousto-optical modulator, resulting in a frequency-shifted CARS signal. Using a voltage controlled oscillator (VCO), an external frequency ($\Omega$) is added to the detected laser repetition rate (around 80 MHz) and applied to the acousto-optical modulator. This $\Omega$ shift is translated to a 2 $\Omega$ shift at the CARS wavelength (two fundamental photons in the CARS process). The beat between the CARS signal and the local oscillator is detected on a photodiode and fed to a lock-in amplifier set to detect at 2 $\Omega$.

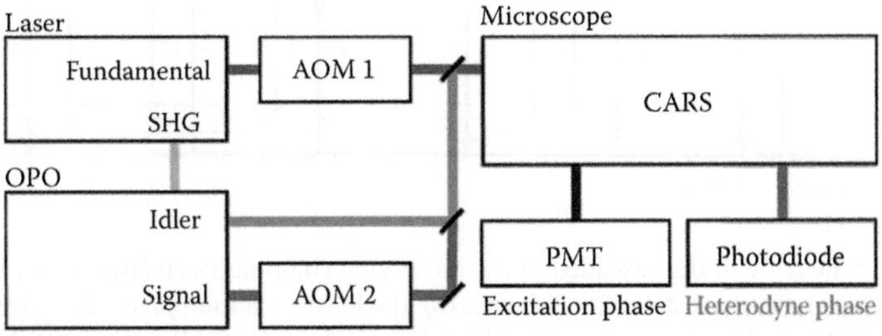


**FIGURE 12.3** Schematic of the VPC-CARS setup, where AOM 1 is phase modulated and AOM 2 is amplitude modulated.

For detection of the local excitation phase (Figure 12.2b), the amplitude of the signal of the OPO must be changed from a few nW to the same level as the fundamental beam to balance the contribution of both pathways, route 1 and route 2. The resulting interference between both routes is probed by another photon (OPO-signal photon) and detected on a PMT (or photodiode) by a second lock-in amplifier set to detect at the same 2 Ω modulation. Switching between the two power levels is done by a second acousto-optical modulator in the OPO signal branch. This acousto-optical modulator is also used in first order (to obtain the necessary high extinction ratio), but driven at the laser repetition rate to prevent frequency modulation. The amplitude modulation frequency (~1 kHz) of AOM 2 is set to match the pixel scan rate and is therefore a multiple of the frequency of the galvano scanners in the microscope.

To overcome interferometric instabilities which are too fast to collect complete images at a steady level, phase detection is done point-by-point. Scanning is done at 1 ms/pixel, alternating between the two different phase detection schemes. First the heterodyne phase is detected on the photodiode, when AOM 2 passes only a few nW, and second the local excitation phase is detected on the PMT, when AOM 2 passes all OPO-signal power, see Figure 12.4. With switching and sample-hold electronics the overload during the excitation phase detection time on the photodiode is removed before the signal enters the lock-in. The detection of the local excitation phase is done by a second lock-in amplifier set to detect at the same 100 kHz modulation. Both lock-in amplifiers are set to 1 ms integration time. Shorter pixel dwell times can be obtained by higher external modulation frequencies and shorter integration times.

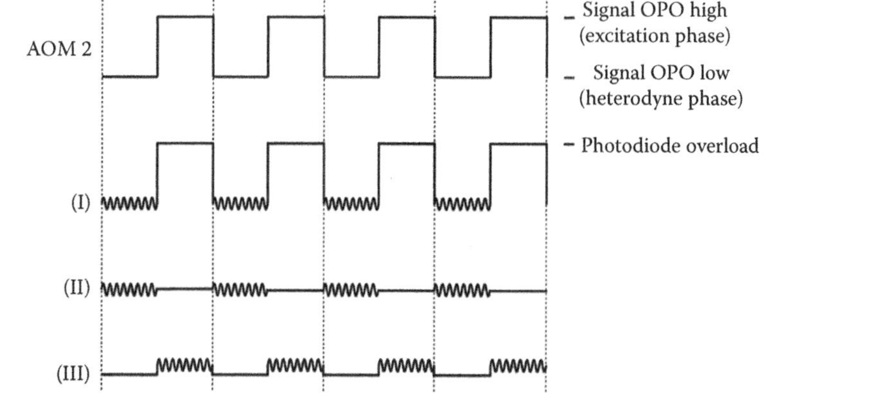

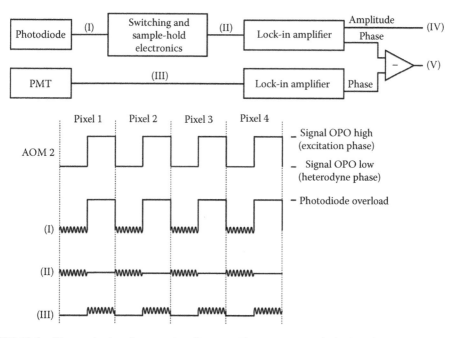


**FIGURE 12.4** Electronic signal processing diagram. The output signals (IV) and (V) are respectively the vibrational amplitude and phase. AOM 2 shows the switching of the signal of the OPO, (I) detected Photodiode signal, (II) Photodiode signal after switching and sample-hold electronics and (III) PMT signal.

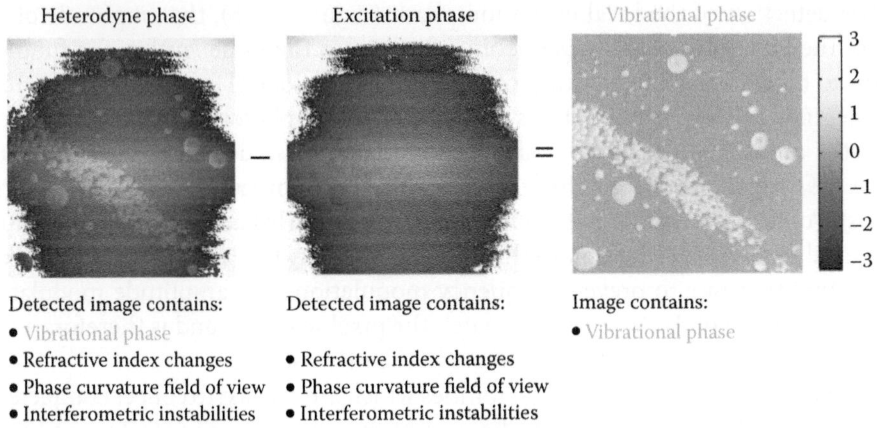


**FIGURE 12.5** A sample of fat globules (mayonnaise) of different sizes distributed in an agarose gel solution imaged at 2865 cm$^{-1}$. The subtraction of the heterodyne phase from the excitation phase shows a clean vibrational phase.

A demonstration of the vibrational phase detection is shown in Figure 12.5, where a sample of fat globules (mayonnaise) of different sizes distributed in agarose gel is imaged. The phase of the heterodyne signal contains the vibrational phase of the sample, but the vibrational phase is hard to distinguish (due to the refractive index changes in the sample, phase curvature over the field of view and interferometric instabilities). The excitation phase also shows the phase disturbances, but not the vibrational phase. The subtraction of the excitation phase from the heterodyne phase yields the vibrational phase of the sample, without disturbances. In the vibrational phase image, phase steps can be seen on the location of the fat globules.

## 12.7 Vibrational Phase Contrast CARS Microscopy

As a second demonstration HeLa cells are imaged. The heterodyne phase, Figure 12.6a, shows that the phase over the field-of-view is not constant, but distorted and noisy. At the location of the lipids in the cell we can see phase steps. These phase steps are mostly due to the difference in phase between the resonant and non-resonant signals. Figure 12.6b shows the local excitation phase detected on the PMT which is also distorted and noisy. Phase steps are again observed at the location of the lipid droplets, but these steps are caused by the differences in refractive index between the lipids and the water in the cell. The amplitude image of the CARS signal is shown in Figure 12.6c. Note that there are some droplets represented by dark spots in the amplitude image. Subtraction of the local excitation phase from the heterodyne phase can be seen in Figure 12.6d. The distortion and noise cancel out, allowing for averaging of this vibrational phase image to increase the precision on the phase value. Taking the sine of the vibrational phase image and multiplying it by the amplitude image, removes the non-resonant signal, which is done in Figure 12.6e. Comparing the amplitude image and the background free image, it can be seen that the internal structure of the cell and the edge of the cell have disappeared and only the resonant lipids are observed. This shows that the cell structure in the amplitude image is non-resonant and should not be confused for the resonant material

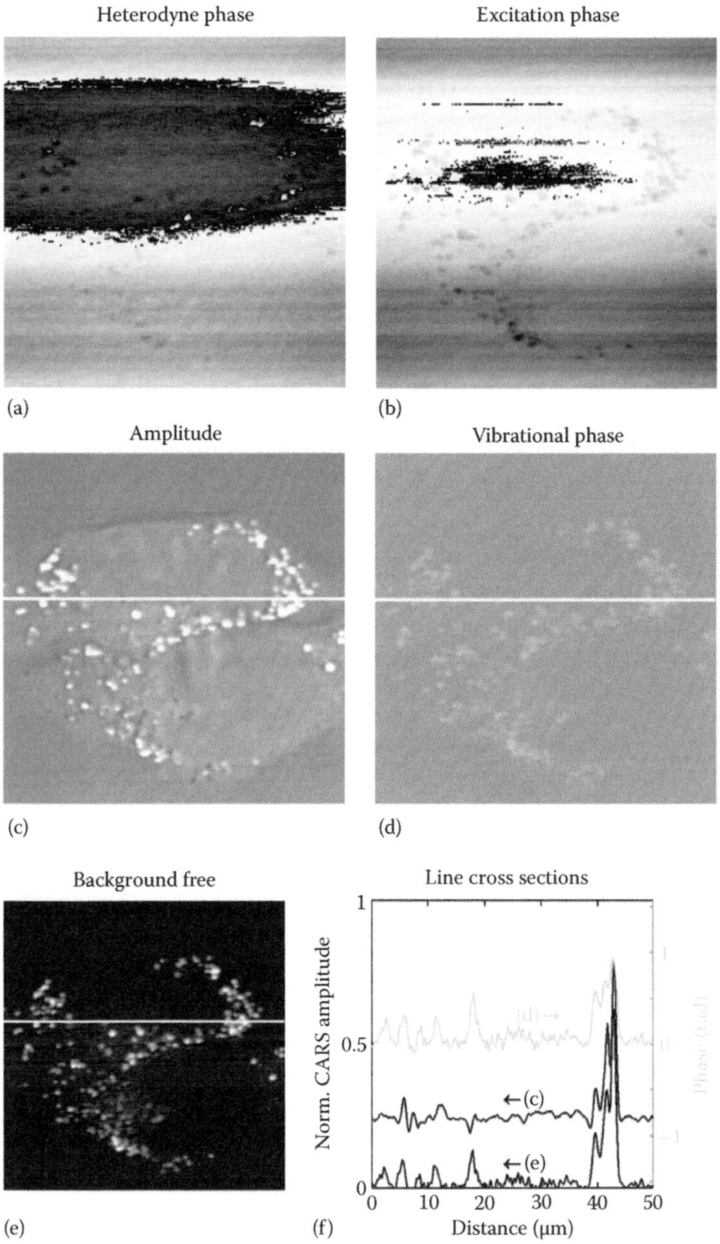


**FIGURE 12.6** Fixed HeLa cells in water, imaged (50 × 50 μm, 262 × 262 pixels) with a 1.2 NA water objective at the vibrational stretch of 2845 cm$^{-1}$. Heterodyne phase (a) and local excitation phase (b). Average of five measured images of CARS amplitude (c) and vibrational phase (d) (subtraction of the local excitation phase from the heterodyne phase). (e) Background free amplitude. (f) Line cross sections of images c, d, and e. For these measurements a frequency modulation of 50 kHz is used on the fundamental and a switching modulation of 1 kHz of the OPO signal.

at the cell boundary. In the line cross section, see Figure 12.6f, the dispersive feature at 18 μm, is also accompanied by a phase step, resulting in a recovered clean amplitude in the background free image.

## 12.8 Visualization within the Complex Plane

The detection of amplitude and phase allows for a different representation of the data. At a fixed driving frequency, the anti-Stokes emission of a molecule in the focal volume will have an amplitude and phase relative to the input fields that can be represented as a point in the complex plane defined by a vector length and angle. A single isolated resonance traces a circle in the complex plane as a function of frequency, while multiple resonances result in additional loops. [9]. The added value of the complex plane representation is most apparent when two or more chemical species with overlapping vibrational bands are present, as illustrated in Figure 12.7. At 2900 cm$^{-1}$ the CARS amplitudes of polyethylene and polystyrene are very similar, but their phases are very different. Plotting the amplitude and phase of each material against the driving frequency constructs a spiral or corkscrew pattern in complex space. The projection along the frequency axis shows the complex plane; along the imaginary axis shows the real component, with the typical dispersion caused by wavelength-dependent refractive index changes; and along the real axis shows the imaginary component, with the resonant (Raman) part of the response. The advantage of this visualization is the ease with which the appropriate driving frequency can be determined to maximize separation of the two components in complex space, thereby optimizing chemical selectivity for dynamic processes as mixing and dilution [10].

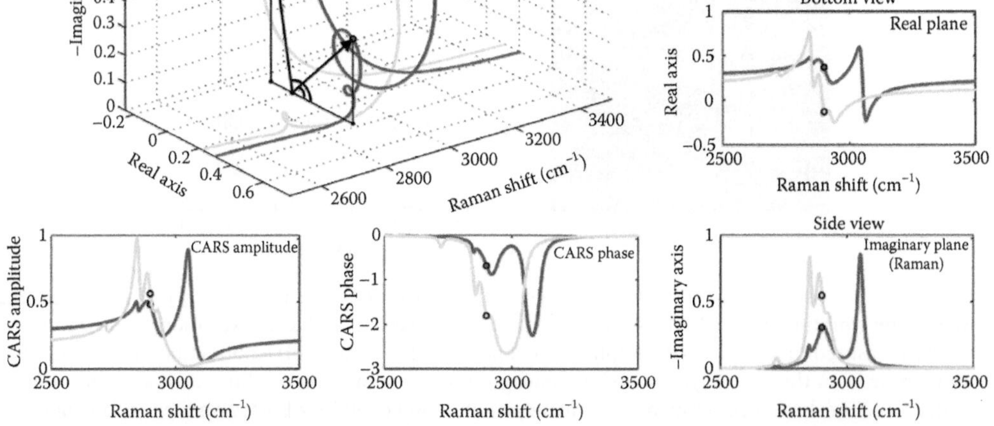

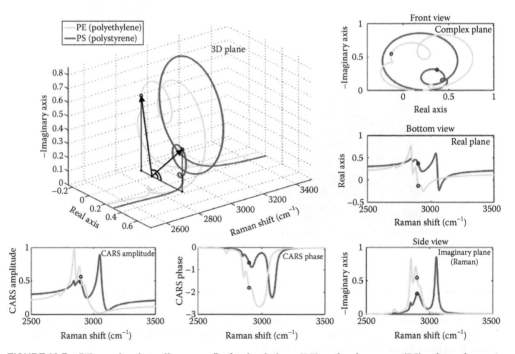


**FIGURE 12.7** "The molecular rollercoaster" of polyethylene (PE) and polystyrene (PS), where the position in the complex plane is plotted against the frequency. Three different projections are on the right side. The amplitude and phase of the CARS signal are plotted in the bottom-left.

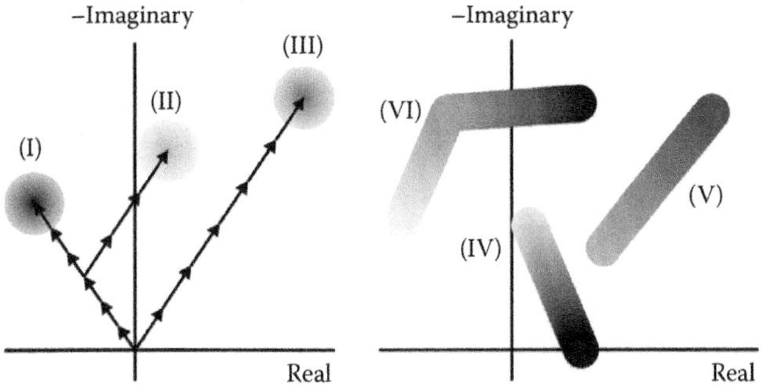


**FIGURE 12.8** Complex plane trajectories obtained at one vibrational resonance. Left: compounds (I) and (III) are constructed contain the same number of molecules but have a different angle and vector length. A 50%/50% molecular mixture between (I) and (III) gives location (II) in the complex plane. Right: (IV) shows the dilution of a resonant compound, (V) shows a mixing or reaction between two resonant compounds and (VI) shows a reaction via an intermediate state.

The signal that is measured for each pixel in an image is the combined result of all the amplitudes and phases of all the molecules (oscillators) in the focal volume for that pixel. Within the focal volume the signal from all these molecules is added coherently. When the focal volume contains only one type of molecule we obtain a total signal such as sketched by (I) or (III) in Figure 12.8, where (III) would represent a stronger vibrational response than (I). A 50%/50% molecular mixture of molecules (I and III) would result in a signal as depicted by (II), constructed from half of the response (I) added to half of the response (III).

Changes in amplitude and phase over time can be used to monitor a sample or system. A dissolution experiment, in which a resonant compound dissolves into a non-resonant compound and disappears, is depicted in Figure 12.8(IV). The start of the experiment is highly resonant (close to the imaginary axis) and the end is nearly non-resonant (near the real axis). When two partially resonant compounds are mixed together we could obtain a figure like 12.8(V). Figure 12.8(V) could also be obtained for a transformation of a molecule (crystallinity, folding). More complex situations such as a reaction between two molecules via an intermediate state might result in something like Figure 12.8(VI). These are just a few theoretical examples of the possibilities of monitoring molecular dynamics in the complex plane.

An example of a real dilution graph is given in Figure 12.9a. Ethanol (resonant) is mixed in different ratios with water (non-resonant) and probed at 2927 $cm^{-1}$. Eleven samples, ranging from pure ethanol to pure water with 10% volume ratio intervals, are flown through a cell. The dilution does not follow a straight line between the pure substances, as would be expected but curves significantly. This curvature is due to the fact that the water and the ethanol interact and it is consistent with the Raman scattering data on the location of the resonance peak of ethanol as a function of the mixing ratio with water, see Figure 12.9b. It can be seen that the peak position shifts toward higher Raman frequencies when the ethanol is diluted with water. This shift results in a shift of the associated vibrational phase, decreasing the angle while diluting. This experiment shows that the interaction dynamics between ethanol and water are more complicated than just a combination of two non-interacting compounds.

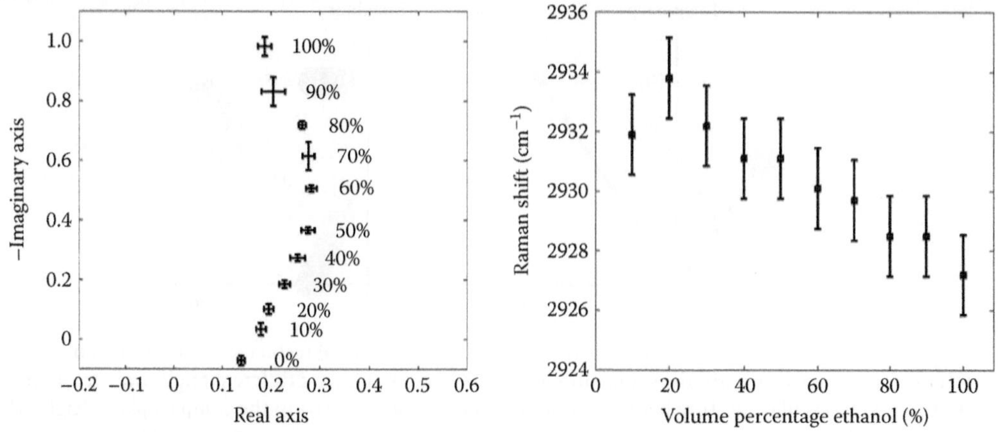


FIGURE 12.9 Dilution of ethanol in water as function of volume percentage.

Mixing of two compounds that are resonant but non-interacting results in straight lines through the complex plane (see Figure 12.8(V)) where the position of the line depends on the frequency that is used to probe the mixture. Based on Raman spectra of the pure substances, the complex response curves of ethanol and methanol are calculated and shown in Figure 12.10. The length of the difference vector as function of frequency is shown in Figure 12.10d. In the experiment ethanol and methanol are mixed in five steps, from pure ethanol to pure methanol with 25% volume percentage mixing steps.

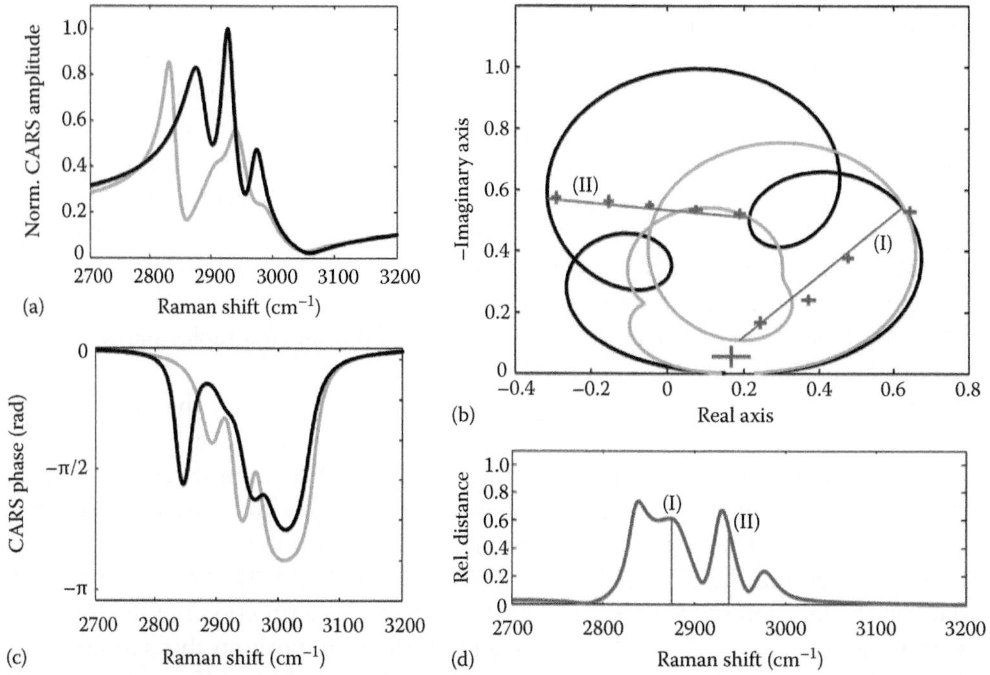


FIGURE 12.10 CARS amplitude (a), phase (b), and complex representation (c) of ethanol (gray) and methanol (black). (d) Shows the distance between the curves as a function of frequency. (I) and (II) represent measurements at 2871 and 2936 cm$^{-1}$. The solid straight lines in (b) represent the calculated response and the crosses indicate the measurements.

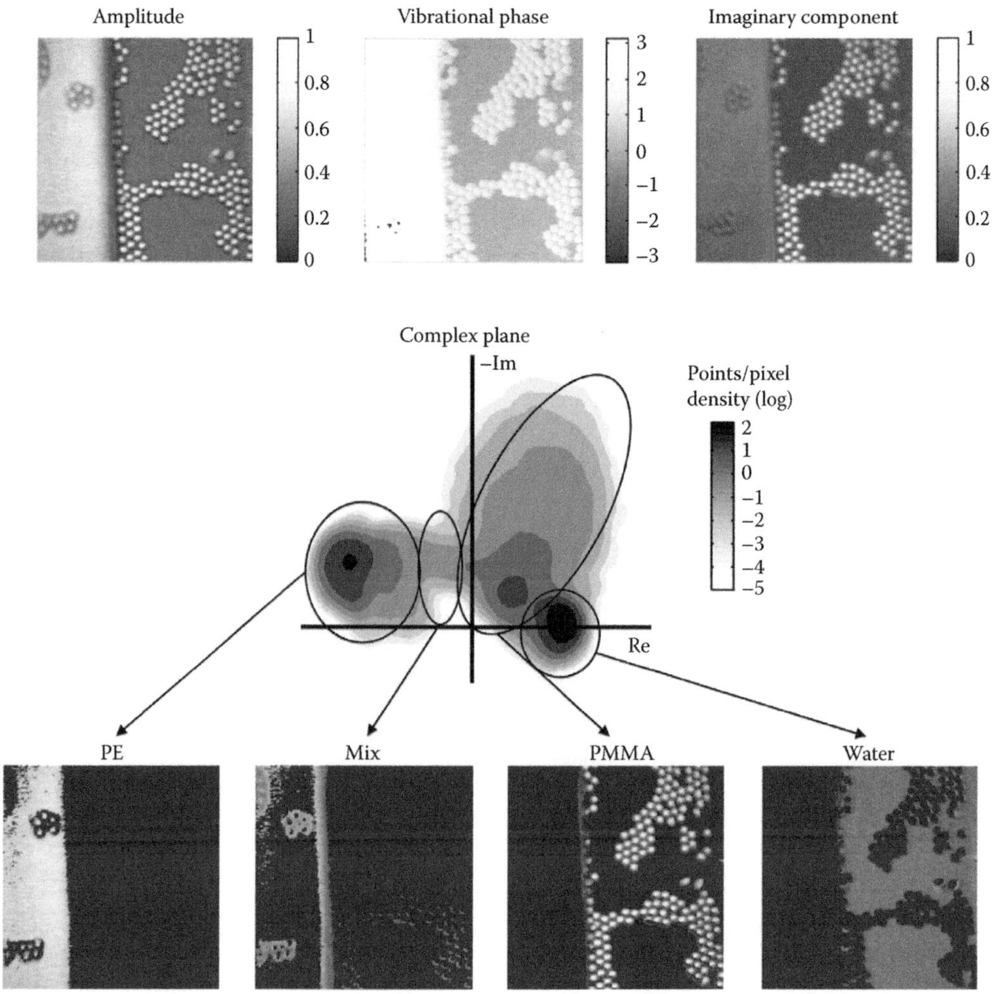


**FIGURE 12.11** Top: amplitude, vibrational phase, and background free image of a sample that contains a sheet of PE, 4 μm PMMA beads and water, imaged at 2940 $cm^{-1}$. Middle: density graph of the projected amplitude and vibrational phase points in the complex plane. Bottom: multicomponent analysis, where the different components are separated by their location in the complex plane.

The mixtures, measured at 2871 (I) and 2936 $cm^{-1}$ (II) are shown as crosses in 12.10(b) and show a good agreement with the theoretical (solid) line.

To demonstrate the imaging selectivity of representation in the complex plane, a heterogeneous mixture of three different materials is shown in Figure 12.11. The sample contains a polyethylene (PE) sheet, 4 μm poly-(methyl)-methacrylate (PMMA) beads, and water. The amplitudes of the three materials are very similar at this frequency. However, the phase image presents a clear difference between the water, which is non-resonant, and PE and PMMA. When the amplitude and phase information of all points in the image are displayed as a density plot in the complex plane, distinct regions emerge. These regions can be used to locate the three pure constituents in the sample. Between the PE, the PMMA, and the water there is a crossing-area that corresponds to those parts in the sample image where the focal volume consists of a mix of compounds.

The reconstructed sample images are all free from background signal; they relate only to the (complex) signal from the substances. This example illustrates how vibrational phase contrast (VPC)-CARS allows for multicomponent separation in a single measurement.

## 12.9 Conclusion

VPC-CARS is a powerful technique to increase both the sensitivity and selectivity of vibrational imaging. With VPC-CARS it is possible to obtain background free images, comparable to the Raman scattering intensity image. The extra dimension provided by direct measurement of the vibrational phase could grant insights into a wide range of scientific questions, such as the temporal dynamics of chemical reactions or the behavior of cells during replication.

## References

1. F. Zernike, Phase contrast, a new method for the microscopic observation of transparent objects, *Physica* 9, 686–698 (1942).
2. C. Evans, E. Potma, and X. Xie, Coherent anti-Stokes Raman scattering spectral interferometry: Determination of the real and imaginary components of nonlinear susceptibility chi(3) for vibrational microscopy, *Opt. Lett.* 29, 2923–2925 (2004).
3. E. Potma, C. Evans, and X. Xie, Heterodyne coherent anti-Stokes Raman scattering (CARS) imaging, *Opt. Lett.* 31, 241–243 (2006).
4. F. Lu, W. Zheng, and Z. Huang, Heterodyne polarization coherent anti-Stokes Raman scattering microscopy, *Appl. Phys. Lett.* 92, 123901 (2008).
5. M. Jurna, J.P. Korterik, C. Otto, and H.L. Offerhaus, Shot noise limited heterodyne detection of CARS signals, *Opt. Express* 15(23), 15207–15213 (2007).
6. R. Boyd, *Nonlinear Optics*, 2nd edn. (Academic Press, San Diego, CA, 2003).
7. M. Jurna, J.P. Korterik, C. Otto, J.L. Herek, and H.L. Offerhaus, Background free CARS imaging by phase sensitive heterodyne CARS, *Opt. Express* 16(20), 15863–15869 (2008).
8. M. Jurna, J.P. Korterik, C. Otto, J.L. Herek, and H.L. Offerhaus, Vibrational phase contrast microscopy by use of coherent anti-Stokes Raman scattering, *Phys. Rev. Lett.* 103(4), 043905 (2009).
9. G. Lucassen, Polarization sensitive coherent Raman spectroscopy on (bio)molecules in solutions, PhD thesis, University of Twente, Enschede, the Netherlands, 1992.
10. M. Jurna, E.T. Garbacik, J.P. Korterik, C. Otto, J.L. Herek, and H.L. Offerhaus, Visualizing resonances in the complex plane with vibrational phase contrast coherent anti-Stokes Raman scattering, *Anal. Chem.* 82(18), 7656–7659 (2010).

# 13. Multiplex CARS Microscopy

**James P.R. Day, Katrin F. Domke, Gianluca Rago, Erik M. Vartiainen, and Mischa Bonn**





*Coherent Raman Scattering Microscopy*. Edited by Ji-Xin Cheng and X. Sunney Xie © 2013 CRC Press/Taylor & Francis Group, LLC. ISBN: 978-1-4398-6765-5.

## 13.1 Introduction

All forms of coherent anti-Stokes Raman scattering (CARS) microscopy generate contrast based on the different vibrations present within a sample. All molecules inherently possess these vibrations and the different functional groups within a molecule vibrate at different frequencies. The frequency range that these vibrations can cover is ~4000 $cm^{-1}$ wide, ranging from the high-frequency O–H stretches in, for example, liquid water, to low-frequency phonon modes in crystalline solids. The concept of multiplex CARS microscopy is to address simultaneously this entire frequency range (or an appreciable subset) and acquire a vibrational spectrum at each place within the sample.

Multiplex CARS microscopy can be compared to spontaneous Raman microscopy as a vibrational spectrum is taken at each pixel in the image (hyperspectral imaging). However, the nonlinear, coherent nature of multiplex CARS gives a signal enhancement of up to several orders of magnitude. Furthermore, its spatial resolution is superior to spontaneous Raman owing to the cubic dependency on the incident laser intensity, it is intrinsically confocal and the output beam is directional. Thus, multiplex CARS does not experience the complication potentially occurring in Raman scattering that signal light originating from elsewhere in the sample is scattered onto the detector. Combined, these advantages mean that generally much shorter acquisition times are required in multiplex CARS experiments than in spontaneous Raman experiments to achieve a comparable signal-to-noise ratio.

The multiplex approach to CARS microscopy currently does not allow such rapid imaging rates as those achievable with single-frequency CARS microscopies, but it does offer two essential advantages. First, there is considerably more information contained throughout an entire spectrum than at just a single frequency within that spectrum. This advantage is particularly useful when complex samples with many overlapping peaks are studied. Second, multiplex CARS microscopy is much more amenable to quantitative analysis of samples (Day 2011). Most microscopic techniques only report upon qualitative variations in the composition of a sample as a function of position, but many questions in the biological and material sciences require a quantitative analysis of a microscopic sample. Multiplex CARS microscopy is able to both distinguish between different chemical species without the need for exogenous labels and quantify local concentrations of those species on the microscopic scale—even in complex mixtures. In this chapter, we will discuss the basis of the multiplex CARS response and the challenges inherent in its interpretation. We then describe the various approaches that have been developed in the literature to acquire multiplex CARS spectra and images. Lastly, we discuss some recent examples from the literature where multiplex CARS microscopy is used to address questions from both the biological and material sciences.

## 13.2 Physical Basis of the Multiplex CARS Response

In this section we describe the origin of the CARS response with particular emphasis on its use in multiplex CARS microscopy for quantitative microscopic imaging. CARS is a four-wave mixing technique that employs a pump field at frequency $\omega_{pu}$, a Stokes field at frequency $\omega_S$, and a probe field at frequency $\omega_{pr}$. Typically, two laser sources are used, one for the Stokes field and one for both the pump and probe fields, such that $\omega_{pu} = \omega_{pr}$.

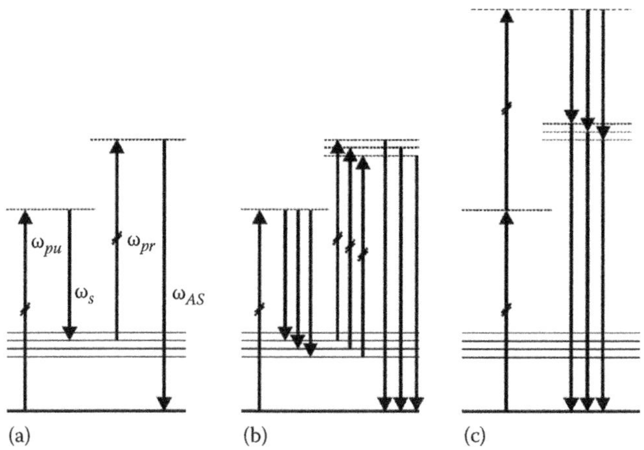


**FIGURE 13.1** CARS energy level diagrams: (a) vibrationally resonant scheme (single-frequency CARS), (b) vibrationally resonant scheme (multiplex CARS), (c) non-resonant scheme. $\omega_{pu}$, $\omega_S$, $\omega_{pr}$, and $\omega_{AS}$ are the pump, Stokes, probe, and anti-Stokes frequencies, respectively.

These incident fields interact with the third-order nonlinear susceptibility of the sample, $\chi^{(3)}$ and the CARS signal is emitted at the anti-Stokes frequency, $\omega_{AS} = 2\omega_{pu} - \omega_S$. $\chi^{(3)}$ describes the properties of the sample and can be separated into two terms:

$$\chi^{(3)} = \chi_R^{(3)} + \chi_{NR}^{(3)} \tag{13.1}$$

The resonant term, $\chi_R^{(3)}$, reports on vibrational resonances and hence contains the quantitative and chemically specific information about the sample. In order to demonstrate the difference between single-frequency and multiplex CARS, we compare their energy-level diagrams in Figure 13.1a and b, respectively. A representative non-resonant transition, $\chi_{NR}^{(3)}$, is shown in Figure 13.1c. $\chi_{NR}^{(3)}$ is not dependent upon the vibrational transitions and only reports on the electronic properties of the material. Away from electronic resonances, $\chi_{NR}^{(3)}$ is strictly real and frequency-invariant, although the magnitude of $\chi_{NR}^{(3)}$ is expected to vary throughout a heterogeneous sample. The CARS signal, $I_{AS}$, depends upon both terms:

$$I_{AS} \propto \left|\chi^{(3)}\right|^2 = \left|\chi_R^{(3)} + \chi_{NR}^{(3)}\right|^2 = \left|\chi_R^{(3)}\right|^2 + 2\chi_{NR}^{(3)}\,\mathrm{Re}\left[\chi_R^{(3)}\right] + \left|\chi_{NR}^{(3)}\right|^2 \tag{13.2}$$

It is evident from Equation 13.2 that both the resonant and non-resonant responses of the sample contribute to the detected signal and that the non-resonant term cannot be retrieved by a simple background correction. Furthermore, the presence of the non-resonant contribution to the overall susceptibility means that there is no simple, linear correlation between $I_{AS}$ and concentration. Although in single-frequency CARS experiments the non-resonant background is a hindrance that limits the detection sensitivity, in multiplex CARS the non-resonant background is actually beneficial and serves to amplify the resonant response through the cross-term in Equation 13.2. Nevertheless, the presence of the non-resonant term does alter the CARS spectrum compared to the spontaneous Raman spectrum, as illustrated in Figure 13.2 for an equimolar mixture of AMP/ADP/ATP in aqueous solution at a total concentration of 500 mM (Vartiainen et al. 2006).

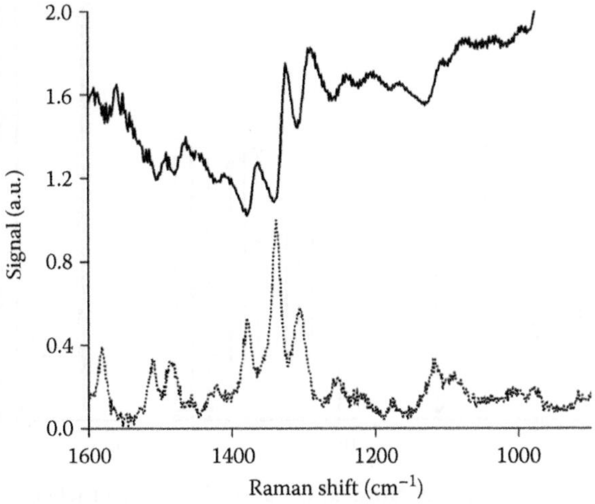


**FIGURE 13.2** Comparison of (dotted) spontaneous Raman and (solid) multiplex CARS spectra of an equimolar mixture of AMP/ADP/ATP in water at a total concentration of 500 mM. (Reproduced from Vartiainen, E.M. et al., *Opt. Express*, 14, 3622, 2006. With permission.)

As mentioned above, multiplex CARS spectra contain quantitative information through the resonant term, $\chi_R^{(3)}$, given by (Shen 1984):

$$\chi_R^{(3)} = \frac{NA}{\Omega - (\omega_{pu} - \omega_S) - i\Gamma} \tag{13.3}$$

where
- $A$, $\Omega$, and $\Gamma$ are the amplitude, frequency, and line width of the vibrational mode
- $N$ is the number of scatterers per unit volume

For comparison, the spontaneous Raman intensity, $I_{Raman}$, is given by

$$I_{Raman} \propto \frac{NA\Gamma}{(\Omega - \omega_0)^2 + \Gamma^2} \tag{13.4}$$

It is important to note that $I_{Raman}$ is directly comparable to the imaginary part of $\chi_R^{(3)}$:

$$\mathrm{Im}\left[\chi_R^{(3)}\right] = \frac{NA\Gamma}{(\Omega - (\omega_{pu} - \omega_S))^2 + \Gamma^2} \tag{13.5}$$

$\mathrm{Im}[\chi_R^{(3)}]$ is both chemically specific and increases linearly with concentration (Rinia et al. 2007). Also, as the $\mathrm{Im}[\chi_R^{(3)}]$ spectrum is comparable to the spontaneous Raman spectrum, reference can be made to correlation charts of known Raman bands. In order to retrieve $\mathrm{Im}[\chi_R^{(3)}]$ from the CARS intensity, it is sufficient if we know the phase of the CARS field, $\theta$, according to

$$\mathrm{Im}[\chi_R^{(3)}] = \left|\chi^{(3)}\right| \sin\theta \propto \sqrt{I_{AS}}\,\sin\theta \tag{13.6}$$

The techniques that are commonly applied to extract the phase are discussed below, following the experimental section.

## 13.3 Experimental Approaches to Multiplex CARS Microscopy

In order to perform multiplex CARS microscopy experiments, an experimental setup needs to fulfill two requirements. First, the setup needs to address a significant portion of the vibrational spectrum simultaneously, which is achieved by employing a Stokes beam with sufficient bandwidth. Second, the setup requires sufficient spectral resolution to resolve bands of interest. The spectral resolution is derived from the bandwidth of the probe beam. In condensed phases, vibrational Raman bands have a bandwidth that is typically >10 $cm^{-1}$, although in some circumstances such high resolution is not required.

There is one additional issue that must be addressed when considering multiplex CARS microscopy. In all cases with nonlinear microscopies such as CARS, one is trying to maximize the signal-to-noise-ratio by using sufficiently intense incident beams, without destroying the sample through linear or nonlinear photodamage. For the same average power, the spectral density in the Stokes beam of a single-frequency CARS experiment is very high compared to multiplex CARS, which results in a stronger signal. The broader the bandwidth of the Stokes beam becomes, the lower the spectral brightness and hence the lower the total signal. However, this reduction in signal does not directly lead to a reduction in the signal-to-noise ratio due to the fact that the nonresonant contribution is not treated as a background term but as an amplifying term in multiplex CARS microscopy.

### 13.3.1 Picosecond Probe, Femtosecond Stokes

Some of the first approaches to multiplex CARS microscopy employed electronically synchronized Ti:sapphire picosecond and femtosecond sources to provide the pump/probe and Stokes beams, respectively (Cheng et al. 2002, Muller and Schins 2002). In these approaches, a ~100 fs Stokes beam was employed to give a frequency window with a usable bandwidth of ~400 $cm^{-1}$, which is sufficient to acquire spectra of, for example, the entire C–H stretching range. Furthermore, by tuning the center wavelength of the Stokes pulse, other regions of the vibrational spectrum can also be investigated, albeit not simultaneously. High spectral resolution of ~2–3 $cm^{-1}$ was achieved by using a ~10 ps source for the pump and probe beam. Although electronic synchronization makes this approach slightly more cumbersome to implement, this combination of ps and fs beams does have the advantage that relatively high spectral densities in the fs beam can be employed for a given average power.

### 13.3.2 Supercontinuum Stokes

Despite the success of employing a ps-probe with a fs-Stokes beam, this approach does not have sufficient spectral bandwidth to deconvolute overlapping peaks in the spectra obtained from more complex samples. Fortunately, the rapid development of photonic crystal fiber-based supercontinuum sources has led to the availability of Stokes beams that can cover the entire 4000 $cm^{-1}$ vibrational spectrum simultaneously. Furthermore, as the beam that is used to generate the supercontinuum can also be used as the pump/probe beam, this approach eliminates the need for electronic synchronization of the pump and Stokes beams. Both femtosecond (Chen et al. 2011, Kano and Hamaguchi 2005a, Kee and Cicerone 2004, Parekh et al. 2010, von Vacano et al. 2007) and nanosecond (Okuno et al. 2008, 2010) sources have been employed to generate the supercontinuum.

### 13.3.3 Pulse Shaping Approaches

Pulse shaping approaches have been employed by researchers to surmount a number of the challenges inherent to multiplex CARS microscopy. Rehbinder et al. used a pulse-shaper to tailor the supercontinuum output from a photonic crystal fiber in such a way as to suppress the intensity in silent regions of the vibrational spectrum (Rehbinder et al. 2010). This approach keeps the spectral density of the Stokes beam high in active regions of the vibrational spectrum while reducing photodamage for a given average incident power. The Lim group have developed Fourier transform spectral interferometry based on pulse shaping techniques to generate ultrabroadband CARS spectra with high spectral resolution at high readout rates (Chen and Lim 2008).

## 13.4 Retrieving Quantitative Information from Multiplex CARS Spectra

As discussed above, there are two key advantages to multiplex CARS microscopy. First, knowledge of the full spectrum is essential when studying complex samples with many overlapping peaks and second, multiplex CARS is a quantitative microscopy. However, for both of these advantages to be realized, it is necessary to extract the $\mathrm{Im}[\chi_R^{(3)}]$ spectrum from the raw CARS spectrum as the $\mathrm{Im}[\chi_R^{(3)}]$ spectrum is comparable to the normal Raman response—unlike the CARS intensity. It is directly proportional to the number of scatterers and the large number of existing Raman references and libraries can be employed for band assignment and chemical identification of the scatterers. There are a number of approaches for this extraction, as discussed in the following sections.

### 13.4.1 Time Gating

The resonant and non-resonant responses can be separated in the time domain. The non-resonant response follows instantaneously the incident field; it vanishes as this driving field vanishes. However, the resonant response exhibits a finite dephasing time: the vibrational coherence that arises from beating between the pump and Stokes beams can last for several picoseconds depending upon the linewidth of the vibration. Thus the information-rich resonant response and the non-resonant response can be separated in the time domain by exploiting this difference in their dephasing times, as shown in Figure 13.3.

At a time delay of zero ps, the CARS spectrum of acetonitrile exhibited typical dispersive resonances, but in this instance a delay of 1 ps removed the effect of the non-resonant contribution. However, to extract quantitative information from time-gated spectra, it is necessary to correct for the different linewidths of different modes. Broader peaks decay

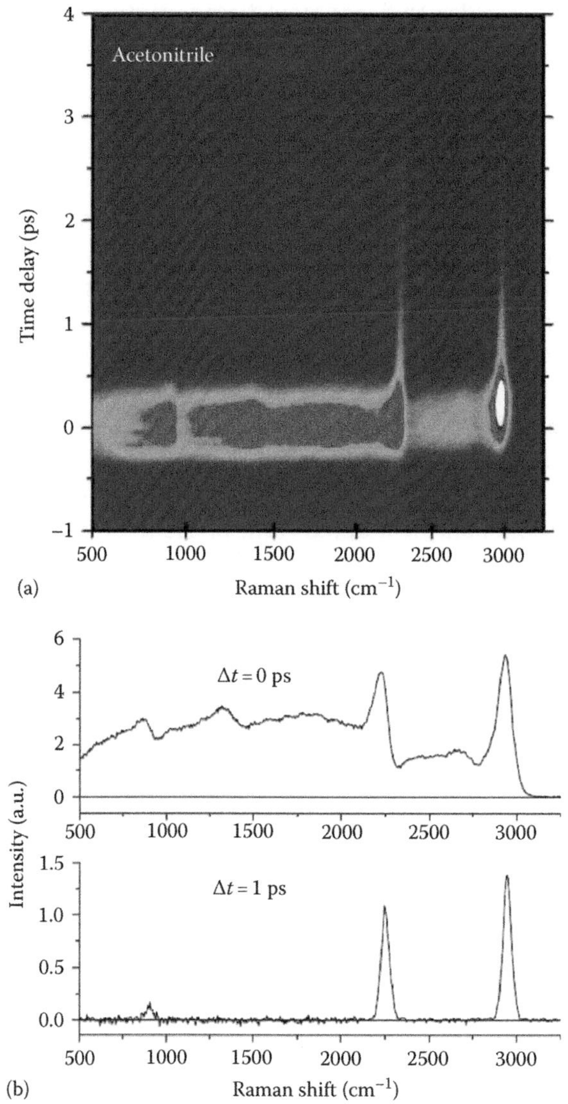


FIGURE 13.3 (a) Time-resolved CARS spectra of acetonitrile. (b) CARS spectra at delay times of 0 and 1 ps. (Reproduced from Lee, Y.J. et al., *Opt. Express*, 18, 4371, 2010. With permission.)

more rapidly, so to obtain quantitative data it is necessary to measure the intensity of the resonant signal as a function of time for finite delay, and extrapolate to zero.

## 13.4.2 Peak Fitting

In this approach, the number of vibrational resonances within a spectrum is estimated or inferred from spontaneous Raman spectra, and the CARS response is modeled according to Equations 13.2 and 13.3 under the assumptions that $\chi_{NR}^{(3)}$ is purely real, positive, and frequency-invariant (Cheng et al. 2002, Muller and Schins 2002). If the spectrum is not too congested, this approach works well. However, the increasing complexity of the samples that are now being studied by CARS means that this approach has been superseded by phase retrieval methods which do not require such *a priori* information.

## 13.4.3 Phase Extraction

The $\mathrm{Im}[\chi_R^{(3)}]$ spectrum can be extracted from the CARS spectrum when the vibrational phase spectrum is known. The non-resonant response has zero phase as it is instantaneous, but the resonant response has a nonzero vibrational phase whose magnitude varies across the spectrum. Weak Raman scatterers ($\chi_R^{(3)} \ll \chi_{NR}^{(3)}$) have a phase that is close to zero, whereas strong Raman scatterers can exhibit a phase shift up to $\pi$ radians.

Although the CCD detectors employed in a multiplex CARS experiment only measure the intensity of the CARS spectrum, a reliable approximation to the vibrational phase can be extracted mathematically from this spectrum. Despite the fact that there are potentially an infinite number of solutions to such a phase extraction problem, in practice there are considerable constraints on the allowed form of the solution (Millane 1990). Two well-established approaches to phase extraction from CARS spectra are the time-domain Kramers–Kronig (TD-KK) transform and the maximum entropy method (MEM).

### 13.4.3.1 Time-Domain Kramers–Kronig Transform Method

The vibrational phase spectrum can be extracted via a Fourier transform into the time domain (Liu et al. 2009):

$$\theta(\omega) = \begin{cases} 2\,\mathrm{Im}\left\{\Im\left(\Im^{-1}\left[\ln\left|I_{AS}(\omega)\right|\right]\right) - \dfrac{\ln\left|I_{AS}(\omega)\right|}{2}\right\}, & t \geq 0 \\ 2\,\mathrm{Im}\left\{\Im\left(\Im^{-1}\left[\ln\left|I_{NR}(\omega)\right|\right]\right) - \dfrac{\ln\left|I_{NR}(\omega)\right|}{2}\right\}, & t < 0 \end{cases} \tag{13.7}$$

where

$\Im$ and $\Im^{-1}$ indicate the Fourier and inverse Fourier transform

$I_{NR}$ is the non-resonant response

Sung et al. have shown that the spectra extracted by this approach are linear in concentration for a series of $Na_2SO_4$ solutions (Sung et al. 2010). Parekh et al. have employed this approach to extract $\mathrm{Im}[\chi_R^{(3)}]$ spectra from mouse fibroblast cells and differentiate and image the nuclear and the cytoplasmic regions of the cells based solely on their vibrational spectra (Parekh et al. 2010).

### 13.4.3.2 Maximum Entropy Method

The maximum entropy method (MEM) extracts the phase spectrum from the CARS spectrum based upon the maximum entropy principle. Typically, a CARS spectrum is only known over a frequency range that is limited by the bandwidth of the Stokes laser, which in the time domain equates to knowing only a limited number of autocorrelations. The MEM enables a fit to a CARS spectrum using this limited number of autocorrelations, in order to retrieve the phase with no prior assumptions and no fitting parameters. This approach has been described in detail in the literature (Rinia et al. 2007, Vartiainen 1992, Vartiainen et al. 1996, 2006).

#### 13.4.3.3 Comparison of the TD-KK and MEM

The MEM and TD-KK methods are related, as the aim of both approaches is to retrieve the vibrational phase spectrum. However, each approach is influenced in different ways to artifacts in the spectrum. Multiplex CARS spectra are usually complicated by the presence of a non-resonant background that is not spectrally flat. This background results from the interaction between the pump and Stokes pulses (Liu et al. 2009) and in the TD-KK approach it must be estimated in the CARS spectrum for phase retrieval to be possible. The background can be estimated by iteration (Liu et al. 2008). In contrast, the MEM does not require such an estimate and the effect of the background appears as a slowly varying background to the phase, termed the error phase. For weak Raman scatterers in the limit $\chi_R^{(3)} < \chi_{NR}^{(3)}$, the phase is zero away from vibrational resonances (Vartiainen et al. 2006):

$$\theta = \tan^{-1}\left(\frac{\mathrm{Im}\left[\chi_R^{(3)}\right]}{\chi_{NR}^{(3)} + \mathrm{Re}\left[\chi_R^{(3)}\right]}\right) \approx 0 \tag{13.8}$$

Here, the error phase can be estimated as a polynomial background correction to the retrieved phase. In practice, the phase retrieved by the TD-KK method must also be corrected due to inaccuracies in the estimation of the non-resonant background. Background correction is common in both Raman and IR spectroscopy, and many approaches have been developed.

Both methods allow for quantitative analysis of CARS spectra as there is little difference in the effectiveness of each phase retrieval method. The MEM approach has the slight advantage that no estimate of the non-constant background is required. Also, high-frequency noise can be removed from the spectrum through use of fewer autocorrelations. This subject is discussed in more detail elsewhere (Day 2011).

## 13.5 Applications of Multiplex CARS Microscopy

To date, the majority of applications of multiplex CARS microscopy have tended toward questions from the field of biological science, focusing in particular on lipids. CARS microscopy is particularly suited to the study of lipids as the Raman scattering cross section of C–H stretching vibrations is typically much stronger than that of other bands in the spectrum. However, recent technological advances now mean that it is becoming easier to study other functional groups, which in turn has sparked interest in the use of multiplex CARS microscopy to answer questions from material science.

### 13.5.1 Biological Science Studies

#### 13.5.1.1 Lipid Membranes and Vesicles

Much work in multiplex CARS microscopy has focused on lipid membranes and vesicles (Cheng et al. 2002, Wurpel et al. 2002). Despite being only two molecules thick, lipid membranes are essential to the proper functioning of cells, which has stimulated so much work

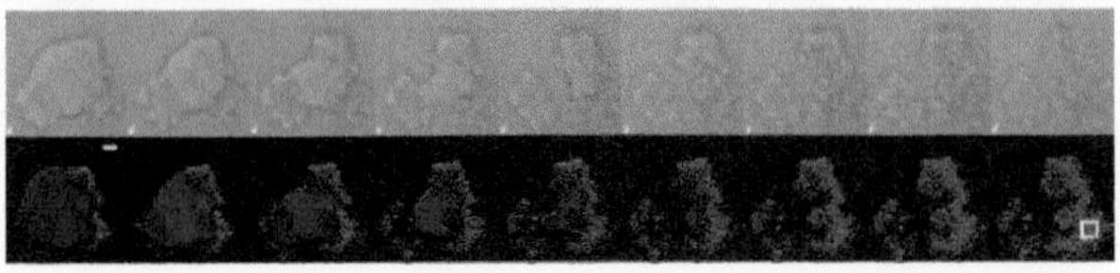


**FIGURE 13.4** Time-resolved images of the digestion of a glyceryl trioleate droplet by porcine pancreatic lipase. (a) Bright-field microscopy images, (b) False-color images obtained by CARS microspectroscopy of glyceryl trioleate (red) and lipolytic products (green). Scale bar = 5 μm, pixel step-size = 1 μm. (Reproduced from Day, J.P.R. et al., *J. Am. Chem. Soc.*, 132, 8433, 2010. With permission.)

on these model systems. Early work employing the peak fitting analysis described above was able to give insight into the fluidity and density of the lipids in vesicles and supported bilayers (Muller and Schins 2002, Rinia et al. 2006, Wurpel and Muller 2006, Wurpel et al. 2004, 2005), as well as the interlamellar water (Wurpel and Muller 2006). More recent work has employed the MEM to investigate quantitatively the lipid phase transition (Rinia et al. 2007).

#### 13.5.1.2 Lipid Droplets

Enzyme-catalyzed hydrolysis of lipids is a key step in their digestion and uptake by the human gut. Quantitative multiplex CARS microscopy has been employed to investigate the *in vitro* digestion of an unsaturated triacylglyceride emulsion. The possibility to discriminate between the digested and undigested species was demonstrated, based solely upon their vibrational spectra as shown in Figure 13.4. Also, the local concentrations of progesterone and vitamin $D_3$ within saturated triacylglyceride emulsion droplets during digestion were quantified (Day et al. 2010). Such information is essential in order to help optimize formulations for the delivery of bioactive ingredients and lipid-soluble drugs.

The fatty acids produced during digestion are adsorbed by fat cells and converted back into triacylglycerides. These triacylglycerides are stored within the cell as lipid droplets. CARS microscopy was used to image the lipid droplets in mouse adipocyte cells incubated with different mixtures of saturated (palmitic acid, PA) and unsaturated (linolenic acid, LA) fatty acids (Rinia et al. 2008). This allowed for the quantitative determination of not only the composition, but also the distributions of the fluidity and unsaturation within the lipid droplets, as shown in Figure 13.5.

#### 13.5.1.3 Cellular Imaging

The imaging of entire cells by multiplex CARS microscopy has great potential, but the vibrational spectra are often weak and the spatial variations in composition are subtle. These issues make the interpretation of the vibrational spectra challenging. The recent advent of ultrabroadband multiplex CARS microscopes has helped greatly to address these challenges. Workers in the Hamaguchi group have performed a series of ultrabroadband multiplex CARS studies on live yeast cells (Kano and Hamaguchi 2005b, 2006). In their most recent study, they acquired spectra of a budding yeast cell spanning the entire fingerprint range (800–1800 $cm^{-1}$). By also using the MEM and reducing noise in the spectrum by the singular value decomposition, they could identify features in the spectra that were used to give simultaneous movies of the behavior of mitochondrial phospholipids, proteins, and a vacuole dancing body within the budding yeast cell, as

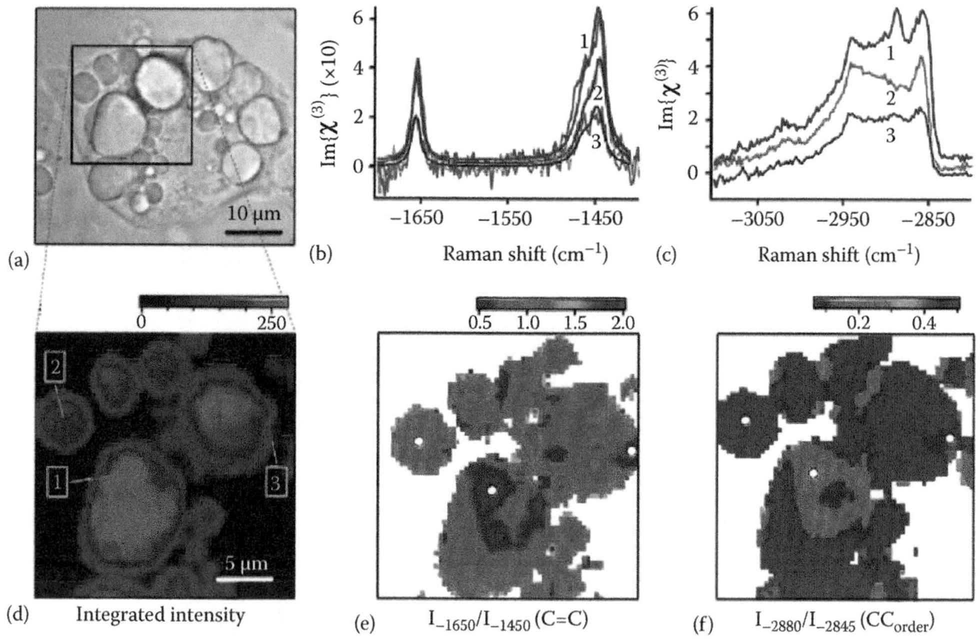


**FIGURE 13.5** (a) Brightfield image of an adipocyte incubated with a 1:3 mixture (mol:mol) of LA and PA. The marked region was imaged with multiplex CARS microscopy. Based on the measured $\mathrm{Im}[\chi_R^{(3)}]$ spectra at every pixel, images are derived with the contrast based on (d) lipid concentration (integrated intensity), (e) C=C concentration derived from the CC-stretch region ($I_{-1650}/I_{-1450}$) and (f) acyl chain order ($I_{-2880}/I_{-2845}$). In (b) and (c), representative $\mathrm{Im}[\chi_R^{(3)}]$ spectra are shown recorded at the locations indicated in (d) for the CC-stretch and CH-stretch spectral region, respectively. The least-squares fit of a sum of three Lorentzians to the data is also shown in (b). The acquisition time was 20 ms for a multiplex CARS spectrum at every pixel position (~1 min/image). (Reproduced from Rinia, H.A. et al., *Biophys. J.*, 95, 4908, 2008. With permission.)

shown in Figure 13.6 (Okuno et al. 2010). The Cicerone group have also been at the forefront of the development of ultrabroadband CARS sources, and they have also recently mapped entire cells and discriminated between the nuclear and cytoplasmic regions (Parekh et al. 2010). Petrov et al. have similarly studied single bacterial endospores and were able to distinguish all the characteristic features of the spores (Petrov et al. 2007).

## 13.5.2 Material Science Studies

### 13.5.2.1 Polymer Films

Multiplex CARS microscopy lends itself well to the study of polymer films, thanks to the intrinsic sectioning capability of CARS and the intensity of the polymer peaks. Von Vacano et al. were able to discriminate between the different polymers in tertiary blends of polystyrene (PS), polymethylmethacrylate (PMMA) and polyethyleneterephthalate (PET) and PS, PMMA, and polyethylene (PE) and hence map their occurrence through each heterogeneous sample, as well as a virtual section of an adhesive and its carrier film (von Vacano et al. 2007). The Jeoung group have also used the sectioning capability of multiplex CARS microscopy to determine quantitatively the thickness of sub-micron polymeric films (Choi et al. 2008, Yahng and Jeoung 2011).

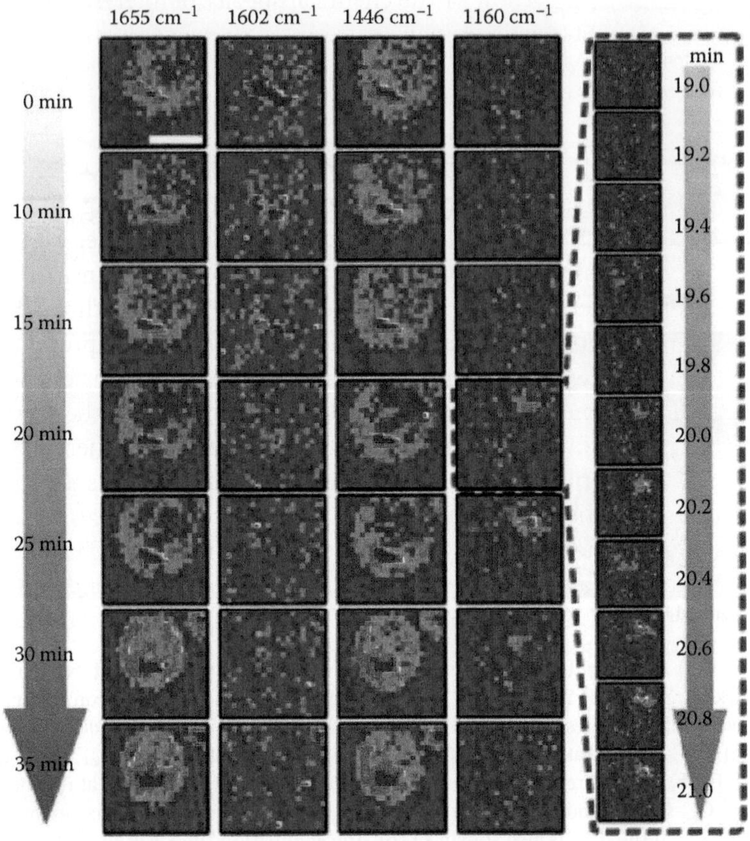


**FIGURE 13.6** Time-resolved CARS images of the Raman bands at 1655, 1602, 1446, and 1160 $cm^{-1}$, respectively. The scale bar inside the image is 5 μm. The color scale for each band is the same for all the times measured. The red frame on the right contains CARS images at 1160 $cm^{-1}$ from 19 to 21 min. Each image is measured every 12 s. Each of the seven rows of images is constructed from Raman intensities retrieved from one CARS spectrum. (Reproduced from Okuno, M. et al., *Angewandte Chemie Int. Ed.*, 49, 6773, 2010. With permission.)

#### 13.5.2.2 Zeolites

Multiplex CARS microscopy has also been used to study reactions in zeolite crystals. CARS is an ideal tool to study the interplay between the catalyst architecture and reactivity and hence gains previously inaccessible information about fundamental processes in catalysts. Specifically, the precursor states of thiophene conversion over individual H-ZSM-5 zeolite crystals have previously been studied (Kox et al. 2009). Maps (Figure 13.7) of the catalysts loaded with thiophene revealed a heterogeneous, diffusion-limited distribution throughout the zeolite, with the analyte accumulating in the center of the crystal and along defect sites.

#### 13.5.2.3 Microfluidic Devices

Microfluidic devices (MFDs) are ideal reaction vessels but to exploit fully their capabilities, quantitative on-chip monitoring of fluid flow, mixing and reaction rates is required. Multiplex CARS microscopy has been used to quantify the local

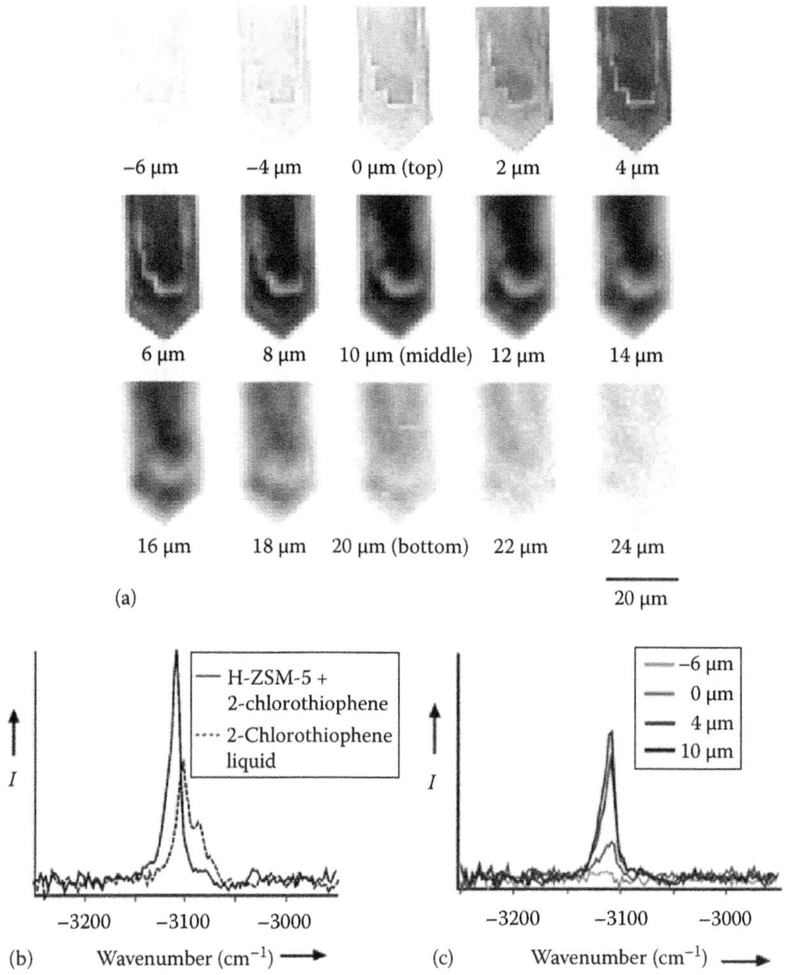


**FIGURE 13.7** (a) CARS images of the normalized concentration of 2-chlorothiophene in an H-ZSM-5 zeolite crystal. (b) $\mathrm{Im}[\chi_R^{(3)}]$ spectra of 2-chlorothiophene and an H-ZSM-5 crystal containing 2-chlorothiophene. (c) $\mathrm{Im}[\chi_R^{(3)}]$ spectra of 2-chlorothiophene at different depths within the crystal. (Reproduced from Kox, M.H.F. et al., *Angewandte Chemie Int. Ed.*, 48, 8990, 2009. With permission.)

concentrations of miscible fluids within an MFD. For unreactive species, this information reports upon local mass transport and can be used to determine diffusion coefficients or fluid flow behavior in shear-sensitive media (Schafer et al. 2009). For reactive chemicals, the local concentration maps can be used to investigate fast reaction kinetics, such as the acid–base reaction between acetic acid and pyrrolidine. As shown in Figure 13.8, this approach enabled the direct determination of reactant and product concentrations throughout the MFD, allowing the resolution of the lower limit for the rate constant of this reaction (Schafer et al. 2008). This technique is also highly suitable to study the physical chemistry of fast, homogeneously catalyzed reactions. Camp et al. have also employed multiplex CARS microscopy in MFDs to perform label-free cytometry (Camp et al. 2009). Although so far only demonstrated with polymer beads, this approach has the potential to greatly expand the discriminatory power of flow cytometers.

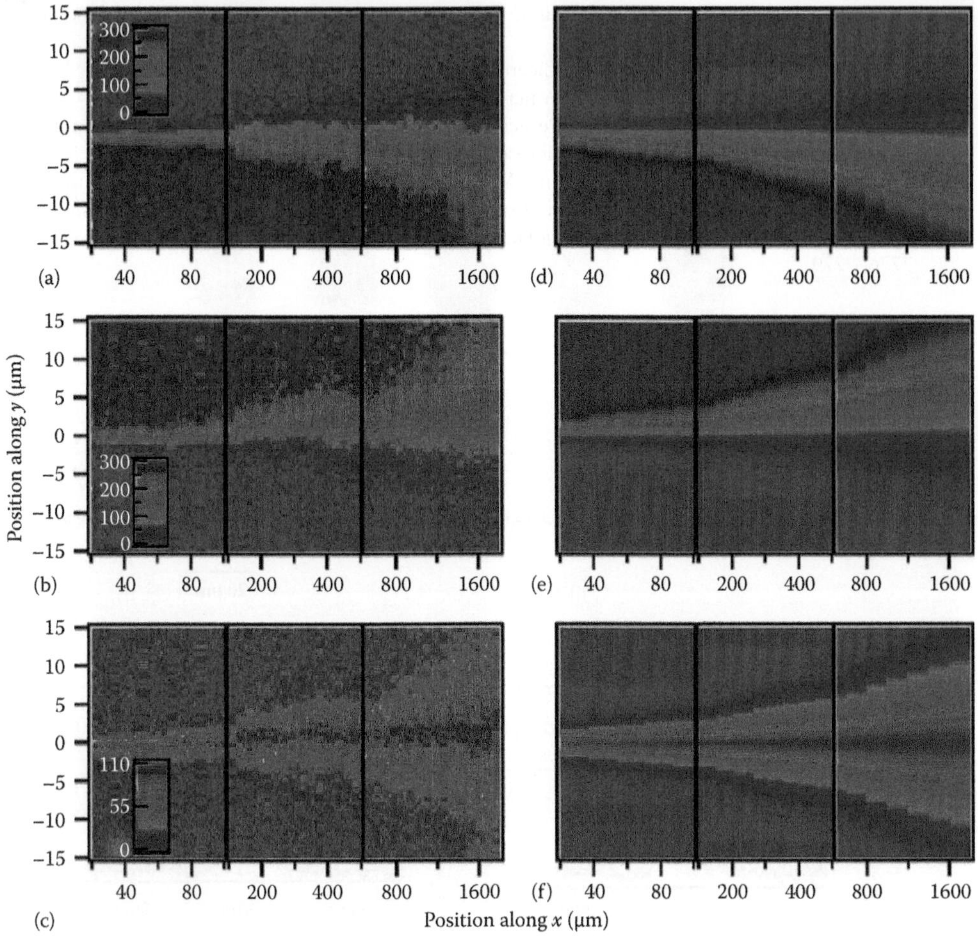


**FIGURE 13.8** Concentration profiles of (a) pyrrolidine, (b) acetic acid and (c) their reaction product within a microfluidic device. Panels (d), (e), and (f) provide the corresponding results of a numerical simulation. (Reproduced from Schafer, D. et al., *J. Am. Chem. Soc.*, 130, 11592, 2008. With permission.)

### 13.5.2.4 Multiplex CARS Tweezers

The ability to manipulate the position and geometry of a microscopic object while at the same time probing its vibrational spectrum is extremely appealing, and has led to the combination of multiplex CARS microscopy with optical tweezers. Shi et al. used the tight focusing of the incident beams in a high numerical aperture (NA) objective to both trap and acquire ultrabroadband CARS spectra of a polystyrene bead (Shi et al. 2007). In an alternative approach, Ulriksen et al. employed a more sophisticated counterpropagating geometry with relatively low NA, long working distance objectives for the trap (Ulriksen et al. 2008). This approach leaves sufficient space close to the sample to further illuminate the sample with a separate objective for CARS. As a result, the trap and incident CARS beams can be moved independently, greatly increasing the versatility of CARS tweezers.

## 13.6 Conclusion

In both the biological and material sciences, it has become possible over the past few years to study complex samples in considerably more detail by CARS microscopy. Multiplex CARS microscopy offers a number of advantages that make it a particularly useful technique to probe such complex samples. The hyperspectral data sets that are generated in a typical multiplex CARS experiment are extremely rich in information. With the advent of phase retrieval techniques such as the MEM and TD-KK transform method, such spectra become amenable to the analysis by multivariate techniques such as principal component analysis and related approaches (de Juan et al. 2009). As a result, different species within a sample can be distinguished in a label-free manner based on their vibrational signatures, and the local concentration of each species can be obtained down to mM concentrations, or perhaps even to μM concentrations with electronic resonant enhancement (Min et al. 2009). For biological samples, the vibrational spectra can be so congested that it is not possible to isolate the response of an individual species. In these instances, techniques like cluster analysis become increasingly useful, where regions within an image that exhibit similar spectral features are grouped together, such as tissue structures or cellular organelles (Lasch et al. 2004). These sophisticated algorithms coupled with the rapid spectral acquisition should allow researchers in the future to answer questions in both biological and material science fields that manifest as only subtle changes in the vibrational spectrum and occur on sub-micron lengthscales and at sub-ms timescales.

## Acknowledgment

This work is part of the research program of the Stichting Fundamenteel Onderzoek der Materie (Foundation for Fundamental Research on Matter) with financial support from the Nederlandse Organisatie voor Wetenschappelijk Onderzoek (Netherlands Organization for the Advancement of Research).

## References

Camp, C. H., S. Yegnanarayanan, A. A. Eftekhar, H. Sridhar, and A. Adibi. 2009. Multiplex coherent anti-Stokes Raman scattering (MCARS) for chemically sensitive, label-free flow cytometry. *Optics Express* 17:22879–22889.

Chen, B.-C. and S.-H. Lim. 2008. Optimal laser pulse shaping for interferometric multiplex coherent anti-Stokes Raman scattering microscopy. *Journal of Physical Chemistry B* 112:3653–3661.

Chen, B.-C., J. Sung, X. Wu, and S. H. Lim. 2011. Chemical imaging and micro-spectroscopy with spectral focusing coherent anti-Stokes Raman scattering (CARS). *Journal of Biomedical Optics* 16:021112.

Cheng, J.-X., A. Volkmer, L. D. Book, and X. S. Xie. 2002. Multiplex coherent anti-Stokes Raman scattering microspectroscopy and study of lipid vesicles. *Journal of Physical Chemistry B* 106:8493–8498.

Choi, D. S., S. C. Jeoung, and B.-H. Chon. 2008. Thickness dependent CARS measurement of polymeric thin films without depth-profiling. *Optics Express* 16:2604–2613.

Day, J. P. R., K. F. Domke, G. Rago, H. Kano, H. Hamaguchi, E. M. Vastiainen, and M. Bonn. 2011. Quantitative coherent anti-Stokes Raman scattering (CARS) microscopy. *Journal of Physical Chemistry B* 115:7713–7725.

Day, J. P. R., G. Rago, K. F. Domke, K. P. Velikov, and M. Bonn. 2010. Label-free imaging of lipophilic bioactive molecules during lipid digestion by multiplex coherent anti-Stokes Raman scattering microspectroscopy. *Journal of American Chemical Society* 132:8433–8439.

de Juan, A., M. Maeder, T. Hancewicz, L. Duponchel, and R. Tauler. 2009. Chemometric tools for image analysis. In *Infrared and Raman Spectroscopic Imaging*, R. Salzer and H. W. Siesler, eds. Wiley-VCH, Weinheim, Germany.

Kano, H. and H. Hamaguchi. 2005a. Ultrabroadband (>2500 cm$^{-1}$) multiplex coherent anti-Stokes Raman scattering microspectroscopy using a supercontinuum generated from a photonic crystal fiber. *Applied Physics Letters* 86:121113.

Kano, H. and H. Hamaguchi. 2005b. Vibrationally resonant imaging of a single living cell by supercontinuum-based multiplex coherent anti-Stokes Raman scattering microspectroscopy. *Optics Express* 13:1322–1327.

Kano, H. and H. Hamaguchi. 2006. In-vivo multi-nonlinear optical imaging of a living cell using a supercontinuum light source generated from a photonic crystal fiber. *Optics Express* 14:2798–2804.

Kee, T. W. and M. T. Cicerone. 2004. Simple approach to one-laser, broadband coherent anti-Stokes Raman scattering microscopy. *Optics Letters* 29:2701–2703.

Kox, M. H. F., K. F. Domke, J. P. R. Day, G. Rago, E. Stavitski, M. Bonn, and B. M. Weckhuysen. 2009. Label-free chemical imaging of catalytic solids by coherent anti-Stokes Raman scattering and synchrotron-based infrared microscopy. *Angewandte Chemie International Edition* 48:8990–8994.

Lasch, P., W. Haensch, and M. Diem. 2004. Imaging of colorectal adenocarcinoma using FT-IR microspectroscopy and cluster analysis. *Biochimica et Biophysica Acta* 1688:176–186.

Lee, Y. J., S. H. Parekh, Y. H. Kim, and M. T. Cicerone. 2010. Optimized continuum from a photonic crystal fiber for broadband time-resolved coherent anti-Stokes Raman scattering. *Optics Express* 18:4371–4379.

Liu, Y., Y. J. Lee, and M. T. Cicerone. 2008. Fast extraction of resonant vibrational response from CARS spectra with arbitrary nonresonant background. *Journal of Raman Spectroscopy* 40:726–731.

Liu, Y., Y. J. Lee, and M. T. Cicerone. 2009. Broadband CARS spectral phase retrieval using a time-domain Kramers–Kronig transform. *Optics Letters* 34:1363–1365.

Millane, R. P. 1990. Phase retrieval in crystallography and optics. *Journal of the Optical Society of America A* 7:394–411.

Min, W., S. Lu, G. R. Holtom, and X. S. Xie. 2009. Triple-resonance coherent anti-Stokes Raman scattering microspectroscopy. *ChemPhysChem* 10:344–347.

Muller, M. and J. M. Schins. 2002. Imaging the thermodynamic state of lipid membranes with multiplex CARS microscopy. *Journal of Physical Chemistry B* 106:3715–3723.

Okuno, M., H. Kano, P. Leproux, V. Couderc, J. P. R. Day, M. Bonn, and H. Hamaguchi. 2010. Quantitative CARS molecular fingerprinting of single living cells with the use of the maximum entropy method. *Angewandte Chemie International Edition* 49:6773–6777.

Okuno, M., H. Kano, P. Leproux, V. Couderc, and H. Hamaguchi. 2008. Ultrabroadband multiplex CARS microspectroscopy and imaging using a subnanosecond supercontinuum light source in the deep near infrared. *Optics Letters* 33:923–925.

Parekh, S. H., Y. J. Lee, K. A. Aamer, and M. T. Cicerone. 2010. Label-free cellular imaging by broadband coherent anti-Stokes Raman scattering microscopy. *Biophysical Journal* 99:2695–2704.

Petrov, G. I., R. Arora, V. V. Yakovlev, X. Wang, A. V. Sokolov, and M. O. Scully. 2007. Comparison of coherent and spontaneous Raman microspectroscopies for noninvasive detection of single bacterial endospores. *Proceedings of the National Academy of Sciences of the United States of America* 104:7776–7779.

Rehbinder, J., C. Pohling, T. Buckup, and M. Motzkus. 2010. Multiplex coherent anti-Stokes Raman microspectroscopy with tailored Stokes spectrum. *Optics Letters* 35:3721–3723.

Rinia, H. A., M. Bonn, and M. Muller. 2006. Quantitative multiplex CARS spectroscopy in congested spectral regions. *Journal of Physical Chemistry B* 110:4472–4479.

Rinia, H. A., M. Bonn, M. Muller, and E. M. Vartiainen. 2007. Quantitative CARS spectroscopy using the maximum entropy method: The main lipid phase transition. *ChemPhysChem* 8:279–287.

Rinia, H. A., K. N. J. Burger, M. Bonn, and M. Muller. 2008. Quantitative label-free imaging of lipid composition and packing of individual cellular lipid droplets using multiplex CARS microscopy. *Biophysical Journal* 95:4908–4914.

Schafer, D., M. Muller, M. Bonn, D. W. M. Marr, J. van Maarseveen, and J. A. Squier. 2009. Coherent anti-Stokes Raman scattering microscopy for quantitative characterization of mixing and flow in microfluidics. *Optics Letters* 34:211–213.

Schafer, D., J. A. Squier, J. van Maarseveen, D. Bonn, M. Bonn, and M. Muller. 2008. In situ quantitative measurement of concentration profiles in a microreactor with submicron resolution using multiplex CARS microscopy. *Journal of the American Chemical Society* 130:11592–11593.

Shen, Y. R. 1984. *The Principles of Nonlinear Optics*. Wiley Interscience, New York.

Shi, K., P. Li, and Z. Liu. 2007. Broadband coherent anti-Stokes Raman scattering Spectroscopy in supercontinuum optical trap. *Applied Physics Letters* 90:141116.
Sung, J., B.-C. Chen, and S.-H. Lim. 2011. Fast three-dimensional chemical imaging by interferometric multiplex coherent anti-Stokes Raman scattering microscopy. *Journal of Raman Spectroscopy*, 42:130–136.
Ulriksen, H.-U., J. Thøgersen, S. R. Keiding, I. R. Perch-Nielsen, J. S. Dam, D. Z. Palima, H. Stapelfeldt, and J. Gluckstad. 2008. Independent trapping, manipulation and characterization by an all-optical biophotonics workstation. *Journal of the European Optical Society* 3:08034.
Vartiainen, E. M. 1992. Phase retrieval approach for coherent anti-Stokes Raman scattering spectrum analysis. *Journal of the Optical Society of America B* 9:1209–1214.
Vartiainen, E. M., K.-E. Peiponen, and T. Asakura. 1996. Phase retrieval in optical spectroscopy: Resolving optical constants from power spectra. *Applied Spectroscopy* 50:1283–1289.
Vartiainen, E. M., H. A. Rinia, M. Muller, and M. Bonn. 2006. Direct extraction of Raman line shapes from congested CARS spectra. *Optics Express* 14:3622–3630.
von Vacano, B., L. Meyer, and M. Motzkus. 2007. Rapid polymer blend imaging with quantitative broadband multiplex CARS microscopy. *Journal of Raman Spectroscopy* 38:916–926.
Wurpel, G. W. H. and M. Muller. 2006. Water confined by lipid bilayers: A multiplex CARS study. *Chemical Physics Letters* 425:336–341.
Wurpel, G. W. H., H. A. Rinia, and M. Muller. 2005. Imaging orientational order and lipid density in multilamellar vesicles with multiplex CARS microscopy. *Journal of Microscopy* 218:37–45.
Wurpel, G. W. H., J. M. Schins, and M. Muller. 2002. Chemical specificity in three-dimensional imaging with multiplex coherent anti-Stokes Raman scattering microscopy. *Optics Letters* 27:1093–1095.
Wurpel, G. W. H., J. M. Schins, and M. Muller. 2004. Direct measurement of chain order in single phospholipid mono- and bilayers with multiplex CARS. *Journal of Physical Chemistry B* 108:3400–3403.
Yahng, J. S. and S. C. Jeoung. 2011. Thickness determination with chemical identification of double-layered polymeric thin film by using multiplex CARS. *Optics and Lasers in Engineering* 49:66–70.

# 14. Interferometric Multiplex CARS

**Sang-Hyun Lim**



## 14.1 Introduction

Conventional CARS microscopy experiments typically employ two synchronized narrowband laser pulses to generate CARS signals at a single vibrational resonance (Evans and Xie 2008). This configuration of CARS microscopy has proven to be an excellent imaging tool for visualizing lipid-rich structures in cells and tissues when it detects CARS signals at 2840 $cm^{-1}$ (Le et al. 2010). Its chemical selectivity is, however, considerably limited; CARS imaging at a single vibrational frequency can at best map out the distribution of one chemical bond, which is of far less chemical information than full vibrational spectrum can provide. CARS microscopy can be configured to measure a vibrational spectrum if a broadband laser pulse is incorporated (Cheng et al. 2002, Müller and Schins 2002). In this multiplex CARS method, broadband laser pulses simultaneously excite multiple vibrational peaks and a narrowband probe pulse generates CARS spectrum, which is then measured by a spectrometer. In principle, multiplex CARS microscopy can provide great chemical information via vibrational spectral analysis at each image pixel.

The non-resonant CARS signal has been a serious issue in CARS microscopy because it obscures the desired vibrationally resonant CARS signal (Cheng and Xie 2004, Evans and Xie 2008). This electronic four-wave-mixing signal not only adds up as a background but also interferes with the resonant CARS signal, making quantitative vibrational peak analysis difficult. It is more problematic in multiplex CARS experiments, because the use of broadband pulses (shorter pulses in the time domain) increases its relative contribution significantly.

*Coherent Raman Scattering Microscopy*. Edited by Ji-Xin Cheng and X. Sunney Xie © 2013 CRC Press/Taylor & Francis Group, LLC. ISBN: 978-1-4398-6765-5.

Various techniques have been developed to either suppress the non-resonant signal or extract the pure vibrational signals from the measurements (Chen and Lim 2008, Chen et al. 2010, Evans and Xie 2008, Ganikhanov et al. 2006, Jurna et al. 2009, Oron et al. 2003, Volkmer et al. 2001).

The resonant and non-resonant CARS signals are coherent and have characteristic phase structures (Dudovich et al. 2003). In the frequency domain, the resonant signal is a complex quantity that can be modeled as a linear sum of complex Lorentzian functions. The imaginary part of the resonant signal corresponds to the spontaneous Raman signals. In contrast, the non-resonant signal is a real quantity that varies smoothly upon excitation frequency due to its instantaneous time response. In typical CARS experiments, the measured signal is a coherent sum of these two quantities and the interference between them generates characteristic CARS spectral response, which can be very different from that of spontaneous Raman scattering (Cheng and Xie 2004, Evans and Xie 2008). On the other hand, this interference can be manipulated if the relative phase between the resonant and non-resonant signals can be controlled (Oron et al. 2002a,b, 2003). In recent years, our group has applied this idea to multiplex CARS microscopy and demonstrated that Raman-equivalent CARS spectrum can be measured by the combination of laser pulse shaping and real-time spectral interferometry (Chen and Lim 2008, Lim et al. 2005, 2006, 2007, Sung et al. 2008). In this chapter, two CARS techniques based on this concept will be introduced and their performance will be discussed in terms of imaging applications (Chen and Lim 2008, Lim et al. 2006).

## 14.2 CARS with Single Broadband Laser Pulses

Coherent anti-Stokes Raman scattering (CARS) is a third order nonlinear optical process, where two laser pulses at the frequencies of $\omega_P$ and $\omega_S$ (pump and Stokes pulses) excite coherent vibrations at $\Omega = \omega_P - \omega_S$, with which a third pulse, $\omega_{Pr}$ (probe pulse) generates anti-Stokes signals at the frequency $\omega_{CARS} = \omega_P - \omega_S + \omega_{Pr}$ (Figure 14.1) (Evans and Xie 2008). The excited vibrational frequency is determined by the frequency difference between the pump and Stokes pulses (i.e., $\omega_P - \omega_S$), which ranges over a few hundred to ~3400 cm$^{-1}$ in typical organic molecules.

The essential mechanism of the CARS methods discussed in this chapter can be understood in the two different nonlinear processes shown in Figure 14.1. Suppose that a combination of spectrally overlapping broadband and narrowband pulses is used to

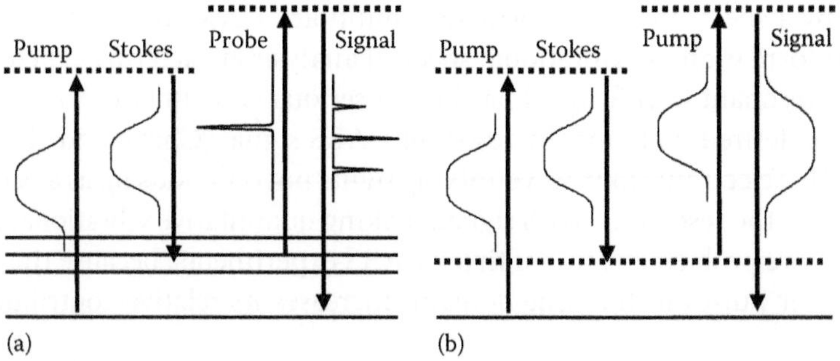


**FIGURE 14.1** (a) Multiplex resonant CARS process. (b) Non-resonant CARS process that accompanies the resonant CARS signal generation.

generate CARS signals. In this scenario, all possible combinations of wavelength components inside the broadband pulses will excite all the vibrational resonances covered within their bandwidth if the pulses are temporally compressed (Dudovich et al. 2003). In the resonant CARS process (Figure 14.1a), the narrowband probe generates frequency-resolved vibrational spectrum. The broadband pump pulses, however, can also act as probe to generate spectrally broad CARS signals (Figure 14.1b). With transform-limited broadband laser pulses, the non-resonant signals typically dominate in this process due to the short temporal durations of the pulses. Note that these two nonlinear signals are coherent and interfere with each other. Also note that the relative phase between the two signals can be controlled if the phase of the probe pulses is manipulated relative to the broadband pump and Stokes pulses.

A single broadband laser can provide both broad- and narrowband pulses discussed earlier (Oron et al. 2002a, 2003). A narrow frequency range inside the laser bandwidth is selected as the probe by a pulse shaper (Figure 14.2) while all the remaining frequency parts are used as the pump and Stokes. We use a 4*f* pulse shaper based on liquid crystal spatial light modulator (LC-SLM) (Weiner 2000). A dual band LC-SLM is preferred since it can control not only the phase but also polarization or amplitude of different frequency components. This pulse shaper setup can be configured into one of two operational modes. The first one can be chosen by removing the polarizer (P2 in Figure 14.2) after the LC-SLM. This configuration can control both the phase and polarization of individual frequency pixels at the Fourier plane. If both polarizers (P1 and P2 in Figure 14.2) are placed around the LC-SLM, the phase and amplitude of each frequency components can be controlled.

To apply the previously mentioned CARS mechanism, it is important to compress the laser pulse at the sample position since the techniques rely on precise phase manipulation of intra-frequency components inside the broadband pulses. In high-resolution CARS microscopy, a high numerical aperture (NA) microscope objective introduces significant high-order dispersions, which cannot be compensated by prism or grating pairs. Fortunately, the pulse shaper can compensate arbitrary pulse dispersion if the spectral phase of the pulse *at the sample position* is known. We have developed two complimentary single beam SPIDER (spectral phase interferometry for direct electric field reconstruction) methods, which can extract the spectral phase with the help of the pulse shaper (Chen and Lim 2007, Sung et al. 2008). In these pulse characterization methods,

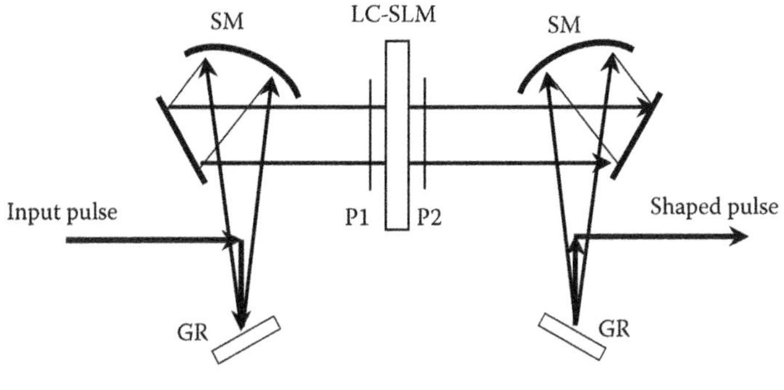


**FIGURE 14.2** Pulse shaper. GR, grating; SM, spherical mirror; P1 and P2, polarizer; LC-SLM, liquid crystal spatial light modulator.

SHG (second harmonic generation) signals are generated from a thin nonlinear crystal at the sample position and its spectrum is measured. Applying phase shifts only at narrowband frequency regions causes interference patterns in SHG spectrum, from which one can extract the spectral phase of the pulse analytically. These methods do not require any reference pulse and its precision is better than 0.1 rad over the FWHM of the laser bandwidth (Sung et al. 2008). By applying a counter phase mask in the SLM, one can create a transform-limited pulse at the microscope sample position.

## 14.3 Double Quadrature Spectral Interferometric CARS

In the first method, single phase- and polarization-controlled broadband laser pulses generate two orthogonally polarized CARS signals, which are measured by an imaging spectrometer (Lim et al. 2005, 2006). The normalized difference between the two spectra corresponds to background-free Raman-equivalent vibrational CARS spectrum. This DQSI (double quadrature spectral interferometric) CARS method will be discussed in this section.

With single broad band pulse excitation, all possible combinations of wavelengths within the bandwidth can be used as pump, Stokes, and probe pulses to generate the CARS signal. In this case, the resonant and non-resonant CARS signals become (Dudovich et al. 2003)

$$P_R(\omega) \propto \int_0^\infty d\Omega \chi_R(\Omega) E(\omega - \Omega) A(\Omega) \tag{14.1}$$

$$P_{NR}(\omega) \propto \int_0^\infty d\Omega \chi_{NR}(\Omega) E(\omega - \Omega) A(\Omega) \tag{14.2}$$

where

$$A(\Omega) = \int_0^\infty d\omega' E^*(\omega') E(\Omega + \omega') \tag{14.3}$$

$P_R$ and $P_{NR}$ are the resonant and non-resonant polarizations, respectively
$E$ is the laser field
$\chi_R$ and $\chi_{NR}$ represent the third-order nonlinear susceptibilities for the resonant and non-resonant CARS processes, respectively

Note that $\chi_{NR}$ is real and $\chi_R$ is complex.

To obtain a multiplex CARS signal, it is necessary to separate a spectrally narrow probe and this can be done by applying phase- and polarization-control techniques (Dudovich et al. 2003, Oron et al. 2002a). If the spectral linewidth of the probe is narrow

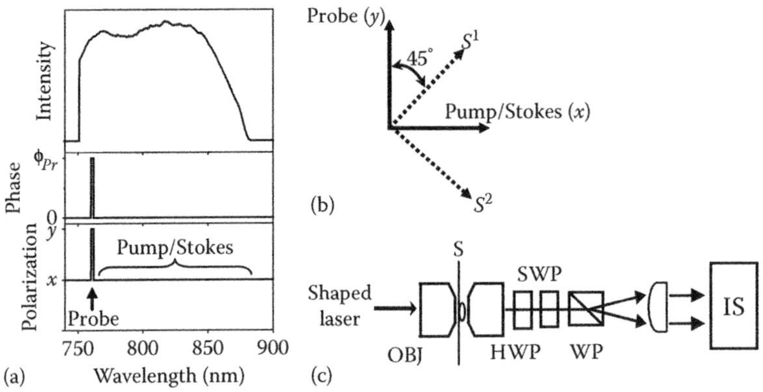


FIGURE 14.3 DQSI-CARS experiment. (a) Intensity, phase, and polarization of the shaped laser pulse. (b) Polarization directions of pump/Stokes, probe, and CARS signal detection (1 and 2). (c) Experimental setup for DQSI-CARS microscopy. OBJ, microscope objective lens; S, sample; HWP, half-wave plate; SWP, short-wave-pass filter; WP, Wollaston prism; IS, imaging spectrometer. (Reprinted in part with permission from Lim, S.-H., Caster, A.G., and Leone, S.R., *Phys. Rev. A*, 72, 041803, 2005. Copyright 2005 by the American Physical Society.)

as shown in the phase and polarization masks of Figure 14.3a, the resonant CARS component becomes (Lim et al. 2005)

$$\begin{aligned} P_R(\omega) &\propto \int_0^\infty d\Omega \chi_R(\Omega) E_{Pr}(\omega-\Omega) A(\Omega) \\ &\approx \int_0^\infty d\Omega \chi_R(\Omega) E_{Pr}^0 \delta(\omega-\Omega-\omega_{Pr}) \exp(i\phi_{Pr}) A(\Omega) \\ &= \chi_R(\omega-\omega_{Pr}) E_{Pr}^0 \exp(i\phi_{Pr}) A(\omega-\omega_{Pr}) \end{aligned} \tag{14.4}$$

where we assume that $E_{Pr}(\omega) \approx E_{Pr}^0 \delta(\omega-\omega_{Pr})\exp(i\phi_{Pr})$, i.e., the probe spectrum is infinitely narrow. $E_{Pr}^0$ is the amplitude of the narrowband probe pulse, $\delta$ is the delta function, $\varphi_{Pr}$ is the relative phase of the probe with respect to the excitation part of the pulse (as shown in Figure 14.3a), and $\omega_{Pr}$ is the frequency of the probe pulse.

Figure 14.3a shows the intensity, phase, and polarization of the laser pulse controlled by the pulse shaper (Lim et al. 2006). Polarizations of the pump/Stokes and the probe pulses are along the $x$- and $y$-directions, respectively, as shown in Figure 14.3b. If we detect CARS signals along the ±45° polarization directions relative to the $x$-axis (1- and 2-directions in Figure 14.3b), the CARS signals in the 1- and 2-polarization directions become

$$\begin{aligned} P_R^1(\omega) &= \frac{1}{\sqrt{2}} \chi_R(\omega-\omega_{Pr}) E_{Pr}^0 \exp(i\phi_{Pr}) A(\omega-\omega_{Pr}) \\ P_R^2(\omega) &= -\frac{1}{\sqrt{2}} \chi_R(\omega-\omega_{Pr}) E_{Pr}^0 \exp(i\phi_{Pr}) A(\omega-\omega_{Pr}) \\ P_{NR}^1(\omega) &= P_{NR}^2(\omega) = \frac{1}{\sqrt{2}} P_{NR}(\omega) \end{aligned} \tag{14.5}$$

The detected CARS signal intensities ($S^1$ and $S^2$) are

$$S^1 \propto |P_R^1 + P_{NR}^1|^2 = |P_R^1|^2 + \frac{1}{2}|P_{NR}|^2 + \sqrt{2}P_{NR}\,\mathrm{Re}[P_R^1]$$
$$S^2 \propto |P_R^2 + P_{NR}^2|^2 = |P_R^2|^2 + \frac{1}{2}|P_{NR}|^2 + \sqrt{2}P_{NR}\,\mathrm{Re}[P_R^2] \tag{14.6}$$

Since only a small portion of the laser spectrum is used for the probe, $P_{NR}$ is primarily parallel to the pump/Stokes direction ($x$-axis in Figure 14.3b). Combining Equations 14.5 and 14.6, we find that the difference, $S^1 - S^2$, becomes

$$\begin{aligned} S^1(\omega) - S^2(\omega) &\propto \sqrt{2}P_{NR}(\omega)\left(\mathrm{Re}\left[P_R^1(\omega)\right] - \mathrm{Re}\left[P_R^2(\omega)\right]\right) \\ &= 2P_{NR}(\omega)E_{Pr}^0 A(\omega-\omega_{Pr})\mathrm{Re}[\chi_R(\omega-\omega_{Pr})\exp(i\phi_{Pr})] \\ &= P_{NR}(\omega)E_{Pr}^0 A(\omega-\omega_{Pr})\{\mathrm{Re}[\chi_R(\omega-\omega_{Pr})]\cos\phi_{Pr} - \mathrm{Im}[\chi_R(\omega-\omega_{Pr})]\sin\phi_{Pr}\} \end{aligned} \tag{14.7}$$

Equation 14.7 predicts that if we can detect $S^1$ and $S^2$ simultaneously, the real or imaginary parts of the vibrationally resonant susceptibility ($\chi_R$) can be extracted simply by controlling the phase of the probe pulse ($\phi_{Pr}$ in Figure 14.3a). Since $S^1 + S^2$ contains mostly non-resonant signals, we can also retrieve $P_R(\omega)$ in Equation 14.7 by the following normalized difference ($D$),

$$\begin{aligned} D(\omega) &= \frac{S^1(\omega) - S^2(\omega)}{\sqrt{S^1(\omega) + S^2(\omega)}} \\ &\approx \frac{P_{NR}(\omega)E_{Pr}^0 A(\omega-\omega_{Pr})\{\mathrm{Re}[\chi_R(\omega-\omega_{Pr})]\cos\phi_{Pr} - \mathrm{Im}[\chi_R(\omega-\omega_{Pr})]\sin\phi_{Pr}\}}{\sqrt{|P_{NR}(\omega)|^2}} \\ &= E_{Pr}^0 A(\omega-\omega_{Pr})\{\mathrm{Re}[\chi_R(\omega-\omega_{Pr})]\cos\phi_{Pr} - \mathrm{Im}[\chi_R(\omega-\omega_{Pr})]\sin\phi_{Pr}\} \end{aligned} \tag{14.8}$$

where we assume that $|P_{NR}|^2 \gg |P_R^1|^2 + |P_R^2|^2$. In this way, we can detect only the resonant signals from samples free from the effect of the non-resonant signals.

In the experimental setup (Figure 14.3c), CARS signals are generated and collected with high NA microscope objective lenses, the polarization of the CARS signal is rotated by 45° with a half wave plate, laser pulses are removed by a short-wave-pass filter, and the 1- and 2-polarization signals are separated by a Wollaston prism (Lim et al. 2006). Directing the split signals (the 1- and 2-polarization signals) into an imaging spectrometer and onto a two-dimensional CCD, we obtain both spectral traces simultaneously.

Figure 14.4 shows the experimental signal spectra from toluene along the 1- and 2-directions (Figure 14.4a and b) and the normalized differences according to Equation 14.8 (Figure 14.4c and d) with two different probe phases (Lim et al. 2005). As predicted by

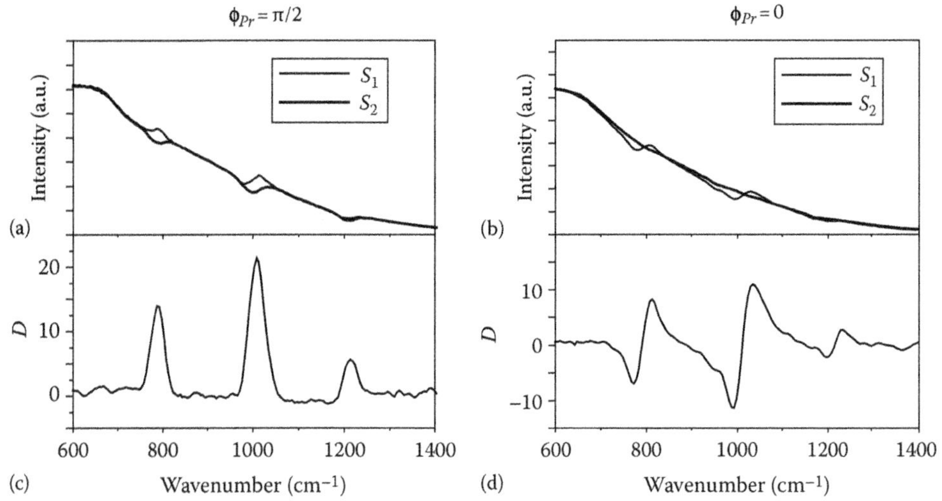


**FIGURE 14.4** DQSI-CARS of toluene: (a) and (b) The spectra in each polarization direction (1 and 2 as in Figure 14.3b) taken simultaneously in the imaging spectrometer. (c) and (d) The normalized difference spectra ($D$) according to Equation 14.8. (a) and (c) are taken with $\phi_{P_r} = \pi/2$. (b) and (d) are taken with $\phi_{P_r} = 0$. (Reprinted in part with permission from Lim, S.-H., Caster, A.G., and Leone, S.R., *Phys. Rev. A*, 72, 041803, 2005. Copyright 2005 by the American Physical Society.)

Equation 14.7, the imaginary or real part of the vibrational spectrum can be obtained by $\phi_{P_r} = \pi/2$ and 0. As noted previously, the imaginary part (Figure 14.4c) corresponds to the spontaneous Raman scattering signal.

Figure 14.5a shows the experimental DQSI-CARS spectra of several common lab solvents, which correspond well to the known Raman spectra (Lim et al. 2005). Detecting the

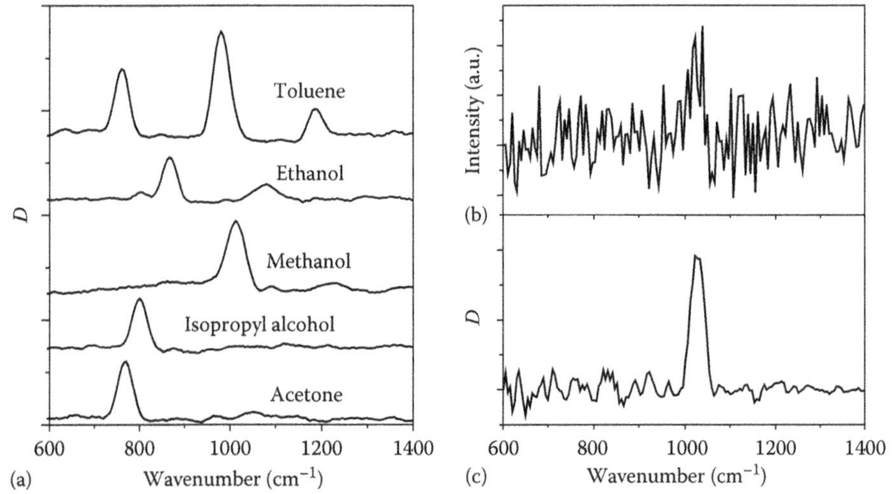


**FIGURE 14.5** (a) Experimental DQSI-CARS spectra of organic solvents. (b) and (c) Comparison of signal levels between the resonant-only CARS method and the homodyne-amplified DQSI-CARS method. The sample is 1 M aqueous $KNO_3$ solution. (b) the phase- and polarization-controlled method according to Oron et al. (2003). (c) DQSI method under the same experimental conditions. (Reprinted in part with permission from Lim, S.-H., Caster, A.G., and Leone, S.R., *Phys. Rev. A*, 72, 041803, 2005. Copyright 2005 by the American Physical Society.)

cross term, $P_{NR}$ Re$[P_R]$ in Equation 14.6, instead of $|P_R|^2$, has two further advantages. First, $P_R$ is amplified with the help of the much bigger $P_{NR}$, which is usually called homodyne amplification. Second, the signal depends linearly on the sample concentration. The linear concentration dependence is an important benefit, since it makes quantitative concentration measurement easier. Figure 14.5b and c demonstrate the homodyne amplification advantage that the DQSI-CARS technique offers. The sample is 1 M $KNO_3$ solution in water and Figure 14.4b shows a CARS spectrum taken in a similar fashion to the phase- and polarization-controlled CARS method developed by Silberberg and co-workers (Oron et al. 2003). In this technique, the non-resonant signal is blocked by a polarizer and only the resonant signal is measured. In Figure 14.4c, the spectrum is taken by the DQSI-CARS technique with exactly the same experimental conditions (i.e., the laser power and CCD exposures are identical). One can clearly see the improvement of signal-to-noise ratio via homodyne amplification (Lim et al. 2005).

## 14.4 Fourier Transform Spectral Interferometric CARS

The DQSI-CARS method discussed in the previous section has a few drawbacks in its microscopy applications; (1) it cannot be applied to birefringent or highly scattering samples since it requires balanced CARS spectra along two polarization directions. If the magnitudes of the two measured spectral traces are significantly different, which can be caused by sample depolarization, the resulting DQSI-CARS spectrum no longer corresponds to the imaginary part of the resonant CARS signals. (2) The CCD should be operated in the so-called multi-track mode, measuring two vertically separated spectral traces simultaneously (Lim et al. 2006). This mode is significantly slower than the full binning mode resulting in slow image acquisition. The fastest CCD operation in the multi-track mode is typically a few 10 ms and it takes several minutes to acquire one image. (3) Since the pulse shaper cannot control the spectral amplitude, the entire frequency components of the laser pulses should be focused to the sample. Note that the SLM-based pulse shaper can be operated in either phase/polarization or phase/amplitude modes. The use of the entire pulse bandwidth increases the nonlinear photodamage significantly due to the shorter pulse duration. If the amplitude pulse shaping is allowed, one can remove the unused frequency components in the laser spectrum and stronger CARS signals can be generated with the same laser power. Accordingly, we have developed a new method that requires a single spectral measurement and can extract Raman-equivalent CARS spectrum via phase/amplitude pulse shaping (Chen and Lim 2008, Lim et al. 2007, Sung et al. 2010).

Fourier transform spectral interferometry (FTSI) is a single-shot spectral interferometry method that extracts a full complex signal quantity via the different time responses between the signal and local oscillator (Lepetit et al. 1995). In CARS experiments, the resonant CARS signal obeys the causality principle, which means that vibrational excitation should not occur before the excitation laser arrives at the sample. This section will show how this causality can be used to extract the imaginary part of the resonant CARS signal (Chen and Lim 2008).

Figure 14.6a shows the amplitude and phase spectra of the shaped laser pulse used in the FTSI-CARS method (Chen and Lim 2008). The laser spectrum has a distinct high intensity frequency region that is used as the probe in CARS signal generation.

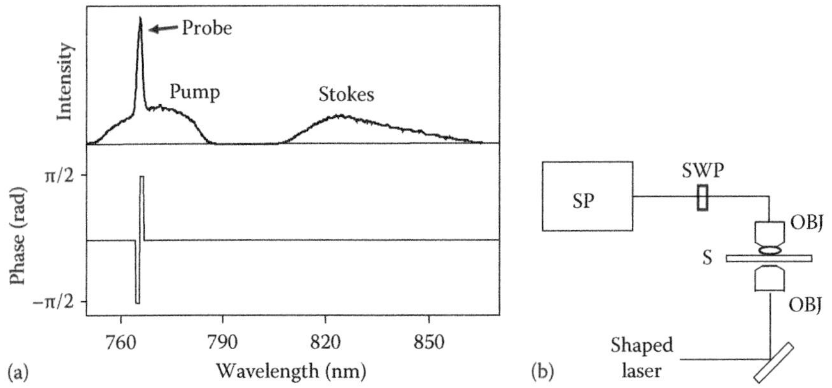


FIGURE 14.6 (a) Intensity and phase of the laser pulse used in FTSI-CARS experiments. (b) Experimental setup for FTSI-CARS. OBJ, microscope objective lens; S, sample; SWP, short-wave-pass filter; SP, spectrometer. (Reprinted in part with permission from Chen, B.-C. and Lim, S.-H., Optimal laser pulse shaping for interferometric multiplex coherent anti-Stokes Raman scattering microscopy, *J. Phys. Chem. B*, 112, 3653–3661, 2008. Copyright 2008 American Chemical Society.)

A $\pi$ step phase is applied only in the narrow probe region as shown in the bottom trace in Figure 14.6a. Figure 14.6b is the schematic experimental setup, which is simpler than that of DQSI-CARS. As in the DQSI-CARS experiments, the phase-shifted probe creates spectral interference between the resonant CARS signals and the smooth non-resonant background, which can be seen in the CARS signal from cyclohexane ($S_\pi$ in Figure 14.7a) (Chen and Lim 2008). If the $\pi$ step phase is removed (i.e., transform-limited laser pulse), the measured spectrum becomes smooth ($S_0$ in Figure 14.7a). Note that the reference signal ($S_0$) is dominated by the non-resonant background due to the short pulse duration used in this experiment.

Comparing the two traces ($S_\pi$ and $S_0$) in Figure 14.7a, one can see that the narrow spectrally resolved CARS signals appear on top of the broad background in $S_\pi$. This is in fact the cross term in the spectral interference between the non-resonant ($P_{NR}(\omega)$) and spectrally resolved resonant ($P_R(\omega)$) signals. Under the approximation of $P_{NR}(\omega) \gg P_R(\omega)$, the measured signal, $S_\pi(\omega)$ becomes

$$S_\pi(\omega) \propto \left|P_{NR}(\omega) + P_R(\omega)\right|^2 \approx P_{NR}(\omega)^2 + 2P_{NR}(\omega)\,\mathrm{Re}[P_R(\omega)] \tag{14.9}$$

The cross term $\mathrm{Re}[P_R(\omega)]$ can be obtained by normalizing the CARS signal ($S_\pi(\omega)$) with the reference signal ($S_0$). The normalized signal becomes

$$\frac{S_\pi(\omega)}{S_0(\omega)} \approx \left|\frac{P_{NR}(\omega)}{P_{NR}^0(\omega)}\right|^2 + \frac{2P_{NR}(\omega)}{\left|P_{NR}^0(\omega)\right|^2}\mathrm{Re}[P_R(\omega)] \tag{14.10}$$

where $P_{NR}^0(\omega)$ are the non-resonant signals by the reference excitation laser pulse (i.e., no $\pi$ step phase in the probe pulse). Figure 14.7b shows the normalized CARS signal ($S_\pi/S_0$) from Figure 14.7a. Since the spectral shape of the non-resonant signal does not depend on the sample and the $\pi$ probe part of the laser pulse does not excite the non-resonant signals, the first term in Equation 14.10 ($\left|P_{NR}(\omega)/P_{NR}^0(\omega)\right|^2$) becomes a constant offset.

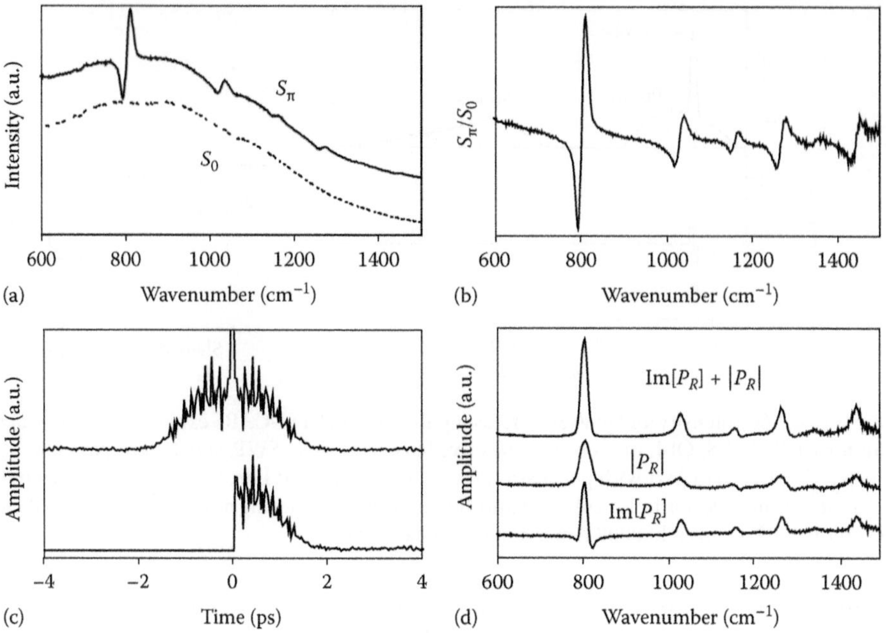


**FIGURE 14.7** FTSI-CARS experiment: (a) Measured CARS signals from cyclohexane by the CARS ($S_\pi$, top) and the reference ($S_0$, bottom) laser pulses. (b) Normalized CARS signal ($S_\pi/S_0$). (c) (top) FT of the normalized signal ($S_\pi/S_0$) and (bottom) the zero-filled time profile before the time zero. (d) Imaginary part, amplitude, and a sum of the imaginary part and amplitude for the inverse FT of the zero-field time profile (the bottom trace in (c)). Traces in (a), (c), and (d) are vertically displaced for clarity. The CCD exposure time is 1 ms. (Reprinted in part with permission from Chen, B.-C. and Lim, S.-H., Optimal laser pulse shaping for interferometric multiplex coherent anti-Stokes Raman scattering microscopy, *J. Phys. Chem. B*, 112, 3653–3661, 2008. Copyright 2008 American Chemical Society.)

The non-resonant contribution in the second term ($|P_{NR}(\omega)/P^0_{NR}(\omega)|^2$) also has a very smooth frequency dependence compared to the resonant signal (Re[$P_R(\omega)$]). As a result, $S_\pi/S_0$ have the spectral line shape of Re[$P_R(\omega)$] on top of a flat background as shown in Figure 14.7b. The complex function of $P_R(\omega)$ can be acquired out of Re[$P_R(\omega)$] by the following method.

So far we have considered CARS signals only in the frequency domain. In the time domain, it has an interesting symmetry property, which can be used to extract the full complex quantity of $P_R(\omega)$ out of Re[$P_R(\omega)$]. Let us consider Fourier transformation (FT) of Figure 14.7b. The upper trace in Figure 14.7c shows FT of the normalized CARS signal in Figure 14.7b. This is essentially FT of the real part of a Lorentzian function. Note that FT of a complex Lorentzian function has zero amplitudes before the time zero and its amplitude decays exponentially after. Mathematically, (Arfken 2001)

$$\begin{aligned}\mathrm{FT}\left(\frac{1}{(\Omega_0-\Omega)+i\Gamma_0}\right) &= \frac{1}{\sqrt{2\pi}}\int_{-\infty}^{\infty} d\omega e^{i\omega t}\frac{1}{(\Omega_0-\Omega)+i\Gamma_0}\\ &= 0 \qquad \text{for } t<0\\ &= \sqrt{2\pi}\exp(-\Gamma_0 t)\exp(i\Omega_0 t-\pi i/2) \quad \text{for } t>0 \end{aligned} \tag{14.11}$$

It is also interesting to compare FT of $P_R(\omega)$ and $P_R(\omega)^*$. A FT of a complex conjugate quantity in the frequency domain is equivalent to an inversion in the time domain. This can be easily proved by (Arfken 2001)

$$\mathrm{FT}[P_R(\omega)^*] = \frac{1}{\sqrt{2\pi}} \int_{-\infty}^{\infty} d\omega e^{i\omega t} P_R(\omega)^* = \left[ \frac{1}{\sqrt{2\pi}} \int_{-\infty}^{\infty} d\omega e^{-i\omega t} P_R(\omega) \right]^* = P_R(-t)^* \tag{14.12}$$

In other words, $P_R(\omega)$ and $P_R(\omega)^*$ are separated in the time domain. If we apply the Heaviside step function, $\theta(t)$, to the FT of the normalized CARS signal of Equation 14.10 (the lower trace in Figure 14.7c), we can select only $P_R(t)$. i.e.,

$$\theta(t)\mathrm{FT}[P_R(\omega) + P_R(\omega)^*] = \mathrm{FT}[P_R(\omega)] \tag{14.13}$$

Then an inverse FT recovers the full complex quantity of $P_R(\omega)$.

$$\mathrm{FT}^{-1}[\theta(t)\mathrm{FT}[P_R(\omega) + P_R(\omega)^*]] = P_R(\omega) \tag{14.14}$$

Figure 14.7d shows the imaginary (bottom), amplitude (middle), and the sum of imaginary and amplitude parts (top) of the retrieved $P_R(\omega)$. Note that the use of the $\pi$ step phase in the probe (instead of the flat phase used in DQSI-CARS) creates dips around the vibrational peaks in the retrieved imaginary part of the CARS signals. We found that $\mathrm{Im}[P_R] + |P_R|$ generates vibrational spectrum that is very similar to the spontaneous Raman scattering.

It is worth to consider the relationship between the FTSI-CARS and the Kramers–Kronig relations. As one may realize, this method transforms the real part of a CARS signal into the imaginary part of it by FTs. In fact, this is one of the Kramers–Kronig relations (Levenson and Kano 1988),

$$\mathrm{Im}[\chi(\omega)] = \frac{1}{\pi} \int_{0}^{\infty} d\omega_0 \frac{\mathrm{Re}[\chi(\omega_0)]}{\omega_0 - \omega} \tag{14.15}$$

where $\chi(\omega)$ is a general susceptibility. The FTSI-CARS method can be reformulated with the convolution theorem (i.e., a multiplication in the time domain is equivalent to a convolution in the frequency domain). Equation 14.14 is then rewritten as

$$\begin{aligned} \mathrm{FT}^{-1}[\theta(t)\mathrm{FT}[P_R(\omega) + P_R(\omega)^*]] &= \mathrm{FT}^{-1}[\theta(t)\mathrm{FT}[2\,\mathrm{Re}[P_R(\omega)]]] \\ &= \frac{1}{\sqrt{2\pi}} \mathrm{FT}^{-1}[\theta(t)] \otimes \mathrm{FT}^{-1}[\mathrm{FT}[2\,\mathrm{Re}[P_R(\omega)]]] \\ &= \frac{1}{\sqrt{2\pi}} \mathrm{FT}^{-1}[\theta(t)] \otimes 2\,\mathrm{Re}[P_R(\omega)] \end{aligned} \tag{14.16}$$

where ⊗ is a convolution operator. With

$$\mathrm{FT}^{-1}[\theta(t)] = \frac{1}{i\sqrt{2\pi}\omega} + \sqrt{\frac{\pi}{2}}\delta(t)$$

Equation 14.16 becomes

$$\mathrm{FT}^{-1}[\theta(t)\mathrm{FT}[P_R(\omega) + P_R(\omega)^*]] = \frac{1}{\sqrt{2\pi}}\left(\frac{1}{i\sqrt{2\pi}\omega} + \sqrt{\frac{\pi}{2}}\delta(\omega)\right) \otimes 2\,\mathrm{Re}[P_R(\omega)]$$

$$= \frac{-i}{\pi}\int_{-\infty}^{\infty} d\omega_0 \frac{\mathrm{Re}[P_R(\omega_0)]}{\omega - \omega_0} + \mathrm{Re}[P_R(\omega)]$$

The real and imaginary parts of the retrieved signal become

$$\mathrm{Re}[\mathrm{FT}^{-1}[\theta(t)\mathrm{FT}[P_R(\omega) + P_R(\omega)^*]]] = \mathrm{Re}[P_R(\omega)]$$

$$\mathrm{Im}[\mathrm{FT}^{-1}[\theta(t)\mathrm{FT}[P_R(\omega) + P_R(\omega)^*]]] = \frac{-1}{\pi}\int_{-\infty}^{\infty} d\omega_0 \frac{\mathrm{Re}[P_R(\omega_0)]}{\omega - \omega_0} = \mathrm{Im}[P_R(\omega)] \tag{14.17}$$

In Equation 14.17, we have the Kramers–Kronig relation (Equation 14.15) with $P_R(\omega)$ instead of $\chi(\omega)$. This implies that our FTSI method is equivalent to the Kramers–Kronig relation (Chen and Lim 2008).

Figure 14.8a shows the FTSI-CARS spectra obtained from cyclohexane and toluene with the CCD exposure time of 20 μs (Chen and Lim 2008). Note that this CARS spectrum is taken with only 40 laser pulses since the repetition rate of the laser is 2 MHz. The signal-to-noise ratio of the CARS spectra in Figure 14.8a is ~130 and ~280 for cyclohexane and toluene, respectively. Also note that the vibrational peaks up to 1400 $\mathrm{cm}^{-1}$

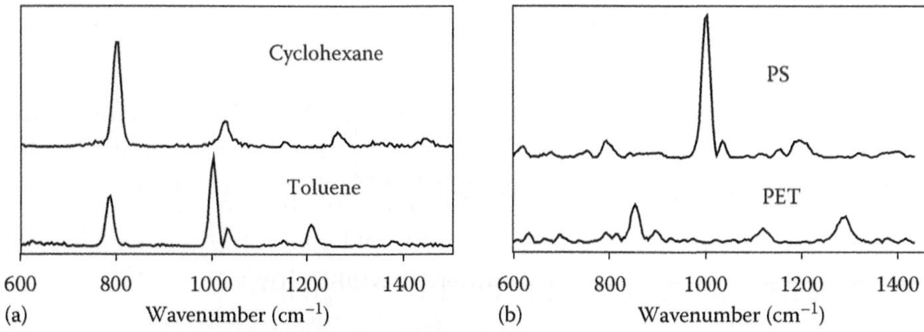


**FIGURE 14.8** (a) FTSI-CARS spectra of cyclohexane and toluene taken under the microscope setup with 20 μs CCD exposure time. (b) FTSI-CARS spectra of polystyrene (PS) and polyethylene terephthalate (PET) with 100 μs CCD exposure time. Each trace is displaced vertically for clarity. (Reprinted in part with permission from Chen, B.-C. and Lim, S.-H., Optimal laser pulse shaping for interferometric multiplex coherent anti-Stokes Raman scattering microscopy, *J. Phys. Chem. B*, 112, 3653–3661, 2008. Copyright 2008 American Chemical Society.)

are clearly visible in Figure 14.8a and b. The pump, Stokes, and probe powers used in Figure 14.8a are 1.2, 0.8, and 0.3 mW, respectively. The laser repetition rate is 2 MHz. The spectral resolution of ~20 $cm^{-1}$ is measured by the full width at the half maximum of the 1001 $cm^{-1}$ peak of toluene. One can see that the two neighboring peaks at 1001 and 1028 $cm^{-1}$ are clearly resolved in this experiment. Figure 14.8b shows the FTSI-CARS spectra of polystyrene (PS) and polyethylene terephthalate (PET) film with the CCD exposure time of 100 μs (Chen and Lim 2008). Since the polymers have lower photodamage thresholds, the pump power is lowered down from 1.2 to 0.5 mW while keeping the same Stokes and probe powers as used in Figure 14.8a. The CARS signals of polymers in Figure 14.8b are noisier than the liquid signals in Figure 14.8a due to the lower pump power, but all the major vibrational peaks of the polymers are clearly visible.

## 14.5 Optimal Laser Pulse Shape and Sensitivity

In the previously mentioned CARS methods, the pump, Stokes, and probe pulses are all selected inside a single broadband laser pulse. In this section, further optimization of the amplitude and phase pulse shape is considered with a goal of generating strong resonant CARS signals over the fingerprint region (600–1800 $cm^{-1}$). Since the sample photodamage limits the maximum allowable laser power, we seek after the optimal laser pulse shape under a given total laser power (Chen and Lim 2008).

It is easier to understand the CARS process by dividing it into the excitation and probing steps. The pump and Stokes pulses generate the coherent vibration (Raman coherence) in the excitation step. Then, the probe pulse scatters off the coherent vibration to generate the CARS signal (Figure 14.1a).

Let us consider the excitation step first. From Equation 14.3, $A(\Omega)$, the coherent vibrational excitation field at the frequency of $\Omega$ depends on the phase of the pulse by

$$A(\Omega) = \int_0^{\infty} d\omega' \left|E_P(\omega')\right| \left|E_S^*(\omega' - \Omega)\right| \exp(i\phi_P(\omega') - i\phi_S(\omega' - \Omega)) \tag{14.18}$$

where

- $E_P$ and $E_S$ are the electric fields of pump and Stokes laser pulses (Dudovich et al. 2003)
- $\phi_P(\omega')$ and $\phi_S(\omega')$ are the phase of pump and Stokes pulses at the frequency of $\omega$, respectively

As one can see in Equation 14.18, $A(\Omega)$ is maximized when $E_P$ and $E_S$ are transform limited (i.e., $\phi_P(\omega') - \phi_S(\omega' - \Omega) = 0$). There is, however, a choice of bandwidth for $E_P$ and $E_S$ in the multiplex CARS experiment. In Figure 14.9a, we compare two different amplitude shapes of pump/Stokes pulse combinations with the same total laser power. Since the Stokes pulse has a broad bandwidth, one can achieve a reasonable vibrational window in both cases. $A(\Omega)$ over 0–600 $cm^{-1}$ is excited by either the two pump or two Stokes pulses. The 600–1800 $cm^{-1}$ region is from the pulse combination of the pump and Stokes pulses. Here we are only interested in optimizing the excitation over the vibrational fingerprint region.

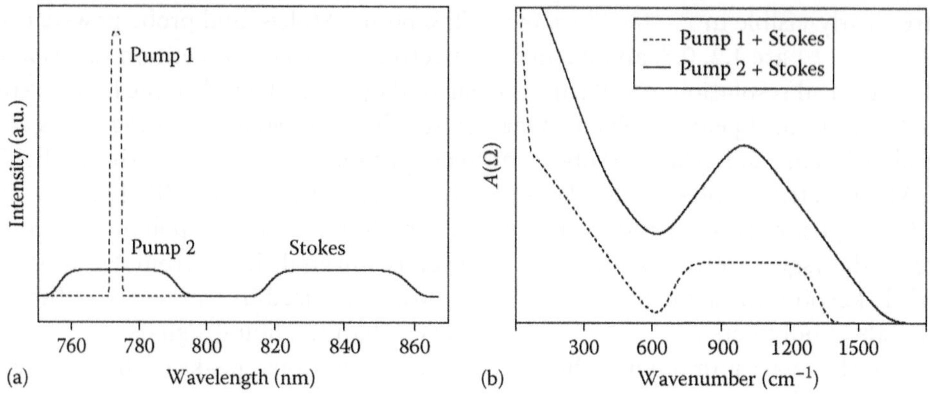


**FIGURE 14.9** (a) Spectra of two different pump/Stokes pulse shapes. (b) The simulations of coherent vibration excitations, $A(\Omega)$ with two different laser spectra in (a). The powers of pump 1 (dotted line) and 2 (solid line) are set to be equal. The Stokes pulse is identical for the both simulations. (Reprinted in part with permission from Chen, B.-C. and Lim, S.-H., Optimal laser pulse shaping for interferometric multiplex coherent anti-Stokes Raman scattering microscopy, *J. Phys. Chem. B*, 112, 3653–3661, 2008. Copyright 2008 American Chemical Society.)

Note that the pump pulse with a narrow bandwidth (Pump 1 in Figure 14.9a) is used in typical multiplex CARS experiments. Figure 14.9b shows the simulated $A(\Omega)$ with the two pump/Stokes pulse configurations in Figure 14.9a and Equation 14.17. Stronger and broader $A(\Omega)$ is generated by the broadband pump pulse (solid line in Figure 14.9b) (Chen and Lim 2008). There is, however, a complicating experimental factor. The sample photodamage is highly nonlinear with femtosecond laser pulses and a shorter pump pulse (broader bandwidth) causes more photodamage than a longer pulse with the same laser power. Accordingly, these two factors should be compromised to generate the optimal signal. In experiments with polymer samples, we find that the pump pulse with ~300 $cm^{-1}$ bandwidth generates the optimum CARS signal.

When the probe pulse scatters off the generated coherent vibration, CARS signal is generated. Since a high spectral resolution is desired, the probe pulse should have a narrow bandwidth. Note that we use separate pump and probe pulses, which is different from typical multiplex CARS experiments where a degenerate pump and probe pulse are used. This type of CARS signal generation is called three-color CARS process as opposed to two-color one (Lee et al. 2007). In the previously mentioned CARS methods, the probe pulse is primarily selected by the phase pulse shaping since the pump and probe pulses are taken in the same frequency region of the laser. Since the resonant CARS signal generated by this narrow phase-shifted probe pulse has a different phase than the non-resonant signal, we can extract the resonant signal with the help of DQSI or FTSI. Then question arises regarding what type of phase structure generates most efficient CARS signals with highest spectral resolution. Here we consider the two different phase masks for the probe pulse discussed earlier (Figures 14.3a and 14.6a). One has a flat phase over the entire probe bandwidth (Figure 14.3a) and the other has the $\pi$ step phase (Figure 14.6a). In the case of the flat phase probe, it is possible that the probe pulse can be used as pump pulse to create coherent vibrational excitation. This is not a serious problem as long as the power of the narrowband probe pulse is small with respect to that of the broadband pump pulse. If the probe pulse has a significant intensity, however, the excitation ($A(\Omega)$) begins to have a non-negligible component generated by the phase-shifted probe pulse and the phase structure of $A(\Omega)$ becomes complicated. This can be avoided by using the $\pi$ step phase pulse (Chen and Lim 2008).

Consider $A(\Omega)$ excited by the $\pi$ phase probe as a pump pulse (Figure 14.6a). Since the Stokes pulse ($E_S(\omega)$) has a much broader bandwidth than the probe pulse, Equation 14.18 becomes

$$A(\Omega) = \int_{\omega_{pr}-\Delta}^{\omega_{pr}} d\omega' E_{Pr}(\omega')\, E_S^*(\omega'-\Omega) + \int_{\omega_{pr}}^{\omega_{pr}+\Delta} d\omega' E_{Pr}(\omega') E_S^*(\omega'-\Omega)$$

$$\approx \left|E_{Pr}(\omega_{Pr})\right| (e^{i\pi/2} + e^{-i\pi/2}) E_S^*(\omega_{Pr}-\Omega)\Delta \tag{14.19}$$

where

$E_{Pr}$ is the electric field of the probe
$\omega_{Pr}$ is the center frequency of the probe pulse
$\Delta$ is the half bandwidth of the probe pulse

Under this condition, the integration in Equation 14.19 cancels. That is,

$$A(\Omega) \approx \left|E_{pr}(\omega_{pr})\right| (i-i) E_S^*(\omega_{pr}-\Omega)\Delta = 0$$

So the $\pi$ probe pulse does not excite coherent vibrations with a broadband Stokes pulse (Chen and Lim 2008). This is an important aspect since one can use a high intensity narrowband probe pulse without generating a complicated phase structure in the resulting CARS spectrum.

The resonant nonlinear susceptibility, $\chi_R$ has the Lorentzian spectral line shape, i.e.,

$$\chi_R(\Omega) \equiv \sum_k \frac{a_k}{(\Omega_k - \Omega) + i\Gamma_k} \tag{14.20}$$

where $a_k$, $\Omega_k$, and $\Gamma_k$ are the relative intensity, frequency, and linewidth of vibrational mode $k$, respectively. Note that $\chi_R$ has a $\pi$ phase shift at every vibrational resonance. The $\pi$ probe pulse compensates the $\pi$ phase shift of $\chi_R$ to maximize the CARS signal generation and also yields a narrow vibrational linewidth. As pointed out by Silberberg and coworkers, the probe phase, $\phi(\omega) = \arctan((\omega - \omega_{pr})/\Gamma)$ is optimal for the resonant CARS signal generation (Dudovich et al. 2002). The closest phase shape which we can produce with a typical pulse shaper is the $\pi$ step phase (Figure 14.6a) due to the limited spectral resolution of LC-SLM. Simulations show that the vibrational linewidth with the $\pi$ probe pulse is about twice narrower than that by the flat phase probe. Accordingly, the $\pi$ phase probe is also a good choice in terms of the spectral resolution.

The sensitivity of FTSI-CARS technique was tested with aqueous $Na_2SO_4$ solution and the detection limit of 970 cm$^{-1}$ vibrational peak of sulfate ions was found to be ~15 mM concentration in the CCD exposure time of 2 s (Sung et al. 2010). In this concentration level, there are approximately $1.2 \times 10^6$ sulfate ions in the focal volume of the laser focus. We also test the detection limit of $1037\ cm^{-1}$ peak of nitrate ions and find a similar sensitivity limit (Sung et al. 2010).

In the case of the homodyne limit (i.e., $P_{NR} \ll P_R$), measured CARS signals $(S(\omega) \propto |P_{NR}(\omega) + P_R(\omega)|^2)$ are dominated by $|P_R|^2$. In this situation, one observes quadratic dependence of the CARS signal. In the heterodyne limit (i.e., $P_{NR} \gg P_R$), essentially the cross-term ($P_{NR}\,\mathrm{Re}[P_R]$) is measured and the retrieved signal magnitude is linearly proportional to the sample concentration. This has been experimentally observed in aqueous sulfate and nitrate solutions (Sung et al. 2010).

The main noise source in the heterodyne regime is the shot noise from non-resonant background ($P_{NR}^2$). The standard deviation ($\sigma$) of the quantum fluctuation of non-resonant background (i.e., shot noise) becomes $P_{NR}$ from the Poison distribution (i.e. $\sigma = P_{NR}$). The previously mentioned CARS methods retrieve the cross-term ($P_{NR} P_R$) and the signal-to-noise ratio (SNR) is estimated as

$$\mathrm{SNR} = \frac{\mathrm{Signal}}{\mathrm{Noise}} = \frac{P_{NR} P_R}{P_{NR}} = P_R \tag{14.21}$$

As such, the DQSI- and FTSI-CARS methods are supposed to be shot-noise-limited. We verified this by estimating the shot-noise from the CCD signal counts. Theoretical shot-noise level was very close to the measured noise floor in the extracted FTSI-CARS signal (Sung et al. 2010).

## 14.6 Spectral Imaging

Both the DQSI- and FTSI-CARS methods have been successfully applied to high-resolution microscopy of condensed-phase samples (Chen and Lim 2008, Lim et al. 2006, Sung et al. 2010). Since both methods can measure the vibrational spectrum in a single shot, scanning of the sample position or excitation beam over the sample plane can generate hyperspectral CARS images. In DQSI-CARS demonstration, the samples were raster scanned since the multi-track operation of the CCD is slow (Lim et al. 2006). In the DQSI-CARS experiments, the maximum speed of CCD exposure plus the data transfer to the computer was 20 ms (10 ms for exposure and another 10 ms for data transfer per spectrum). FTSI-CARS can be performed significantly faster since it can utilize the full-binning CCD mode (Sung et al. 2010). To apply this faster CCD operation to multiplex CARS microscopy experiments, a new scanning mechanism was developed, which is explained in the following text.

In this new scanning mechanism, the laser beam is scanned in the sample plane along one axis by a scanning mirror while the sample stage is translated along the orthogonal direction (Chen and Lim 2008). The laser beam is scanned such that the collected signal is imaged along the entrance slit of the imaging spectrometer. The beam scanning is synchronized with the CCD operation to ensure that the scanning laser spot position is correlated with the measured CCD spectrum. After acquiring one line of the image, the sample is moved in the perpendicular direction, and the next line of the image is measured by the laser beam scanning. In this way, the maximum scan speed is limited by the CCD data transfer rate, not by the slow sample translation speed since the laser beam can be scanned faster than the sample stage. Note that modern scientific CCD camera can measure full vertical binned spectra with an exposure time as short as a few 10 μs without a mechanical shutter. However, its data transfer speed is much slower; the maximum data transfer speed of the CCD used in FTSI-CARS demonstrations was ~2 ms per a spectrum. After measuring the CARS spectra over the entire sample positions, the data are processed with the FTSI method. Due to the efficient fast Fourier transform algorithm, the data operation can be done very fast (a few seconds with a decent personal computer).

Figure 14.10a and b show chemically selective microscope images of a PS/PET mixture film with the FTSI-CARS method. Note that we take the entire vibrational spectra at every spatial point. We construct the images by integrating 970–1040 $cm^{-1}$ for PS and 1060–1140 $cm^{-1}$ for PET in the 10,000 vibrational spectra. One can clearly see that circular PS domains are surrounded by a PET domain in Figure 14.10. An exposure time of 1 ms per pixel is used and the total image acquisition time was 43 s. Note that the actual CCD exposure time for 10,000 spectra is 20 s but the data transfer takes additional 23 s.

The data transfer speed of CCD can be improved further via the so-called crop mode operation (Sung et al. 2010). In this mode, the CCD was set to read only 128 rows out of the original 512 rows to maximize CCD readout speed. In this condition, the maximum CCD speed of 250 μs per spectrum has been achieved with 200 μs exposure time per image pixel. This improved scanning mechanism allowed us to obtain a 100 × 100 pixel image in 3.4 s, which includes the CCD exposure, data transfer, and FTSI numerical process (Sung et al. 2010). Although this speed is slower than typical beam scanning CARS microscopy experiments by one order of magnitude, the rapid advance in CCD and CMOS detectors may overcome this speed limit.

In many cases, the image acquisition speed of multiplex CARS microscopy experiments is limited by the sample photodamage since longer CCD exposure time has to be used to obtain enough signals to create decent images. With a short laser pulse with 20 fs pulse duration, the power density at the focus of 1.2 NA objective can reach $5 \times 10^{12}$ $W/cm^2$ with 1 mW of laser power at a 2 MHz repetition rate. This is a power density high enough to cause multiphoton absorption, plasma generation, and even optical breakdown. So the maximum sensitivity of the CARS methods is mostly limited by

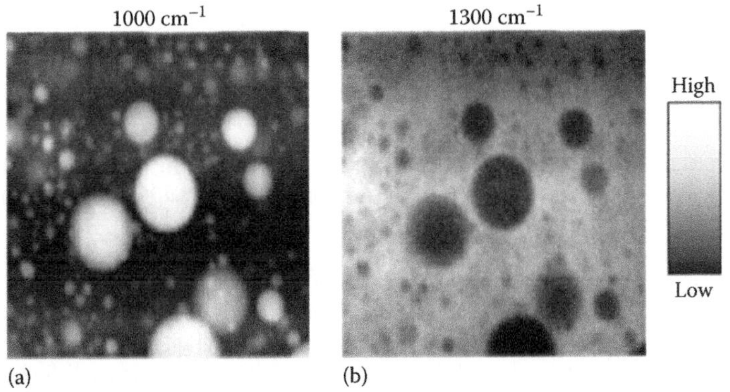


FIGURE 14.10 Chemical imaging of polystyrene (PS) and polyethylene terephthalate (PET) mixture film. Images are constructed with peaks at (a) 1000 $cm^{-1}$ (PS), and (b) 1200 $cm^{-1}$ (PET), respectively.

the nonlinear photodamage of the sample. Removing the laser frequency components from 780 to 810 nm (as shown in Figure 14.6a) makes the pulse duration significantly longer while keeping the same CARS sensitivity over the vibrational window from 600 to 1400 $cm^{-1}$. Among the three CARS excitation laser pulses (i.e., pump, Stokes, and probe), however, the probe pulse has much longer pulse duration (~ps) and its contribution to the photodamage is minimal comparing to the other two short pulses (pump and Stokes). We also find that the photodamage is more severe with the pump laser pulse due to the short wavelength of the pump beam. It is worth to compare the nonlinearity of power dependence for the CARS and sample photodamage. Photodamage nonlinearity is reported from 2.5 (Hopt and Neher 2001) to 1.1 (Fu et al. 2006), which is lower than 3, the nonlinearity of CARS. If the nonlinearity of sample photodamage is lower than that of CARS, a laser with a lower repetition rate and higher peak power is preferred in the CARS experiment since it can generate more signals than the laser with a higher repetition rate and lower peak power. We have compared the CARS signals with two different repetition rates (2 and 90 MHz) to find that the laser pulse with a lower repetition rate and higher peak power (2 MHz) has a better CARS sensitivity by comparing the CARS signals with the maximum allowable laser power determined by the photodamage of polymers (Chen and Lim 2008).

## 14.7 Chapter Summary

- Single broadband laser pulse can generate a CARS spectrum over the entire vibrational fingerprint region in a single measurement.
- Combination of the laser pulse shaping and spectral interferometry allows the retrieval of background-free Raman-equivalent vibrational spectrum.
- The key requirement in these CARS methods is the precise manipulation of the relative phase between the different frequency components inside the original broadband pulse.
- Detection sensitivity is currently around 10 mM concentration of sulfate and nitrate ions.
- Multiplex CARS imaging can be performed in a fast speed by scanning both the sample and laser beam. An image acquisition time of a few seconds can be achieved.
- Interferometric multiplex CARS offers allow full vibrational analysis to identify chemical species in inhomogeneous sample.

## References


Arfken, G. B. 2001. *Mathematical Methods for Physicists*. San Diego, CA: Academic Press.

Chen, B. and S.-H. Lim. 2007. Characterization of a broadband pulse for phase controlled multiphoton microscopy by single beam SPIDER. *Opt. Lett.* 32:2411–2413.

Chen, B.-C. and S.-H. Lim. 2008. Optimal laser pulse shaping for interferometric multiplex coherent anti-Stokes Raman scattering microscopy. *J. Phys. Chem. B* 112:3653–3661.

Chen, B. C., J. Sung, and S.-H. Lim. 2010. Chemical imaging with frequency modulation coherent anti-Stokes Raman scattering microscopy at the vibrational fingerprint region. *J. Phys. Chem. B* 114:16871–16880.

Cheng, J. X., A. Volkmer, L. D. Book, and X. S. Xie. 2002. Multiplex coherent anti-Stokes Raman scattering microspectroscopy and study of lipid vesicles. *J. Phys. Chem. B* 106:8493–8498.

Cheng, J. X. and X. S. Xie. 2004. Coherent anti-Stokes Raman scattering microscopy: Instrumentation, theory, and applications. *J. Phys. Chem. B* 108:827–840.

Dudovich, N., D. Oron, and Y. Silberberg. 2002. Single-pulse coherently controlled nonlinear Raman spectroscopy and microscopy. *Nature* 418:512–514.

Dudovich, N., D. Oron, and Y. Silberberg. 2003. Single-pulse coherent anti-Stokes Raman spectroscopy in the fingerprint spectral region. *J. Chem. Phys.* 118:9208–9215.

Evans, C. and X. S. Xie. 2008. Coherent anti-Stokes Raman scattering microscopy: Chemical imaging for biology and medicine. *Annu. Rev. Anal. Chem.* 1:883–909.

Fu, Y., H. Wang, R. Shi, and J. X. Chen. 2006. Characterization of photodamage in coherent anti-Stokes Raman scattering microscopy. *Opt. Express* 14:3942–3951.

Ganikhanov, F., C. L. Evans, B. G. Saar, and X. S. Xie. 2006. High-sensitivity vibrational imaging with frequency modulation coherent anti-Stokes Raman scattering (FM CARS) microscopy. *Opt. Lett.* 31:1872–1874.

Hopt, A. and E. Neher. 2001. Highly nonlinear photodamage in two-photon fluorescence microscopy. *Biophys. J.* 80:2029–2036.

Jurna, M., J. P. Korterik, C. Otto, J. L. Herek, and H. L. Offerhaus. 2009. Vibrational phase contrast microscopy by use of coherent anti-Stokes Raman scattering. *Phys. Rev. Lett.* 103:043905.

Le, T. T., S. Yue, and J. X. Cheng. 2010. Shedding new light on lipid biology with coherent anti-Stokes Raman scattering microscopy. *J. Lipid Res.* 51:3091–3102.

Lee, Y. J., Y. Liu, and M. T. Cicerone. 2007. Characterization of three-color CARS in a two-pulse broadband CARS spectrum. *Opt. Lett.* 32:3370–3372.

Lepetit, L., G. Chériaux, and M. Joffre. 1995. Linear techniques of phase measurement by femtosecond spectral interferometry for applications in spectroscopy. *J. Opt. Soc. Am. B-Opt. Phys.* 12:2467–2474.

Levenson, M. D. and S. S. Kano. 1988. *Introduction to Nonlinear Spectroscopy*. San Diego, CA: Academic Press.

Lim, S.-H., A. G. Caster, and S. R. Leone. 2005. Single-pulse phase-control interferometric coherent anti-Stokes Raman scattering spectroscopy. *Phys. Rev. A* 72:041803.

Lim, S.-H., A. G. Caster, and S. R. Leone. 2006. Chemical imaging by single pulse interferometric coherent anti-Stokes Raman scattering microscopy. *J. Phys. Chem. B* 110:5196–5204.

Lim, S.-H., A. Caster, and S. R. Leone. 2007. Fourier transform spectral interferometric coherent anti-Stokes Raman scattering (FTSI-CARS) spectroscopy. *Opt. Lett.* 32:1332–1334.

Müller, M. and J. M. Schins. 2002. Imaging the thermodynamic state of lipid membranes with multiplex CARS microscopy. *J. Phys. Chem. B* 106:3715–3723.

Oron, D., N. Dudovich, and Y. Silberberg. 2002a. Single-pulse phase-contrast nonlinear Raman spectroscopy. *Phys. Rev. Lett.* 89:273001.

Oron, D., N. Dudovich, and Y. Silberberg. 2003. Femtosecond phase-and-polarization control for background-free coherent anti-Stokes Raman spectroscopy. *Phys. Rev. Lett.* 90:213902.

Oron, D., N. Dudovich, D. Yelin, and Y. Silberberg. 2002b. Narrow-band coherent anti-Stokes Raman signals from broad-band pulses. *Phys. Rev. Lett.* 88:63004–63001.

Sung, J., B. Chen, and S.-H. Lim. 2008. Single-beam homodyne SPIDER for multiphoton microscopy. *Opt. Lett.* 33:1404–1406.

Sung, J., B. Chen, and S.-H. Lim. 2010. Fast three-dimensional chemical imaging by interferometric multiplex coherent anti-Stokes Raman scattering microscopy. *J. Raman Spectrosc.* (Early View):DOI 10.1002/jrs.2647

Volkmer, A., J. X. Cheng, and X. S. Xie. 2001. Vibrational imaging with high sensitivity via epidetected coherent anti-Stokes Raman scattering microscopy. *Phys. Rev. Lett.* 87:023901.

Weiner, A. M. 2000. Femtosecond pulse shaping using spatial light modulators. *Rev. Sci. Instrum.* 71:1929–1960.

# 15. Photonic Crystal Fiber–Based Broadband CARS Microscopy

**Marcus T. Cicerone, Young Jong Lee, Sapun H. Parekh, and Khaled A. Aamer**



## 15.1 Background

Broadband coherent anti-Stokes Raman (B-CARS) microscopy combines imaging with the label-free chemical specificity of vibrational spectroscopy. Imaging has long been a workhorse of biological and clinical investigation because it is intuitively information rich. Methods such as brightfield and phase contrast can convey significant information to a trained eye, and they are completely noninvasive. On the other hand, they provide minimal biochemical specificity and no molecular-level information. Fluorescence microscopy can provide real-time, spatially resolved information about biological structures and biochemical species. It has consequently become an indispensable tool for cell biology and physiology studies. However, fluorescence imaging presents potential complications from fluorescent probes, affecting molecular function and cytotoxic freeradical generation. In addition, fluorescent probes are subject to bleaching, limiting their ultimate signal generation.

Single-frequency CARS microscopy (Duncan et al., 1982; Zumbusch et al., 1999), described in detail elsewhere in this book, can provide high-resolution label-free images at video rates without bleaching but has limited chemical resolving power. Usually signal is observed from one or a small number of strong vibrational peaks such as the CH-stretch region of the spectrum (2840–3000 cm$^{-1}$). Thus, while it is a very desirable method for

*Coherent Raman Scattering Microscopy*. Edited by Ji-Xin Cheng and X. Sunney Xie © 2013 CRC Press/Taylor & Francis Group, LLC. ISBN: 978-1-4398-6765-5.

high-resolution spatio-chemical imaging, it is not ideal for distinguishing components with subtle differences in chemical composition that, for example, may be apparent only through a ratio of multiple peak intensities. This point has been described in detail in the preceding chapter.

In contrast to the rapid imaging modalities discussed earlier, spontaneous Raman scattering microscopy is relatively slow but presents an abundant amount of chemical information with no need for labeling. For example, Raman imaging and spectroscopy have been used to identify chemical changes in subcellular components and track changes in balance of mRNA and protein during stem cell differentiation (Chan et al., 2006, 2009; Notingher et al., 2004a,b), to discriminate very subtle cell types (Schut et al., 2002), and to discriminate various types of cancerous and normal tissues with sensitivity and specificity greatly exceeding that obtainable through traditional methods (Chan and Lieu, 2009; Kast et al., 2008). This, and similar chemical information, is usually obtained from changes in the "fingerprint" (500–1800 $cm^{-1}$) region of the Raman spectra (Chan et al., 2008; Matthaus et al., 2007; Uzunbajakava et al., 2003).

While advantageous in terms of label-free imaging and characterization, spontaneous Raman microspectroscopy has significant limitations due to intrinsically weak Raman scattering. Even when the laser power is applied near the physical damage threshold, spectral acquisition requires times ranging from seconds to minutes per spatial pixel (Chan et al., 2008, 2009; Uzunbajakava et al., 2003). The long acquisition time makes routine imaging applications impractical, and the high laser powers make truly noninvasive cellular imaging nearly impossible (Chernenko et al., 2009). A further complication of spontaneous Raman spectroscopy is the presence of a spectrally overlapping autofluorescence from the sample and substrate, the shot noise of which can dominate the weaker Raman scattering. Because CARS derives its signal from the Raman susceptibilities and it is a nonlinear process, it overcomes the signal-level limitations of spontaneous Raman scattering described earlier. Furthermore, CARS emission is blue-shifted relative to all excitation beams and is thus spectrally removed from any linear fluorescence emission. Expanding on the ideas presented in the preceding chapter, B-CARS provides spectral bandwidth sufficient to cover the fingerprint as well as the CH-stretch region and thus offers superb chemical resolving power with improved acquisition rates.

As previously discussed, in CARS a vibrational coherence is generated when a pair of photons (pump and Stokes) interact with the sample to excite a vibrationally resonant mode at frequency $\omega_{vib} = \omega_{pump} - \omega_{Stokes}$. A third (probe) photon is inelastically scattered off this coherent excitation, and anti-Stokes light ($\omega_{as} = \omega_{pump} - \omega_{Stokes} + \omega_{probe}$) is emitted from the sample. The third-order polarization ($P^{(3)}$) generated in the sample is given by

$$P_i^{(3)} = \sum_{j,k,l} \chi^{(3)}_{i,j,k,l}(\omega_p + \omega_{pr} - \omega_S) : \mathbf{E}_p^j(\omega_p)\mathbf{E}_{pr}^k(\omega_{pr})\mathbf{E}\sum_S^{*l}(\omega_S) \quad (15.1)$$

where

**E** are the electric field vectors

$\chi^{(3)}$ is the third-order nonlinear susceptibility

The subscripts *p*, *pr*, and *S* indicate pump, probe, and Stokes light, respectively

The CARS intensity $I_{CARS} \sim |P^{(3)}|^2$, and each element of $\chi^{(3)}$ has a frequency-independent nonresonant component ($\chi_{NR}^{(3)}$), which is entirely real, and a frequency-dependent resonant component, which is composed of a real ($\chi_{R'}^{(3)}$) and an imaginary part ($\chi_{R''}^{(3)}$),

$$\chi^{(3)} - \chi_{NR}^{(3)} + \chi_{R'}^{(3)} + i\chi_{R''}^{(3)} \tag{15.2}$$

$$\chi_R = \frac{\sigma}{\Omega - \Omega_R + i\Gamma} = \chi_{R'} + i\chi_{R''} \tag{15.3}$$

where
- $\sigma$ is the Raman cross section
- $\Omega$ is the excitation frequency
- $\Omega_R$ is the Raman resonance frequency
- $\Gamma$ is the linewidth

The imaginary part ($\chi_{R''}^{(3)}$) contains elements with the same bandshape as the spontaneous Raman signal, and it is the component of interest to us. This component, however, is not directly attainable from the raw CARS signal since the resonant and nonresonant signals are coherent, and the real and imaginary parts are convolved.

The resonant part of the CARS signal can be orders of magnitude larger than spontaneous Raman scattering under the right conditions (Cui et al., 2009; Tolles and Turner, 1977). These conditions are easily met for Raman peaks with large cross section and high oscillator concentrations, such as for CH-stretch in bulk lipids or in bilayer membranes. However, these conditions cannot be met for the resonances from the fingerprint region of most biological systems (Cui et al., 2009). This point will be discussed in detail under Section 15.2, but the bottom line is that without amplifying the resonant CARS signal, there are very few options for acquiring fingerprint CARS from cells and other biological systems at a rate faster than can be done for spontaneous confocal Raman microspectroscopy (CRMS).

The remainder of this chapter focuses primarily on technical challenges associated with implementing photonic crystal fiber (PCF)–based B-CARS and its resolution thus far. A major challenge is that the B-CARS signal is spread over a broad spectral range and there is a phototoxicity-induced limit on impinging light flux, thus efficient use of excitation light in signal generation of paramount importance. There are several signal generation mechanisms available to B-CARS, and these will be discussed in Section 15.2. Another challenge has been generation of a well-behaved continuum pulse, as this is crucial for generating B-CARS signal. We briefly discuss recent trends in the field of continuum generation and their implications for B-CARS in Section 15.3. Further issues are associated with coherence of the signal as can be inferred directly from Equations 15.1 through 15.3. First, because there are separate linear and quadratic terms involving the resonant contribution, the raw CARS signal amplitude scales linearly in the limit of low analyte concentration and quadratically in the limit of high analyte concentration. Secondly, shot noise associated with the nonresonant contribution can mask weak resonant signals when the nonresonant term is sufficiently large. The resolution of both these issues is found in resonant signal phase retrieval, discussed in Section 15.4. Phase retrieval allows us to exploit the intrinsic heterodyning in the CARS signal, as expressed by the second term of Equation 15.10, amplifying the weaker signals, and linearizing the resonant signal. Finally, in Section 15.5 we briefly review milestones achieved with nonlinear fiber–based CARS microscopy and provide some perspective on the potential for B-CARS for biological and materials imaging in Section 15.6.

## 15.2 Signal Generation

Three light fields are required to generate CARS signal. Most often at least one of these is degenerate, so two pulsed sources can be used if they are synchronized (intrinsically or extrinsically). In B-CARS signal generation, one of the two pulses is narrowband and the other is broadband. CARS signal is generated by a "2-color" mechanism (Lee et al., 2007) as shown in Figure 15.1 when the pump and probe transitions are degenerately induced by the narrowband pulse, and the Stokes transition by the continuum pulse. In the "3-color" CARS mechanism, two different frequency components within the continuum pulse induce the pump and Stokes transitions and the narrowband pulse induces the probe transition as shown in Figure 15.1.

The CARS nonlinear polarizations for the 2- and 3-color mechanisms are given by

$$P^{(3)}_{3\text{-}color}(\omega_n+\Omega) \propto A(\Omega,\Omega_R,\Gamma)E_n(\omega_n)\int_{-\infty}^{\infty} E_c^{\bullet}(\omega)E_c(\omega+\Omega)d\omega \tag{15.4}$$

$$P^{(3)}_{2\text{-}color}(\omega_n+\Omega) \propto A(\Omega,\Omega_R,\Gamma)E_n(\omega_n)E_n^{\bullet}(\omega_n)E_c(\omega_n+\Omega) \tag{15.5}$$

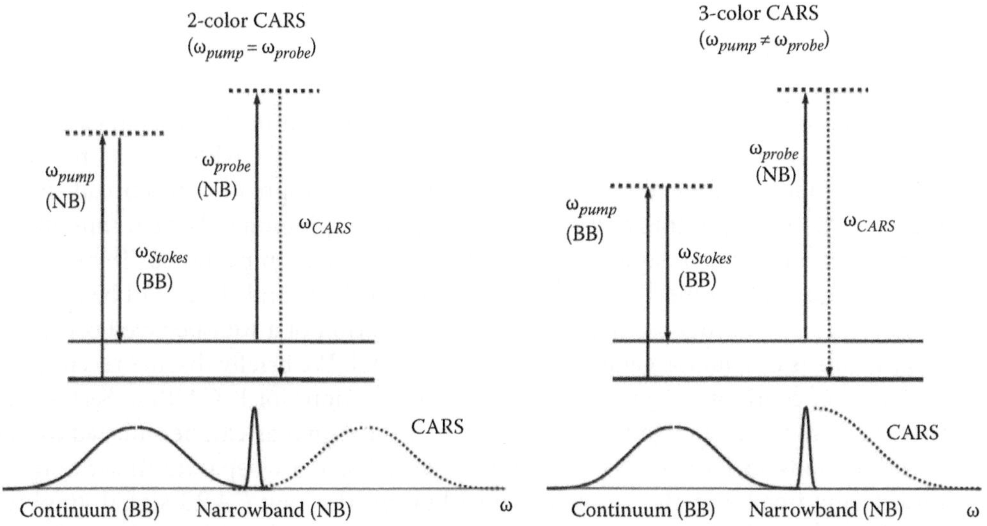


**FIGURE 15.1** Energy diagrams of the 2-color and 3-color CARS generation schemes, where the solid arrows indicate the transitions induced by the continuum pulse; the dotted arrows, by the narrowband pulse; and the dashed arrows indicate the CARS emission.

where
$E_n$ and $E_c$ are the narrowband and continuum laser fields, respectively
$\omega_n$ is the frequency of the narrowband laser
$\Omega$ is the Raman shift frequency
$\Omega_R$ is the frequency of the Raman-active mode in the medium
$\Gamma$ is the spontaneous Raman linewidth

$$A = [(\Omega - \Omega_R) + i\Gamma]^{-1}$$

Equations 15.4 and 15.5 indicate that the 2- and 3-color CARS signals have different dependencies on narrowband and continuum laser pulse powers. As indicated in Figure 15.1, the 3-color signal dominates at lower anti-Stokes shifts, and the 2-color signal dominates at higher shifts. By measuring CARS signal at an anti-Stokes shift of 900 $cm^{-1}$, we have demonstrated the expected quadratic dependence of the 3-color CARS signal on the continuum power and linear dependence on the narrowband pulse power. The converse power dependence is observed at 3000 $cm^{-1}$, showing the 2- and 3-color CARS signals generation scheme at the high and low frequency range, respectively (Lee et al., 2007).

Three-color CARS generation has several advantages over 2-color generation for the imaging of biological systems; these advantages are associated with its quadratic dependence on the peak power of the continuum pulse, which is red-shifted with respect to the probe. The "water window" for imaging aqueous-based systems extends out to wavelengths of roughly 1.3 μm. Living biological systems will tolerate much higher light flux toward the red end of this window than even slightly bluer wavelengths around 800 nm (Yakovlev, 2003). Because of the quadratic dependence on the continuum power, 3-color signal generation takes better advantage of this red light tolerance. Also, living systems appear to have slightly higher tolerance for increased peak power of shorter pulses (Fu et al., 2006; Hopt and Neher, 2001; Yakovlev, 2003). The photodamage rate for biological systems appears to scale as $P^{2.5}/\tau^{1.5}$ (König, 2000), where $P$ is the time-averaged power and $\tau$ is the laser pulse width. Here also, the 3-color CARS signal is increased much more by reduction of temporal pulse width than is the 2-color signal. If we assume transform-limited pulses, the decreased pulse width is associated with increased bandwidth. Figure 15.2a shows that, at constant total power for transform-limited pulses, the 3-color CARS signal exhibits significant increase in both the signal intensity and the measurable spectral range for the spectrally broader continuum pulse. Figure 15.2b shows that the 2-color CARS signal intensity is reduced, but its bandwidth is increased with the spectrally broader continuum pulse. Figure 15.2c shows that the 3-color CARS signal is also very sensitive to pulse chirp, with significant reduction of the low-frequency signal ($\omega_{as}$ < 1100 $cm^{-1}$) resulting from even a small amount of chirp in the continuum pulse (Lee et al., 2007).

The 3-color CARS approach can also be used to measure the dynamics of vibrationally excited states by controlling the time delay between the continuum and narrowband pulses. In this way, the 2-pulse 3-color microscopy approach can provide vibrational dephasing time images (Lee and Cicerone, 2008), which are a potential imaging modality

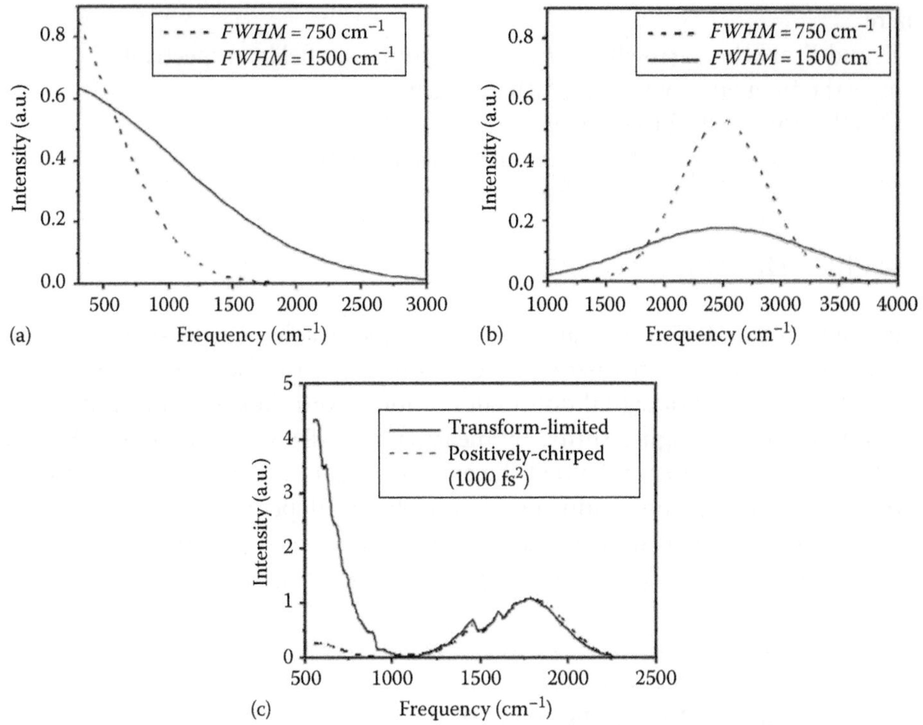


**FIGURE 15.2** Simulated pulse width effects of a transform-limited broadband pulse on 3-color (a) and 2-color (b) CARS spectra. The time-averaged power $P$ is adjusted as $P^{2.5}/\tau^{1.5}$ to simulate a constant photodamage rate. (c) Experimental chirp effects on 2-color (1000–2500 cm$^{-1}$) and 3-color (500–1000 cm$^{-1}$) CARS spectra of 0.5 M benzonitrile in ethanol. The laser powers are 4 and 8 mW at the sample position for the narrowband and the femtosecond pulses, respectively.

of the local molecular environment of vibrational modes. Figure 15.3 shows that time-resolved CARS results from neat benzonitrile using the optimized continuum pulse in the 3-color arrangement. Panel (a) shows clear differences in the measured CARS spectra of $\Delta t = 0$ and 1 ps. At $\Delta t = 0$ ps, nonresonant background (NRB) of both 2- and 3-color CARS interferes with resonant signal and produces dispersive line shapes overlaid on a broad baseline. At $\Delta t = 1$ ps, the NRB contribution diminishes and only 3-color resonant CARS signal remains. The frequency range shown in the time-resolved spectra is noteworthy. In previously reported NRB-free CARS spectra using similar time-resolved techniques, Raman peaks are barely analyzable at frequencies greater than 2300 cm$^{-1}$ (Lee and Cicerone, 2008). This higher-frequency region includes CH-stretch resonances, whose dephasing time will be impacted by physiologically important phenomena, such as membrane phase changes and protein insertion. The optimized time-resolved broadband CARS technique presented here provides access to vibrational dephasing times of multiple Raman modes from 500 to 3100 cm$^{-1}$ in a single time delay scan without any additional laser tuning. Figure 15.3b and d shows time evolution of low-, medium-, and high-frequency Raman modes. The instrumental response functions (IRFs) are shown by dotted lines and are measured in the absence of resonant features from a glass cover slip (anonresonant medium) under the same measurement condition as the sample liquids.

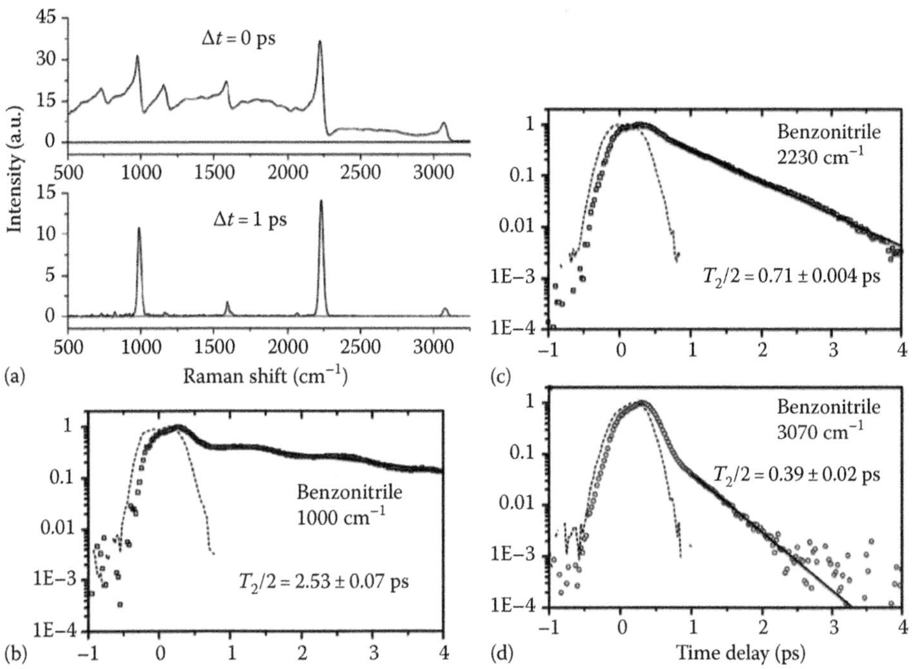


**FIGURE 15.3** Time-resolved CARS spectroscopy of neat benzonitrile. (a) CARS spectra at probe delay $\Delta t = 0$ and 1 ps. The average power of the continuum pulse was constant at 14 mW, and the average power of the narrowband pulse was 1 mW at the sample position. For each time scan, the time-independent baseline was measured at $\Delta t = -2$ ps and subtracted from the total CARS spectrum data. The baseline was 10%–20% of the peak signal. (b–d) Dephasing time measurements of various Raman modes in benzonitrile. The dotted lines indicate time profiles of the nonresonant background signal from a glass coverslip at the same Raman shift in each figure. The data of time delay ($\Delta t$) later than 1 ps are fitted to a single exponential function, $I(t) = \exp(-2\Delta t/T_2)$, where $T_2$ is the vibrational dephasing time. The average powers of the continuum and narrowband pulses were 14 and 1 mW, respectively, at the sample position. The exposure time was 600 ms for all spectra.

The IRF full-width-half-maximum (FWHM) ranges between 500 and 550 fs. The bandwidth of the narrowband pulse can be easily adjusted to give the desired temporal and spectral resolution based on the dephasing time of a mode and its frequency proximity to adjacent Raman modes. Taking into account the finite pulse duration of the probe and the IRF, data only after $\Delta t > 1$ ps are used for analysis to avoid the interference from NRB contribution. The time profiles are fitted to single exponential functions, $I(\Delta t) = \exp(-2\Delta t/T_2)$, where $T_2$ is the vibrational dephasing time (Lee et al., 2010a).

## 15.3 Continuum Generation

PCF-based supercontinuum is at the heart of a B-CARS system. Until recently, a major roadblock preventing full-spectral B-CARS microscopy in biological samples was a lack of an appropriate broadband continuum source that provides sufficient power and spectral breadth. We recently demonstrated an approach to supercontinuum generation using a Ti:Sapphire oscillator that was sufficient to overcome this limitation (Lee et al., 2010b).

The ideal continuum pulse for broadband CARS imaging should be (1) maximally coherent (i.e., transform-limited), (2) sufficiently broad in frequency (>3500 $cm^{-1}$ on the long wavelength side of the narrowband pulse), and (3) sufficiently powerful (>10 mW at the sample position for 80 MHz repetition rate). Extensive theoretical and experimental studies have been performed to control the characteristics of continuum pulses (Dudley et al., 2006). A continuum pulse can be generated in a highly nonlinear fiber via various mechanisms depending on both the medium (fiber length, nonlinearity, and dispersion curve) and the input pulse (wavelength, pulsewidth, pulse energy, polarization, and chirping). It remains extremely challenging to generate continuum that simultaneously satisfies the three requirements of coherence, bandwidth, and power using an unamplified (*ca.* 10 nJ) input pulse near 800 nm characteristic of Ti:Sapphire oscillator output. However, Selm et al. have recently demonstrated the use of a continuum generated from 1550 nm light that seems to meet all these criteria (Selm et al., 2010).

In Figures 15.4 and 15.5, we present an experimental optimization of continuum generation in a commercially available PCF that produces a sufficiently broad, coherent, and energetic pulse to generate a strong 3-color CARS signal with bandwidth to cover the fingerprint and CH regions (Lee et al., 2010b). Figures 15.4 and 15.5 both show time-resolved nonresonant CARS spectra from a glass coverslip where the continuum

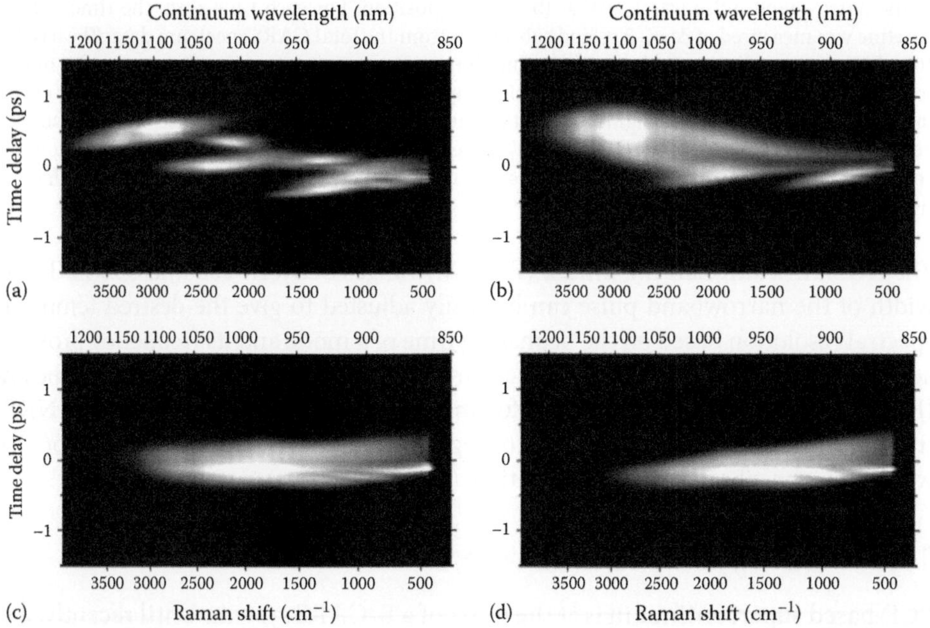


**FIGURE 15.4** Time-resolved CARS spectra generated in a glass coverslip. The continuum pulse is generated from a PCF pumped by a pulse with different group delay dispersion (GDD) of (a) 4,000 $fs^2$, (b) 12,000 $fs^2$, (c) 18,000 $fs^2$, and (d) 21,000 $fs^2$, where the uncertainty of GDD is 500 $fs^2$. For each spectrum, the phase of the narrowband pulse is optimized to the transform limit at the sample position. The continuum-generating pulse is centered at 830 nm, and its average power is 350 mW before the fiber. Average powers of the continuum and narrowband pulses are 2 and 24 mW, respectively, at the sample position. The exposure time was 100 ms.

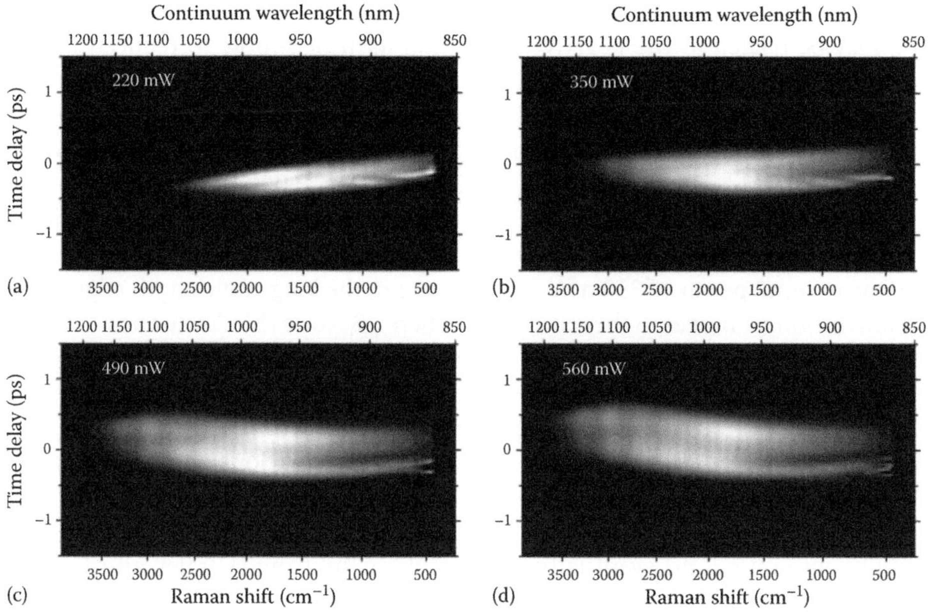


**FIGURE 15.5** Time-resolved CARS spectra generated in a glass coverslip by continuum pulses generated by an input pulse with different average powers of (a) 220 mW, (b) 350 mW, (c) 490 mW, and (d) 560 mW. At the sample position, the average powers of the continuum and narrowband pulses were controlled at constant values of 2 and 25 mW, respectively. The exposure time was 100 ms. The GDD of the input pulse is $-19{,}000\ fs^2$.

power was reduced sufficiently so that CARS is detected only when the narrowband and continuum are temporally overlapped. Under these conditions, the probe-delay time dependence of the NRB mirrors the temporal profile of the continuum pulse (analogous to the 2-color scheme described earlier). We use these spectrograms as diagnostics to highlight the influence of input pulse dispersion and power on the continuum bandwidth and temporal profile.

We tested several parameters of the input pulse, including wavelength, dispersion, and power. Input pulse wavelengths tested for continuum generation are 765, 830, and 860 nm, all of which are in the anomalous region of the fiber dispersion curve. The long wavelength end of the generated continuum spectrum is not sensitive to the input pulse wavelength and remains ~1200 nm at an average input power of 350 mW (4.4 nJ/pulse). For B-CARS, the short wavelength end of the continuum at the sample position is determined by the cutoff wavelength of the longpass filter combining the continuum and narrowband beams. We used an 830 nm narrowband pulse and an 850 nm longpass filter for the data acquisition reported in the remainder of this report.

In Figure 15.4, we show the effects of input pulse group delay dispersion (GDD) on continuum time profiles at constant average input power. We observe that an input pulse with a small amount of negative GDD generates a continuum with broad bandwidth, but also multiple solitons having significant temporal separation (Figure 14.4a). As the input pulse becomes more negatively chirped, the soliton splitting is greatly mitigated and the pulse becomes more stable, but as the magnitude of the negative GDD increases further, (Figure 15.4d) the bandwidth begins to narrow. The weakening of soliton splitting may be related to the negative dispersion itself or simply to reduced peak power due to pulse broadening (Dudley et al., 2006). The 70 fs transform-limited pulse will be temporally broadened to 170 and 800 fs by GDD of −4,000 and −21,000 $fs^2$, respectively. For efficient CARS signal generation, we are interested in the most stable continuum with the broadest bandwidth and least chirp (Lee et al., 2007).

In Figure 15.5, we show the effect of input pulse power on continuum generation at fixed GDD. Increasing input pulse power increases the bandwidth of the continuum until at an input power of 350 mW for this system, the long-wavelength edge of the continuum no longer moves to the red (1200 nm here). Beyond this "saturation" input power, additional power only leads to more complex structure in the time domain (Figure 15.5c and d) and results in less efficient CARS generation. Theoretical and numerical studies have shown that the majority of spectral broadening occurs within the initial stages of propagation. The length of the initial stage depends on various parameters of laser and fiber, including laser power. For a fixed fiber length, higher peak power will cause the initial broadening to occur earlier, imparting greater fiber-induced dispersion to the output continuum, and eventually lead to soliton splitting (Dudley and Coen, 2002). This scenario is consistent with the data of Figure 15.4 as well as Figure 15.5. At a higher power (Figure 15.5c and d), the average GDD of the continuum is −200 $fs^2$, while that at a lower power (Figure 15.5a) is +500 $fs^2$ at the sample plane (note that the dispersion of the fiber is anomalous, so its dispersion has the opposite sign of typical glass). From the average dispersion of the fiber (~70 ps/(km · nm)) in the continuum wavelength range, we can equate this difference in chirp of output continuum to a difference in the propagation within in the fiber of ~2 cm; at higher power, the pulse seems to be generated ~2 cm earlier than at low power. Thus, by tuning both chirp and power of the input pulse, we have some control over the continuum chirp, and we can pre-compensate for the dispersion of an objective lens, yielding a mean continuum GDD of near-zero at the sample position, which is ideal for CARS. The GDD of a continuum output can be varied to a value as large as −4500 $fs^2$, depending on where the continuum broadening occurs in the 12 cm fiber while the GDD of typical high numerical aperture objective lenses are ~2000–6500 $fs^2$ (Guild et al., 1997). This fortuitous GDD cancellation removes the necessity of additional pulse compression or pulse shaping for the continuum pulse, allowing maximum peak power delivery for the continuum pulse into the sample position even though the higher-order dispersion is still not compensated.

Coupling of the input pulse into the PCF can be easily affected by thermal drift. We have stabilized the supercontinuum output by locating the focusing lens of the input pulse on a feedback-controlled three-axis stage. Without the active stabilization, the power and spectrum of the continuum output slowly drifts after about 1 h in our experimental environment. The stability of the coupling will depend on individual experimental conditions. We find it useful to periodically manually align the focusing lens for the input pulse using the feedback-controlled stage; however, we usually run CARS imaging experiments with the feedback turned off to avoid additional high-frequency jitter in our continuum.

The broad bandwidth in NIR and the large numerical aperture from the fiber make it challenging to collimate the continuum output using a refractive lens without chromatic aberration and temporal dispersion. Either of these would result in significant reduction in signal intensity and spectral bandwidth of measured CARS spectra. A reflective lens could remove both of the complications, but a spherical mirror cannot collimate the beam with a large numerical aperture without significant spherical aberration and astigmatism. Therefore, an off-axis Au-coated parabolic mirror was used to collimate the continuum output despite its complicated alignment. To minimize spatial mode distortion of the reflected beam by the diamond turning fabrication of the parabolic mirror, the distance between the mirror and the focusing objective lens was reduced as much as possible.

## 15.4 Data Analysis

B-CARS images are acquired as a hyperspectral data cube, with vibrational spectra, usually in the range of 300–3500 $cm^{-1}$, at each spatial point. Several processing steps are taken to convert the raw B-CARS data into useful Raman-like spectra and further into spatiochemical information. Among these are measures to subtract detector baseline and random noise. Additionally, one must extract the resonant component from the overall CARS signal. In this section, we discuss treatment of random noise and extraction of resonant signal.

For most B-CARS applications, a charge-coupled device (CCD) detector is used to collect the signal with some amount of vertical binning so that the output is a linear array of non-negative integers representing signal amplitudes as a function of wavenumber. The combined data from all spatial pixels are formatted into 3D matrix $(X, Y, W)$ where at every wavenumber $W$, there exists an $XY$ image, or conversely, at every point $XY$ there is a full spectrum. A spectral baseline is collected just before acquiring the raw B-CARS data and is subtracted from each spectrum in the image.

Random noise from sources such as CCD reading and laser intensity fluctuations comprise a significant portion of total the noise associated with the raw B-CARS signal.

Random noise such as is found in the B-CARS signal can be efficiently removed by singular value decomposition (SVD) when there are at least two nondegenerate dimensions to the data. Because the data dimensions must be nondegenerate, spatial dimensions $X$ and $Y$ could be used as the required two dimensions only if they contain distinct and independent information. SVD is a generalized approach to finding a set of basis vectors (eigenvectors), the linear superposition of which fully describe the data. In our case the independent dimensions are spatial and spectral. The method generates basis vectors for reconstruction of data along both of these dimensions, but we will discuss the process in terms of spatial reconstruction because we will rely on spatial patterns in the reconstructed data to discriminate signal from noise.

SVD is performed by first formatting the data into a 2D rectangular matrix $A_{mn}(M \times N)$, where $M$ rows contain the spectral amplitudes as a function of wavenumber and $N$ columns contain the $XY$ spatial data. The matrix $A_{mn}$ is decomposed into the product of three matrices—$U_{mm}$: a square orthogonal matrix; $S_{mn}$: a rectangular diagonal matrix; and $V_{nn}^{\mathrm{T}}$: the transpose of a square orthogonal matrix $V_{nn}$—as follows:

$$A_{mm} = U_{mm} S_{mn} V_{nn}^{\mathrm{T}} \tag{15.6}$$

Both $U$ and $V$ are orthonormal matrices where the columns of $U$ are orthonormal eigenvectors of $AA^{\mathrm{T}}$ and the columns of $V$ are orthonormal eigenvectors of $A^{\mathrm{T}}A$: $U^{\mathrm{T}}U = V^{\mathrm{T}}V = I$ where $I$ is the identity matrix.

The matrix $S$ contains singular values (SVs) along the diagonal, that is, square roots of eigenvalues shared by $U$ and $V$. These are arranged in descending value order, so that the amplitude of the SVs decreases with increasing SV index; $S_{nm} \geq S_{nm+1}$. The $U$ square matrix contains the spectral eigenvectors, and the $V$ square matrix contains the spatial eigenvectors. A subset of the spatial and spectral eigenvectors will be sufficient to reconstruct the primary features of the CARS signal while others will represent primarily noise.

The key to successful filtering is to correctly determine the eigenvectors that represent signal and to exclude those that do not. Once this decision is made, the selected eigenfunctions are used to reconstruct the de-noised data. The reconstructed data are generated as the product of a modified $S$ matrix ($S_{mn}^{\text{mod}}$) matrix along with the $U$ and $V^{\text{T}}$ matrices:

$$A_{mm} = U_{mm} S_{mn}^{\text{mod}} V_{nn}^{T} \tag{15.7}$$

where the elements of $S_{mn}^{\text{mod}}$ corresponding to undesired eigenvectors are set to zero.

Selecting eigenvectors for inclusion in the reconstructed data set can be done manually or in an automated fashion. The manual method relies on the fact that spatial eigenvectors representing noise will have spectral counterparts that also represent noise. Whether or not a spatial eigenvector represents noise can be determined simply by plotting its scalar elements in a way that spatially corresponds to the image from which the original data were obtained.

Figure 15.6a shows a panel of 15 spatial eigenvectors from an image of two L929 fibroblast cells. The eigenvector elements are plotted according to their $x$–$y$ coordinates in the original image. These eigenvector plots are arranged by SV index (1–15) in order of decreasing SV amplitudes. It is clear from inspection that at least the first six eigenvectors contain image-related features. Eigenvector plots with higher SV indices are associated with smaller SVs, so they contribute less amplitude to the overall image. Indeed, many of these seem to entirely represent noise, although some of the eigenvector maps at higher index appear to have image-related information (such as SV12 in Figure 15.6a).

There are several common automated methods for selecting eigenvectors for reconstruction. Three such methods are "intersecting lines," "first-order autocorrelation," and "residual plots." The intersecting lines method takes advantage of the fact that there are generally two regimes for values of the quantity $\Delta S_{mn}/\Delta n$. At low index, this quantity takes on relatively large values, but these become asymptotically small at higher index. In the intersecting lines method, the SV threshold is set at the breakpoint between these two regimes, which is determined as the index value at which linear extrapolations of the high-index and low-index regimes intersect.

The residual plots method focuses on variation of texture between the eigenvector plots and the original image. Here the SV threshold is determined from the difference ($D_{mn}$) between the original image and images reconstructed from an individual spatial eigenvector:

$$D_{mn} = A_{mn} - \left(U_{mm} S_{mn}^{\text{mod}} V_{nn}^{\text{T}}\right) \tag{15.8}$$

where all but one element of $S_{mn}^{\text{mod}}$ is set to zero. If the eigenvector in question significantly contributes to the image, the elements of $D_{mn}$ should be randomly distributed about zero.

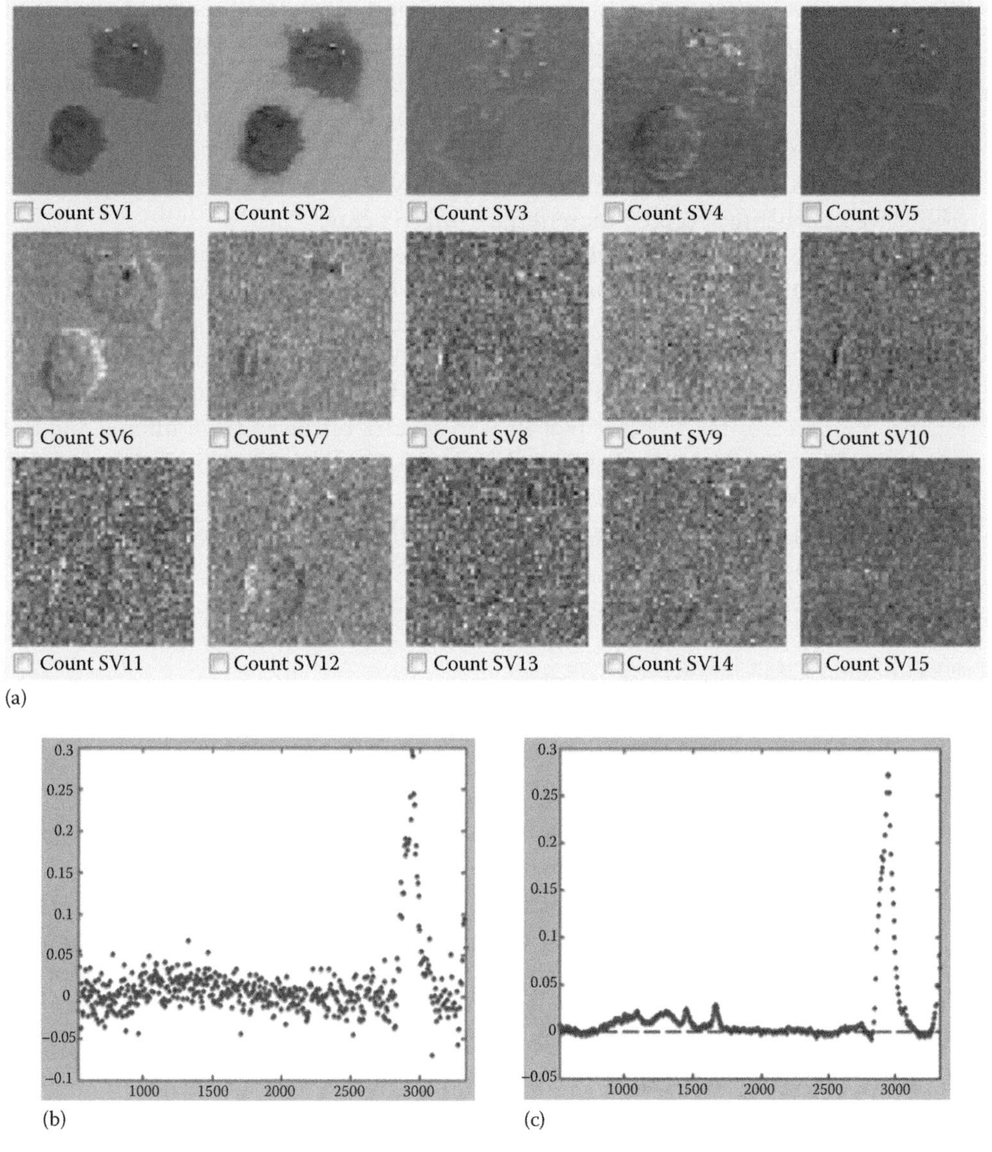


**FIGURE 15.6** (a) Spatial arrangement of eigenvectors from first 15 SVs of a spectral cell image plotted in $x$–$y$ coordinates corresponding spatial arrangement of original image. Eigenvector plots bearing resemblance to original image are expected to contain information. Others are expected to represent only noise. (b) Raman spectrum retrieved from CARS before de-noising by SVD. (c) Same spectrum after de-noising.

In the first autocorrelation function method, the first autocorrelation function for the columns of $U$ or $V$ matrices is calculated as

$$C(x_i) = \sum x_{i,j} \times x_{i,j+1} \tag{15.9}$$

where $x_{i,j}$ and $x_{i,j+1}$ are the $j$th and $(j + 1)$th row elements of column $i$ from either the $U$ or $V$ matrix. The correlation value, $C(x_i)$, represents how adjacent $U$ or $V$ row elements (eigenvectors) vary. $C(x_i)$ takes values between 1 and −1 since all rows of both $U$ and $V$ are normal vectors. Values close to −1 indicate anti-correlation and signal that the eigenvector represents primarily noise. The threshold for the acceptance of an eigenvector is typically set to $C(x_i) \geq 0.8$ since this represents a signal-to-noise ratio of 1 (Haq et al., 1997). We have applied this method to using both $U$ and $V$ matrices and find consistently that the autocorrelations from the $V$ matrix are slightly higher, resulting in a larger number of eigenvectors being accepted than in the $U$ matrix case.

Of the three automated methods discussed here, we favor the first-order autocorrelation functions of the $V$ matrix, as we find it to be the most liberal (i.e., least likely to exclude important eigenvectors). However, each of the automated approaches tends to neglect some eigenvectors with slightly smaller SV index, even though they may have notable morphological features that can be identified with the image, as with SV12 of Figure 15.6a. In Figure 15.6b, we plot Raman spectra retrieved from filtered and unfiltered raw CARS data from a single pixel of the L929 cells depicted in Figure 15.6a. The reduction in noise level is notable.

Once the random noise has been filtered from the raw B-CARS data, we are ready to extract the resonant component. The overall CARS signal is composed of a resonant and nonresonant contribution as indicated in Equation 15.10:

$$I_{CARS} \propto |\chi|^2 = |\chi_{NR}|^2 + 2\chi_{NR}\chi_R' + |\chi_R|^2 \tag{15.10}$$

The presence of the NRB distorts the spectral shape and makes quantitative spectral interpretation impossible without deconvolving the resonant and nonresonant parts. Experimental approaches to suppress the NRB have been successful for high concentration analytes in solution where the resonant field (third term in Equation 15.10) is relatively strong, and the detected intensity is quite large. However, in cellular samples, signals not associated with CH vibrations are relatively very weak, and the heterodyne term in Equation 15.10 can be used to substantially increase the CARS signal of interest as well as linearize the response to analyte concentration, making for easier signal interpretation.

Computational methods from our group (Liu et al., 2009) as well as others (Kano and Hamaguchi, 2005; Vartiainen et al., 2006) have been demonstrated to extract the resonant signal from raw CARS spectra. Two computational methods, a modified Kramers–Kronig (KK) transform and a maximum entropy approach, have been used to retrieve the resonant signal of interest from the NRB in M-CARS and B-CARS spectra (Liu et al., 2009; Vartiainen et al., 2006). Here we briefly review the KK method and its use.

The modified KK transform used for retrieval of CARS spectra is described in detail by Liu et al. (2009). It uses the principle of causality embodied in a KK transform and makes accommodation for a temporally finite excitation pulse. Practically, we obtain raw CARS signal and an NRB from the same location Fourier-transform both into the time domain as free induction decays (FIDs). The NRB carries envelope and background phase information which must be subtracted from the raw CARS signal in order to extract the spectral phase of interest. The envelope and background phase subtraction is accomplished simply by replacing the negative-time CARS FID with the negative-time NRB FID amplitude. This combined FID is back-transformed, and spectral phase is extracted. Operationally this is accomplished using Equations 15.15 and 15.16. A detailed mathematical description is given in the following.

We start with a KK expression for spectral phase.

$$\phi(\omega) = -\frac{P}{\pi}\int_{-\infty}^{\infty}\frac{\ln|\chi(\omega')|}{\omega'-\omega}d\omega' \tag{15.11}$$

where

$\chi(\omega) = |\chi(\omega)|\exp[i\varphi(\omega)]$

$P$ is the Cauchy principle value

For convenience, we define an operator

$$\psi(f(\omega)) = F\left[u(t)F^{-1}[f(\omega)]\right] \tag{15.12}$$

where

$F$ and $F^{-1}$ denote the Fourier and inverse Fourier transforms, respectively

$u(t)$ is the Heaviside function

Using the convolution theorem, we rewrite this operator as

$$\psi(f(\omega)) = \frac{1}{2}\left[\frac{-i}{\pi}P\int_{-\infty}^{\infty}\frac{f(\omega')}{\omega'-\omega}d\omega' + f(\omega)\right] \tag{15.13}$$

By comparison with Equation 15.11, it is clear that applying the operator $\psi$ is closely related to calculating a spectral phase using a conventional KK transform. In fact, when the operator is applied to $\ln|\chi(\omega)|$, we obtain spectral phase with minor manipulation:

$$\phi(\omega) = 2\,\mathrm{Im}\left\{\psi\left(\ln|\chi(\omega)|\right) - \frac{\ln|\chi(\omega)|}{2}\right\} \tag{15.14}$$

We modify the conventional KK transform by replacing the Heaviside function with the FID of the NRB, which contains the envelope and background phase to be subtracted from the CARS signal. To do so, we replace the product $u(t)\,F^{-1}[f(\omega)]$ in Equation 15.12 with

$$\eta(t; f(\omega)) = \begin{cases} F^{-1}[f(\omega)] & t \geq 0 \\ F^{-1}[f_{NR}(\omega)] & t < 0 \end{cases} \tag{15.15}$$

where $f_{NR}(\omega)$ is the NRB of the CARS spectrum. Thus, the spectral phase is given as

$$\phi(\omega) = 2\,\mathrm{Im}\left\{F\left[\eta\left(t;\ln|\chi(\omega)|\right)\right] - \frac{\ln|\chi(\omega)|}{2}\right\} \tag{15.16}$$

Using a desktop computer, extraction takes on the order of 1 ms for each spatial pixel in the image, so resonant B-CARS hyperspectral data can be retrieved in about 1 min for 151 × 151 spatial pixels with ~600 spectral points at each spatial location. Once processed for resonant signal extraction, the spectrum from each individual pixel no longer has the dominant NRB envelope that masks the sharp vibrationally resonant features. Instead, the spectra resemble data that are obtained from spontaneous Raman spectroscopy.

Obtaining an accurate representation of the NRB at each spatial pixel is crucial to reliable resonant extraction. It is important to obtain the NRB under conditions as similar as possible to those under which the raw CARS spectrum was obtained. When individual cells are being imaged, we have determined the following to be an acceptable approach. First, spatial regions of the B-CARS image containing cells are masked based on physical contrast (changes in the NRB level), and the inverse of the mask is used to select the cell-free areas of the image to calculate an averaged NRB. Spectra from pixels outside the mask are averaged line by line to obtain a line-averaged NRB signal. The resonant signal from each pixel within the mask is then extracted using the line-averaged NRB from the same line. After resonant extraction, any residual, slowly varying spectral baseline due to variance of the line-averaged NRB signal from the actual NRB at a given pixel is removed by fitting the overall baseline shape to two disparate regions of the extracted spectrum (~500–800 $cm^{-1}$ and 1700–2700 $cm^{-1}$) with a third-order polynomial. The fitted curve is subtracted from the extracted resonant spectrum on a pixel-by-pixel basis to detrend the data. This detrending is necessary for reliable qualitative or quantitative analysis of the spectral images and is similar to processes commonly used in spontaneous Raman and M-CARS spectral analyses (Lieber and Mahadevan-Jansen, 2003; Rinia et al., 2007).

## 15.5 Applications

B-CARS microscopy and spectroscopy have been used in materials characterization and imaging and solution chemistry and are now seeing increased utility in biological applications. In this section, we present examples from PCF-based CARS imaging and spectroscopy systems. The examples presented in the following represent typical applications in biology and materials characterization with current PCF-based technology. Spectral imaging (Kano and Hamaguchi, 2005, 2006; Kee and Cicerone, 2004; Parekh et al., 2010) and spectroscopy (Ikeda and Uosaki, 2009; Lee and Cicerone, 2008; Petrov et al., 2005, 2007) are two such applications where PCF continuum generation provides improved acquisition rates or additional information as compared to CRMS. Though the breadth of applications presented here is limited, rapidly improving continuum generation and CARS signal detection technology is poised to open new avenues for PCF-based CARS spectral techniques moving forward.

Spectral CARS microscopy with a PCF continuum light source was first demonstrated by Kee and Cicerone on tertiary polymer blends of polystyrene, poly(methylmethacrylate), and poly(ethylene terephthalate) (Kee and Cicerone, 2004). This demonstration used a Ti:Sapphire laser tuned to $\lambda = 785$ nm with a 150 fs pulse at 76 MHz to pump a PCF for continuum generation. The fundamental laser was used as the pump and probe, and the continuum was used as the Stokes beam (2-color scheme). This technique was used to acquire B-CARS spectra, including both fingerprint 500–1600 $cm^{-1}$ and CH-stretch 2840–3000 $cm^{-1}$ spectral regions at each image pixel. By pseudo-coloring each pixel according to the spectral signatures associated with it, the authors were able to generate phase maps showing discrete domains in the polymer film (Figure 15.7).

After this initial B-CARS microscopy demonstration on polymer blends, Kano and Hamaguchi used a similar laser and PCF system to produce the first spectral CARS images from live biological samples. They acquired spectra in the CH-stretch region from live Schizosaccharomyces pombe fission yeast (Kano and Hamaguchi, 2005). In these experiments, a $\lambda = 800$ nm Ti:Sapphire laser was used to pump the PCF, which produced a continuum that allowed imaging of individual live yeast cells using spectral intensities from 2000 to 4500 $cm^{-1}$. From these spectra, the authors reconstructed resonant vibrational images based on CARS signal intensities in the CH-stretch

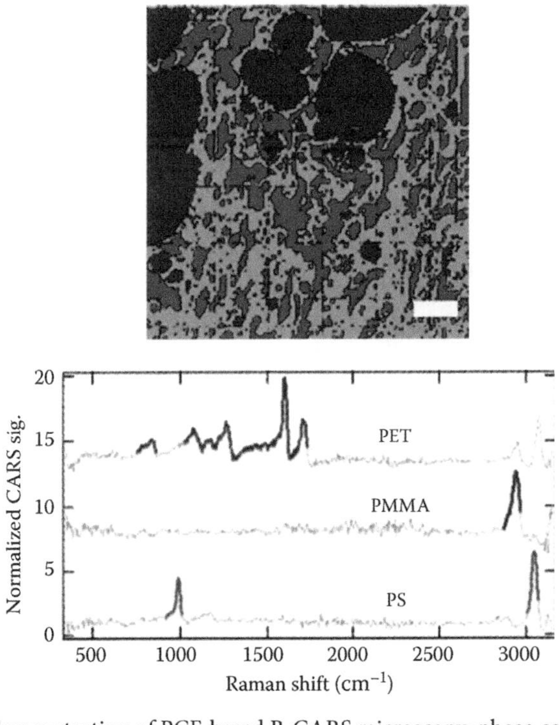


**FIGURE 15.7** Initial demonstration of PCF-based B-CARS microscopy: phase-separated polymer blend consisting of equal parts PMMA, PS, and PET. The pseudo-color image (top) is 150 × 150 pixels; the colors red, green, and blue correspond to PMMA, PS, and PET, respectively. Reference spectra from each of the individual polymer components (bottom) were offset vertically for clarity. The bold, colored line segments indicate spectral regions that were used for identification of different blend components in the spectrum at each spatial pixel. Scale bar = 100 μm. (Images adapted from Kee, T.W. and Cicerone, M.T., *Opt. Lett.*, 29, 2701, 2004.)

(Figure 15.8a and b). Hamaguchi and colleagues have since published numerous studies demonstrating the use of PCF-based multiplex CARS (M-CARS) microscopy (with spectra covering the CH-stretch region) for imaging live and fixed biological samples. Significantly, they have shown that different organelles in HeLa cells have distinct spectral shapes in the CH-stretch region, which can be seen in raw CARS spectra from these locations (Kano and Hamaguchi, 2005) (Figure 15.8c and d).

As mentioned in Sections 15.3 and 15.4, recent advances in continuum generation (Lee et al., 2010b) and signal processing (Liu et al., 2009) have permitted simultaneous

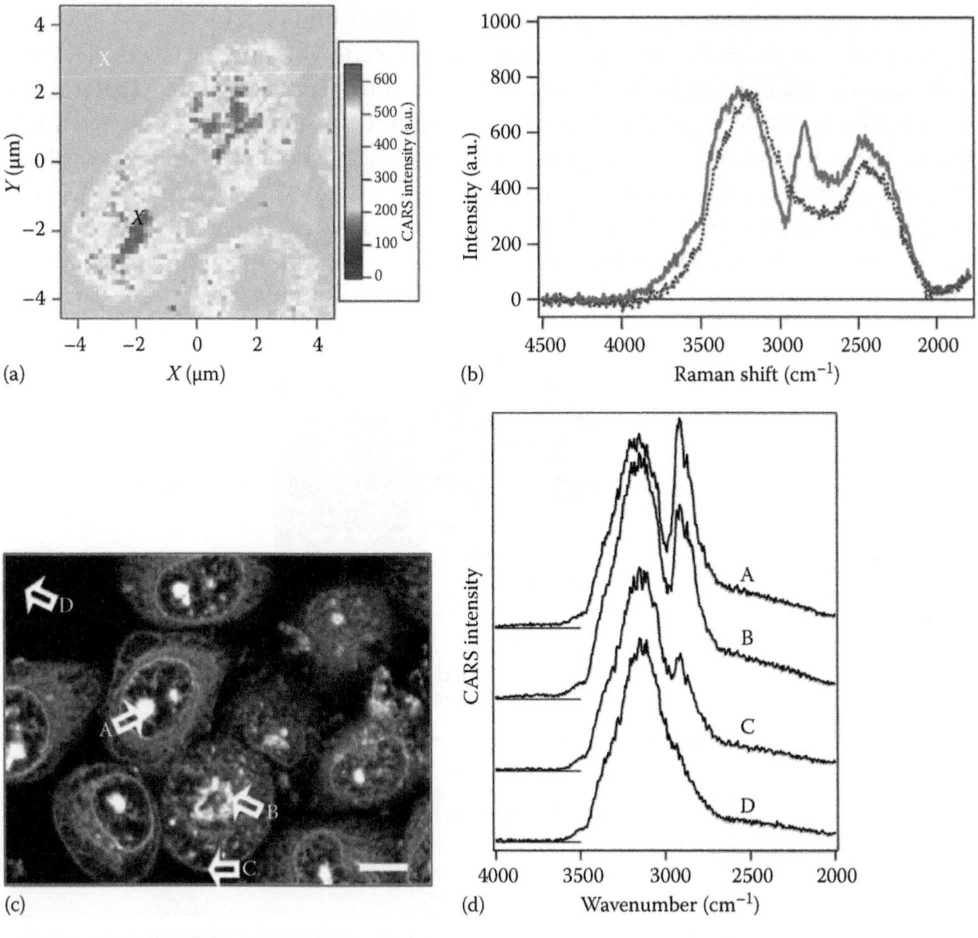


**FIGURE 15.8** PCF-based M-CARS imaging of biological samples. (a) Spectral imaging of live *Schizosaccharomyces pombe* cells. An *X–Y* image of a single cell showing pixel intensity values from raw CARS spectra at 2856 $cm^{-1}$. (b) Solid and dotted spectra are from the positions marked by the black and white *X*, respectively, in (a). (Images adapted from Kano, H. and Hamaguchi, H., Vibrationally resonant imaging of a single living cell by supercontinuum-based multiplex coherent anti-Stokes Raman scattering microspectroscopy, *Opt. Express*, 13, 1322–1327. Copyright 2005 Optical Society of America.) (c) CH-stretch CARS images of HeLa cells constructed from intensities at 2870 $cm^{-1}$. (d) CARS spectral from cellular organelles marked by letters in (c): nucleolus A, chromosome B, cell membrane C, and background D. Scale bar = 10 μm. (Images adapted from Kano, H.: Molecular vibrational imaging of a human cell by multiplex coherent anti-Stokes Raman scattering microscopy using a supercontinuum light source. *J. Raman Spectrosc.*, 2008. 39(11). 1649–1652. Copyright Wiley-VCH Verlag GmbH & Co. KGaA. With permission.)

fingerprint and CH-stretch imaging of single cells with B-CARS microscopy (Parekh et al., 2010). Parekh et al. acquired spectra from individual fixed L929 fibroblast cells on glass coverslips at fivefold faster speeds than CRMS. The quality of these spectra was sufficient for spectral discrimination of nuclear and cytoplasmic regions that colocalized very well with fluorescently stained images (Figure 15.9). This is the only demonstration to date that has shown simultaneous fingerprint and CH-stretch spectral imaging with CARS microscopy on single cells.

In addition to enabling broadband spectral detection, a PCF-generated continuum (Stokes) and a second spectrally broad (~130 cm$^{-1}$) pulse (pump/probe) can be used in combination with a single-point detector such as a photo-multiplier tube (PMT) for rapid imaging. In one demonstration, the wavelengths of the pulses were tuned such that they would detect CARS signal from CH-stretch spectral region (Murugkar et al., 2007),

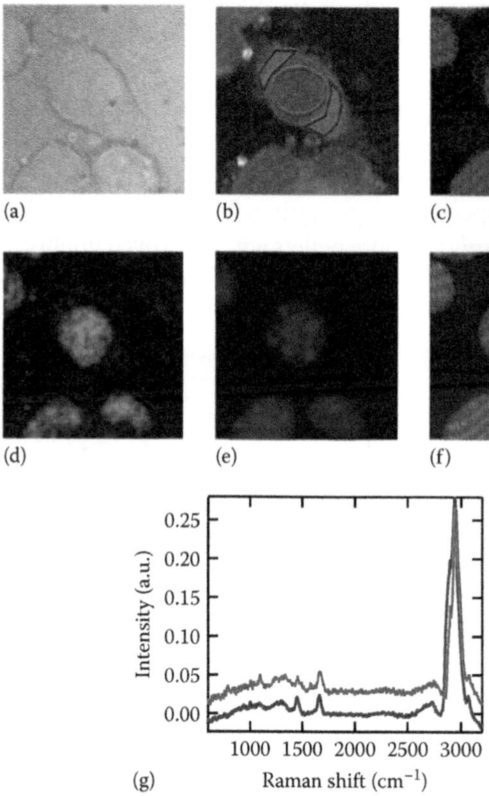

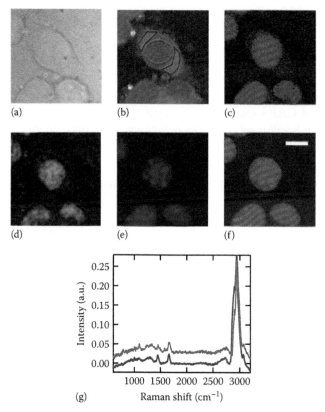

**FIGURE 15.9** B-CARS imaging of cell nuclei covering both fingerprint and CH-stretch spectra regions. (a) Bright-field image of L929 cells. (b) Raman image constructed from resonant-extracted B-CARS signal at 2900 cm$^{-1}$. (c) Image constructed from difference in B-CARS resonant-extracted signal intensity between 3003 and 2853 cm$^{-1}$ generates chemical contrast highlighting the nuclear regions of the cell. (d) Image of cell nuclei (green) and cytosol (blue) generated by principal component analysis of resonant-extracted B-CARS data. (e) Fluorescent image of Hoechst-stained DNA in the same cells. (f) Overlay of images from panels (c) and (e) demonstrates accurate nuclear identification using B-CARS microscopy. (g) Spectra generated by averaging extracted spectra from $n$ = 751 and 731 pixels in nuclear and cytoplasmic regions of the cell, respectively. Regions of interest in panel (b) are color-coded traces, and traces are vertically offset for clarity. Scale bar = 10 μm. (Images adapted from Parekh, S.H. et al., *Biophys. J.*, 99, 2695, 2010. With permission.)

which provides strong signal for biological samples. This approach offers some attractive features compared to traditional single-frequency CARS having simpler instrumentation and reduced optical setup cost while offering somewhat comparable imaging speed. Murugkar et al. recently demonstrated 84 μs positional dwell times when imaging adipocytes and fatty tissue with a 500 $cm^{-1}$ bandwidth PCF-generated continuum and a 130 $cm^{-1}$ Ti:Sapphire pump laser (Figure 15.10). Compared to the video-rate single-frequency CARS (Evans et al., 2005), this technique is an order of magnitude slower, but its results demonstrate the possibility of using PCF continua for high-speed CARS imaging. In addition to single-element CARS imaging, Stolow and colleagues have shown that a similar PCF continuum laser system is useful for multimodal optical imaging of biological samples. They obtained second harmonic generation, two-photon excitation fluorescence, and CARS with the same instrument (Pegoraro et al., 2009). These imaging capabilities demonstrate the potential for using cheaper PCF-based laser systems to accomplish similar functions as more expensive parametric oscillator and laser oscillator combinations used for traditional single-frequency CARS and multimodal microscopy (Cheng et al., 2002).

In addition to imaging applications, PCF-generated continua have also been used for M-CARS and B-CARS spectroscopy in pathogen detection and materials characterization, respectively. Petrov et al. have shown that a $GeO_2$ fiber continuum (Petrov et al., 2005) used for M-CARS spectroscopy acquires fingerprint vibrational spectra (400 $cm^{-1}$ bandwidth) from *Bacillus subtilis* spore pellets with improved fidelity as compared to spontaneous Raman spectroscopy (Petrov et al., 2007). A follow-up to this study compared the

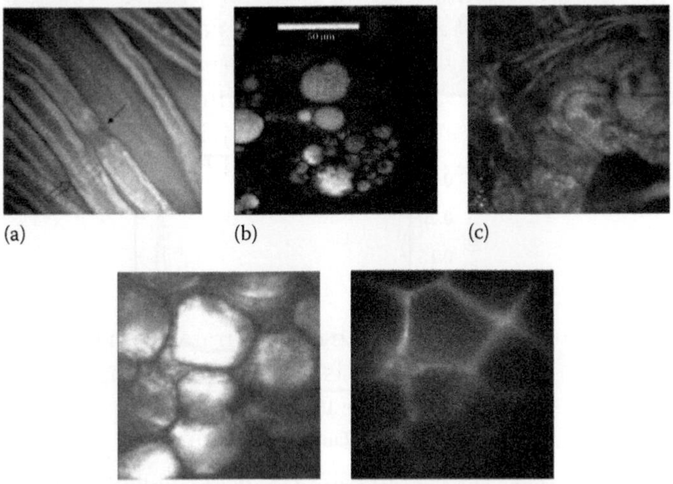

(a) (b) (c) (d) (e)

**FIGURE 15.10** High-speed (84 μs dwell time) PCF-based, single-element CARS images of biological samples in the CH-stretch region (2840–3000 $cm^{-1}$). CARS intensity from the entire 200 $cm^{-1}$ is integrated in a single PMT. (a) Isolated unstained live rat dorsal root axons, (b) lipid droplets in unlabeled 3T3 L1 adipocyte cell culture, (c) sebaceous gland (in pseudo-color) in a mouse ear at a depth of ~40 μm, (d) adipocyte cells at a depth of 100 μm in a mouse ear imaged in the forward mode through several hundred microns of tissue and (e) same as in (d), but with Stokes beam blocked. Scale bar = 50 μm. (Images adapted from Murugkar, S., Brideau, C., Ridsdale, A., Naji, M., Stys, P.K., and Anis, H., Coherent anti-Stokes Raman scattering microscopy using photonic crystal fiber with two closely lying zero dispersion wavelengths, *Opt. Express*, 15, 14028–14037. Copyright 2007 Optical Society of America.)

quality of the retrieved Raman spectra from M-CARS spectroscopy of individual *B. subtilis* spores generated using a continuum from either a PCF or $GeO_2$ fiber. The fingerprint spectrum 800–1800 $cm^{-1}$ quality was superior in spectra acquired using a $GeO_2$ fiber relative to the PCF (Petrov et al., 2005). The authors noted that although the PCF continuum bandwidth exceeds that of the $GeO_2$ continuum, the spectral fluctuations in the PCF-generated continuum were especially problematic and resulted in reduced spectral quality. However, as described in Section 15.3, recent work shows that PCF-generated continua can be made quite stable under specific excitation conditions.

PCF continua have also found use in time-resolved CARS spectroscopy and microscopy. Two recent studies have used such an approach to obtain picosecond relaxation processes in single-walled carbon nanotubes (SWCNTs) (Ikeda and Uosaki, 2009; Lee et al., 2010a). In both cases, the continuum was used for intrapulse excitation (3-color CARS mechanism) for investigating the temporal dynamics of electron–phonon coupling in these materials. These dynamics studies are a growing area of use for continuum-based CARS spectroscopy where large bandwidth and reasonably good compressibility of the continuum are critical for accuracy of these measurements. Recent improvements in continuum properties using dispersive wave (blue-shifted) continuum generated at 1550 nm (Selm et al., 2010) will make such time-resolved measurements even more feasible.

The examples presented earlier show a sample of recent work using PCF continua technology in CARS imaging and spectroscopy. Rapid progress in continuum generation and application will inevitably lead to increasingly widespread application of these approaches. The following section provides a more specific outlook for the future of continuum-based CARS in vibrational imaging.

## 15.6 Future Prospects

As we have seen, continuum-based CARS microscopy shows promise in both narrowband and broadband imaging modes, as well as providing a simple approach to time-resolved spectroscopy. Until now, however, broadband CARS microscopy has been largely a research endeavor. That is, most effort has gone into technique development rather than application. Although conceptually simple, there are many technical challenges associated with making broadband CARS work sufficiently well for application in biology. For example, the signal quality depends critically on noise characteristics of the continuum, the spectral chirp, and higher-order structure. Additionally, working with such broad pulses necessitates finding ways to ameliorate chromatic aberrations, and having one's signal spread out over a broad spectral band means that the average signal level is relatively low.

Fortunately, there are mitigating factors that currently allow signal detection at rates considerably faster than spontaneous Raman, which can be further leveraged to increase signal detection rates by perhaps orders of magnitude. For example, broadband spectral detection allows one to rather straightforwardly use phase retrieval methods and thus take advantage of intrinsic heterodyning in the CARS signal. Generation of hyperspectral images also permits the use of powerful statistical methods such as SVD to filter random noise. Furthermore, recent advances in continuum generation and CMOS detector technology portend significant improvements in broadband CARS performance and ease of implementation.

Ultimately our motivation for doing broadband CARS is to make the application of broadband vibrational microscopy practical and convenient. For many years now, spontaneous Raman spectroscopy has been used to detect the biochemical signatures of single cells without exogenous labels. As a result of decades of work in this field, researchers can now use the information-rich fingerprint region of the vibrational spectrum to discriminate with very high confidence between different cell types such as stem cells and phenotypically committed cells (Chan et al., 2009) and between cancerous and normal cells (Gniadecka et al., 2004) and thus to detect malignant tissue (Kast et al., 2008). However, spontaneous Raman microscopy is quite slow (often seconds or minutes/spectrum); thus, imaging applications have not become widespread. B-CARS microscopy has already been shown to allow acquisition of the same vibrational spectra but at much higher acquisition rates, making high-resolution imaging of large numbers of samples a possibility. (Parekh et al., 2010). Already, the ability to quickly generate high-resolution chemical maps of cells in tissues afforded by B-CARS is likely to pave the way for tremendous advances in our knowledge of important biological and disease processes. Furthermore, we look forward to continued improvement in speed and increased signal-to-noise ratios in upcoming instrumentation innovations.

## References

Chan, J., Fore, S., Wachsman-Hogiu, S., and Huser, T. 2008. Raman spectroscopy and microscopy of individual cells and cellular components. *Laser and Photonics Reviews*, 2, 325–349.

Chan, J. W. and Lieu, D. K. 2009. Label-free biochemical characterization of stem cells using vibrational spectroscopy. *Journal of Biophotonics*, 2, 656–668.

Chan, J. W., Lieu, D. K., Huser, T., and Li, R. A. 2009. Label-free separation of human embryonic stem cells and their cardiac derivatives using Raman spectroscopy. *Analytical Chemistry*, 81, 1324–1331.

Chan, J. W., Taylor, D. S., Zwerdling, T., Lane, S. M., Ihara, K., and Huser, T. 2006. Micro-Raman spectroscopy detects individual neoplastic and normal hematopoietic cells. *Biophysical Journal*, 90, 648–656.

Cheng, J. X., Jia, Y. K., Zheng, G. F., and Xie, X. S. 2002. Laser-scanning coherent anti-Stokes Raman scattering microscopy and applications to cell biology. *Biophysical Journal*, 83, 502–509.

Chernenko, T., Matthaus, C., Milane, L., Quintero, L., Amiji, M., and Diem, M. 2009. Label-free Raman spectral imaging of intracellular delivery and degradation of polymeric nanoparticle systems. *ACS Nano*, 3, 3552–3559.

Cui, M., Bachler, B. R., and Ogilvie, J. P. 2009. Comparing coherent and spontaneous Raman scattering under biological imaging conditions. *Optics Letters*, 34, 773–775.

Dudley, J. M. and Coen, S. 2002. Numerical simulations and coherence properties of supercontinuum generation in photonic crystal and tapered optical fibers. *IEEE Journal of Selected Topics in Quantum Electronics*, 8, 651–659.

Dudley, J. M., Genty, G., and Coen, S. 2006. Supercontinuum generation in photonic crystal fiber. *Reviews of Modern Physics*, 78, 1135–1184.

Duncan, M. D., Reintjes, J., and Manuccia, T. J. 1982. Scanning coherent anti-Stokes Raman microscope. *Optics Letters*, 7, 350–352.

Evans, C. L., Potma, E. O., Puoris'haag, M., Cote, D., Lin, C. P., and Xie, X. S. 2005. Chemical imaging of tissue in vivo with video-rate coherent anti-Stokes Raman scattering microscopy. *Proceedings of the National Academy of Sciences of the United States of America*, 102, 16807–16812.

Fu, Y., Wang, H., Shi, R., and Cheng, J. X. 2006. Characterization of photodamage in coherent anti-Stokes Raman scattering microscopy. *Optics Express*, 14, 3942–3951.

Gniadecka, M., Philipsen, P. A., Sigurdsson, S., Wessel, S., Nielsen, O. F., Christensen, D. H., Hercogova, J., Rossen, K., Thomsen, H. K., Gniadecki, R., Hansen, L. K., and Wulf, H. C. 2004. Melanoma diagnosis by Raman spectroscopy and neural networks: Structure alterations in proteins and lipids in intact cancer tissue. *Journal of Investigative Dermatology*, 122, 443–449.

Guild, J. B., Xu, C., and Webb, W. W. 1997. Measurement of group delay dispersion of high numerical aperture objective lenses using two-photon excited fluorescence. *Applied Optics*, 36, 397–401.
Haq, I., Chowdhry, B. Z., and Chaires, J. B. 1997. Singular value decomposition of 3-D DNA melting curves reveals complexity in the melting process. *European Biophysics Journal*, 26, 419–426.
Hopt, A. and Neher, E. 2001. Highly nonlinear photodamage in two-photon fluorescence microscopy. *Biophysical Journal*, 80, 2029–2036.
Ikeda, K. and Uosaki, K. 2009. Coherent phonon dynamics in single-walled carbon nanotubes studied by time-frequency two-dimensional coherent anti-Stokes Raman scattering spectroscopy. *Nano Letters*, 9, 1378–1381.
Kano, H. 2008. Molecular vibrational imaging of a human cell by multiplex coherent anti-Stokes Raman scattering microscopy using a supercontinuum light source. *Journal of Raman Spectroscopy*, 39(11), 1649–1652.
Kano, H. and Hamaguchi, H. 2005. Vibrationally resonant imaging of a single living cell by supercontinuum-based multiplex coherent anti-Stokes Raman scattering microspectroscopy. *Optics Express*, 13, 1322–1327.
Kano, H. and Hamaguchi, H. 2006. In-vivo multi-nonlinear optical imaging of a living cell using a supercontinuum light source generated from a photonic crystal fiber. *Optics Express*, 14, 2798–2804.
Kast, R. E., Serhatkulu, G. K., Cao, A., Pandya, A. K., Dai, H., Thakur, J. S., Naik, V. M., Naik, R., Klein, M. D., Auner, G. W., and Rabah, R. 2008. Raman spectroscopy can differentiate malignant tumors from normal breast tissue and detect early neoplastic changes in a mouse model. *Biopolymers*, 89, 235–241.
Kee, T. W. and Cicerone, M. T. 2004. Simple approach to one-laser, broadband coherent anti-Stokes Raman scattering microscopy. *Optics Letters*, 29, 2701–2703.
König, K. 2000. Multiphoton microscopy in life sciences. *Journal of Microscopy*, 200, 83–104.
Lee, Y. J. and Cicerone, M. T. 2008. Vibrational dephasing time imaging by time-resolved broadband coherent anti-Stokes Raman scattering microscopy. *Applied Physics Letters*, 92, 041108.
Lee, Y. J., Liu, Y., and Cicerone, M. T. 2007. Characterization of three-color CARS in a two-pulse broadband CARS spectrum. *Optics Letters*, 32, 3370–3372.
Lee, Y. J., Parekh, S. H., Fagan, J. A., and Cicerone, M. T. 2010a. Phonon dephasing and population decay dynamics of the G-band of semiconducting single-wall carbon nanotubes. *Physical Review B*, 82, 5432.
Lee, Y. J., Parekh, S. H., Kim, Y. H., and Cicerone, M. T. 2010b. Optimized continuum from a photonic crystal fiber for broadband time-resolved coherent anti-Stokes Raman scattering. *Optics Express*, 18, 4371–4379.
Lieber, C. A. and Mahadevan-Jansen, A. 2003. Automated method for subtraction of fluorescence from biological Raman spectra. *Applied Spectroscopy*, 57, 1363–1367.
Liu, Y. X., Lee, Y. J., and Cicerone, M. T. 2009. Broadband CARS spectral phase retrieval using a time-domain Kramers–Kronig transform. *Optics Letters*, 34, 1363–1365.
Matthaus, C., Chernenko, T., Newmark, J. A., Warner, C. M., and Diem, M. 2007. Label-free detection of mitochondrial distribution in cells by nonresonant Raman microspectroscopy. *Biophysical Journal*, 93, 668–673.
Murugkar, S., Brideau, C., Ridsdale, A., Naji, M., Stys, P. K., and Anis, H. 2007. Coherent anti-Stokes Raman scattering microscopy using photonic crystal fiber with two closely lying zero dispersion wavelengths. *Optics Express*, 15, 14028–14037.
Notingher, I., Bisson, I., Bishop, A. E., Randle, W. L., Polak, J. M. P., and Hench, L. L. 2004a. in situ spectral monitoring of mRNA translation in embryonic stem cells during differentiation in vitro. *Analytical Chemistry*, 76, 3185–3193.
Notingher, I., Jell, G., Lohbauer, U., Salih, V., and Hench, L. L. 2004b. In situ non-invasive spectral discrimination between bone cell phenotypes used in tissue engineering. *Journal of Cellular Biochemistry*, 92, 1180–1192.
Parekh, S. H., Lee, Y. J., Aamer, K. A., and Cicerone, M. T. 2010. Label-free cellular imaging by broadband coherent anti-Stokes Raman scattering microscopy. *Biophysical Journal*, 99, 2695–2704.
Pegoraro, A. F., Ridsdale, A., Moffatt, D. J., Pezacki, J. P., Thomas, B. K., Fu, L. B., Dong, L., Fermann, M. E., and Stolow, A. 2009. All-fiber CARS microscopy of live cells. *Optics Express*, 17, 20700–20706.
Petrov, G. I., Arora, R., Yakovlev, V. V., Wang, X., Sokolov, A. V., and Scully, M. O. 2007. Comparison of coherent and spontaneous Raman microspectroscopies for noninvasive detection of single bacterial endospores. *Proceedings of the National Academy of Sciences of the United States of America*, 104, 7776–7779.

Petrov, G. I., Yakovlev, V. V., Sokolov, A. V., and Scully, M. O. 2005. Detection of *Bacillus subtilis* spores in water by means of broadband coherent anti-Stokes Raman spectroscopy. *Optics Express*, 13, 9537–9542.
Rinia, H. A., Bonn, M., Muller, M., and Vartiainen, E. M. 2007. Quantitative CARS spectroscopy using the maximum entropy method: The main lipid phase transition. *Chemphyschem*, 8, 279–287.
Schut, T. C. B., Wolthuis, R., Caspers, P. J., and Puppels, G. J. 2002. Real-time tissue characterization on the basis of in vivo Raman spectra. *Journal of Raman Spectroscopy*, 33, 580–585.
Selm, R., Winterhalder, M., Zumbusch, A., Krauss, G., Hanke, T., Sell, A., and Leitenstorfer, A. 2010. Ultrabroadband background-free coherent anti-Stokes Raman scattering microscopy based on a compact Er:fiber laser system. *Optics Letters*, 35, 3282–3284.
Tolles, W. M. and Turner, R. D. 1977. Comparative analysis of analytical capabilities of coherent anti-Stokes Raman-spectroscopy (CARS) relative to Raman-scattering and absorption spectroscopy. *Applied Spectroscopy*, 31, 96–103.
Uzunbajakava, N., Lenferink, A., Kraan, Y., Volokhina, E., Vrensen, G., Greve, J., and Otto, C. 2003. Nonresonant confocal Raman imaging of DNA and protein distribution in apoptotic cells. *Biophysical Journal*, 84, 3968–3981.
Vartiainen, E. M., Rinia, H. A., Muller, M., and Bonn, M. 2006. Direct extraction of Raman line-shapes from congested CARS spectra. *Optics Express*, 14, 3622–3630.
Yakovlev, V. V. 2003. Advanced instrumentation for non-linear Raman microscopy. *Journal of Raman Spectroscopy*, 34, 957–964.
Zumbusch, A., Holtom, G. R., and Xie, X. S. 1999. Three-dimensional vibrational imaging by coherent anti-Stokes Raman scattering. *Physical Review Letters*, 82, 4142–4145.

# 16. Multiplex Stimulated Raman Scattering Microscopy

Dan Fu and Xiaoliang Sunney Xie



## 16.1 Principles of Multiplex Stimulated Raman Scattering

Coherent Raman microscopy is a powerful tool in imaging samples with diffraction-limited resolution and chemical specificity at high speed up to video rate (Cheng and Xie, 2004; Evans and Xie, 2008; Freudiger et al., 2008; Saar et al., 2010a). Because it relies on molecules' intrinsic vibration contrast, it has the unique advantage of label-free detection and is an attractive alternative to fluorescent labeling. An additional advantage of vibrational contrast is that it is not subject to photobleaching, allowing long-term monitoring. Capitalizing on these advantages, coherent Raman microscopy

*Coherent Raman Scattering Microscopy.* Edited by Ji-Xin Cheng and X. Sunney Xie © 2013 CRC Press/Taylor & Francis Group, LLC. ISBN: 978-1-4398-6765-5.

has been applied to many different areas of studies such as lipidomics, drug delivery, medical imaging, and biomass energy production, etc. as will be discussed in later chapters (Wang et al., 2005, 2011; Bégin et al., 2009; Kim et al., 2010; Saar et al., 2010b; Pezacki et al., 2011). While some of these problems could also be studied with confocal Raman microscopy, its slow imaging speed is a major deterrent to its widespread use, especially for live cells or samples that change with time. However, confocal Raman microscopy has a distinctive advantage over current narrowband implementations of coherent Raman microscopy in that it provides complete and reliable Raman spectra. This is important for analyzing complex systems such as biological cells because in most cases, multiple components have major overlapping Raman contributions (Puppels et al., 1990; Feofanov et al., 2000; Anita, 2003; Ellis and Goodacre, 2006; Yamakoshi et al., 2011). In order to extract quantitative information of any particular component, at least acquisition at several Raman bands or a portion of the full spectrum, followed by multivariate data analysis is required (Enejder et al., 2005; Haka et al., 2005). One approach to overcome the conflicting requirement of imaging speed and information content is to combine Raman spectroscopy with a fast imaging technique (Caspers et al., 2003; Slipchenko et al., 2009). Spectroscopy acquisition is only performed at certain points of interest in the images. With very limited sampling, these techniques could not provide chemical maps at high imaging speeds.

Coherent Raman microscopy models such as coherent anti-Stokes Raman scattering (CARS) and stimulated Raman scattering (SRS) are well suited for fast imaging, but are rather limited in acquiring spectroscopic information. In the simplest implementation, both techniques use synchronized picosecond laser pulses (typically a laser oscillator and an optical parametric oscillator (OPO)) for excitation. This allows high spectral resolution imaging at a single predetermined Raman band. Imaging at a different band requires tuning of one of the lasers to a new wavelength. That process is typically slow and changes optical power. Nonetheless, tuning across a limited spectral range is possible and has been used to acquire CARS spectra as well as SRS spectra (Freudiger et al., 2008; Begin et al., 2011; Potma et al., 2011). The more popular approach to obtain spectroscopic information is to start with a broadband femtosecond laser and a narrowband picosecond laser. The spectral resolution is then determined by the narrowband laser while the spectral coverage is determined by the broadband laser. This approach has been implemented by many groups collectively termed as multiplex CARS microscopy. As discussed in detail in Chapters 13 through 15, multiplex CARS combines the advantage of spectral detection and high-resolution imaging, offering the potential for detailed chemical analysis at subcellular resolution (Cheng et al., 2002; Wurpel et al., 2002; Kano and Hamaguchi, 2005).

Multiplex CARS is simpler to implement than multiplex SRS because CARS detects new wavelengths that can be spectrally dispersed and measured by parallel detection (for example using a CCD), while for SRS, a lock-in amplifier (LIA) array is also needed for demodulation (Freudiger et al., 2008) (**Figure 16.1**). As will be discussed in detail in Section 16.2, Ploetz et al. used a software demodulation scheme in the first implementation of SRS microscopy, which works when the stimulated Raman gain or loss is very large (Ploetz et al., 2007). Typically high excitation power from a laser amplifier is required, which significantly lowers imaging speed and increases sample damage concern. In most other

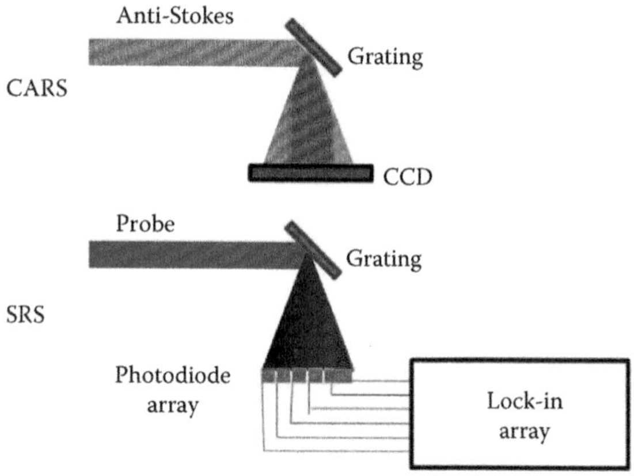


**FIGURE 16.1** Schematic diagram comparing multiplex CARS and SRS using spectral dispersion detection.

cases where femtosecond oscillators are used for excitation, the SRS signal has a modulation depth of less than $10^{-4}$. Therefore, demodulation has to be done through hardware LIA detection. Constructing a LIA array could be both time-consuming and prohibitive in complexity or cost.

Nonetheless, as has been discussed extensively in Chapter 4, SRS imaging has significant advantages over CARS imaging owing to the complete removal of non-resonant background that plagues all CARS imaging techniques. For quantitative multicomponent chemical analysis, it is important to avoid any complications arising from spectral distortion and the nonlinear concentration dependence of the signal. These common problems associated with multiplex CARS are limiting factors, despite the simplicity in implementation. A few mathematical algorithms have been suggested to recover the resonant part of the CARS spectrum (Wurpel et al., 2004; Rinia et al., 2007; Liu et al., 2009; Okuno et al., 2010), but they all involve complicated post-image data processing, which may fail when the data quality is not good, especially at high speed imaging conditions.

In contrast, simple linear algebra operations can be used for multiplex SRS data analysis. This is similar in concept to the well-established fluorescence spectral imaging (Tsurui et al., 2000). Here we describe a simple analytic tool for SRS chemometric analysis of an inhomogeneous sample (Fu et al., 2012). Assume that we have $M$ molecule species in a given sample and they are inhomogeneously distributed. SRS signals for a total of $K$ Raman bands at each sample point on an $N \times N$ spatial grid are measured. At a representative pixel location $(i, j)$, the concentration of species $m$ is $C_m$ (for simplicity, we dropped the subscript $ij$). If the Raman cross-section for each species at a certain Raman band $k$ is $\sigma_{k,m}$, we can write the SRS signal at all Raman band as a linear sum of all the individual contributors:

$$SRS_k = \sum_m C_m \sigma_{k,m} P_k \tag{16.1}$$

where $P_k$ comprises the multiplication of pump and Stokes power corresponding to the Raman band $k$ as well as the scaling factor related to detection efficiency. The Raman

Chapter 16

cross-section $\sigma_{m,k}$ together with the power factor $P_k$ can be directly measured using pure solutions of each species at known concentration, resulting in a calibration matrix $A$:

$$A = \begin{bmatrix} \sigma_{1,1}P_1 & \cdots & \sigma_{1,M}P_M \\ \vdots & \ddots & \vdots \\ \sigma_{K,1}P_1 & \cdots & \sigma_{K,M}P_M \end{bmatrix} \tag{16.2}$$

Therefore, the SRS signal for the $k$ Raman channels can be written in matrix form:

$$SRS = A * C \tag{16.3}$$

where
$SRS$ is a column vector with $K$ elements
$C$ is a column vector with $M$ elements

For each pixel, the concentration of each species can be obtained by solving a set of $K$ linear equations. In order to solve them, the number of Raman bands $K$ has to be larger than or equal to the number of species $M$. In the case $K = M$, $A$ is a square matrix. When the $K$ bands are appropriately chosen (no two species have the same vector $[\sigma_1, \sigma_2, \ldots, \sigma_K]$, which is usually the case), the $K$ sets of equations are nondegenerate. In other words, the matrix $A$ is invertible:

$$C = A^{-1} * SRS \tag{16.4}$$

On the basis of this equation, the concentration of each species can be readily calculated based on the acquired $K$ channel SRS signals. In the case that $K > M$, the system is over-determined and the matrix is not invertible. A solution can be obtained by the following matrix operation:

$$C = (A^T * A)^{-1} * A^T * SRS \tag{16.5}$$

where $A^T$ is the transpose matrix of the calibration matrix $A$. $(A^T * A)^{-1} * A^T$ is also called the pseudo-inverse of $A$. Having a greater number of channels improves the accuracy of concentration determination. In practice, there is always noise associated with SRS measurement at each Raman channel:

$$SRS = A * C + \delta \tag{16.6}$$

where $\delta$ is a vector of uncorrelated noise in each of the $K$ channels. The solution $C$ could be obtained through the ordinary least square operation:

$$\min \| A * C - SRS \| \tag{16.7}$$

which is equivalent to Equation 16.5. Because chemical concentration cannot be negative, there is an added constraint to Equation 16.7:

$$C_m \geq 0 \quad \text{for all } m \tag{16.8}$$

This is known as the nonnegative least square solution. For imaging, this analysis is repeated over the entire $N * N$ grid. We note that calibration matrix is the same for all pixels, therefore only the least square analysis is needed for each pixel. The general principle described here applies to all forms of multiplex SRS that will be discussed later.

In this chapter, in addition to Ploetz's multiplex SRS by parallel detection, we will present three different spectral detection approaches that overcome the difficulties in multiplex SRS detection. Contrary to spontaneous Raman, multiplex CARS, and multiplex SRS by parallel detection, the spectroscopic information is obtained from the excitation side instead of the emission side. The first approach utilizes pulse shaping to modify excitation spectrum and therefore achieves molecular specificity. The second approach applies modulation multiplexing to the femtosecond excitation beam and demodulates at several different Raman bands simultaneously using the fast Fourier transform (FFT). The third approach uses spectral focusing to quickly scan through the whole Raman spectrum that falls within the bandwidth of the femtosecond excitation beam. All three approaches use distinctive technologies and have their advantages and disadvantages, but they share the common feature with the simpler single band SRS imaging: detecting the stimulated Raman gain or loss on a single beam using high-frequency modulation and demodulation on a single photodetector. Sample complexity and movement may dictate the most appropriate method for imaging. We will present operating principles, technical details, and experimental demonstrations for each of the three techniques.

## 16.2 Multiplex SRS Microscopy by Parallel Detection

### 16.2.1 Instrumentation

The simplest approach to multiplex SRS microscopy is to perform parallel detection, similar to multiplex CARS. Such a detection scheme requires a multichannel detector and a multichannel lock-in as shown in Figure 16.1. Due to the difficulty of building a multichannel lock-in, parallel detection is achieved through software demodulation at low modulation frequencies. Because digital data acquisition has limited dynamic range, it requires that the detected modulation depth is higher than $10^{-4}$.

Figure 16.2 shows the experimental setup of parallel detection based multiplex SRS (Ploetz et al., 2007). The Stokes beam is from a 1 kHz amplified Ti:sapphire femtosecond laser system (Clark CPA). The spectrum bandwidth is narrowed down to 25 cm$^{-1}$ by passing it through a bandpass filter. The pump beam is derived from a white light continuum generated by focusing the output of the amplified laser onto a sapphire plate. The energy for the pump and Stokes beam is 10 and 270 nJ, respectively. A Cassegrainian microscope objective (Ealing, 36×, NA = 0.5) is used to focus the combined beam onto the sample. The effective NA of the objective is actually lower because of its central obstruction. The light after the sample is imaged onto the entrance slit of a spectrograph, which was then detected by a 512 element diode array (Hamamatsu S3902-512Q). The output

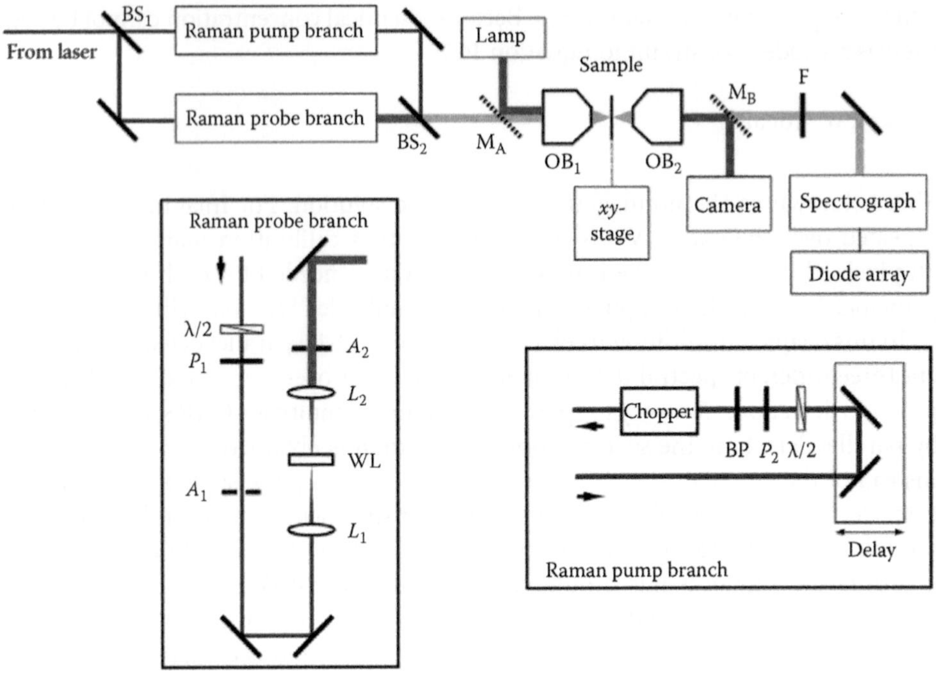


**FIGURE 16.2** Schematic of the femtosecond SRS microscopy setup. The pump branch (Stokes beam) is from an amplified laser system. The probe branch (pump beam) is derived from white light generation from a sapphire disk, which contains a large bandwidth for simultaneous excitation of the full Raman spectrum. The parallel detection by the spectrograph and diode array allows simultaneous acquisition of the entire spectrum. A, apertures; WL, white light generation; BS, beamsplitter; $OB_1$, Cassegrainian microscope objective; $OB_2$, refractive microscope objective; F, notch filter; BP, band pass filter; $M_{A,B}$, removable mirrors. (From Ploetz, E., et al., *Appl. Phys. B* 87, 389–393, 2007.)

of the diode array is read out at 1 kHz, which is twice the frequency of the amplitude modulation of the pump through a chopper wheel. Therefore, direct demodulation can be achieved by subtracting the pump "on" signal with pump "off" signal.

## 16.2.2 Imaging Example

Multiplex SRS microscopy of polystyrene was demonstrated based on parallel detection. A key advantage of parallel detection is that the whole spectrum at a pixel is obtained simultaneously. To obtain an image, the sample is raster scanned in *xy* by a motorized stage. Based on the spectral information, an image can be reconstructed at any Raman peak of interest. Because the supercontinuum generation is a noisy process, the Raman spectrum bears pixel-to-pixel fluctuation which is removed by a Savitzky–Golay filter. After filtering, the Raman spectrum is subtracted by a smooth offset using spline fitting. Figure 16.3a through c shows the image of 30 μm polystyrene beads at three different SRS shifts after all the corrections: (1) aliphatic and aromatic stretch vibrations from 2830 to 3120 $cm^{-1}$; (2) ring breathe mode at 1000 $cm^{-1}$; (3) water Raman band from 3260 to 3550 $cm^{-1}$. Figure 16.3d shows example SRS spectra at two locations in the image, one on the beads and the other in water. Even though at worse signal-to-noise ratio, the SRS spectra reflect the same features as their spontaneous Raman counterpart. The C–H

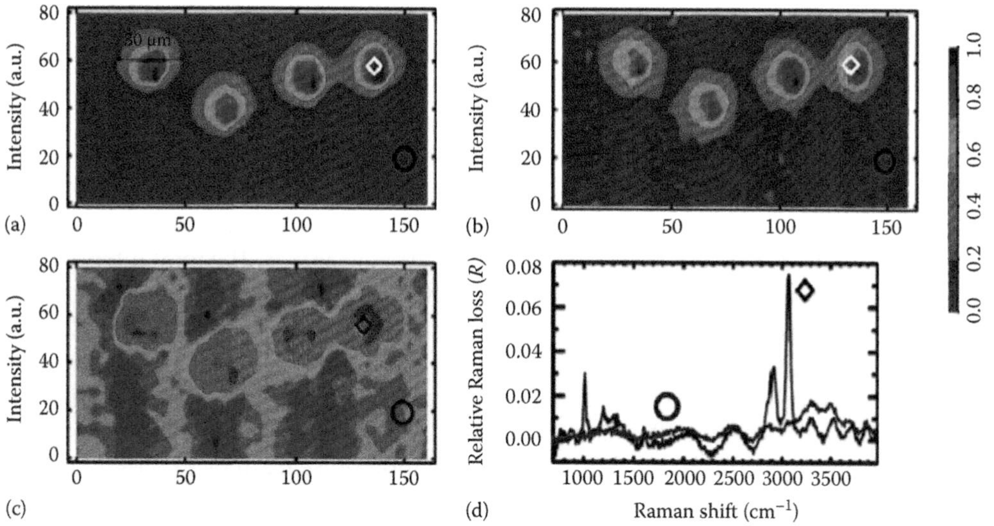


**FIGURE 16.3** SRS images of polystyrene beads (diameter 30 μm) dispersed in water. (a) Integral covering the C–H stretch vibrations of polystyrene. (b) Integral covering the ring breathing mode of polystyrene. (c) Integral covering the O–H stretch vibrations of water. (d) Raman spectra of the locations indicated by the circles and square. (From Ploetz, E., et al., *Appl. Phys. B* 87, 389–393, 2007.)

stretch vibrations around 3000 cm$^{-1}$, the aromatic C–C stretch vibration, and the ring breathe mode at 1000 cm$^{-1}$ are present in the polystyrene SRS spectrum. The beads show negative contrast when the image is reconstructed at the water Raman band as expected.

In order to be able to obtain a good spectrum, the optical power has to be very high to generate modulation signals that are larger than $10^{-4}$ so that it can be detected without a LIA. Clearly such an approach is unsuitable for biological imaging, where the samples have much lower tolerance for high power illumination. Another major drawback is that the low modulation frequency (due to both the laser repetition rate and the modulator) means high laser fluctuation noise, which is made even worse by using a supercontinuum source as one of the excitation beams, thereby significantly reducing the signal-to-noise ratio in the obtained spectrum. In principle, these limitations could be overcome by using lower optical power and higher modulation frequency. The real bottleneck still lies in the multichannel lock-in amplification, which could potentially be overcome in the future.

## 16.3 Multiplex SRS Microscopy by Spectrally-Tailored Excitation

### 16.3.1 Theoretical Basis of Spectrally-Tailored Excitation

The high frequency lock-in detection limits the number of SRS channels that can be simultaneously acquired. However, there is no limit on the excitation side to how many Raman bands can be excited at the same time. One could either sweep the SRS excitation band or engineer the excitation spectrum to target one species at a time. In this section, we will discuss how spectrally-tailored multiplex excitation enables the utilization of Raman spectroscopy information to distinguish and quantify chemical species with overlapping Raman bands, one at a time.

The general idea is to use a spectrally shaped broadband pump together with a narrowband Stokes beam and high-sensitivity detection of SRS to perform Raman spectroscopy on the excitation side rather than the emission side. The broadband pump can probe vibrational resonances of many species simultaneously. For each species, the excitation efficiency is a sum of all the probed vibrational resonances. Altering the relative spectral intensity in the pump $P_k$ will change the relative Raman excitation efficiency of different species. The key is to shape the excitation pump spectrum such that summed excitation is zero for every species except one that is our target. The total intensity of the narrowband Stokes beam is then detected with a single photodiode by high frequency lock-in detection to extract the overall SRS signal (sum of all excitation channels) due to the tailored excitation of the target species.

In order to achieve zero excitation for interfering species, "negative" excitation must be introduced. Because SRS uses phase-sensitive detection, "negative" excitation is achieved through shifting the phase of the excitation pulse train by exactly 180°, which gives rise to a negative SRS signal. The net SRS is a result of the sum of "positive" and "negative" excitations of all species.

$$SRS = \sum_m \sum_k \left[ C_m \sigma_{k,m}^{+} P_k^{+} + C_m \sigma_{k,m}^{-} P_k^{-} \right] \tag{16.9}$$

where $P_k^+$ and $P_k^-$ represent "positive" and "negative" excitation, respectively. In other words, the $K$ channel excitation spectrum is split into two portions. When the following condition for the excitation spectrum is satisfied, SRS detection of only one species $x$ is achieved.

$$\begin{aligned} &\sum_k \left[ \sigma_{k,m} P_k^{+} + \sigma_{k,m} P_k^{-} \right] = 0 \quad \text{for all } m \neq x \\ &\sum_k \left[ \sigma_{k,m} P_k^{+} + \sigma_{k,m} P_k^{-} \right] \neq 0 \quad \text{for } m = x \end{aligned} \tag{16.10}$$

We note that the excitation spectrum is independent of the concentrations of chemical species and applies to all pixels in the image. The total SRS signal has linear dependence on the concentration of species $x$:

$$SRS = \sum_k \left[ \sigma_{k,x}^{+} P_k^{+} + \sigma_{k,x}^{-} P_k^{-} \right] C_x \tag{16.11}$$

To shape the excitation spectrum correctly, $\sigma_{k,m}$ of all existing species have to be measured. They can be obtained through spontaneous Raman measurement at the selected spectral range. The best excitation spectrum is then determined by solving for a set of linear equations, for example, Equation 16.10, with the additional constraint

$$\max \sum_k \left[ \sigma_{k,x}^{+} P_k^{+} + \sigma_{k,x}^{-} P_k^{-} \right] \tag{16.12}$$

We note that this whole process is analogous to Equations 16.3 through 16.7, because excitation and emission are formally identical.

## 16.3.2 Instrumentation

The basic setup for spectrally-tailored excitation SRS (STE-SRS) is very similar to narrowband SRS setup except for the addition of a pulse shaping module. Figure 16.4 shows the schematic diagram of the experimental setup. The narrowband Stokes laser is a Nd:$YVO_4$ picosecond laser (High Q, center wavelength at 1064 nm with 0.3 nm bandwidth), while the broadband pump laser is a Ti:sapphire femtosecond laser (Mira, Coherent, center wavelength at 800 nm with 7 nm bandwidth). The two lasers are electronically synchronized by a Synchrolock unit from Coherent. This system allows SRS imaging in the CH stretching region spanning from 2800 to 3050 $cm^{-1}$. The femtosecond pump beam is chirped by a long glass rod (not shown in figure) to picosecond duration to minimize nonlinear photodamage to the sample. The Stokes beam and pump beam are combined on a dichroic beamsplitter and sent to a confocal laser scanning microscope (Olympus BX61/FV300) for imaging.

The essential element that distinguishes STE-SRS from regular SRS is that a pulse-shaper (Biophotonic Solutions Inc.) is introduced into the pump beam to manipulate the excitation spectrum (Freudiger et al., 2011). The pulse-shaper consists of a grating (to disperse the broadband pulse) and a curved mirror (1 m focal length to achieve a spectral resolution of 0.1 nm) to focus the spectral components onto a polarization SLM in a 4f geometry in reflection mode. The throughput of the pulse-shaper was 55% ± 5%. Figure 16.5 shows schematically STE-SRS excitation compared to broadband SRS excitation in both frequency domain and time domain pictures. In broadband excitation, the number of photons increases in the Stokes beam and decreases in the pump beam due to the SRS processes at different Raman resonance frequencies. The photon number increment for the Stokes component is equivalent to the sum of decrements of pump photons at

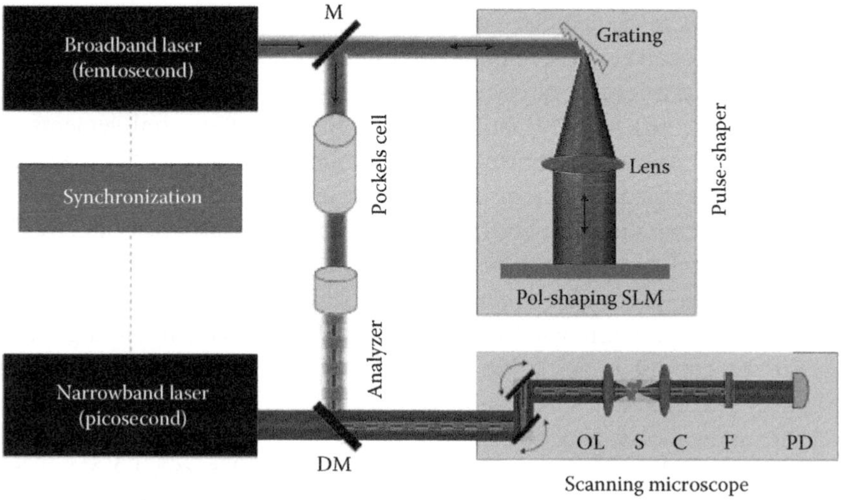


FIGURE 16.4 Schematic diagram of spectrally-tailored excitation. The broadband Ti:sapphire laser is spectrally modified by a spatial light modulator (SLM) based pulse shaper to contain "positive" and "negative" SRS excitation components, allowing for selective imaging of one particular molecular target. (From Freudiger, C.W., et al., *Nat. Photon.* 5, 103–109, 2011. With permission.)

Chapter 16

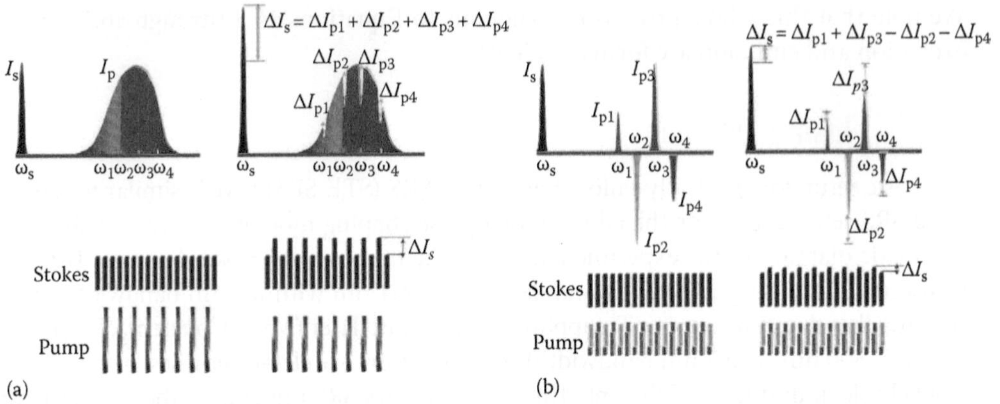


**FIGURE 16.5** Comparison between (a) broadband excitation and (b) spectrally-tailored excitation. The top panels show the frequency domain picture and the bottom panels show the time domain picture, before and after excitation. In both cases, the multiband excitation is the same, but by spectrally-tailoring the pump spectrum, the total excitation could be modified to selectively excite one species (From Freudiger, C.W., et al., *Nat. Photon.* 5, 103–109, 2011. With permission.)

all Raman resonance frequencies. To introduce "negative" excitation, polarization pulse-shaping is used. It works by splitting the spectrum into two portions with orthogonal polarization using the SLM. For illustration purposes, we only show two peaks for each polarization. The EOM and polarizer together perform amplitude modulation at 4 MHz for linear input polarization. After passing these two, the + and – portions of the broadband beam appear on opposite half-cycles of the modulation with the same polarization, meaning that the two portions have exactly 180° phase shift. As a result, the net gain in Stokes photons is the subtraction of "positive" excitations and "negative" excitations. As described in the previous chapter, for sample with multicomponents, the excitation spectrum is designed such that the "positive" excitations are cancelled out by the "negative" excitations for all except one species of interest. In doing so, quantitative imaging of one chosen species can be carried out without interference from all other species. The excitation spectrum can be altered as required for imaging different species.

The number of species that can be distinguished is limited to the number of spectral elements that can be independently manipulated. In this case, 80 spectral elements are used for pulse shaping. In imaging experiments, normally the total number of interfering species in sample is much less than 80. As a result, Equation 16.10 is underdetermined, meaning that many different excitation spectra can be used for imaging a target species. We determined the spectrum by calculating the Moore–Penrose pseudo-inverse of $\sigma_{k,m}P_k$. This procedure is similar to the chemometric method known as classical least squares (CLS). It effectively projects the target spectrum onto the subspace orthogonal to all interfering spectra.

One the detection side, the Stokes beam is detected by a large-area InGaS photodiode (New England Photoconductors, I5-3-5, biased at 12 V), whose output was bandpass filtered around the modulation rate of 4 MHz and then demodulated by a high-frequency lock-in amplifier (Stanford Research Systems, SR844RF). We used a ×60, 1.2 NA water immersion lens (Olympus, UPlanApo/IR) as the excitation objective and the light was collected in transmission with a 1.4 NA oil condenser (Nikon). For imaging of biological samples, the average power was 15 mW for the pump and 120 mW for the Stokes. A time constant of 10 μs ("no filter" mode) was used for imaging and 1 s for the solution spectroscopy studies.

### 16.3.3 Performance Characterization

Figure 16.6 shows binary and tertiary mixture solution measurements based on STE-SRS. First, the spontaneous Raman spectra (Figure 16.6a) of all the individual species are measured with a commercial confocal Raman unit (Horiba Jobin Yvon). Next, the pump spectrum is recorded with a spectrometer (HR4000, Ocean Optics). The multiplication of these two results in SRS spectra for this particular broadband pump spectrum. Based on oleic acid and cholesterol SRS spectra, an excitation mask that selectively targets cholesterol is designed as shown in Figure 16.6c. For this excitation mask the SRS difference signal is shown to be linear with the concentration of cholesterol and

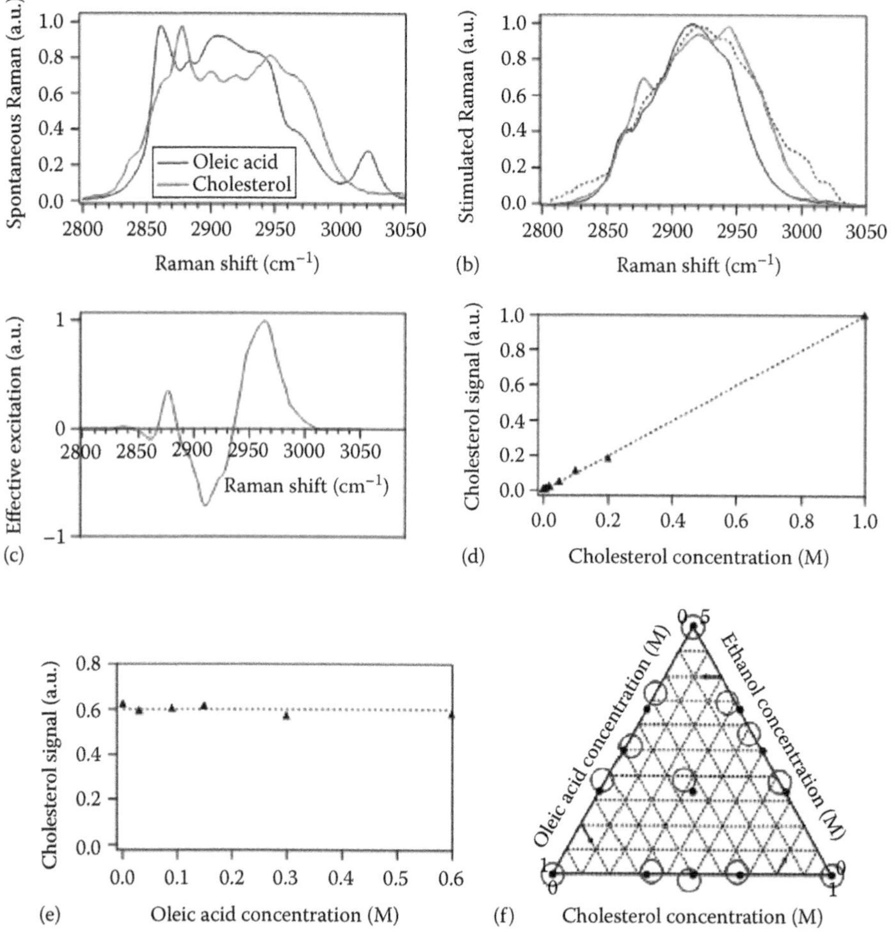


**FIGURE 16.6** (a) Spontaneous Raman spectra of cholesterol (green) and oleic acid (red). (b) Expected SRS spectra for cholesterol (green) and oleic acid (red) calculated by normalizing the spontaneous Raman spectra with the laser excitation spectrum (black dotted). (c) Excitation mask for selective detection of cholesterol (target species) in the presence of oleic acid (interfering species) generated from the SRS spectra shown in (b). (d) Linear dependence of the STE-SRS signal on concentrations of cholesterol allows for straightforward signal quantification. (e) STE-SRS allows interference-free quantification of cholesterol at varying oleic acid concentrations. (f) Ternary plot of mixtures of cholesterol, oleic acid, and ethanol solutions in deuterated chloroform. Solid dots show actual concentration of the mixtures and red circles show measurement with STE-SRS. (From Freudiger, C.W., et al., *Nat. Photon.* 5, 103–109, 2011. With permission.)

independent of the concentration of oleic acid (Figure 16.6d and e). STE-SRS allows suppression against signals from interfering species that are ~2000-fold more concentrated. When three different excitation spectra were used for selectively measuring oleic acid, cholesterol, and ethanol, their concentrations were quantitatively determined by STE-SRS measurements as shown in Figure 16.6f.

### 16.3.4 Imaging Examples

A phantom sample is first imaged to show the suppression effect of STE-SRS. The sample contains a mixture of protein, oleic acid, and stearic acid. Three different excitation masks are designed based on their SRS spectra for selective targeting of one species at a time. Figure 16.7 shows the STE-SRS imaging results of each species with minimal interference from other species.

We further demonstrated the selective imaging of different types of fatty acids *in vivo*. A model organism *Caenorhabditis elegans*, commonly used in lipid research, is used for our imaging. Both CARS and SRS microscopy have been applied to study lipid storage in *C. elegans* before (Hellerer et al., 2007; Wang et al., 2011). They overcome the shortcomings of lipid staining techniques as will be discussed in detail in Chapters 18 and 24. However, there are many different kinds of lipids that may have different functions in

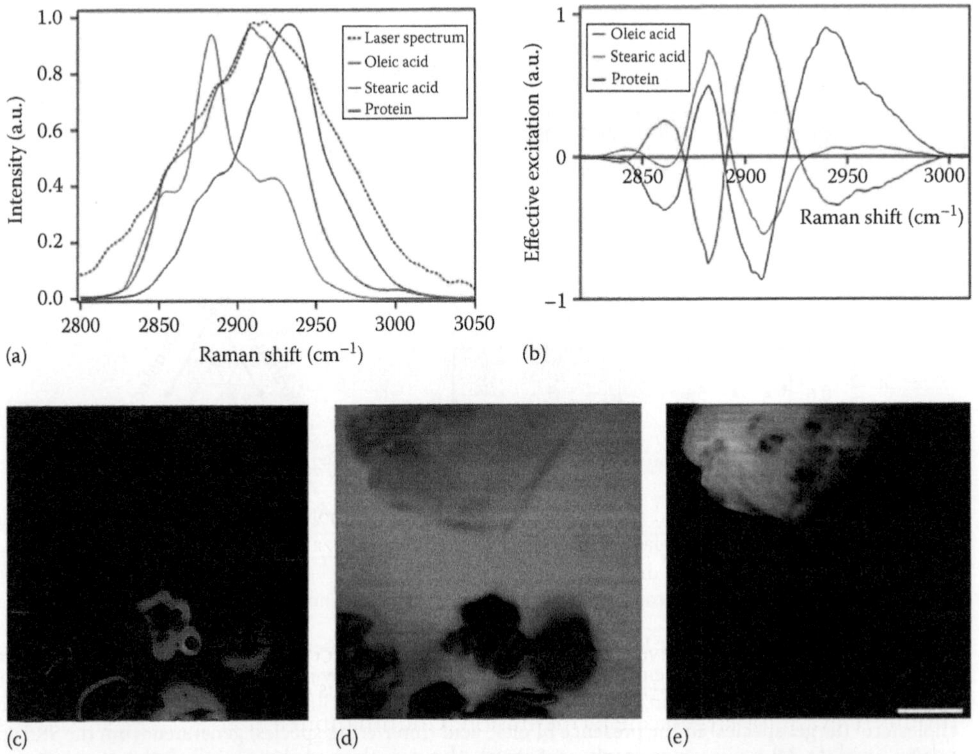


FIGURE 16.7 (a) Spontaneous Raman spectra of protein, oleic acid, and stearic acid, and the laser spectrum (in dashed line). (b) Excitation mask for each of the three species. (c–e) Images show the same area of a mixture containing three species using STE-SRS for three different excitation masks: protein (c), oleic acid (d) and stearic acid (e). Scale bar: 25 μm. (From Freudiger, C.W., et al., *Nat. Photon.* 5, 103–109, 2011. With permission.)

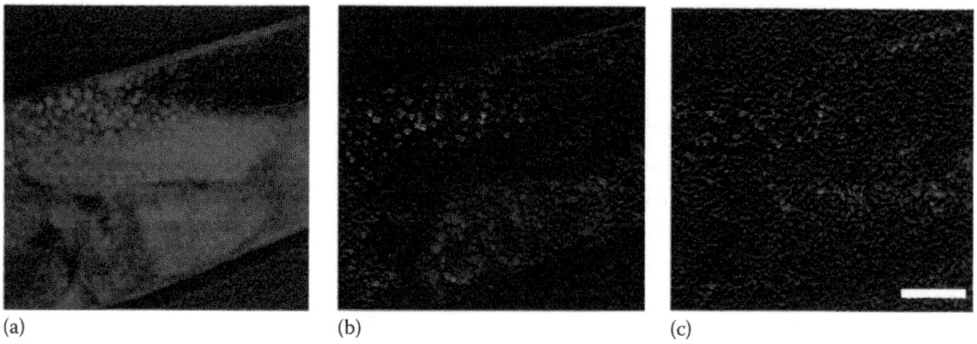

FIGURE 16.8 STE-SRS imaging of lipid storage in *C. elegans* by applying spectral masks for protein (a), oleic acid (b) and stearic acid (c). Comparison of (b) and (c) shows that oleic and stearic acid deposits co-localize and together they co-localize with protein aggregates as well. Scale bar: 25 μm. (From Freudiger, C.W., et al., *Nat. Photon.* 5, 103–109, 2011. With permission.)

the nematode. These lipid species do not have isolated vibrational bands, thus preventing direct imaging by narrowband SRS. With STE-SRS, we have already shown that different lipid species can be distinguished owing to their minute variations in spectrum in the C–H stretching region. In order to distinguish saturated and unsaturated fattyacids, oleic and stearic acid were chosen as their representative spectroscopic samples. The same STE-SRS excitation masks as used previously are applied here for imaging protein, saturated lipids, and unsaturated lipids distribution in *C. elegans*. Figure 16.8 shows the imaging results of the worm. Based on these chemical maps, we notice that both types of lipids co-localize with no isolated storage areas that contain a single species only. We also note that fat deposits co-localize with areas of increased protein aggregation.

In conclusion, STE-SRS makes use of spectroscopy information to image one target species at a time without the interference of other species. Therefore, it combines fast imaging speed with high chemical selectivity. Unlike serial spectral acquisition, the spectroscopy information is built into the data acquisition process such that spectroscopy information at every single pixel is obtained in a short time. Therefore, STE-SRS imaging speed can be orders of magnitude faster than multiplex-CARS. Moreover, it has the advantage that no data processing is needed because SRS signal is linearly proportional to the concentration of one target species. In principle, the number of chemical species that can be distinguished by STE-SRS is only limited by the number of spectral channels on the SLM. The practical limitation comes from laser spectral fluctuation and diminished SNR with an increasing number of species due to increased constraints in excitation mask design. Another limitation comes from the fact that STE-SRS requires that all the chemical species in the sample have to be known a priori and their Raman spectra measured for calculating the excitation masks.

## 16.4 Modulation Multiplex SRS Microscopy

### 16.4.1 Instrumentation

In the previous chapter, we described how we can engineer the excitation spectrum based on the spectroscopy of all species to selectively target one species at a time. In many cases, it is desirable to achieve quantitative information about multiple species

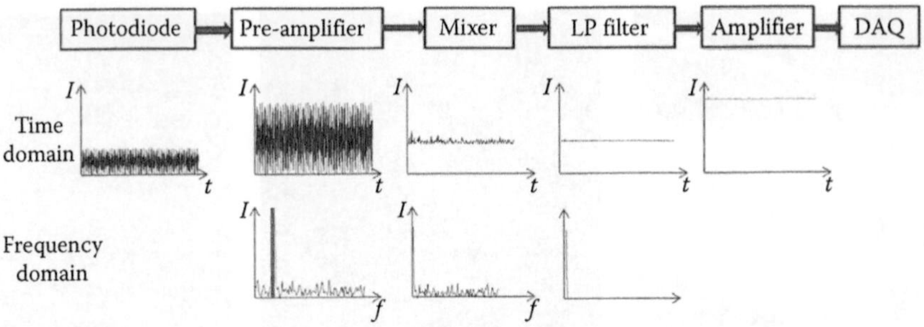


**FIGURE 16.9** Illustration showing how a lock-in amplifier (LIA) works, in both time domain and frequency domain.

at the same time. How can we obtain multiple channel signals without resorting to a multichannel lock-in array? LIA allows extraction of signals at a specific frequency amid a noisy background. A LIA can be thought of as a voltage amplifier combined with a bandpass filter. Figure 16.9 shows a flowchart explaining the LIA. The mixer shifts the frequency of modulated signal to DC and the low pass filter removes all the noise above and below the modulation frequency. This whole process is analog until the analog to digital conversion for data read out (DAQ). The digital equivalent operation of the lock-in is FFT, which allows sampling at all frequencies simultaneously. This can be done after the preamplification stage, using fast DAQ to directly digitize the data and perform a numerical Fourier transform. Therefore, we can design the experiment so that different Raman excitations are modulated at different frequencies and use the FFT to demodulate several Raman bands simultaneously (Fu et al., 2012).

Figure 16.10 shows our experimental design. We use the same laser system as shown in the previous section. A picosecond laser at 1064 nm is electronically locked to a femtosecond laser centered at 805 nm with a bandwidth of 15 nm. To apply different modulations to different Raman excitations, an acousto-optical tunable filter (AOTF, Crystal Technology) is used to modulate the pump beam. The AOTF is not only designed to filter broadband light with an arbitrary combination of colors, but also has the built-in function of applying a digital or analog amplitude modulation. We use an eight-channel digital driver together with a home-built four-channel digital TTL pulse generator to drive the AOTF. Up to four Raman excitation bands within the femtosecond laser bandwidth can be diffracted and amplitude-modulated simultaneously. The diffracted order of the pump beam was used to minimize unwanted laser power illuminated on the sample. The diffracted laser frequency is linearly dependent on the input RF frequency, while its amplitude modulation frequency can be independently controlled between 0 and 200 kHz. To ensure maximal use of the modulation bandwidth and minimal cross-talk between channels, the amplitude modulation frequencies are chosen to be evenly distributed between 50 and 200 kHz.

Because SRS signal detection is based on lock-in amplification, it is important to consider the laser noise spectrum. The laser noise has $1/f$ characteristic at frequencies below 1 MHz. Therefore, to avoid laser fluctuation induced noise, the modulation and demodulation is typically done at a few up to 20 MHz. However, the modulation of AOTF is limited to well below 1 MHz due to the long interaction length of acoustic wave and femtosecond pulses in the crystal. In order to avoid laser fluctuation noise, we used

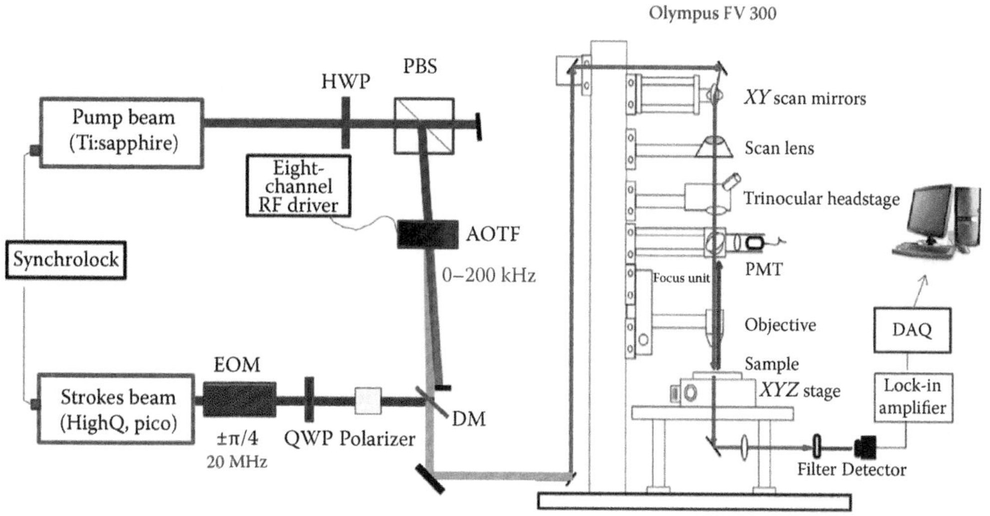


**FIGURE 16.10** Schematic diagram of our modulation multiplex SRS setup. Double modulation and double demodulation are used to detect a few SRS bands simultaneously with the same performance as their narrowband counterparts. The Stokes is modulated at high frequency 20 MHz, and the pump is modulated at low frequency from 0 to 200 kHz. The number of SRS channels is determined by the number of RF channels in the EOM, electro-optical modulator; QWP, quarter-wave plate; DM, dichroic mirror; AOTF, acousto-optical tunable filter; HWP, half-wave plate; PBS, polarizing beamsplitter; PMT, photomultiplier tube; DAQ, data acquisition card.

double modulation by introducing a 20 MHz modulation on the Stokes beam, which effectively shifts the detection away from noisy laser region. On the detection side, we first demodulate at 20 MHz with 1 MHz bandwidth using a home-built fast LIA and then demodulate the low-frequency modulation using fast data acquisition and FFT (Saar et al., 2010a). Because the second step is done digitally in the software, all frequencies are read out simultaneously, allowing for multiplex SRS detection.

We note that using AOTF as the multiplex device has a drawback because the diffracted laser has side-lobe frequencies. Consequently, even though each diffracted band has a FWHM bandwidth of 2.3 nm and could be separated for the chosen Raman excitation bands, the side-lobe of one diffraction band could still overlap with another diffraction band, resulting in cross-talk among the multiple channels. Due to the limitation of this channel cross-talk, we typically limit the number of channels to three to minimize cross-talk. The three channels are modulated at 125, 100, and 75 kHz, respectively.

## 16.4.2 Performance Characterization

The data analysis on modulation multiplex SRS is quite straightforward. As shown in Equations 16.2 through 16.4, we only need to obtain the calibration matrix before imaging a multicomponent system. The calibration matrix can be determined from measurements on pure solutions or even from *in situ* measurements in the multichannel images. As a demonstration showing the quantitative analysis capability of modulation multiplex SRS, we first determined the concentrations of three different species in binary and tertiary mixtures.

The three species used were oleic acid, cholesterol, and cyclohexane. They are dissolved in deuterated chloroform. We measured multiplex SRS signals at three different Raman resonances simultaneously 2850, 2900, and 2960 $cm^{-1}$ (shown in Figure 16.11a, together with the spontaneous Raman spectra of the three species). The solutions of each species are measured individually to construct the calibration matrix (Figure 16.11b). Based on Equation 16.4, when the three-channel measurement on any binary or tertiary mixture is multiplied by the inverse matrix of the calibration matrix, concentrations of the three species are obtained. Figure 16.11c shows the ternary plot of the result of concentration measurement based on modulation multiplex SRS. As can be seen, the accuracy of measurement is relatively good. We believe it can be significantly improved if the channel cross-talk problem is eliminated with better modulation multiplexing technology. An additional source of error is the timing jitter between the picosecond and femtosecond pulse, which could also be eliminated with alternative laser sources such as synchronously pumped lasers or fiber wavelength conversion.

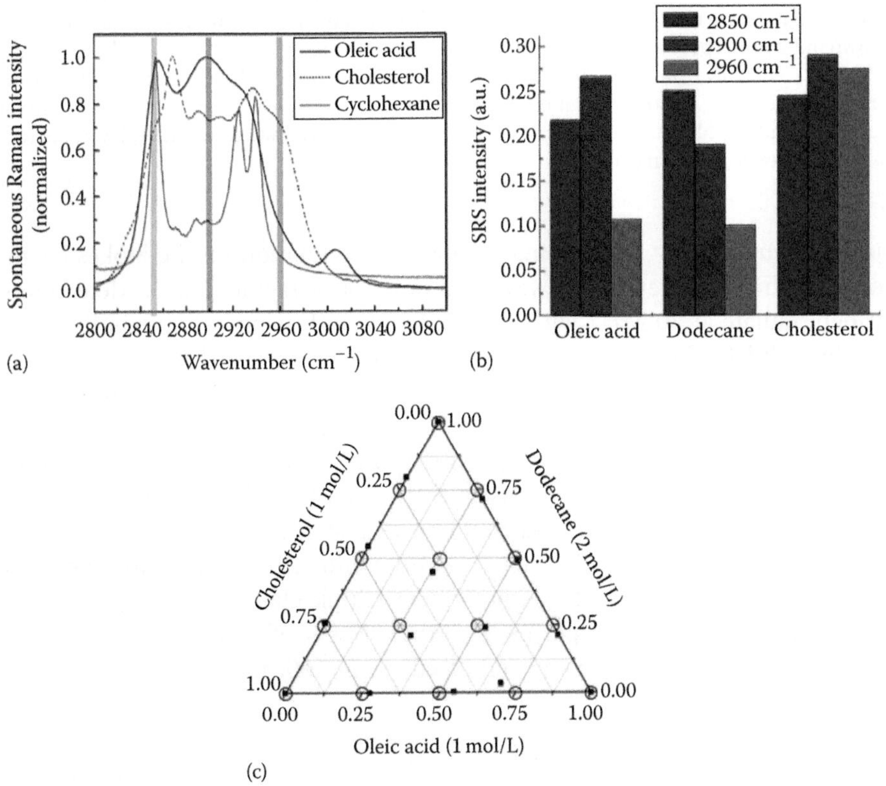


**FIGURE 16.11** (a) Spontaneous Raman spectra of three compounds: oleic acid, cholesterol, and cyclohexane. They have overlapping Raman bands in the C–H stretching region. Colored vertical bars indicate the chosen Raman band for modulation multiplex SRS. (b) Calibration matrix obtained though multiplex SRS measurement on pure solutions of the three compounds. Each color bar corresponds to one Raman band, with the total SRS intensity normalized to 1. (c) Quantitative measurements on binary and tertiary mixture of the three compounds are shown in ternary plot. The dots show the measured concentrations while the red circles show the expected values. (From Fu, D., et al., *J. Am. Chem. Soc.* 134, 3623–3626, 2012. With permission.)

### 16.4.3 Application Examples

In this section we show that imaging based on modulation multiplex SRS allows visualization of multiple species simultaneously with one image scan. Two different application examples are demonstrated: multiplex SRS of microalgae and multiplex SRS of skin tissue. In both systems, we used simple three-component models to describe their biochemical compositions.

Microalgae are of great interest in recent years due to their potential in renewable energy production, mainly through conversion of their lipid contents to diesel fuel. They have much faster growth rates than terrestrial crops and higher per-acre yield. Besides, algae biofuel contains no sulfur, is non-toxic, and is highly biodegradable. Some species of algae are ideally suited to biodiesel production due to their high oil content, as much as 50% by dry weight. There are a number of factors that affect the oil producing capability of algae, including environmental conditions and strains of algae. To evaluate how these factors affect the oil producing capacity, it is critical to have a fast and robust method that can quantitatively measure the oil content in algae. Most recently, Raman microspectroscopy has been used in studying algal lipid composition but with very low throughput (Wu et al., 2011). CARS and SRS are excellent label-free alternatives to study lipid contents because lipids are usually concentrated in droplets and have very strong signal (Nan et al., 2006; Wang et al., 2011). Compared to CARS, SRS is more quantitative because it has a linear concentration dependence as opposed to a nonlinear dependence in CARS (Freudiger et al., 2008).

Raman imaging of microalgae presents significant challenge due to the abundance of chlorophylls that are used for photosynthesis. Chlorophylls are fluorescent, masking any spontaneous Raman signal when excited in the visible. SRS imaging is able to overcome the fluorescence problem because fluorescence is incoherent and therefore has much smaller signal size compared to the heterodyne amplified stimulated Raman signal. However, another related process called two-photon absorption (TPA) is strongly interfering with SRS measurement. Because SRS detection uses high-frequency modulation transfer, other two-color two-photon processes such as TPA, excited state absorption, ground state depletion, and stimulated emission will also contribute to the signal (Fu et al., 2007a,b). Previously TPA of hemoglobin and melanin has already been shown to have strong signal in the NIR when using pulsed lasers as excitation sources. These signals were used as unique contrasts for label-free tissue imaging. However, in this particular case of algae lipid imaging, TPA of chlorophyll is mixed with SRS of lipids and protein, presenting difficulty in extracting the signal from lipids. The advantage of multiplex detection allows us to distinguish different contributions based on their difference in spectral shape. Unlike SRS, TPA has a broad spectral shape due to the manifolds of vibrational and rotational energy levels that are involved in an electronic transition. With three channels in modulation multiplexing, we can distinguish chlorophyll TPA, protein SRS, and lipids SRS, respectively. Figure 16.12 shows an example of multiplex SRS image of microalgae samples. The raw three channel multiplex SRS signals (Figure 16.12a through c) reveal differences in contributions from chlorophyll, lipids, and protein. These contributions are mixed together, making image interpretation difficult. In order to unmix the three major components, we constructed the calibration matrix using multiplex SRS measurement of lipids (oleic acid) and protein

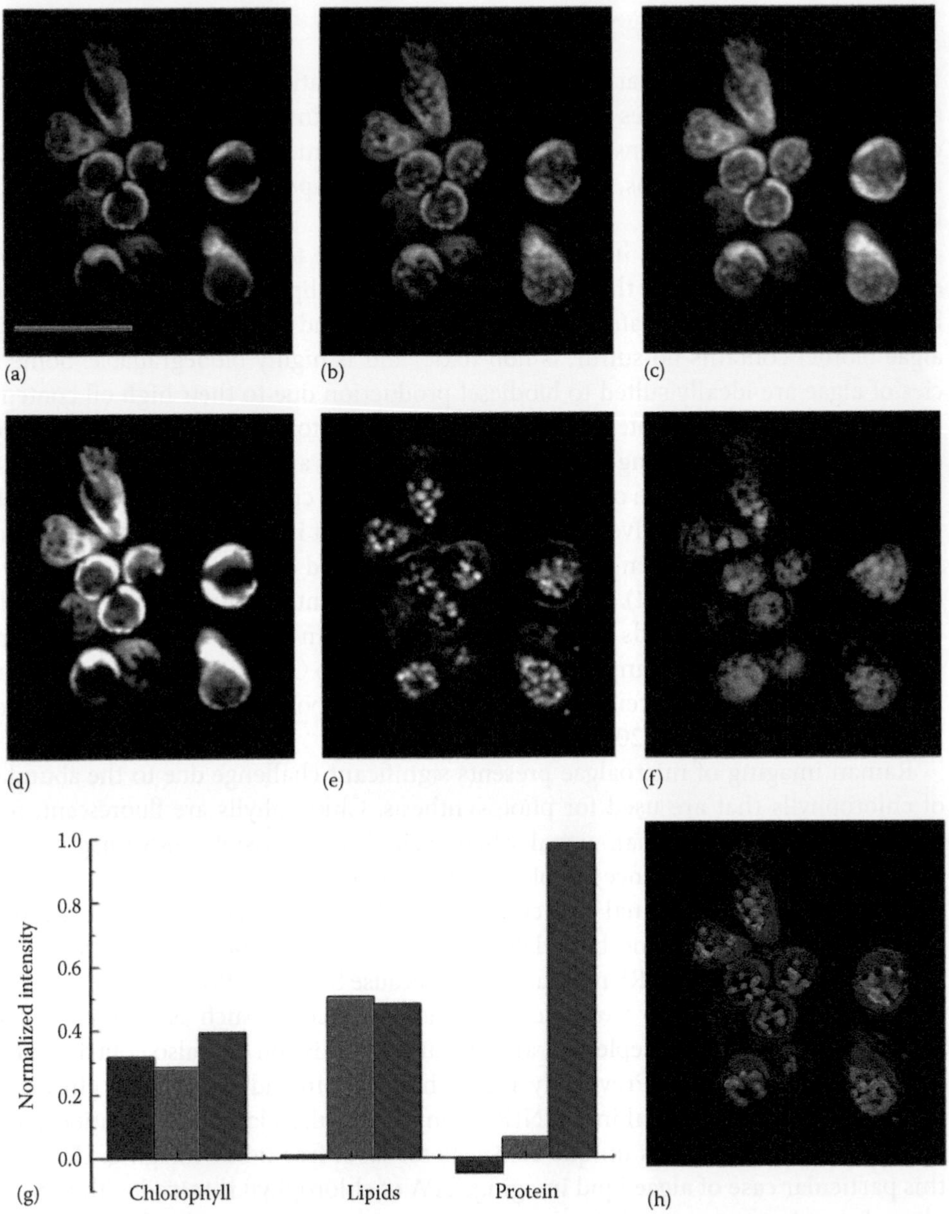


**FIGURE 16.12** (a–c) Multiplex SRS images at 2780, 2850, and 2950 cm$^{-1}$, respectively. Scale bar: 20 μm. (d–f) Chlorophyll, lipids, and protein map based on matrix inversion operation on the three multiplex images. (g) The calibration matrix for the three major components. (h) Composite image of the three chemical maps: red—chlorophyll; green—lipids; blue—protein.

(bovine serum albumin) solutions. The chlorophyll calibration is based on the image regions containing high chlorophyll concentration. From the SRS images we can see that chlorophylls (from TPA signal in 2780 cm$^{-1}$ channel) are highly concentrated in the periphery of the cells forming a clamshell shape. We used signals in these regions to calibrate chlorophyll. The resulting calibration matrix is shown in Figure 16.12g. As can been seen, chlorophyll TPA signal has a more uniform distribution in the three channels while protein has a dominant signal at 2950 cm$^{-1}$ and lipids have contributions from both 2850 and 2950 cm$^{-1}$. The small negative contribution of protein at 2780 cm$^{-1}$ is an artifact either from cross-talk between 2780 cm$^{-1}$ and another channel or from cross-phase modulation (Fu et al., 2007b). In either case, we used the matrix as it is because the sample itself will contain the same artifact. Using Equation 16.4, the inverse matrix was calculated and then used for decomposition of the three-channel SRS signals into respective chemical contributions for every single pixel in the image, resulting in high-resolution chemical maps of the three major components of interests. Figure 16.12d through f shows the decomposed chemical images and Figure 16.12h shows the composite image of the three components. Apparently, lipids appear as oil droplets, which reside entirely in the center of microalgal cells.

We further quantified the total lipid contents of the cells. Microalgae cells (*Botryococcus braunii* algae in modified bold 3N medium, both are from UTEX) grown under two different conditions are compared: continuous illumination versus a 12:12 h light:dark cycle. We imaged a total of 15 frames of each sample. The total number of cells is about 100–150. Figure 16.13 shows the composite image of the three components for both conditions. It is observed that chlorophyll and lipid concentrations are higher in cells under continuous illumination. To quantify the concentration change, we used image

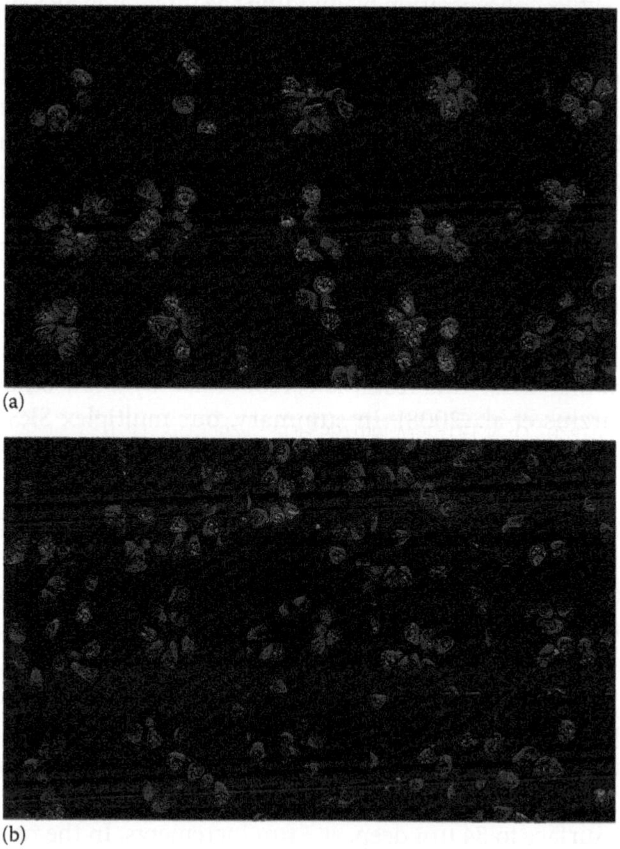

**FIGURE 16.13** Composite images of chlorophyll, lipids, and protein for (a) microalgae under continuous illumination (b) microalgae under 12:12 h light:dark cycle. (From Fu, D., et al., *J. Am. Chem. Soc.* 134, 3623–3626, 2012. With permission.)

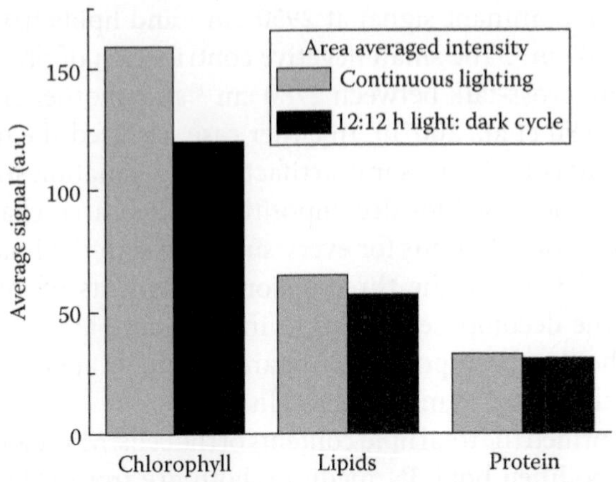


**FIGURE 16.14** Comparison of chlorophyll, lipids, and protein concentrations for microalgae gown under continuous illumination and microalgae grown under 12:12 h light:dark cycle. Cells under continuous illumination show accumulation of larger biomass.

thresholding method to calculate the area of the cells. To a first approximation, we can calculate an effective concentration by dividing the integrated SRS signal by the cell area (shown in bar plot in Figure 16.14). There is an increase in cellular concentration for all three components when algae are grown under continuous illumination. In particular, chlorophyll increases by 25%, lipids by 13%, and protein by 5.5%. It is known that algal cells under intense illumination could attain higher biomass compared with cells that had been adapted to low-level irradiance (Banerjee et al., 2002). Interestingly, the pigment content increases most. This is most likely due to the adaptation of algal cells to light to increase photosynthesis. Another interesting effect we noticed is that cell grows larger on average when they are continuously illuminated. The most important parameter is the oil content, which can be approximately gauged by the fraction of lipids mass in the total lipids plus protein mass. Clearly under continuous lighting, the lipid/protein ratio in the cell increases. This result is corroborated by a few other studies (Fabregas et al., 2002; Darzins et al., 2008). In summary, our multiplex SRS imaging approach provides a quantitative and noninvasive way of measuring a number of key parameters for evaluating the oil content of algae under different environmental conditions.

In the second application, we show that skin tissue structure can be delineated using modulation multiplex SRS. We chose the same three channels to perform multiplex SRS measurement on *ex vivo* mouse ear tissue. In this case, the 2780 $cm^{-1}$ band allows imaging of blood vessels based on TPA of hemoglobin, while the 2850 $cm^{-1}$ and 2950 $cm^{-1}$ band provide information on lipids and proteins. Because SRS is a nonlinear process, it has the same inherent optical sectioning capability as multiphoton fluorescence. We demonstrate this by performing high-resolution three-dimensional optical imaging on the skin tissue. Figure 16.15 shows both the raw composite three-channel images and reconstructed composite chemical images of blood (red), lipid (green), and protein (blue) at increasing imaging depth from the surface to 54 μm deep, at 3 μm increments. In the raw composite images, blood vessels show up in all three channels and lipids show up in the 2850 and 2950 $cm^{-1}$ channels. Multiplex measurement and linear algebra analysis allow us to disentangle these mixed contributions and recover concentration maps of individual components.

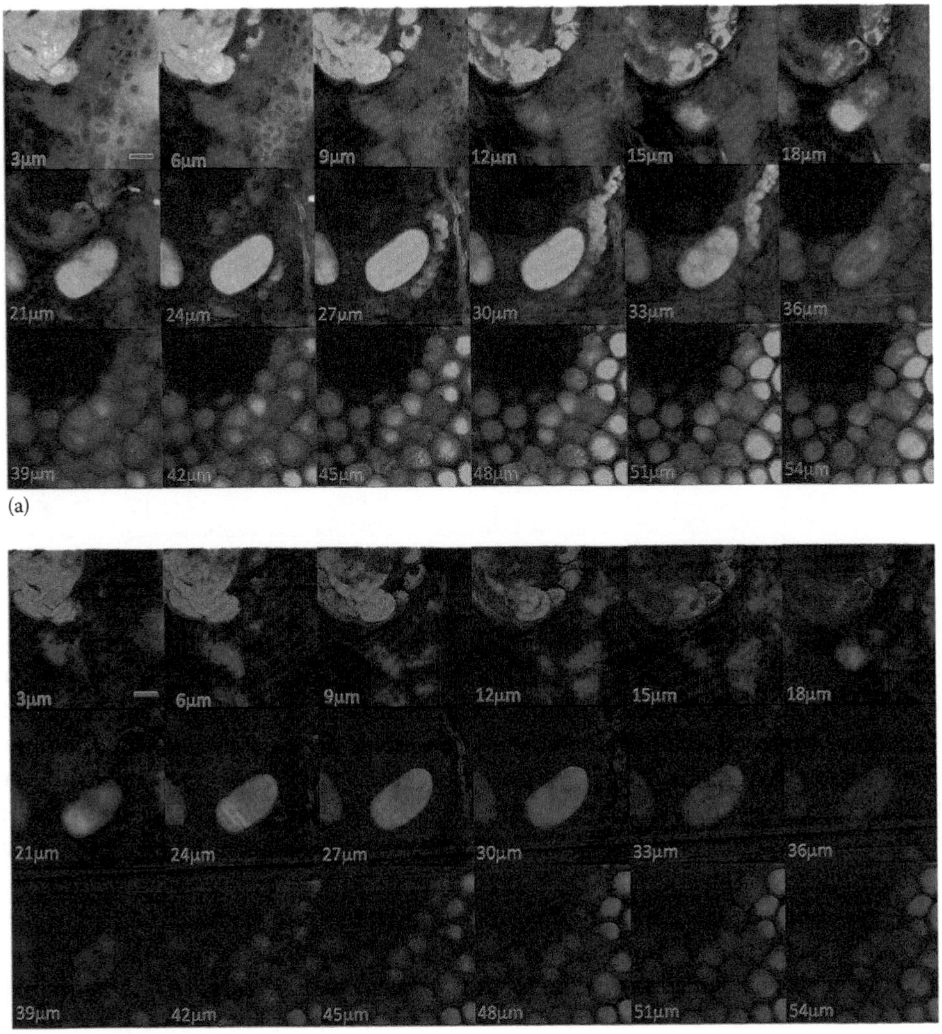


(a)

(b)

FIGURE 16.15 (a) Depth-resolved composite images of multiplex SRS imaging at 2780, 2850, and 2950 cm$^{-1}$. The layered structure of the skin can be delineated based on the varied SRS contrasts. Scale bar: 20 μm. (b) Depth-resolved composite images of calculated chemical maps: blood (red), lipids (green), and protein (blue). Scale bar: 20 μm.

We can observe many interesting signatures in the obtained chemical maps. At the very top of the stratum corneum, there is a layer of lipids which serves as a barrier to keep water inside and keep the bacteria out (Norlen, 2001). At slightly deeper penetration, we can see the characteristic shape of the sebaceous gland on the upper left. Sebaceous glands contain a lot of lipids. It surrounds hair shaft and serves the function of lubricating skin and hair by secreting an oily, waxy substance called sebum. Sebum is primarily composed of triglycerides (TAG), wax esters, and squalene (Thody and Shuster, 1989). Surrounding the sebaceous gland is a few layers of cells that are not rich in lipids. We see clearly the nuclei of epidermal and dermal cells, which show up as dark regions. The nuclear size becomes smaller at deeper skin layers.

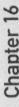

Blood vessels start to appear in the dermis layer at a depth of 20–40 μm. At 50 μm depth, we are already seeing the subcutaneous layer with many fat cells. Each fat cell has a dominant unilocular droplet. It is interesting to note that some fat cells have finer structure: there are many small droplets coated on the surface. The connective regions between different fat cells are concentrated in protein. Some of these features are not easily detectable under a single band SRS imaging experiment. We believe with the multiplex capability, SRS permits detailed structure imaging of skin tissue. When combined with other multiphoton techniques such as multiphoton fluorescence and second harmonic generation, it could be an important tool in monitoring tissue structure alteration during pathological changes such as cancer development. It is also possible to use multiplex SRS as an *in vivo* diagnosis tool because it allows simultaneous imaging of nuclear structure (based on $CH_2$–$CH_3$ contrast) and cytoplasm, offering the label-free equivalent of histology.

## 16.5 SRS Spectral Imaging with Chirped Pulse Excitation

In the previous two sections, we described two completely different approaches in embedding spectroscopy information into SRS imaging. These approaches work well for systems that have simple and known composition. When the system under study becomes more complicated or there is an increasing need for richer chemical information, it is desirable to obtain spectroscopic information directly. This return to spectroscopy mode seems straightforward but practically it is difficult to implement. The most obvious obstacle is limitation in lasers. Unlike spontaneous Raman which can work with any type of narrowband lasers, SRS requires two perfectly synchronized pulsed lasers. In order to obtain spectroscopy information, at least one of the pulsed lasers must have a large spectral bandwidth that is commensurate with the desired Raman spectral range. Even though electronically synchronized picosecond and femtosecond lasers allow us to obtain some spectroscopy information, the inherent timing jitter and limitation in multiplex detection still prevent the acquisition of SRS spectra. One additional drawback that is more subtle is that SRS intensity depends linearly on the optical power in both pump and Stokes beam. If only one of them has a large bandwidth, the power that is actually utilized for excitation of any particular Raman band is quite low, reducing the signal-to-noise ratio.

In this chapter, we present a SRS spectral imaging approach that utilizes two chirped femtosecond pulse trains to achieve simultaneous high spectral resolution and large Raman spectral coverage in SRS imaging. By serially sweeping the Raman resonance, SRS spectra can be acquired for every individual pixel in the SRS image. Similar approaches have been used in CARS imaging, first demonstrated by Zumbusch and co-workers (Hellerer et al., 2004). Later on, other groups have adopted this technique for CARS spectral imaging based on the use of fiber wavelength shifting sources (Pegoraro et al., 2009; Chen et al., 2011). However, the majority of demonstrations are still limited to imaging of samples with strong signals, such as polymer beads or lipid mixtures. Even in those cases, the CARS spectrum is still distorted due to the non-resonant background contribution. Complicated phase retrieval techniques such as the maximum entropy method have to be employed to recover the resonant part of the multiplex CARS spectrum (Hilde et al., 2006; Rinia et al., 2007).

### 16.5.1 Theoretical Basis of Chirped Pulse Excitation

To resolve the conflict of narrow spectra resolution and large spectra bandwidth, chirped pulse excitation, known as spectral focusing, has been developed for CARS microscopy (Hellerer et al., 2004; Knutsen et al., 2004; Pegoraro et al., 2009). Recently, attempts have also been made to use spectral focusing for SRS microscopy (Andresen et al., 2011; Beier et al., 2011). Here we use a simple harmonic oscillator model to derive the analytical formula for chirped pulse excitation SRS.

Raman spectroscopy probes the vibration of chemical bond, which can be described by a harmonic oscillator driven by optical frequencies (Boyd, 2003):

$$\frac{d^2q}{dt^2}+2\gamma\frac{dq}{dt}+\omega_v^2 q=\frac{\alpha E}{m} \tag{16.13}$$

where

$\alpha$ is the polarizability of the vibrational band
$E$ is the input electrical field
$m$ is the mass parameter of the two atoms

The input electric field for two chirped Gaussian femtosecond pulses can be expresses as:

$$E=\frac{A_1 e^{i(k_1 z-\omega_1 t)}}{\sqrt{\tau}e^{-t^2/2\tau^2}e^{-ibt^2}}+\frac{A_2 e^{i(k_2 z-\omega_2 t)}}{\sqrt{\tau}e^{-t^2/2\tau^2}e^{-ibt^2}}+\text{c.c.} \tag{16.14}$$

Here we assume both pulses are linearly chirped with the same chirp parameter $b$. $A$, $k$, $\omega$, and $\tau$ represent the amplitude, wavevector, wavenumber, and pulse duration of the propagating pulses, respectively. Subscripts 1 and 2 denote pump and Stokes pulses, respectively. c.c. represents complex conjugate. Equation 16.13 can then be solved by a Fourier transform:

$$q^*(z,\Omega)=\frac{\sqrt{\pi}(\alpha/m)A_1^* A_2 e^{-(\Omega+\omega_1-\omega_2)\tau^2/4}e^{-iKz}}{\omega_v^2-\Omega^2-2i\gamma\Omega} \tag{16.15}$$

where $K=k_1-k_2$, and we only care about the driving force at the difference frequency. It has the same form as that of excitation with unchirped pulses of the same pulse duration because the slowly varying phase of the two pulses was exactly canceled. When the induced polarization $N\alpha q^* E_1$ is heterodyne mixed with the input electric field $E_2$, we obtain the SRS modulation signal

$$\Delta I_{SRS}\propto\frac{\alpha^2|A_1A_2|^2}{\sqrt{1/4\tau^2+b^2\tau^2}}\int_{-\infty}^{\infty}d\omega\int_{-\infty}^{\infty}d\bar{\omega}\frac{e^{-(\bar{\omega}+\omega_1-\omega_2)^2\tau^2/4}e^{-(\omega-\bar{\omega}-\omega_1)^2/(2/\tau^2+i4b)}e^{-(\omega-\omega_2)^2/(2/\tau^2-i4b)}}{\omega_v^2-\bar{\omega}^2-2i\gamma\bar{\omega}} \tag{16.16}$$

Chapter 16

Numerical simulation shows that the signal is independent of the chirp parameter for pulses with fixed pulse duration, suggesting that picosecond narrowband SRS would have the same signal size as femtosecond SRS when the pulses are chirped to the same pulse duration. Intuitively this result makes sense because at any instant in time the molecular bond is driven at the same frequency difference and amplitude by the two pulses, independent of their respective frequency, and the total signal is a result of the integration of the interaction time of the two pulses.

The advantage of using chirped pulse excitation comes in when an adjustable time delay $t_0$ is introduced. The input electric field in Equation 16.2 can be rewritten as:

$$E = \frac{A_1 e^{i(k_1 z - \omega_1 t + \omega_1 t_0)}}{\sqrt{\tau} e^{-(t-t_0)^2/2\tau^2} e^{-ib(t-t_0)^2}} + \frac{A_2 e^{i(k_2 z - \omega_2 t)}}{\sqrt{\tau} e^{-t^2/2\tau^2} e^{-ibt^2}} + \text{c.c.} \tag{16.17}$$

We can see that the translation in time of one pulse effects a change in oscillation frequency from $\omega_1 - \omega_2$ to $\omega_1 - \omega_2 - 2bt_0$ with the accompanying signal decrease of $\exp[-(t_0/2\tau)^2]$. The resolution of SRS spectroscopy using chirped pulses is determined by the actual pulse duration $\Delta\tau$ (at FWHM), which is directly related to the chirp parameter $b$ for pulses with known optical bandwidth $\Delta\lambda$ (full width at half maximum, FWHM). Therefore, by sweeping the pulse delay, we are effectively probing a sequence of Raman shifts, allowing us to obtain the full spectroscopy information allowed by the spectral bandwidth of the two pulses.

## 16.5.2 Instrumentation

Two synchronized femtosecond laser sources are used (Figure 16.16a). One is a home-built Yb:KGW oscillator mode locked by a semiconductor saturable absorber mirror (SESAM). It is operated at 1040 nm center wavelength with 10 nm bandwidth and 7 W output power. About 1 W of the oscillator output is used as the Stokes beam and the

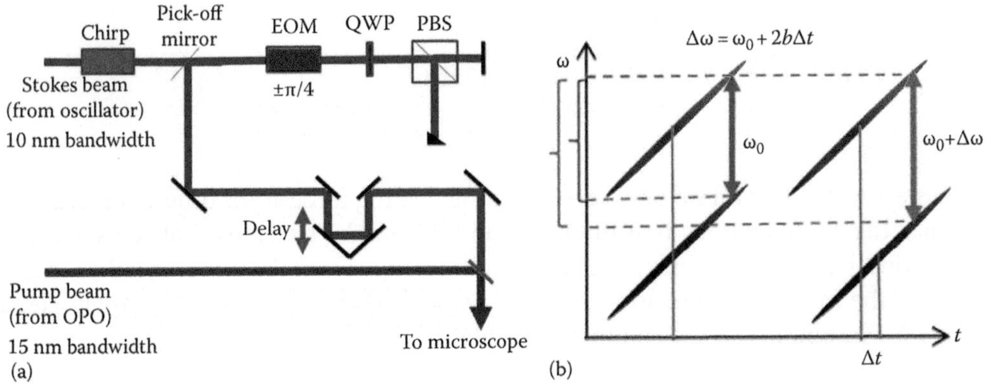


FIGURE 16.16 (a) Schematic diagram of SRS spectral imaging setup based on chirped femtosecond lasers. Both pump and Stokes beams are chirped using long glass rods. The Stokes beam is amplitude modulated at high frequency. EOM, electro-optical modulator; QWP, quarter-wave plate; PBS, polarizing beam splitter. (b) Relationship between time delay and probed Raman frequency shift for two linearly chirped femtosecond pulses.

rest is frequency doubled to 520 nm by a temperature-stabilized LBO crystal. The green light is then used to pump a home-built OPO. The OPO output is used as the pump beam. It is tunable from 750 to 980 nm by adjusting the cavity length and crystal temperature. Output power is about 300–400 mW. The bandwidth of the OPO output is also adjustable by changing cavity dispersion. For C–H stretching imaging, we typically tune the center wavelength to 796 nm and adjust for a bandwidth of 15 nm. The Stokes beam passes through an electro-optical modulator (EOM, Thorlabs) operating at 18.7 MHz. In combination with a quarter waveplate and a polarizing beamsplitter, close to 100% amplitude modulation is achieved. We double pass the Stokes beam through the EOM and then combine it with the pump beam on a long pass dichroic filter (CARS-LP, Chroma). Both beams are chirped with long NSF57 glass blocks (Casix). A total of 36 and 43 cm of NSF57 are used for the Stokes and pump beam, respectively. A motorized delay stage (Newport) is placed in the Stokes beam path to adjust the interpulse delay between pump and Stokes beam.

After the two beams are overlapped spatially and temporally, they are sent into a laser scanning confocal microscope (Olympus IX61/FV300). A water immersion objective (UPLSAPO 60XW, Olympus) is used to focus the light onto the sample. After the sample, the transmitted beams are collected by a high NA condenser (NA = 1.4, Nikon), and then filtered by a bandpass filter (CARS 890/220 m, Chroma) to reject the Stokes beam. The pump beam is detected by a $7 \times 6$ mm$^2$ photodiode (Advanced Photonix) biased at 50 V. The photocurrent is filtered by a 20 MHz bandpass filter (Minicircuits) and then fed into a home-built LIA (Saar et al., 2010a). The output of the LIA is directly read out by the Olympus data acquisition board and displayed as an image.

The major difference between this setup and a narrowband SRS setup is the addition of a delay stage. It allows scanning of the pump and Stokes pulse delay, which is linearly proportional to the probed Raman frequency as described in the previous section. Figure 16.16b shows a diagram depicting the chirped pulse excitation. The frequency difference between the linearly chirped pump and Stokes pulses is concentrated in a band much narrower than the bandwidth of either pulse. This frequency difference corresponds to the Raman band that is excited by the two pulses and it changes with the interpulse delay at a fixed ratio that is related to the chirp parameter $b$. Therefore, by linearly sweeping the time delay and acquiring SRS images synchronously, a three-dimensional dataset can be obtained, with the first two dimensions representing the spatial domain and the third dimension representing the spectral domain.

In order to obtain the SRS spectrum, a calibration is needed to obtain the ratio between time delay and Raman shift. We used samples with known Raman peak positions to obtain the calibration curve. In addition, because both pump and Stokes beam have limited spectral bandwidth, the SRS signal at larger $|\Delta t|$ will be decreased due to the diminished overlap of the two pulses. As a result, the obtained SRS spectrum will be attenuated at its two ends compared to the center. In principle, we can also cancel this effect by dividing the SRS spectrum by a calibration spectrum that is linearly proportional to the pulse overlap. Practically we used a two-photon absorbing sample to obtain this calibration spectrum. Our assumption is that TPA spectra are broader than our spectral range, which is reasonable for a small spectral range of 200–400 cm$^{-1}$ that we are targeting. Figure 16.17 shows the time delay calibration and spectral intensity calibration. We chose a 50%:50% mixture of toluene and cyclohexane as the calibration sample. The two peaks of cyclohexane at

Chapter 16

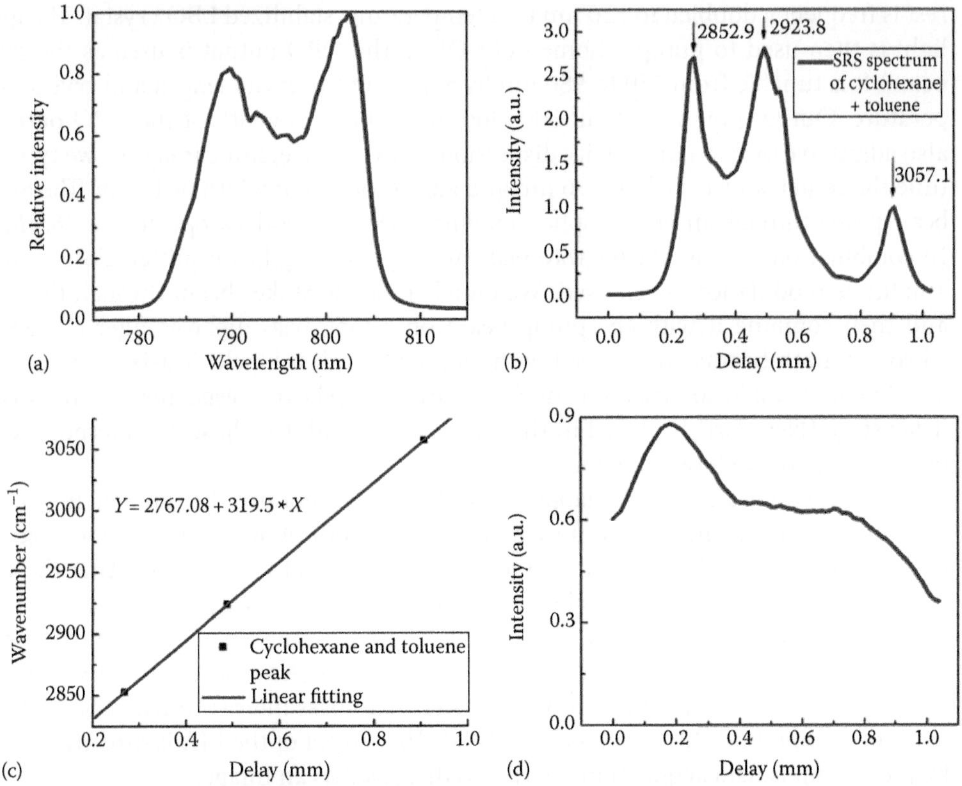


**FIGURE 16.17** (a) Pump beam spectrum from the OPO laser with a full-width half maximum bandwidth of 15–20 nm. (b) Measured SRS signal of 50%:50% cyclohexane/toluene mixture as a function of interpulse delay. The three peak positions are obtained from spontaneous Raman of cyclohexane (2852.9 and 2923.8 $cm^{-1}$) and toluene (3057.1 $cm^{-1}$). These peaks serve as landmarks for Raman shifts. (c) Linear fitting of Raman shift to interpulse delay based on the three landmark Raman peaks. (d) Two-photon absorption (TPA) spectrum of R6G using the same chirped laser pulses for excitation. The TPA spectrum is very broad. The tailing off at early and late delays reflect the smaller overlap of the two pulses at these delays. SRS spectral normalization is done by dividing measured SRS spectrum with the TPA spectrum.

2852.9 and 2923.8 $cm^{-1}$, and the peak of toluene at 3057.1 $cm^{-1}$ are used as landmarks for Raman shift frequency. By linearly fitting these Raman shifts against delay stage position, we were able to obtain a calibration curve that gives the Raman frequency for any delay position (Figure 16.17c). For spectral intensity calibration, we used the TPA of R6G as a measure of pulse overlap (Figure 16.17d). The corrected SRS spectra can be obtained by dividing any measured SRS spectra by this TPA spectrum.

### 16.5.3 Performance Characterization

We first determined the accuracy of Raman spectra obtained by SRS spectral imaging. As demonstrated in earlier SRS development through sequential wavelength tuning, SRS spectra are identical to corresponding spontaneous Raman spectra (Freudiger et al., 2008). Our new approach presents an alternative way of acquiring SRS spectra at much faster speed and wider spectral range, but it should have the same linear dependence on spontaneous Raman. Figure 16.18 shows the SRS spectra of several species measured in

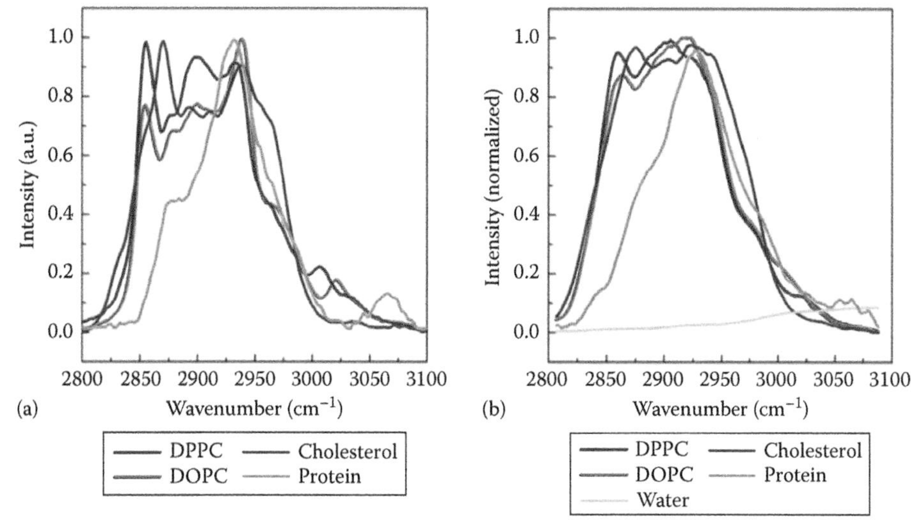


**FIGURE 16.18** (a) Spontaneous Raman spectra of four biomolecules in solutions: DPPC, DOPC, cholesterol, and protein. DPPC, DOPC, and cholesterol are dissolved in deuterated chloroform. Protein is dissolved in deuterated water. (b) SRS spectra measured by spectral focusing of the same four biomolecules and water.

the high wavenumber region. They include 1,2-dipalmitoyl-sn-glycero-3-phosphocholine (DPPC, representing saturated lipids), 1,2-dioleoyl-sn-glycero-3-phosphocholine (DOPC, representing unsaturated lipids), cholesterol (representing cholesterol ester), bovine serum albumin (representing protein). The corresponding spontaneous Raman spectra are measured with a commercial confocal Raman unit.

As can be seen, SRS spectra posses the same features of their spontaneous counterparts, except at reduced resolution. This loss of resolution is due to the deviation from linear chirp for the two excitation lasers. When higher order chirp is present, the frequency difference between pump and Stokes is broadened and changes as interpulse delay is varied. Nonetheless, because SRS calibration spectra can be acquired on the same platform for SRS imaging, this reduced resolution will not affect its capability in quantitative spectral decomposition.

To demonstrate the quantitative spectral decomposition of binary mixtures to their individual components, we prepared binary mixtures of two compounds with strongly overlapping Raman bands: oleic acid and cholesterol. Both are dissolved in deuterated chloroform. The oleic acid concentration was fixed at 0.2 mol/L while the cholesterol concentration varies between 0.004 mmol/L and 0.2 mol/L. Figure 16.19a shows the measured SRS spectra for a total of six different mixtures, as well as two pure solutions. Using the two pure solution SRS spectra as the calibration matrix and applying nonnegative least square analysis as described in Equation 16.7, the concentrations of the two compounds in each mixture are obtained. Because chloroform is highly volatile and the volumes of solutions are small, the concentration could change as we prepare the sample. Therefore, the obtained cholesterol concentration is divided by the oleic acid concentration to cancel out this artifact. The resulting cholesterol concentration is plotted out in Figure 16.19b. We can accurately determine the concentration of cholesterol down to 0.004 mmol/L even in the presence of oleic acid that has 50 times higher concentration, proving the robustness of SRS spectra measurement by spectral focusing.

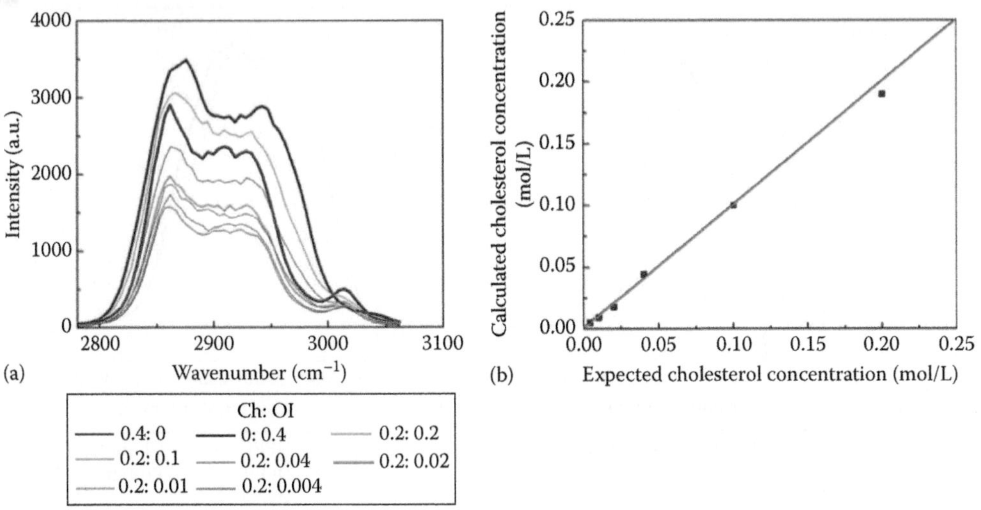


**FIGURE 16.19** (a) Measured SRS spectra of eight solutions comprising of cholesterol (Ch) and Oleic acid (Ol) dissolved in deuterated chloroform. (b) Calculated cholesterol concentration based on spectral decomposition and normalization to oleic acid concentration, plotted against expected cholesterol concentration.

## 16.5.4 Application Examples

The spectroscopic capability of SRS is particularly useful in distinguishing and quantifying different species in an inhomogeneous sample such as a biological cell. Here we demonstrate the spectral imaging capability of SRS using chirped pulse excitation, first in a polymer bead mixture and then in mammalian cells.

A polymer bead mixture is prepared in agar gel. It contains polystyrene microspheres (2 μm diameter), melamine beads (3 μm diameter), and polymethylmethacrylate (PMMA, 1–10 μm diameter). The mixture is smeared onto a cover-slide and then sandwiched by a coverslip. We imaged the beads at sequential Raman shifts by synchronizing the delay stage movement to the frame trigger of the laser scanning microscope. A total of 80 SRS imaging frames spanning 300 cm$^{-1}$ were acquired at about 4 cm$^{-1}$ intervals. Figure 16.20a shows a few selected frames at different Raman shifts. Obviously, the three kinds of beads appear in different frames. We can manually pick out the individual beads and plot their SRS spectra. The spectra are shown in Figure 16.20b and they are very similar to the spontaneous spectra shown in Figure 16.20c. Using the three SRS spectra as three basis spectra and nonnegative least square method to solve for the concentration of the three components, a composite image is created with red, green, and blue representing melamine, polystyrene, and PMMA, respectively (Figure 16.20d). With the spectroscopic information, different chemical components are easily distinguished.

Next, we show SRS spectral imaging of eukaryotic cells and their spectral decomposition. HeLa cells grown in DMEM medium are fixed in 4% formaldehyde and then imaged immediately afterward. A total of 60 SRS frames at about 5 cm$^{-1}$ spacing were acquired for each sample. Each frame takes 1.12 s, resulting in a total acquisition time of less than 70 s. The optical power used is about 40 mW for the pump and 58 mW for

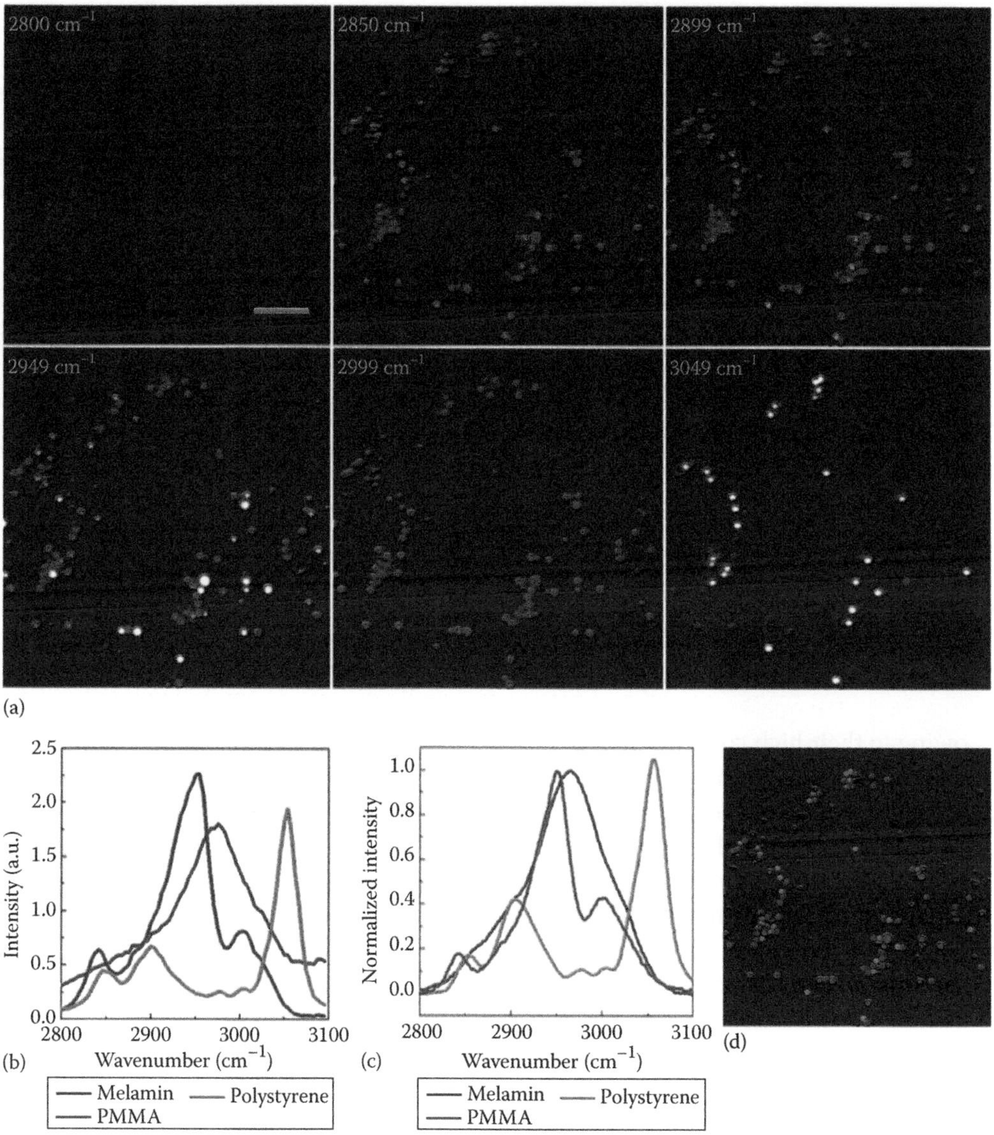


**FIGURE 16.20** (a) Six SRS images acquired at different Raman shifts show spatially varying contrast, revealing the difference in spectra information for the three different kinds of beads. Scale bar: 20 μm. (b) SRS spectra from each kind of beads. (c) Spontaneous Raman spectra of the three beads. (d) Reconstructed chemical image based on the nonnegative least square analysis of the 3D SRS spectral imaging data. Red—melamine; green—polystyrene; blue—PMMA.

the Stokes. Figure 16.21a shows a few representative frames. In order to perform spectral decomposition, we used the calibration spectra of DOPC, DPPC, cholesterol, protein, and water as the basis spectra (shown in Figure 16.18). After nonnegative least-square analysis pixel by pixel, we can reconstruct the chemical maps for the five components, which are shown in Figure 16.21b.

We can observe prominent features distinguishing cytoplasm and nucleus in unsaturated lipids, cholesterol, and protein channels. The nucleoli are also visible in the protein map

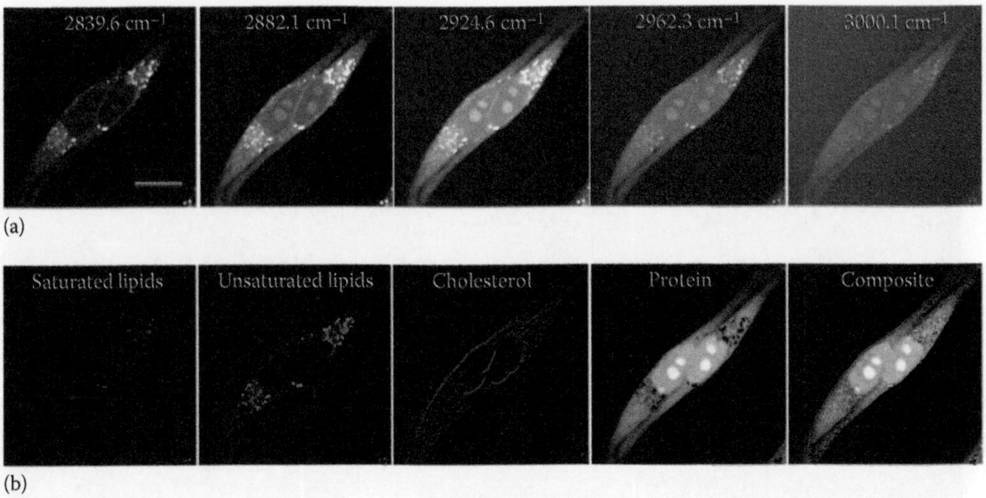


**FIGURE 16.21** (a) SRS spectral imaging at a few selected Raman shifts. The total number of SRS frames is 60 with a frame spacing of 4.1 $cm^{-1}$. (b) Reconstructed chemical maps of saturated lipids, unsaturated lipids, cholesterol, and protein based on spectral decomposition of SRS spectral imaging data. The last image shows the color-coded composite image of the four components. Both lipids and cholesterol are devoid within the nucleus. The saturated lipids and unsaturated lipids are largely localized in droplets with saturated lipids tending to populate the center of the droplets. Scale bar: 20 μm.

owing to their high concentration of protein and nucleic acids. We are not distinguishing protein and nucleic acids here because they have similar Raman spectra in the high wave number region. Another prominent feature is that saturated and unsaturated lipids mostly co-localize in the droplet, which are devoid of protein. Those droplets also contain less cholesterol compared to other cytoplasmic regions. These characteristic features can be only revealed with reliable and complete spectroscopic information.

## 16.6 Conclusion

Multiplex SRS combines the advantage of fast imaging with spectroscopy information. It is an important tool in disentangling the many different chemical components in complex environments such as biological cells or tissue. Without such spectroscopic capability, SRS is limited in its ability for quantitative analysis due to the strongly overlapping nature of Raman spectra. Because SRS is free of the nonresonant background, unlike another commonly used technique multiplex CARS, there is no need for complicated data processing to recover the real Raman spectrum. Thus, it is more reliable and more sensitive. Multiplex SRS typically has two to three orders of magnitude faster imaging speed than either multiplex CARS or confocal Raman.

The three different approaches described in this chapter are distinctive in terms of the technology that is employed, thus offering different benefits and drawbacks. Spectrally-tailored excitation potentially has the fastest imaging speed because it makes use of the full spectroscopic information of all chemical components to selectively target one species. But its output is limited to a single species. Therefore, it could be particularly useful in imaging one species of interest, provided that all species are known a priori.

When information of multiple components is needed, modulation multiplex SRS is a better compromise between speed and spectral information. One additional advantage is that the spectra information is obtained in a parallel fashion. Therefore it is more immune to dynamic motion of the sample. It is particularly useful if the ratio of two compounds is desired. However, due to practical limitations in modulation and demodulation technology, the spectral decomposition is suboptimal and the speed is compromised. The third approach provides the most complete information and it does not require any a priori information of the system. It is most useful for complex systems or unknown systems. But because spectra are measured sequentially, it is slower compared to the other techniques.

## References


Andresen E. R., Berto P., and Rigneault H. 2011. Stimulated Raman scattering microscopy by spectral focusing and fiber-generated soliton as Stokes pulse. *Opt. Lett.* **36**, 2387–2389.

Anita M.-J. 2003. *Biomedical Photonics Handbook*. Boca Raton, FL: CRC Press.

Banerjee A., Sharma R., Chisti Y., and Banerjee U. C. 2002. *Botryococcus braunii*: A renewable source of hydrocarbons and other chemicals. *Crit. Rev. Biotechnol.* **22**, 245–279.

Bégin S., Bélanger E., Laffray S., Vallée R., and Côté D. 2009. In vivo optical monitoring of tissue pathologies and diseases with vibrational contrast. *J. Biophotonics* **2**, 632–642.

Begin S., Burgoyne B., Mercier V., Villeneuve A., Vallee R., and Cote D. 2011. Coherent anti-Stokes Raman scattering hyperspectral tissue imaging with a wavelength-swept system. *Biomed. Opt. Express* **2**, 1296–1306.

Beier H. T., Noojin G. D., and Rockwell B. A. 2011. Stimulated Raman scattering using a single femtosecond oscillator with flexibility for imaging and spectral applications. *Opt. Express* **19**, 18885–18892.

Boyd R. W. 2003. *Nonlinear Optics*. San Diego, CA: Academic Press.

Caspers P. J., Lucassen G. W., and Puppels G. J. 2003. Combined in vivo confocal Raman Spectroscopy and confocal microscopy of human skin. *Biophys. J.* **85**, 572–580.

Chen B.-C., Sung J., Wu X., and Lim S.-H. 2011. Chemical imaging and microspectroscopy with spectral focusing coherent anti-Stokes Raman scattering. *J. Biomed. Opt.* **16**, 021112–021118.

Cheng J.-x., Volkmer A., Book L. D., and Xie X. S. 2002. Multiplex coherent anti-Stokes Raman scattering microspectroscopy and study of lipid vesicles. *J. Phys. Chem. B* **106**, 8493–8498.

Cheng J. X. and Xie X. S. 2004. Coherent anti-Stokes Raman scattering microscopy: Instrumentation, theory, and applications. *J. Phys. Chem. B* **108**, 827–840.

Darzins A., Hu Q., Sommerfeld M., Jarvis E., Ghirardi M., Posewitz M., and Seibert M. 2008. Microalgal triacylglycerols as feedstocks for biofuel production: Perspectives and advances. *Plant J.* **54**, 621–639.

Ellis D. I. and Goodacre R. 2006. Metabolic fingerprinting in disease diagnosis: Biomedical applications of infrared and Raman spectroscopy. *Analyst* **131**, 875–885.

Enejder A. M. K., Scecina T. G., Oh J., Hunter M., Shih W. C., Sasic S., Horowitz G. L., and Feld M. S. 2005. Raman spectroscopy for noninvasive glucose measurements. *J. Biomed. Opt.* **10**, 031114.

Evans C. L. and Xie X. S. 2008. Coherent anti-Stokes Raman scattering microscopy: Chemical imaging for biology and medicine. *Annu. Rev. Anal. Chem.* **1**, 883–909.

Fabregas J., Maseda A., Dominguez A., Ferreira M., and Otero A. 2002. Changes in the cell composition of the marine microalga, *Nannochloropsis gaditana*, during a light: Dark cycle. *Biotechnol. Lett.* **24**, 1699–1703.

Feofanov A. V., Grichine A. I., Shitova L. A., Karmakova T. A., Yakubovskaya R. I., Egret-Charlier M., and Vigny P. 2000. Confocal Raman microspectroscopy and imaging study of theraphthal in living cancer cells. *Biophys. J.* **78**, 499–512.

Freudiger C. W., Min W., Holtom G. R., Xu B., Dantus M., and Sunney Xie X. 2011. Highly specific label-free molecular imaging with spectrally tailored excitation-stimulated Raman scattering (STE-SRS) microscopy. *Nat. Photon.* **5**, 103–109.

Freudiger C. W., Min W., Saar B. G., Lu S., Holtom G. R., He C., Tsai J. C., Kang J. X., and Xie X. S. 2008. Label-free biomedical imaging with high sensitivity by simulated Raman scattering microscopy. *Science* **322**, 1857–1861.

Fu D., Lu F., Zhang X., Freudiger C., Pernik D. R., Holtom G., and Xie, X. S. 2012. Quantitative chemical imaging with multiplex stimulated Raman scattering microscopy. *J. Am. Chem. Soc.* **134**, 3623–3226.

Fu D., Ye T., Matthews T. E., Chen B. J., Yurtserver G., and Warren W. S. 2007a. High-resolution in vivo imaging of blood vessels without labeling. *Opt. Lett.* **32**, 2641–2643.

Fu D., Ye T., Matthews T. E., Yurtsever G., and Warren W. S. 2007b. Two-color, two-photon, and excited-state absorption microscopy. *J. Biomed. Opt.* **12**, 054004.

Haka A. S., Shafer-Peltier K. E., Fitzmaurice M., Crowe J., Dasari R. R., and Feld M. S. 2005. Diagnosing breast cancer by using Raman spectroscopy. *Proc. Natl. Acad. Sci. USA* **102,** 12371–12376.

Hellerer T., Axang C., Brackmann C., Hillertz P., Pilon M., and Enejder A. 2007. Monitoring of lipid storage in *Caenorhabditis elegans* using coherent anti-Stokes Raman scattering (CARS) microscopy. *Proc. Natl Acad. Sci. USA* **104**, 14658–14663.

Hellerer T., Enejder A. M. K., and Zumbusch A. 2004. Spectral focusing: High spectral resolution spectroscopy with broad-bandwidth laser pulses. *Appl. Phys. Lett.* **85**, 25–27.

Hilde A. R., Mischa B., Erik M. V., Chris B. S., and Michiel M. 2006. Spectroscopic analysis of the oxygenation state of hemoglobin using coherent anti-Stokes Raman scattering. *J. Biomed. Opt.* **11**, 050502.

Kano H. and Hamaguchi H. 2005. Ultrabroadband (>2500 $cm^{-1}$) multiplex coherent anti-Stokes Raman scattering microspectroscopy using a supercontinuum generated from a photonic crystal fiber. *Appl. Phys. Lett.* **86**, 121113.

Kim S.-H., Lee E.-S., Lee J. Y., Lee E. S., Lee B.-S., Park J. E., and Moon D. W. 2010. Multiplex coherent anti-Stokes Raman spectroscopy images intact atheromatous lesions and concomitantly identifies distinct chemical profiles of atherosclerotic lipids. *Circ. Res.* **106**, 1332–13341.

Knutsen K. P., Johnson J. C., Miller A. E., Petersen P. B., and Saykally R. J. 2004. High spectral resolution multiplex CARS spectroscopy using chirped pulses. *Chem. Phys. Lett.* **387**, 436–441.

Liu Y., Lee Y. J., and Cicerone M. T. 2009. Broadband CARS spectral phase retrieval using a time-domain Kramers–Kronig transform. *Opt. Lett.* **34**, 1363–1365.

Nan X., Potma E. O., and Xie X. S. 2006. Nonperturbative chemical imaging of organelle transport in living cells with coherent anti-Stokes Raman scattering microscopy. *Biophys. J.* **91**, 728–735.

Norlen L. 2001. Skin barrier structure and function: The single gel phase model. *J. Investig. Dermatol.* **117**, 830–836.

Okuno M., Kano H., Leproux P., Couderc V., Day J. P. R., Bonn M., and Hamaguchi H.-o. 2010. Quantitative CARS molecular fingerprinting of single living cells with the use of the maximum entropy method. *Angew. Chem. Int. Ed.* **49**, 6773–6777.

Pegoraro A. F., Ridsdale A., Moffatt D. J., Jia Y., Pezacki J. P., and Stolow A. 2009. Optimally chirped multimodal CARS microscopy based on a single Ti:sapphire oscillator. *Opt. Express* **17**, 2984–2996.

Pezacki J. P., Blake J. A., Danielson D. C., Kennedy D. C., Lyn R. K., and Singaravelu R. 2011. Chemical contrast for imaging living systems: Molecular vibrations drive CARS microscopy. *Nat. Chem. Biol.* **7**, 137–145.

Ploetz E., Laimgruber S., Berner S., Zinth W., and Gilch P. 2007. Femtosecond stimulated Raman microscopy. *Appl. Phys. B* **87**, 389–393.

Potma E. O., Lin C. Y., Suhalim J. L., Nien C. L., Miljkovic M. D., Diem M., and Jester J. V. 2011. Picosecond spectral coherent anti-Stokes Raman scattering imaging with principal component analysis of meibomian glands. *J. Biomed. Opt.* **16**, 021104.

Puppels G. J., de Mul F. F. M., Otto C., Greve J., Robert-Nicoud M., Arndt-Jovin D. J., and Jovin T. M. 1990. Studying single living cells and chromosomes by confocal Raman microspectroscopy. *Nature* **347**, 301–303.

Rinia H. A., Bonn M., Müller M., and Vartiainen E. M. 2007. Quantitative CARS spectroscopy using the maximum entropy method: The main lipid phase transition. *ChemPhysChem* **8**, 279–287.

Saar B. G., Freudiger C. W., Reichman J., Stanley C. M., Holtom G. R., and Xie X. S. 2010a. Video-rate molecular imaging in vivo with stimulated Raman scattering. *Science* **330**, 1368–1370.

Saar B. G., Zeng Y., Freudiger C. W., Liu Y.-S., Himmel M. E., Xie X. S., and Ding S.-Y. 2010b. Label-free, real-time monitoring of biomass processing with stimulated Raman scattering microscopy. *Angew. Chem.* **122**, 5608–5611.

Slipchenko M. N., Le T. T., Chen H., and Cheng J.-X. 2009. High-speed vibrational imaging and spectral analysis of lipid bodies by compound Raman microscopy. *J. Phys. Chem. B* **113**, 7681–7686.

Thody A. J. and Shuster S. 1989. Control and function of sebaceous glands. *Physiol. Rev.* **69**, 383–416.

Tsurui H., Nishimura H., Hattori S., Hirose S., Okumura K., and Shirai T. 2000. Seven-color fluorescence imaging of tissue samples based on Fourier spectroscopy and singular value decomposition. *J. Histochem. Cytochem.* **48**, 653–662.

Wang H., Fu Y., Zickmund P., Shi R., and Cheng J.-X. 2005. Coherent anti-Stokes Raman scattering imaging of axonal myelin in live spinal tissues. *Biophys. J.* **89**, 581–591.

Wang M. C., Min W., Freudiger C. W., Ruvkun G., and Xie X. S. 2011. RNAi screening for fat regulatory genes with SRS microscopy. *Nat. Methods* **8**, 135–138.

Wu H., Volponi J. V., Oliver A. E., Parikh A. N., Simmons B. A., and Singh S. 2011. In vivo lipidomics using single-cell Raman spectroscopy. *Proc. Natl Acad. Sci. USA* **108**, 3809–3814.

Wurpel G. W. H., Schins J. M., and Muller M. 2002. Chemical specificity in three-dimensional imaging with multiplex coherent anti-Stokes Raman scattering microscopy. *Opt. Lett.* **27**, 1093–1095.

Wurpel G. W. H., Schins J. M., and Muller M. 2004. Direct measurement of chain order in single phospholipid mono- and bilayers with multiplex CARS. *J. Phys. Chem. B* **108**, 3400–3403.

Yamakoshi H., Dodo K., Okada M., Ando J., Palonpon A., Fujita K., Kawata S., and Sodeoka M. 2011. Imaging of EdU, an alkyne-tagged cell proliferation probe, by Raman microscopy. *J. Am. Chem. Soc.* **133**, 6102–6105.

# III

# Applications

# 17. Imaging Myelin Sheath *Ex Vivo* and *In Vivo* by CARS Microscopy

Yan Fu, Yunzhou (Sophia) Shi, and Ji-Xin Cheng





*Coherent Raman Scattering Microscopy*. Edited by Ji-Xin Cheng and X. Sunney Xie © 2013 CRC Press/Taylor & Francis Group, LLC. ISBN: 978-1-4398-6765-5.

## 17.1 Introduction

Myelin sheath is a multiple-layer membrane wrapping around axons and is critical for rapid and efficient transmission of impulses along the nerve. Myelin sheath is formed by oligodendrocytes in the central nervous system and Schwann cells in the peripheral nervous system. The degradation of myelin sheath, known as demyelination, leads to various degrees of axonal conduction impairment and debilitating functional deficits seen in diseases and trauma such as multiple sclerosis (MS), Guillain–Barrè syndrome, and spinal cord injury (SCI) (Blight 1985; Smith et al. 1999; Compston and Coles 2002). The term "demyelination" dates back to the nineteenth century when researchers found demyelination of nerves in brain and spinal cord as the characteristic pathological lesion of MS. Today, the understanding of the molecular mechanisms underlying demyelination still remains limited, partially due to the lack of microscopic tools for high-resolution imaging of myelin in its natural state. The lack of an effective readout of the disease has also hindered the development of effective therapeutic interventions to remyelination.

The current methods of imaging myelin include electron microscopy (EM), histology, immunohistochemistry, two-photon excited fluorescence (TPEF), magnetic resonance imaging (MRI), and positron emission tomography (PET). EM has the highest resolution that resolves individual myelin layer. Histology and immunohistochemistry allow detection of demyelinating lesions at cellular level on stained tissue slices. These three methods require fixed tissues, staining, and complicated sample preparation that prevents *in vivo* observation. TPEF allows *in vivo* imaging of demyelination on rodents with fluorescent dye–labeled myelin or green fluorescent protein (GFP)-labeled glial cells. The image quality is limited by fluorescence labeling efficiency. MRI detects the demyelinating lesions based on the difference in apparent diffusion coefficients of water between normal and demyelinating tissue. Sometimes an imaging contrast agent, gadolinium, is used to facilitate imaging. PET monitors the radiolabeled dyes, which accumulate at lesion sites. MRI and PET are being used on patients due to their noninvasiveness and long-term monitoring capability, but their resolution is limited to 100 μm. To monitor cellular activities during demyelination and remyelination *in vivo*, an imaging tool with minimal invasiveness, intrinsic contrast, high speed, and good resolution is needed.

Coherent anti-Stokes Raman scattering (CARS) microscopy has become a powerful tool for studying demyelination and remyelination in fresh tissues and live animals for its noninvasiveness, inherent three-dimensional (3D) submicron resolution, high imaging speed, and multimodality. As one of Raman scattering imaging methods, CARS detects intrinsic vibrations of chemical bonds without any labeling. To image myelin, the frequency difference of the two lasers is tuned to 2840 $cm^{-1}$ to match the symmetric $CH_2$ stretching vibration. The nonlinear dependence on excitation intensity allows CARS signal to be generated only at the focal plane, which gives inherent 3D submicron resolution to observe fine structures such as nodes of Ranvier. Moreover, laser scanning facilitates real-time imaging of fresh tissues and live animals at a speed of tens of frames per second (Misgeld and Kerschensteiner 2006). Also, as a nonlinear optical (NLO) modality, CARS microscopy utilizes near IR excitation, which provides both superior optical penetration into tissues as well as reduced photodamage (Helmchen and Denk 2002). Furthermore, other NLO imaging modalities can be integrated onto a CARS microscope for probing different cellular structures. Over the past few years,

CARS microscopy has been applied to monitoring myelin in various chemical and animal models of demyelination.

In this chapter, we first introduce the CARS setup, and then describe the method for imaging myelin *ex vivo*, followed by the method for *in vivo* myelin imaging. Finally, we describe mechanistic study of demyelination enabled by CARS microscopy.

## 17.2 Multimodal CARS Microscope

### 17.2.1 CARS-Based Multimodal Imaging

The CARS setup generally utilizes two laser beams, the pump beam ($\omega_p$) and the Stokes beam ($\omega_s$). Several electromagnetic fields are generated at the same time as the CARS process at $\omega_p + \omega_p - \omega_s$, which produce multiple NLO signals known as sum frequency generation (SFG) at $\omega_p + \omega_s$, second harmonic generation (SHG) at $2\omega_p$ or $2\omega_s$, stimulated Raman gain (SRG) at $\omega_p - \omega_p + \omega_s$, and stimulated Raman loss (SRL) at $\omega_s - \omega_s + \omega_p$. Using a multimodal CARS microscope, Fu et al. first observed astroglial filaments arising from SFG and SHG (Fu et al. 2007b). The microtubule filaments inside the axon at the nodal area were also clearly visualized by SFG (Huff et al. 2008). In addition, TPEF is used for imaging vasculature tracers; fluorescent probe-labeled ions or proteins such as dextran-FITC; $Ca^{2+}$ and $K^+$ ion indicators; and GFP-labeled oligodendrocyte precursor cells, Schwann cells, and axons. To perform multimodal imaging of nervous system, the frequency of the pump or Stokes laser is selected according to TPEF excitation spectrum. The frequency difference of the two lasers is tuned to 2840 $cm^{-1}$ to match the symmetric $CH_2$ stretching vibration. By using suitable band-pass filters at different detection channels, CARS, TPEF, SFG, and SHG signals can be collected simultaneously.

### 17.2.2 Dual-Scanning CARS Microscope

To increase the speed and area of imaging, the Cheng lab built a dual-scanning CARS microscope (Fu et al. 2008). In this microscope, the combination of polygon mirror with a galvanometer provides laser scanning at a speed of 20 frames per second. A sample stage driven by two stepper motors fulfills large-area mosaic mapping. The two beams are collinearly combined on a dichroic mirror and directed to a multifaceted polygon mirror, which provides fast laser scanning along the x-axis (Rajadhyaksha et al. 1999). The scanned beams pass through a telescope to be focused onto a plane galvanometer, which provides the y-axis scan. The beams are then expanded by a second telescope to fill the back aperture of a 40× water immersion objective. Considering that the width of the mouse spinal cord is about 1.2–1.5 mm, the entire lumbar region will take approximately 60 frames. A whole horizontal mouse brain slice (Figure 17.1) was mapped by computer-controlled stitching of 9579 partially overlapped CARS images acquired with a 20× air objective.

### 17.2.3 Integration of CARS Imaging with Compound Action Potential Recording

To correlate the physiological properties of nervous tissue with morphological changes, CARS imaging was combined with electrophysiological measurement. As shown in Figure 17.2, compound action potential (CAP) is measured using a double

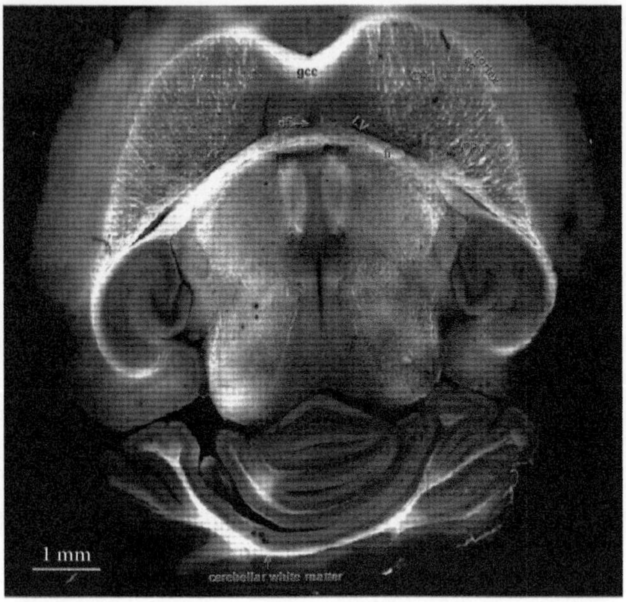


**FIGURE 17.1** Mosaic CARS image of a horizontal mouse brain slice. The whole image contains about 9579 partially overlapped images acquired with a 20× objective in a dual-scanning CARS microscope. The gross anatomy of mouse brain was depicted. gcc, genu of corpus callosum; CPu, caudate putamen; ec, external capsule; dfi, dorsal fornix; fi, fimbria hippocampus; LV, lateral ventricle. (From Fu, Y., Huff, T.B. et al., *Opt. Express* 16, 19396–19409, 2008. With permission of Optical Society of America.)

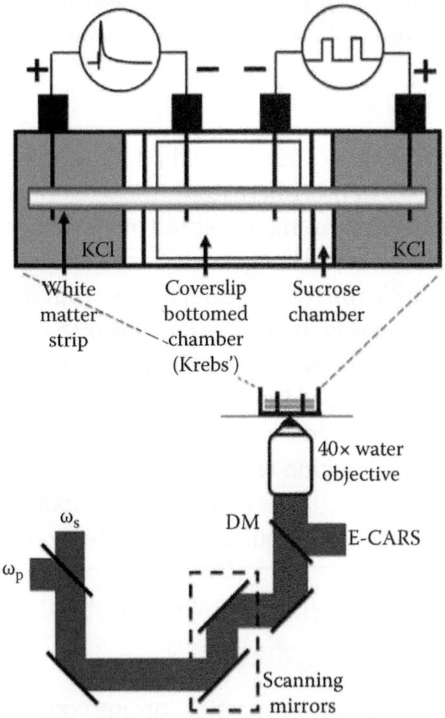


**FIGURE 17.2** Experimental setup for simultaneous E-CARS imaging, high-frequency stimulation, and compound action potential recording. DM, dichroic mirror.

sucrose gap chamber (Shi et al. 1999). The center chamber is specially designed with a glass coverslip bottom for optical penetration to allow simultaneous imaging of the same sample by CARS. The center chamber is filled with Krebs' solution enriched with oxygen. The white matter strip is draped across the electrodes in the two end chambers and allowed to rest on the glass coverslip bottom of the center chamber. The two end chambers are filled with 120 mM KCl. The white matter strip can be maintained at 37°C either by circulating a preheated Krebs' solution to the center chamber at a flow rate of 2 mL/min or by embedding a heating coil in the wall of the chamber. A constant infusion of 95%$O_2$/5%$CO_2$ into the Krebs' solution is needed to maintain the physiological condition during an experiment. The electrophysiological measurement is controlled by a LabView™ program and can be synchronized with the optical imaging by utilizing the same Transistor-Transistor Logic (TTL) signal that triggers the optical scanning.

## 17.3 Method for *Ex Vivo* CARS Imaging of Myelin

### 17.3.1 Sample Preparation

CARS imaging of myelin has been successfully implemented on fixed tissue slices, frozen sections, cultured tissue slices, and fresh tissues. Due to tissue scattering, the optical penetration depth is limited to about 100 μm. Therefore, to image myelin structures inside brain, spinal cord, or sciatic nerves, samples are sliced into slices of a few hundred micron thickness using vibrotome or cryotome. Fresh tissues are placed on a glass coverslip bottom cell culture dish maintained in 37°C oxygenated Krebs' solution. Fixed or frozen tissues are usually sandwiched by two glass coverslips filled with medium in between to avoid dehydration. The medium used is usually phosphate buffer solution. If other media are used, chemicals with rich C–H groups should be avoided as the CARS signal from myelin is based on C–H stretch vibration.

### 17.3.2 Characterization of Axonal Myelin

*Forward versus epi-detected CARS imaging*: For tissue slices or isolated thin samples such as spinal cord and sciatic nerve, either epi-detected CARS (E-CARS) or forward-detected CARS (F-CARS) can be used. If the laser beams are focused at the equatorial plane of the axon, then the CARS radiation from myelin sheath predominantly goes forward. In this case, E-CARS signal mainly arises from the back-reflected F-CARS signal by the myelin/fluid interface. The F-CARS signal is partially scattered by the thick tissue. For an isolated guinea pig spinal tissue, the average intensity of F-CARS is 18.9 times that of E-CARS (Figure 17.3) according to Wang et al. (2005). In contrast, if the laser beams are focused at the top or bottom of the myelin fibers, then the average F- to E-CARS intensity ratio becomes 2.9, much smaller than the ratio for the equatorial area. For imaging thick tissue such as the whole brain, epi-detected CARS should be used.

*CARS spectral profile*: The CARS signal of axonal myelin (Figure 17.4A) arising from symmetric $CH_2$ stretch vibration appears at 2840 $cm^{-1}$ with a resonant signal to nonresonant background ratio of 10:1 in CARS spectra (Figure 17.4B) (Wang et al. 2005). A weak

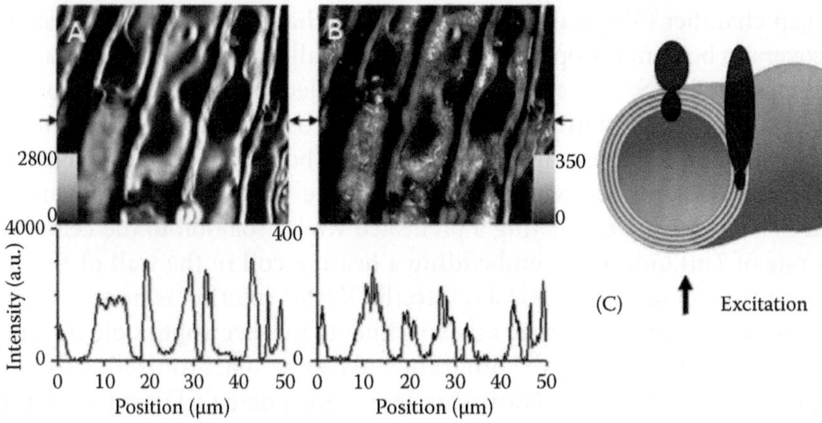


**FIGURE 17.3** (A) Forward-detected CARS (F-CARS) and (B) epi-detected CARS (E-CARS) images of axons in a spinal cord sample. The intensity profiles along the lines indicated by arrows are shown below the images. (C) Schematic radiation profile of CARS from the top and equatorial part of a myelin fiber. (Reprinted from *Biophys. J.* 89, Wang, H., Fu, Y., Zickmund, P., Shi, R. and Cheng, J.-X., Coherent anti-Stokes Raman scattering imaging of axonal myelin in live spinal tissues 581–591. Copyright 2005, with permission from Elsevier.)

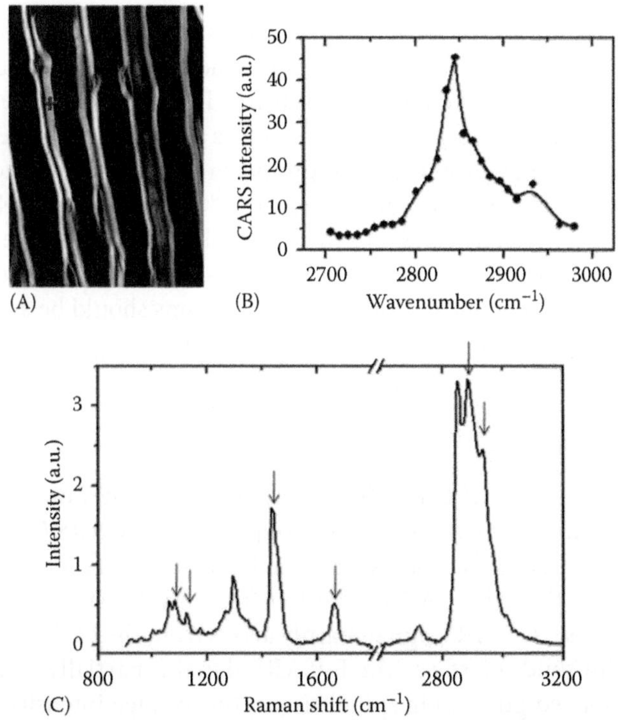


**FIGURE 17.4** (A) CARS image of myelin in a fresh spinal tissue. Red mark indicates the position where spectra are recorded. (B) CARS spectra of myelin. (Reprinted from *Biophys. J.* 89, Wang, H., Fu, Y., Zickmund, P., Shi, R. and Cheng, J.-X., Coherent anti-Stokes Raman scattering imaging of axonal myelin in live spinal tissues 581–591. Copyright 2005, with permission from Elsevier.) (C) Confocal Raman spectra of myelin lipid. The peaks at 1076, 1122, 1445, 1650, 2885, and 2930 $cm^{-1}$ from left to right are indicated by black arrows.

peak at 2930 $cm^{-1}$ observed in both spinal cord tissue and myelinated sciatic nerves is considered to be the $CH_3$ stretch mode of proteins, which constitute 30% weight of myelin sheath. The ratios among signal from symmetric $CH_2$, asymmetric $CH_2$ (2885 $cm^{-1}$), and $CH_3$ stretching vibrations indicate the lipid packing condition related to the diseased and healthy stages of myelin (Fu et al. 2011). Confocal Raman spectral analysis on the CARS platform (Slipchenko et al. 2009) reveals the conformational and/or chemical changes of hydrocarbon chains of lipids (Figure 17.4C). Especially, the ratio of $I_{2930}/I_{2885}$ reflects the intermolecular chain disorder. The ratio of $I_{1650}/I_{1445}$ represents the ratio of C=C bonds to C–H bonds in lipid acyl chains, indicating the degree of lipid unsaturation. The ratio of $I_{1122}/I_{1076}$ is related to the ratio of the *trans* and *cis* conformation populations of C–C bonds, examining the intramolecular chain ordering in lipid hydrocarbon chain. These ratios help identify myelin status during demyelination and remyelination (Fu et al. 2011).

*Polarization characterization*: The compact multilamellar structure of myelin sheath and highly ordered $CH_2$ groups in lipid membranes give rise to polarization-sensitive CARS signal. By changing the polarization of excitation beams with respect to the orientation of $CH_2$ groups in the lipid membranes, one can control the relative amplitudes and phases of CARS signal (Wang et al. 2005; Fu et al. 2008). For quantitative analysis of myelin amount, the polarization modulation effect can be minimized by reconstruction of two complementary images with perpendicular excitation polarizations, as shown in Figure 17.5 (Fu et al. 2008), or by using circularly polarized excitation beams (Bélanger et al. 2009). Apart from $CH_2$ groups, when tuned to 3200 $cm^{-1}$, water molecules can be probed through O–H stretch vibration. It has been shown that the intramyelin water molecules are partially ordered with the symmetry axis perpendicular to that of the $CH_2$ groups (Wang et al. 2005).

*Measurement of g ratio*: The g ratio is defined as the ratio between inner diameter of the axon and outer diameter of the myelin wrapping the axon and is widely used to quantify the degrees of myelination. The g ratio can be measured by EM observation of fixed tissues. CARS microscopy allows measurement of g ratio of myelin in its natural state. The laser beams are focused onto the equatorial plane and the CARS intensity profile is used to determine the inner and outer diameters of the myelin fiber. This method was used to monitor lysophosphatidylcholine-induced demyelination in real time (Fu et al. 2007a).

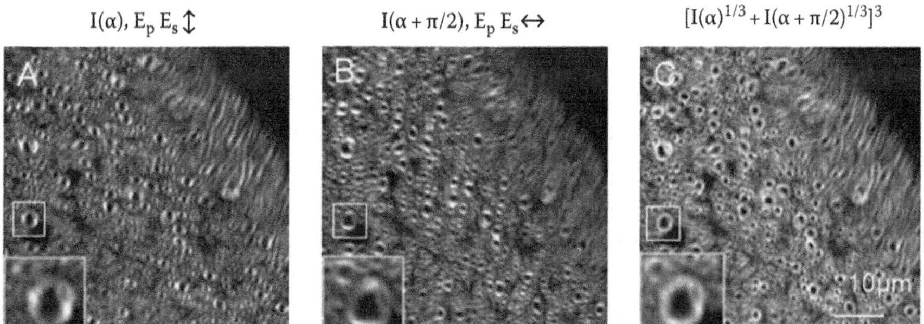


**FIGURE 17.5** Dependence of CARS intensity on excitation polarization. (A) CARS image with vertically polarized excitations showing stronger CARS signal from vertically oriented membranes and fibers. (B) CARS image with horizontally polarized excitations showing stronger CARS signal from horizontally oriented membranes and fibers. (C) Reconstructed CARS images showing independence of CARS signal on excitation polarization. Bar = 10 μm. (From Fu, Y., et al., *Opt. Express* 16, 19396–19409, 2008. With permission of Optical Society of America.)

## 17.4 Method for *In Vivo* CARS Imaging of Myelin in Sciatic Nerve, Spinal Cord, and Brain

Visualization of myelin degradation and regeneration on an individual axon longitudinally is key to understanding the demyelination and remyelination process. We summarize techniques that were used by the Cheng lab to enhance the optical penetration and to permit repetitive imaging of the same location in the following.

### 17.4.1 Enhancement of Optical Penetration

In explanted fresh spinal cord and brain tissue, animals are usually perfused with Krebs' solution to wash away red blood cells; the arachnoid mater, dura mater, and pia mater lose the cerebrospinal fluid in subarachnoid space and reduce the spatial separation; and physical barriers such as bone, muscle, and fat are completely removed. In the *in vivo* environment, all these factors will increase the scattering of both the incident light and the generated signals. To enhance the penetration depth *in vivo*, the following physical and optical methods have been used.

*Miniature lens*: A miniature lens usually has a needle-like tip that encloses objective lens inside and is able to penetrate into the soft tissue to approach the target tissue. For instance, the gradient index (GRIN) lens is available from 0.5 to 2 mm in diameter. The 20× MicroProbe Objective (MPO) (Olympus Inc., Tempe, AZ) lens has a 1.3 mm outer diameter with a working distance of 200 μm. Inserting a miniature lens into a soft sample provides an effective approach to imaging deep tissues with minimal surgery. The GRIN lens has been applied to fluorescence imaging of deep brain tissues (Flusberg et al. 2005), and the MPO has been applied to CARS imaging in spinal cord (Wang et al. 2007; Shi et al. 2011b).

*Use of adaptive optics*: As the laser beams penetrate through live biological tissues, the effect of wave-front distortions greatly reduces the performance of CARS system. Adaptive optics, first envisioned by Horace W. Babcock in 1953, has later been used in astronomical telescopes and laser communication systems to remove the effects of atmospheric distortion (Tyson 1998), as well as in retinal imaging systems to reduce the impact of optical aberrations (Wright et al. 2005). By measuring the distortions in a wave front and compensating with a device such as a deformable membrane mirror or a liquid crystal array, and combining a random search optimization algorithm to improve signal to noise ratio at large sample depths, adaptive optics-combined CARS system has been shown to increase the signal by an average factor of 3 for test samples at a depth of 700 μm and 6 for muscle tissue at a depth of 260 μm (Wright et al. 2007).

*Lysis of red blood cells*: After surgery to remove the bone and tissue above the spinal cord, a layer of red blood cells (RBCs) will deposit at the surgical site, which reduces the optical penetration depth. As RBCs have a different refractive index compared with the lipid membranes we are interested in, the refractive index mismatch can be compensated by using red blood cell lysis buffer. RBC lysis buffer is composed of 8.3 g $NH_4Cl$, 1.0 g $KHCO_3$, and 1.8 mL of 5% ethylenediaminetetracetic acid (EDTA) in 1 L de-ionized (DI) water. After incubating the imaging area with RBC lysis buffer for 5 min after surgery, the RBCs burst due to imbalance of intra- and extracellular ion strength. The buffer will then be removed and the tissue will be rinsed several times with sterile saline. RBC lysis buffer is usually applied to local surface of surgical area, thus minimal toxic effect to animals is expected.

### 17.4.2 Prevention of Scar Formation

As surgical procedures in *in vivo* imaging preparation cause scar formation at the exposure, site, repetitive imaging of the same site becomes extremely difficult over a period of several weeks. One method to reduce scar formation is to apply a layer of sterile 3% agarose gel on top of the surgical area after each imaging session. Agarose gel is a hydrogel with well-established biocompatibility; thus, it has been used as an index-matching material for imaging GFP-labeled neurons in mouse brain cortex and has been recently employed for TPEF imaging of the spinal cord blood vessels to facilitate reimaging after several days.

### 17.4.3 Using Blood Vessels as Landmark

As the field of view of a multiphoton image is limited to a few hundred microns square, thus a "roadmap" of region of interest is needed for quick navigation. A common practice is to utilize anatomical features as landmarks. Blood vessels have been utilized as anatomical markers for repeat imaging due to their unique network morphology (Davalos et al. 2008). During diseased stage, however, the use of a vascular "road map" may be precluded by neovascularization following the injury, which can make identification of the same region suspect unless alternative methods are employed (Dray et al. 2009).

### 17.4.4 *In Vivo* Imaging of Mouse Sciatic Nerve

The sciatic nerve is a perfect candidate for *in vivo* CARS imaging since it is the largest single nerve in mammals and located in the peripheral nervous system isolated from the movement induced by breathing and heartbeat. Compared to the brain and spinal cord, it is relatively easy to access the sciatic nerve with minimal surgery. The sciatic nerve is visible through musculature after opening a small window in the skin. The exposed sciatic nerve is placed on the coverslip chamber to facilitate imaging on an inverted microscope (Huff and Cheng 2007). Tissue hydration is maintained during the imaging process. CARS, SHG, and/or TPEF signal (Figure 17.6) is detected at the

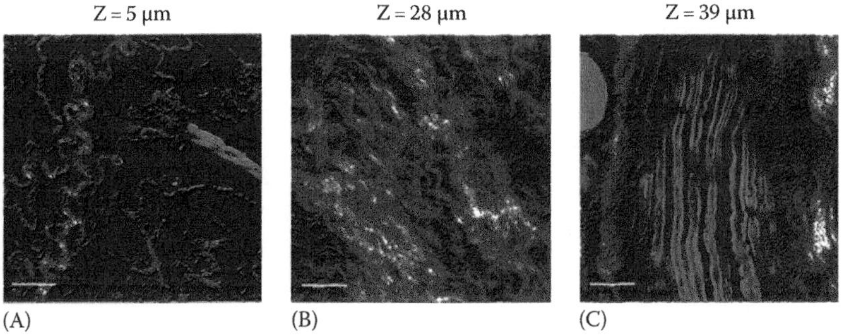


FIGURE 17.6 E-CARS (green) and SHG (red-orange) images of mouse sciatic nerve tissue at varying depths. (A) Z = 5 μm, close to skin. (B) Z = 28 μm. (C) Z = 39 μm. E-CARS probes the fat cells and myelinated axons whereas SHG probes the collagen fibrils surrounding sciatic nerve. Bar = 20 μm. (From Huff, T.B., et al., In vivo coherent anti-Stokes Raman scattering imaging of sciatic nerve tissue, *J. Microsc.*, 2007, 225, 175–182. Copyright Wiley-VCH Verlag GmbH & Co. KGaA. Reproduced with permission.)

Chapter 17

back port of the microscope. By 3D imaging of fat cells that surround the nerve, the underlying contrast mechanisms of *in vivo* CARS are found to arise from interfaces as well as back reflection of forward CARS signal. Unavoidable animal movement is corrected by image stabilization algorithms (Henry et al. 2009; Bélanger et al. 2011).

## 17.4.5 *In Vivo* Imaging of Spinal Cord

Longitudinal *in vivo* CARS imaging of spinal cord has been demonstrated through a careful experimental design to overcome complexities induced by the laminectomy surgery (Shi et al. 2011b). To reduce the amount of tissue to be removed and improve the survivability and recovery time of the animals, the spine was exposed by making an incision through the skin and muscle tissue at T10, where the natural curvature of the spinal cord makes this region more superficial to the skin than other locations. Following exposure of the spinal cord, the spine was needed to be stabilized by a

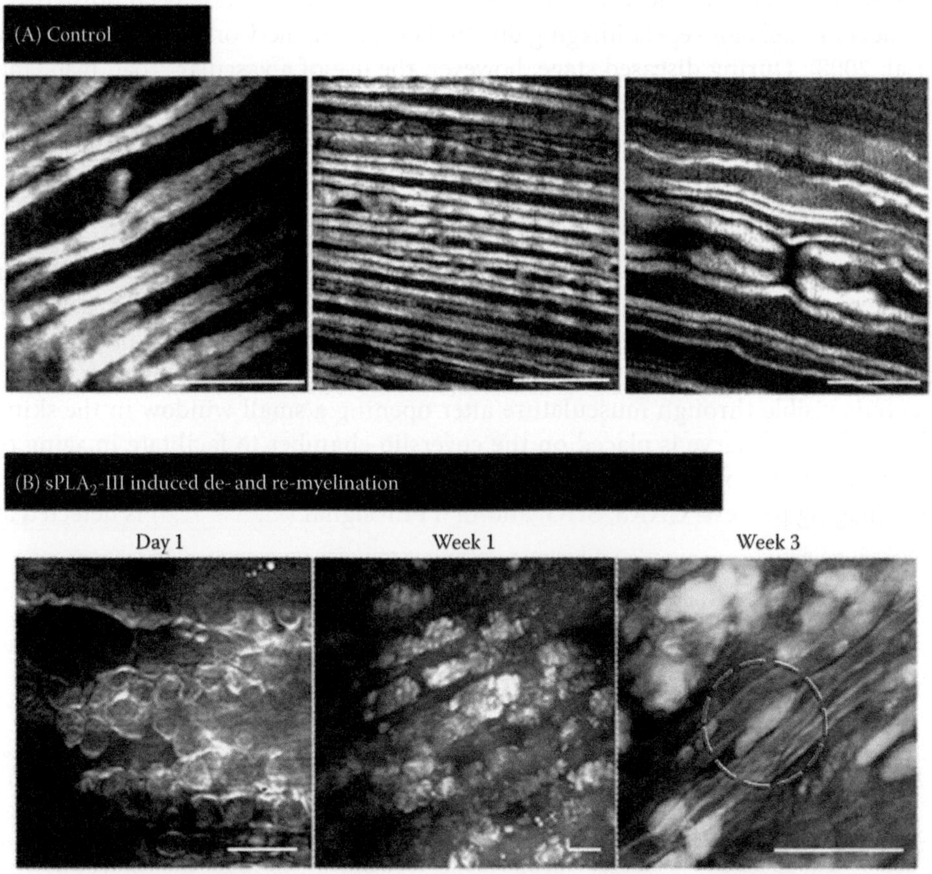


**FIGURE 17.7** Longitudinal *in vivo* CARS imaging of a rat spinal cord. (A) CARS images of healthy myelin in a live rat. (B) Longitudinal CARS images of spinal cord in a rat injected with secretory phospholipase $A_2$ III in the spinal cord. Myelin degradation was observed in 24 h by formation of myelin vesicles. These vesicles appear to be digested by macrophages/microglia 1 week after injection. By 3 weeks postinjection, signs of Schwann cell–mediated remyelination were visible, with elongated cell nuclei (dashed circle) adjacent to myelinated axons at the injection site. Bar = 20 μm.

clamping system, which allowed the animal to breathe freely and at the same time avoided the motion-induced image distortion. To repetitively image the same location at the spinal cord, a cushion of agarose gel was needed to be deposited above the surgical site. Subsequent CARS imaging would be performed by reopening the laminectomy site and carefully removing the agarose gel and a thin layer of red blood cells with RBC lysis buffer. The application of an agarose gel and lysis of red blood cells permitted longitudinal imaging of the same region of the spinal cord white matter for a period of 3 weeks (Figure 17.7).

### 17.4.6 *In Vivo* Imaging of Mouse Brain

*In vivo* imaging of a mouse brain has been demonstrated with invasive surgery (Fu et al. 2008), in which an upright microscope was used with a dipping mode water objective. In the field of craniotomy of brain, only a few myelinated axons with random orientations can be found in the cortex since gray matter is made up of neuronal cell bodies. Because CARS microscopy has limited optical penetration depth approximately 30 μm in both white matter and gray matter, to access the white matter containing bundles of myelinated axons, the gray matter layer in the cortex had to be removed. The exposed subcortex white matter displayed the bundled myelinated axons along one direction without any labeling. The invasiveness limits the application of CARS microscopy in *in vivo* brain imaging. This difficulty will be resolved by dramatic improvement of penetration depth of CARS imaging or development of miniature lens with longer working distance.

## 17.5 Demyelination Studied by CARS Microscopy

Demyelination, the loss of normal myelin sheath, is responsible for long-term neurologic disability (Waxman 1998; Lazzarini 2004). Myelin degradation under clinical and experimental conditions has been extensively studied by EM and immunofluorescence. Myelin vesiculation was observed in the acute lesions of MS and experimental autoimmune encephalomyelitis (EAE) (Genain et al. 1999). However, the need for fixed tissue samples by these techniques prevents real-time monitoring of enzymatic activities and demyelination dynamics. Given the high-speed imaging, 3D submicron resolution, and multimodal capability, label-free technique CARS microscopy has been applied to the mechanical study of demyelination in fresh nerve tissues and live animals.

In CARS imaging of nerve tissues, demyelination is characterized by changes of myelin morphology, reduction of CARS contrast, and loss of polarization dependence of CARS signal. The g ratio and spectral profiles also help to identify demyelination. Submicron resolution of CARS imaging allows examination of detailed structural changes at the single axon such as paranodal myelin. Moreover, the capability of probing myelin in the fresh tissue environment allows examination of enzymatic and cellular activities involved in demyelination. It was found that demyelination was initiated at nodes of Ranvier in glutamate excitotoxicity (Fu et al. 2009) and under high-frequency stimulation (Huff et al. 2011). Fu et al. showed that myelin swelling occurred at internodes in the lysophosphatidylcholine-induced demyelination model (Fu et al. 2007a). The underlying mechanisms were investigated. In all cases, a calpain-dependent pathway is involved.

### 17.5.1 Myelin Retraction and Splitting at Nodes of Ranvier

The node of Ranvier appearing as a periodic gap on a myelinated axon is critical for rapid saltatory conduction of nerve impulses. Adjacent to the nodes are paranodes where axolemma and the lateral borders of myelin sheath are connected through the adhesion junctions. The integrity of paranodal domains is vital to action potential conduction along a myelinated axon. Real-time CARS imaging of fresh spinal cord tissue revealed paranodal myelin splitting and retraction (Figure 17.8A) induced by glutamate excitotoxicity (Fu et al. 2009), a lipid peroxidation product (Shi et al. 2011a), and high-frequency stimulation (Huff et al. 2011). Paranodal myelin retraction was characterized by the increase of ratio of nodal length to nodal diameter (Figure 17.8B and C). Combining CARS and two-photon immunofluorescence showed that paranodal myelin retraction caused exposure of juxtaparanodal $K^+$ channels. In all cases, paranodal myelin injury was demonstrated to be a $Ca^{2+}$-dependent process.

### 17.5.2 Internodal Demyelination

Lysophosphatidylcholine (lyso-PtdCho) was widely used to induce acute local demyelination in pathological and therapeutical study of demyelination and remyelination. Fu et al. applied CARS imaging to monitor lyso-PtdCho-induced demyelination in a fresh spinal cord tissue (Fu et al. 2007a) and in mouse sciatic nerve *in vivo* (Huff and Cheng 2007). Demyelination characterized by reduction of CARS intensity and loss of dependence on excitation polarization was found to start from the outer surface of myelin sheath (Figure 17.9). The g ratio of the compact myelin region in a swollen myelin was observed to increase over time (Figure 17.9C). Real-time CARS imaging and

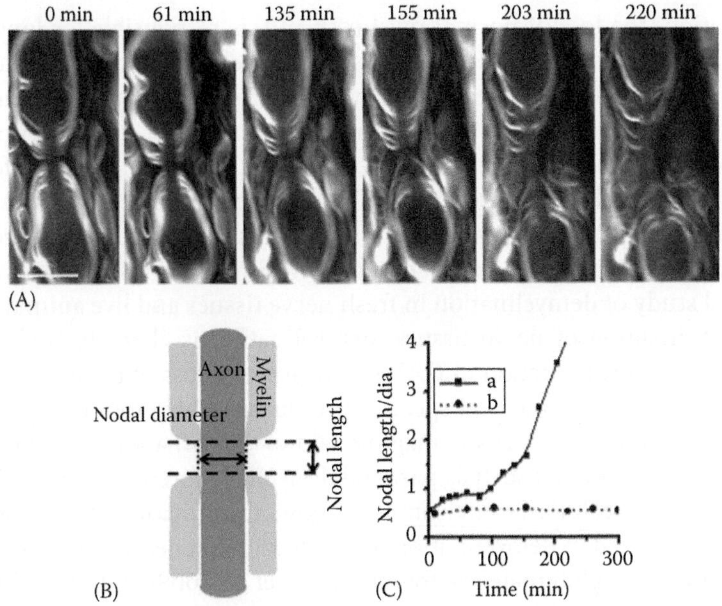


**FIGURE 17.8** Real-time CARS imaging of paranodal myelin splitting and retraction in an isolated fresh spinal tissue. (A) CARS images of paranodal myelin after different periods of 1 mM glutamate treatment *ex vivo*. Bar = 10 µm. (B) Schematic nodal length and nodal diameter. (C) Changes of the nodal length-to-diameter ratio with time.

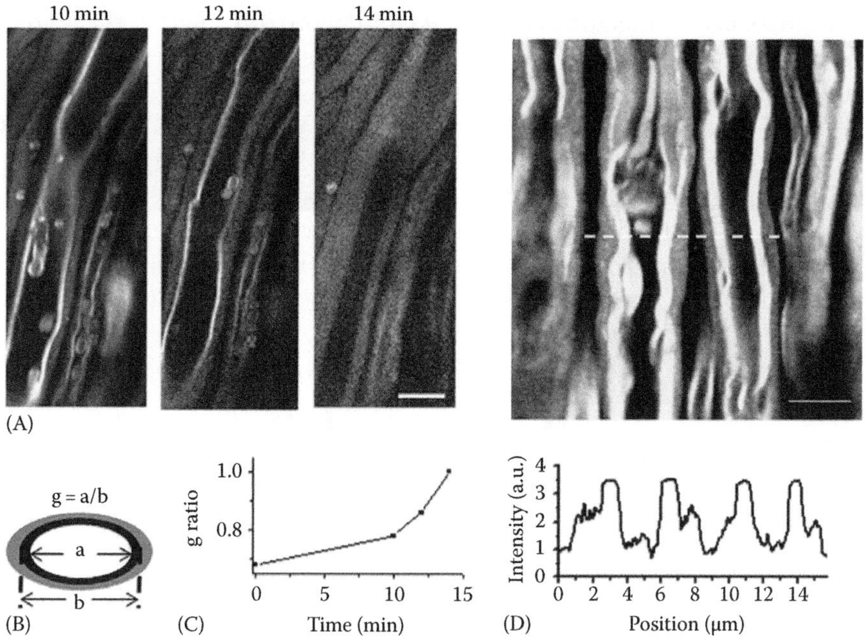


**FIGURE 17.9** *Ex vivo* and *in vivo* CARS imaging of myelin degradation induced by lyso-PtdCho. (A) Time-lapse CARS images of myelin swelling in the spinal tissue incubated with a Krebs' solution containing 10 mg/mL lyso-PtdCho. Bar = 10 μm. (B) Diagram of measuring the g ratio of a partially swollen myelin fiber based on the remaining compact region. (C) The increase of g ratio during the process of myelin swelling. (D) *In vivo* CARS image of partially swollen myelin in a mouse sciatic nerve. Intensity profile along the dash line in (D) is shown below the image. Bar = 5 μm. (From Fu, Y., et al., Coherent anti-Stokes Raman scattering imaging of myelin degradation reveals a calcium-dependent pathway in lyso-PtdCho-induced demyelination, *J Neurosci Res.*, 2007, 85, 2870–2881. Copyright Wiley-VCH Verlag GmbH & Co. KGaA. Reproduced with permission.)

electrophysiological experiments suggest that lyso-PtdCho induces myelin degradation via $Ca^{2+}$ influx into myelin and subsequent activation of $cPLA_2$ and calpain that break down myelin lipids and proteins (Fu et al. 2007a).

Demyelination in a controlled sciatic nerve crush injury was examined by video-rate *in vivo* CARS imaging (Bélanger et al. 2011). Following a crushed nerve injury, minor myelin swelling was observed with diminished effect over time. A large amount of myelin debris with very few myelinated axons was observed after 2 weeks postinjury. *Ex vivo* imaging examination revealed a decrease of the myelin thickness 1 week postinjury.

In a spinal cord injected with secretory phospholipase $A_2$-III, repetitive *in vivo* CARS imaging of injection site revealed myelin changes in a sequence of myelin vesiculation, a large amount of myelin debris, and spontaneous remyelination by Schwann cells (Figure 17.7B) over a 3 week period (Shi et al. 2011b).

## 17.5.3 Demyelination in the EAE Model

EAE is a widely used animal model that reproduces specific features of the histopathology and neurobiology of MS. Accepted as a T cell-mediated autoimmune disease in the central nervous system (Martin and Dhib-Jalbut 2000), EAE model allows for investigation of the mechanism of myelin loss and assessment of remyelination efficacy

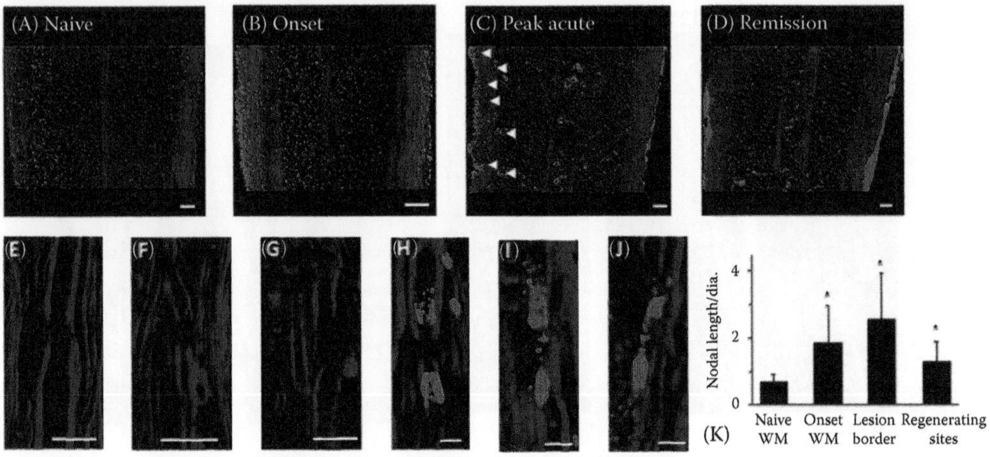


**FIGURE 17.10** Characterization of demyelination and remyelination of mouse spinal cord in a relapsing EAE model. (A–D) CARS images of lumbar spinal cord slices from a naïve mouse, relapsing EAE mice at disease onset, peak acute stage, and remission stage, respectively. Red: myelin imaged by CARS microscopy; Green: Hoechst-labeled cell nuclei imaged by TPEF. Bar = 100 μm. (E–G) Typical CARS images of myelinated nodes in the white matter of a naïve mouse, EAE mice at disease onset, and within borders of demyelinating lesions at the peak acute EAE, respectively. Bar = 10 μm. (H) A typical image of paired Kv1.2 channels (green: immunolabels imaged by TPEF) in the naive white matter, which are located at the juxtaparanodes and concealed beneath the compact myelin. (I and J) Exposure and displacement of juxtaparanodal Kv1.2 channels into the paranodes and nodes in the white matter of EAE onset and at the border of demyelination lesions. Bar = 5 μm in (H–J). (K) Comparison of ratios of nodal length to nodal diameter at different stages of EAE.

of therapeutic interventions. Fu et al. employed CARS microscopy to examine myelin integrity in the lumbar spinal cord tissue isolated from naïve SJL mice, and mice at the onset, peak acute, and remission stages of relapsing EAE (Figure 17.10A through D) (Fu et al. 2011). Myelin morphology, density, and thickness were recorded and compared at different stages of relapsing EAE. At the onset of EAE, mice manifested weak tail, and parallel myelin fibers were present in the lumbar tissue. No significant difference in the myelin density and thickness was found between the onset stage and the naïve stage. At single axon level, retraction of paranodal myelin was observed both at the onset of disease and at the borders of acute demyelinating lesions without loss of internodal myelin (Figure 17.10F and G). The disruption of paranodal myelin subsequently exposed Kv1.2 channels at the juxtaparanodes and led to the displacement of Kv1.2 channels to the paranodal and nodal domains (Figure 17.10I and J). These observations were consistent with the function deficit. At the remission stage, paranodal myelin was found to be partially restored indicating spontaneous myelin regeneration (Figure 17.10K). In another study that combined video-rate reflectance, fluorescence confocal imaging, and CARS microscopy (Imitola et al. 2011), local areas of severe loss of CARS signal from myelin and loss of the reflectance signal from axons were observed in the corpus callosum and spinal cord slices of EAE animals. Increased expression of GFP-coded microglia was found to correlate with the diffusive decrease of CARS signal and regional loss of reflectance. Collectively, label-free CARS imaging provides an efficient readout for demyelination and remyelination in animal models of multiple sclerosis.

## 17.6 Summary

This chapter discussed the use of CARS for visualization and characterization of myelin in *ex vivo* tissues and live animals. CARS provides advantages of high-speed imaging, 3D submicron resolution, and selectivity due to its excellent contrast from myelin lipid content. The applications include imaging of sciatic nerve, spinal cord, and brain. The examples of neurodegenerative disease were highlighted in EAE, crush injury of sciatic nerve, and spinal cord injury. Moreover, CARS microscopy allowed examination of dynamic activities of myelin sheath in the paranodal areas. Other techniques such as two-photon fluorescence, sum frequency generation, and video-rate reflectance can be incorporated into CARS microscopy platform allowing examination of various cellular component activities. CARS imaging holds great potential for longitudinal *in vivo* imaging of myelin at single axon level in animal models of neurological disorders. To this purpose, technical improvement for increase of penetration depth and decrease of image distortion due to the animal movement is desired.

## References


Bélanger, E., S. Bégin et al. (2009). Quantitative myelin imaging with coherent anti-Stokes Raman scattering microscopy: Alleviating the excitation polarization dependence with circularly polarized laser beams. *Opt Express* **17**(21): 18419–18432.

Bélanger, E., F. P. Henry et al. (2011). In vivo evaluation of demyelination and remyelination in a nerve crush injury model. *Biomed Opt Express* **2**: 2698–2708.

Blight, A. R. (1985). Delayed demyelination and macrophage invasion: A candidate for secondary cell damage in spinal cord injury. *Cent Nerv Syst Trauma* **2**: 299–315.

Compston, A. and A. Coles (2002). Multiple sclerosis. *Lancet* **359**: 1221–1231.

Davalos, D., J. K. Lee et al. (2008). Stable in vivo imaging of densely populated glia, axons and blood vessels in the mouse spinal cord using two-photon microscopy. *J Neurosci Methods* **169**: 1–7.

Dray, C., G. Rougon et al. (2009). Quantitative analysis by in vivo imaging of the dynamics of vascular and axonal networks in injured mouse spinal cord. *Proc Natl Acad Sci USA* **108**: 6282–6287.

Flusberg, B. A., E. D. Cocker et al. (2005). Fiber-optic fluorescence imaging. *Nat Methods* **2**: 941–950.

Fu, Y., T. J. Frederick et al. (2011). Paranodal myelin retraction proceeds demyelination in relapsing experimental autoimmune encephalomyelitis. *J Biomed Opt* **16**: 106006.

Fu, Y., T. B. Huff et al. (2008). Ex vivo and in vivo imaging of myelin fibers in mouse brain by coherent anti-Stokes Raman scattering microscopy. *Opt Express* **16**: 19396–19409.

Fu, Y., W. Sun et al. (2009). Glutamate excitotoxicity inflicts paranodal myelin splitting and retraction. *PLoS ONE* **4**(8): e6705.

Fu, Y., H. Wang et al. (2007a). Coherent anti-Stokes Raman scattering imaging of myelin degradation reveals a calcium dependent pathway in lyso-PtdCho induced demyelination. *J Neurosci Res* **85**: 2870–2881.

Fu, Y., H. F. Wang et al. (2007b). Second harmonic and sum frequency generation imaging of fibrous astroglial filaments in ex vivo spinal tissues. *Biophys J* **92**(9): 3251–3259.

Genain, C. P., B. Cannella et al. (1999). Identification of autoantibodies associated with myelin damage in multiple sclerosis. *Nat Med* **5**(2): 170–175.

Helmchen, F. and W. Denk (2002). New developments in multiphoton microscopy. *Curr Opin Neurobiol* **12**: 593–601.

Henry, F. P., D. Côté et al. (2009). Real-time in vivo assessment of the nerve microenvironment with coherent anti-Stokes Raman scattering microscopy. *Plast Reconstr Surg* **123**: 123S–130S.

Huff, T. B. and J. X. Cheng (2007). In vivo coherent anti-Stokes Raman scattering imaging of sciatic nerve tissue. *J Microsc* **225**(2): 175–182.

Huff, T. B., Y. Shi et al. (2008). Multimodal nonlinear optical microscopy and applications to central nervous system imaging. *IEEE J Sel Topics Quantum Electron* **14**: 4–9.

Huff, T. B., Y. Shi et al. (2011). Real-time CARS imaging reveals a calpain-dependent pathway for paranodal myelin retraction during high-frequency stimulation. *PLoS ONE* **6**: e17176.

Imitola, J., D. Côté et al. (2011). Multimodal coherent anti-Stokes Raman scattering microscopy reveals microglia-associated myelin and axonal dysfunction in multiple sclerosis-like lesions in mice. *J Biomed Opt* **16**: 021109.

Lazzarini, R. A., Ed. (2004). *Myelin Biology and Its Disorders.* San Diego, CA: Elsevier Academic Press.

Martin, R. and S. Dhib-Jalbut (2000). *Multiple Sclerosis: Diagnosis, Medical Management, and Rehabilitation.* New York: Demos Medical Publishing, Inc.

Misgeld, T. and M. Kerschensteiner (2006). In vivo imaging of the diseased nervous system. *Nat Rev Neurosci* **7**: 449–463.

Rajadhyaksha, M., R. R. Anderson et al. (1999). Video-rate confocal scanning laser microscope for imaging human tissues in vivo. *Appl Opt* **38**: 2105–2115.

Shi, R., R. B. Borgens et al. (1999). Functional reconnection of severed mammalian spinal cord axons with polyethylene glycol. *J Neurotrauma* **16**(8): 727–738.

Shi, Y., W. Sun et al. (2011a). Acrolein induces myelin damage in mammalian spinal cord. *J Neurochem* **117**: 554–564.

Shi, Y., D. Zhang et al. (2011b). Longitudinal in vivo CARS imaging of demyelination and remyelination in injured spinal cord. *J Biomed Opt* **16**: 106012.

Slipchenko, M., T. T. Le et al. (2009). Compound Raman microscopy for high-speed vibrational imaging and spectral analysis of lipid bodies. *J Phys Chem B* **113**: 7681–7686.

Smith, K. J., R. Kapoor et al. (1999). Demyelination: The role of reactive oxygen and nitrogen species. *Brain Pathol* **9**: 69–92.

Tyson, R. K. (1998). *Principles of Adaptive Optics.* Boston, MA: Academic Press.

Wang, H., Y. Fu et al. (2005). Coherent anti-Stokes Raman scattering imaging of live spinal tissues. *Biophys J* **89**: 581–591.

Wang, H., T. B. Huff et al. (2007). Increasing the imaging depth of coherent anti-Stokes Raman scattering microscopy with a miniature microscope objective. *Opt Lett* **32**: 2212–2214.

Waxman, S. G. (1998). Demyelinating diseases—New pathological insights, new therapeutic targets. *N Engl J Med* **338**(5): 323–325.

Wright, A. J., B. A. Patterson et al. (2005). Dynamic closed-loop system for focus tracking using a spatial light modulator and a deformable membrane mirror. *Opt Express* **14**: 222–228.

Wright, A. J., S. P. Poland et al. (2007). Adaptive optics for enhanced signal in CARS microscopy. *Opt Express* **15**: 18209–18219.

# 18. Imaging Lipid Metabolism in *Caenorhabditis elegans* and Other Model Organisms

**Helen Fink, Christian Brackmann, and Annika Enejder**



## 18.1 Introduction

This chapter describes recent advances in imaging lipids with coherent anti-Stokes Raman scattering (CARS) microscopy. Applications of the technique on different model organisms used in lipid research are highlighted. Some models have already found broad interest and are widely used, while others have just recently been introduced in lipid research.

The increase in prevalence of obesity worldwide has intensified the search to identify both the genes and metabolic regulating pathways that control the storage of fat in mammalian tissues. Energy homeostasis is maintained by an extremely complex system of signaling pathways involved in uptake of nutrients, storage, and feeding behavior. Impairments in any of

*Coherent Raman Scattering Microscopy*. Edited by Ji-Xin Cheng and X. Sunney Xie © 2013 CRC Press/ Taylor & Francis Group, LLC. ISBN: 978-1-4398-6765-5.

these functions can lead to systemic imbalance resulting in excess energy storage as lipid bodies. The health consequences are significant since obesity is linked to many widespread diseases such as type-2 diabetes and various cardiovascular diseases that in turn are associated with neurodegenerative conditions like Alzheimer's disease (Luchsinger and Gustafson 2009). Understanding the pathways of lipid synthesis, storage, and metabolism, including the regulation thereof, are therefore of immense importance.

For these purposes suitable model systems have been developed, ranging from single-cell cultures (Alberts et al. 1974; Brasaemle et al. 2004) to whole animal models, predominantly rodents (Moitra et al. 1998; Haemmerle et al. 2006; Speakman et al. 2008).

Nevertheless, studies of relatively simple organisms such as yeast cells, worms, and zebrafish are valuable since they serve as model organisms that together with several others have contributed to extensive knowledge into the development of organisms in general. Although the key regulators of lipid metabolism have so far been identified using mammalian models, invertebrate model systems including *Caenorhabditis elegans* have accelerated the discovery of new genes that are important in lipid hemostasis.

Over the past two decades, it has become increasingly clear that many important pathways are preserved and that knowledge from one organism can provide remarkable insight when combined with that obtained from other organisms. Whereas the whole-organism models allow for full systemic studies, single-cell models such as yeast are useful to gain in-depth information on specific cellular metabolic pathways. The yeast cell has for a long time proven to be highly useful in understanding the biology of the cell, also for higher organisms.

## 18.2 Model Organisms for Lipid Research

All organisms must balance their energy input and output in order to survive, and lipids are a vital component of energy storage throughout eukaryotes. Many different model organisms have been deployed to study lipid storage and metabolism in order to answer important questions concerning obesity, including rodents and recently nonmammalian organisms such as fruit flies (Gronke et al. 2005; Welte et al. 2005), *C. elegans* (Ashrafi et al. 2003), and yeast (Kurat et al. 2006) of which the last two together with zebrafish are described in this chapter. Their main advantage is to allow high-throughput screening in both single-cell and whole-animal context.

## 18.3 *Caenorhabditis elegans*

Research in molecular and developmental biology began in the early 1970s and *C. elegans* has since then been used extensively as a model organism. It was the first multicellular organism to have its whole genome sequenced and it became apparent that 30% of the identified genes have human homologues, although this organism is otherwise very distinct from any mammalian organisms or humans. This nematode has many organ systems similar to other animals and is one of the simplest organisms with a nervous system. Thus, it is easily appreciated why this organism is so well studied and has found enormous popularity.

*C. elegans* is a transparent free-living nematode (roundworm) that lives in temperate soil and is about 1 mm in length with a maximum diameter around 80 μm (Figure 18.1). As a model organism, *C. elegans* offers many experimental advantages. *C. elegans* is very easy

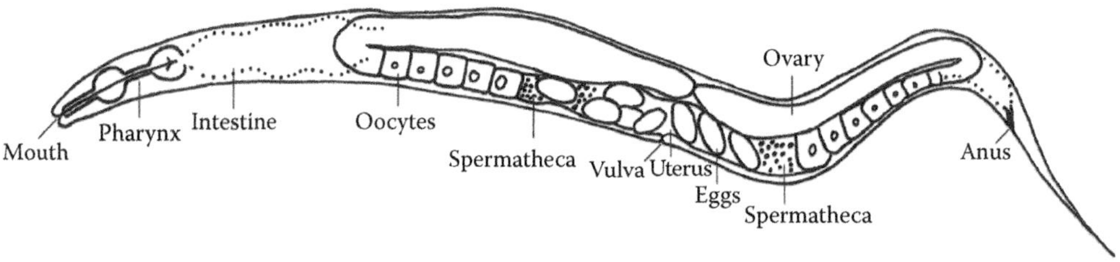


**FIGURE 18.1** Schematic illustration of the anatomy of the *C. elegans* nematode.

to grow, has an average life span of 2–3 weeks, has a short generation time of approximately 4 days, and produces a great number of progeny (>300) by self-fertilization. After hatching, *C. elegans* larvae pass through four larval stages (L1–L4). In dense populations or in the absence of nutrition, they can enter an alternative larval stage called the dauer stage. In this stasis-like stage, dauer larvae are stress resistant and can survive harsh conditions due to accumulated lipid stores during development (Wood 1988).

*C. elegans* was introduced as a model to study developmental and neuronal biology (Brenner 1974), but today it is used to study a variety of biological processes such as aging (Kenyon et al. 1993), cell cycle (MacQueen et al. 2005), cell signaling (Kaletsky and Murphy 2010), apoptosis (Vlachos and Tavernarakis 2010), gene regulation (Mooijaart et al. 2005), and lipid metabolism (Spanier et al. 2009). Many key discoveries, both in basic biology and medically relevant areas, were first done in this nematode. Indeed, following comparisons of the human and *C. elegans* genomes have confirmed that the majority of human disease genes and disease pathways are present in *C. elegans*. For example, the first presenilin was discovered in *C. elegans* (Sundaram and Greenwald 1993), and 2 years later, mutations in the human presenilin-1 gene were associated with early-onset familial Alzheimer's disease. A remarkably functional conservation between *C. elegans* and humans could be demonstrated (Sherrington et al. 1995). Another example is that genetic studies in *C. elegans* identified negative regulators of the insulin signaling pathway, and one of these genes, *daf-16*, encodes the *C. elegans* ortholog of the transcription factor FOXO. Five years later, FOXO loss-of-function was found to rescue the diabetic phenotype of insulin-resistant mice (Nakae et al. 2002).

### 18.3.1 Lipid Metabolism in *C. elegans*

Like other disease genes and pathways, the metabolic pathways, including biochemical pathways of fat, carbohydrates, and protein synthesis, are highly conserved from *C. elegans* to humans. Thus, this organism has become an excellent model for exploring the genetic basis of fatty-acid synthesis and regulation of lipid storage, as well as processes involved in associated diseases such as obesity and diabetes.

Fatty acids build up triacylglycerides, which are stored in lipid droplets and yolk. In *C. elegans*, lipids are stored in the form of small bodies or lipid droplets that consist of a highly hydrophobic core of neutral lipids, mainly triacylglycerides and/or steryl esters, which is surrounded by a phospholipid monolayer, with a small proportion of embedded proteins. These lipid stores are a vital energy source during embryogenesis, periods of low food availability, and the nonfeeding dauer stage. The triacylglyceride fat stores make up approximately 40%–55% of the total lipids depending on diet and growth stage (Ashrafi 2006).

The fat regulation in *C. elegans* differs from mammals in some aspects. Since *C. elegans* has no mesoderm-derived adipose cells dedicated for lipid storage, this nematode stores its fat primarily in intestinal and hypodermal (skin-like) cells. In *C. elegans*, lipid droplets are confined within a bilayer of phospholipids rather than a monolayer that is the case in mammals and is stored in lysosome-related organelles. The nematode is also auxotrophic for sterols such as cholesterol, which means it is dependent on endogenous uptake through their nutrition (Hieb and Rothstein 1968). Despite these differences, the highly evolutionary conservation of key metabolic pathways and their regulators suggest that investigations into lipid storage mechanisms should be of great value. Indeed several newly identified regulatory pathways have been found to have analogous mammalian pathways that regulate lipid storage (McKay et al. 2003).

In mammals, a number of transcription factors have been identified and shown to play important roles in fatty-acid and lipid metabolism, including peroxisome proliferator–activated receptors, CCAAT/enhancer-binding proteins (C/EBPs), and sterol regulatory element-binding proteins (SREBPs) (Hansen and Connolly 2008). Similarly, key transcription factors have been identified to orchestrate transcription of several genes, whose gene products are involved in either catabolism or anabolism of lipids in *C. elegans*.

The SREBP family exemplifies the remarkable conservation of metabolic regulators throughout eukaryotes. The SREBP homologues are crucial for metabolism in yeast, worms, and flies (Dobrosotskaya et al. 2002; Hughes et al. 2005). In mammals, SREBP proteins regulate genes involved in the de novo synthesis of monounsaturated and polyunsaturated fatty acids and their subsequent incorporation into triglycerides and phospholipids (Raghow et al. 2008). *C. elegans* expresses only one ortholog of SREBP called SBP-1 (McKay et al. 2003). Accordingly, genes that have been shown to be direct SREBP targets in mammalian cells such as acyl-CoA carboxylase, fatty-acid synthase, stearoyl-CoA desaturases, fatty-acid elongases, glycerol 3-phosphate acyltransferase, malic enzyme, and ATP citrate lyase have all been shown to be regulated by SBP-1 in *C. elegans* (Kniazeva et al. 2004). This shared origin of SREBP function is further underscored by the observation that the activation domain of SBP-1 was found to stimulate transcription in humans (Yang et al. 2006).

Many of the adverse health effects of excess fat accumulation in humans are unlikely to occur in *C. elegans*. Nevertheless, the studies that have been reported so far already reveal remarkable similarities between molecular components of *C. elegans* and mammalian fat pathways that extend to disease-associated genes.

### 18.3.2 Imaging of Lipid Metabolism in *C. elegans*

Historically, a number of techniques have been used to visualize the lipid stores in *C. elegans* by staining with a number of different fat-soluble dyes such as Sudan black, oil red, and Nile red dyes or with BODIPY-labeled fatty acids. An early method to evaluate fat content was the use of Sudan Black, a lipophilic dye, on fixed animals. This allowed, for the first time, visual analysis of lipid storage. Since this method requires fixation, it is time consuming and has also shown inconsistent staining; therefore, it is usually used to confirm genetic interactions that have already been established by other phenotypes (Kimura et al. 1997) or to examine dramatic phenotypes (Greer et al. 2008). The major advance was the use of vital dye staining, which allows examination of fat content in intact individuals. Over the last decade, the most commonly used fluorescent dyes have been Nile red and fatty-acid-conjugated BODYPI. Generally these dyes are quite sensitive, enabling detection of subtle differences in fat levels. Visualization is, however, dependent upon endogenous uptake and transport pathways, as well as the physical properties of the dyes themselves. For example, the fluorescence yield of Nile red is highly dependent on the lipid phase, where liquid phase is less favorable than more ordered gel phase, and Nile red does not stain hypodermal stores, while BODIPY-labeled fatty acids do. Both dyes mainly stain intestinal fat stores, but in certain cases visualization of these stores is not possible (Schroeder et al. 2007). Thus, the validity of these lipid-staining methods has recently come into question.

In a study by O'Rourke et al., the lipid stores of different mutants have been investigated comparing fluorescent (Nile red and BODIPY) or nonfluorescent (Sudan black and oil red O) labeling methods with biochemical analysis of triglyceride/phospholipid levels. Mutants with altered Nile red or BODIPY phenotypes typically do not exhibit any lipid phenotypic differences or even an opposite phenotype to that indicated by biochemical measurements of lipid content. The insulin-reporter-like mutant, *daf-2*, for instance, shows increased lipid storage when measured with biochemical analysis but the opposite when analyzed with Nile red or BODIPY–fatty acid staining. Together with further inconsistencies between Nile red fluorescence and triglyceride levels found in other mutants, it is getting apparent that these stains are not accurate for visualizing major lipid stores in *C. elegans*. Furthermore, the study by O'Rourke et al. shows that the major fat is stored in distinct cellular compartments that are not stained by Nile red. Label-free imaging by CARS microscopy provides a method to avoid these limitations and has been used to visualize fat deposits in living intact animals (Figure 18.2).

In the first quantitative application of CARS microscopy to *C. elegans*, Hellerer et al. demonstrated the major benefits of imaging without any reporter molecule by comparing fluorescence and CARS microscopy (Hellerer et al. 2007). In wild-type nematodes, intestinal fat stores visualized by Nile red correspond quite well to CARS microscopy measurements.

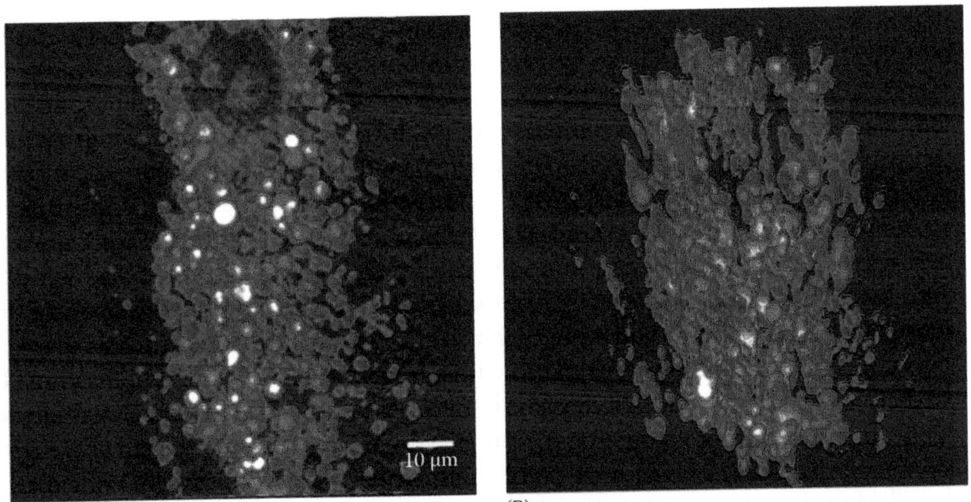


(A) (B)

**FIGURE 18.2** CARS microscopy image of lipid droplets in *C. elegans daf-4* dauer (A) and corresponding volume view reconstructed from a z-stack of images (B).

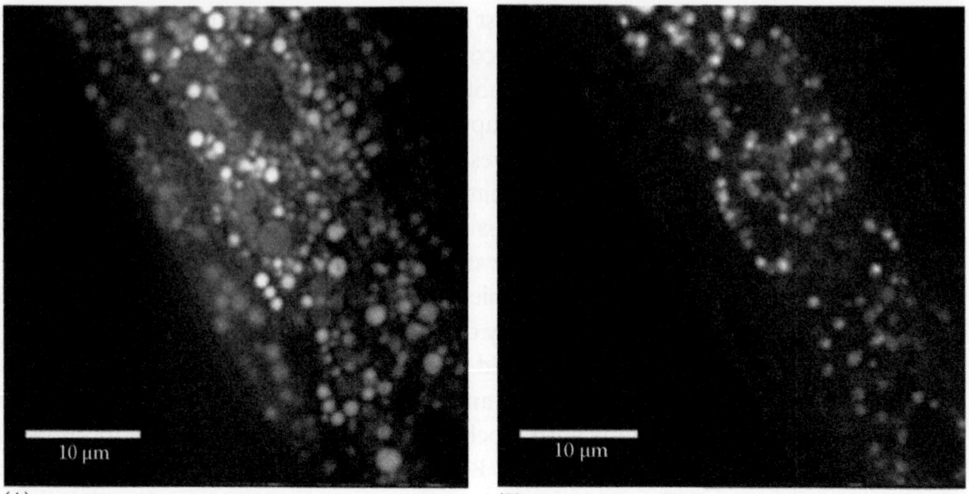


**FIGURE 18.3** Comparison of CARS and fluorescence microscopy in *C. elegans.* (A) CARS and (B) two-photon fluorescence images of Nile red–stained *daf-4* mutant arrested in its dauer stage. While the CARS image shows the full distribution of lipid droplets, also the hypodermal fraction, the Nile red fluorescence merely shows the lipids in the intestine.

However, while CARS microscopy was shown to be able to visualize the full distribution of lipid droplets, also the hypodermal accumulations, Nile red fluorescence fail in that aspect (Figure 18.3). In wild-type nematodes, this hypodermal fraction is very small making differences in Nile red staining less obvious. In other mutants, however, with genetic variations in the insulin and transforming growth factor (IGF and TGF-β) signaling pathway (*daf-2, daf-4 dauer*), this fraction is much larger. In the *daf-2* mutants, for example, the total accumulation of lipid stores is much higher than for the wild type, as should be expected, and significantly lower for the mutants with genetic deletions of proteins involved in the feeding behavior (*pha-3*). In these mutants, the relative hypodermal lipid storage is substantially larger than that for the wild type, hosting as much as 40% of the total lipid volume in contrast to around 9% for the wild-type larvae. In consequence, the hypodermal energy storage pool might be underestimated or even neglected in studies that rely solely on Nile red staining. This study shows that CARS microscopy has the potential to become a sensitive and important tool for studies of lipid storage mechanisms, improving our understanding of phenomena underlying metabolic disorders.

Additional inconsistencies in Nile red staining were demonstrated in a study by Mörck et al., where the impact of cholesterol-lowering drugs on lipid storage was evaluated (Mörck et al. 2009). Statin treatment reduces the level of Nile red staining but does not affect neutral lipid storage as monitored with CARS microscopy.

In more recent studies, Le et al. identified two distinct subgroups of lipid storage compartments, one that can only be visualized with CARS and another that can be visualized both with CARS and autofluorescence (Le et al. 2010). They confirmed the visualization capability of CARS microscopy for neutral lipids and investigated different mutants, *daf-4*, as described earlier, and mutants with deletions in genes encoding lipid desaturation enzymes (*fat5–7*), antioxidant enzymes (*sod-1*), and lipid synthesis transcription factors (*sbp-1*) and demonstrated the value of this method in monitoring changes in lipid metabolism.

In addition to differences in neutral lipid storage, differences in the amount of autofluorescent lipid compartments, corresponding to lipofuscin-containing gut granules, could be observed between mutants and wild-type nematodes (Le et al. 2010). Several other studies have also identified these fluorescent bodies (Clokey and Jacobson 1986; Hosokawa et al. 1994).

When Yen et al. compared dye-labeled imaging to label-free imaging with two-photon fluorescence and CARS microscopy, respectively, it was further confirmed that feeding worms with vital dyes does not lead to staining of major fat stores but rather the autofluorescent lysosome-related organelles and that mutants in the insulin/IGF-1 signaling exhibit distinct phenotypes (Yen et al. 2010). In a separate study, it was also found that the vital dye Nile red stains a subset of granules known as gut granules, which do not correlate with lipid content, by means of comparing Nile red staining with fixative dyes (O'Rourke et al. 2009). It has previously been reported that vital dyes are endocytosed and transported to degradative compartments (Clokey and Jacobson 1986). This is confirmed by investigation of mutants with deficiency in lysosomal biogenesis that exhibit reduced Nile red staining (Schroeder et al. 2007) and perfect overlap of simultaneous staining of vital dyes and lysosome-specific tracker molecules. Collectively, these studies suggest that Nile red staining might not be a reliable measure of lipid content in *C. elegans*.

Although CARS signals are generated from all lipid-rich structures including cell membranes and cellular organelles, it is possible to distinguish lipid droplets based on their CARS signal, which is quadratically dependent on the concentration of $CH_2$ groups. Consequently, the CARS signal from the lipid droplets in intestinal and hypodermal fat stores is enhanced due to a high density and high local concentration and therefore discriminates significantly from the background. The strong resonant response and selectivity for lipid droplets, probed at the symmetric stretch vibration of frequency 2840 $cm^{-1}$, are demonstrated by two CARS images measured in the intestinal cells of wild-type worms. Lipid droplets are clearly visible probing the resonance of the symmetric $CH_2$ vibration whereas these structures are only vaguely visualized off resonance (Figure 18.4).

Raman spectroscopy can be used to determine the order of polymethylene chains (Snyder et al. 1978) and phospholipids (Gaber and Peticolas 1977) by calculating the ratio between the signal intensities of the asymmetric and symmetric $CH_2$ vibrations. A higher ratio is characteristic for a gel phase of highly ordered methyl chains as compared to the liquid phase of less ordered chains. This has also been confirmed for lipid droplets measured by CARS microspectroscopy (Wurpel et al. 2002). Correspondingly, CARS spectra measured in *C. elegans* nematodes suggest that the shift in metabolism to increased lipid storage observed for the stressed *daf-2* and *daf-4* (dauer) mutants is accompanied by a shift in the ordering of the lipids from gel (high ordering) to liquid phase (low ordering) (Hellerer et al. 2007). From this it can be concluded that the degree of hypodermal lipid storage and the lipid phase can be used as markers of lipid metabolism shift. The CARS spectra can be further improved by subtracting the background especially that from water when measuring lipids (Hagmar et al. 2008; Li et al. 2010).

Furthermore, coupling CARS microscopy imaging with spectral information from pointwise confocal Raman microspectroscopy provides the capability of detecting lipid-chain unsaturation. Le et al. measured lipid-chain unsaturation levels in wild-type and mutant *C. elegans* using this combination, highlighting the capability for analysis

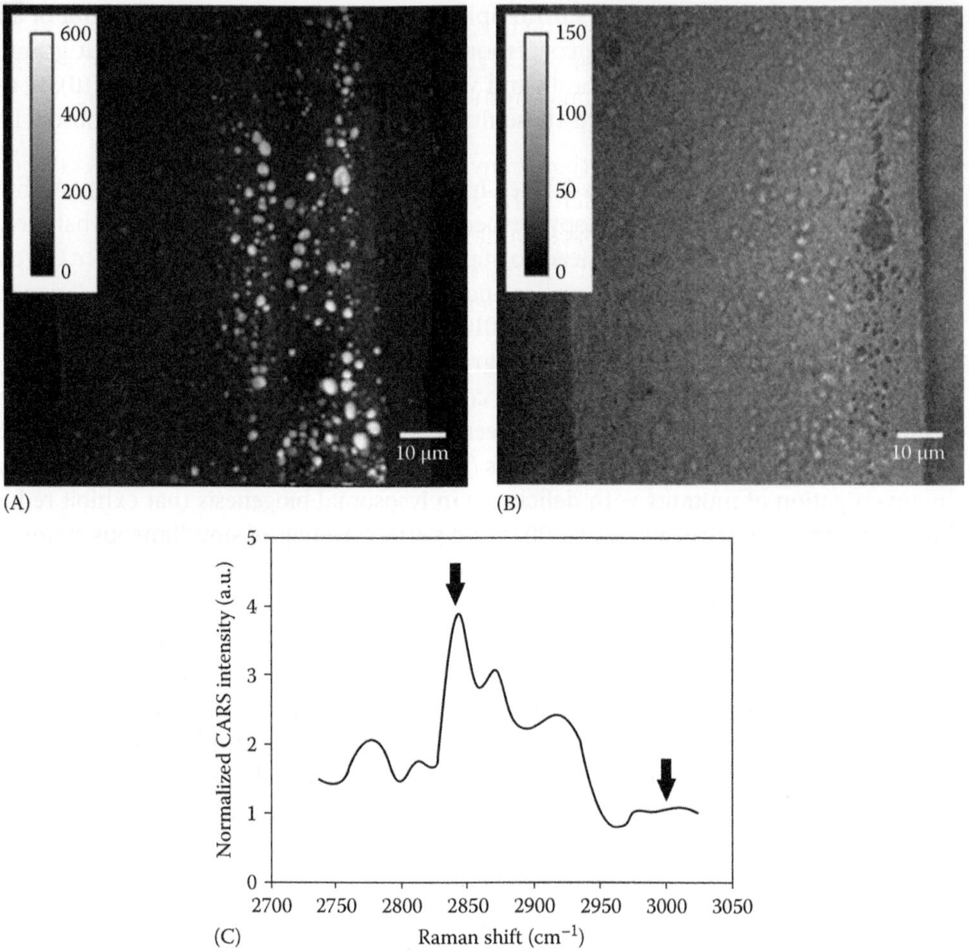


**FIGURE 18.4** CARS images (normalized) of lipid droplets in *C. elegans* showing (A) resonance (2845 $cm^{-1}$) and (B) off-resonance (3023 $cm^{-1}$). Lipid droplets are visible at the resonance of $CH_2$ vibrations, whereas structures can only vaguely be discerned off resonance. (C) CARS spectra of *C. elegans* obtained from series of CARS images collected *in vivo*. Corresponding reference images measured in glass were used for normalization of lipid droplet intensities (arrows indicate frequencies corresponding to images A and B). The lipid droplets exhibit peaks at 2845 and 2870 $cm^{-1}$, corresponding to the symmetric and asymmetric $CH_2$ stretch vibrations, respectively.

of complex genotype–phenotype relationships between lipid storage, peroxidation, and desaturation in single living *C. elegans* (Le et al. 2010).

CARS microscopy has also been used to evaluate the impact of cholesterol-lowering drugs on lipid storage. Statins are compounds prescribed to lower blood cholesterol in millions of patients worldwide. They act by inhibiting HMG-CoA reductase, the enzyme responsible for the rate-limiting step in the mevalonate pathway, which splits into two branches: one leading to cholesterol synthesis and the other to synthesis of farnesyl pyrophosphate. The nematode *C. elegans* possesses a mevalonate pathway that lacks the branch leading to cholesterol synthesis, and thus represents an ideal organism to specifically study the noncholesterol roles of the pathway. The lipid stores and fatty-acid composition were unaffected in statin-treated worms (Mörck et al. 2009).

CARS imaging provides a distinct visualization measure of total lipid stores with greater accuracy than previous visualization methods and in more detail. For example, not only the amount of lipid stores can be assessed but also a good estimation of the size and number as well as saturation of individual lipid droplets can be provided. Indeed, CARS imaging is able to give quantification similar to biochemical triglyceride measurements (Yen et al. 2010).

Together, these initial studies showcase the versatility of CARS microscopy for noninvasive and label-free visualization and compositional analysis of lipid droplets in *C. elegans*. When combined with *C. elegans* genetics and high-throughput screening, CARS microscopy is likely to rapidly facilitate functional studies of lipid metabolism at the genomic scale and advance our understanding of the roles of lipids in human diseases.

## 18.4 Baker's Yeast (*Saccharomyces cerevisiae*)

*Saccharomyces cerevisiae* (*S. cerevisiae*) is one of the most intensively studied eukaryotic model organisms in molecular and cell biology. As a single-cell organism, *S. cerevisiae* is small and cells are round to ovoid, 5–10 μm in diameter. It reproduces by a division process known as budding with a short generation time (doubling time 1.5–2 h at 30°C) and can be easily cultured. These are all positive characteristics in that they allow for swift production and maintenance of multiple specimen lines at low cost. *S. cerevisiae* was the first eukaryotic organism to have its entire genome sequenced (Goffeau et al. 1996) and has for a long time proven to be highly useful in understanding the biology of cells as well as higher organisms. In addition, *S. cerevisiae* exhibits high evolutionary conservation in complex processes such as regulatory mechanisms of cell cycle (Hartwell 2002) and fundamental metabolic pathways (Kurat et al. 2006; Czabany et al. 2007; Rajakumari et al. 2008). Thus, this organism could provide important information on metabolic processes involving lipid storage.

### 18.4.1 Lipid Metabolism in *S. cerevisiae*

As for *C. elegans*, the major neutral lipids in *S. cerevisiae* are triacylglycerols and steryl esters, which lack charged groups and are therefore not suited as components of membranes. Consequently, they are also accumulated into lipid droplets. In yeast, lipid droplets are formed by budding from the endoplasmic reticulum and contain most enzymes for sterol synthesis and triacylglycerol formation (Athenstaedt et al. 1999). Moreover, the enzymes involved in the synthesis of triacylglycerols and steryl esters are remarkably conserved in *S. cerevisiae*.

### 18.4.2 Imaging of Lipid Metabolism in *S. cerevisiae*

Previously, evaluation of lipid storage and droplet morphology have relied on biochemical analysis of whole-cell content or fluorescent imaging using lipid-specific staining or green fluorescent protein (GFP)-tagged proteins associated with lipid droplets (Greenspan et al. 1985; Szymanski et al. 2007). However, as described previously in this chapter, staining is an indirect and invasive method. The dyes are required to enter the cells and bind to lipids, with potential side effects such as altered chemical and

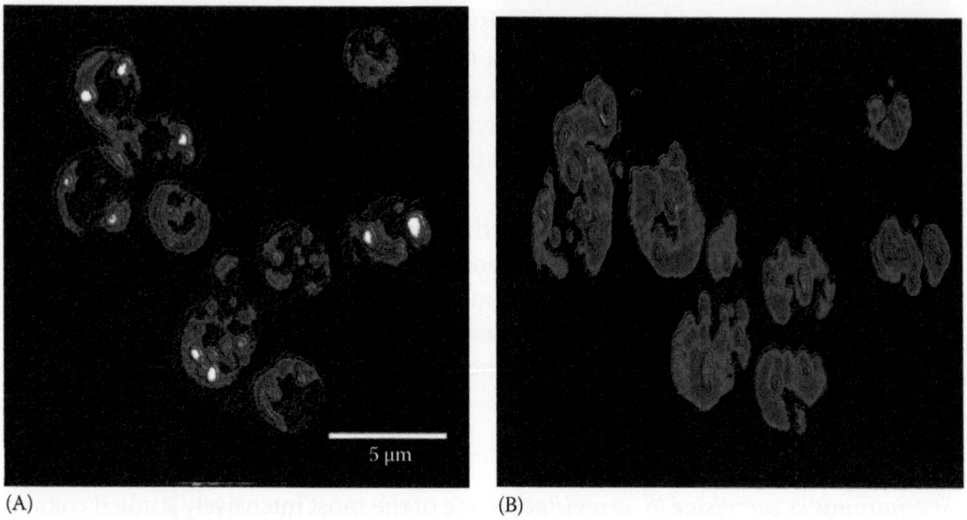


**FIGURE 18.5** (A) CARS microscopy image of *S. cerevisiae* yeast cells, collected probing 2845 cm$^{-1}$, with intracellular lipid droplets clearly visible. (B) tree-dimensional reconstruction from z-stack of CARS images giving a volume view of the lipid droplets in an yeast cell.

physiological properties of the lipids and uncertainties in the labeling—problems previously described for Nile red and staining of lipids in *C. elegans* (Fukumoto and Fujimoto 2002; Hellerer et al. 2007; Yen et al. 2010).

CARS microscopy has been evaluated as a tool to study lipid storage *in vivo* in yeast (Brackmann et al. 2009). Lipid droplets in yeast cells can clearly be visualized as shown in Figure 18.5 and confirmed by CARS microspectroscopy, comparative studies with GFP-labeled lipid droplets, and a negative control on cells of a strain unable to synthesize lipids. From the CARS images, quantitative data of size, shape, number, and intracellular location of the neutral lipid stores can be deduced with sub-micrometer three-dimensional resolution.

In proof-of-principle measurement on fission yeast (*Schizosaccharomyces pombe*), it was hypothesized that the strong intracellular CARS signal is derived from mitochondria and the endoplasmatic reticulum (Kano and Hamaguchi 2007). Based on the lack of co-localization between GFP-tagged mitochondria imaged with two-photon fluorescence microscopy and lipid droplets visualized by simultaneous CARS microscopy (Figure 18.6), Brackmann et al. instead suggest that the images by Kano et al. show lipid droplets.

Brackmann et al. observed that long-term starvation induces a significant accumulation of lipid droplets (Brackmann et al. 2009). Furthermore, a mutant (bcyl) unable to store carbohydrates showed a twofold increase in stored neutral lipids compared to wild-type cells. It is also notable that there is a significant cell-to-cell variability in neutral lipid storage in general, that is, there is a correspondence to the noise found for gene expression also in lipidomics. This further exemplifies the strength and usability of this method for two cases: the impact on lipid storage of the nutritional condition (starvation and type of carbon source available) as well as of genetic modification of two fundamental metabolic regulation pathways involving carbohydrate and lipid storage (BCY1 and DGA1, LRO1, ARE1/2 deletions), respectively.

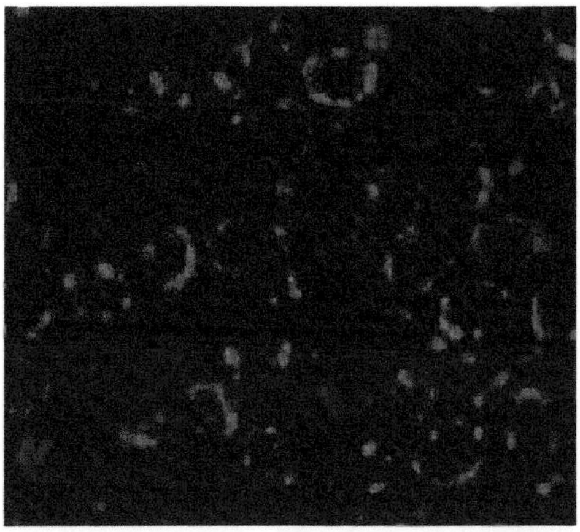

**FIGURE 18.6** CARS (red) and simultaneously collected two-photon fluorescence (green) microscopy images of yeast cells with GFP-labeled mitochondria. No colocalization is observed between fluorescent mitochondria (green) and lipid droplets imaged with CARS (red).

Chumnanpuen et al. have by means of CARS microscopy monitored the neutral lipid accumulation throughout the different metabolic phases of yeast under nutritional conditions relevant for large-scale fermentation (Chumnanpuen et al. 2011). They show that while lipid droplet sizes are fairly constant, the number of droplets is a dynamic parameter depending on whether the cell metabolism has to rely on glucose or ethanol. The lowest number of lipid droplets was observed in the transition phase between glucose and ethanol fermentation, followed by a buildup during the ethanol phase. The surplus of accumulated lipids was then mobilized during the glucose and ethanol starvation in the subsequent stationary phase. The highest amount of lipids was found in the ethanol phase and quantified to approximately 0.3 fL/cell. This is important information with the increasing interest for large-scale biosynthesis of nutritional lipids using microorganisms. In addition, this study demonstrates that lipid accumulation is a highly dynamic process and that CARS microscopy is able to follow and give new insights into these processes *in vivo* at single-cell level.

More detailed chemical information is obtained with spectroscopic measurements, and CARS spectroscopy in living budding yeast has been compared to spontaneous Raman spectroscopy. Compared to the required exposure time of 300 s for Raman spectroscopy, CARS spectra were acquired during 50 ms. Thus, CARS spectroscopy represents a much less time-consuming method than traditional Raman spectroscopy (Okuno et al. 2010). Faster measurements are clearly advantageous when studying metabolic processes in yeast cells and other organisms that have a very short generation time.

## 18.5 Zebrafish (*Danio rerio*)

Zebrafish is a small freshwater fish, a cyprinoid teleost that lives in shallow water at temperatures ranging from 24°C to 39°C. It has become a popular organism for several lines of biomedical research including developmental biology, genetics, toxicology, oncology, and neurobiology. It emerged as a model organism for modern biological investigation after the pioneering work of Streisinger et al. (1981).

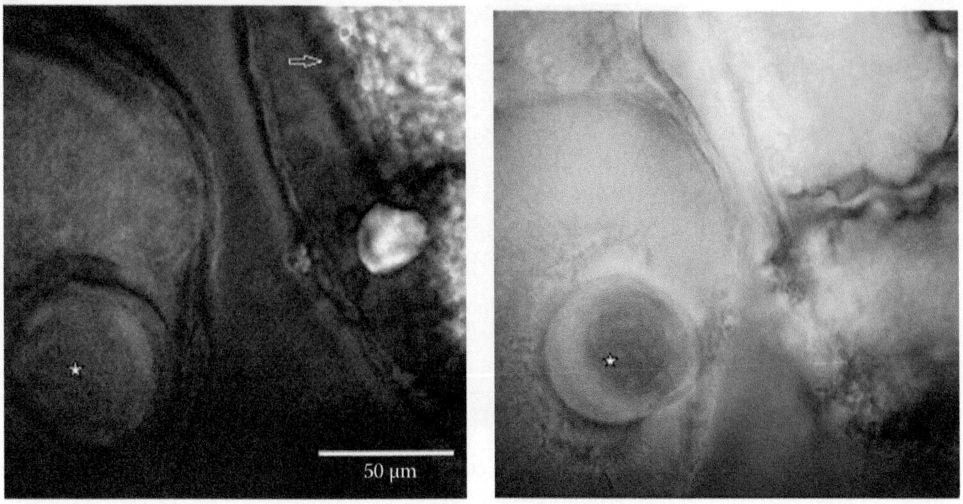


**FIGURE 18.7** CARS microscopy image measured in a zebrafish embryo (A) and corresponding bright-field image (B). The lipid-rich yolk gives a strong CARS signal as indicated by the white arrow. The eye of the embryo is indicated by a star.

The development and metabolic processes of zebrafish are very similar to higher vertebrates including humans. They also show many similarities with terrestrial vertebrates in terms of their genome, brain patterning, and structure and function of neurochemical and behavioral systems. Zebrafish have a short generation time of 3–5 months and produce a large number of progeny, which makes large-scale screens possible. Studies of mutants are easier because of the gene duplication resulting in extended survival of embryos with genetic defects. In laboratory environment, embryos hatch after 3 days and start to eat after 5 days after fertilization.

### 18.5.1 Imaging Lipid Metabolism in Zebrafish

Zebrafish is still new as a model system in lipid research and so far only proof-of-principle of CARS measurements on lipids has been performed. The high lipid content in the zebrafish yolk, consisting of neutral lipids and polar phospholipids, generates a strong signal in CARS microscopy as seen in Figure 18.7 (Holtta-Vuori et al. 2010). Although zebrafish has so far not been systematically evaluated, this organism shows great promise as a model system for lipid research, and combined with noninvasive visualization of lipids with CARS microscopy, this opens up for many future possibilities.

## 18.6 Algae

Algae are a large and diverse group of organisms, ranging from unicellular to multicellular forms. Microalgae are microscopic unicellular species, which exist individually or in chains or groups. Depending on the species, their size can range from a few up to hundreds of micrometers. Nearly all algae have photosynthetic machinery and they produce approximately half of the atmospheric oxygen while simultaneously consuming carbon dioxide to grow photoautotrophically.

## 18.6.1 Lipid Metabolism in Algae

The interests in the lipid metabolism of algae lie not in understanding the pathways leading to obesity but instead the balance between carbohydrates and lipids for efficient development of biofuels. Microalgae are considered one of the most promising feedstocks for biofuels since their productivity in converting carbon dioxide into energy-rich compounds, such as triacylglycerol and starch, greatly exceeds that of agricultural crops, without competing for arable land. Produced carbohydrates and lipids can be utilized for the production of several biofuels, including ethanol and biodiesel. Therefore, currently global research efforts aim at increasing and modifying the accumulations of lipids, carbohydrates, and other energy storage compounds through genetic engineering to enhance biofuel production (Radakovits et al. 2010).

Algae have also become interesting for the food and health industry as a source of omega-3 fatty acids. While fish oil has become famous for its omega-3 fatty-acid content, fish do not actually produce omega-3 acids, instead they accumulate omega-3 reserves by consuming microalgae. Therefore, microalgae oil has been considered as a vegetarian source for the omega-3 fatty acids: eicosapentaenoic acid (EPA) and docosahexaenoic acid (DHA) (Doughman et al. 2007).

## 18.6.2 Imaging Lipid Metabolism in Algae

CARS microscopy would constitute an excellent method to monitor changes in the amount and composition of lipids in different genetically modified algae species. Initial CARS imaging shown in Figure 18.8 demonstrates that lipid droplets can be clearly visualized in different algae.

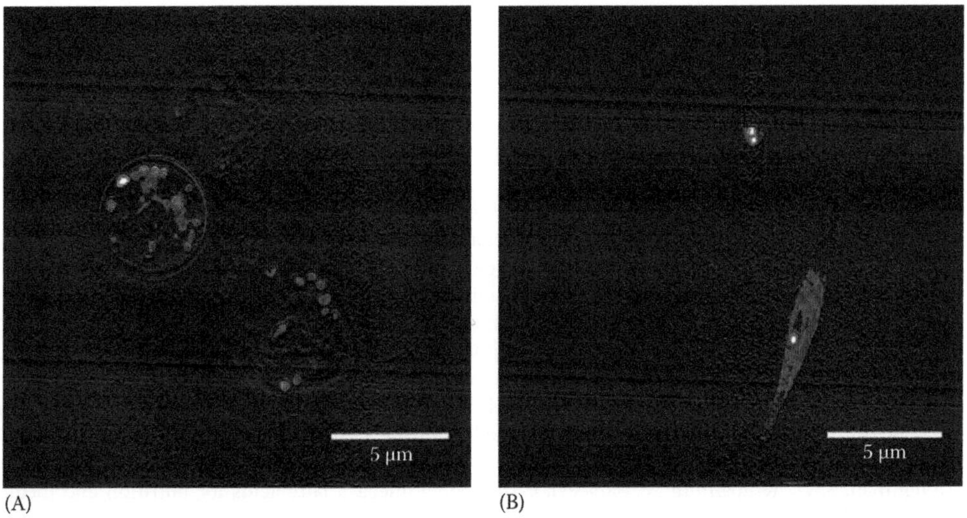


**FIGURE 18.8** CARS microscopy images of different microalgae: (A) tetraselmis and (B) phaeodactylum. Lipid droplets are clearly visible in both algae.

## 18.7 Conclusions

Current invasive techniques for lipid studies prevent characterization of the dynamic interactions between various lipid metabolism pathways. Label-free CARS imaging has proven to be a superior method for assessment of lipid storage in different model organisms and will contribute to the understanding of the complex processes involved in lipid synthesis, storage, and metabolism, and the regulation thereof. The technique provides a distinct measure of total lipid stores with greater accuracy and higher detail than previous visualization methods. Not only the amount of stored lipids but also the size and number of lipid droplets can be assessed. Additionally, the composition of individual lipid droplets can be analyzed.

Thus, CARS microscopy presents an exceptional opportunity to study lipid metabolism in living organisms like *C. elegans* and *S. cerevisiae* as well as other, not yet investigated, model systems and will in the future undoubtedly be invaluable in the strive to answer important questions in lipid research.

## Acknowledgments

Our studies of the lipid metabolism in *C. elegans* and other animal models using CARS microscopy would not have been possible without the collaboration with Professors Marc Pilon, Lena Gustafsson, Eva Albers, Jens Nielsen, Elina İkonen, and colleagues.

## References

Alberts, A. W., Ferguson, K., Hennessy, S., Vagelos, P. R. 1974. Regulation of lipid synthesis in cultured animal cells. *Journal of Biological Chemistry* 249: 5241–5249.

Ashrafi, K. 2006. Mapping out starvation responses. *Cell Metabolism* 3: 235–236.

Ashrafi, K., Chang, F. Y., Watts, J. L. et al. 2003. Genome-wide RNAi analysis of *Caenorhabditis elegans* fat regulatory genes. *Nature* 421: 268–272.

Athenstaedt, K., Zweytick, D., Jandrositz, A., Kohlwein, S. D., Daum, G. 1999. Identification and characterization of major lipid particle proteins of the yeast *Saccharomyces cerevisiae. Journal of Bacteriology* 181: 6441–6448.

Brackmann, C., Norbeck, J., Åkeson, M. et al. 2009. CARS microscopy of lipid stores in yeast: The impact of nutritional state and genetic background. *Journal of Raman Spectroscopy* 40: 748–756.

Brasaemle, D. L., Dolios, G., Shapiro, L., Wang, R. 2004. Proteomic analysis of proteins associated with lipid droplets of basal and lipolytically stimulated 3T3-L1 adipocytes. *Journal of Biological Chemistry* 279: 46835–46842.

Brenner, S. 1974. The genetics of *Caenorhabditis elegans. Genetics* 77: 71–94.

Chumnanpuen, P., Brackmann, C., Nandy, S. K. et al. 2011. Lipid biosynthesis monitored at the single-cell level in *Saccharomyces cerevisiae. Biotechnology Journal* 7(5): 594–601. doi 10.1002/biot.201000386

Clokey, G. V., Jacobson, L. A. 1986. The autofluorescent "lipofuscin granules" in the intestinal cells of *Caenorhabditis elegans* are secondary lysosomes. *Mechanisms of Ageing and Development* 35: 79–94.

Czabany, T., Athenstaedt, K., Daum, G. 2007. Synthesis, storage and degradation of neutral lipids in yeast. *Biochimica et Biophysica Acta* 1771: 299–309.

Dobrosotskaya, I. Y., Seegmiller, A. C., Brown, M. S., Goldstein, J. L., Rawson, R. B. 2002. Regulation of SREBP processing and membrane lipid production by phospholipids in Drosophila. *Science* 296: 879–883.

Doughman, S. D., Krupanidhi, S., Sanjeevi, C. B. 2007. Omega-3 fatty acids for nutrition and medicine: Considering microalgae oil as a vegetarian source of EPA and DHA. *Current Diabetes Review* 3: 198–203.

Fukumoto, S., Fujimoto, T. 2002. Deformation of lipid droplets in fixed samples. *Histochemistry and Cell Biology* 118: 423–428.

Gaber, B. P., Peticolas, W. L. 1977. On the quantitative interpretation of biomembrane structure by Raman spectroscopy. *Biochimica et Biophysica Acta* 465: 260–274.
Goffeau, A., Barrell, B. G., Bussey, H. et al. 1996. Life with 6000 genes. *Science* 274: 546, 563–547.
Greenspan, P., Mayer, E. P., Fowler, S. D. 1985. Nile red: A selective fluorescent stain for intracellular lipid droplets. *Journal of Cell Biology* 100: 965–973.
Greer, E. R., Perez, C. L., Van Gilst, M. R., Lee, B. H., Ashrafi, K. 2008. Neural and molecular dissection of a *C. elegans* sensory circuit that regulates fat and feeding. *Cell Metabolism* 8: 118–131.
Gronke, S., Mildner, A., Fellert, S. et al. 2005. Brummer lipase is an evolutionary conserved fat storage regulator in *Drosophila. Cell Metabolism* 1: 323–330.
Haemmerle, G., Lass, A., Zimmermann, R. et al. 2006. Defective lipolysis and altered energy metabolism in mice lacking adipose triglyceride lipase. *Science* 312: 734–737.
Hagmar, J., Brackmann, C., Gustavsson, T., Enejder, A. 2008. Image analysis in nonlinear microscopy. *Journal of the Optical Society of America A* 25: 2195–2206.
Hansen, M. K., Connolly, T. M. 2008. Nuclear receptors as drug targets in obesity, dyslipidemia and atherosclerosis. *Current Opinion in Investigational Drugs* 9: 247–255.
Hartwell, L. H. 2002. Nobel lecture: Yeast and cancer. *Bioscience Reports* 22: 373–394.
Hellerer, T., Axang, C., Brackmann, C. et al. 2007. Monitoring of lipid storage in *Caenorhabditis elegans* using coherent anti-Stokes Raman scattering (CARS) microscopy. *Proceedings of the National Academy of Sciences of the United States of America* 104: 14658–14663.
Hieb, W. F., Rothstein, M. 1968. Sterol requirement for reproduction of a free-living nematode. *Science* 160: 778–780.
Holtta-Vuori, M., Salo, V. T., Nyberg, L. et al. 2010. Zebrafish: Gaining popularity in lipid research. *Biochemical Journal* 429: 235–242.
Hosokawa, H., Ishii, N., Ishida, H. et al. 1994. Rapid accumulation of fluorescent material with aging in an oxygen-sensitive mutant mev-1 of *Caenorhabditis elegans. Mechanisms of Ageing and Development* 74: 161–170.
Hughes, A. L., Todd, B. L., Espenshade, P. J. 2005. SREBP pathway responds to sterols and functions as an oxygen sensor in fission yeast. *Cell* 120: 831–842.
Kaletsky, R., Murphy, C. T. 2010. The role of insulin/IGF-like signaling in *C. elegans* longevity and aging. *Disease Models and Mechanisms* 3: 415–419.
Kano, H., Hamaguchi, H. O. 2007. Supercontinuum dynamically visualizes a dividing single cell. *Analytical Chemistry* 79: 8967–8973.
Kenyon, C., Chang, J., Gensch, E., Rudner, A., Tabtiang, R. 1993. A *C. elegans* mutant that lives twice as long as wild type. *Nature* 366: 461–464.
Kimura, K. D., Tissenbaum, H. A., Liu, Y., Ruvkun, G. 1997. daf-2, an insulin receptor-like gene that regulates longevity and diapause in *Caenorhabditis elegans. Science* 277: 942–946.
Kniazeva, M., Crawford, Q. T., Seiber, M., Wang, C. Y., Han, M. 2004. Monomethyl branched-chain fatty acids play an essential role in *Caenorhabditis elegans* development. *PLoS Biol* 2: E257.
Kurat, C. F., Natter, K., Petschnigg, J. et al. 2006. Obese yeast: Triglyceride lipolysis is functionally conserved from mammals to yeast. *Journal of Biological Chemistry* 281: 491–500.
Le, T. T., Duren, H. M., Slipchenko, M. N., Hu, C. D., Cheng, J. X. 2010. Label-free quantitative analysis of lipid metabolism in living *Caenorhabditis elegans. Journal of Lipid Research* 51: 672–677.
Li, D., Zheng, W., Zeng, Y., Qu, J. Y. 2010. In vivo and simultaneous multimodal imaging: Integrated multiplex coherent anti-Stokes Raman scattering and two-photon microscopy. *Applied Physics Letters* 97: 223702–223703.
Luchsinger, J. A., Gustafson, D. R. 2009. Adiposity and Alzheimer's disease. *Current Opinion in Clinical Nutrition and Metabolic Care* 12: 15–21.
MacQueen, A. J., Phillips, C. M., Bhalla, N. et al. 2005. Chromosome sites play dual roles to establish homologous synapsis during meiosis in *C. elegans. Cell* 123: 1037–1050.
McKay, R. M., McKay, J. P., Avery, L., Graff, J. M. 2003. *C. elegans*: A model for exploring the genetics of fat storage. *Developmental Cell* 4: 131–142.
Moitra, J., Mason, M. M., Olive, M. et al. 1998. Life without white fat: A transgenic mouse. *Genes and Development* 12: 3168–3181.
Mooijaart, S. P., Brandt, B. W., Baldal, E. A. et al. 2005. *C. elegans* DAF-12, nuclear hormone receptors and human longevity and disease at old age. *Ageing Research Reviews* 4: 351–371.
Mörck, C., Olsen, L., Kurth, C. et al. 2009. Statins inhibit protein lipidation and induce the unfolded protein response in the non-sterol producing nematode *Caenorhabditis elegans. Proceedings of the National Academy of Sciences of the United States of America* 106: 18285–18290.

Nakae, J., Biggs, W. H., 3rd, Kitamura, T. et al. 2002. Regulation of insulin action and pancreatic beta-cell function by mutated alleles of the gene encoding forkhead transcription factor Foxo1. *Nature Genetics* 32: 245–253.

O'Rourke, E. J., Soukas, A. A., Carr, C. E., Ruvkun, G. 2009. *C. elegans* major fats are stored in vesicles distinct from lysosome-related organelles. *Cell Metabolism* 10: 430–435.

Okuno, M., Kano, H., Leproux, P. et al. 2010. Quantitative CARS molecular fingerprinting of single living cells with the use of the maximum entropy method. *Angewandte Chemie International Edition in English* 49: 6773–6777.

Radakovits, R., Jinkerson, R. E., Darzins, A., Posewitz, M. C. 2010. Genetic engineering of algae for enhanced biofuel production. *Eukaryotic Cell* 9: 486–501.

Raghow, R., Yellaturu, C., Deng, X., Park, E. A., Elam, M. B. 2008. SREBPs: The crossroads of physiological and pathological lipid homeostasis. *Trends in Endocrinology and Metabolism* 19: 65–73.

Rajakumari, S., Grillitsch, K., Daum, G. 2008. Synthesis and turnover of non-polar lipids in yeast. *Progress in Lipid Research* 47: 157–171.

Schroeder, L. K., Kremer, S., Kramer, M. J. et al. 2007. Function of the *Caenorhabditis elegans* ABC transporter PGP-2 in the biogenesis of a lysosome-related fat storage organelle. *Molecular Biology of the Cell* 18: 995–1008.

Sherrington, R., Rogaev, E. I., Liang, Y. et al. 1995. Cloning of a gene bearing missense mutations in early-onset familial Alzheimer's disease. *Nature* 375: 754–760.

Snyder, R. G., Hsu, S. L., Krimm, S. 1978. Vibrational spectra in the C–H stretching region and the structure of the polymethylene chain. *Spectrochimica Acta Part A: Molecular Spectroscopy* 34: 395–406.

Spanier, B., Lasch, K., Marsch, S. et al. 2009. How the intestinal peptide transporter PEPT-1 contributes to an obesity phenotype in *Caenorhabditits elegans. PLoS One* 4: e6279.

Speakman, J., Hambly, C., Mitchell, S., Krol, E. 2008. The contribution of animal models to the study of obesity. *Laboratory Animals* 42: 413–432.

Streisinger, G., Walker, C., Dower, N., Knauber, D., Singer, F. 1981. Production of clones of homozygous diploid zebra fish (*Brachydanio rerio*). *Nature* 291: 293–296.

Sundaram, M., Greenwald, I. 1993. Suppressors of a lin-12 hypomorph define genes that interact with both lin-12 and glp-1 in *Caenorhabditis elegans. Genetics* 135: 765–783.

Szymanski, K. M., Binns, D., Bartz, R. et al. 2007. The lipodystrophy protein seipin is found at endoplasmic reticulum lipid droplet junctions and is important for droplet morphology. *Proceedings of the National Academy of Sciences of the United States of America* 104: 20890–20895.

Vlachos, M., Tavernarakis, N. 2010. Non-apoptotic cell death in *Caenorhabditis elegans. Developmental Dynamics* 239: 1337–1351.

Welte, M. A., Cermelli, S., Griner, J. et al. 2005. Regulation of lipid-droplet transport by the perilipin homolog LSD2. *Current Biology* 15: 1266–1275.

Wood, W. B. 1988. *The Nematode Caenorhabditis elegans*. Cold Spring Harbor, NY: Cold Spring Harbor Laboratory.

Wurpel, G. W., Schins, J. M., Muller, M. 2002. Chemical specificity in three-dimensional imaging with multiplex coherent anti-Stokes Raman scattering microscopy. *Optics Letters* 27: 1093–1095.

Yang, F., Vought, B. W., Satterlee, J. S. et al. 2006. An ARC/mediator subunit required for SREBP control of cholesterol and lipid homeostasis. *Nature* 442: 700–704.

Yen, K., Le, T. T., Bansal, A. et al. 2010. A comparative study of fat storage quantitation in nematode *Caenorhabditis elegans* using label and label-free methods. *PLoS One* 5: e12810.

# 19. Lipid-Droplet Biology and Obesity-Related Health Risks

**Thuc T. Le**



## 19.1 Introduction

Discovered in the nineteenth century (Altmann 1890; Wilson 1896) and ignored for most of the twentieth century, lipid droplets are experiencing a renaissance in the twenty-first century due to the rise of the obesity epidemic and the prevalence of obesity-related health risks in human populations (Martin and Parton 2006; Farese and Walther 2009). Excessive lipid-droplet accumulation in adipocytes, macrophages, hepatocytes, and cancer cells is strongly linked to obesity and type 2 diabetes, cardiovascular diseases, fatty liver diseases, and aggressive cancer behaviors, respectively (Lusis 2000; Libby 2002; Calle and Kaaks 2004; Despres and Lemieux 2006; Hotamisligil 2006; Reddy and Rao 2006; Hotamisligil and Erbay 2008). Yet it remains vaguely understood the roles of lipid droplets in these major human diseases. Lipid droplets are dynamic organelles that expand or shrink according to the cellular metabolic states (Guo et al. 2009; Walther and Farese 2009). Yet lipid-droplet biology has been studied using mostly static measurement techniques that require cell fixation, lipid staining, or lipid extraction. The revival of coherent anti-Stokes

*Coherent Raman Scattering Microscopy*. Edited by Ji-Xin Cheng and X. Sunney Xie © 2013 CRC Press/ Taylor & Francis Group, LLC. ISBN: 978-1-4398-6765-5.

Raman scattering (CARS) microscopy in the last decade (Cheng and Xie 2004; Evans and Xie 2008), which allows label-free visualization of lipid droplets, is opening up exciting opportunities to examine the roles of lipid-droplet dynamics in disease development. This book chapter aims to provide a brief summary of lipid-droplet biology and to explore several critical research areas where CARS imaging of lipid-droplet dynamics can shed new light on the mechanisms underlying the correlation between lipid-droplet accumulation and the development of obesity-related health risks.

## 19.2 Lipid Droplets as a Conserved Energy Storage

Lipid droplets are cytoplasmic bodies comprising mainly of triacylglycerols (TAGs) and sterol esters (Martin and Parton 2006). TAG is an evolutionarily conserved source of energy formed by the ester bonds between a glycerol and three fatty acids. The hydrophobic packing of the TAG fatty acid chains provides a compact and anhydrous means of energy storage. The energy yield of fatty acid catabolism is 9 kcal/g, which is more than twice the energy yield of 4 kcal/g of carbohydrates, the next high-energy molecule in biological systems (Voet and Voet 2010). In addition, the packing of anhydrous fatty acids provides more energy per storage mass as compared to hydrous packing. For example, 1 g of glycogen, the carbohydrate energy storage form, can be hydrated with 2 g of water, which reduces the energy yield per gram to 1.33 kcal. Thus, for the same storage mass, fatty acids provide greater than sixfolds more energy than carbohydrates. The efficient means of energy storage is conserved from yeast to human where excess energy is converted to fatty acids and TAG and stored as cytoplasmic lipid droplets.

## 19.3 Lipid Droplets as a Dynamic Organelle

As energy storage depots, lipid droplets represent a dynamic pool of TAG that expands with nutritional excess and shrinks with starvation. However, since their initial description in the nineteenth century, lipid droplets were mostly viewed as simple inert and immobile energy depots whose structures and functions were sparsely characterized. In recent decades, lipid droplets are increasingly recognized as a dynamic organelle that participates in critical cellular processes including energy metabolism, vesicle trafficking, and signaling (Guo et al. 2009). This emerging dynamic view of lipid droplets is supported by many discoveries including (a) surface-coated proteins such as PAT proteins (perilipin, adipophilin/adipose differentiation-related protein, TIP47, and other related proteins), Rab18, caveolin-1 and -2, and acyl CoA:diacylglycerol acyltransferase 2 (DGAT2), (b) highly regulated lipolysis process that involves phosphorylation by protein kinase A (PKA) and translocation of lipase proteins to the surface of lipid droplets such as hormone-sensitive lipase (HSL) and adipose tissue triglyceride lipase (ATGL), (c) highly mobile pools of lipid droplets that interact with other cellular organelles such as the mitochondria and endoplasmic reticulum (ER), and (d) interaction with viral replication and assembly machinery (Martin and Parton 2006; Farese and Walther 2009; Guo et al. 2009; Walther and Farese 2009). Yet significant details of lipid-droplet biology remain unknown including the dynamics of their formation, growth, fusion, trafficking, and mobilization. The complex but vaguely understood biogenesis and functional roles of lipid droplets are becoming a new research frontier in cell biology (Farese and Walther 2009).

## 19.4 Lipid Droplets and Metabolic Diseases

Lipid droplets are a strong physiological indicator of metabolic diseases. Although the largest default energy storage site in mammals is the white adipose tissue (WAT), significant ectopic energy storage is also found in skeletal and cardiac muscles, as well as liver and pancreatic tissues (Meex et al. 2009). Increase in the concentration of lipid droplets of WAT as in obesity is directly associated with chronic inflammation, hyperlipidemia, and increased risks for type 2 diabetes and cardiovascular diseases (Hotamisligil 2006; Hotamisligil and Erbay 2008). Excess lipid droplets, or steatosis, in the muscle is associated with insulin resistance, which hinders the uptake of glucose leading to hyperglycemia and the development of type 2 diabetes (Kopelman 2000). Excess accumulation of lipid droplets in foam cells of the arterial wall is linked to the development of atherosclerosis and increased risks of thrombosis and stroke (Lusis 2000). In liver steatosis, or fatty liver, excess lipid-droplet accumulation can lead to insulin resistance of the hepatocytes, which reduces glycogen synthesis and increases glucose production and release into the blood, thus elevating the risk of type 2 diabetes development (Browning and Horton 2004; Reddy and Rao 2006). Lipid droplets are strongly correlated to human metabolic diseases and have been used extensively for pathological analysis. Nonetheless, how lipid droplets contribute to the onset and progression of metabolic diseases is not clearly understood. Consequently, targeting lipid droplets to treat metabolic diseases is a novel but unrealized concept due to the lack of mechanistic understanding of lipid-droplet dynamics (Farese and Walther 2009).

## 19.5 Lipid Droplets and Systems Biology

Biological studies of the last several centuries have been following a reductionist approach where complex organisms are characterized by their molecular composition. This reductionist approach culminated with the completion of the Human Genome Project in 2003 when the first complete human genetic blueprint was decoded (McPherson et al. 2001; Venter et al. 2001; Collins et al. 2003). However, it became clear that the degree of biological complexity is not linearly dependent on the size of the genetic blueprint or the number of coding genes (Claverie 2001). For example, the human genomes have 2.9 billion DNA base pairs, which are only 14% longer than the mouse genome with 2.5 billion DNA base pairs. Both human and mouse have approximately the same number of coding genes of between 20,000 and 25,000 genes. Furthermore, while some genetic diseases such as sickle cell anemia, cystic fibrosis, and Tay–Sachs disease are caused by mutations in single genes, the contribution of single gene mutations to common diseases including diabetes and cardiovascular diseases is relatively rare. It has been proposed that biological complexity arises from the dynamics of post-genomic events such as the control of gene and protein expression by noncoding RNAs, RNA splicing, histone modifications, and interactions between proteins, RNAs, and DNAs (Carninci et al. 2005; Katayama et al. 2005; Mattick 2007). Similarly, the causes of major human diseases have likely arisen from the accumulation of disruptive post-genomic events (Badano and Katsanis 2002). Accordingly, biological studies in the post-genomic era are focusing on the dynamic spatial–temporal interactions of the molecular components within complex organisms. The systemic approach to understand the architecture of biological networks, the mechanisms that govern decision-making processes such as cell differentiation and cell death, or the multifactorial causes of human diseases is aptly described as systems biology (Ideker et al. 2001; Kitano 2002; Weston and Hood 2004).

The roles of lipid-droplet dynamics in metabolic diseases can be investigated using systems biology approach. Lipid droplets serve as an excellent morphological readout for cellular metabolic states. An increase in the size or number of lipid droplets indicates an upregulation of lipid metabolism genes and an increase in the activity of lipid synthesis enzymes. Conversely, a reduction in lipid droplets indicates an increase in energy mobilization characterized by a suppression of lipid synthesis and an increase in lipid catabolism. The formation of lipid droplets involves synthesis of fatty acids and sterols in the cytosol, import of fatty acyl-CoA and sterols into the ER for TAG and phospholipids synthesis, export of lipid droplets from the ER, and fusion of the lipid droplets (Wakil et al. 1983; Guo et al. 2009). The mobilization of lipid droplets involves cAMP signaling, PKA phosphorylation of lipases, release of glycerol and free fatty acids to the microenvironment and bloodstream, transport of fatty acyl-CoA into the mitochondria for β-oxidation and energy generation, or diffusion of very long chain fatty acids into peroxisomes for incomplete β-oxidation where oxidation products are transported to mitochondria for further degradation (Walther and Farese 2009). In healthy state, lipid-droplet dynamics is tightly regulated to maintain the balance between energy storage and mobilization. In diseased states, impaired lipid synthesis, transport, oxidation, and peroxidation are key contributors to the accumulation and release of reactive oxygen species (ROS) and inflammatory lipid molecules leading to metabolic diseases and central nervous system injuries (Hotamisligil 2006; Adibhatla and Hatcher 2010). Such impairments clearly affect lipid-droplet dynamics; however, the impacts have not been described. A systems biology approach to study the correlation between lipid metabolism activities, lipid-droplet dynamics, and disease development promises to shed new understanding on the roles of lipid-droplet dynamics in health and diseases.

## 19.6 Lipid-Droplet Dynamics and CARS Microscopy

The studies of lipid-droplet dynamics and its roles in diseases require noninvasive visualization methods that allow monitoring of lipid droplets as a function of disease onset and progression. However, this requirement is challenging with standard light and fluorescent imaging techniques. In cell cultures, large lipid droplets can be visualized with light microscopy due to high optical diffraction, but the optical diffraction of small lipid droplets of less than 1 μm in diameter cannot be distinguished from that of cellular organelles. Lipid droplets are alternatively visualized with hydrophobic dyes using fixative or vital staining methods. Fixation facilitates the diffusion of hydrophobic dyes across the cell membrane to intercalate with cytoplasmic lipid droplets but prevents live cell studies. On the other hand, vital staining is inefficient to report lipid-droplet dynamics because its concentration is constant but the size and number of lipid droplets are variable in live cells. Vital staining with Nile red has been shown to label cellular structures other than lipid droplets, which further undermines the use of this method for lipid-droplet studies (O'Rourke et al. 2009; Yen et al. 2010). In addition, the incorporation of hydrophobic dyes onto the membrane can cause phase separation or induce lipid-raft formation and perturb cellular signaling (Palmantier et al. 2001; Ghosh et al. 2002;

Baumgart et al. 2007). The challenges facing lipid-droplet visualization and quantitation are major hindrances to the studies of lipid-droplet dynamics.

The powerful capabilities of CARS microscopy are opening up exciting new possibilities for the studies of lipid-droplet dynamics. In CARS microscopy, the contrast mechanism arises from the intrinsic molecular vibration of the probed molecules; thus, lipid-rich structures can be visualized without labeling using the abundance of $CH_2$ stretch vibration at 2845 $cm^{-1}$ (Cheng and Xie 2004). CARS microscopy is a multiphoton process that utilizes pulsed lasers at near infrared wavelengths; thus, it is capable of three-dimensional resolution and high penetration depth due to the lack of endogenous absorber in tissues at long wavelengths (Evans and Xie 2008). Furthermore, the quadratic dependence of CARS signal intensity on the oscillator concentration renders CARS microscopy highly sensitive to densely packed $CH_2$ chains of lipid-rich structures and permits video-rate image acquisition (Evans et al. 2005). With diffraction-limited optical resolution of ~300 nm, CARS microscopy is capable of resolving fine structures such as nascent lipid droplets (Nan et al. 2003) or the nodes of Ranvier and the Schmidt–Lanterman incisures of the myelin sheaths (Wang et al. 2005). In the last decade, CARS microscopy has been employed for label-free visualization of a wide range of lipid-rich structures in cell cultures, tissue explants, and live animals (Cheng 2007; Muller and Zumbusch 2007; Krafft et al. 2009). Especially, CARS microscopy has been used to study lipid-droplet synthesis (Nan et al. 2003; Le and Cheng 2009), trafficking (Nan et al. 2006), lipolysis (Yamaguchi et al. 2007), and interaction with the viral replication machinery (Lyn et al. 2009).

Most importantly, the multifunctionality of a CARS microscope renders it ideally suited for the studies of lipid-droplet dynamics in health and diseases. The pulsed lasers used in CARS microscopy can be employed for other nonlinear optical imaging modalities such as two-photon fluorescence (TPF), second harmonic generation (SHG), sum frequency generation (SFG), and third harmonic generation (THG) (Wang et al. 2009a). The signal detection scheme can be configured such that multiple imaging modalities can be performed simultaneously. Indeed, multimodal imaging has been employed to describe tumor microenvironment (Le et al. 2007b), atherosclerotic plaques (Le et al. 2007a; Wang et al. 2009b; Kim et al. 2010; Ko et al. 2010; Lim et al. 2010), spinal cord injuries (Fu et al. 2007, 2009; Shi et al. 2010), lipid droplets' interaction with cellular organelles (Nan et al. 2006; Lyn et al. 2009), and visualization of lipid-droplet synthesis together with cellular activities including insulin signaling and lipid metabolism gene expression (Le and Cheng 2009). Furthermore, spectral scanning CARS, multiplex CARS, or CARS with integrated spontaneous Raman spectroscopy allows Raman spectrum analysis of the composition of the interested lipid structures (Hellerer et al. 2007; Rinia et al. 2008; Slipchenko et al. 2009; Wu et al. 2009). The lipid composition analysis capability has been used to evaluate lipid-droplet packing order, lipid-chain unsaturation, trafficking of deuterated fatty acids, activity of desaturation enzymes, and the effects of drugs on lipid composition of foam macrophages (Rinia et al. 2008; Slipchenko et al. 2009; Kim et al. 2010; Le et al. 2010). The multifunctionality of CARS microscopy allows the visualization and analysis of lipid droplets together with other cellular activities (Le and Cheng 2009; Lyn et al. 2009). This capability is critical to the investigation of the roles of lipid-droplet dynamics in diseases using systems biology approach.

## 19.7 Lipid Droplets as Readout for Lipid Metabolism Activities

As a product of lipid metabolism, lipid droplets serve as a reliable readout for cellular energetic states. Recent applications of multimodal CARS microscopy to image living nematode *Caenorhabditis elegans* highlight the use of lipid droplets to study the dynamics of lipid metabolism (Hellerer et al. 2007; Morck et al. 2009; Le et al. 2010; Yen et al. 2010). With simultaneous CARS and TPF imaging, two dynamic lipid-droplet pools are observed (Le et al. 2010). The neutral lipid pool resides in both the intestinal and hypodermal cells whose signal shows up only with CARS imaging, whereas the fluorescent lipid pool resides exclusively in the intestinal cells whose signal shows up in both CARS and TPF imaging. Several independent studies point to the locations of fluorescent lipid droplets as the lysosome-related organelles (Clokey and Jacobson 1986; O'Rourke et al. 2009; Yen et al. 2010). These observations suggest that the neutral lipid pool represents the energy storage reservoir, whereas the fluorescent lipid pool represents the lipids being oxidized in lysosome-related organelles.

Multimodal CARS imaging of *C. elegans* reveals that the expression level and morphology of the lipid-droplet pools can be used to assay lipid metabolism activities (see Figure 19.1). In worms with genetic defect in sterol regulatory element binding protein 1 (SBP-1) (Yang et al. 2006), a nuclear transcription factor that controls the expression of lipid metabolism genes, the expression of both neutral and fluorescent lipid droplets are suppressed. In worms with genetic defect in insulin-like receptor DAF-2 (Kimura et al. 1997), the expression of neutral lipid droplets increases and fluorescent lipid droplets decreases, pointing to a shift toward lipid storage and away from oxidation. Interestingly, mutations in DAF-2 have been shown to double the lifespan of worms, suggesting a role of lipid oxidation in the lifespan of *C. elegans*. In worms with

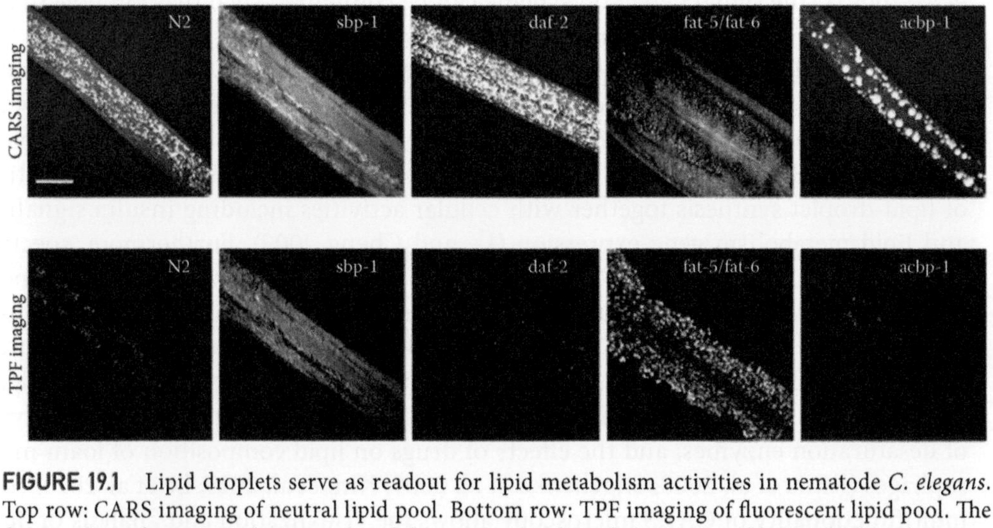


**FIGURE 19.1** Lipid droplets serve as readout for lipid metabolism activities in nematode *C. elegans*. Top row: CARS imaging of neutral lipid pool. Bottom row: TPF imaging of fluorescent lipid pool. The figures show changes in the neutral and fluorescent lipid pools between wild-type (N2) and mutant worms with mutations in the genes encoding for the sterol regulatory element binding protein 1 (sbp-1), insulin-like growth factor (daf-2), Δ9 desaturases (fat-5/fat-6), acyl-CoA binding protein 1 (acbp-1). Scale bar: 25 μm.

genetic defects in Δ9 desaturases (Brock et al. 2007), the level of neutral lipid droplets decreases while fluorescent lipid droplets increases, suggesting the importance of lipid desaturation for neutral lipid-droplet synthesis and the lack of lipid desaturation activity could drive lipids toward oxidation in the lysosome-related organelles. In worms with genetic defect in acyl-CoA binding protein 1 (ACBP-1) (Kragelund et al. 1999), a transporter protein that shuttles fatty acyl-CoA to the ER for TAG and phospholipid synthesis or to the mitochondria for β-oxidation, the average size of neutral lipid droplets increases more than threefolds from an average of 1.5 μm in wild-type worms to over 5 μm in diameter in ACBP-1 mutant worms. This observation suggests possible impairment in phospholipid biosynthesis, which is a key determinant of lipid-droplet size and number, in ACBP-1 mutant worms. Taken together, the genotype–phenotype screening of lipid droplets in wild-type and mutant worms with CARS microscopy clearly shows the potential use of lipid droplets as a readout for lipid metabolism activity. Because lipid metabolism pathways are conserved from worms to humans, it is conceivable that CARS imaging of lipid droplets could also be used to evaluate lipid metabolism activities in human metabolic diseases.

## 19.8 Lipid Droplets and the Metabolism–Immunity Interface

The immune response and metabolic regulation are highly integrated and evolutionarily conserved. From the evolutionary perspective, metabolic and immune organs in higher organisms evolve from the same ancestral structure. In *Drosophila melanogaster*, a single functional unit known as the "fat body" carries out critical functions such as nutrient sensing and immune response. In mammals, the functions of the "fat body" are performed with distinct and specialized organs of adipose tissues, liver, and hematopoietic system (Hotamisligil 2006). Nonetheless, the regulation of metabolic and immune response of higher organisms remains highly conserved and coordinated. Nuclear receptors such as the peroxisome proliferator–activated receptors (PPARs) and liver X receptors (LXRs), which are activated by free fatty acids and cholesterol metabolites, exert regulatory control over the expression of a wide range of both metabolic and inflammatory genes (Bensinger and Tontonoz 2008). Both adipocytes and macrophages have a large capacity to store intracellular lipids, have extensive overlapping in the expression of metabolic and inflammatory genes, and secrete similar cytokines. The architecture of the adipose tissues and liver still resemble that of the "fat body," where the metabolic cells such as adipocytes or hepatocytes are within proximity of the immune cells such as macrophages, Kupffer cells, and lymphocytes. The conserved organization and regulation of metabolic and immune cells support their functional dependence and help to explain the prevalence of chronic inflammation in metabolic diseases.

Obesity is commonly associated with chronic adipose tissue inflammation due to increased accumulation of adipose tissue macrophages or ATMs. The exact function of ATMs remains unknown. However, ATMs exhibit functional and morphological polarization that parallel with the progression of obesity (Lumeng et al. 2007a, 2008). In lean adipose tissues, ATMs are mostly in the M2 state and express anti-inflammatory markers such as arginase and interleukin 10 (IL-10). On the contrary, in obese adipose tissues,

ATMs are mostly in the M1 state and express pro-inflammatory markers such as TNFα, IL-6, and inducible nitric oxide synthase (iNOS). M1 state ATMs also exhibit increased intracellular lipid accumulation and increased expression of lipid metabolism genes, including those encoding for peroxisome proliferator–activated receptor-γ (PPARγ) and adipocyte differentiation–related protein (ADFP) (Hotamisligil 2006; Hotamisligil and Erbay 2008; Tontonoz and Spiegelman 2008). It has been proposed that the paracrine loop between adipocytes and ATMs and the ratio of certain regulatory T cells regulate ATMs' phenotypic switch (Lumeng et al. 2007a; Nishimura et al. 2009). Recently, immunotherapy with CD4+ T lymphocytes, which prevents obesity-associated insulin resistance in mice, shows an association with reversed ATMs' polarization from M1 toward the M2 state (Winer et al. 2009). In addition, a growing body of literature points to the significance of ATMs' phenotypic switch in the progression of obesity-associated diseases and supports the interference with ATMs' polarization as a viable therapeutic strategy for insulin resistance and type 2 diabetes (Hotamisligil 2006; Lumeng et al. 2007a; Odegaard et al. 2007).

Critical for the functional integration of the metabolic and immune systems is the dynamic communication between adipocytes and ATMs. Yet adipocyte-ATM communication is one of the least studied systems. Is increase in the M1-to-M2 ATM ratio in obesity due to the recruitment of M1 macrophages or due to the proliferation of resident M2 ATMs and the activation of resident M2 into M1 ATMs? If hypertrophic adipocytes can activate the phenotypic switch of ATMs as proposed, are adipokines and lipokines sufficient or the malfunction of regulatory T cells must be a corequisite? How does the phenotypic switch affect the reverse cholesterol transport of ATMs? How do M1 ATMs affect adipocytes' metabolic activities such as insulin sensitivity, glucose uptake and utilization, lipid transport, lipogenesis, and lipolysis? Do M1 ATMs contribute to the development of obesity by affecting the preadipocyte–adipocyte transition? Do M1 ATMs affect angiogenesis and extracellular matrix remodeling in adipose tissues? These questions are important to the understanding of the immunity and metabolic interface. Yet they remain vaguely understood. Clearly, further in-depth studies of adipocyte–ATM communication and its roles in obesity-associated diseases are critically needed. Understanding the molecular signals that control adipocyte–ATM communication could provide specific targets for the intervention of obesity-associated immunity and metabolic diseases.

CARS microscopy is an ideal tool to probe the metabolism–immunity interface by examining the adipocyte–macrophage communication. Previously, CARS microscopy has been used to examine subcutaneous and mammary adipocytes using signals emanating from the monocular lipid droplets. CARS microscopy has also been used to detect foam macrophages of atherosclerotic plaques using signals arising from the cytoplasmic lipid droplets. In the visceral adipose tissues of diet-induced obesity mice, CARS imaging reveals cohabitation of adipocytes and foam cells (see Figure 19.2). These foam cells have previously been identified as pro-inflammatory M1 state ATMs (Lumeng et al. 2007a,b, 2008). Yet it remains unknown the relationship between lipid-droplet accumulation in ATMs and the increase in the production of the pro-inflammatory signaling molecules. On the other hand, adipocytes secrete soluble bioactive peptides and lipids, known as adipokines and lipokines, which exert endocrine functions to regulate systemic energy homeostasis (Cao et al. 2008). In obesity, the secretion

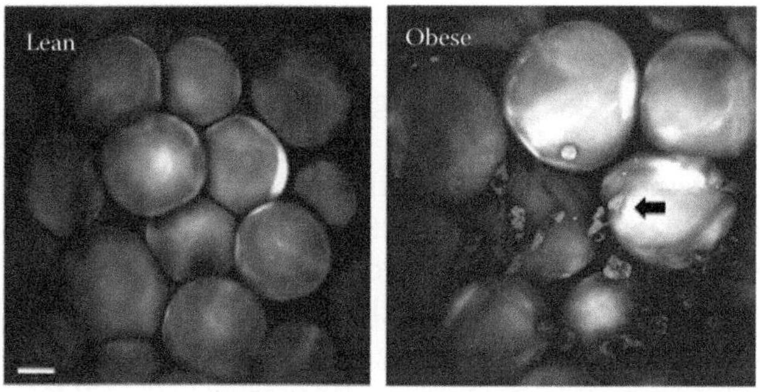


FIGURE 19.2 CARS visualization of lipid droplets of lean (left panel) and obese visceral adipose tissues (right panel). Arrow points to one of many foam cells residing in the obese visceral adipose tissue. Scale bar: 25 μm.

profiles of adipokines and lipokines are altered leading to altered neuronal control of satiety, hyperlipidemia, and local and systemic inflammation. However, it is not clear how pro-inflammatory macrophages perturb the secretion profiles of adipocytes. To study adipocyte–macrophage communication, CARS microscopy can be applied to monitor lipid-droplet dynamics of macrophages and adipocytes to study how perturbations to lipid metabolism affect the proliferation, differentiation, and function of the immunity and metabolic cells. Multifunctional CARS microscopy can be deployed to measure insulin signaling, glucose transport, and metabolic or inflammatory gene activities. In addition, *in vivo* imaging with multifunctional CARS microscopy should allow the examination of extracellular matrix remodeling or recruitment of immune cells as a function of adipocyte–macrophage communication in lean and obese adipose tissues. It is foreseeable that CARS microscopy will serve as an indispensible tool for the studies of inflammation in obesity-related health risks.

## 19.9 Lipid Droplets and Fatty Liver Disease

The liver plays a central role in whole-body energy metabolism by regulating the plasma glucose and lipid levels (Browning and Horton 2004; Adams et al. 2005). In times of energy excess, plasma glucose enters the hepatocytes to be stored as glycogen or broken down into acetyl-CoA for de novo fatty acid synthesis, and free fatty acids are imported to be converted into TAG for storage. In times of energy demand, glycogen is broken down to glucose and TAG is converted into very low density lipoprotein (VLDL) for secretion back into the bloodstream. Excessive alcohol consumption, obesity, or metabolic and genetic conditions that affect fatty acid metabolism can disrupt the liver function and lead to chronic liver condition such as fatty liver disease (Anderson and Borlak 2008). Currently, fatty liver disease is affecting up to 25% of the general adult population and over 70% of obese or diabetic individuals in the United States (Reddy and Rao 2006). The persistence of fatty liver disease invariably leads to the development of steatohepatitis or inflammation of fatty liver, fibrosis, cirrhosis, liver cancer, and ultimately liver failure. A defining hallmark of fatty liver disease is the excessive accumulation of lipid droplets in hepatocytes. Yet the role of lipid droplets in the pathogenesis of fatty liver disease is not clearly understood. Understanding lipid-droplet biology of hepatocytes is of tremendous interest to the studies of fatty liver disease.

Lipid-droplet accumulation in hepatocytes is dependent on the balance between lipid synthesis and lipid oxidation. In the liver, de novo fatty acid synthesis is regulated independently by insulin and glucose through the activation of the transcription factors sterol regulatory element–binding protein-1c (SREBP-1c) and carbohydrate response element–binding protein (ChREBP), respectively (Browning and Horton 2004; Anderson and Borlak 2008). These transcription factors promote the conversion of excess glucose into fatty acids by turning on genes and activating proteins necessary for glycolysis and lipogenesis. In addition, the activation of the transcription factor PPARγ is also observed for the expression of lipogenesis genes required for fatty acid and lipid-droplet synthesis. Endogenous and exogenous fatty acids are transported to the ER for TAG and phospholipid synthesis and conversion into lipid droplets. On the other hand, lipid oxidation in the liver is mediated by the nuclear receptor PPARα, which serves as a lipid sensor (Reddy and Rao 2006). In responding to fatty acid influx, PPARα induces the activities of all genes encoding for the mitochondrial and peroxisomal β-oxidation systems and microsomal ω-oxidation system. The dynamic balance between the sensing of insulin, glucose, and fatty acids of the transcription factors and between lipid synthesis and oxidation activity is critical to the health of the liver. Impairments in the sensing mechanism or lipid metabolism are the main cause of hepatic steatosis (Browning and Horton 2004; Glass and Ogawa 2006; Reddy and Rao 2006).

Hepatic steatosis is a prerequisite for the development of steatohepatitis and other end-stage liver diseases. While the mechanisms underlying liver disease progression are yet to be determined, extensive data point to lipid-induced cellular injuries due to lipid oxidation, lipid peroxidation, and oxidative stress as the critical causes (Browning and Horton 2004; Adams et al. 2005; Reddy and Rao 2006; Anderson and Borlak 2008). The main site for lipid β-oxidation occurs in the mitochondria using the mitochondrial respiratory chain (MRC). Impairment of the MRC can disrupt the electron flow and transfer the electron to oxygen molecules forming superoxide anions and hydrogen peroxide or ROS. Mitochondrial dysfunction or excessive cytosolic fatty acids promotes alternative β-oxidation pathway in peroxisomes and ω-oxidation in microsomes leading to further increase in ROS. In addition, a product of microsomal fatty acid ω-oxidation, dicarboxylic acids, can impair mitochondrial function by uncoupling oxidative-phosphorylation and causing further oxidative stress (Tonsgard and Getz 1985; Hermesh et al. 1998). While ROS are relatively short lived, they cause peroxidation of polyunsaturated fatty acids and form stable aldehyde by-products such as *trans*-4-hydroxy-2-nonenal (HNE) and malondialdehyde (MDA) (Esterbauer et al. 1991). Aldehydes HNE and MDA are cytotoxic, which cause ATP and NAD depletion, DNA and protein damage, and glutathione depletion leading to eventual cell death. HNE and MDA can diffuse into the extracellular space and induce inflammation by attracting neutrophils. Furthermore, ROS and lipid peroxidation products can induce hepatic stellate cells to synthesize collagen, which can lead to liver fibrosis (Browning and Horton 2004). Taken together, lipid overloading of hepatocytes appears to be a major factor in the impairment of lipid oxidation machinery and the induction of oxidative stress and inflammation of the liver.

The multifunctionality of a CARS microscope renders it an ideal diagnostic and research imaging tool to study fatty liver disease. Combined CARS, SHG, and TPF imaging of the liver allows label-free detection of lipid droplets of hepatocytes and Kupffer cells, collagen fibrils, and nicotinamide adenine dinucleotide phosphate (NADPH) autofluorescence, respectively (Chen et al. 2009). Integrated spontaneous Raman spectroscopy further allows spectral analysis of lipid-droplet composition (Wu et al. 2009) and the detection of flavin adenine dinucleotide (FAD) (see Figure 19.3). These capabilities should be of tremendous significance for clinical diagnosis of hepatic steatosis, inflammation, fibrosis, and metabolic activities based on the molecular concentration of NADPH and FAD and the degree of lipid-chain unsaturation. In addition, multifunctional CARS microscopy has been used to study various aspects of lipid metabolism including lipid-droplet dynamics in mammalian cells (Nan et al. 2006) and interactions between lipid storage, desaturation, and peroxidation in *C. elegans* (Le et al. 2010). It is conceivable that similar applications of multifunctional CARS microscopy to study lipid metabolism in the liver should shed light on the mechanisms underlying hepatic steatosis and the progression toward end-stage liver diseases. The ability to assay various aspects of lipid metabolism should also allow multifunctional CARS microscopy to be an effective tool to screen drugs targeting lipid metabolism for the treatment of fatty liver disease.

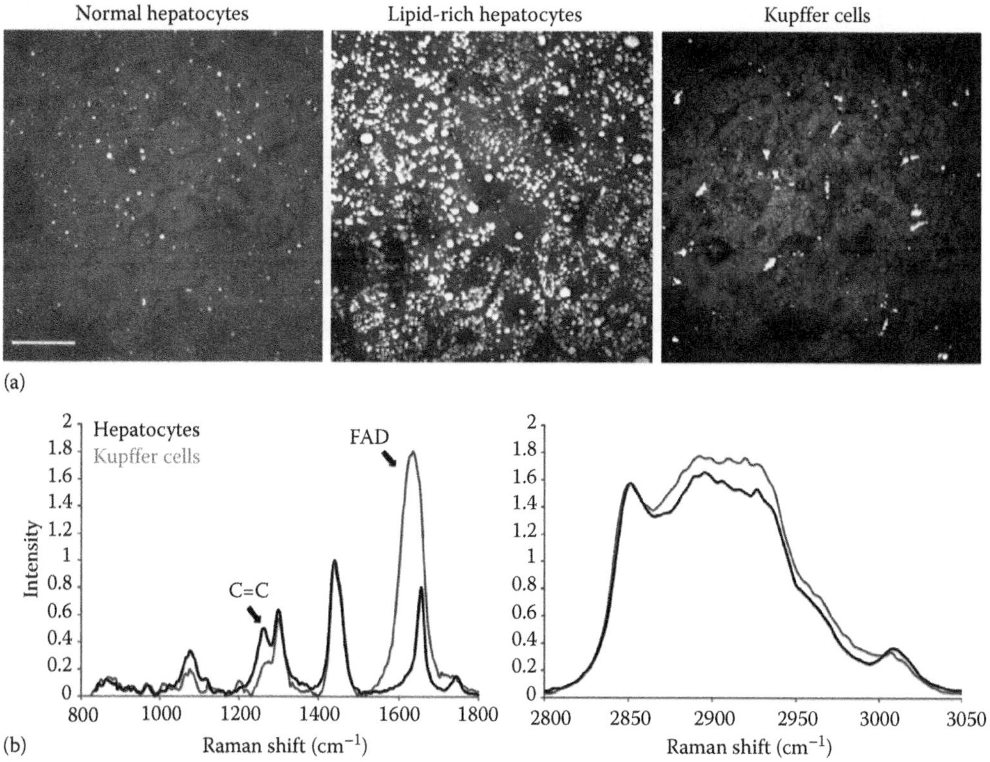


**FIGURE 19.3** Label-free imaging and composition analysis of lipid droplets in liver tissues. (a) CARS imaging of a normal liver tissue (left panel), a liver tissue with hepatic steatosis (middle panel), and a liver tissue with high Kupffer cells infiltration (right panel). Kupffer cells are small lipid-rich cells that wedge in the interstitial space between hepatocytes. Scale bars: 25 μm. (b) Spontaneous Raman spectral analysis of lipid droplets in hepatocytes (black) and Kupffer cells (gray). Lipid droplets of Kupffer cells exhibit more saturated lipid chain and a prominent peak at 1612 $cm^{-1}$, which indicates the presence of flavin adenine dinucleotide (FAD). FAD serves as a prosthetic group for inducible nitric oxide synthase (iNOS), an enzyme that participates in the inflammatory response of Kupffer cells.

## 19.10 Lipid Droplets and Cancer Cell Metabolism

Cytoplasmic lipid droplets have been observed in many types of cancers (Ramos and Taylor 1973; Sijens et al. 1996; Metser et al. 2006). Early clinical studies in the 1970s linked lipid-rich carcinoma of mammary glands to high incidence of cancer mortality, metastatic tumors, and aggressive clinical behavior (Ramos and Taylor 1973). In recent years, increased expression of lipid metabolism genes (Menendez and Lupu 2007) has been found in aggressive brain (Seyfried and Mukherjee 2005), mammary (Alli et al. 2005), and prostate cancer (Migita et al. 2009). More specifically, prostate tumors overexpressing fatty acid synthase (FAS) exhibit more aggressive behavior compared to tumors with normal FAS levels (Baron et al. 2004). Overexpression of fatty acid binding protein 7 (FABP-7) in human melanoma (Slipicevic et al. 2008) and FABP-5 in human prostate and breast cancer cells (Jing et al. 2000) are associated with increased invasion and proliferation potential of cancer cells. Together, these observations show a strong relationship between lipid metabolism and cancer aggressiveness (Menendez and Lupu 2007). However, limited understanding of the roles of lipids in cancer development hinders their use as a factor in cancer diagnosis.

A key hallmark of cancer is the altered energy metabolism of cancer cells (Warburg 1956; Gatenby and Gillies 2004; Hsu and Sabatini 2008). Described in the 1920s by Otto Warburg, cancer cells prefer to metabolize glucose by glycolysis, even in the abundance of oxygen, despite the significantly less energy yield than oxidative phosphorylation. The reasons for the preferred anaerobic respiration of cancer cells remain vaguely understood. However, it has been proposed that cancer cell metabolism is reflective of proliferative metabolism where the increases in biomass of nucleotides, amino acids, and lipids are more critical for the production of new cells than energy efficiency (Vander Heiden et al. 2009). This proposal lends further support to the roles of lipid metabolism in cancer development. Indeed, in addition to altered glucose metabolism, studies in human cancer patients reveal an increase in free fatty acid turnover, oxidation, and clearance.

CARS imaging of lipid droplets in cancer cells can be used for cancer diagnosis and for the studies of lipid metabolism in cancer development. In recent years, CARS microscopy has been applied to tumor imaging in tissue biopsies based on lipid ($CH_2$) and protein (amide) vibrational contrast with relative success (Evans et al. 2007; Le et al. 2007b, 2009; Chowdary et al. 2010). Particularly, CARS imaging reveals lipid-droplet accumulation in invading, circulating, and metastasized cancer cells (Le et al. 2009). These observations suggest the potential use of CARS microscopy for label-free intravital imaging of lipid-rich cancer cell migration and intravital flow cytometry to enumerate lipid-rich circulating tumor cells. They also strongly support the correlation between increased lipid metabolism and aggressive cancer cell behaviors. Future in-depth applications of CARS microscopy to study lipid-droplet biology in cancer cells should further the understanding of the roles of lipids in cancer cell metabolism and cancer aggressiveness (see Figure 19.4). As lipid metabolism proteins such as FAS and FABP are becoming therapeutic targets for cancer treatment (Baron et al. 2004; Furuhashi and Hotamisligil 2008), CARS microscopy will conceivably find increasing applications in cancer research.

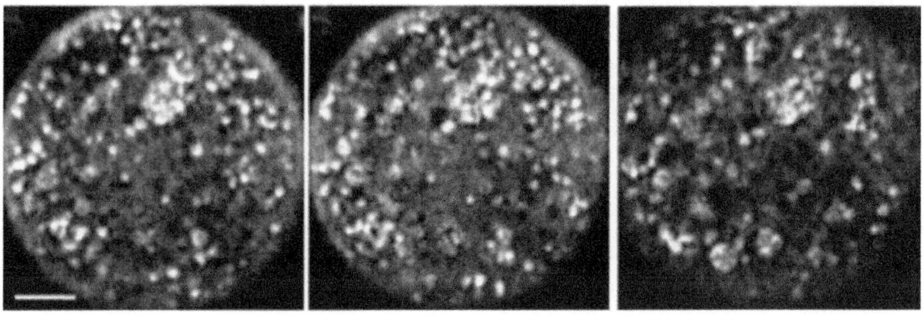


FIGURE 19.4 Dynamics of lipid droplets in cancer cells visualized with CARS microscopy. Lipid droplets of a metastatic lung cancer cell line M109 exhibit high mobility as a function of time. Scale bar: 5 μm.

## 19.11 Summary

Obesity has reached an epidemic proportion in the United States and many developed nations (Kopelman 2000). The rise of the obesity epidemic is accompanied by the increased frequency of metabolic diseases and risks for many type of cancers among the affected human populations (Calle and Kaaks 2004; Despres and Lemieux 2006). The obesity-related health risks are currently the greatest threat to human welfare globally. A common feature in all obesity-related illnesses is the excessive accumulation of intracellular lipid droplets in affected tissues. Yet lipid droplets are the least studied cellular organelle. Consequently, how lipid droplets contribute to the development and progression of obesity-related diseases is vaguely understood. As the main energy source for cell metabolism, lipid droplets are constantly being modified for growth or shrinkage according to cellular energy demand (Guo et al. 2009). However, lipid droplets have been characterized mostly with methods that require fixation or extraction. Clearly, new technologies that allow the studies of lipid-droplet dynamics are critically needed to further our understanding of this important organelle.

The recent development of multifunctional CARS microscopy is removing the technological barrier and opening up exciting opportunities for the studies of lipid-droplet dynamics. CARS microscopy allows label-free visualization of lipid droplets in living cells with high sensitivity and resolution (Cheng 2007; Evans and Xie 2008). More importantly, multifunctional CARS microscopy allows multiple biological structures and cellular events to be monitored simultaneously with lipid-droplet dynamics. The latter capability enables the studies of lipid-droplet dynamics in disease development using systems biology approach. Excessive lipid-droplet accumulation in visceral adipose tissue alters its endocrine function and induces chronic inflammation leading to altered energy metabolism and the development of type 2 diabetes (Hotamisligil 2006; Hotamisligil and Erbay 2008). Excessive lipid-droplet accumulation in macrophages residing in the arterial walls is associated with the development of atherosclerotic plaques and the risks of thrombosis and stroke (Lusis 2000; Libby 2002). Excessive lipid-droplet accumulation in hepatocytes causes hepatic steatosis, which the persistence invariably leads to end-stage liver diseases (Reddy and Rao 2006). Excessive lipid-droplet accumulation in cancer cells promotes aggressive behaviors by inducing tissue invasion and metastasis (Le et al. 2009). While the list of lipid droplet–associated diseases can expand further, the few diseases aforementioned highlight the importance of lipid droplets in human disease development. By enabling the visualization of lipid-droplet dynamics, CARS microscopy is poised to become an indispensible imaging tool for the studies of lipid-droplet biology in obesity-related health risks.

## References

Adams, L. A., P. Angulo et al. (2005). Nonalcoholic fatty liver disease. *CMAJ* **172**(7): 899–905.

Adibhatla, R. M. and J. F. Hatcher (2010). Lipid oxidation and peroxidation in CNS health and disease: From molecular mechanisms to therapeutic opportunities. *Antioxidants and Redox Signaling* **12**(1): 125–169.

Alli, P. M., M. L. Pinn et al. (2005). Fatty acid synthase inhibitors are chemopreventive for mammary cancer in neu-N transgenic mice. *Oncogene* **24**(1): 39–46.

Altmann, R. (1890). *Die Elementarorganisem und ihre Beziehungen zu den Zellen.* (Leipzig, Germany: Veit).

Anderson, N. and J. Borlak (2008). Molecular mechanisms and therapeutic targets in steatosis and steatohepatitis. *Pharmacological Reviews* **60**(3): 311–357.

Badano, J. L. and N. Katsanis (2002). Beyond Mendel: An evolving view of human genetic disease transmission. *Nature Reviews Genetics* **3**(10): 779–789.

Baron, A., T. Migita et al. (2004). Fatty acid synthase: A metabolic oncogene in prostate cancer? *Journal of Cellular Biochemistry* **91**(1): 47–53.

Baumgart, T., A. T. Hammond et al. (2007). Large-scale fluid/fluid phase separation of proteins and lipids in giant plasma membrane vesicles. *Proceedings of the National Academy of Sciences of the United States of America* **104**(9): 3165–3170.

Bensinger, S. J. and P. Tontonoz (2008). Integration of metabolism and inflammation by lipid-activated nuclear receptors. *Nature* **454**(7203): 470–477.

Brock, T. J., J. Browse et al. (2007). Fatty acid desaturation and the regulation of adiposity in *Caenorhabditis elegans. Genetics* **176**(2): 865–875.

Browning, J. D. and J. D. Horton (2004). Molecular mediators of hepatic steatosis and liver injury. *Journal of Clinical Investigation* **114**(2): 147–152.

Calle, E. E. and R. Kaaks (2004). Overweight, obesity and cancer: Epidemiological evidence and proposed mechanisms. *Nature Reviews Cancer* **4**(8): 579–591.

Cao, H. M., K. Gerhold et al. (2008). Identification of a lipokine, a lipid hormone linking adipose tissue to systemic metabolism. *Cell* **134**(6): 933–944.

Carninci, P., T. Kasukawa et al. (2005). The transcriptional landscape of the mammalian genome. *Science* **309**(5740): 1559–1563.

Chen, H. T., H. F. Wang et al. (2009). A multimodal platform for nonlinear optical microscopy and microspectroscopy. *Optics Express* **17**(3): 1282–1290.

Cheng, J. X. (2007). Coherent anti-Stokes Raman scattering microscopy. *Applied Spectroscopy* **61**(9): 197a–208a.

Cheng, J. X. and X. S. Xie (2004). Coherent anti-Stokes Raman scattering microscopy: Instrumentation, theory, and applications. *Journal of Physical Chemistry B* **108**(3): 827–840.

Chowdary, P. D., Z. Jiang et al. (2010). Molecular histopathology by spectrally reconstructed nonlinear interferometric vibrational imaging. *Cancer Research* **70**(23): 9562–9569.

Claverie, J. M. (2001). Gene number. What if there are only 30,000 human genes? *Science* **291**(5507): 1255–1257.

Clokey, G. V. and L. A. Jacobson (1986). The autofluorescent lipofuscin granules in the intestinal-cells of *Caenorhabditis-elegans* are secondary lysosomes. *Mechanisms of Ageing and Development* **35**(1): 79–94.

Collins, F. S., M. Morgan et al. (2003). The human genome project: Lessons from large-scale biology. *Science* **300**(5617): 286–290.

Despres, J. P. and I. Lemieux (2006). Abdominal obesity and metabolic syndrome. *Nature* **444**(7121): 881–887.

Esterbauer, H., R. J. Schaur et al. (1991). Chemistry and biochemistry of 4-hydroxynonenal, malonaldehyde and related aldehydes. *Free Radical Biology and Medicine* **11**(1): 81–128.

Evans, C. L., E. O. Potma et al. (2005). Chemical imaging of tissue in vivo with video-rate coherent anti-Stokes Raman scattering microscopy. *Proceedings of the National Academy of Sciences of the United States of America* **102**(46): 16807–16812.

Evans, C. L. and X. S. Xie (2008). Coherent anti-Stokes Raman scattering microscopy: Chemically selective imaging for biology and medicine. *Annual Review of Analytical Chemistry* **1**(1): 883–909.
Evans, C. L., X. Y. Xu et al. (2007). Chemically-selective imaging of brain structures with CARS microscopy. *Optics Express* **15**(19): 12076–12087.
Farese, R. V. and T. C. Walther (2009). Lipid droplets finally get a little R-E-S-P-E-C-T. *Cell* **139**(5): 855–860.
Fu, Y., W. Sun et al. (2009). Glutamate excitotoxicity inflicts paranodal myelin splitting and retraction. *PLoS One* **4**(8): e6705.
Fu, Y., H. F. Wang et al. (2007). Coherent anti-Stokes Raman scattering imaging of myelin degradation reveals a calcium-dependent pathway in lyso-PtdCho-induced demyelination. *Journal of Neuroscience Research* **85**(13): 2870–2881.
Furuhashi, M. and G. S. Hotamisligil (2008). Fatty acid-binding proteins: Role in metabolic diseases and potential as drug targets. *Nature Reviews Drug Discovery* **7**(6): 489–503.
Gatenby, R. A. and R. J. Gillies (2004). Why do cancers have high aerobic glycolysis? *Nature Reviews Cancer* **4**(11): 891–899.
Ghosh, P. K., A. Vasanji et al. (2002). Membrane microviscosity regulates endothelial cell motility. *Nature Cell Biology* **4**(11): 894–900.
Glass, C. K. and S. Ogawa (2006). Combinatorial roles of nuclear receptors in inflammation and immunity. *Nature Reviews Immunology* **6**(1): 44–55.
Guo, Y., K. R. Cordes et al. (2009). Lipid droplets at a glance. *Journal of Cell Science* **122**(6): 749–752.
Hellerer, T., C. Axang et al. (2007). Monitoring of lipid storage in *Caenorhabditis elegans* using coherent anti-Stokes Raman scattering (CARS) microscopy. *Proceedings of the National Academy of Sciences of the United States of America* **104**(37): 14658–14663.
Hermesh, O., B. Kalderon et al. (1998). Mitochondria uncoupling by a long chain fatty acyl analogue. *Journal of Biological Chemistry* **273**(7): 3937–3942.
Hotamisligil, G. S. (2006). Inflammation and metabolic disorders. *Nature* **444**(7121): 860–867.
Hotamisligil, G. S. and E. Erbay (2008). Nutrient sensing and inflammation in metabolic diseases. *Nature Reviews Immunology* **8**(12): 923–934.
Hsu, P. P. and D. M. Sabatini (2008). Cancer cell metabolism: Warburg and beyond. *Cell* **134**(5): 703–707.
Ideker, T., T. Galitski et al. (2001). A new approach to decoding life: Systems biology. *Annual Review of Genomics and Human Genetics* **2**: 343–372.
Jing, C., C. Beesley et al. (2000). Identification of the messenger RNA for human cutaneous fatty acid-binding protein as a metastasis inducer. *Cancer Research* **60**(9): 2390–2398.
Katayama, S., Y. Tomaru et al. (2005). Antisense transcription in the mammalian transcriptome. *Science* **309**(5740): 1564–1566.
Kim, S. H., E. S. Lee et al. (2010). Multiplex coherent anti-Stokes Raman spectroscopy images intact atheromatous lesions and concomitantly identifies distinct chemical profiles of atherosclerotic lipids. *Circulation Research* **106**(8): 1332–1341.
Kimura, K. D., H. A. Tissenbaum et al. (1997). daf-2, an insulin receptor-like gene that regulates longevity and diapause in *Caenorhabditis elegans*. *Science* **277**(5328): 942–946.
Kitano, H. (2002). Systems biology: A brief overview. *Science* **295**(5560): 1662–1664.
Ko, A. C., A. Ridsdale et al. (2010). Multimodal nonlinear optical imaging of atherosclerotic plaque development in myocardial infarction-prone rabbits. *Journal of Biomedical Optics* **15**(2): 020501.
Kopelman, P. G. (2000). Obesity as a medical problem. *Nature* **404**(6778): 635–643.
Krafft, C., B. Dietzek et al. (2009). Raman and CARS microspectroscopy of cells and tissues. *Analyst* **134**(6): 1046–1057.
Kragelund, B. B., K. Poulsen et al. (1999). Conserved residues and their role in the structure, function, and stability of acyl-coenzyme A binding protein. *Biochemistry* **38**(8): 2386–2394.
Le, T. T. and J. X. Cheng (2009). Single-cell profiling reveals the origin of phenotypic variability in adipogenesis. *PLoS One* **4**(4): e5189.
Le, T. T., H. M. Duren et al. (2010). Label-free quantitative analysis of lipid metabolism in living *Caenorhabditis elegans*. *Journal of Lipid Research* **51**(3): 672–677.
Le, T. T., T. B. Huff et al. (2009). Coherent anti-Stokes Raman scattering imaging of lipids in cancer metastasis. *BMC Cancer* **9**: 42.
Le, T. T., I. M. Langohr et al. (2007a). Label-free molecular imaging of atherosclerotic lesions using multimodal nonlinear optical microscopy. *Journal of Biomedical Optics* **12**(5): 054007.
Le, T. T., C. W. Rehrer et al. (2007b). Nonlinear optical imaging to evaluate the impact of obesity on mammary gland and tumor stroma. *Molecular Imaging* **6**(3): 205–211.

Libby, P. (2002). Inflammation in atherosclerosis. *Nature* **420**(6917): 868–874.
Lim, R. S., A. Kratzer et al. (2010). Multimodal CARS microscopy determination of the impact of diet on macrophage infiltration and lipid accumulation on plaque formation in ApoE-deficient mice. *Journal of Lipid Research* **51**(7): 1729–1737.
Lumeng, C. N., J. L. Bodzin et al. (2007a). Obesity induces a phenotypic switch in adipose tissue macrophage polarization. *Journal of Clinical Investigation* **117**(1): 175–184.
Lumeng, C. N., J. B. DelProposto et al. (2008). Phenotypic switching of adipose tissue macrophages with obesity is generated by spatiotemporal differences in macrophage subtypes. *Diabetes* **57**(12): 3239–3246.
Lumeng, C. N., S. M. Deyoung et al. (2007b). Increased inflammatory properties of adipose tissue macrophages recruited during diet-induced obesity. *Diabetes* **56**(1): 16–23.
Lusis, A. J. (2000). Atherosclerosis. *Nature* **407**(6801): 233–241.
Lyn, R. K., D. C. Kennedy et al. (2009). Direct imaging of the disruption of hepatitis C virus replication complexes by inhibitors of lipid metabolism. *Virology* **394**(1): 130–142.
Martin, S. and R. G. Parton (2006). Lipid droplets: A unified view of a dynamic organelle. *Nature Reviews Molecular Cell Biology* **7**(5): 373–378.
Mattick, J. S. (2007). A new paradigm for developmental biology. *Journal of Experimental Biology* **210**(9): 1526–1547.
McPherson, J. D., M. Marra et al. (2001). A physical map of the human genome. *Nature* **409**(6822): 934–941.
Meex, R. C., P. Schrauwen et al. (2009). Modulation of myocellular fat stores: Lipid droplet dynamics in health and disease. *American Journal of Physiology—Regulatory, Integrative and Comparative Physiology* **297**(4): R913–R924.
Menendez, J. A. and R. Lupu (2007). Fatty acid synthase and the lipogenic phenotype in cancer pathogenesis. *Nature Reviews Cancer* **7**(10): 763–777.
Metser, U., E. Miller et al. (2006). F-18-FDG PET/CT in the evaluation of adrenal masses. *Journal of Nuclear Medicine* **47**(1): 32–37.
Migita, T., S. Ruiz et al. (2009). Fatty acid synthase: A metabolic enzyme and candidate oncogene in prostate cancer. *Journal of the National Cancer Institute* **101**(7): 519–532.
Morck, C., L. Olsen et al. (2009). Statins inhibit protein lipidation and induce the unfolded protein response in the non-sterol producing nematode *Caenorhabditis elegans. Proceedings of the National Academy of Sciences of the United States of America* **106**(43): 18285–18290.
Muller, M. and A. Zumbusch (2007). Coherent anti-Stokes Raman scattering microscopy. *ChemPhysChem* **8**(15): 2156–2170.
Nan, X. L., J. X. Cheng et al. (2003). Vibrational imaging of lipid droplets in live fibroblast cells with coherent anti-Stokes Raman scattering microscopy. *Journal of Lipid Research* **44**(11): 2202–2208.
Nan, X. L., E. O. Potma et al. (2006). Nonperturbative chemical imaging of organelle transport in living cells with coherent anti-Stokes Raman scattering microscopy. *Biophysical Journal* **91**(2): 728–735.
Nishimura, S., I. Manabe et al. (2009). CD8+ effector T cells contribute to macrophage recruitment and adipose tissue inflammation in obesity. *Nature Medicine* **15**(8): 914–920.
Odegaard, J. I., R. R. Ricardo-Gonzalez et al. (2007). Macrophage-specific PPARgamma controls alternative activation and improves insulin resistance. *Nature* **447**(7148): 1116–1120.
O'Rourke, E. J., A. A. Soukas et al. (2009). *C. elegans* major fats are stored in vesicles distinct from lysosome-related organelles. *Cell Metabolism* **10**(5): 430–435.
Palmantier, R., M. D. George et al. (2001). Cis-polyunsaturated fatty acids stimulate beta(1) integrin-mediated adhesion of human breast carcinoma cells to type IV collagen by activating protein kinases C-epsilon and -mu. *Cancer Research* **61**(6): 2445–2452.
Ramos, C. V. and H. B. Taylor (1973). Lipid-rich carcinoma of the breast: A clinicopathologic analysis of 13 examples. *Cancer* **33**(3): 812–819.
Reddy, J. K. and M. S. Rao (2006). Lipid metabolism and liver inflammation. II. Fatty liver disease and fatty acid oxidation. *American Journal of Physiology—Gastrointestinal and Liver Physiology* **290**(5): G852–G858.
Rinia, H. A., K. N. J. Burger et al. (2008). Quantitative label-free imaging of lipid composition and packing of individual cellular lipid droplets using multiplex CARS microscopy. *Biophysical Journal* **95**(10): 4908–4914.
Seyfried, T. N. and P. Mukherjee (2005). Targeting energy metabolism in brain cancer: Review and hypothesis. *Nutrition and Metabolism (London)* **2**: 30.
Shi, Y., S. Kim et al. (2010). Effective repair of traumatically injured spinal cord by nanoscale block copolymer micelles. *Nature Nanotechnology* **5**(1): 80–87.

Sijens, P. E., P. C. Levendag et al. (1996). H-1 MR spectroscopy detection of lipids and lactate in metastatic brain tumors. *NMR in Biomedicine* **9**(2): 65–71.
Slipchenko, M. N., T. T. Le et al. (2009). High-speed vibrational imaging and spectral analysis of lipid bodies by compound Raman microscopy. *Journal of Physical Chemistry B* **113**: 7681–7686.
Slipicevic, A., K. Jorgensen et al. (2008). The fatty acid binding protein 7 (FABP7) is involved in proliferation and invasion of melanoma cells. *BMC Cancer* **8**: 276.
Tonsgard, J. H. and G. S. Getz (1985). Effect of Reye's syndrome serum on isolated chinchilla liver mitochondria. *Journal of Clinical Investigation* **76**(2): 816–825.
Tontonoz, P. and B. M. Spiegelman (2008). Fat and beyond: The diverse biology of PPARgamma. *Annual Review of Biochemistry* **77**: 289–312.
Vander Heiden, M. G., L. C. Cantley et al. (2009). Understanding the Warburg effect: The metabolic requirements of cell proliferation. *Science* **234**: 1029–1033.
Venter, J. C., M. D. Adams et al. (2001). The sequence of the human genome. *Science* **291**(5507): 1304–1351.
Voet, D. and J. G. Voet (2010). *Biochemistry*. (New York: Wiley).
Wakil, S. J., J. K. Stoops et al. (1983). Fatty-acid synthesis and its regulation. *Annual Review of Biochemistry* **52**: 537–579.
Walther, T. C. and R. V. Farese (2009). The life of lipid droplets. *Biochimica Et Biophysica Acta-Molecular and Cell Biology of Lipids* **1791**(6): 459–466.
Wang, H. F., Y. Fu et al. (2005). Coherent anti-Stokes Raman scattering imaging of axonal myelin in live spinal tissues. *Biophysical Journal* **89**(1): 581–591.
Wang, H. W., Y. Fu et al. (2009a). Chasing lipids in health and diseases by coherent anti-Stokes Raman scattering microscopy. *Vibrational Spectroscopy* **50**(1): 160–167.
Wang, H. W., I. M. Langohr et al. (2009b). Imaging and quantitative analysis of atherosclerotic lesions by CARS-based multimodal nonlinear optical microscopy. *Arteriosclerosis, Thrombosis, and Vascular Biology* **29**(9): 1342–1348.
Warburg, O. (1956). Respiratory impairment in cancer cells. *Science* **124**(3215): 269–270.
Weston, A. D. and L. Hood (2004). Systems biology, proteomics, and the future of health care: Toward predictive, preventative, and personalized medicine. *Journal of Proteome Research* **3**(2): 179–196.
Wilson, E. (1896). *The Cell in Development and Inheritance*. (New York: Macmillan).
Winer, S., Y. Chan et al. (2009). Normalization of obesity-associated insulin resistance through immunotherapy. *Nature Medicine* **15**(8): 921–929.
Wu, Y. M., H. C. Chen et al. (2009). Quantitative assessment of hepatic fat of intact liver tissues with coherent anti-Stokes Raman scattering microscopy. *Analytical Chemistry* **81**(4): 1496–1504.
Yamaguchi, T., N. Omatsu et al. (2007). CGI-58 facilitates lipolysis on lipid droplets but is not involved in the vesiculation of lipid droplets caused by hormonal stimulation. *Journal of Lipid Research* **48**(5): 1078–1089.
Yang, F. J., B. W. Vought et al. (2006). An ARC/mediator subunit required for SREBP control of cholesterol and lipid homeostasis. *Nature* **442**(7103): 700–704.
Yen, K., T. T. Le et al. (2010). A comparative study of fat storage quantitation in nematode *Caenorhabditis elegans* using label and label-free methods. *PLoS One* **5**(9): e12810.

# 20. White Matter Injury

## Cellular-Level Myelin Damage Quantification in Live Animals

**Erik Bélanger, F.P. Henry, R. Vallée, M.A. Randolph, I.E. Kochevar, J.M. Winograd, Charles P. Lin, and Daniel Côté**



*Coherent Raman Scattering Microscopy*. Edited by Ji-Xin Cheng and X. Sunney Xie © 2013 CRC Press/ Taylor & Francis Group, LLC. ISBN: 978-1-4398-6765-5.

## 20.1 Introduction

Peripheral nerve pathology, whether through direct injury (trauma) or chronic disease (carpal tunnel compressive neuropathy), results in a disability that has widespread repercussions for both personal and occupational rehabilitation. While complete division of a nerve can be directly observed and managed appropriately with standard microsurgery, the spectrum of injury related to crushed nerves or chronic neuropathies cannot be accurately assessed. The ability to observe the microarchitecture of the nerve, in particular its level of myelination, is limited at present to destructive histological techniques that are not appropriate in a clinical setting and therefore currently there is no method that can help a clinician assess nerve health *in vivo* following injury. Electrical studies, although a useful diagnostic technique, cannot in the early phase of injury distinguish between nerves with minor or severe internal disruption. Furthermore, only after recovery has occurred over a period of several months can such a determination be made empirically. This "wait and see" approach prolongs the period of muscle denervation distally, lengthens the time to ultimate recovery if surgical reconstruction is required, and ultimately hastens the time after which meaningful reconstruction, particularly of motor neuron lesions, is no longer possible. A technique that would provide information to allow for grading of the nerve injury would be a valuable clinical tool in terms of both a diagnostic and prognostic indication of functional recovery.

Following neural injury, axonal demyelination coupled with subsequent remyelination over time may be used as an indicator of both severity of injury and degree of neural recovery expected. The degree of axonal remyelination observed following initial insult corresponds directly to the level of functional recovery achieved. Current methods of assessing axonal myelination rely on destructive histological techniques (*ex vivo*, transverse slice, mechanically cut), which are not suitable in the clinical setting. This chapter focuses on coherent anti-Stokes Raman scattering (CARS) microscopy, which has been demonstrated to be sensitive to lipid-rich cells such as adipocytes [1], Schwann cells [2,3], and oligodendrocytes [4–6] without the use of exogenous labeling. This technique utilizes the $CH_2$ symmetric stretch vibration that is present in lipids by exciting it with two lasers tuned to different wavelengths such that their frequency difference corresponds to the frequency of the vibration. Through a vibrationally resonant third-order nonlinear process, a photon at a higher frequency is emitted to provide lipid-based contrast. By virtue of being a nonlinear process, it benefits from three-dimensional (3D) sectioning and high spatial resolution similar to two-photon excitation fluorescence (TPEF) [7–10]. When combined with distortion-free images (without breathing and heartbeat artifacts) obtained with *in vivo* video-rate microscopy, CARS imaging is ideally positioned to quantify peripheral nerve myelination in live animals over an extended period of time. This longitudinal assessment of neural injury provides unique histology without compromising the nerve itself, that is, in its native state. Standard histology is usually done *ex vivo* on transverse sections of fixed nerve slices, a preparation not suitable for *in vivo* evaluation due to its intrinsically destructive nature. Reconstructions from *z*-stacks can provide similar information when CARS contrast is used [5]. Yet, even with video-rate imaging, it is challenging to obtain 3D data that allow for high quality *z*-stack reconstruction. For this reason, we propose a new procedure compatible with live animal imaging. Histomorphometry based on CARS images is extracted from frames recorded from the coronal plane of the sciatic nerve.

In this chapter, we discuss the equipment needed to perform live animal imaging and demonstrate its use on a model of nerve degeneration. The rat sciatic nerve is used as a model of Wallerian degeneration in which we perform myelin histomorphometry on images obtained with CARS imaging tuned to lipid-based contrast. After a controlled crush injury, the nerves are surgically exposed and imaged *in vivo* and *ex vivo* in the coronal plane with confocal reflectance and CARS microscopy at different time points and at different locations on the sciatic nerve. We quantify demyelination proximal and remyelination distal to the crush site *ex vivo* and *in vivo*, respectively. From this, we show that CARS microscopy may be used as a reliable, nondestructive, *in vivo* technique with sufficient accuracy to assess axonal myelination of normal and injured peripheral nerves.

## 20.2 Multimodal Video-Rate Microscope

### 20.2.1 Microscope Hardware

While many groups have successfully adapted commercial microscopes for intravital imaging, most of these systems are not optimized for live animal studies. Before describing applications of CARS microscopy in live animals, we describe our multimodality laser scanning microscope platform that is designed specifically for live animal imaging, with the following design considerations:

1. Combination of high spatial resolution with large field of view (FOV). High spatial resolution and optical sectioning capability are required to visualize cells in deep tissue. At the same time having a large FOV, ideally millimeters, will allow us to place the high-resolution image into the proper context of a larger surrounding tissue environment.
2. Combination of high imaging speed, 30 frames per second (fps) full frame, with image stabilization algorithm for extended acquisition and signal averaging. Video-rate imaging is needed in general to visualize fast cellular movements such as leukocyte rolling on vascular endothelial walls or to measure fast dynamical processes such as neuronal activity; even higher imaging speeds are needed to detect free-flow cells in the blood stream. At the same time, fast scanning is extremely useful for minimizing motion artifacts, even in situations where the events being imaged do not require high imaging speed per se (see below).
3. Combination of multiple imaging modalities to take advantage of intrinsic CARS contrast but also exogenous labels and molecular reporters. Although CARS imaging is an essential component of the system, the goal is to track one or multiple cell populations and to visualize the interactions among the cells and with surrounding tissue structures.
4. Easy integration in biologists' workflow with software for real-time feedback to the microscope user.

The heart of the system is the fast laser scanning mechanism and its associated software. Many different strategies may be used to scan laser beams in microscopes. The scanning platform must achieve an acquisition speed of 30 fps while maintaining a relatively high resolution, at least 500 × 500 scanned points. The use of two closed-loop galvanometer scanners limits the scanning speed for large areas, while the use of a resonant scanner complicates the data acquisition because of the sinusoidal dependence of the position with time. Systems using acousto-optic deflectors (AODs) recently have been constructed and have very high scan rates in addition to having random-access capability, but they require careful attention to pulse dispersion and wavelength dependence of the deflection [11]. The chosen approach for the scanning platform, based on the original design of Rajadhyaksha et al. [12–14], consists of a galvanometer-mounted mirror and a spinning polygonal mirror to produce a unidirectional raster scan pattern. This combination has the advantage of producing a linear scan (contrary to a resonant scanner) and avoids the complications of AODs. A consequence of this linear scan is that video frame grabbers can be used for acquisition, which greatly simplifies integration with software. One disadvantage of polygonal mirrors is that zooming along the horizontal direction cannot be performed easily as the sweep angle is fixed by the polygon and telescopes. See Ref. [14] for all the technical details and circuit description.

The microscope design is such that it can perform confocal and multiphoton detection simultaneously. In confocal mode (reflectance or fluorescence), the reemitted light must be descanned and focused to a pinhole size detector. Multiphoton excited CARS or fluorescence signals do not need such a confocal detection given their inherent small excitation volume. In these modes, the reemitted light is reflected off the long-pass dichroic and sent to the photomultiplier tube immediately after the dichroic mirror. A 3 cm focal length lens positioned after the dichroic filter and 6 cm away from the back of the objective focuses the light onto the photomultiplier tubes (PMTs) to maximize the amount of collected light. All PMTs are extended multi-alkali photocathodes (R3896, Hamamatsu), with wavelength coverage up to approximately 850 nm and a large active area (approximately 1 $cm^2$). The large active area is necessary to detect a maximum number of photons, including those that have scattered multiple times. Note that in the case of nonlinear excitation, it is possible to obtain both the nonlinear signal and the reflectance simultaneously (two-color acquisition), as the reflectance signal is always descanned. Water immersion objectives are used to match the index of refraction of tissue and minimize aberrations. At the focal plane of the objective, the sample is mounted on a user- or computer-controlled *xyz* stage (Sutter Instruments, ROE-200 with MPC-385) to allow micron scale precision displacements over millimeter distances. Small animals such as mice or rats can easily be mounted directly onto the stage for imaging.

## 20.2.2 Laser Sources

The video-rate scanning platform can be adapted to many imaging modalities by selecting different laser sources. For CARS microscopy, the Stokes pulse is provided by a 10 W, 7 ps Nd:Vanadate pulsed laser (picoTRAIN, High Q Laser) operating at 1064 nm, while the pump pulse, at a wavelength of 816.8 nm, is obtained from a synchronously pumped OPO (Levante Emerald ps, APE) generating approximately 400 mW at 80 MHz. Both beams are recombined spatially and temporally with a dichroic mirror mounted on a delay line, after their divergences have been corrected independently using two independent telescopes. It is usually possible to simultaneously acquire multiple modalities. For instance, reflectance imaging can always be performed with other modalities

(CARS and TPEF), which makes reflectance perfect for guiding the user in tissue. Typically, powers incident on the sample are on the order of 50–100 mW for CARS imaging. Because of the fast scanning rate and short pixel dwell time, no damage to the tissue or cells is observed.

### 20.2.3 Video Acquisition and Processing

For acquiring images, the analog signal from the detectors and the synchronization signals from the polygon (horizontal synchronization) and the galvanometer (vertical synchronization) are sent to a video acquisition board (Snapper-24, Active Silicon), which generates the pixel clock internally. Typically, a pixel clock of 10 MHz is used, corresponding to 133 ns of pixel dwell time. The software used for acquiring and processing is written in-house to allow for specific needs to be addressed, such as still frame averaging, movie acquisition, movement correction, automated stage control for *xy* maps and *z*-stacks, and live streaming to disk for extended movies. The software for the video acquisition board runs on a PowerMac G5 with Mac OS X 10.5 and makes use of libraries such as Cocoa, OpenGL, Core Image, vImage and vDSP for real-time processing and display of the images, and Quicktime for archiving images and movies in different standard file formats. One notable feature of the acquisition software is its real-time movement correction capability that enables extended integration times with instant visual feedback to the user, as discussed in the next section.

### 20.2.4 Stabilization Strategies

A critical issue when performing live animal imaging is movement of the specimen. Three categories of solutions have already been examined: (1) elimination of the source of movement, (2) mechanical contention of the tissue being imaged, and (3) computer-assisted methods of post-acquisition image gating. Methods aimed at eliminating the source of movement can be extremely invasive, for example, interruption of the animal respiration during image acquisition [15,16] and cardiopulmonary bypass for extracorporal blood oxygenation [17,18]. Contention methods have limited efficiency and are generally applicable to restricted conditions or regions with limited movement [19,20]. We make use of three different strategies: software-based *xy* correction, software-based image similarity gating in *z*, and active *z* movement compensation with dedicated hardware, schematically shown in Figure 20.1.

#### 20.2.4.1 *xy* Software Correction

The *xy* correction is performed as follows. When acquisition starts, a differential translation vector $\Delta\mathbf{T}$ is set to the null vector. Every time an image is acquired, a user-defined region of interest (ROI) is sampled in the acquired channel. The two-dimensional (2D) normalized cross-correlation with the previous uncorrected image is computed, and the position of its maximum is determined, yielding the displacement $\mathbf{T}$ between these two images. The current image is then translated by an amount corresponding to the subtraction of differential translation vector $\Delta\mathbf{T}$ and the current correction $\mathbf{T}$, after which the differential translation vector is updated to $\Delta\mathbf{T} \leftarrow \mathbf{T} - \Delta\mathbf{T}$. This sequence is repeated for each subsequent frame.

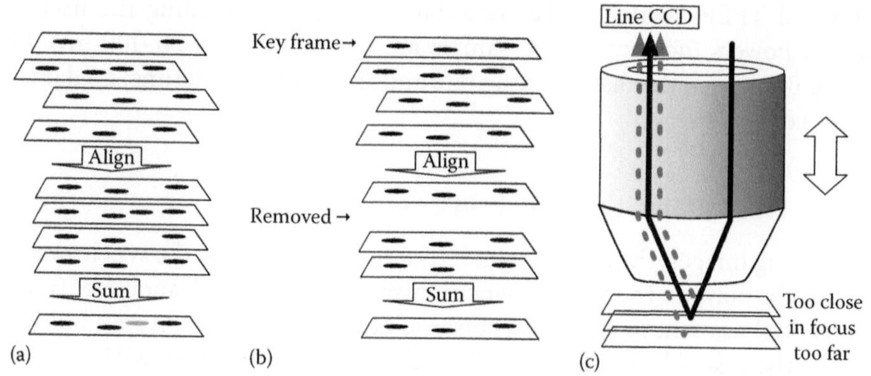


**FIGURE 20.1** Three different strategies for stabilization of images before averaging: (a) software coregistration of all images before averaging. This corrects only for *xy* movement since images are still averaged even if the image plane has changed. (b) Software coregistration and removal of images where similarity to a key frame is insufficient (the maximum value of the 2D normalized cross-correlation with the key frame is below a user-set threshold). (c) Hardware adjustment of the objective position during imaging based on the position of the top surface of the tissue, followed by software coregistration of images.

#### 20.2.4.2 Similarity-Gated Averages

The image similarity gating strategy is an addition to the previous correction: frames are averaged only if the maximum of the normalized cross-correlation is superior to a user-defined similarity threshold, typically 0.75. The calculation can also be performed with nonnormalized cross-correlations, but the similarity parameter then depends on the intensity of the images, and the setting of the similarity threshold requires user intervention to be set properly.

#### 20.2.4.3 *z* Movement Compensation

Finally, a hardware solution for active *z* movement compensation has been developed [21] and is discussed briefly here for completeness, although it was not used in this work. The device is a highly modified sensor (ATF-4, WDI) that estimates the position of the surface of the animal and uses this position signal to control a piezo-mounted microscope objective in a closed feedback loop. The device emits a hemicircular light beam at 785 nm, and an integrated line CCD sensor operating at 7 kHz is used to monitor the position of the reflected beam. The laser beam at 785 nm is injected off-axis in the microscope via a dichroic (LPD01-785RS-25, Semrock) after the tube lens and before the microscope. With its own collimation telescope, this beam can be adjusted to focus on a different surface than the imaging beam and is typically adjusted to focus onto the surface of the tissue. After being reflected off the surface, the beam is returned to the sensor where the line CCD images its profile. This is used to determine the *z* position of the surface based on the lateral position of the profile: An internal processor uses a signal processing algorithm involving a rolling window average (down to 60 Hz) and digital filtering coupled with a peak fitting procedure and linear interpolation to determine the best current position of the imaging plane of interest. The predicted *z* position for each data point is generated by subtracting the system's self-motion from the data points before averaging. The correction is then computed based on the best linear fit to extrapolate to current position. The signal is sent to the nanopositioner for correction. More information about the system can be found elsewhere [21].

## 20.3 Methodology

### 20.3.1 Animal Model

We have used the sciatic nerve of Sprague Dawley rats as a model for this study. Approval was obtained from our local institutional animal care and use committee. Surgical exposure of the sciatic nerve was achieved by means of a dorsolateral muscle splitting incision. A standardized demyelinating crush injury was reproduced in all animals [14] by means of a #5 jeweler's forceps held closed across the nerve for 30 s. Functional assessment of the sciatic nerve was carried out by means of a walking track analysis. Following sciatic nerve injury and during subsequent recovery, the hindprint of the rat undergoes several morphological changes from which a sciatic function index (SFI) can be calculated [22]. Prints were obtained by coating the animal hindpaws in ink and allowing it to walk down a paper-lined track. The resulting SFI allows for a longitudinal functional assessment of sciatic nerve regeneration. For the purpose of imaging, the animal was anaesthetized (intraperitoneal injection of pentobarbital) and surgery was performed to expose the sciatic nerve. The animal is then mounted on an adapted stage that is incorporated into the custom-built upright multimodal video-rate microscope. Imaging with confocal reflectance (detecting the pump beam) and CARS is performed simultaneously.

### 20.3.2 Image Acquisition

Confocal reflectance is used as a guidance modality to confirm position and nature of the tissue in question prior to imaging with CARS. During the imaging session, animal breathing leads to lateral shifts between subsequent frames. Axial movement of the animal (up/down) is reduced by mechanically restricting leg movements and with additional $xy$ correction performed using the software as described in Section 20.2.4 to permit long integration times. An image is a frame recorded in the coronal plane (inset of Figure 20.2b) of the sciatic nerve as opposed to a transverse image (inset of Figure 20.2a) that is rendered from a $z$-stack of multiple images recorded in the coronal plane. When axial movement is minimized, $z$-stacks (60 images, 1 μm apart) are acquired to reconstruct transverse images corresponding to the standard histology paradigm. After the imaging session, the animal is sacrificed to permit *ex vivo* CARS imaging.

### 20.3.3 Image Processing for Histology in the Transverse Plane

Image analysis was performed with an in-house software [5] implemented in MATLAB® (Mathworks, Natick, Massachusetts) and with ImageJ (NIH, Bethesda, Maryland). To assess the myelin health, $z$-stacks of images were acquired and transverse views were rendered with the freely accessible Volume Viewer plugin. Then, ROIs that include a myelinated axon were manually cropped from the whole dataset of transverse images. All ROIs were resampled by a factor of 4 with a bicubic algorithm. Each individual axon was thresholded with an automatic adaptive Niblack algorithm [23] to avoid any user bias. Local window size was set to 21 pixels and weight of the standard deviation

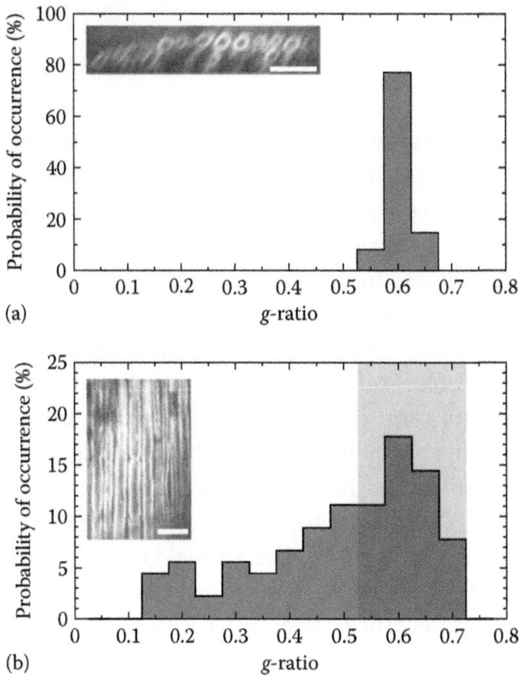


**FIGURE 20.2** CARS histomorphometry in the transverse and coronal planes. Probability histogram of *g*-ratio measurements extracted from (a) reconstructed transverse planes and from (b) the *z*-stack of coronal planes. (a, inset) Typical CARS image in the transverse plane. (b, inset) Typical CARS image in the coronal plane. All scale bars are 25 μm.

contribution to 0.55 (i.e., the *k*-value). The threshold parameters were set empirically on a subset of images of a control animal and kept constant for all subsequent measurements. At that time, an automated function determined the inner and outer boundaries of the circular myelin sheath on the binary image, from which the diameter of the axon and that of the fiber are mathematically computed based on the area of their respective regions. Finally, the g-ratio (ratio of the axon to the fiber diameter) is calculated by dividing the two previous measurements.

## 20.3.4 Image Processing for Histology in the Coronal Plane

Image analysis was performed with MATLAB and ImageJ. Morphology based on CARS images was extracted from frames recorded from the coronal plane of the sciatic nerve. The proposed approach is an analogy of the laser beam size measurement method making use of a scanning edge [24]. First, individual myelinated axons were manually cropped from the original dataset. Great care is taken to make sure that rectangular ROIs were parallel to their respective axons. Then, all ROIs were resampled by a factor of 4 with a bicubic algorithm. At this time, all the individual line profiles coming from a particular ROI were averaged to obtain a single representative line profile of the fiber. At this point, a linear baseline subtraction was performed in order to level the representative line profile on a straight line. The leveled representative line profile was then normalized by dividing every of his points by its maximum value. A cumulative integral was numerically computed on the leveled representative line profile. The fiber and axon

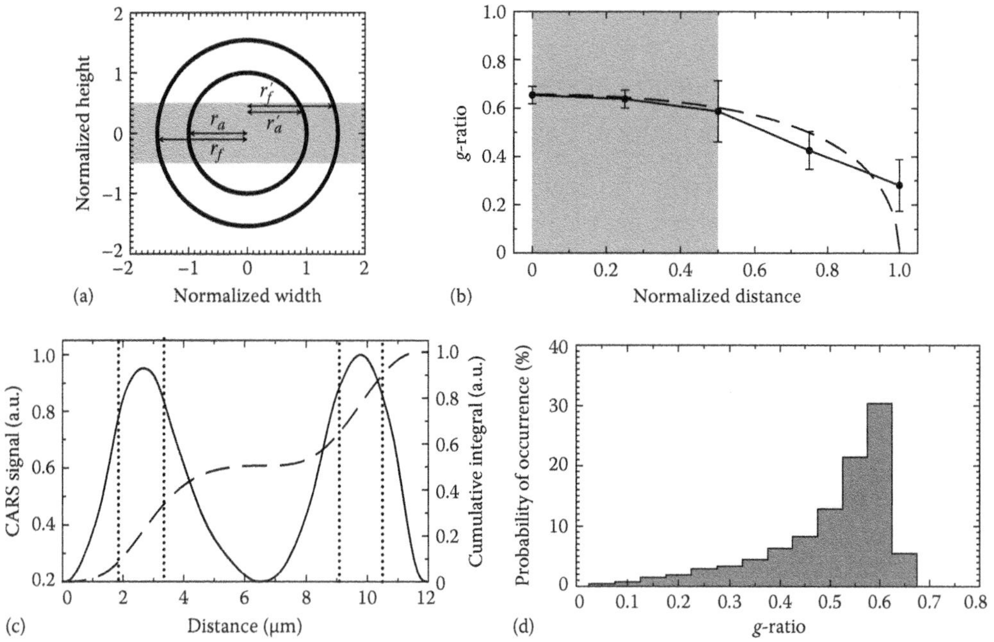


**FIGURE 20.3** (a) Scheme of the imaging plane relative to the fiber and axon location ($r_a$: axon radius, $r_f$: fiber radius, $r_a'$: biased axon radius, and $r_f'$: biased fiber radius). (b) Measured *g*-ratio versus the normalized displacement from the center of the axon (solid: experimental and dashed: simulated). (c) Leveled line profile (solid) overlaid with its cumulative integral (dashed) and the edges of the fiber and axon diameters (dotted). (d) Theoretical biased probability histogram of *g*-ratio measurements.

diameters are determined by the position of the edges of the fiber and axon diameters. The edges' position corresponds to a width of half of the contributed area of each peak of the line profile to the cumulative integral (i.e., roughly 12.5%, 37.5%, 62.5%, and 87.5%). The limits were set empirically on a subset of images of a control animal and kept constant for all subsequent measurements. Figure 20.3c shows a typical leveled line profile (solid line) overlaid with its cumulative integral (dashed line). The dotted lines highlight the positions that were automatically determined by the cumulative integral to be the edges of the fiber and axon diameters. The g-ratio was calculated by dividing the axon by the fiber diameter. Myelin thickness was computed by subtracting the axon from the fiber diameter and then dividing by the factor 2.

## 20.3.5 Statistical Analysis

Values are expressed in terms of means ± standard error of the mean (SEM) unless otherwise stated. Statistical analysis of the data was performed using MATLAB. Values of $p \leq 0.05$ were considered statistically significant (*), $p \leq 0.01$ were considered highly statistically significant (**), and $p \leq 0.001$ were considered extremely statistically significant (***). Mann–Whitney *U*-test was used to evaluate the differences in medians of g-ratio distributions and myelin thickness prevalence at various position along a crushed nerve (Figure 20.5d) and at different time points postcrush (Figure 20.7b), respectively.

## 20.4 Results

### 20.4.1 Myelin Histomorphometry in the Transverse and Coronal Plane of CARS Images

The objective of this section is to demonstrate that it is possible to provide accurate estimates of myelin histomorphometric parameters from CARS images of live uncut tissue, the two most important parameters being the *g*-ratio and the myelin thickness. With this method, preparation and cutting artifacts inherent to standard histology are avoided. Histomorphometric analysis can be performed on transverse images obtained by 3D reconstruction of a *z*-stack (Figure 20.2a) or on coronal images (Figure 20.2b). The *g*-ratio histogram extracted from transverse images of sciatic nerves of Figure 20.2a is obtained from 61 morphometric measurements (1 animal, 70 transverse images) and reveals a mean *g*-ratio of 0.60 with a standard deviation of 0.02. The *g*-ratio histogram for the same nerve section, but instead analyzed on coronal slices, is shown on Figure 20.2b. The g-ratio histogram shows 90 measurements (1 animal, 60 images) and illustrates that under a random sampling of the position of the imaging plane throughout the axon diameter, the distribution is skewed toward smaller values. We hypothesize that the upper half of the *g*-ratio distribution, that is, every measurement over the median, is minimally affected and is suitable for histomorphometry (region highlighted in light gray in Figure 20.2b). In the present case, the mean g-ratio of every measurement over the 50th percentile is $0.62 \pm 0.04$.

As is shown in Figure 20.3a and b, the randomly positioned imaging plane leads to measured *g*-ratio within 15% of the nominal *g*-ratio value 50% of the time. Importantly, the value is always underestimated and never overestimated. A simple model illustrated in Figure 20.3a reveals the trend in *g*-ratio bias in Figure 20.3b (dashed line: calculation, solid line: experimental data points from Figure 20.2b). The normalized displacement is defined as the distance from the axon center divided by its diameter. Myelin parameters on coronal images are extracted with the procedure described in Section 20.3.4 (see Figure 20.3c). The predicted asymmetrical resulting histogram, for a fiber distribution similar to Figure 20.2a, is shown in Figure 20.3d and confirms that every measurement over the 50th percentile is minimally affected. Therefore, with a random sampling of imaging plane positions and a large number of measurements, this technique is satisfactory for studying neuropathies in which a single population of hypomyelinated axons is expected.

### 20.4.2 Myelin Histomorphometry Reveals Demyelination of *Ex Vivo* Crushed Sciatic Nerve

Figure 20.4 shows a large-scale and high-resolution CARS map of an *ex vivo* crushed sciatic nerve at the millimeter scale at 1 week postinjury. This mosaic is a collection of images (i.e., tiles) manually stitched together. Each individual tile measures 435 μm × 435 μm. This map measures approximately 0.5 mm width by 4 mm long. The vertical scale indicates the approximate transition from the healthy to the proximal and from the proximal to the distal part of the lesion. Well aligned and globally organized myelin sheaths are clearly observed in the healthy region. In the proximal region, myelin

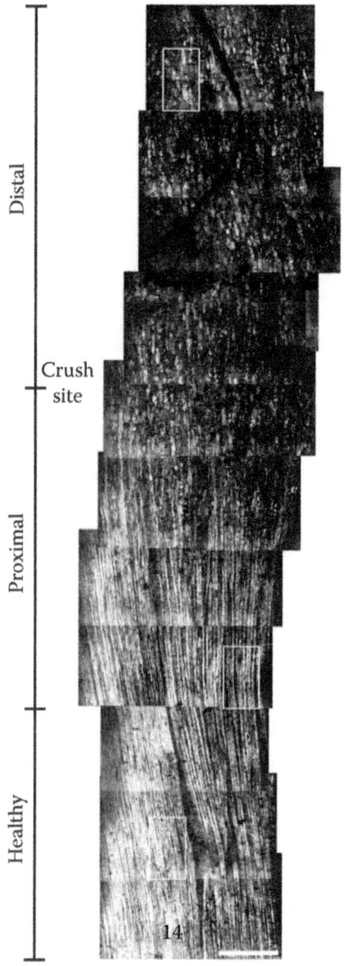


FIGURE 20.4 Large-scale and high-resolution CARS map of a crushed sciatic nerve at the millimeter scale at 1 week postinjury. Three ROIs depicting healthy, proximal, and distal regions of the nerve are highlighted. Scale bar is 250 μm.

swelling and spheroids appear. The distal region is mostly filled with myelin debris with no distinguishable sign of myelinated axons.

Figure 20.5a through c shows typical snapshots (corresponding to the three ROIs of Figure 20.4) of the healthy, proximal, and distal regions of the lesion at 1 week postcrush, respectively. The g-ratio histogram is measured at different locations along the crushed nerve, and its asymmetrical distribution is used to extract the minimally biased *g*-ratios. The mean g-ratio increases from approximately 0.58–0.60 (1 animal; 6 images; 19, 21, and 23 measurements) in the healthy region to 0.63–0.65 (1 animal; 12 images; 25, 23, 20, and 13 measurements) proximal to the lesion site, as shown in Figure 20.5d. All mean g-ratios in the proximal zone were compared (Mann–Whitney *U*-test) with the central point of the healthy region (indicated by a light gray square). This analysis reveals a highly statistically significant (**) to extremely statistically significant increases (***) between the different locations along the crushed nerve suggesting that healthy and proximal regions have different axonal myelination.

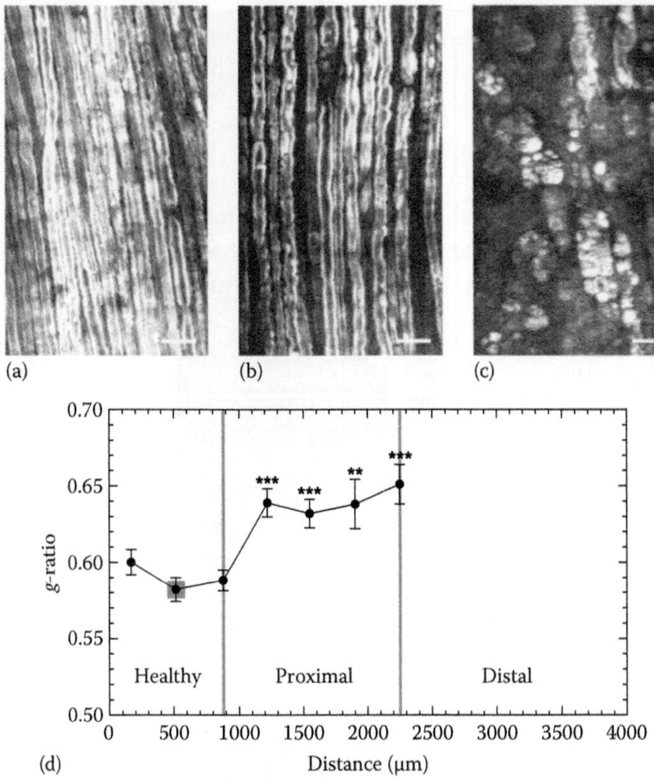


**FIGURE 20.5** Histomorphometry on *ex vivo* crushed sciatic nerve 1 week postinjury. (a–c) Snapshots of the healthy, proximal, and distal regions of the lesion, respectively, corresponding to the three ROIs of Figure 20.4. (d) The *g*-ratio versus the position along the crushed sciatic nerve. All scale bars are 25 μm.

### 20.4.3 Myelin Histomorphometry Reveals Remyelination of *In Vivo* Crushed Sciatic Nerve

Following a crushed nerve injury, axonal damage and demyelination are expected to occur mostly distal to the point of injury. The nerve also demyelinates to the proximal node of Ranvier [25]. The superficial layers of the sciatic nerves were imaged at different time points of recovery after injury, as shown in Figure 20.6. The control image (Figure 20.6d) shows myelin sheaths with normal myelin thickness, as expected from healthy nerves. Figure 20.6a through c shows myelin sheaths proximal to the crush site at 2–4 weeks postinjury, respectively. Only modest changes are observed to the myelinated axons architecture, as expected for the proximal part of the lesion. In fact, minor myelin swelling is observed with diminished effect over time. However, the situation is very different distal to the lesion site. Figure 20.6e and f reveals a large amount of myelin debris with very few myelinated axons at 2 and 4 weeks postcrush, respectively. Myelin debris are organized in a cell-like shape, suggesting that they are phagocytosed by either Schwann cells or macrophages [26].

Loss of function following injury is seen to partially recover over the course of 4 weeks. While the timeframe for remyelination can vary between animals and injuries it is expected to begin roughly 1 week following neural insult and upon cessation of the

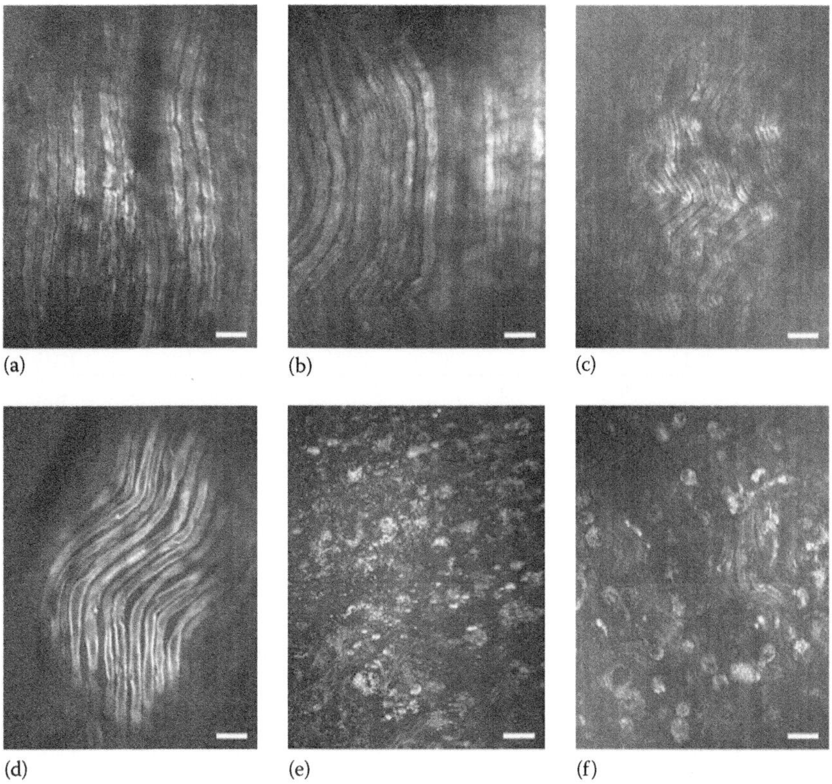


**FIGURE 20.6** *In vivo* CARS images of crushed sciatic nerves at different time points of recovery. (a–c) Myelin sheaths proximal to the crush site at 2–4 weeks postinjury, respectively. (d) Myelin sheaths of a control sciatic nerve. (e) and (f) Myelin sheaths distal to the crush site at 2 and 4 weeks postcrush, respectively. All scale bars are 25 μm.

initial inflammatory response [25]. This slow process is mirrored by the return of motor function as demonstrated in Figure 20.7a by a constant increase of the SFI starting at 1 week postcrush (14 animals). In order to address remyelination *in vivo* at the cellular level, as opposed to behavioral assessment, the myelin thickness (circles) and the *g*-ratio (squares) have been measured distal to the lesion at 2 and 4 weeks postinjury and are illustrated in Figure 20.7a and b, respectively. The myelin thickness decreases from 1.4 to 1.1 μm (***) from the control (7 animals, 412 images, 233 measures) to 2 weeks postcrush (1 animal, 94 images, 15 measures), unraveling a significant demyelination. At 4 weeks (1 animal, 100 images, 6 measures), myelin thickness augments up to 1.2 μm (*), showing a partial remyelination. However, the level of myelination is still lower than in the control (*).

## 20.5 Discussion

The gold standard for myelin histomorphometry is to compute morphological parameters on transverse images on ultrathin slices of fixed tissue. With myelin CARS imaging, it is possible to use this approach *in vivo* with optical sectioning and reconstructed *z*-stacks to obtain reliable metrics as shown in Figure 20.2a. However, this paradigm is often difficult in a live animal due to animal movement (breathing and

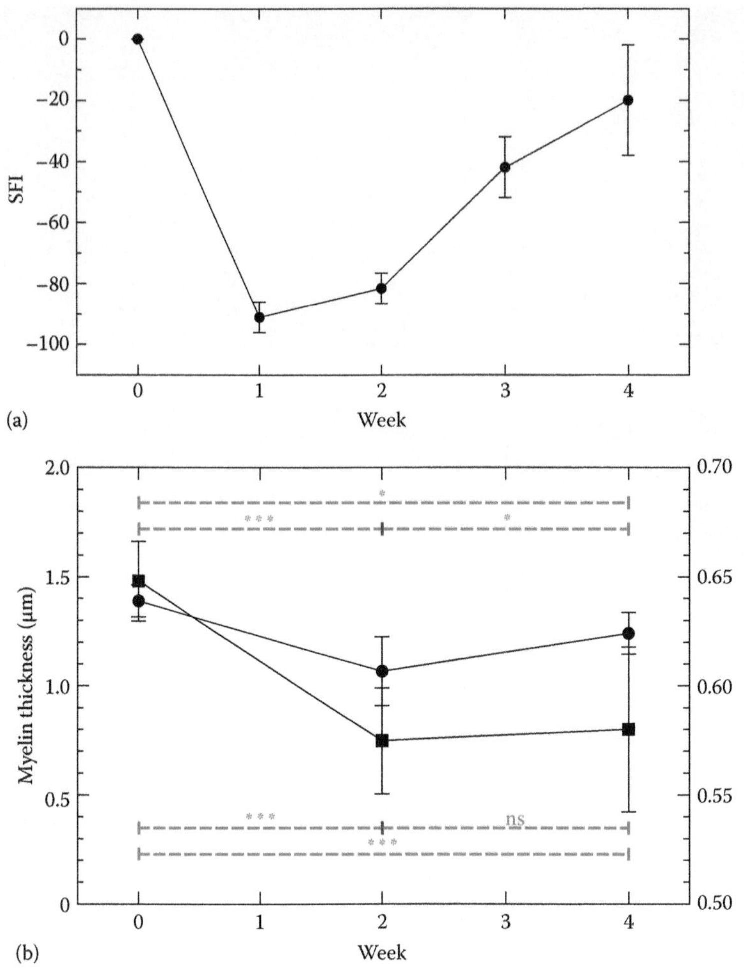


**FIGURE 20.7** Behavioral assessment and histomorphometry on *in vivo* CARS images of crushed sciatic nerves distal to the lesion at different time points. (a) SFI and (b) myelin thickness (circles) and the *g*-ratio (squares) versus time.

heartbeat): *z*-stack recording requires a stable tissue volume to avoid any distortion for the duration of the acquisition, which can be several seconds. On the other hand, a single image at video rate can easily be acquired distortion free. Histomorphometry in the coronal plane is fully compatible with live animal imaging since it requires solely a collection of single images randomly positioned inside the axon diameter rather than a well registered *z*-stack. This results in a *g*-ratio distribution that is often underestimated; a single bias-free *g*-ratio can be calculated with transverse plane images (Figure 20.2a), whereas on coronal images, a distribution of *g*-ratios is obtained (as shown in Figures 20.2b and 20.3d).

We applied this method *in vivo* for studying sciatic nerve crush injuries. Figure 20.5d shows an increase of the *g*-ratio (i.e., a thinning of myelin) in the proximal region of the lesion 1 week postinjury. Similar findings have also been reported recently in an *in vivo* model of acute axonal degeneration in the optic nerve [27]. Figure 20.7b shows a decrease of the myelin thickness and of the *g*-ratio simultaneously distal to the lesion over time. This behavior has also been pointed out by Ref. [27] and explained it as axons possessing a condensed axoplasm.

Images postcrush (especially at 2 weeks) are difficult to acquire because of the large inflammation in the vicinity of the lesion. A thickening of the epinerium and an increased vascularization surrounding the lesion site limit the penetration depth, both by lowering excitation and detection of the back-scattered photons. This results in a limited amount of measures in the pathological states compared to the control, limiting generalization of our results. However, demyelination arising from chronic compression injuries do not result in dramatic inflammatory responses, and in these cases, CARS microscopy would be very appropriate.

The *g*-ratio and the myelin thickness were not totally adequate to describe the axons morphology. For example, the whole distal part of the large-scale and high-resolution CARS map in Figure 20.4 cannot be quantified with these standard metrics. Also, only a few myelinated axons have been found in Figure 20.6c and f, suggesting that the use of new metrics such as texture analysis [28] could complement the usual morphological parameters in depicting the complete structural changes arising from demyelinating pathologies.

With the rapid emergence of hand-held probes [29] and microendoscopes [30], CARS microscopy has the potential to become an experimental and clinical tool in the field of peripheral nerve pathologies. It would provide nondestructive, real-time, and label-free information to allow for both diagnostic and prognostic at the cellular level.

## 20.6 Conclusion

In conclusion, video-rate CARS microscopy was used to image the microenvironment of the sciatic nerve *ex vivo* and *in vivo*. Thanks to the fast acquisition rate, image blur is minimized within one frame. This permits the use of stabilization and movement correction strategies, critical for live animal imaging, to extract high-quality images for analysis. Histomorphometry in the coronal plane, an approach compatible with live animal imaging, has been used to assess axonal myelination. Demyelination and subsequent remyelination have been observed, quantified, and correlated with behavioral assessment. This nondestructive, real-time, *in vivo* microscopy may be useful in the near future as an experimental and clinical tool in the field of peripheral nerve pathologies.

## References


1. C. L. Evans, E. O. Potma, M. Puoris'haag, D. Côté, C. P. Lin, and X. S. Xie, Chemical imaging of tissue in vivo with video-rate coherent anti-Stokes Raman scattering microscopy. *Proceedings of the National Academy of Sciences of the United States of America* **102**(46), 16807–16812 (2005).
2. T. B. Huff and J.-X. Cheng, In vivo coherent anti-Stokes Raman scattering imaging of sciatic nerve tissue. *Journal of Microscopy* **225**(Pt 2), 175–182 (2007).
3. F. P. Henry, D. Côté, M. A. Randolph, E. A. Z. Rust, R. W. Redmond, I. E. Kochevar, C. P. Lin, and J. M. Winograd, Real-time in vivo assessment of the nerve microenvironment with coherent anti-Stokes Raman scattering microscopy. *Plastic and Reconstructive Surgery* **123**(2 Suppl), 123S–30S (2009).

4. H. Wang, Y. Fu, P. Zickmund, R. Shi, and J.-X. Cheng, Coherent anti-Stokes Raman scattering imaging of axonal myelin in live spinal tissues. *Biophysical Journal* **89**(1), 581–591 (2005).
5. E. Bélanger, S. Bégin, S. Laffray, Y. De Koninck, R. Vallée, and D. Côté, Quantitative myelin imaging with coherent anti-Stokes Raman scattering microscopy: Alleviating the excitation polarization dependence with circularly polarized laser beams. *Optics Express* **17**(21), 18419–18432 (2009).
6. J. Imitola, D. Côté, S. Rasmussen, X. S. Xie, Y. Liu, T. Chitnis, R. L. Sidman, C. P. Lin, and S. J. Khoury, Multimodal coherent anti-Stokes Raman scattering microscopy reveals microglia-associated myelin and axonal dysfunction in multiple sclerosis-like lesions in mice. *Journal of Biomedical Optics* **16**(2), 021109 (2011).
7. C. L. Evans and X. S. Xie, Coherent anti-Stokes Raman scattering microscopy: Chemical imaging for biology and medicine. *Annual Review of Analytical Chemistry* **1**(1), 883–909 (2008).
8. S. Bégin, E. Bélanger, S. Laffray, R. Vallée, and D. Côté, In vivo optical monitoring of tissue pathologies and diseases with vibrational contrast. *Journal of Biophotonics* **2**(11), 632–642 (2009).
9. T. T. Le, S. Yue, and J.-X. Cheng, Shedding new light on lipid biology with coherent anti-Stokes Raman scattering microscopy. *Journal of Lipid Research* **51**(11), 3091–3102 (2010).
10. J. P. Pezacki, J. A. Blake, D. C. Danielson, D. C. Kennedy, R. K. Lyn, and R. Singaravelu, Chemical contrast for imaging living systems: Molecular vibrations drive CARS microscopy. *Nature Chemical Biology* **7**(3), 137–145 (2011).
11. R. Salomé, Y. Kremer, S. Dieudonné, J.-F. Léger, O. Krichevsky, C. Wyart, D. Chatenay, and L. Bourdieu, Ultrafast random-access scanning in two-photon microscopy using acousto-optic deflectors. *Journal of Neuroscience Methods* **154**(1–2), 161–174 (2006).
12. M. Rajadhyaksha, M. Grossman, D. Esterowitz, R. H. Webb, and R. R. Anderson, In vivo confocal scanning laser microscopy of human skin: Melanin provides strong contrast. *Journal of Investigative Dermatology* **104**(6), 946–952 (1995).
13. M. Rajadhyaksha, S. González, J. M. Zavislan, R. R. Anderson, and R. H. Webb, In vivo confocal scanning laser microscopy of human skin II: Advances in instrumentation and comparison with histology. *Journal of Investigative Dermatology* **113**(3), 293–303 (1999).
14. I. Veilleux, J. Spencer, D. Biss, D. Côté, and C. P. Lin, In vivo cell tracking with video rate multimodality laser scanning microscopy. *IEEE Journal of Selected Topics in Quantum Electronics* **14**(1), 10–18 (2008).
15. M. Kerschensteiner, M. E. Schwab, J. W. Lichtman, and T. Misgeld, In vivo imaging of axonal degeneration and regeneration in the injured spinal cord. *Nature Medicine* **11**(5), 572–577 (2005).
16. T. Misgeld, I. Nikić, and M. Kerschensteiner, In vivo imaging of single axons in the mouse spinal cord. *Nature Protocols* **2**(2), 263–268 (2007).
17. R. Drdla, M. Gassner, E. Gingl, and J. Sandkühler, Induction of synaptic long-term potentiation after opioid withdrawal. *Science* **325**(5937), 207–210 (2009).
18. H. Ikeda, J. Stark, H. Fischer, M. Wagner, R. Drdla, T. Jäger, and J. Sandkühler, Synaptic amplifier of inflammatory pain in the spinal dorsal horn. *Science* **312**(5780), 1659–1662 (2006).
19. D. Davalos, J. K. Lee, W. B. Smith, B. Brinkman, M. H. Ellisman, B. Zheng, and K. Akassoglou, Stable in vivo imaging of densely populated glia, axons and blood vessels in the mouse spinal cord using two-photon microscopy. *Journal of Neuroscience Methods* **169**(1), 1–7 (2008).
20. E. Chaigneau, P. Tiret, J. Lecoq, M. Ducros, T. Knöpfel, and S. Charpak, The relationship between blood flow and neuronal activity in the rodent olfactory bulb. *Journal of Neuroscience* **27**(24), 6452–6460 (2007).
21. S. Laffray, S. Pagès, H. Dufour, P. De Koninck, Y. De Koninck, and D. Côté, Adaptive movement compensation for in vivo imaging of fast cellular dynamics within a moving tissue. *PLoS ONE* **6**(5), e19928 (2011).
22. J. R. Bain, S. E. Mackinnon, and D. A. Hunter, Functional evaluation of complete sciatic, peroneal, and posterior tibial nerve lesions in the rat. *Plastic and Reconstructive Surgery* **83**(1), 129–138 (1989).
23. W. Niblack, *An Introduction to Digital Image Processing* (Prentice-Hall International, Englewood Cliffs, NJ, 1986).
24. G. Veshapidze, M. L. Trachy, M. H. Shah, and B. D. DePaola, Reducing the uncertainty in laser beam size measurement with a scanning edge method. *Applied Optics* **45**(32), 8197–8199 (2006).
25. M. Coleman, Axon degeneration mechanisms: Commonality amid diversity. *Nature Reviews Neuroscience* **6**(11), 889–898 (2005).
26. K. Hirata and M. Kawabuchi, Myelin phagocytosis by macrophages and nonmacrophages during Wallerian degeneration. *Microscopy Research and Technique* **57**(6), 541–547 (2002).

27. J. Knöferle, J. C. Koch, T. Ostendorf, U. Michel, V. Planchamp, P. Vutova, L. Tönges, C. Stadelmann, W. Brück, M. Bähr, and P. Lingor, Mechanisms of acute axonal degeneration in the optic nerve in vivo. *Proceedings of the National Academy of Sciences* **107**(13), 6064–6069 (2010).
28. R. Haralick, K. Shanmuga, and I. Dinstein, Textural features for image classification. *IEEE Transactions on Systems Man and Cybernetics* **SMC3**(6), 610–621 (1973).
29. M. Balu, G. Liu, Z. Chen, B. J. Tromberg, and E. O. Potma, Fiber delivered probe for efficient CARS imaging of tissues. *Optics Express* **18**(3), 2380–2388 (2010).
30. S. Murugkar, B. Smith, P. Srivastava, A. Moica, M. Naji, C. Brideau, P. K. Stys, and H. Anis, Miniaturized multimodal CARS microscope based on MEMS scanning and a single laser source. *Optics Express* **18**(23), 23796–23804 (2010).

# 21. CARS Microscopy Study of Liquid Crystals

**Heung-Shik Park and Oleg D. Lavrentovich**



## 21.1 Introduction

In the preface to his book *The Physics of Liquid Crystals*, P.G. de Gennes presented the importance of imaging molecular arrangements in liquid crystal systems: "the study of liquid crystals is complicated because it involves several different scientific disciplines… and

*Coherent Raman Scattering Microscopy*. Edited by Ji-Xin Cheng and X. Sunney Xie © 2013 CRC Press/Taylor & Francis Group, LLC. ISBN: 978-1-4398-6765-5.

also a certain sense of vision in three-dimensional space in order to visualize complex molecular arrangements." As a distinctive feature, liquid crystals (LCs) exhibit a long-range orientational order and a complete or partial absence of positional order in their building units, which can be individual molecules or their aggregates. Observing the behavior of these ordered systems and how they respond to various stimuli, such as temperature, pressure, applied electric and magnetic fields, etc., is critical for understanding their physical properties and for development of relevant applications (de Gennes and Prost 1993).

Many optical imaging tools have been used to understand the orientational order in LCs. The most widely used technique is polarizing light microscopy (PM). PM tests the orientation of the optic axes that are related to the pattern of orientational order (Hartshorn 1974, Meeten 1986, Bellare et al. 1990, Kleman and Lavrentovich 2003). Unfortunately, a PM image carries only two-dimensional (2D) information, as the three-dimensional (3D) pattern of birefringence-induced retardation is integrated over the path of light (Bellare et al. 1990). It is hard to get information about the orientational order along the direction of observation. Another technique that has been recently presented, scanning transmission x-ray microscopy (STXM), offers higher spatial resolution than PM but is still limited to 2D information (Kaznacheev and Hegmann 2007).

Three-dimensional imaging of orientational order, both in the plane of observation and along the direction of observation, has been achieved by combining the complementary capabilities of PM and fluorescence confocal microscopy into what has been called a fluorescence confocal polarizing microscopy (FCPM) (Smalyukh et al. 2001, Shiyanovskii et al. 2001, Lavrentovich 2003, Smalyukh and Lavrentovich 2006). In FCPM, the LC host is doped with anisometric fluorescent dye molecules that are aligned by the ordered matrix. The sample is scanned with a focused polarized light and the average local orientations of molecules represented by the director ($\hat{\mathbf{n}}$) is deduced from the intensity of fluorescence that depends on the mutual orientation of the polarization of probing light and the transition dipole **d** (related to the director of the orientational order, usually as $\mathbf{d}||\hat{\mathbf{n}}$) of the fluorescent dye. Development of fast scanning microscopes allows one to increase the rate of FCPM imaging to ~1000 frames per second by using a Nipkow disc (Tanaami et al. 2002). The FCPM technique, however, requires doping the sample with a specially selected dye, though typically the needed dye concentrations are very small (0.01 wt.%). Reduction of the fluorescence intensity either by photobleaching or quenching, frequently observed during microscopic measurements, is another problem of fluorescence-based microscopy.

Vibrational microscopies, such as infrared (IR) and Raman microscopy, offer intrinsic chemical selectivity, as different molecules have specific vibrational frequencies. Though IR microscopy has been applied to study the polymer dispersed liquid crystal film (Rafferty et al. 2002), it is limited by a number of difficulties, including low sensitivity due to non-background-free detection and low spatial resolution associated with long infrared wavelengths (Rodriguez et al. 2006). Confocal Raman microscopy allows one to achieve 3D imaging that is both chemically selective and noninvasive. In Raman microspectroscopy, the chemical information is acquired through the inelastically scattered light. The frequency shifts from the monochromatic pump wavelength correspond to the natural frequencies of vibration of the individual chemical groups. For example, Ofuji et al. (2006) used Raman microscopy to image solid films of orientationally ordered fibrils in polyacetylene films with a contrast based on the C=C bond stretching frequency, while Buyuktanir et al. (2008) used the cyano (CN) group stretching frequency to image defects in LC samples. Blach et al. (2006) developed a theory of polarized Raman microscopy in a birefringent medium and probed the director of LC samples by using the CN stretching vibrational mode at 2226 $cm^{-1}$. Camorani and Fontana (2007) used polarization sensitive confocal Raman microscopy to characterize liquid crystal alignment at a micropatterned substrate. Raman microscopy offers a wealth of chemical information, but it requires high laser power and relatively long integration times (100–200 ms per pixel), since the Raman effect is extremely weak (typical photon conversion efficiencies are lower than 1 in $10^{16}$) (Potma and Xie 2008).

As ultrafast laser technology has progressed, new microscopy techniques to image LCs have been developed, employing nonlinear optical processes, such as multiharmonic generation (Yelin et al. 1999, Yoshiki et al. 2005, Pillai et al. 2006), multiphoton excitation fluorescence (Xie and Higgins 2004), and coherent anti-Stokes Raman scattering (CARS) microscopy (Kachynski et al. 2007, 2008, Saar et al. 2007, Chen and Lim 2009). To view director distortions induced by colloidal particles in a liquid crystal, Pillai et al. (2006) used third-harmonic generation (THG) microscopy, building on the previous work of Yelin et al. (1999). In THG imaging, the contrast is based both on changes in the material's nonlinear susceptibility in the focus of the laser beam and on the local orientation of the director relative to the laser polarization, but it does not offer any direct chemical information. THG has been shown to be sensitive to the birefringence induced by the liquid crystalline materials, but the weak signal strength requires relatively long pixel dwell times (>50 ms at the average powers used in the study of Pillai et al. [2006]), making characterization of samples in a wide range of conditions a time-consuming process and making fast dynamics impossible to monitor. Multiphoton fluorescence from LCs has also been used to study polymer dispersed liquid crystal films (Xie and Higgins 2004), but multiphoton excitation fluorescence microscopy does not provide information on chemical composition.

An ideal optical imaging technique for viewing liquid crystalline samples would have the following properties:

1. Three-dimensional imaging of the orientational order
2. No perturbation of orientational order by labeling molecules/agents or by the probing beams
3. Large penetration depth allowing for imaging in thick samples
4. Rapid imaging to allow either visualization of sample dynamics or characterization under a range of experimental conditions in a reasonable amount of time

In this review, we demonstrate that many of the aforementioned requirements for 3D imaging of the orientational order of LC systems can be in principle satisfied by using laser-scanning coherent anti-Stokes Raman scattering (CARS) microscopy, which does not require doping the LC with fluorescent dyes and offers chemical selectivity and imaging the orientation of preselected chemical bonds.

## 21.2 Liquid Crystals

LCs, that are often called mesophases, show properties intermediate between those of solid crystals and isotropic fluids. They flow like an isotropic liquid yet possess a long-range orientational order and a complete or partial absence of positional order of building units, which can be individual molecules or their aggregates (Kleman and Lavrentovich 2003). There are two types of LCs, thermotropic LCs and lyotropic LCs. Thermotropic LC phases exist in a certain temperature range, typically between the high-temperature isotropic fluid phase and a low-temperature crystalline phase. Their basic building units are usually individual molecules with pronounced shape anisotropy, such as rods, disks, etc. Thermotropic LCs are used in the flat panel displays. Lyotropic LCs are formed when the mesogenic material is dissolved in a solvent (usually water), and thus the mesophases are controlled mostly by concentration. The distinct feature of lyotropic LCs is self-assembly of molecules into supermolecular structures such as micelles or aggregates (Petrov 1999). The typical lyotropic LCs are formed by water and surfactants (amphiphiles), such as soaps, synthetic detergents, and lipids. Surfactant molecules contain a hydrophilic part chemically bound to a hydrophobic part, such as a flexible hydrocarbon chain.

### 21.2.1 Thermotropic LCs

The most common type of thermotropic LC molecules are rod-like molecules that can produce a number of different mesophases. The nematic (N) phase is the least ordered LC phase in which the molecules show a long-range orientational order and no long-range positional order (Figure 21.1a). The direction of average molecular orientation is

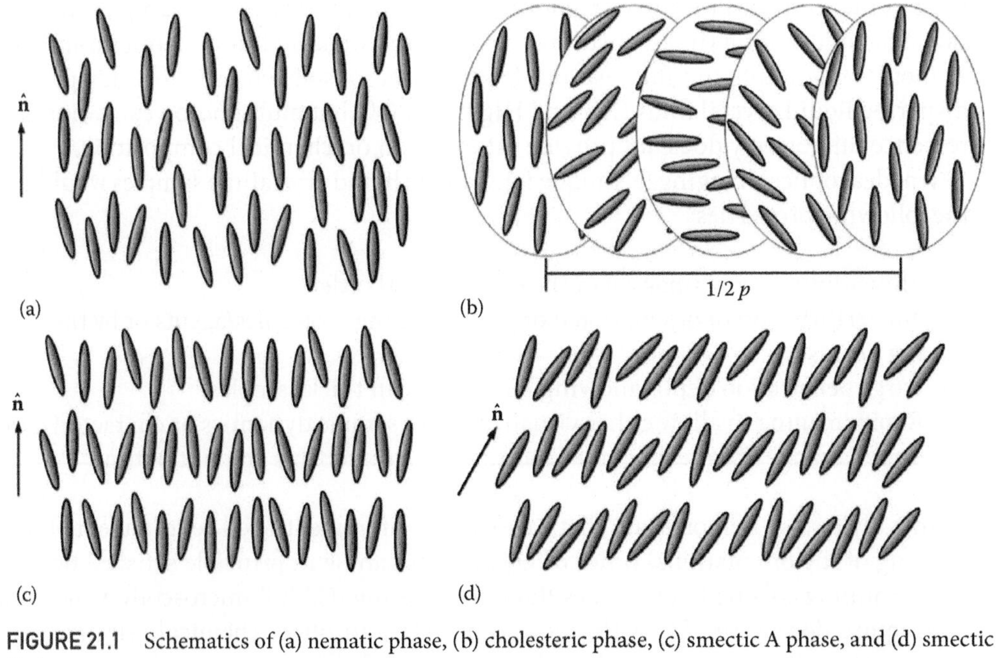


FIGURE 21.1 Schematics of (a) nematic phase, (b) cholesteric phase, (c) smectic A phase, and (d) smectic C phase.

called the director $\hat{\mathbf{n}}$. The two opposite directions are equivalent, $\hat{\mathbf{n}} \equiv -\hat{\mathbf{n}}$, as the nematic order is not polar. The N phase shows an optically positive uniaxial behavior with the local optical axis coinciding with $\hat{\mathbf{n}}$. When the molecules are chiral, the material shows a cholesteric (Ch) phase, in which the director rotates in helical fashion about an axis perpendicular to the director. The pitch, $p$, of the chiral phase is defined as the distance it takes for the director to rotate one full turn in the helix (Figure 21.1b). The shortest cholesteric pitch is typically a few hundreds of nanometers. In the smectic (Sm) phases, the molecules show some translational order not present in the N phase. Smectics are layered phases with a quasi-long-range one-dimensional order of centers of molecules in a direction normal to the layers. In the Sm A phase, the director is perpendicular to the smectic layers (Figure 21.1c). In the Sm C phase, the director is tilted. In Sm A and Sm C, there is no long-range positional order within the layer (Figure 21.1d).

## 21.2.2 Director Field and Its Distortions

An aligned monodomain of an N phase is characterized by a coordinate-independent director $\hat{\mathbf{n}}$ = const. However, in general, the director varies through space, because of conflicting orientation at bounding surfaces or because of the presence of defects or because of the action of the external fields. The three basic types of director distortions are called splay, twist, and bend (Figure 21.2), with the associated free energy density

$$f = \frac{1}{2}K_1[\nabla \cdot \hat{\mathbf{n}}]^2 + \frac{1}{2}K_2\left[\hat{\mathbf{n}} \cdot (\nabla \times \hat{\mathbf{n}})\right]^2 + \frac{1}{2}K_3\left[\hat{\mathbf{n}} \times (\nabla \times \hat{\mathbf{n}})\right]^2 \tag{21.1}$$

featuring the elastic constants of splay ($K_1$), twist ($K_2$), and bend ($K_3$). The typical values for these constants are about $10^{-11}$ N.

A misaligned LC sample usually contains defects in the director field. The defects can be points, lines, or walls. Line defects, called disclinations, are the most common in nematic LCs confined between two parallel plates of glass. Under the PM, they produce the characteristic Schlieren texture when the director is parallel to the glass plates. The disclinations are typically joining the two opposite plates and

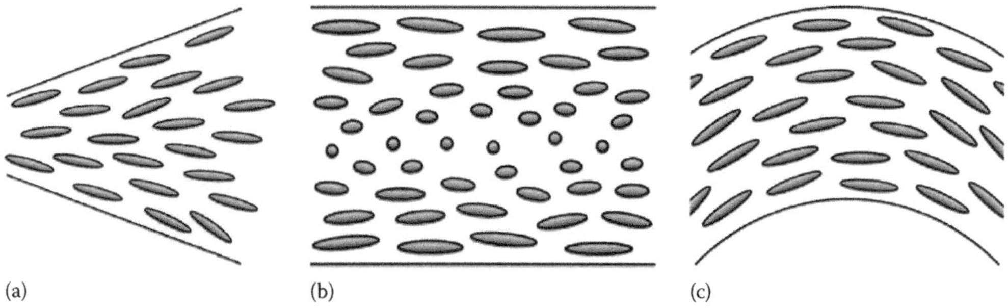

FIGURE 21.2 Deformation of (a) splay, (b) twist, and (c) bend.

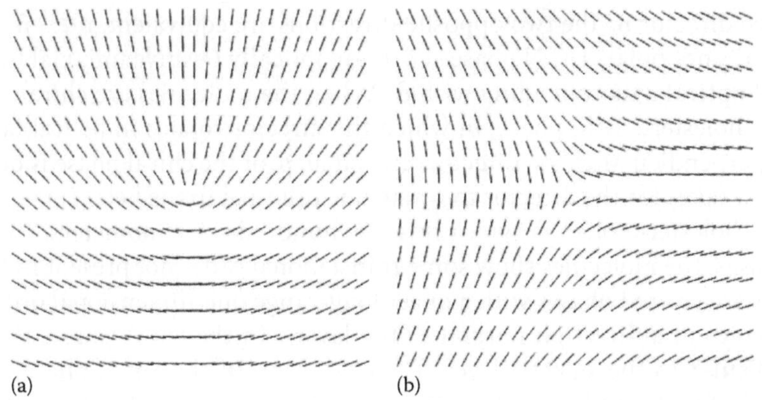


**FIGURE 21.3** Director configurations around an axial disclination of different strength, $k = +1/2$ (a) and $k = -1/2$ (b).

arrange along the normal to the plates to minimize the length (and energy). They are seen as singular points. Each point gives rise to two broad dark bands of light extinction. These bands mark the regions of the texture in which $\hat{\mathbf{n}}$ is parallel to either the polarizer or analyzer. Figure 21.3 shows two possible director configurations characterized by a "strength" of disclination, which assumes either the value of $k = +1/2$ or $k = -1/2$.

The SmA samples in which the layers are not uniformly parallel to the substrates often show "focal conic domain" (FCD) structures. The layers adopt the shape of Dupin cyclides, i.e., surfaces whose lines of curvature are circles (Figure 21.4). A family of Dupin cyclides is associated with a pair of defect lines, a confocal hyperbola and an ellipse, located in two mutually perpendicular planes (Kleman and Lavrentovich 2003). The layers fold around the ellipse and the hyperbola in such a way that they remain equidistant to each other. FCDs allow one to smoothly fill the space with layered structure, satisfying (often conflicting) boundary conditions and creating only linear defects (as opposed to wall defects).

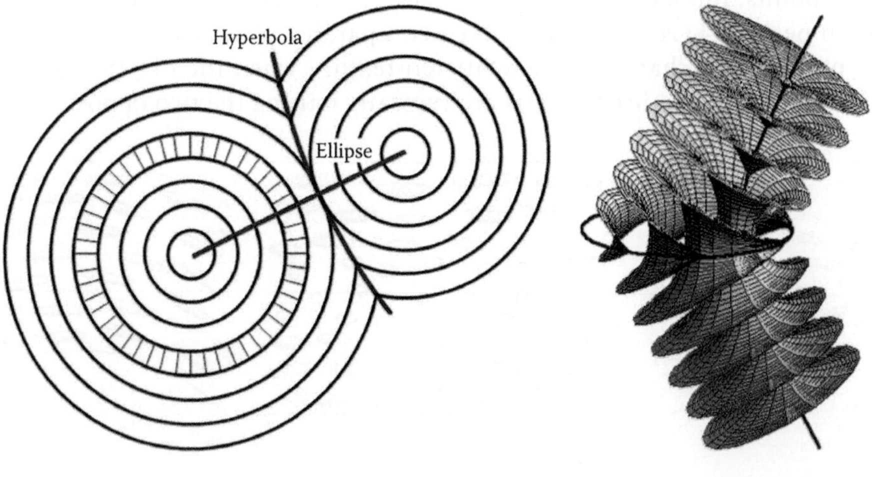


**FIGURE 21.4** Focal conic domain.

### 21.2.3 Electric Field–Induced Director Reorientation

The orientation order of LCs leads to the anisotropy of their elastic, electric, magnetic, and optical properties. The anisotropic dielectric susceptibility allows one to use the electric field in realigning the director, a so-called Frederiks effect that is the basis for liquid crystal display technologies.

To study the dielectric reorientation of the director, the LC is confined between two glass plates coated with thin layers of indium tin oxide, serving as transparent electrodes. The substrates are typically covered with a polymer aligning layer that can be mechanically rubbed to set a preferred orientation of the director at the surface. This direction is called an easy axis. Suppose that the LC is of a positive dielectric anisotropy, $\Delta\varepsilon = \varepsilon_{par} - \varepsilon_{perp} > 0$, where $\varepsilon_{par}$ and $\varepsilon_{perp}$ are the dielectric susceptibilities measured for the electric field parallel to the director and perpendicular to it, respectively. When the electric field is applied normal to the surface, the LC director will tend to align along the field. For the geometry shown in Figure 21.5, the realignment takes place at the field values exceeding some threshold $E_{th} = \pi/d\sqrt{(K_1/\varepsilon_0\Delta\varepsilon)}$, determined by the cell thickness $d$ and the materials parameters $K_1$ and $\Delta\varepsilon$. The reorientation at moderate values of $E > E_{th}$ would not be complete because of the surface anchoring at the substrates. In this particular geometry, the deformations involve bend and splay of the director. When the field is removed, the director relaxes back into the initial orientation (parallel to the bounding plates). Since the director reorientation implies a modification of the optical properties of the cell, the Frederiks effect is widely used in display applications. The cell is typically viewed between the crossed polarizers. The principle of operation is similar to what is described previously, but the geometry of director configuration might be different. For example, in a patterned vertical alignment mode, one uses a vertical orientation of the director, parallel to the direction of the field, but the LC material has a negative dielectric anisotropy, so that the field causes reorientation of the LC molecules toward the substrates. The field-off state is dark when viewed between the crossed polarizers, while the field-on state is transparent.

To conclude, the LC materials show a variety of director configurations related to their phases and structures. Imaging the 3D director structures of LCs is thus important for both fundamental and practical reasons. The following section gives a brief account of how CARS microscopy can be applied to 3D imaging of LCs.

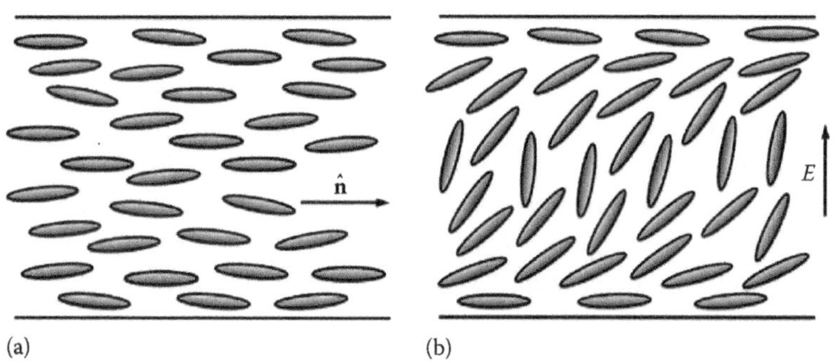


**FIGURE 21.5** Illustration of the electric field induced Fredericks transition, when $E = 0$ (a) and $E > E_{th}$ (b).

## 21.3 Principles of CARS Microscopy of Orientationally Ordered Media

### 21.3.1 Four Wave Mixing Process

CARS is a four wave mixing process involving a "pump" beam at frequency $\omega_p$ and a "Stokes" beam at frequency $\omega_s$. When the beat frequency $\omega_p - \omega_s$ is tuned onto resonance with a selected vibrational mode of LC molecule, an enhanced CARS signal is generated at the "anti-Stokes" frequency, $\omega_{as} = 2\omega_p - \omega_s$, through a four wave mixing process (Cheng and Xie 2004). This is used as the contrast mechanism for microscopic imaging. The CARS signal generated by a given material is related to its third-order nonlinear susceptibility $\chi^{(3)}$. $\chi^{(3)}_{apps}$ is a fourth rank tensor, whose four indices represent the polarization directions of the *a*nti-Stokes, *p*robe (the same as the pump in our case), *p*ump, and *S*tokes beams, respectively. In the general case, $\chi^{(3)}$ can have 81 elements, of which a few are independent in system with symmetry. For example, there are four nonvanishing elements in an isotropic material: $\chi^{(3)}_{1111}$, $\chi^{(3)}_{1122}$, $\chi^{(3)}_{1212}$, and $\chi^{(3)}_{1221}$ (Levenson 1982). The coherent pump and Stokes fields induce a nonlinear polarization at the anti-Stokes frequency:

$$\mathbf{P}^{(3)} \propto \chi^{(3)}_{apps}(-\omega_a, \omega_p, \omega_p, \omega_s)\mathbf{E}_p\mathbf{E}_p\mathbf{E}_s^* \tag{21.2}$$

where

$\mathbf{E}_p$ and $\mathbf{E}_s^*$ are the pump and Stokes fields, respectively

the asterisk (*) indicates a complex conjugate that corresponds to the emission of a photon

The nonlinear susceptibility contains two contributions, a vibrationally resonant term $\chi_r^{(3)}$ and nonresonant term $\chi_{nr}^{(3)}$; $\chi^{(3)} = \chi_r^{(3)} + \chi_{nr}^{(3)}$. Nonresonant contributions include the virtual electronic resonances from molecular states at energies different from the excitation beams and two-photon interactions (Jones 1988). Nonresonant contributions are responsible for the nonzero background signal that decreases the contrast of images. The CARS signal intensity is proportional to the square of the polarization at the anti-Stokes frequency:

$$I_{CARS} \propto \left|\chi_r^{(3)}\right|^2 I_p^2 I_s \tag{21.3}$$

where $I_p$ and $I_s$ are the intensity of the pump and Stokes beams, respectively.

A nonlinear process can be efficiently generated only near the focal volume using short pulsed lasers, since most nonlinear processes have low transition probabilities compared with the linear processes. To generate sufficient CARS signal, the pump and Stokes pulse have to be combined in time and space and focused through a high numerical aperture microscope objective. Cheng et al. (2002) showed the intensity distribution in the focal region of a Gaussian beam at wavelength $\lambda$. When a beam propagating along the $+z$ axis is focused by an objective lens of NA = 1.4, the lateral intensity profile at $z = 0$ exhibits a full width at half maximum (FWHM) of $0.4\lambda$ and the axial intensity profile at $x = y = 0$ has a

FWHM of 1.0λ. The CARS signal is generated from a point-like excitation volume where the pump and Stokes beams are focused, thus producing confocal sectioning without the use of confocal pinholes. The CARS microscopy also provides 3D imaging even in strongly scattering medium such as a tissue because of the localization of signal generation. For a concise overview of CARS microscopy, the reader is also referred to review papers in literature (Cheng et al. 2002, Cheng and Xie 2004, Volkmer 2005, Evans and Xie 2008) and other chapters in this book.

## 21.3.2 Chemical Selectivity

All molecules, including the LC molecules, have multiple resonant frequencies corresponding to the chemical bonds, ranging from low frequency (~100 cm$^{-1}$) for twisting and bending modes to high frequency (~3500 cm$^{-1}$) for stretching modes. Understanding of the vibrational spectrum of a given material is a crucial step in interpreting CARS images (Wurpel et al. 2002). A typical Raman spectrum of the thermotropic LC shows several sharp and isolated features, as demonstrated by the pure compound octylcyanobiphenyl (8CB) and a mixture E7 (Figure 21.6). The spectra show the CN stretching (2215 cm$^{-1}$) and aryl stretching (1600 cm$^{-1}$) vibrations, which can be easily selected for probing CARS because their peaks are strong and located away from that of the other vibrational modes (Figure 21.6). In some biological samples, the nonresonant background can be overwhelmingly strong, reducing the proper contrast. In the thermotropic LC samples, however, the vibrational resonance enhancement is substantial and allows for high-contrast chemical imaging without taking extra measures for background suppression.

Recent advances in CARS microscopy have allowed it to be used for new applications in materials science (Potma et al. 2004, Kano and Hamaguchi 2006, Kim et al. 2008) as well as in the biological and biomedical sciences (Evans et al. 2005, Nan et al. 2006, Huff and Cheng 2007). Several studies have demonstrated the applicability of CARS microscopy to imaging liquid crystals showing nematic, cholesteric, and smectic A textures, such as planar monodomain N samples, electrically induced reorientation in the N bulk with depth-dependent director distortions, Schlieren textures formed by line and point defects in the N phase (Saar et al. 2007), cholesteric finger print texture (Kachynski et al. 2008), and SmA textures with FCDs (Kachynski et al. 2007, 2008, Saar et al. 2007, Chen and Lim 2009).

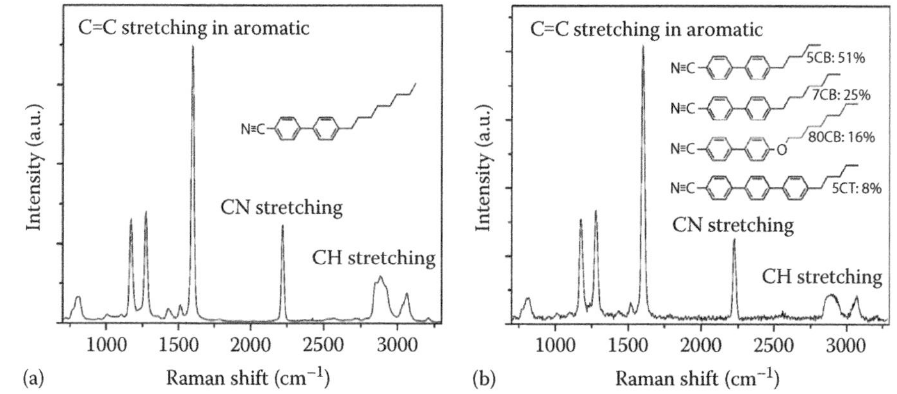


**FIGURE 21.6** Raman scattering spectra and molecular structure of liquid crystal 8CB (a) and E7 (b).

### 21.3.3 Orientational Dependence of CARS Signal

When the pump and probe beams are derived from the same laser beam and the Stokes is polarized parallel to the pump and probe polarization in the *x–y* plane of the image, the imaging of orientational order is detected primarily by the difference between $\chi^{(3)}_{xxxxx}$ and $\chi^{(3)}_{yyyy}$. The images obtained from the different polarization states of probing beams allow one to reconstruct the 3D pattern of $\hat{\mathbf{n}}$. The intensity of CARS signal depends on the angle θ between the collinear polarization of the excitation beams and the chemical bond orientations at the point of focus (Figure 21.7). In the case of CARS utilizing the third-order nonlinear process, the signal intensity depends on the sixth power of the cosθ (Kachynski et al. 2008, Sun et al. 2008):

$$I_{CARS} \propto \cos^6\theta \tag{21.4}$$

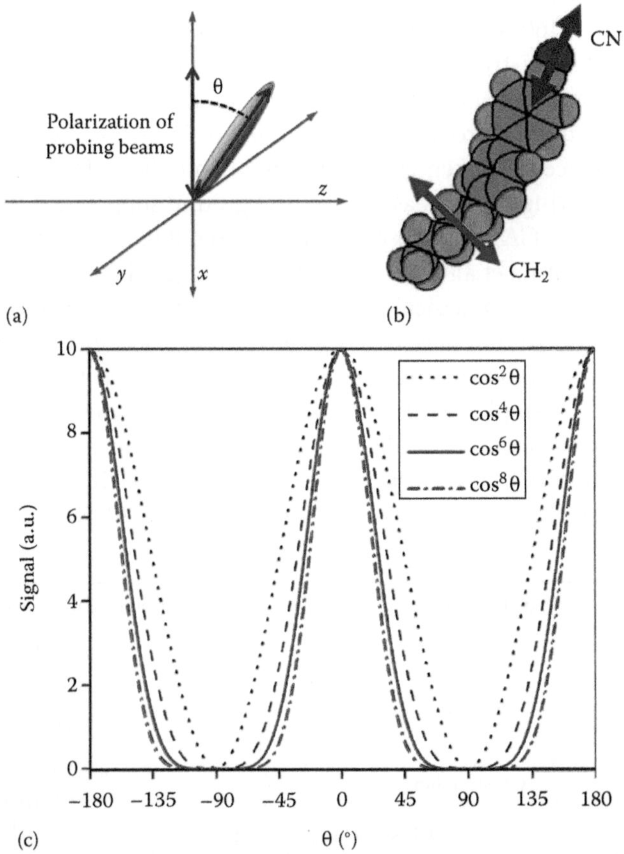


**FIGURE 21.7** (a) Illustration of the angle θ between the collinear polarization of the excitation beams and the chemical bond orientation. (b) Molecular structure of the typical LC molecule pentylcyanobiphenyl (5CB) with the directions of CN and $CH_2$ bonds. (c) Angular dependence of the signals from one-photon ($\cos^2\theta$) and multi-photon ($\cos^4\theta$, $\cos^6\theta$, $\cos^8\theta$) processes. θ is the angle between the collinear polarization of the excitation beams and the bond/molecular orientations.

The intensity is the highest when the polarization directions of the beams are parallel to the orientation of the vibrational mode of the LC molecules. Note that in the LC material 8CB or E7 the orientation of $\hat{\mathbf{n}}$ coincides with the average orientation of CN bonds, which are along the long axes of the molecules. If another polarizer, parallel with the polarization direction of the excitation beam, is placed in the detection channel, the CARS intensity scales as cos $\theta^8$ (Wurpel et al. 2005, Kachynski et al. 2008, Sun et al. 2008). The strong angular dependence of the CARS intensity allows one to decipher the spatial configuration of the specific vibrational mode of the molecules. The signal intensity shows the maximum at $\theta = 0$ and drops rapidly as $\theta$ changes from $\theta = 0$, which provides the image with sharper contrast near $\theta = 0$ than that from the one- or two-photon processes. Note, however, that the dependence on high power of cos $\theta$ also implies worsening of contrast in the range of angles close to $\theta = 90$ (Figure 21.7).

By imaging the sample at the same vibrational frequency using two different polarization geometries ($I_x$ and $I_y$, the subscript denotes the polarization of the pump, probe, and Stokes, which are all parallel to each other), the director orientation at each point can be deduced according to arctan($I_x/I_y$) and the local degree of order can be calculated using $(I_x - I_y)/(I_x + I_y)$ (Saar et al. 2007).

# 21.4 CARS Microscopy Measurement

## 21.4.1 Optical Setup

A CARS imaging experiment requires one to have two pulsed lasers at two different wavelengths that are overlapped in space and synchronized in time. A laser system setup is demonstrated in Figure 21.8, which closely follows the one described in Saar et al. (2007); detailed instructions are available elsewhere (Cheng et al. 2002, Cheng and Xie 2004, Evans et al. 2005, Evans and Xie 2008). A passively mode-locked Nd:$YVO_4$ oscillator (HighQ Laser, PicoTrain) that delivers a ~6 ps pulse trains at 1064 and 532 nm with a repetition rate of 76 MHz. The 532 nm (~4 W) light is used to synchronously pump an optical parametric oscillator (OPO, APE GmbH, Levante Emerald) based on

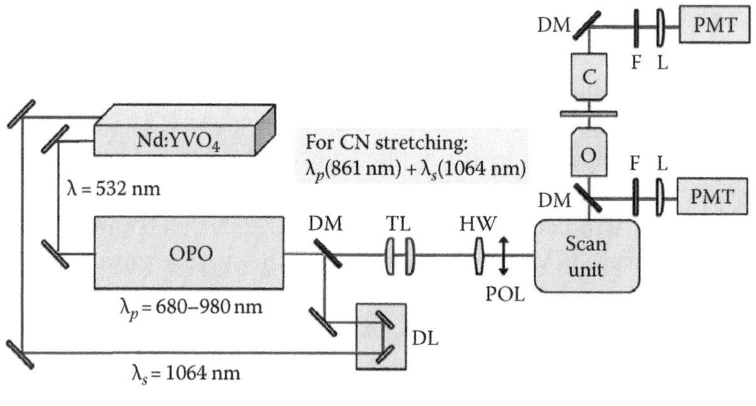


FIGURE 21.8 Schematic diagram of the CARS setup. Nd:$YVO_4$: 532 and 1064 nm laser; OPO, optical parametric oscillator; DM, dichroic mirror; TL, telescope lenses; HW, broadband half-wave plate; POL, polarizer; O, objective lens; C, condenser; F, short-pass filter; L, collecting lens; PMT, photon multiplier tube.

Chapter 21

a noncritically phase-matched lithium triborate (LBO, $LiB_3O_5$) crystal. The signal wavelength of the OPO is tunable in the range 680–980 nm by adjusting the LBO temperature and is used as the pump wavelength for CARS. The pump beam is combined with the 1064 nm Stokes beam from the Nd:$YVO_4$ laser on a dichroic mirror. The time delay is inserted in the path of the Stokes beam prior to mixing with the pump beam to allow compensation for the differences in optical path and refractive index differences of the optics at the two wavelengths. The combined beams are linearly polarized. The direction of polarization can be rotated using an achromatic zero order $\lambda/2$ plate in the beam path and attenuated to the desired power using a variable neutral density filter. The beams are then coupled into a modified laser-scanning microscope (Olympus, FV300-IX71) and focused on the sample using a water immersion objective (Olympus UPLSAPO, NA1.2, X60). The signal can be simultaneously detected in forward and backward (epi) directions. A condenser lens is used to collect the forward CARS signal. The CARS signal is separated from the excitation beams using optical filter sets and detected in the forward direction on a non-descanned detector utilizing a photomultiplier tube (Hamamatsu, R9876). For epi-detection, a dichroic beam splitter is used to reflect the excitation beams and transmit the signal. Forward-detection is suitable for imaging objects of a size comparable to or larger than the excitation wavelength, while epi-detection provides a sensitive means of imaging objects smaller than the excitation wavelength (Cheng and Xie 2004). Scan speeds of ~4 μs per pixel are routinely used, though imaging at video rate (~100 ns pixel dwell time) has also been demonstrated for CARS (Evans et al. 2005).

## 21.4.2 Laser-Induced LC Realignment

CARS is a third-order nonlinear process in which the intensity of CARS signal is dependent on the intensity of pump and Stokes beams: $I_{CARS} \propto I_{pump}^2 I_{Stokes}$. The high-intensity pump and Stokes beams can allow enough CARS signals during the short integration times needed for the studies of dynamic processes; however, high-power laser beams can cause director distortions through the optical Fredericks effect (Khoo 1995). Laser-induced LC molecular realignment usually occurs at intensities above a certain threshold. The threshold intensity increases with a decrease in the optical anisotropy of LC materials and with an increase in the elastic constants. LC materials in the smectic and columnar phase show higher threshold intensity than materials in the nematic phase (Durbin et al. 1981, Khoo et al. 1987, Santamato et al. 1987, Khoo 1995). To use a laser beam for optical imaging of liquid crystalline samples, it is important to establish whether the torque imparted by the applied electric field of the laser beam can reorient the sample, perturbing the images. As shown by Enikeeva et al. (2005), short-pulsed (femtosecond) laser irradiation causes director reorientation comparable to that caused by the continuous wave (CW) irradiation at the same average power, suggesting that the pulse width is not the most critical parameter for this effect.

To address this issue, Kachynski et al. (2008) explored the effects of director realignment by using laser powers higher than usually needed for imaging of stationary director structure. They tested the effect of director realignment of 5CB by changing the scanning area and changing laser power while keeping the constant image area. They showed that when the scanning area at constant laser power (~350 mW) is decreased to a certain threshold value (~512 μm$^2$) the laser-induced realignment is observed. Similarly, when the excitation area is kept constant but the laser power is varied, the laser-induced realignment is observed at a certain intensity that depends on the scanning area. The threshold excitation intensity needed for the LC realignment increases with increasing the size of the scanning area. They also showed that the threshold powers of the realignment in the planar cells are larger than those in the homeotropic cells. Saar et al. (2007) did not observe any optically induced director distortion of the E7 cell under their experimental conditions (average excitation power: 16–64 mW, pulse width: ~6 ps, and repetition rate: 80 MHz). Compared with other techniques such as THG (Pillai et al. 2006), short dwell time per pixel in CARS microscopy experiment ($\tau_{dwell} = 4\,\mu s$) can help to mitigate the possible director distortion.

## 21.5 In-Plane CARS Imaging of Flat Samples

The CARS signal intensity depends strongly on the angle $\theta$ between the collinear polarization **P** of the excitation beams and the chemical bond orientations. This dependency $I_{CARS}(\theta)$ can be established by using a uniformly aligned LC cell. Saar et al. (2007) demonstrated the large dependence of $I_{CARS}(\theta)$ measured for $\theta = 0$ and for $\theta = \pi/2$. The CARS signal strength of the aligned E7 slab imaged at 2215 cm$^{-1}$ corresponding to the CN group strongly depends on the polarization **P** of the excitation beams, reaching maximum when $\hat{\mathbf{n}} \parallel \mathbf{P}$ and minimum when $\hat{\mathbf{n}} \perp \mathbf{P}$. The ratio of the maximum signal to the minimum signal is ~10, indicating that the CN groups are well aligned along the director as expected, because these groups are on average parallel to the long axes of LC molecules, as illustrated in Figure 21.7 for 5CB (one of the components of E7) (Saar et al. 2007). The entire dependence of $I_{CARS}(\theta)$ was tested using 5CB (Chen and Lim 2009) and resulted in the relationship, $I_{CARS}(\theta) \propto \cos^6\theta$. Kachynski et al. (2008) showed that the angular dependence of the CARS signal caused by the CN bond in 5CB becomes stronger, $I_{CARS}(\theta) \propto \cos^8\theta$, when an additional polarizer is placed in the detection channel collinear with the polarizations of excitation beams.

Figure 21.9 shows typical intensity maps of the texture for the CN group (oscillation at 2215 cm$^{-1}$) of E7 in a flat cell with untreated bounding plates that provide a tangential orientation of the director, $\hat{\mathbf{n}} = (n_x, n_y, 0)$ without setting a preferred orientation in the $XY$ plane. These samples, when viewed in a PM with two crossed polarizers, give rise to the so-called Schlieren textures, caused by linear defect disclinations connecting the opposite plates and point defects (boojums) located at the bounding plates. The boojums can be considered as the remnants of the disclination lines of an integer strength in which the director "escaped" along the vertical direction, see (Chiccoli et al. 2002). When one circumnavigates the defect core once (in the plane of the substrate), the director reorients by an angle $\pi$ if the defect is a disclination line and by $2\pi$ when the defect is a boojum point. Figure 21.9a through c illustrates the horizontal slices of an N sample with vertical disclination (one is shown inside a circle in Figure 21.9a) whose two ends are located underneath each other at the opposite plates. The texture also shows point defects boojums located at the surfaces of the sample (one is enclosed in a rectangle in Figure 21.9a) around which the director rotates by $2\pi$. By rotating the linear polarization of the pump and Stokes beams together and making a map of the regions with $\hat{\mathbf{n}} \parallel \mathbf{P}$, one can reconstruct the entire director configuration. The reconstructed in-plane director field around the boojum is shown in Figure 21.9d. In PM, a texture would show extinction whenever $\hat{\mathbf{n}}$ is along the polarizer or analyzer direction (Hartshorn 1974, Bellare et al. 1990). Because of this degeneracy, PM does not distinguish readily two complementary director structures that are mutually

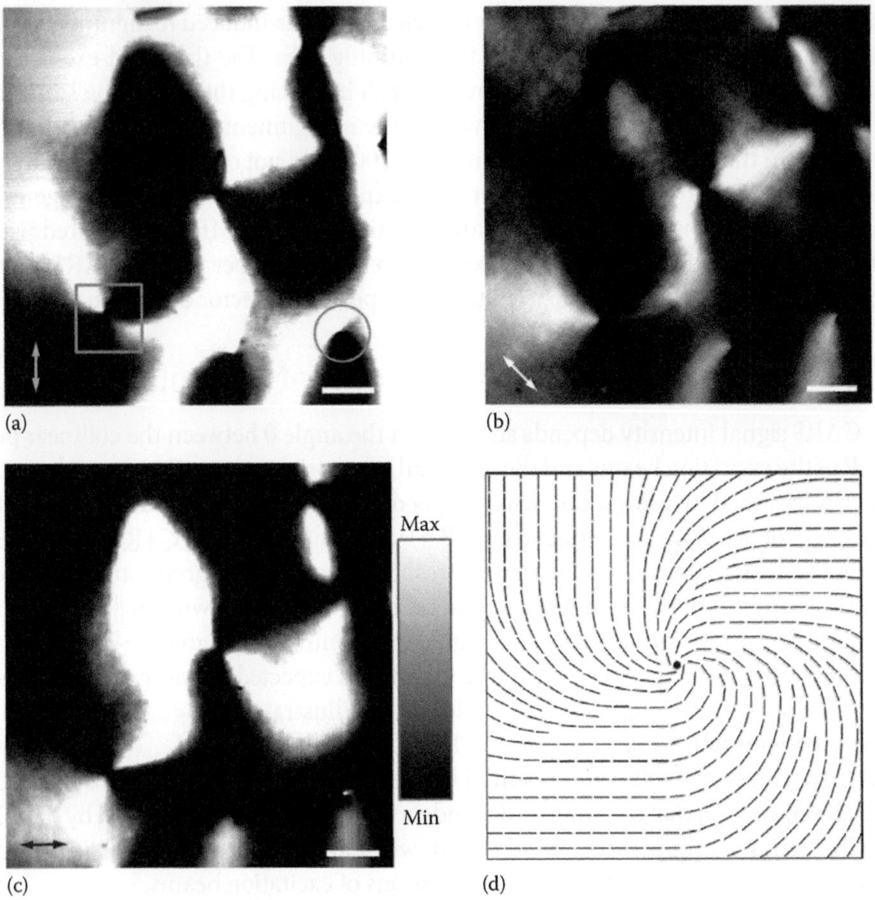


**FIGURE 21.9** CARS images of the Schlieren texture of E7 imaged at 2215 cm$^{-1}$ corresponding to CN stretching, with the pump and Stokes beams polarized along the direction indicated by the double-headed arrow. The square embraces a surface point defect—boojum, while the circle embraces the center of a disclination line of strength 1/2. The maximum CARS signal corresponds to the regions with $\hat{\mathbf{n}} \| \mathbf{P}$ and minimum to $\hat{\mathbf{n}} \perp \mathbf{P}$. Scale bar is 20 μm. (d) Sketch of the director field around a point defect in a square in (a), based on data from (a–c).

orthogonal (for that task, an optical compensator is required [Kleman and Lavrentovich 2003]). CARS microscopy does not suffer from this type of degeneracy. Figure 21.9d shows the in-plane director configuration within the area limited by a rectangle in Figure 21.9a, around the point defect with a core marked by a dot.

## 21.6 CARS Images of the Vertical Cross-Section of LC Cells

### 21.6.1 Imaging Fredericks Transition (Field-Induced Distortion)

Liquid crystals such as E7 are dielectrically anisotropic; consequently, a sufficiently strong external electric field can reorient $\hat{\mathbf{n}}$, which is called the Fredericks transition, see Section 21.2.3. Using a CARS microscope, Saar et al. (2007) imaged a Fredericks transition of the homogeneously aligned E7, placed between two glass plates coated with transparent indium tin oxide (ITO) electrodes. The electric field reorients the director vertically toward the state $\hat{\mathbf{n}} = (0, 0, 1)$ when the voltage is high, since the dielectric anisotropy of E7 is positive. At small voltages, below the well-defined threshold, the structure remains uniform, $\hat{\mathbf{n}} = (1, 0, 0)$. Above this threshold, the electric field becomes strong enough to overcome the anchoring forces at the bounding plates and elasticity of the N sample to reorient $\hat{\mathbf{n}}$ from its strictly horizontal orientation toward the *Z*-axis. The maximum deviation was observed in the middle of the cell, as predicted by theory (Blinov and Chigrinov 1994). Figure 21.10 shows the vertical cross-section of the cell imaged at a series of voltages, 0, 2.5, 5, and 8 V (Saar et al. 2007). The *XZ* profiles at the CN stretching frequency (2215 cm$^{-1}$) demonstrate the typical signature of a Fredericks transition: at 0 V, the director lies along *X*, and the sample shows a constant signal along the *Z* direction within the N phase and fades away once the probing beams move into the glass substrates. At 2.5 V, the molecules in the center of the sample are reoriented along *Z* and thus generate a decreased amount of CARS signal, yielding bands of intense signal only near the electrode boundaries, where $\hat{\mathbf{n}}$ remains horizontal or only slightly tilted. At even higher fields, the absolute signal level decreases as the molecules are tilted forward *Z* and, therefore, generate weaker signals upon the polarized excitation along *X*. The images at the $CH_2$ stretching frequency (2845 cm$^{-1}$) are quite different from the images at the CN stretching since the $CH_2$ groups are not parallel to the CN groups and have a lower degree of orientational order.

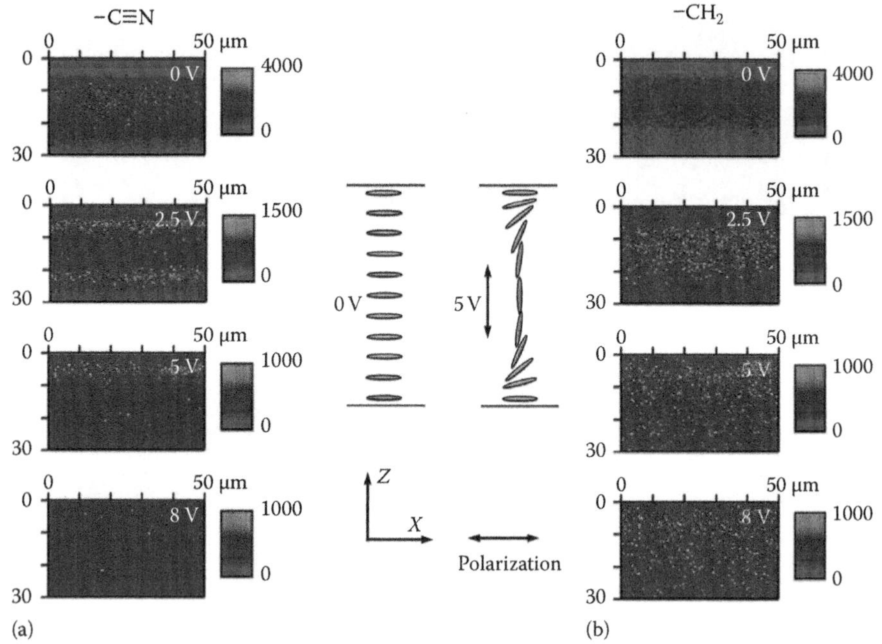


**FIGURE 21.10** Visualizing the Fredericks transition with CARS. CARS depth profiles of the E7 cell tuned into the CN stretch at 2215 cm$^{-1}$ at various voltages (a). The loss of signal in the center of the sample at elevated voltages is characteristic of a Frederiks transition and demonstrates that the CN groups (and the director) reorient vertically. CARS images of the same sample at the $CH_2$ stretching frequency of 2845 cm$^{-1}$ (b).

### 21.6.2 Imaging FCDs in a Smectic A Phase

Figure 21.11 shows the CARS images of the toric focal conic domains (TFCD) of 8CB (octylcyanobiphenyl) between boundaries with different orientation, planar and homeotropic plates, imaged at the CN stretching frequency of 2215 cm$^{-1}$. 8CB has a Sm A phase at room temperature. The strongest CARS signal is obtained at the spatial locations where $\hat{\mathbf{n}}||\mathbf{P}$ and smectic layers are perpendicular to **P** (Figure 21.11). Chen and Lim (2009) have shown the vertically sectioned image of an FCD in a smectic A phase of 8CB in a 50 μm thick sandwich cell (Figure 21.12).

## 21.7 Imaging Cholesteric LCs

The CARS images visualize the so-called fingerprint structure of a cholesteric LC, which contains 2π director twist along a helical axis tilted with respect to the cell normal. Figure 21.13 shows the fingerprint texture of cholesteric LC (a mixture of nematic host E7 and a chiral agent CB15, EM industries, the United States) with pitch $p = 5\,\mu m$ between two glass plates treated for the homeotropic LC alignment. In the twisted N cell or in cholesteric

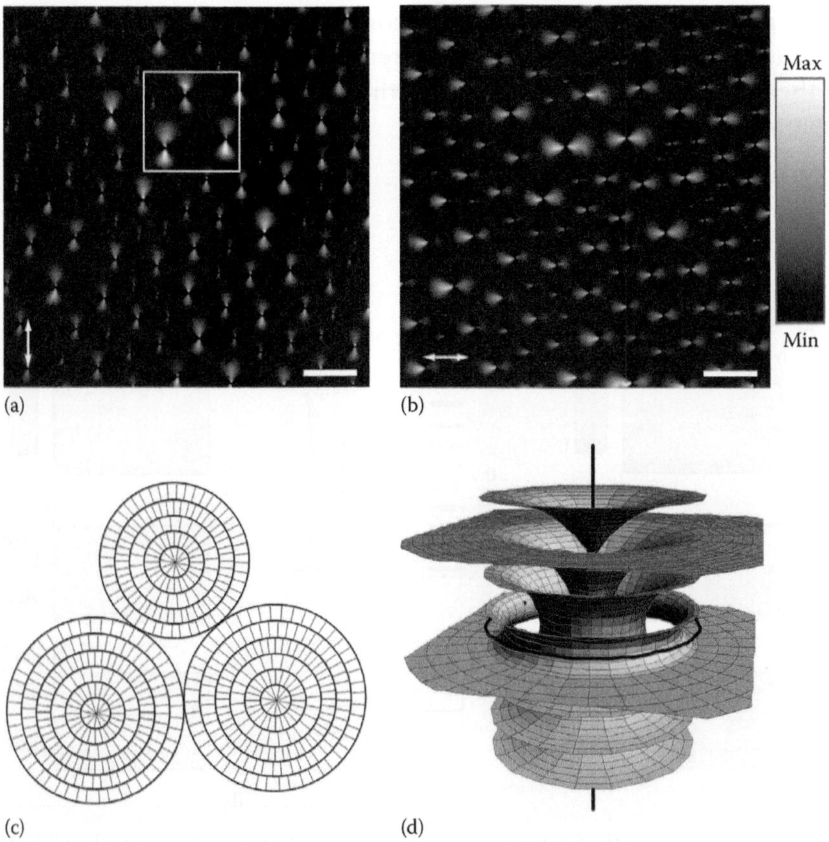


**FIGURE 21.11** CARS images of TFCDs of 8CB between boundaries with different orientation, planar and homeotropic plates, imaged at 2215 cm$^{-1}$ corresponding to CN stretching, with the pump and Stokes beams polarized along the direction indicated by double-headed arrow (a) and (b). Scale bar is 20 μm. (c) Sketch of the director field in a square in (a). (d) TFCD in a stack of parallel smectic layers.

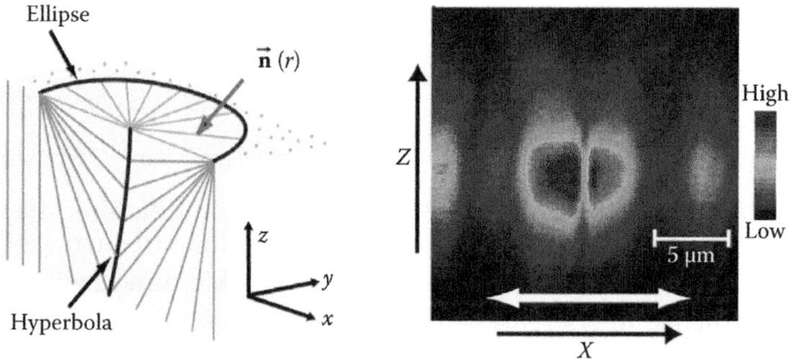


**FIGURE 21.12** Vertical cross-sectioned image of an FCD in an 8CB film. Double-headed arrow represents the laser polarization directions. (Reprinted in part with permission from Chen, B.-C. and Lim, S.-H., Three-dimensional imaging of director field orientations in liquid crystals by polarized four-wave mixing microscopy, *Appl. Phys. Lett.*, 94, 171911. Copyright 2009, American Institute of Physics.)

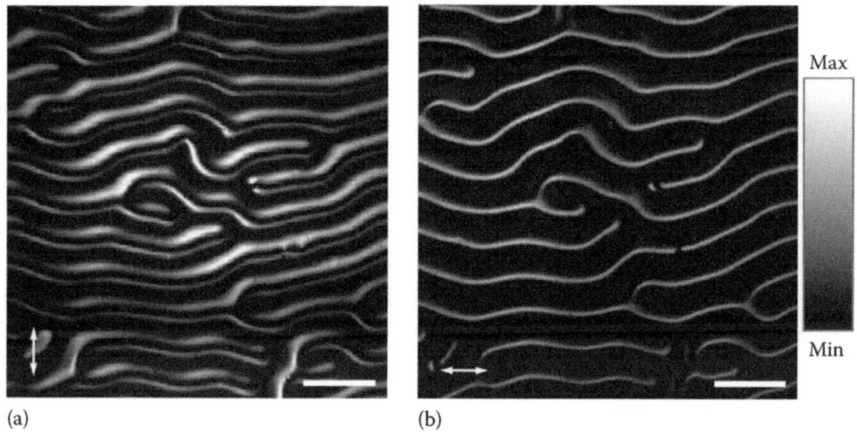


**FIGURE 21.13** (a) and (b) Fingerprint textures in a homeotropic cell with the E7 doped with chiral additive CB15, imaged at the CN stretching frequency of 2215 cm$^{-1}$. The double-headed arrow indicates the polarization of the pump and Stokes beams. Scale bar is 5 μm.

LCs, polarization of both ordinary and extraordinary waves follows the twist of the director (the so-called Mauguin regime) (Kleman and Lavrentovich 2003). This is quantified by the Mauguin parameter, $Ma = p\Delta n/2\lambda$, where $p$ is the cholesteric pitch and $\Delta n$ is the difference of the extraordinary and ordinary refractive indices. When $Ma \gg 1$, the polarization of probing beams follows the director twist, which makes it difficult to image the twisted director. Compared with the single-photon microscopy, CARS microscopy mitigates this issue by using the longer wavelength excitation beams that can reduce the Mauguin parameter.

Recently Lee et al. (2010) demonstrated the feasibility of simultaneous CARS and other nonlinear optical techniques, such as multiphoton excitation fluorescence microscopy and second-harmonic generation microscopy. They showed the vertical cross-sectional images of ~30 μm thick cells that have the planar ground state cholesteric structure. The two-photon excitation fluorescence (2PF) image showed the cholesteric layer structure via fluorescence of the dye *N*,*N*′-bis(2,5-di-*tert*-butylphenyl)-3,4,9,10-perylenedicarboximide (BTBP), while three-photon excitation fluorescence (3PF) and CARS images were obtained without the use of a dye (Lee et al. 2010).

# 21.8 CARS Images of the Heterogeneous and Lyotropic Samples

## 21.8.1 Imaging Sm A LC and Glycerol Mixtures

Figure 21.14 shows the forward-detected CARS image (a) and epi-detected CARS image (b) of an 8CB droplet in glycerol at the CN stretching frequency of 2215 $cm^{-1}$. The forward-detection seen in Figure 21.14a shows the brighter image with a stronger CARS signal by constructive interference (Evans and Xie 2008). The glycerol region does not show a polarization preference for excitation, consistent with an isotropic nature of this medium. This study also demonstrates that CARS microscopy is not limited to visualizing molecular orientation patterns in homogeneous LC materials. CARS microscopy can also be used for the study of a composite system of anisotropic LCs and isotropic materials.

## 21.8.2 Imaging Lyotropic LCs

Compared with the traditional optical microscopy, a technique using a linear absorption process, CARS microscopy provides higher resolution for the lyotropic LC system with chemical selectivity and polarization sensitivity by using near-infrared excitation beams that penetrate deeper into the scattering medium. Potma and Xie (2003) have visualized the $CH_2$ stretching mode at the frequency of 2845 $cm^{-1}$ in membranes of erythrocyte ghost vesicles using CARS microscopy. The CARS signal is maximized when the excitation polarization is parallel to the $CH_2$ group symmetry axis that is perpendicular to the lipid hydrocarbon chain. The image reveals that $CH_2$ groups are on average parallel to the surface of the bilayer (Figure 21.15).

# 21.9 Conclusions

Imaging 3D spatial patterns of molecular orientations is important in the studies of modern materials such as LCs and polymers. CARS microscopy has been demonstrated to be an excellent tool of nondestructive label-free chemical imaging. Practical applications of

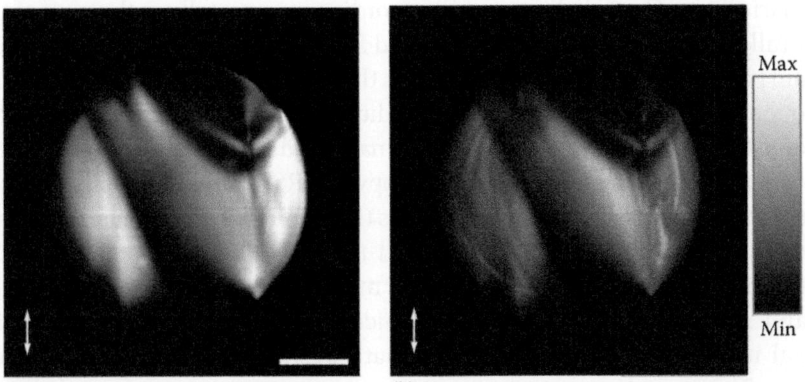


(a) (b)

**FIGURE 21.14** Forward-detected (a) and epi-detected (b) CARS images of an 8CB droplet in glycerol at the CN stretching frequency of 2215 $cm^{-1}$. A double-headed arrow indicates the polarization of the pump and Stokes beams. Scale bar is 5 μm.

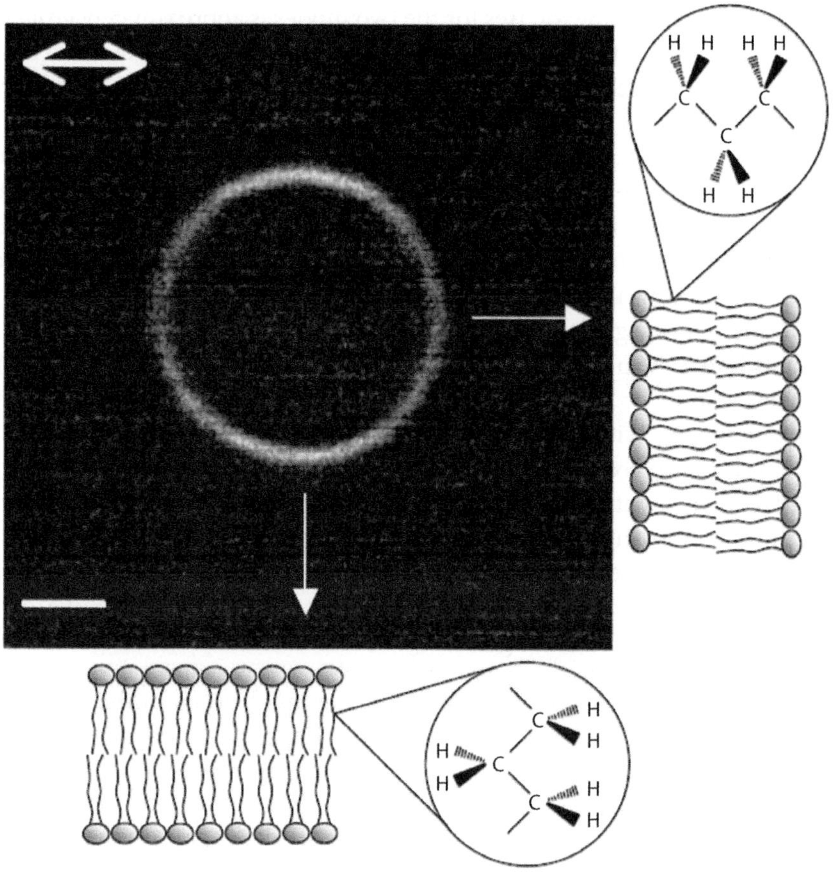


FIGURE 21.15 CARS images of an erythrocyte ghost vesicle at the $CH_2$ stretching frequency of 2845 $cm^{-1}$. The average powers of the pump and the Stokes beam were 0.3 and 0.1 mW, respectively, at a repetition of 250 kHz. A double-headed arrow indicates the polarization of the pump and Stokes beams. Scale bar is 2 μm. (From Potma, E.O. and Xie, X.S., *J. Raman Spectrosc.*, 34, 642, 2003. With permission.)

CARS microscopy require some care to avoid aberrations and other side effects. There are issues of common importance for both optically isotropic and anisotropic media, as well as issues specific to ordered materials, as specified in the following:

1. Light-induced director realignment. A focused laser beam of a sufficient power can cause director reorientation in the medium. In FCPM, the problem is not significant, since the power is usually less than 1 μW (Smalyukh et al. 2001). In CARS, the average power might be of the order of 1–100 mW, and, in principle, the issue of director reorientation needs to be addressed. However, because of the very short dwell time, the director distortions are not significant. Continuous instrumentation improvements will allow for faster image acquisition and shorter signal integration times, which can solve this issue and also enable the study of the dynamics of the director in LCs.
2. Optics of anisotropic medium. Focusing the laser beams in a birefringent medium, even if the director is not perturbed, is more complicated than in an isotropic medium since there are two propagating modes with different indices of refraction. To reduce the aberrations one can use a liquid crystal host with a low birefringence, $\Delta n$: the

spatial separation of the focuses for the two modes is approximately $g\Delta nh/n^{-}$, where $n^{-}$ is the average refractive index, $h$ is the distance between the entry and focus in the liquid crystal cell, and $g$ is the number of the order of unity (dependent on the sample orientation) (Smalyukh et al. 2001).

3. Finite absorption. Figure 21.10 clearly illustrates that the top and the bottom of the vertical slices of the CARS images are slightly asymmetric, due to losses across the sample, associated with absorption of the material and scattering. The effect can be verified by turning the sample upside down. Reducing the sample thickness or depth of scanning all help to mitigate the problem.
4. Adiabatic following of polarization. The well-known effect in the optics of liquid crystals is that the polarization of light follows the director twist (Mauguin regime). This effect must be taken into account while interpreting the confocal images for nematic samples with twist deformations, especially when the twist scale is supramicron and light propagates along the twist axis.
5. Polarization geometry. We described only one configuration of light polarization, in which the pump and Stokes light are linearly polarized in one direction and these polarization directions are rotated together. Depending on the need, one can use many other polarization geometries, such as polarization sensitive detection to reduce the nonresonant background in CARS microscopy (Cheng et al. 2001).

In summary, CARS microscopy is a versatile tool for rapid, nonperturbative 3D imaging of media with orientational order, supplemented with a specific chemical information. CARS microscopy offers a number of advantages over the existing modes of observation, such as PM, FCPM, THG, and spontaneous Raman microscopy. CARS microscopy can also be combined with other imaging techniques, such as multiphoton excitation fluorescence microscopy and multiharmonic generation microscopy, to provide a multimodal platform for the study of liquid crystals (Lee et al. 2010). Given the large number of possible detection schemes and experimental conditions, CARS will likely prove to be an invaluable tool for studying orientationally ordered materials, such as liquid crystals, for years to come.

## Acknowledgment

The work was supported by NSF grant DMR 1104850 and 11212878.

## References

Bellare, J. R., H. T. Davis, W. G. Miller, and L. E. Scriven. 1990. Polarized optical microscopy of anisotropic media: Imaging theory and simulation. *J. Colloid Interface Sci.* 136: 305–326.

Blach, J. F., M. Warenghem, and D. Bormann. 2006. Probing thick uniaxial birefringent medium in confined geometry: A polarised confocal micro-Raman approach. *Vib. Spectrosc.* 41: 48–58.

Blinov, L. M. and V. G. Chigrinov. 1994. *Electrooptic Effects in Liquid Crystal Materials.* New York: Springer.

Buyuktanir, E. A., K. Zhang, A. Gericke, and J. L. 2008. Raman imaging of nematic and smectic liquid crystals. *Mol. Cryst. Liq. Cryst.* 487: 39–51.

Camorani, P. and M. P. Fontana. 2007. Local three-dimensional characterization of a micro-patterned liquid crystalline cell by confocal Raman microscopy. *Mol. Cryst. Liq. Cryst.* 465: 143–152.

Chen, B.-C. and S.-H. Lim. 2009. Three-dimensional imaging of director field orientations in liquid crystals by polarized four-wave mixing microscopy. *Appl. Phys. Lett.* 94: 171911.

Cheng, J. X., L. D. Book, and X. S. Xie. 2001. Polarization coherent anti-Stokes Raman scattering microscopy. *Opt. Lett.* 26: 1341–1343.
Cheng, J. X., A. Volkmer, and X. S. Xie. 2002. Theoretical and experimental characterization of coherent anti-Stokes Raman scattering microscopy. *J. Opt. Soc. Am. B* 19: 1363–1375.
Cheng, J. X. and X. S. Xie. 2004. Coherent anti-Stokes Raman scattering microscopy: Instrumentation, theory, and applications. *J. Phys. Chem. B* 108: 827–840.
Chiccoli, C., I. Feruli, O. D. Lavrentovich, P. Pasini, S. V. Shiyanovskii, and C. Zannoni. 2002. Topological defects in Schlieren textures of biaxial and uniaxial nematics. *Phys. Rev. E* 66: 030701.
Durbin, S. D., S. M. Arakelian, and Y. R. Shen. 1981. Optical-field-induced birefringence and Freedericksz transition in a nematic liquid crystal. *Phys. Rev. Lett.* 47: 1411–1414.
Enikeeva, V. A., V. A. Makarov, I. A. Ozheredov, A. P. Shkurinov, I. A. Budagovsky, V. F. Kitaeva, A. S. Zolot'ko, and M. I. Barnik. 2005. Orienting influence of femtosecond pulses on nematic liquid crystals. *Mol. Cryst. Liq. Cryst.* 442: 1–18.
Evans, C. L., E. O. Potma, M. Puoris'haag, D. Cote, C. P. Lin, and X. S. Xie. 2005. Chemical imaging of tissue in vivo with video-rate coherent anti-Stokes Raman scattering microscopy. *Proc. Natl Acad. Sci. USA* 102: 16807–16812.
Evans, C. L. and X. S. Xie. 2008. Coherent anti-Stokes Raman scattering microscopy: Chemical imaging for biology and medicine. *Annu. Rev. Anal. Chem.* 1: 883–909.
de Gennes, P. G. and J. Prost. 1993. *The Physics of Liquid Crystals*. Oxford, U.K.: Oxford University Press.
Hartshorn, N. H. 1974. *Microscopy of Liquid Crystals*. London, U.K.: Microscope Publications.
Huff, T. B. and J. X. Cheng. 2007. In vivo coherent anti-Stokes Raman scattering imaging of sciatic nerve tissue. *J. Microsc.* 225: 175–182.
Jones, W. J. 1988. Raman amplification spectroscopy. In *Advances in Non-Linear Spectroscopy*, eds. R. J. H. Clark and R. E. Hester, pp. 51–96. New York: Wiley.
Kachynski, A. V., A. N. Kuzmin, P. N. Prasad, and I. I. Smalyukh. 2007. Coherent anti-Stokes Raman scattering polarized microscopy of three-dimensional director structures in liquid crystals. *Appl. Phys. Lett.* 91: 151905.
Kachynski, A. V., A. N. Kuzmin, P. N. Prasad, and I. I. Smalyukh. 2008. Realignment-enhanced coherent anti-Stokes Raman scattering and three-dimensional imaging in anisotropic fluids. *Opt. Express* 16: 10617–10632.
Kano, H. and H. Hamaguchi. 2006. Three-dimensional vibrational imaging of a microcrystalline J-aggregate using supercontinuum-based ultra-broadband multiplex coherent anti-Stokes Raman scattering microscopy. *J. Phys. Chem. B* 110: 3120–3126.
Kaznacheev, K. and T. Hegmann. 2007. Molecular ordering in biaxial smectic-A phase studied by scanning transmission x-ray microscopy (STXM). *Phys. Chem. Chem. Phys.* 9: 1705–1712.
Khoo, I. C. 1995. *Liquid Crystals: Physical Properties and Nonlinear Optical Phenomena*. New York: Wiley.
Khoo, I. C., T. H. Liu, and P. Y. Yan. 1987. Nonlocal radial dependence of laser-induced molecular reorientation in a nematic liquid crystal: Theory and experiment. *J. Opt. Soc. Am. B* 4: 115–120.
Kim, H., D. K. Taggart, C. Xiang, R. M. Penner, and E. O. Potma. 2008. Spatial control of coherent anti-Stokes emission with height-modulated gold zig-zag nanowires. *Nano Lett.* 8: 2373–2377.
Kleman, M. and O. D. Lavrentovich. 2003. *Soft Matter Physics: An Introduction*. New York: Springer-Verlag.
Lavrentovich, O. D. 2003. Fluorescence confocal polarizing microscopy: Three-dimensional imaging of the director. In *Liquid Crystals and Other Soft Materials*, ed. B. K. Sadashiva, pp. 373–384. Bangalore, India: Indian Academy of Sciences, reprinted from *Pramana—J. Phys.* 61(2): 373–384.
Lee, T., R. P. Trivedi, and I. I. Smalyukh. 2010. Multimodal nonlinear optical polarizing microscopy of long-range molecular order in liquid crystals. *Opt. Lett.* 35: 3447–3449.
Levenson, M. D. 1982. *Introduction to Nonlinear Laser Spectroscopy*. New York: Academic Press.
Meeten, G. H. 1986. *Optical Properties of Polymers*. Essex, U.K.: Elsevier.
Nan, X., E. O. Potma, and X. S. Xie. 2006. Nonperturbative chemical imaging of organelle transport in living cells with coherent anti-Stokes Raman scattering microscopy. *Biophys. J.* 91: 728–735.
Ofuji, M., Y. Takano, Y. Houkawa, Y. Takanishi, K. Ishikawa, H. Takezoe, T. Mori, M. Goh, S. Guo, and K. Akagi. 2006. Microscopic orientational order of polymer chains in helical polyacetylene thin films studied by confocal laser Raman microscopy. *Jpn. J. Appl. Phys.* 45: 1710–1713.
Petrov, A. G. 1999. *The Lyotropic State of Matter: Molecular Physics and Living Matter Physics*. Amsterdam, the Netherlands: Gordon and Breach Science.
Pillai, R. S., M. Oh-e, H. Yokoyama, G. J. Brakenhoff, and M. Müller. 2006. Imaging colloidal particle induced topological defects in a nematic liquid crystal using third harmonic generation microscopy. *Opt. Express* 14: 12976–12983.

Potma, E. O. and X. S. Xie. 2003. Detection of single lipid bilayers with coherent anti-Stokes Raman scattering (CARS) microscopy. *J. Raman Spectrosc.* 34: 642–650.

Potma, E. O. and X. S. Xie. 2008. Theory of spontaneous and coherent Raman scattering. In *Handbook of Biomedical Nonlinear Optical Microscopy*, eds. B. R. Masters and P. T. So, pp. 164–185. Oxford, U.K.: Oxford University Press.

Potma, E. O., X. S. Xie, L. Muntean, J. Preusser, D. Jones, J. Ye, S. R. Leone, W. D. Hinsberg, and W. Schade. 2004. Chemical imaging of photoresists with coherent anti-Stokes Raman scattering (CARS) microscopy. *J. Phys. Chem. B* 108: 1296–1301.

Rafferty, D., J. L. Koening, G. Magyar, and J. L. West. 2002. FT-IR imaging of nematic liquid crystals. *Appl. Spectrosc.* 56: 284–287.

Rodriguez, L. G., S. J. Lockett, and G. R. Holtom. 2006. Coherent anti-Stokes Raman scattering microscopy: A biological review. *Cytom. Part A* 69A: 779–791.

Saar, B. G., H.-S. Park, X. S. Xie, and O. D. Lavrentovich. 2007. Three-dimensional imaging of chemical bond orientation in liquid crystals by coherent anti-Stokes Raman scattering microscopy. *Opt. Express* 15: 13585–13596.

Santamato, E., G. Abbate, P. Maddalena, and Y. R. Shen. 1987. Optically induced twist Fréedericksz transitions in planar-aligned nematic liquid crystals. *Phys. Rev. A* 36: 2389–2392.

Shiyanovskii, S. V., I. I. Smalyukh, and O. D. Lavrentovich. 2001. Computer simulations and fluorescence confocal polarizing microscopy of structures in cholesteric liquid crystals. In *Defects in Liquid Crystals: Computer Simulations, Theory and Experiment*, Vol. 43: NATO Science Series, II. Mathematics, Physics and Chemistry, eds. O. D. Lavrentovich, P. Pasini, C. Zannoni, and S. Zumer, pp. 229–270. Dordrecht, the Netherlands: Kluwer Academic Publishers.

Smalyukh, I. I. and O. D. Lavrentovich. 2006. Defects, surface anchoring, and tree-dimensional director fields in the lamellar structure of cholesteric liquid crystals as studied by fluorescence confocal polarizing microscopy. In *Topology in Condensed Matter*, ed. M. Monastyrsky, pp. 205–250. New York: Springer.

Smalyukh, I. I., S. V. Shiyanovskii, and O. D. Lavrentovich. 2001. Three-dimensional imaging of orientational order by fluorescence confocal polarizing microscopy. *Chem. Phys. Lett.* 336: 88–96.

Sun, Y., W. Lo, J. W. Su, S.-J. Lin, S.-H. Jee, and C.-Y. Dong. 2008. Multiphoton polarization microscopy. In *Handbook of Biomedical Nonlinear Optical Microscopy*, eds. B. R. Masters and P. T. So, pp. 484–556. Oxford, U.K.: Oxford University Press.

Tanaami, T., S. Otsuki, N. Tomosada, Y. Kosugi, M. Shimizu, and H. Ishida. 2002. High-speed 1-frame/ms scanning confocal microscope with a microlens and Nipkow disks. *Appl. Opt.* 41: 4704–4708.

Volkmer, A. 2005. Vibrational imaging and microspectroscopies based on coherent anti-Stokes Raman scattering microscopy. *J. Phys. D Appl. Phys.* 38: R59–R81.

Wurpel, G. W. H., H. A. Rinia, and M. Müller. 2005. Imaging orientational order and lipid density in multilamellar vesicles with multiplex CARS microscopy. *J. Microsc.* 218: 37–45.

Wurpel, G. W. H., J. M. Schins, and M. Müller. 2002. Chemical specificity in three-dimensional imaging with multiplex coherent anti-Stokes Raman scattering microscopy. *Opt. Lett.* 27: 1093–1095.

Xie, A. and D. A. Higgins. 2004. Electric-field-induced dynamics in radial liquid crystal droplets studied by multiphoton-excited fluorescence microscopy. *Appl. Phys. Lett.* 84: 4014.

Yelin, D., Y. Silberberg, Y. Barad, and J. S. Patel. 1999. Phase-matched third-harmonic generation in a nematic liquid crystal cell. *Phys. Rev. Lett.* 82: 3046.

Yoshiki, K., M. Hashimoto, and T. Araki. 2005. Second-harmonic-generation microscopy using excitation beam with controlled polarization pattern to determine three-dimensional molecular orientation. *Jpn. J. Appl. Phys.* 44: L1066.

# 22. Live Cell Imaging by Multiplex CARS Microspectroscopy

**Hideaki Kano**



White-light laser source is applied to coherent anti-Stokes Raman scattering (CARS) microspectroscopy. Owing to the ultrabroadband spectral profile of the white-light laser, a wide range of vibrational resonance can be investigated. Using the white-light laser source, we visualized living cells with multiple vibrational contrast.



*Coherent Raman Scattering Microscopy*. Edited by Ji-Xin Cheng and X. Sunney Xie © 2013 CRC Press/ Taylor & Francis Group, LLC. ISBN: 978-1-4398-6765-5.

## 22.1 Introduction

CARS microscopy is applied widely to cell and tissue imaging *ex vivo* and *in vivo* [1]. Fascinating results are obtained, and are described from various aspects in other chapters. CARS microscopy, however, provides only a monochromatic image using a single vibrational frequency such as $CH_2$ stretch vibrational mode. In general, the signal at the single frequency does not correspond directly to the pure vibrationally resonant signal due to a nonresonant background (NRB) component. This is critical in particular in the fingerprint region, where a large number of vibrational bands with various bandwidths are congested. In our experience of live cell CARS imaging, the vibrationally resonant signal intensity in the fingerprint region is in the similar order to that of the NRB. The spectral profile in the fingerprint region thus shows complicated lineshape, from which it is difficult to obtain "vibrationally resonant" CARS signal intensity. Several methods such as polarization CARS and time-delayed CARS are proposed and demonstrated to suppress the NRB [2,3]. However, at the expense of the NRB suppression, the vibrationally resonant CARS signal is significantly decreased. Besides the difficulty of differentiation of the CARS signal from the NRB, there is an intrinsic drawback in monochromatic (single-frequency) imaging by fixed laser frequencies of $\omega_1$ and $\omega_2$. Monochromatic imaging using a single frequency ($\omega_1 - \omega_2$) cannot distinguish a change of the number of the molecule, $\Delta N$, from its spectral change due to a conformational change, $\Delta\Omega$. Here $\Delta N$ and $\Delta\Omega$ correspond to the variations of the number of molecule and its vibrational frequency in the focal volume, respectively. In other words, the image contrast depends on two values ($\Delta N$ and $\Delta\Omega$), and these two factors cannot be differentiated simply by using the CARS signal intensity at a fixed frequency. Since vibrational spectroscopy is sensitive to such structural or conformational changes of a molecule, $\Delta N$ and $\Delta\Omega$ can be evaluated independently from the band intensity and its bandshape (or simply its peak position), respectively. Therefore, it is indispensible to measure a spectral profile of the CARS signal to distinguish $\Delta N$ from $\Delta\Omega$. Recently, alternative methods to CARS microscopy such as stimulated Raman scattering (SRS) microscopy have been reported [4,5]. Although SRS microscopy does not suffer from the NRB, this problem still remains in the fingerprint region, because SRS microscopy also exploits monochromatic (single-frequency) detection scheme.

In order to (1) obtain structural information of molecules through multiple vibrational bands and (2) retrieve pure vibrationally resonant spectra, spectroscopic imaging methods such as multiplex CARS microspectroscopy are superior to other techniques. In the multiplex CARS process, we use narrowband pump/probe and broadband Stokes laser pulses to drive multiple vibrational modes simultaneously. As a result, we can obtain full spectral information of the CARS signal. As alternatives of CARS spectroscopic imaging, SRS spectroscopic imaging [6] and spontaneous Raman spectroscopic imaging are reported. Since typical SRS imaging needs enhancement of signal sensitivity using a modulation method such as lock-in detection [4], multi-channel detection of the SRS signal with high signal-to-noise ratio requires a complicated apparatus such as a multi-channel lock-in amplifier. Spontaneous Raman microspectroscopy also exploits multi-channel detection. Recent improvement of spontaneous Raman microspectroscopy enables us to obtain Raman spectra and image with high speed using multi-confocal [7] and line confocal [8] techniques, those of which can be implemented to multiplex CARS microspectroscopy.

In the present chapter, I review recent studies on live cell imaging using a multiplex CARS technique. In order to perform live cell imaging, several difficulties should be overcome to obtain CARS spectra with appreciable signal-to-noise ratio, because the CARS signal of a live cell in the fingerprint region is weak in general, biased and deformed by the NRB. Moreover, since various bands are closely located in the fingerprint region, the spectral profile is easily smeared by poor spectral resolution to show structureless profile. In order to identify each vibrational band in the spectral domain, we have to take great care of a spectral detection. These technical points will also be discussed.

This chapter is organized in the following. First, I overview recent studies on live cell multiplex CARS imaging. Second, I review our recent studies on live cell multiplex CARS imaging. Depending on the laser source, we can obtain not only multiplex CARS but also TPEF, second and third harmonics (SH and TH) simultaneously. Nonlinear multi-modal imaging such as combinations of CARS, TPEF, SH, and TH shows intriguing overlaid images manifested by unique properties of their own nonlinear optical processes. In this chapter, two setups, namely multiplex CARS setups using a femtosecond Ti:sapphire laser [9] and sub-nanosecond Nd:YAG laser [10] will be introduced. In both setups, a white-light laser source is used for the ultrabroadband Stokes pulse.

## 22.2 Brief Review of Cell Imaging by Multiplex CARS

There is a trend in multiplex CARS studies; from X–H stretch imaging (X corresponds to C, N, and O) to the fingerprint imaging. After the reports of cell imaging by single-frequency CARS microscopy [11,12], multiplex CARS microspectroscopy has been demonstrated for live cell imaging [13,14]. Cheng et al. performed live cell imaging and visualized lipids inside of the NIH 3T3 cells by polarization multiplex CARS [13]. We visualized living fission yeast cells (*Schizosaccharomyces pombe*) by both multiplex CARS and two-photon excitation fluorescence (TPEF) as described in detail in the following section [15]. Rinia et al. visualized cellular lipid droplets, not only by C–H stretch but also by *cis* C=C vibrational modes [16]. Isobe et al. reported C–H and O–H imaging by single laser pulses with octave spanning spectra [17].

Based on the peak position of the C–H stretch vibrational frequency, we can roughly distinguish lipids from proteins using $CH_2$ (~2840 $cm^{-1}$) and $CH_3$ (~2940 $cm^{-1}$) stretch vibrational modes, because the former and the latter bands give intense signal in aliphatic chains of lipids and in terminal methyl groups of proteins, respectively. For plant and yeast cells, polysaccharides also contribute to the signal at the C–H stretch vibrational modes. The C–H stretch bands are isolated in comparison with other bands in the fingerprint region, and the signal intensities of the C–H stretch bands are stronger than those of the fingerprint Raman bands. However, they are less sensitive to structural information such as secondary structure of protein than the fingerprint Raman bands.

Cell imaging in the fingerprint region is still challenging [18–23]. Thanks to the recent progress of the state-of-the-art laser technology, ultrabroadband multiplex CARS imaging is demonstrated simply by using a laser oscillator [9,18,24]. The laser sources for CARS microspectroscopy are classified into two; the one is the system using a supercontinuum light source (hereafter we call it “white-light laser source”) typically generated from a photonic crystal fiber (PCF), and the other is the system using a direct output from a Ti:sapphire laser oscillator with ultrabroadband spectral bandwidth [17,23]. Petrov et al. employ the first strategy using a picosecond mode-locked laser oscillator, and measure a single bacterial spore. Based on the averaged CARS spectrum of bacterial spores and spectral data analysis with maximum entropy method (MEM), they report Im[$\chi^{(3)}$] spectra of 2,6-pyridinedicarboxylic acid (or dipicolinic acid [DPA]), which is one of the major components of bacteria [18]. Since the size of bacteria is smaller than spatial resolution of CARS microspectroscopy, it is difficult to obtain molecular distribution inside of bacteria. Intra-cellular imaging has been recently performed in the fingerprint region [21–23]. Parekh et al. performed CARS spectral imaging of a fixed L929 cell. Based on the spectral analysis using time-domain Kramers–Kronig transformation, they demonstrated to retrieve Im[$\chi^{(3)}$] spectra from distorted CARS spectra. Chen et al. reported a frequency modulation CARS method combined with chirped broadband pulses, and visualized cellulose and lignin in a plant cell's walls [23]. In these studies, a femtosecond or picosecond laser system is required to drive and/or detect a vibrational coherence to generate a CARS signal. On the other hand, we develop a novel multiplex CARS system using a *sub-nanosecond* microchip Nd:YAG laser [10,21]. By seeding the nanosecond laser pulse into a PCF, white-light laser is also obtainable. Using this system and spectral analysis including MEM, dynamical behavior of a living cell is visualized for the first time in the fingerprint region (*molecular fingerprinting*) [21].

## 22.3 Live Cell Imaging by Multiplex CARS Microspectroscopy Using a Ti:Sapphire Laser

### 22.3.1 Experimental Setup

Figure 22.1 shows a schematic of our multiplex CARS microspectrometer [9]. An unamplified mode-locked Ti:sapphire laser oscillator (Coherent, Vitesse-800) is used for a master laser. Typical temporal duration, center wavelength, pulse energy, and repetition rate are 100 fs, 800 nm, 12 nJ, and 80 MHz, respectively. A portion of the output from the oscillator is coupled into a PCF (Crystal Fibre, NL-PM-750) to generate white-light laser pulses. The input pulse energy for the white-light laser generation is less than 2.3 nJ. As shown in Figure 22.1, the fundamental of the Ti:sapphire laser and the white-light laser pulses are used for the pump ($\omega_1$) and Stokes ($\omega_2$) laser pulses, respectively. In order to obtain CARS spectrum with high-frequency resolution, the pump laser pulses are spectrally filtered by a narrow band-pass filter. The bandwidth was measured to be about 20 cm$^{-1}$. Because the pump laser is in the near infra-red (NIR) region (800 nm), the Stokes laser must also be in the NIR. The visible component in the white-light laser is thus blocked by a long-wavelength pass filter. Two laser pulses are superimposed collinearly using an 800 nm notch filter, and then tightly focused onto the sample with a microscope objective (40×/NA 0.9). The forward-propagating CARS signal is collected by a microscope objective (40×/NA 0.6). Finally, the CARS signal is guided to a polychromator (Acton, SpectraPro-300i) through an optical fiber, and is detected by a CCD camera (Roper Scientific, PIXIS 100BR). The multiplex CARS images are reconstructed after point-by-point acquisition of the CARS spectrum. The sample is scanned by piezo-driven *xyz* translators (MadCity, Nano-LP-200).

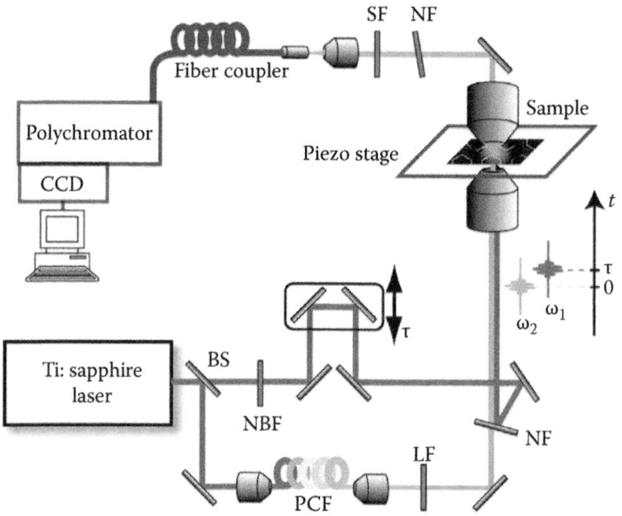


**FIGURE 22.1** Experimental setup for multiplex CARS microspectroscopy; PCF, photonic crystal fiber; BS, beam splitter; NBF, narrow band-pass filter; LF, long-wavelength pass filter; NF, Notch filter; SF, short-wavelength pass filter.

## 22.3.2 Results and Discussion

### 22.3.2.1 Single Pollen Grains

Figure 22.2 shows the results of CARS spectroscopic imaging of a single pollen grain of *Prunus x yedoensise* (cherry) [25]. The exposure time at each spatial point is 100 ms. Since pollen emits strong auto-fluorescence, it is difficult to measure spontaneous Raman spectra. For this reason, pretreatment such as photo-bleaching by laser irradiation [26] or application of surface-enhanced Raman scattering (SERS) [27] was reported previously. In comparison with previous studies, CARS does not suffer from auto-fluorescence, because the signal wavelength is blue-shifted from those of the incident lasers. As a result, CARS spectra of a single pollen grain are obtained without any pretreatment. The multiplex CARS spectra shown in Figure 22.2a consist of several peaks due to vibrational resonances. It is noted that the spectral profile differs from each position of a grain. In particular, three peaks at around 1150, 1520, and 2850 $cm^{-1}$ are prominent in Figure 22.2a in black. On the other hand, the CARS spectrum in gray shows other resonance at around 1600 $cm^{-1}$. Figure 22.2b through e shows the CARS images at around 1150, 1520, 1600, and 2850 $cm^{-1}$, respectively. It is clear that Figure 22.2b and c shows the same image with three intense spots at the granular parts, indicating that these vibrational modes originate from the same molecular species. In contrast, Figure 22.2d and e shows different images from Figure 22.2b and c. Figure 22.2e shows the whole shape of the pollen grain with three intense spots as discussed in Figure 22.2b and c. The groove-like structures are probably germ pores or germ slits. Figure 22.2d shows partial overlap with Figure 22.2e, but no contrast at the three spots. Taking into account that the spectral profile in black in Figure 22.2a is similar to that of β-carotene in the fingerprint region [25], the bands at around 1150 and 1520 $cm^{-1}$ are ascribed to C–C and C=C stretch vibrational modes due to carotenoid derivatives such as sporopollenin. Sporopollenin is a complex polymer of carotenoids and carotenoid esters, which protect pollens from chemical and

Chapter 22

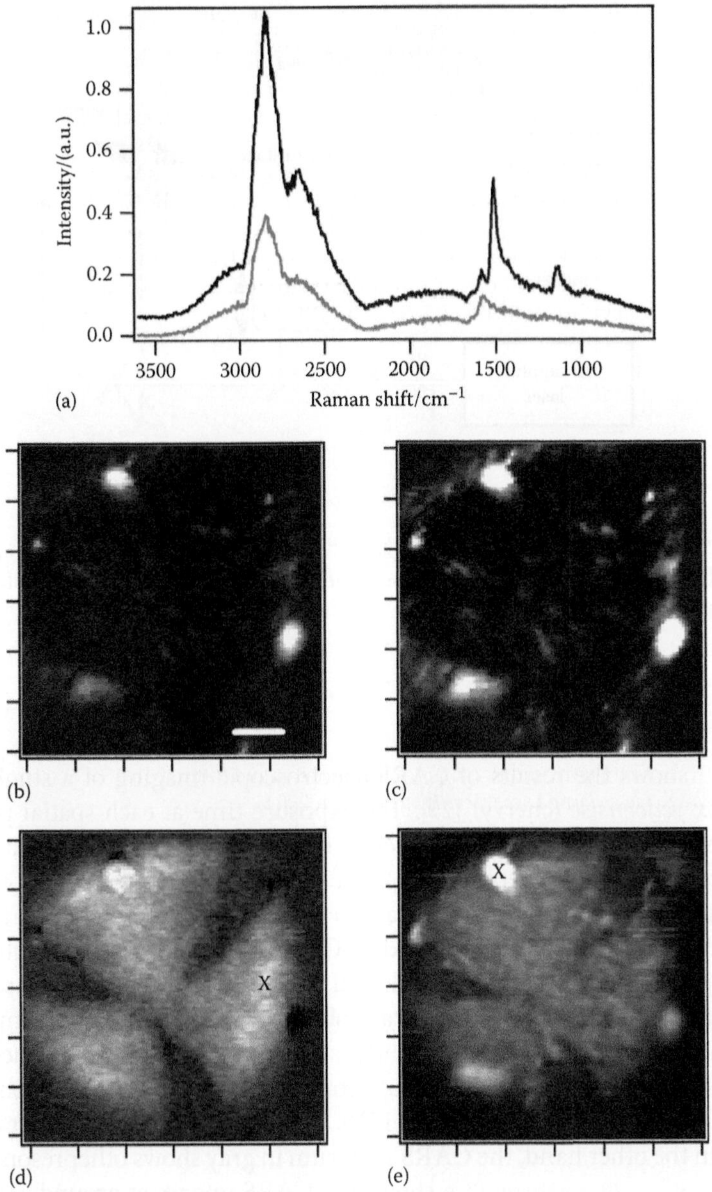


**FIGURE 22.2** (a) CARS spectra of a single pollen grain (intensity uncorrected). The black and gray spectra are obtained at black crosses in Figure 22.2e and d, respectively. The spectrum is offset for clarity; CARS images of a pollen grain at C–C stretch (b), C=C stretch (c), 1600 cm$^{-1}$ (d), and C–H stretch (e) vibrational modes. The scale is 5 μm.

biological invasions [26]. The band at around 2850 cm$^{-1}$ is attributable to C–H stretch vibrational modes. Based on the peak position of the signal, this band is attributable to the $CH_2$ stretch vibrational mode originating from lipid molecules. Interestingly, spatial overlaps are found at the granular parts in Figure 22.2b, c, and e. It means that both carotenoid derivatives and lipids coexist in high concentration in these granular parts. In contrast, the localization of the molecular species corresponding to Figure 22.2d is exclusive to those of carotenoid derivatives. The molecular species for Figure 22.2d have not been assigned yet, but one of the possible candidates is flavonoid derivatives [28,29]. As clearly shown, CARS spectroscopic imaging based on the multiplex CARS process is capable of distinguishing molecular species by spectral information.

#### 22.3.2.2 Living Yeast Cells

Figure 22.3a shows a typical spectral profile of the CARS signal of a living yeast cell [15]. The sample is a fission yeast cell (*S. pombe*). Figure 22.3a shows a strong signal at around 2850 cm$^{-1}$. This band originates mainly from $CH_2$ stretch vibrational mode. It visualizes mainly lipid molecules which are found in membranous organelles such as mitochondria. In this experimental setup, it is hard to find vibrational resonance in the fingerprint region. It is partly explained by poor spectral resolution mainly due to the spectral bandwidth of the pump laser (20 cm$^{-1}$). Figure 22.3b shows the CARS image at $CH_2$ stretch mode. In Figure 22.3b, living yeast cells at various cell-cycle stages are clearly visualized. In particular, a septum is visualized in the cell around the center. The septum is composed of polysaccharides, which also give the strong CARS signal at the C–H stretch vibrational mode.

Thanks to the three-dimensional sectioning capability, we can obtain not only lateral but also axial slices of a living yeast cell. Figure 22.4a and b shows lateral and axial CARS images of a yeast cell, respectively. The CARS signal is weaker at the top part than at the bottom. This is caused by imperfect focusing of the pump and Stokes laser pulses due to the propagation in the spatially heterogeneous cell. Figure 22.4c shows optical sectioning of a yeast cell at different depth positions indicated at the top in a micrometer scale.

The white-light laser source can also be used as an excitation light source for the TPEF [30–32]. Owing to the broadband spectral profile of the white-light laser, two-photon allowed electronic states can be accidentally excited efficiently in comparison with TPEF microscopy using a conventional Ti:sapphire oscillator. Figure 22.5a shows typical CARS and TPEF spectra of a living yeast cell at the G1 (gap-1) phase, when a septum is formed at the middle of the cell [33]. We used yeast cells expressing

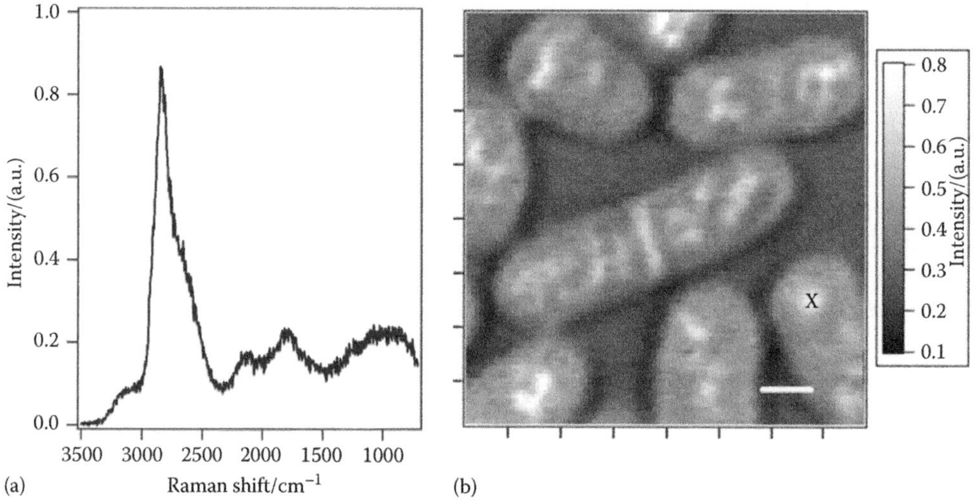


**FIGURE 22.3** (a) CARS spectrum of a living yeast cell (intensity uncorrected); (b) CARS image of living yeast cells at the $CH_2$ stretch vibrational mode. The scale is 2 μm. The CARS spectrum in Figure 22.3a is obtained at the position of a cross in Figure 22.3b.

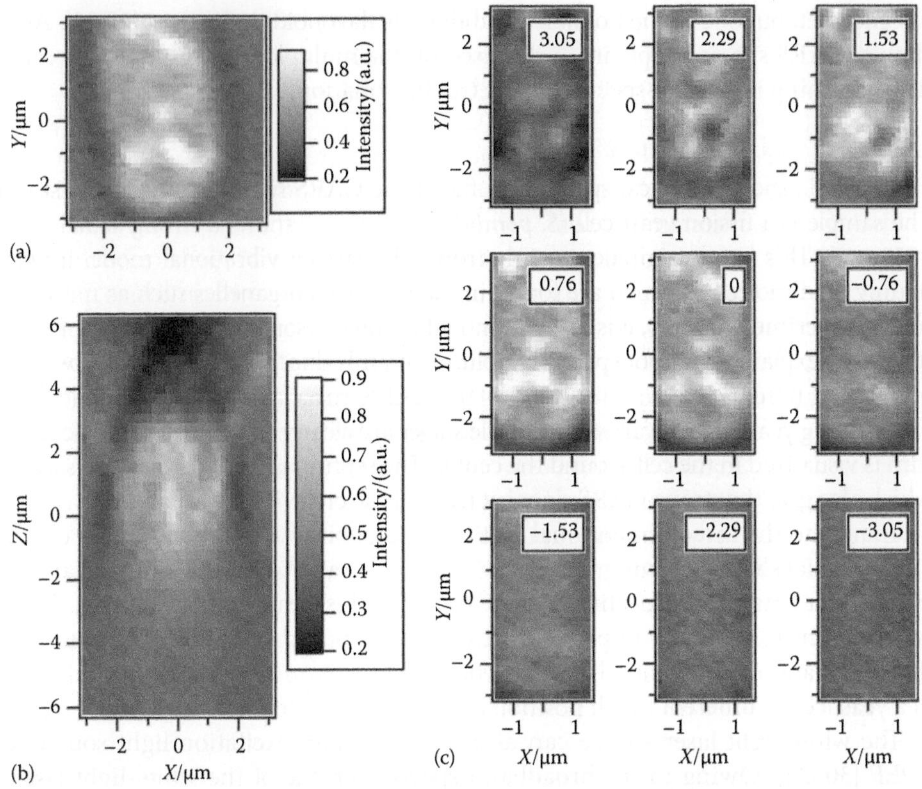


**FIGURE 22.4** Lateral (a) and axial (b) CARS images of a living yeast cell at the $CH_2$ stretch vibrational mode. Axial CARS image is obtained at $y = 0$; (c) Axial slices of the yeast cell at different depth positions indicated at the top in a micrometer scale. (Reproduced from Kano, H. and Hamaguchi, H., *Opt. Express*, 14, 2798, 2006.)

mitochondria-targeted green fluorescent protein (GFP). Each spectrum is obtained at the position of septum (red), mitochondria (green), and surrounding water (black). All spectra are spatially averaged in the region of $(0.3\,\mu m)^2$. The CARS spectra in red and green are similar to that in Figure 22.3, showing strong vibrational resonance at around $2850\,cm^{-1}$. In the present condition, the temporal overlap with the pump and Stokes laser pulses is optimized both for CARS and TPEF. It changes the spectral profile of the CARS signal in the fingerprint region in comparison with that in Figure 22.3.

In addition to the CARS signal at the wavelength longer than 620 nm, a broad band is observed at around 510 nm. From the peak position and spectral profile, this band is safely assigned to TPEF from GFP. Since mitochondria of the yeast cell are labeled by GFP, the TPEF image corresponds to the image of mitochondria.

Figure 22.5b and c shows the result of multi-modal imaging of a living yeast cell at the G1 phase by CARS (b) and TPEF (c). It is emphasized that both of the signals are obtained simultaneously. CARS at the $CH_2$ stretch vibrational mode is indicated in Figure 22.5b. As discussed in Figure 22.3, the CARS image at $CH_2$ stretch vibrational mode shows high contrast not only at membranous organelles but also at septum.

Figure 22.5c indicates the TPEF image, which corresponds to the distribution of mitochondria. It is clear that no GFP signal is observed at the position of septum. In order

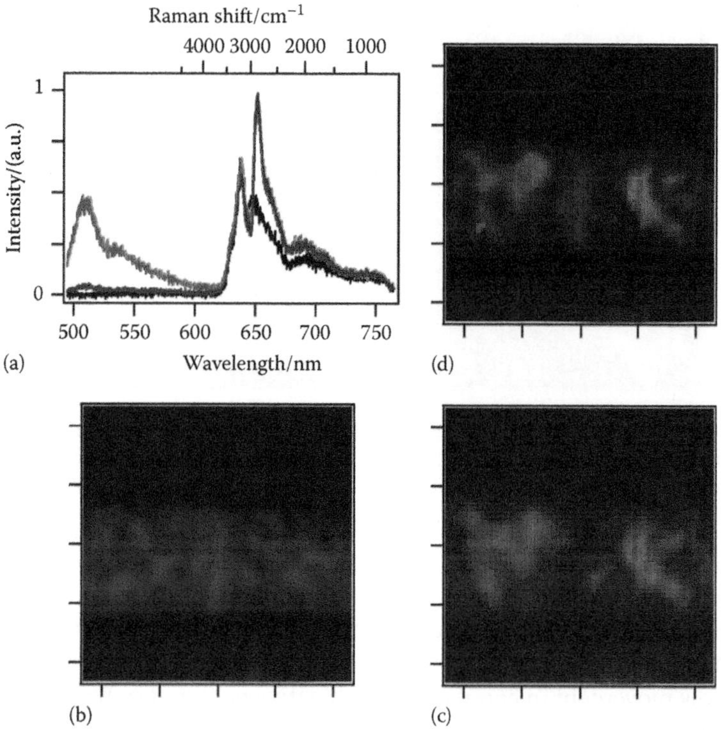


**FIGURE 22.5** (a) CARS and TPEF spectra of a living yeast cell at G1 (gap-1) phase (intensity uncorrected). Each spectrum is obtained at the position of septum (red), mitochondria (green), and surrounding water (black). All spectra are spatially averaged in the region of $(0.3\,\mu m)^2$. Exposure time is 200 ms. Apparent peaks at the Raman shift of approximately 3200 cm$^{-1}$ are due to the spectral profile of the Stokes laser; (b and c) Nonlinear multi-modal imaging of a living yeast cell at G1 phase. CARS at the C–H stretch vibrational mode (b), TPEF (c), and merged (d) images are indicated. The scale bar corresponds to 2 μm. (Reproduced from Kano, H. and Hamaguchi, H., *Anal. Chem.*, 79, 8967, 2007.)

to examine both CARS and TPEF images in more detail, these images are merged (Figure 22.5d). Yellow color in Figure 22.5d indicates that both CARS and TPEF signals are strong at the corresponding area. The CARS images are in almost accord with the TPEF images, but they are not perfectly matched with each other. It means that the CARS signal at the $CH_2$ stretch vibrational mode probes not only mitochondria but also other organelles. Concerning organelles with high contrast for the $CH_2$ stretch signal, Brackmann et al. assigned them to lipid droplets [34], because their CARS image of budding yeast cells at the $CH_2$ stretch vibrational mode is similar (but not perfectly overlapped) to the TPEF image of lipid droplets labeled by GFP via ERG6. In fact, there are several differences between their results and ours. First, the shape of the organelle is different. The organelles they reported look particles, whereas those we reported distribute widely inside of a cell. Second, the appearance of lipid droplets depends on a growth-curve phase and cultural conditions. Further work is required to assign these unidentified organelles.*

* Recently, we have developed a new multimodal molecular imaging system that combines CARS, SHG, THG, and multiplex TSFG (third-order sum frequency generation) using a subnanosecond white-light laser source [35]. We found particle-like structures in the cells probably due to lipid droplets using both CARS and TSFG.

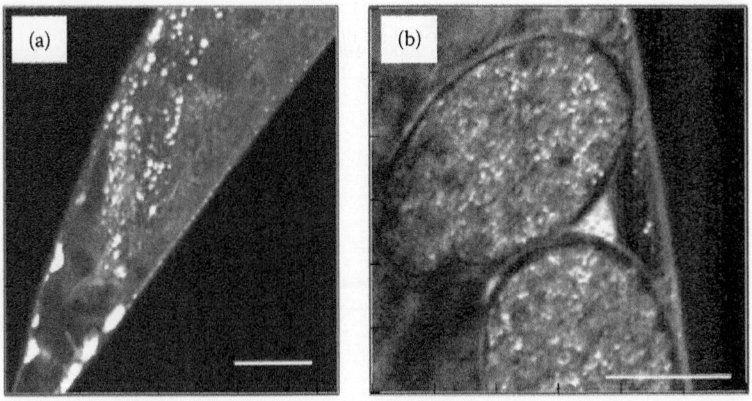


FIGURE 22.6 (a) CARS images at the $CH_2$ stretch vibrational mode of a tail region (a) and an abdominal region (b) of a nematode. Exposure time at each spatial point is 50 ms. The short bar measures 20 μm. (Reproduced from Kano, H., *Bull. Chem. Soc. Jpn.*, 83, 735, 2010.)

#### 22.3.2.3 Living Nematodes

We applied our multiplex CARS technique to a living nematode (*Caenorhabditis elegans*), which is a well-known biological model organism [36]. The sample is unc-119 mutants, showing slow movement in comparison with native nematodes. Before measurement, the sample specimens were exposed to levamisole to suppress their movement. Figure 22.6 shows the vibrationally resonant CARS images of a nematode at the $CH_2$ stretch vibrational mode. A tail region (a) and an abdominal region (b) are visualized. In this measurement, the experimental setup is optimized for C–H stretch region. At the expense, the CARS signal in the fingerprint region becomes weak, and thus does not have enough signal-to-noise ratio to find any vibrational resonance in the fingerprint region. Exposure time at each spatial point is 50 ms. The CARS images show various organs inside of the nematode. In particular, many small and bright spots are distributed inside of the body as shown in Figure 22.6a. Based on the previous study [37], these structures are assigned mainly to lipid droplets. On the other hand, large and round eggs are clearly observed in Figure 22.6b. It is noted that several dark spots are observable inside of the eggs. They are probably nuclei, showing that the eggs contain multiple cells.

## 22.4 Cell Imaging by Multiplex CARS Microspectroscopy Using a Sub-Nanosecond Microchip Nd:YAG Laser

### 22.4.1 From Femtosecond to Nanosecond Laser Source

As shown in the previous section, a femtosecond laser source combined with a PCF for broadband Stokes radiation can be used for the light source of multiplex CARS imaging. It suffers, however, from limited spectral resolution of the pump/probe laser and temporal chirp of the Stokes laser [38]. Although there are several reports to overcome these issues, sophisticated techniques such as coherent control or interferometric methods are required [39,40]. In this section, I describe our recent approach to overcome these issues; multiplex CARS imaging by a sub-nanosecond microchip laser source.

In comparison with the white-light laser generation using a femto- or pico-second laser system, nanosecond white-light laser is almost free from the problem on the temporal chirp of the white-light laser, because the temporal duration of the pump/probe laser is long enough to be overlapped with the Stokes laser pulses. In other words, we do not need to compress the white-light laser pulse for the broadband multiplex CARS process. Moreover, the spectral bandwidth of the pump laser pulse is narrow enough ($<0.1\ cm^{-1}$ in our setup) for the multiplex CARS process.

## 22.4.2 Experimental Setup

Figure 22.7a shows the experimental setup. The master laser source is a passively Q-switched 1064 nm microchip Nd:YAG laser, delivering <1 ns pulses at 33 kHz repetition rate (~10 kW peak power; ~300 mW average power). The laser beam is equally divided into two parts with a beam splitter. One part is directly used for the pump ($\omega_1$)

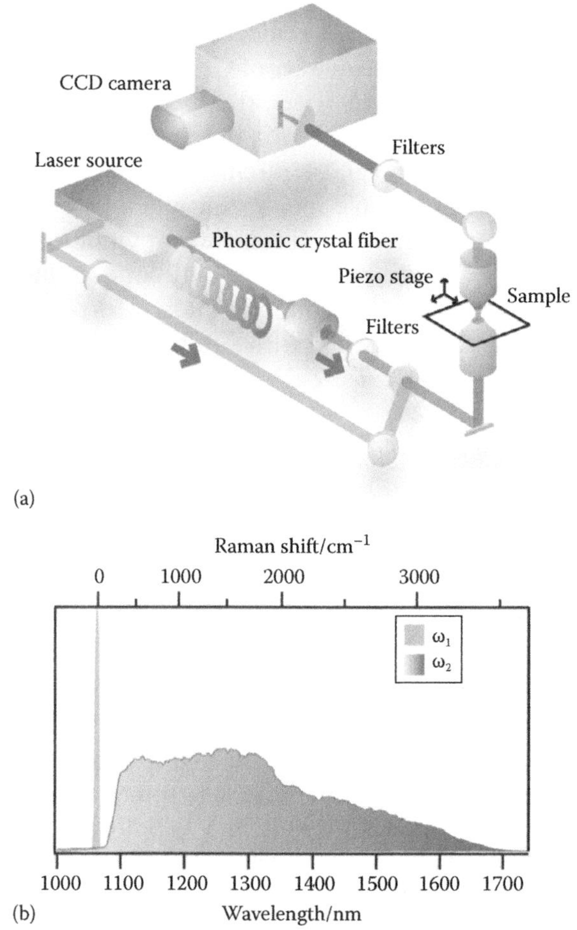


**FIGURE 22.7** (a) Experimental setup of ultrabroadband multiplex CARS microspectrometer for molecular fingerprinting (b) the spectra of the $\omega_1$ and $\omega_2$ radiations (intensity uncorrected). (Reproduced from Okuno, M., Kano, H. et al.: Quantitative CARS molecular fingerprinting of single living cells with the use of the maximum entropy method. *Angew. Chem. Int. Ed.* 2010. 49. 6773–6777, S6773/6771–S6773/6776. Copyright Wiley-VCH Verlag GmbH & Co. KGaA.)

radiation (~5 kW peak power; ~150 mW average power; <0.1 cm$^{-1}$ spectral bandwidth) of the CARS process after adjusting its power with a variable neutral density filter. The other part is seeded into a 6-m-long air-silica PCF, characterized by a 2.5 μm hole diameter and a 4 μm hole-to-hole spacing, which results in a zero-dispersion wavelength of 1040 nm for the fundamental guided mode of the structure. The white-light laser emission is obtained at the fiber output, with >100 μW/nm spectral power density from 1.05 to 1.6 μm, for the Stokes laser pulse. Figure 22.7b shows the spectral profile of the pump and Stokes laser pulses. The pump and Stokes laser pulses pass through several filters to eliminate anti-Stokes spectral components, and are superimposed by a notch filter. The co-propagating pump and Stokes laser pulses are introduced into the modified inverted microscope (Nikon: ECLIPSE Ti). The pump and Stokes pulses are tightly focused onto the sample with an objective (Nikon: Plan Fluor 100×/NA 1.3). The CARS signal generated by the sample is collected by another objective (Nikon: Apo NIR 60x/NA 1.0) and guided into a spectrometer (Princeton Instruments: SpectraPro2300i or LS785 and PIXIS 100BR). The CARS signal was first guided to the spectrometer through an optical fiber, but changed to direct coupling in free space in order to maintain the quality of CARS spectra (*vide infra*). Owing to the narrow spectral line width of the $\omega_1$ laser, the spectral resolution is determined not by the laser line width but by the dispersion of the spectrometer. The spectral resolution and a spectral coverage are 5 and 1200 cm$^{-1}$ with the use of a $f$ = 30 cm polychromator (SpectraPro2300i). The whole fingerprint region is covered with a resolution good enough to resolve congested spectra. Note that vibrational bands from biological samples tend to be broadened to more than 10 cm$^{-1}$ so that a spectral resolution higher than 5 cm$^{-1}$ is generally not required. The spatial resolution is 470 nm in the lateral and 4.5 μm in the axial direction with a 100×/NA 1.3 microscope objective. The sample is scanned with a three-axis PZT stage (Mad City Labs LP-200).

## 22.4.3 Results and Discussion

### 22.4.3.1 Living Tobacco BY2 Cells

Figure 22.8a shows typical CARS spectrum and image from *Nicotiana tabacum* L. cv. Bright Yellow 2 (BY2) cells. BY2 cells are spread in water on a slide-glass and sandwiched with a cover-glass. The CARS spectrum in Figure 22.8a is obtained at a nucleus. The intensity correction of the CARS spectrum has been carried out using the NRB of water surrounding the cells measured with the same experimental condition. In this experimental condition, no vibrationally resonant CARS signals are clearly observed in the fingerprint region. It is partially due to the detection system; we employed first an optical fiber to guide the CARS field into the spectrometer. We found, however, that it gives rise to extra noise in the CARS spectrum. It is probably due to the interference of the signal in a multi-mode optical fiber. As shown in the next section, the quality of the CARS spectrum is significantly improved after we change the configuration from fiber-guiding to free-space guiding.

Figure 22.8b shows a vibrationally resonant CARS image at the Raman shift of 2902 cm$^{-1}$ due to the C–H stretch vibrational mode. The exposure time for each pixel is 300 ms. BY2 cells are clearly visualized with a high vibrational contrast. Inside of the BY2 cell, strong CARS signals are observed at cell walls and in nucleus.

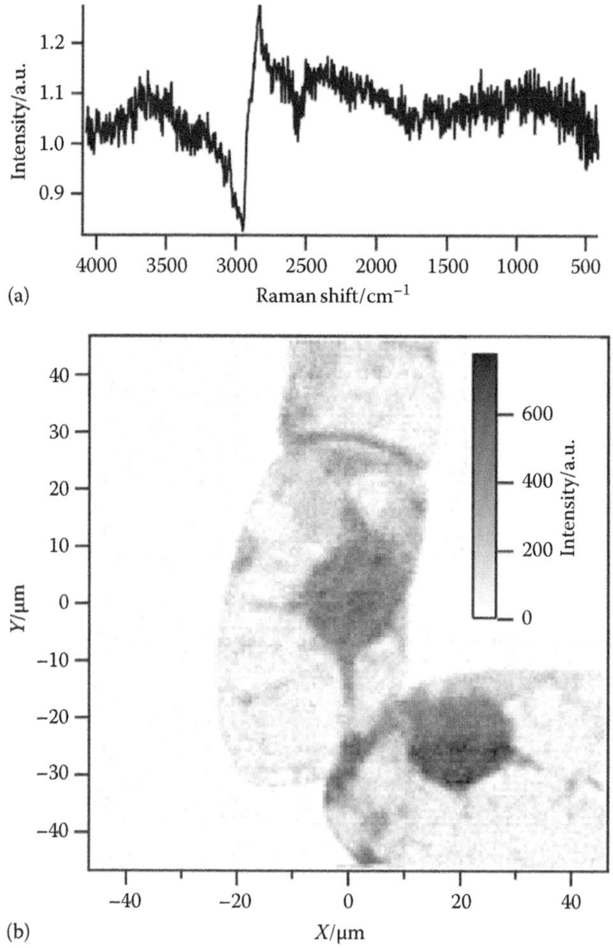


**FIGURE 22.8** (a) Intensity-corrected multiplex CARS spectrum of BY2 cells; (b) CARS image of BY2 cells at a Raman shift around 2900 cm$^{-1}$. (Reproduced from Okuno, M. et al., *Opt. Lett.*, 33, 923, 2008.)

Cytoplasm is also visualized by the CARS signal with moderate intensity. At the positions of vacuole, which is one of the major organelles in BY2 cells, the CARS signal is only weakly found.

### 22.4.3.2 Living Yeast Cells

We applied our technique intensively to living yeast cells. Figure 22.9a and b shows two fingerprint region CARS spectra obtained from two different positions in a living yeast cell [21]. The laser power of both the pump and Stokes lasers is set to 10 mW. The exposure time is 50 ms. A sample is tetraploid strains of budding yeast (zygote of *Saccharomyces cerevisiae* and *Saccharomyces bayanus*). Yeast cultures were maintained in YPD medium. Before measurement, cells are dispersed in water and immobilized on a ConA-coated glass bottom dish.

By comparison with the known Raman spectra of yeast organelles, it is clear that the spectrum in Figure 22.9a is measured from mitochondria and Figure 22.9b from

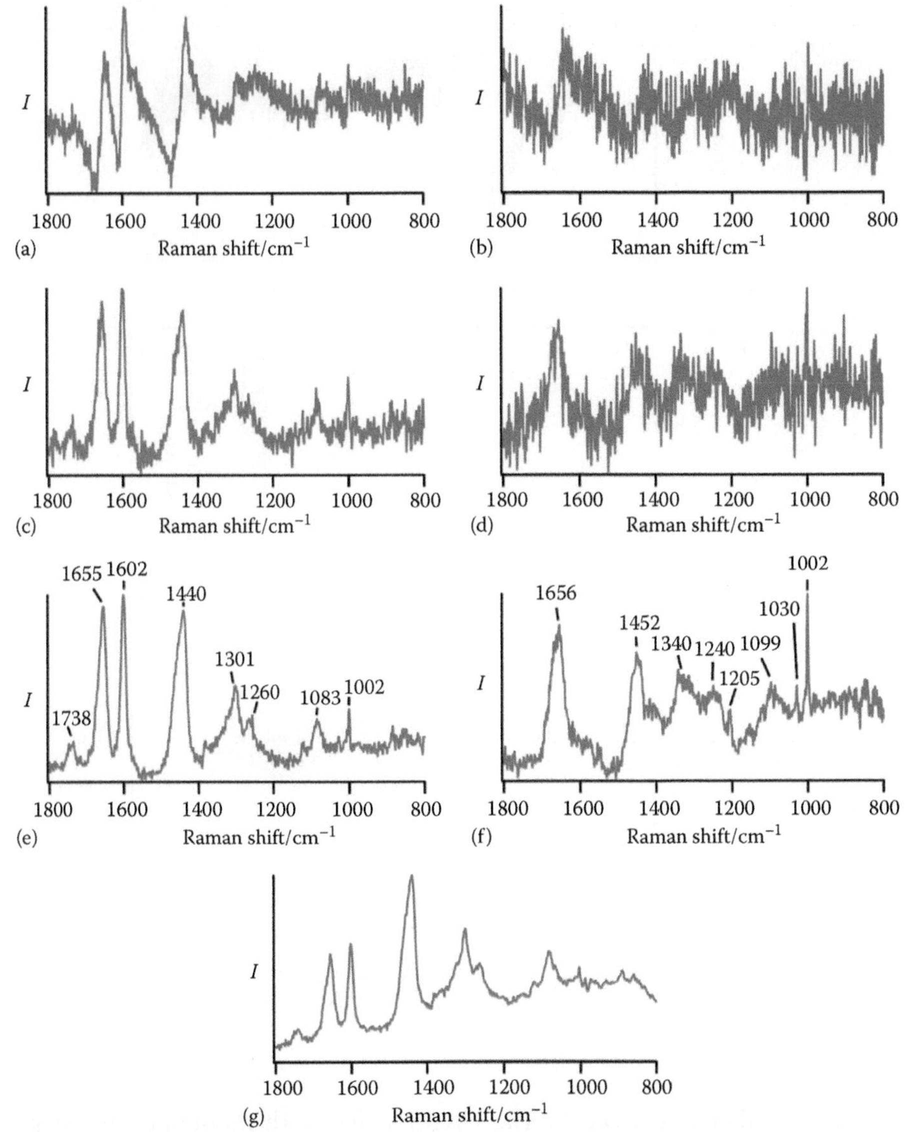


**FIGURE 22.9** Two typical CARS spectra in the fingerprint region of a living budding yeast cell (a) and (b); Im [$\chi^{(3)}$] spectra obtained by using MEM (c) and (d); Im [$\chi^{(3)}$] spectra after a SVD analysis (e) and (f); spontaneous Raman spectrum (g). The exposure time for measuring CARS and spontaneous Raman spectra are 50 ms and 300 s, respectively. (Reproduced from Okuno, M., Kano, H. et al.: Quantitative CARS molecular fingerprinting of single living cells with the use of the maximum entropy method. *Angew. Chem. Int. Ed.* 2010. 49. 6773–6777, S6773/6771–S6773/6776. Copyright Wiley-VCH Verlag GmbH & Co. KGaA.)

cytoplasm/nucleus [41]. The fact that a spectrum can be measured in tens of ms allows us to follow the dynamics of cellular processes, as will be shown in the following text. All the sharp peaks in Figure 22.9a and b originate from vibrational resonances. They show dispersive band shapes due to the interference with the NRB. This coherent nature of CARS enables us to extract the vibrationally resonant components by using the NRB as a local oscillator. We employ MEM to extract the amplitude and phase of vibrational resonances from the obtained multiplex CARS spectra [42]. As discussed in the previous chapter, MEM does not require any a priori knowledge of the vibrational bands contained but still is able to retrieve the phase information on the third-order nonlinear susceptibility $\chi^{(3)}$, whose imaginary part corresponds to ordinary (spontaneous) Raman spectra. Therefore, the amplitude of the retrieved band, Im[$\chi^{(3)}$], is proportional to the molecular concentration. In other words, each band can be evaluated *quantitatively.* In the MEM procedure, the NRB signal plays an important role to enhance the separation capability through the heterodyne amplification of the vibrationally resonant components. Figure 22.9c and d shows the Im[$\chi^{(3)}$] spectra converted by MEM from the CARS spectra in Figure 22.9a and b, respectively. It is clear that dispersive band shapes in Figure 22.9a and b are converted by MEM to normal band shapes (Im[$\chi^{(3)}$] spectra). The wide applicability of MEM for retrieving vibrational resonances has already been demonstrated previously [19,43]. In the present study, we further performed a singular value decomposition (SVD) analysis in order to reduce the noise in the retrieved Im[$\chi^{(3)}$] spectra [44]. The results are shown in Figure 22.9e and f. The signal-to-noise ratio is markedly improved on going from Figure 22.9c and d to Figure 22.9e and f. It is emphasized that such SVD analysis cannot be applied to raw CARS spectra, because of their nonlinear nature. In order to perform noise filtering by SVD, retrieval of Im[$\chi^{(3)}$] spectra by MEM is indispensable. For comparison, Figure 22.9g shows a typical spontaneous Raman spectrum of yeast mitochondria in the same fingerprint region. It is evident that the MEM procedure correctly retrieves the Raman resonant components from the seemingly complex CARS spectrum.

We now focus on the spectrum of mitochondrion in Figure 22.9e. Three prominent bands are observed in the fingerprint region at 1655, 1602, and 1440 cm$^{-1}$. These bands are already assigned by spontaneous Raman spectroscopy [41,45]. The band at 1655 cm$^{-1}$ is assigned to the superposition of the C=C stretch of the *cis* –CH=CH-linkage of unsaturated lipid chains and the amide I mode of proteins. The band at 1440 cm$^{-1}$ is assignable to the C–H bend mode. This peak shows a slightly asymmetric band shape, because two C–H bend modes, $CH_2$ scissors and $CH_3$ degenerate deformation, are overlapped. As shown later, this band is decomposed into two bands originating from $CH_2$ and $CH_3$ in the present high-resolution CARS measurements. The band at 1602 cm$^{-1}$ shows unique properties that have been found in our previous study [41,45]. This Raman band originates exclusively from mitochondria as shown by a spontaneous Raman imaging experiment of a yeast cell expressing mitochondria-targeted GFP. The intensity of this band is correlated with the metabolic activity of mitochondria in a living cell. Addition of a respiration inhibitor such as KCN causes a rapid decrease and subsequent disappearance of this band. We therefore call this band at 1602 cm$^{-1}$ the "Raman spectroscopic signature of life" [45]. Figure 22.9e contains further detailed vibrational information. The Raman bands at 1738, 1301, 1260, 1083 cm$^{-1}$ are assigned to the C=O stretch of the ester linkage, in-phase $CH_2$ twisting, C=C–H in-plane bend of the *cis* –CH=CH-linkage, and the skeletal C–C stretch mode of the *gauche* conformation, respectively. These bands originate from the lipid molecules such as phospholipids, which constitute mitochondria membranes. The sharp band at 1002 cm$^{-1}$ is assigned to the phenylalanine residues in proteins, which is accompanied with the weak bands at 1205 and 1030 cm$^{-1}$. All these features in Figure 22.9e agree excellently with those in the corresponding spontaneous Raman spectrum (Figure 22.9g). MEM is thus proven to be a powerful technique to retrieve the correct Im[$\chi^{(3)}$] spectra from congested CARS spectra over the entire spectral range of the fingerprint region. It enables us to use CARS spectra for quantitative molecular fingerprinting.

Next, we discuss the spectrum of the cytoplasm/nucleus in Figure 22.9f. This spectrum is totally different from that of mitochondrion in Figure 22.9e. It has already been discussed that the band at 1002 $cm^{-1}$ is assigned to the phenylalanine residues in proteins [41]. The high intensity of this band in the spectrum in Figure 22.9f indicates that protein molecules are predominantly located in the cytoplasm/nucleus. Accordingly, the bands at 1656, 1452, 1240 are assigned to amide I, C–H bend, and amide III (C–N stretch and in-plane N–H bending) of proteins. It is noted that the peak position of the C–H bend band in Figure 22.9f (1452 $cm^{-1}$) is different from that (1440 $cm^{-1}$) in Figure 22.9e. This frequency difference indicates that the contribution of $CH_3$ is larger than that of $CH_2$ in the C–H bend band in Figure 22.9f. This is consistent with proteins having a larger $CH_3/CH_2$ ratio than lipid molecules. In addition to the band due to proteins, a nucleic acid band (symmetric stretch of $PO_2^-$) is also observed as a broad feature at 1099 $cm^{-1}$.

Based on the vibrational spectral information, we can perform molecular *fingerprinting* of a living cell. Figure 22.10a through l shows label-free and multi-mode (14 Raman shifts) $Im[\chi^{(3)}]$ images of a living budding yeast cell at the Raman shift of 1738, 1656, 1602, 1456, 1439, 1340, 1301, 1260, 1240, 1205, 1030, and 1002 $cm^{-1}$, respectively [21]. Peak intensities are used for reconstructing these images. The exposure time for each pixel is 50 ms, and each image consists of 21 × 21 pixels. Overall measurement time is 22 s. These images (Figure 22.10a through l) are obtained simultaneously in one scan of the sample. As discussed, the C–H bend band is successfully decomposed into two, the $CH_3$ degenerate deformation (d, 1456 $cm^{-1}$) and $CH_2$ scissors (e, 1439 $cm^{-1}$), respectively. For references, CARS images at the Raman shift of 2930 $cm^{-1}$ (m) and 2850 $cm^{-1}$ (n) are also shown. These images are also simultaneously obtained just by adjusting the center wavelength of spectrometer to the C–H stretch region. We can classify all the $Im[\chi^{(3)}]$ images in Figure 22.10 into three groups. The first one is the images Figure 22.10a,c,e,g,h, and n, namely those obtained at 1738, 1602, 1439, 1301, 1260, and 2850 $cm^{-1}$, respectively. They show localized and intense signal inside the cell. The second group is the images Figure 22.10f,i,j,k, and l, namely those obtained at 1340, 1240, 1205, 1030, and 1002 $cm^{-1}$, respectively. These images show less heterogeneous molecular distributions inside the cell. The third group is Figure 22.10b,d, and m, namely, at 1656, 1456, and 2930 $cm^{-1}$, respectively. They look like the sum of the first and second categories. The images of the first group except that of 1602 $cm^{-1}$ are ascribed to lipid molecules, which are the main molecular species constituting membranous organelles such as mitochondria. The image at 1602 $cm^{-1}$ in Figure 22.10c coincides with the images of the lipids, reflecting that the "Raman spectroscopic signature of life" is localized at mitochondria. In contrast, the images of the second group are assigned to proteins that distribute rather uniformly in the cell. The images of the third group correspond to the sum of images due to lipids and proteins. It is reasonable because the band at 1656 $cm^{-1}$ is the superposition of the band due to the C=C stretch of lipids and the amide I bands of proteins. The Raman bands at 1456 and 2930 $cm^{-1}$ originate from the C–H bend and stretch modes of the $CH_3$ groups. Since both of lipids and proteins contain $CH_3$ groups, it is also reasonable that the images due to $CH_3$ give the sum of the images of the first and second group. Interestingly, the images in Figure 22.10d and e (C–H bend) agree well with those in Figure 22.10m and n (C–H stretch), respectively. There is therefore no need to acquire

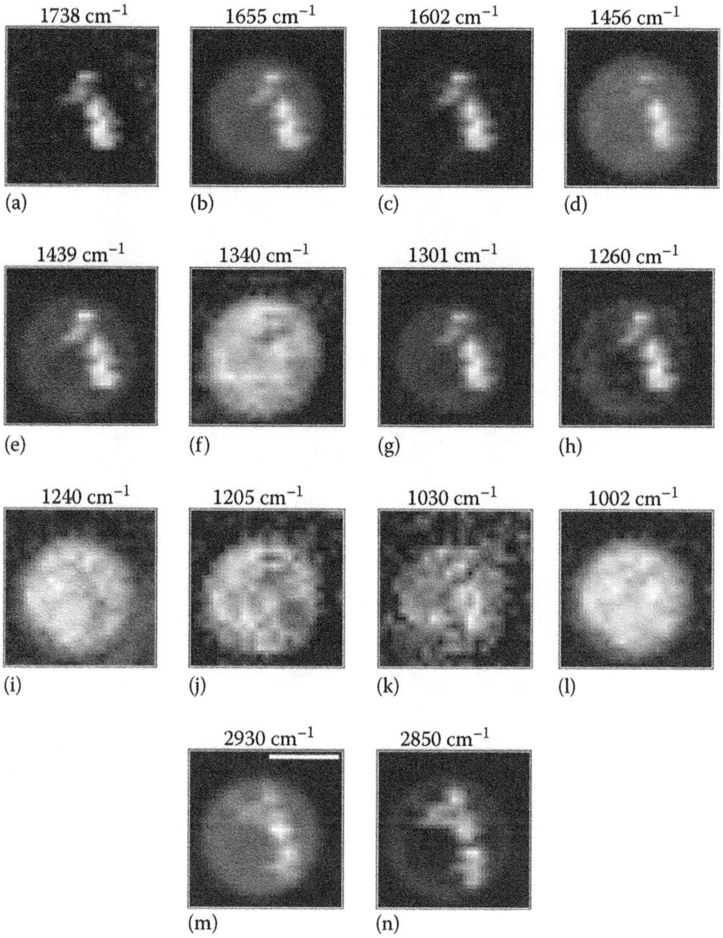


**FIGURE 22.10** Label-free and multi-frequency (14 frequencies) Im[$\chi^{(3)}$] images of a living budding yeast cell at the Raman shift of 1738, 1655, 1602, 1456, 1439, 1340, 1301, 1260, 1240, 1205, 1030, and 1002 $cm^{-1}$ (a–l). The scale bar in (m) is 4 μm. CARS images at the Raman shift of 2930 $cm^{-1}$ (m) and 2850 $cm^{-1}$ (n) are also indicated. (Reproduced from Okuno, M., Kano, H. et al.: Quantitative CARS molecular fingerprinting of single living cells with the use of the maximum entropy method. *Angew. Chem. Int. Ed.* 2010. 49. 6773–6777, S6773/6771–S6773/6776. Copyright Wiley-VCH Verlag GmbH & Co. KGaA.)

CARS images in the C–H stretch region, which only contains part of the vibrational information also contained in the fingerprint region.

CARS molecular fingerprinting enables us to obtain label-free and multi-mode images with high speed. Figure 22.11 shows the result of the real-time CARS study of a budding yeast cell over a time span of 35 min. Four series of time-resolved images at the Raman shifts of, 1655, 1602, 1445, and 1160 $cm^{-1}$ are presented. The band at 1445 $cm^{-1}$ corresponds to the overlap of the bands due to C–H bend originating from $CH_2$ and $CH_3$. The band at 1160 $cm^{-1}$ is obtained from a "dancing body" (DB) in a vacuole [46]. DBs are known to occasionally appear and move vigorously in a vacuole of a starving yeast cell. In our previous spontaneous Raman study [46], it is found that DBs contain high concentration of polyphosphate, which gives a Raman band at 1160 $cm^{-1}$ assigned to the $PO_2^-$ symmetric stretch mode. In Figure 22.11, we trace the

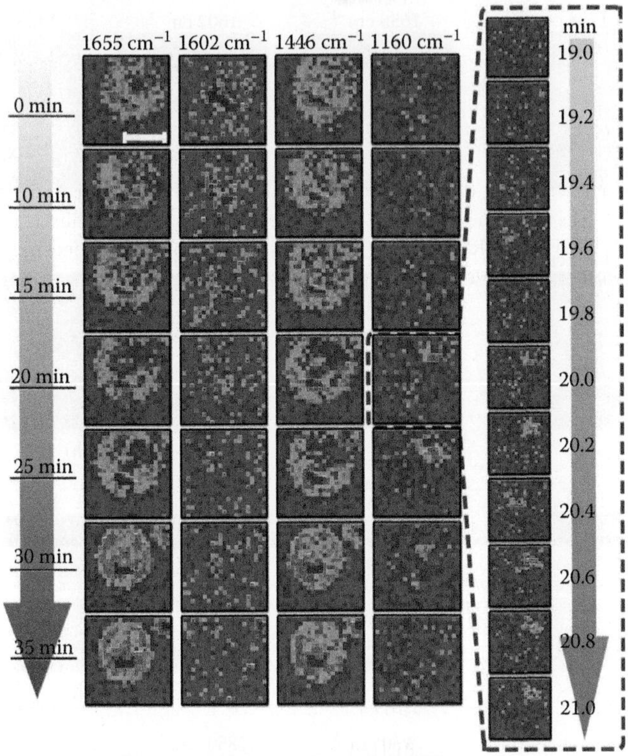


**FIGURE 22.11** Time-resolved CARS images of the Raman bands at 1655, 1602, 1446, and 1160 cm$^{-1}$, respectively. The scale bar inside the image is 5 μm. The color scale for each band is the same for all the times measured. The red frame on the right contains CARS images at 1160 cm$^{-1}$ from 19 to 21 min. Each image is measured every 12 s. Each of the seven rows of images is constructed from Raman intensities retrieved from one CARS spectrum. (Reproduced from Okuno, M., Kano, H. et al.: Quantitative CARS molecular fingerprinting of single living cells with the use of the maximum entropy method. *Angew. Chem. Int. Ed.* 2010. 49. 6773–6777, S6773/6771–S6773/6776. Copyright Wiley-VCH Verlag GmbH & Co. KGaA.)

appearance of a DB and subsequent changes of the cell. At 0 min, the cell seems to be in a normal condition having active mitochondria shown by the bright 1602 cm$^{-1}$ image. The intensity of the 1602 cm$^{-1}$ band gradually decreases with time, though the intensities of the other bands do not show any appreciable changes between 0 and 15 min. At 20 min, the 1602 cm$^{-1}$ band has almost disappeared, indicating that the metabolic activity in mitochondria has become minimal. Concomitantly, the polyphosphate band at 1160 cm$^{-1}$ suddenly appears inside the vacuole. In order to trace this process more closely, time-resolved CARS images between 19 and 21 min are also shown in Figure 22.11 (highlighted by a red frame). The 1160 cm$^{-1}$ band intensity gradually increases with time, showing the appearance of the DB in real time. At 21 min, the DB stops moving and stays at the upper part of the vacuole. At 30 min, the cell shrinks, and the vacuole disappears at the same time. The overall dynamic behavior inside the cell is interpreted in the following way. First, the metabolic activity in mitochondria is gradually lowered. At 20 min, the metabolic activity in mitochondria almost ceases and subsequently a DB appears inside the vacuole. Finally, the DB is disrupted and the cell shrinks at the same time. The cell is dead at this stage.

Multiplex CARS microspectroscopy combined with MEM has realized label-free, high-speed and quantitative molecular mapping of a single living cell in the fingerprint region (*CARS molecular fingerprinting*). Although the fingerprint region is spectrally congested in a living yeast cell, we are able to extract more than 10 vibrationally resonant Im[$\chi^{(3)}$] images by MEM. This method has been applied to a time-resolved study of dynamical processes in a single living yeast cell. It has been found that the laser irradiation with about 10 mW power causes lowering and subsequent ceasing of the metabolic activity of mitochondria, followed by the appearance of a DB, and the eventual cell death in 35 min.

Rapid cell death such as that observed in the present study was never observed in our previous spontaneous Raman experiment in which much lower laser power (4 mW) was used. In harmony with our previous observation of the starving death process, however, a clear correlation of the disappearance of the "Raman spectroscopic signature of life" to the appearance of a DB is found. Figure 22.12 shows two time-resolved multiplex CARS spectra measured from two different spots inside the cell, corresponding to mitochondria and vacuole (DB), respectively. In Figure 22.12a, the intensity of the 1602 cm$^{-1}$ band decreases gradually from 0 min and completely disappears at 20 min. On the other hand, the polyphosphate band suddenly arises around 20 min, lasts until 25 min and disappears at 30 min (Figure 22.12b). Thus, the present CARS experiment has shown clearly that the loss of the "Raman spectroscopic signature of life" precedes the formation of a dancing body. It is highly likely that the loss of metabolic activity in mitochondria causes, through an unknown mitochondria-vacuolar cross talk, the formation of a DB in a vacuole. Our previous spontaneous Raman study was unable to address such details because of its low time resolution. We are now able to discuss quantitatively the life and death of a single cell at the molecular level.

In order to examine the laser irradiation effect, we have conducted the experiment at lower laser power than that in Figure 22.11. Figure 22.13 shows the result of the real-time CARS study of a budding yeast cell over a time span of 35 min. Dynamical changes of the distribution of organelles inside the cell are clearly observed but no perturbed cell behavior is noticeable with the 5 mW pump and 10 mW Stokes laser power. Since the Stokes laser is not tightly focused onto the cell due to chromatic aberration, the irradiation of the Stokes laser is less effective than that of the pump laser. The exposure time at each pixel and image acquisition time is 60 ms and 25 s.

#### 22.4.3.3 Living Nematodes

We applied the molecular fingerprinting technique to a multi-cellular organism. Figure 22.14a shows a spectral profile of multiplex CARS signal of a nematode. Exposure time for each point is 100 ms. Since an optical fiber is used to collect the CARS signal in this experiment, the signal-to-noise ratio of the spectral profile is poorer than that of Figure 22.9. A strong signal is observed at around 2830 cm$^{-1}$, which corresponds to $CH_2$ stretch vibrational mode. In contrast, several dispersive patterns are observed in the fingerprint region. These bands at 1670 and 1470 cm$^{-1}$ are assigned to C=C stretch and C–H bend vibrational modes, respectively.

Figure 22.14b shows three CARS images at $CH_2$ stretch, C=C stretch and C–H bend vibrational modes. Image size is 201 pixel × 201 pixels. It should be noticed that all of the

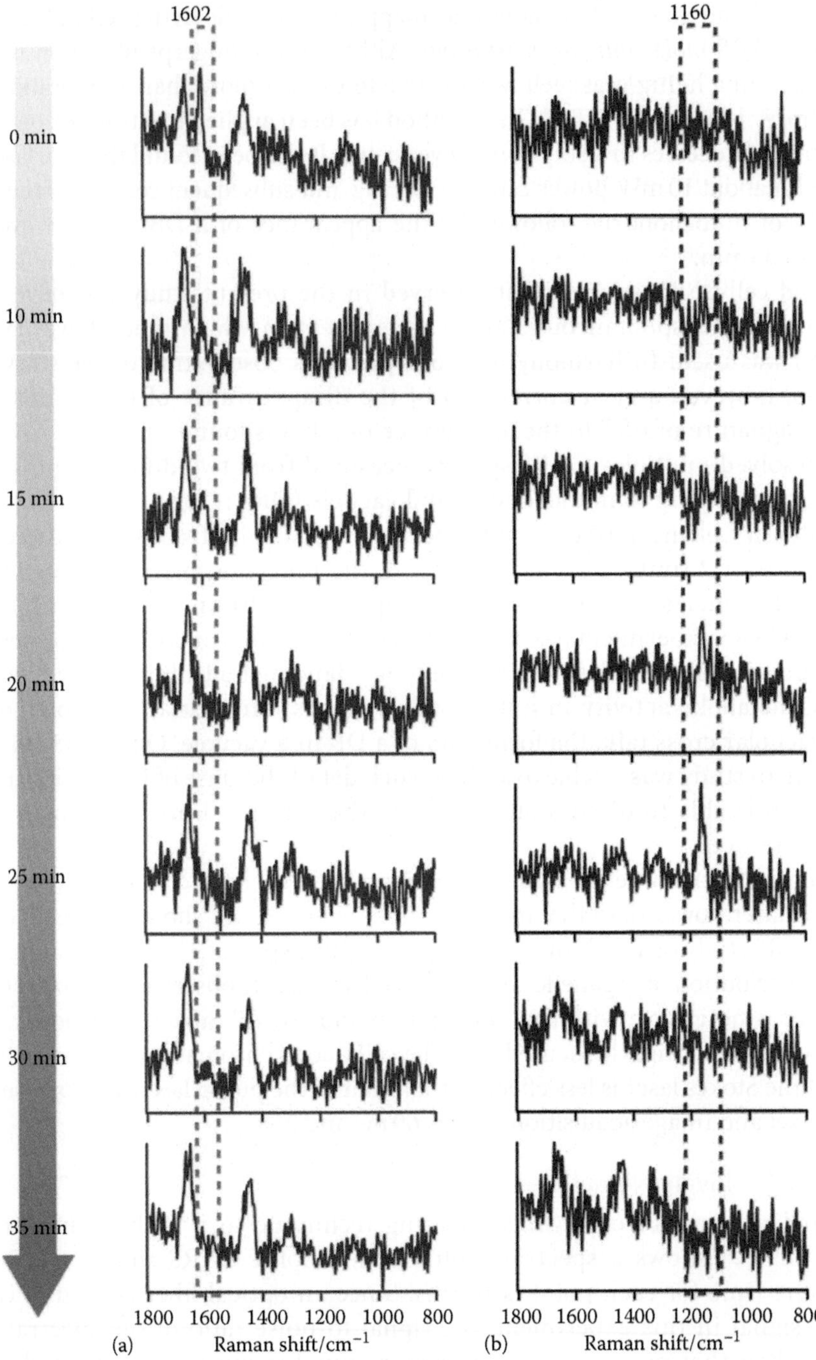


**FIGURE 22.12** Time-resolved multiplex CARS spectra of a budding yeast cell at mitochondria (a), and a dancing body (b). The exposure time for each spectrum is 30 ms. (Reproduced from Okuno, M., Kano, H. et al.: Quantitative CARS molecular fingerprinting of single living cells with the use of the maximum entropy method. *Angew. Chem. Int. Ed.* 2010. 49. 6773–6777, S6773/6771–S6773/6776. Copyright Wiley-VCH Verlag GmbH & Co. KGaA.)

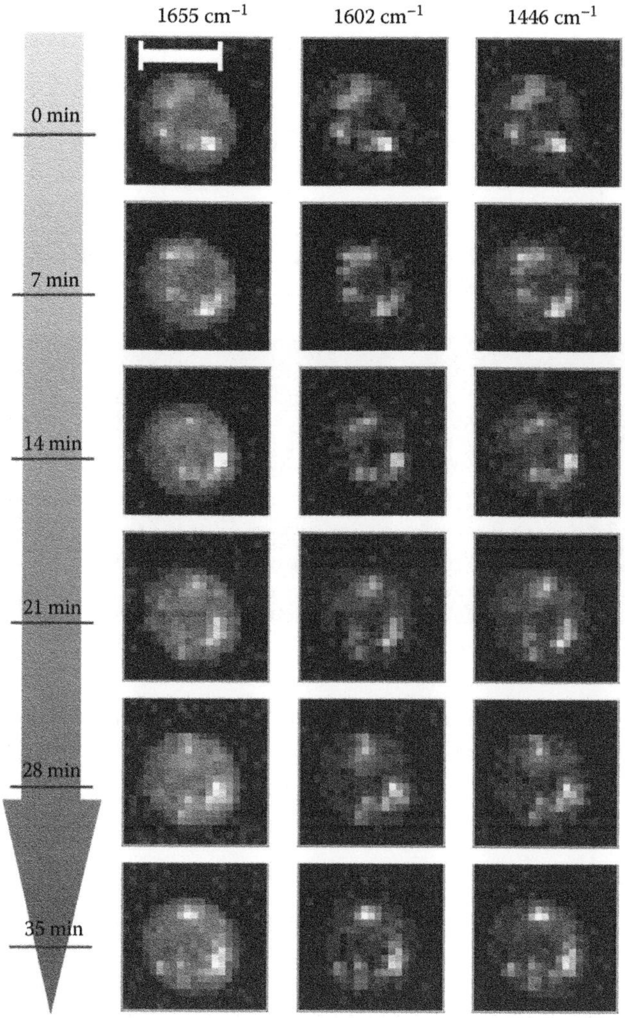


**FIGURE 22.13** Time-resolved CARS images of the Raman bands at 1655, 1602, and 1446 $cm^{-1}$, respectively. The scale bar inside the image is 5 μm. The intensity scale for each band is the same for all the times measured. Each image is measured every 25 s. Each of the six rows of images is constructed from Raman intensities retrieved from one CARS spectrum by MEM and SVD analysis. (Reproduced from Okuno, M., Kano, H. et al.: Quantitative CARS molecular fingerprinting of single living cells with the use of the maximum entropy method. *Angew. Chem. Int. Ed.* 2010. 49. 6773–6777, S6773/6771–S6773/6776. Copyright Wiley-VCH Verlag GmbH & Co. KGaA.)

images exhibit small spots inside of the nematode, and are characterized by the same vibrational contrast. It means that they visualize the same molecular species. Taking account of the vibrational information, they originate from unsaturated lipids, which are stored as lipid droplets inside of the nematodes. Although CARS imaging of nematodes has been performed previously [37], our CARS molecular fingerprinting provides unique information not only on the localization of lipid molecules, but also on their degree of unsaturation simultaneously.

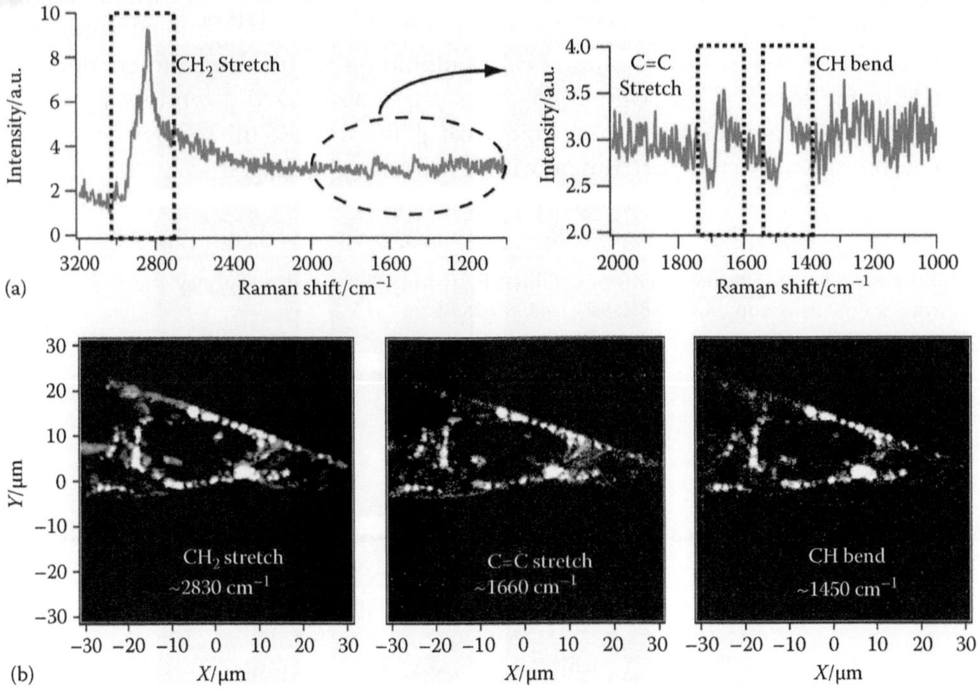


**FIGURE 22.14** (a) CARS spectrum of a nematode. Exposure time for each point is 100 ms; (b) three CARS images at $CH_2$ stretch, C=C stretch and C=H bend vibrational modes. Image size is 201 pixel × 201 pixels.

## 22.5 Conclusions

Owing to its inherent molecular specificity, multiplex CARS microspectroscopy is one of the most promising techniques for molecular imaging, since it provides multi-mode images with rich spectral information on molecular composition and its structure. In comparison with conventional fluorescence imaging, which can visualize only known molecules, CARS spectroscopic imaging can provide information on as-yet-unrecognized molecules through spectral analysis. As a result, CARS spectroscopic imaging enables us to explore such "hidden" molecules and their structures. Combined analysis of CARS with other nonlinear optical processes such as SH, TH, TPEF, and TSFG enables us to perform unique multi-modal nonlinear optical imaging to gain deep insight into a living system [35]. Among them, hyper-Raman spectroscopy [47] is a unique nonlinear Raman process, because infrared active vibrational modes can be probed. Combination of CARS and hyper-Raman provides unique high-spatial-resolution vibrational microspectroscopy that is not restricted by the selection rules.

Molecular spectroscopic imaging featuring nonlinear Raman processes will open up a new scope for the visualization of living cells and organisms *in vivo* and *in situ*, and will be an indispensable tool for medical, life, and material sciences.

## Acknowledgments

It is a pleasure to compile our recent studies on multiplex CARS microspectroscopy of living cells carried out in Hamaguchi laboratory at the University of Tokyo. *I would like to express my sincere gratitude to Professor Hiro-o Hamaguchi* for his guidance, support, discussion, and encouragement throughout the present study. I would like to thank all collaborators in the Hamaguchi laboratory for their dedicated efforts in contributing to this study. In particular, I would like to thank Dr. M. Okuno for development of CARS microspectroscopy using nanosecond white-light laser source. I would also like to thank Dr. C. Onogi for providing the spontaneous Raman spectrum of yeast mitochondria. I would gratefully acknowledge Dr. T. Karashima, Professor M. Yamamoto (the University of Tokyo) and Dr. F. Okura, Ms. H. Yomo (Suntory Co., Ltd.) for their collaboration. I would like to thank Leukos and Horus Laser companies for their technical support. Last but not least, I would gratefully acknowledge J. Ukon (HORIBA, Ltd.) for assisting collaboration between the Japanese and French groups.

## References

1. Evans C. L. and X. S. Xie. 2008. Coherent anti-Stokes Raman scattering microscopy: Chemical imaging for biology and medicine. *Annu. Rev. Anal. Chem.* 1:883–909.
2. Cheng J.-X. and X. S. Xie. 2004. Coherent anti-Stokes Raman scattering microscopy: Instrumentation, theory, and applications. *J. Phys. Chem. B* 108:827–840.
3. Volkmer A. 2005. Vibrational imaging and microspectroscopies based on coherent anti-Stokes Raman scattering microscopy. *J. Phys. D Appl. Phys.* 38:R59–R81.
4. Freudiger C. W., W. Min et al. 2008. Label-free biomedical imaging with high sensitivity by stimulated Raman scattering microscopy. *Science* 322:1857–1861.
5. Saar B. G., C. W. Freudiger et al. 2010. Video-rate molecular imaging in vivo with stimulated Raman scattering. *Science* 330:1368–1370.
6. Ploetz E., S. Laimgruber et al. 2007. Femtosecond stimulated Raman microscopy. *Appl. Phys. B* 87:389–393.
7. Okuno M. and H. Hamaguchi. 2010. Multifocus confocal Raman microspectroscopy for fast multimode vibrational imaging of living cells. *Opt. Lett.* 35:4096–4098.
8. Hamada K., K. Fujita et al. 2008. Raman microscopy for dynamic molecular imaging of living cells. *J. Biomed. Opt.* 13:044027/044021–044027/044024.
9. Kano H. and H. Hamaguchi. 2005. Ultrabroadband (>2500 $cm^{-1}$) Multiplex coherent anti-Stokes Raman scattering microspectroscopy using a supercontinuum generated from a photonic crystal fiber. *Appl. Phys. Lett.* 86:121113/121111–121113.
10. Okuno M., H. Kano et al. 2008. Ultrabroadband multiplex CARS microspectroscopy and imaging using a subnanosecond supercontinuum light source in the deep near infrared. *Opt. Lett.* 33:923–925.
11. Hashimoto M., T. Araki, and S. Kawata. 2000. Molecular vibration imaging in the fingerprint region by use of coherent anti-Stokes Raman scattering microscopy with a collinear configuration. *Opt. Lett.* 25:1768–1770.
12. Zumbusch A., G. R. Holtom, and X. S. Xie. 1999. Three-dimensional vibrational imaging by coherent anti-Stokes Raman scattering. *Phys. Rev. Lett.* 82:4142–4145.
13. Cheng J.-X., A. Volkmer et al. 2002. Multiplex coherent anti-Stokes Raman scattering microspectroscopy and study of lipid vesicles. *J. Phys. Chem. B* 106:8493–8498.
14. Kano H. and H. Hamaguchi. 2005. Vibrationally resonant imaging of a single living cell by supercontinuum-based multiplex coherent anti-Stokes Raman scattering microspectroscopy. *Opt. Express* 13:1322–1327.

15. Kano H. and H. Hamaguchi. 2006. In-vivo multi-nonlinear optical imaging of a living cell using a supercontinuum light source generated from a photonic crystal fiber. *Opt. Express* 14:2798–2804.
16. Rinia H. A., K. N. J. Burger et al. 2008. Quantitative label-free imaging of lipid composition and packing of individual cellular lipid droplets using multiplex CARS microscopy. *Biophys. J.* 95:4908–4914.
17. Isobe K., A. Suda et al. 2009. Single-pulse coherent anti-Stokes Raman scattering microscopy employing an octave spanning pulse. *Opt. Express* 17:11259–11266.
18. Petrov G., V. Yakovlev et al. 2005. Detection of *Bacillus subtilis* spores in water by means of broadband coherent anti-Stokes Raman spectroscopy. *Opt. Express* 13:9537–9542.
19. Petrov G. I., R. Arora et al. 2007. Comparison of coherent and spontaneous Raman microspectroscopies for noninvasive detection of single bacteria endospores. *Proc. Natl Acad. Sci. USA* 104:7776–7779.
20. Pestov D., R. K. Murawski et al. 2007. Optimizing the laser-pulse configuration for coherent roman spectroscopy, *Science* 316:265–268.
21. Okuno M., H. Kano et al. 2010. Quantitative CARS molecular fingerprinting of single living cells with the use of the maximum entropy method. *Angew. Chem. Int. Ed.* 49:6773–6777, S6773/6771–S6773/6776.
22. Parekh S. H., Y. J. Lee et al. 2010. Label-free cellular imaging by broadband coherent anti-Stokes Raman scattering microscopy. *Biophys. J.* 99:2695–2704.
23. Chen B.-C., J. Sung, and S.-H. Lim. 2010. Chemical imaging with frequency modulation coherent anti-Stokes Raman scattering microscopy at the vibrational fingerprint region. *J. Phys. Chem. B* 114:16871–16880.
24. Kee T. W. and M. T. Cicerone. 2004. Simple approach to one-laser, broadband coherent anti-Stokes Raman scattering microscopy. *Opt. Lett.* 29:2701–2703.
25. Kano H. and H. Hamaguchi. 2006. Vibrational imaging of a single pollen grain by ultrabroadband multiplex coherent anti-Stokes Raman scattering microspectroscopy. *Chem. Lett.* 35:1124–1125.
26. Ivleva N. P., R. Niessner, and U. Panne. 2005. Characterization and discrimination of pollen by Raman microscopy. *Anal. Bioanal. Chem.* 381:261–267.
27. Sengupta A., M. L. Laucks, and E. J. Davis. 2005. Surface-enhanced Raman spectroscopy of bacteria and pollen. *Appl. Spectrosc.* 59:1016–1023.
28. Wiermann R. and K. Vieth. 1983. Outer pollen wall, an important accumulation site for flavonoids. *Protoplasma* 118:230–233.
29. Vrielynck L., J. P. Cornard, and J. C. Merlin. 1994. Semi-empirical and vibrational studies of flavone and some deuterated analogues. *Spectrochim. Acta* 50A:2177–2188.
30. McConnell G. and E. Riis. 2004. Photonic crystal fibre enables short-wavelength two-photon laser scanning fluorescence microscopy with fura-2. *Phys. Med. Biol.* 49:4757–4763.
31. Isobe K., W. Watanabe et al. 2005. Multi-spectral two-photon excited fluorescence microscopy using supercontinuum light source. *Jpn. J. Appl. Phys., Part 2* 44:L167–L169.
32. Palero J. A., V. O. Boer et al. 2005. Short-wavelength two-photon excitation fluorescence microscopy of tryptophan with a photonic crystal fiber based light source. *Opt. Express* 13:5363–5368.
33. Kano H. and H. Hamaguchi. 2007. Supercontinuum dynamically visualizes a dividing single cell. *Anal. Chem.* 79:8967–8973.
34. Brackmann C., J. Norbeck et al. 2009. CARS microscopy of lipid stores in yeast: The impact of nutritional state and genetic background. *J. Raman Spectrosc.* 40:748–756.
35. Segawa, H., M. Okuno et al. 2012. Label-free tetra-modal molecular imaging of living cells with CARS, SHG, THG and TSFG (coherent anti-Stokes Raman scattering, second harmonic generation, third harmonic generation and third-order sum frequency generation). *Opt. Express*, 20:9551–9557.
36. Kano H. 2010. Molecular spectroscopic imaging using a white-light laser source. *Bull. Chem. Soc. Jpn.* 83:735–743.
37. Hellerer T., C. Axaeng et al. 2007. Monitoring of lipid storage in *Caenorhabditis elegans* using coherent anti-Stokes Raman scattering (CARS) microscopy. *Proc. Natl Acad. Sci. USA* 104:14658–14663.
38. Kano H. and H. Hamaguchi. 2006. Dispersion-compensated supercontinuum generation for ultrabroadband multiplex coherent anti-Stokes Raman scattering spectroscopy. *J. Raman Spectrosc.* 37:411–415.
39. Lim S.-H., A. G. Caster et al. 2006. Chemical imaging by single pulse interferometric coherent anti-Stokes Raman scattering microscopy. *J. Phys. Chem. B* 110:5196–5204.
40. von Vacano B., W. Wohlleben, and M. Motzkus. 2006. Single-beam CARS spectroscopy applied to low-wavenumber vibrational modes. *J. Raman Spectrosc.* 37:404–410.
41. Huang Y.-S., T. Karashima et al. 2005. Molecular-level investigation of the structure, transformation, and bioactivity of single living fission yeast cells by time- and space-resolved Raman spectroscopy. *Biochemistry* 44:10009–10019.

42. Vartiainen E. M., K. E. Peiponen, and T. Tsuboi. 1990. Analysis of coherent Raman spectra. *J. Opt. Soc. Am. B* 7:722–725.
43. Vartiainen E. M., H. A. Rinia et al. 2006. Direct extraction of Raman line-shapes from congested CARS spectra. *Opt. Express* 14:3622–3630.
44. Van Manen H.-J., Y. M. Kraan et al. 2004. Intracellular chemical imaging of heme-containing enzymes involved in innate immunity using resonance Raman microscopy. *J. Phys. Chem. B* 108:18762–18771.
45. Huang Y.-S., T. Karashima et al. 2004. Raman spectroscopic signature of life in a living yeast cell. *J. Raman Spectrosc.* 35:525–526.
46. Naito Y., A. Toh-e, and H. Hamaguchi. 2005. In vivo time-resolved Raman imaging of a spontaneous death process of a single budding yeast cell. *J. Raman Spectrosc.* 36:837–839.
47. Shimada R., H. Kano, and H. Hamaguchi. 2006. Hyper-Raman microspectroscopy: A new approach to completing vibrational spectral and imaging information under a microscope. *Opt. Lett.* 31:320–322.

# 23. Coherent Raman Scattering Imaging of Drug Delivery Systems

**Ling Tong and Ji-Xin Cheng**



## 23.1 Introduction

Drug delivery is the method or process of administering a pharmaceutical compound to achieve a therapeutic effect in humans or animals. Drug delivery systems are patent protection formulations that carry the drugs and modify drug release profile, absorption, distribution, and elimination for the benefit of improving product efficacy and safety, as well as patient convenience and compliance. Common routes of administration include oral, topical, transmucosal, and inhalation routes. Nanotechnology and polymer technology are designed for drug delivery carriers and devices, such as liposomes, polymer nanoparticles, polymer–drug conjugates, hydrogel and polymer films, and have displayed lots of advantages fighting against a range of diseases. A better understanding of the performance of drug delivery systems in the biological environment by directly visualizing the biodistribution in the body, mass transport among the compartmental barriers, and drug release profile, is desired to improve the therapeutic efficacy and rational designing (Sanhai et al. 2008).

Various microscopy techniques have been developed to study drug delivery systems *in vitro* and *in vivo*. For the drug carriers, fluorescence or radionuclide labeling is most widely used to track the micro- or nano-particles. Although these methods are valuable,

*Coherent Raman Scattering Microscopy*. Edited by Ji-Xin Cheng and X. Sunney Xie © 2013 CRC Press/Taylor & Francis Group, LLC. ISBN: 978-1-4398-6765-5.

there remain challenges. Firstly, the external fluorescent or radionuclide tags might gradually dissociate from the carriers (Chen et al. 2008; Xu et al. 2008); Secondly, the labeling might perturb the inherent properties of the vesicles (Csiszar et al. 2010; Szeto et al. 2005); Thirdly, photobleaching or losing activity impedes long-time monitoring (Tong et al. 2007); Finally, some drugs or particles are difficult to label. One way to overcome some of these challenges is through label-free chemically selective imaging based on signals from inherent molecular vibration.

Vibrational microscopes based on spontaneous Raman scattering (Klueva et al.; Ranade 2010) and near infrared (NIR) (Reich 2005) or infrared (IR) absorption (Levin and Bhargava 2005) have been used for chemical imaging of unstained samples in pharmaceutics. Nevertheless, the applications of these techniques are limited by their shortcomings. For NIR imaging, the spectral features of the components are often heavily overlapped and multivariate analysis is needed to deconvolute spectral information. For IR microscopy, strong water absorption impedes the measurement in aqueous environment. Both NIR and IR imaging are challenged by poor spatial resolution. Raman microscopy provides higher resolution and avoids water absorption because water is a weaker Raman scatterer. Nevertheless, Raman scattering has an extremely small cross section ($\sim10^{-30}$ cm$^2$/sr) compared to fluorescence ($\sim10^{-16}$ cm$^2$/sr). As a result, it usually takes hours to acquire a Raman image. Such slow speed hinders the application of Raman microscopy for the study of highly dynamic systems.

Coherent Raman scattering (CRS) microscopy overcomes the above difficulties by providing a signal level and imaging speed comparable to fluorescence microscopy and 3D sectioning capability with submicron spatial resolution. The potential of coherent anti-Stokes Raman scattering (CARS) and stimulated Raman scattering (SRS) microscopy in drug delivery research has been demonstrated in several recent studies, as summarized in Table 23.1.

**Table 23.1** Application of Coherent Raman Scattering Microscopy to Drug Delivery Systems

| Technology | Application |
|---|---|
| CARS microscopy | Receptor-mediated endocytosis with a polymeric nanoparticle-based CARS probe (Tong et al. 2007) |
| CARS microscopy | Cellular uptake of PLGA nanoparticles (Xu et al. 2008) |
| CARS microscopy | Drug (paclitaxel) distribution and release in polymer films (Kang et al. 2006, 2007) |
| CARS microscopy | Solid-state properties of lipid-based oral dosage forms upon drug (theophylline anhydrate) dissolution (Windbergs et al. 2009) |
| SRS microscopy | Cellular uptake of deuterated liposomes |
| SRS microscopy | Drug distribution in pharmaceutical tablets (amlodipine besylate) (Slipchenko et al. 2010) |
| SRS microscopy | Drug (ibuprofen and ketoprofen) penetration through skin (Saar et al. 2011) |

*Sources:* Tong et al., *J. Phys. Chem. B*, 111(33), 9980, 2007; Xu et al., *Mol Pharmcol.*, 6(1), 190, 2008; Kang et al., *Anal Chem.*, 78(23), 8036, 2006; Kang et al., *J. Control Release*, 122(3), 261, 2007; Windbergs et al., *Anal Chem.*, 81(6), 2085, 2009; Slipchenko et al., *Analyst*, 135, 2613, 2010; Saar et al., *Mol Pharm.*, 8(3), 969, 2011.

## 23.2 Application of CARS Microscopy

### 23.2.1 Imaging Receptor-Mediated Endocytosis with CARS Probe

With the development of colorful fluorescent proteins and probes, fluorescence microscopy has become a powerful tool for biology study. However, most of the fluorophores are prone to photobleaching, which makes continuous observation of biological process difficult. For example, after continuous imaging of a single cell binding with folate-targeted liposomes for 2.7 s, the fluorescent intensity of calcein encapsulated in the liposomes is reduced by 80%, which hindered real-time monitoring of folate receptor-mediated endocytosis and intracellular trafficking by fluorescent microscopy (Tong et al. 2007).

Tong et al. developed a polymeric nanoparticle-based CARS probe for monitoring receptor-medicated endocytosis without photobleaching (Tong et al. 2007). In this study, 200 nm polystyrene particles were loaded into folate-targeted liposomes as probes to monitor the folate receptor-medicated activities in KB cells. Two tightly synchronized 2.5 ps laser beams generated by Ti:sapphire oscillators (Coherent Inc., Santa Clara, CA) were used as pump and Stokes. The laser beams were collinearly combined and sent to a laser-scanning microscope (FV300/IX70, Olympus America Inc., San Jose, CA). Epi-CARS signals were collected by a 60× water immersion objective, spectrally separated from the excitation source by a dichroic mirror and detected by a photomultiplier tube. By turning ($\omega_p - \omega_s$) to 3045 $cm^{-1}$, which corresponds to the aromatic C–H stretching vibration, the epi-detected CARS signal from cellular organelles was cancelled by the destructive interference between the resonant contribution from aliphatic C–H vibration and the nonresonant contribution (Figure 23.1A through C), while the polystyrene nanoparticles with a high density of aromatic C–H bonds were selectively visualized with a high signal-to-noise ratio (Figure 23.1D through F). Without neither photobleaching nor photodamage, the polymeric nanoparticle-based CARS probe allowed real-time imaging of receptor-medicated endocytosis and long-time monitoring of intracellular transport. The intracellular movement of CARS probes was analyzed by single particle tracking. The nanoparticle exhibited a bidirectional active transport motion along microtubules toward the nucleus (negative speed) and cell membrane (positive speed). The mean squared displacements were fully consistent with a directed motion model, with diffusion constant $D = 460\,nm^2/s$ and average speed $\upsilon = 11.4\,nm/s$. No CARS signal reduction was observed during the 25 s period. These results showed the advantages and potential of using a CARS probe and CARS microscopy to study dynamic cellular processes.

### 23.2.2 Imaging Cellular Uptake of Poly(Lactic-Co-Glycolic Acid) Nanoparticle

Poly(lactic-co-glycolic acid) (PLGA) is a widely used polymer for fabricating nanoparticles because of biocompatibility, long-standing track record in biomedical applications, and well-documented utility for sustained drug release. Based on fluorescence imaging of PLGA nanoparticles encapsulating a fluorescent probe and measurement of the intracellular fluorescent level of the probe, many studies demonstrated rapid and efficient cellular uptake of PLGA nanoparticles through endocytosis

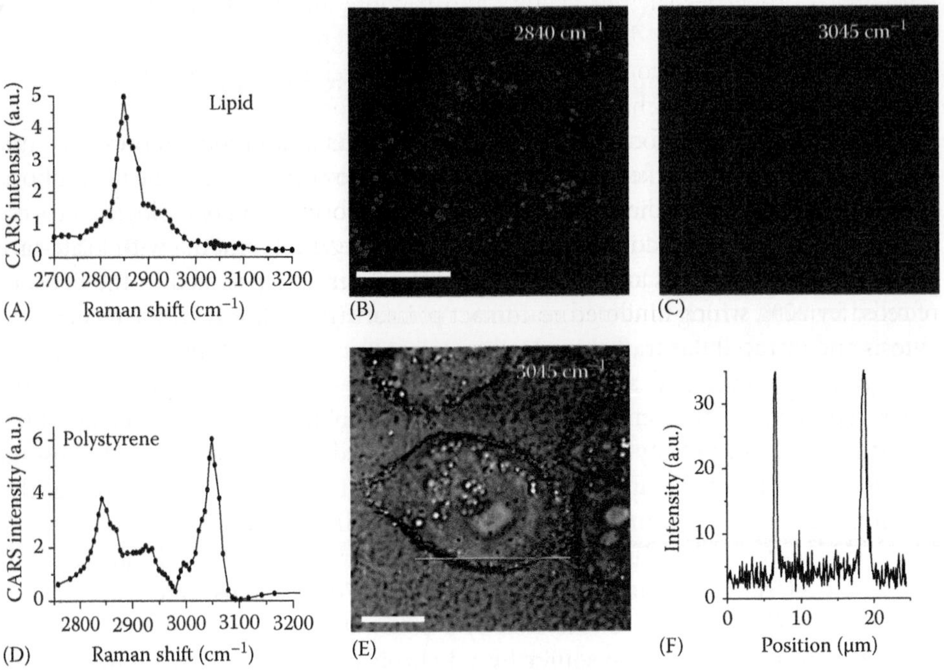


**FIGURE 23.1** Selective imaging of polystyrene beads attached to KB cells by E-CARS microscopy. (A) CARS spectrum of a single lipid droplet inside a KB cell. (B) E-CARS signal (red) from lipid droplets in a KB cell with $(\omega_p - \omega_s)$ at 2840 cm$^{-1}$. (C) No E-CARS signal was probed from organelles in the same KB cell as shown in (B) with $(\omega_p - \omega_s)$ at 3045 cm$^{-1}$. (D) CARS spectrum of polystyrene. (E) After 2 h incubation, 200 nm polystyrene nanoparticles encapsulated in folate-targeted liposomes strongly bound to KB cell surface, probed by E-CARS microscopy. Gray: transmission image. Red: E-CARS signal from polystyrene nanoparticles with $(\omega_p - \omega_s)$ at 3045 cm$^{-1}$. (F) E-CARS intensity profile along a linescan in (E) showed a high signal-to-noise ratio for 200 nm polystyrene nanoparticles. (From Tong, L. et al., *J. Phys. Chem. B*, 111(33), 9980, 2007. With permission.)

and intracellular release of drug (Panyam and Labhasetwar 2003; Sahoo et al. 2002; Vicari et al. 2008). In contrast, an earlier study argued that the lipophilic markers could transfer from PLGA nanoparticles to cell membrane, leading to an overestimation of cellular uptake of PLGA nanoparticles (Pietzonka et al. 2002). To clarify the cellular uptake pathway of PLGA nanoparticles, it is essential to directly visualize the PLGA polymers without labeling.

To address this issue, Xu et al. re-examined the cellular drug delivery mechanism of PLGA nanoparticles by label-free CARS microscopy (Xu et al. 2008). Pure PLGA nanoparticles (NP) and PLGA nanoparticles encapsulated with Nile red (NR/NP) of the same size of 300 nm in diameter were prepared and imaged by confocal fluorescence and CARS microscopy. First, NR/$NP_{300}$ were incubated in the mesothelial cells for 1, 30, 180 min and imaged by confocal microscopy immediately after replacing the particle suspension in the medium with fresh particle-free medium. Red fluorescence was seen inside the cells, especially in the perinuclear region as early as 1 min after incubation with NR/NPs. Fluorescence intensity in the cells increased with incubation time. Similar results were observed in KB cells. The internalization rate was striking for such big NPs of 300 nm in diameter. To examine whether the observed intracellular signals were real PLGA NPs or the dye separated from NPs and transferred to cells, pure PLGA NPs with the same size were incubated in KB cells and imaged with label-free CARS microscopy. The pump and Stokes beams were generated by two synchronized Ti:sapphire oscillators with a pulse width of 5 ps (Spectra-Physics, Santa Clara, CA), collinearly combined and directed into a laser-scanning microscope (FV300/IX71, Olympus America, San Jose, CA). A 60× water immersion objective was used to focus the excitation beams on the sample. The forward-CARS signal was collected by an air condenser and detected with a photomultiplier tube. PLGA exhibits a peak at 2940 $cm^{-1}$ arising from the $CH_3$ stretch vibration, which could be distinguished from the lipid bodies showing a peak at 2840 $cm^{-1}$ from the $CH_2$ stretch vibration in the aliphatic carbon chains (Figure 23.2A). After incubation for 3 h, the PLGA nanoparticles (visible in the 2940 $cm^{-1}$ image but not in the 2840 $cm^{-1}$) were only found in the extracellular space (Figure 23.2B and C), but not inside KB cells. For parallel comparison, confocal microscopy and CARS microscopy were performed on the same KB cells incubated with NR/$NP_{300}$ for 3 h (Figure 23.2D and F). The red intracellular signals observed in confocal microscopy were colocalized with lipid bodies (2840 $cm^{-1}$), whereas the extracellular signals were detected as PLGA (2940 $cm^{-1}$). Note that the extracellular signals do not exactly colocalize because the extracellular NPs were in constant motion. The CARS observation supports that PLGA NPs were not readily taken up by cells, at least in 3 h. The red fluorescence observed in confocal imaging was from Nile red which separated from NPs, transferred to cells, and labeled the lipid bodies. The conclusions were confirmed by examination of PLGA NPs chemically conjugated with fluoresceinamine. Taken together, these findings indicate that the studies of cellular uptake of fluorescence-labeled nanocarriers should be interpreted with caution. Label-free CRS microscopy is a reliable tool for understanding the mechanism of drug delivery system by directly tracking the nanocarriers.

### 23.2.3 Imaging Solid-State Properties of Tablets upon Drug Dissolution

Dissolution behavior is a critical factor for the therapeutic efficacy of oral solid dosage forms. The dissolution testing is generally performed by analyzing the concentration of released drug in the flowing dissolution medium at defined time intervals using UV spectroscopy or HPLC, however, no direct information on the changing of dosage form phenomena is provided in such analysis. A need to monitor the solid-state properties with spatially resolved information during dissolution test is desired. Many techniques have been applied to characterize this process, such as scanning electron microscopy (Aaltonen et al. 2006), x-ray powder diffraction (Debnath et al. 2004), NIR, IR, terahertz and Raman imaging (Reich 2005; Slobodan 2007; Zeitler et al. 2007), as well as secondary ion mass spectrometry (Belu et al. 2000). Although these methods could image physical and/or chemical changes after dissolution testing *ex situ*, they all have weaknesses regarding *in situ* imaging. For example, NIR, IR, and terahertz imaging is restricted by water absorption in the aqueous environment and low spatial resolution. Secondary ion mass spectrometry cannot be performed in a dissolution medium. Raman imaging is limited by the long data acquisition time and possible interference from fluorescence. For *in situ* analysis, it requires: (1) no dosage form destruction; (2) setup with dissolution medium flow; (3) sufficient temporal and spatial resolution.

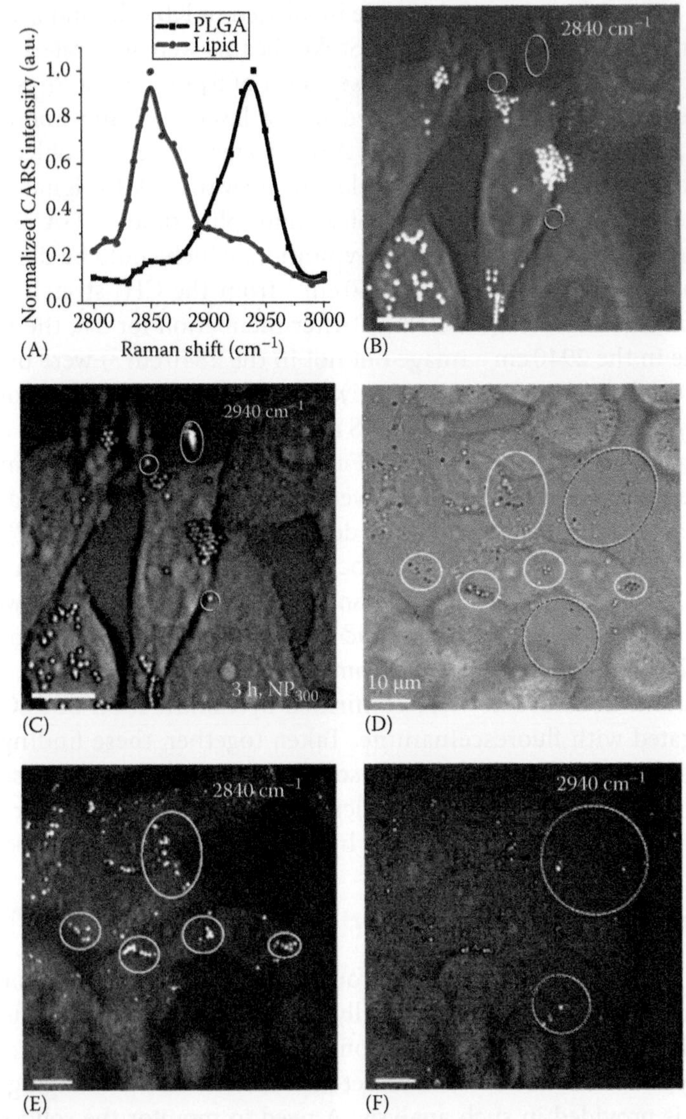


**FIGURE 23.2** Chemically selective imaging of intracellular lipid bodies and PLGA particles by CARS microcopy. (A) CARS spectra of a PLGA film and lipid bodies in KB cells. The two peaks at 2930 and 2840 cm$^{-1}$ arise from $CH_3$ stretch vibration in PLGA and $CH_2$ symmetric stretch vibration in lipid bodies, respectively. (B) CARS image with ($\omega_p - \omega_s$) at 2840 cm$^{-1}$. (C) CARS image with ($\omega_p - \omega_s$) at 2940 cm$^{-1}$. KB cells were incubated with unlabeled PLGA nanoparticles for 3 h. The PLGA nanoparticles (in yellow circles) could only be seen at 2940 cm$^{-1}$ and did not enter KB cells. (D–F) Parallel comparison of confocal microscopy and CARS of KB cells incubated with NR/$NP_{300}$ for 3 h. (D) Confocal image combining the transmission channel and the red channel (Nile red). (E) CARS image at 2840 cm$^{-1}$. (F) CARS image at 2940 cm$^{-1}$. The red intracellular signals (white circle) observed in confocal microscopy were co-localized with lipid bodies (2840 cm$^{-1}$), whereas the extracellular signals (yellow circle) were seen as PLGA (2940 cm$^{-1}$) (Note: the extracellular signals do not exactly co-localize because the extracellular nanoparticles were in constant motion.) The CARS signals were detected in the forward direction. (From Tong, L. et al., *J. Phys. Chem. B*, 111(33), 9980, 2007. With permission.)

CARS microscopy is a suitable tool for the *in situ* dissolution analysis of dosage forms. Kang et al. have reported successfully the mapping of drug distribution and release in polymer film-based form by CARS microscopy (Kang et al. 2006, 2007). More recently Windbergs et al. (2009) employed CARS microscopy for real-time visualization of the solid-state properties of lipid-based oral dosage forms containing the model drug theophylline anhydrate during dissolution. The dissolution testing was performed in a flow-through cell at room temperature with purified water as medium. Tablets or extrudates composed with tripalmitin and theophylline anhydrate or theophylline monohydrate (50/50 w/w) were tested. For the CARS microscope, the fundamental beam (1064 nm, 80 MHz, >15 ps) of a coherent paladin Nd:YAG laser was used as Stokes and the signal beam (700–1000 nm, spectral width of 0.2 nm) from the OPO (APE Levante Emerald) was used as pump. The beams were focused on the sample using a 20× objective (NA 0.5) and the CARS signal was detected in the backward direction.

Firstly, Raman spectra from the pure powdered substances were recorded to determine the suitable vibrational bands for the specific components. The peaks at 2880 $cm^{-1}$ ($CH_3$ symmetric stretching) and at 3109 $cm^{-1}$ (CH stretching associated with the imidazole ring) were identified for qualitative CARS imaging of tripalmitin and theophylline, respectively. A potential way for quantitative measurement based on CARS images was discussed. With the identified peaks, the distribution of drug (green) and lipid (red) in the tablets and extrudates was mapped by CARS imaging with a spatial resolution of 1.5 μm. The images showed that theophylline particles were randomly distributed at the surface of lipid matrix, with theophylline monohydrate showing thin needles whereas theophylline anhydrate exhibiting anisometric particles.

After characterization of the CARS signals from all the components, the drug release and solid-state transformations were monitored during dissolution testing by immersing the dosage forms in 500 mL purified water and comparing the images before and after immersion for 30 and 180 min. After 30 min of immersion, the theophylline monohydrate was still visible whereas no drug was observed on the surface of tablet after 180 min. The lipid matrix remained intact during dissolution. With these results, the dissolution process could be divided in several stages: first, the theophylline anhydrate dissolves, forming a supersaturated solution and creating theophylline monohydrate; then a transformation phase is followed in which monohydrate crystallizes; in the end dissolution of the two forms happens. The impact of extrusion process on the release behavior of drug was studied with the same method. No monohydrate needle was formed on the surface of tablet. A few very small monohydrate needles were observed inside the pores of extrudate at the depth of 50 μm. Then, the solid-state transformation and drug release in real-time were visualized by CARS microscope. The tablet was put into a dissolution flow-through cell and the dissolution process was recorded. These results demonstrated that the method used to combine lipid and drug into a sustained release dosage form can influence the physicochemical behavior of the drug during dissolution. CARS microscopy shows a high potential in characterization of the change of oral solid dosage forms during dissolution and development of efficient dosage forms.

## 23.3 Application of SRS Microscopy

### 23.3.1 Deuterated Liposomes

Liposomes are one of the most widely used drug delivery carriers for both hydrophilic and hydrophobic molecules (Torchilin 2005). Many liposomal drugs, such as liposomal doxorubicin, have been used for clinics (Jarvis 2009). Currently liposome accounts for 26% siRNA delivery (Wu and Mcmillan 2009). Extensive studies have been performed in order to design liposomes for efficient uptake and release of drugs in targeted diseases. A grand challenge facing liposomal drug delivery is to promote the endosomal release before degradation by lysosome. Addressing such a challenge requires a profound understanding of the behavior of liposomes inside cells. Although fluorescence labels have been used to probe the location and activity of liposomes, the potential risks of the external tags including dissociation from liposomes and perturbation the properties of liposomes may cause misleading conclusions. For example, aromatic lipids, which are commonly used as fluorescent markers of liposomes, have been shown to change the cellular uptake process of liposomes by facilitating membrane fusion (Csiszar et al. 2010). Therefore, label-free technology to directly visualize liposomes is critical to circumvent these problems.

Using 2 ps pulse excitation, Cheng group imaged folate-targeted deuterated liposomes in the live CHO-β cells (CHO cells stably transfected with folate receptor β) at $CD_2$ stretching vibration band of 2100 $cm^{-1}$. The liposomes were prepared by polycarbonate membrane extrusion method using d62-DPPC, cholesterol, mPEG-DSPE, and folate-PEG-cholesterol at a molar ratio of 65:30:4.5:0.5, with a diameter of 108.4 ± 29.4 nm. After incubated in CHO-β cells for 10 h, deuterated liposomes were visualized in the perinuclear area by stimulated Raman loss imaging at $CD_2$ stretching vibration of 2100 $cm^{-1}$ with 2 ps lasers (CoherentInc., Santa Clara, CA). No signal was detected when turned away from C–D vibration (Figure 23.3A through C). Raman spectra were measured from the intracellular liposome and cytoplasm, with a peak around 2103 $cm^{-1}$ observed only from liposomes but not cytoplasm (Figure 23.3D). The low intensity of the peak indicated a low concentration of deuterated lipid in the focal volume. The time constant was 1.0 ms and scanning speed was 2 ms/pixel for the current SRS imaging of liposomes. Further improvement of SRS imaging sensitivity could help to increase the imaging speed.

### 23.3.2 Pharmaceutical Tablets

Tablet is a kind of pharmaceutical dosage forms made from a mixture of an active pharmaceutical ingredient (API) and excipients compacted into a solid form. An inhomogeneous distribution of active components and excipients within a tablet may alter the release profile and affect the therapeutic efficacy. Therefore, mapping the distribution of API and excipients in a tablet is critical for the production of high-quality formulations. Slipchenko et al. applied epi-detected SRS microscopy to image API and excipients in amlodipine besylate (AB) tablets provided by six manufactures (Slipchenko et al. 2010). AB is a widely used drug for lowering the blood pressure.

Raman spectra were measured from pure components including microcrystalline cellulose (MCC), dibasic calcium phosphate anhydrous (DCPA), sodium starch glycolate (SSG),

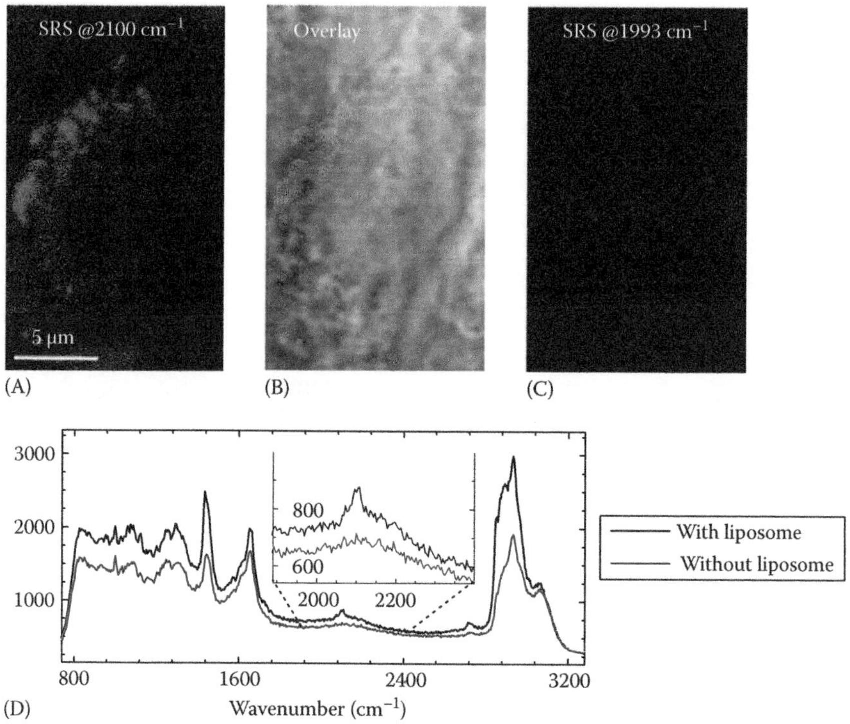


**FIGURE 23.3** SRS imaging of 100 nm deuterated liposomes in CHO-β cell. (A) SRS image of liposomes at 2100 cm$^{-1}$ corresponding to C–D stretching vibration (green). (B) Transmission image (gray) overlaid with SRS image (green). (C) SRS image at 1993 cm$^{-1}$. (D) Confocal Raman spectra from cytoplasm with liposomes (black) and without liposomes (red). Inset: Spectral region around C–D vibration.

and magnesium stearate (MS). The C–C stretching band of AB at 1650 cm$^{-1}$ and P–O stretching band of DCPA at 985 cm$^{-1}$ were directly used for SRS imaging. Since MCC, SSG, and MS have overlapping bands at the C–H stretching region around 2900 cm$^{-1}$, the SRS imaging was performed based on the Raman intensities. Compared with the image at 2900 cm$^{-1}$, the image obtained at 2850 cm$^{-1}$ had 1.4 times higher signal of MS, 2.3 times lower signal of MCC and 5.5 times lower signal of SSG. To separate the signal of SSG and MS, they divided pixel intensity of SRS image at 2900 cm$^{-1}$ by 2.3 and subtracted it from the SRS image at 2850 cm$^{-1}$. With this method, the image had zero intensity at the position of MCC, negative intensity at the position of SSG and positive intensity at the position of MS. SSG was also imaged at 840 cm$^{-1}$ which was isolated from other components. With a 60× water immersion objective, the lateral resolution is about 0.62 μm. A small penetration depth of 10 μm was characterized by depth scanning with 60× water immersion objective. The small penetration depth indicates that multiple particles in the solid dosage forms effectively scatter photons, and these backscattered photons are effectively collected in the epi detection by the same objective, producing a strong epi-SRS signal.

With the method characterized earlier, large area imaging was performed with lower numerical aperture objective (10× objective). Tablets from six different manufactures were imaged (Figure 23.4). No obvious variability of drug particle distribution was observed among the six tablets. On the other hand, the amount and distribution of excipients are different among these companies. Even tablets made from the same

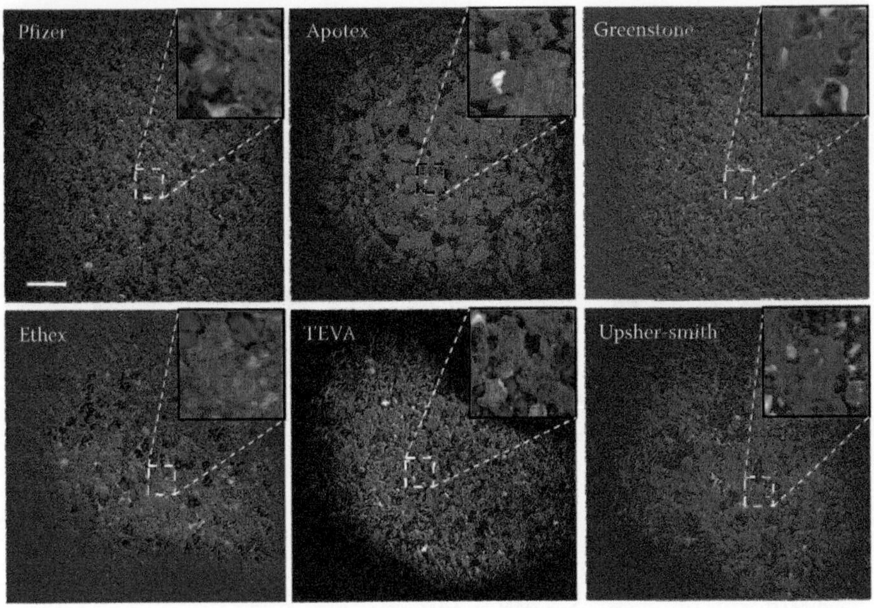


**FIGURE 23.4** Large area SRS imaging of tablets from different manufacturers. MCC, DCPA, AB, SSG and MS were indicated by green, blue, red, yellow/orange, and magenta, respectively. In case of tablet from Apotex the yellow color corresponds to lactose monohydrate and corn starch (Lactose monohydrate and corn starch were not able to be distinguished due to similar Raman intensity ratio between 2900 and 2850 $cm^{-1}$. Lactose has a distinct peak at 356 $cm^{-1}$ that could be used for SRS imaging). Inset: four times magnified areas of the images indicated by dashed squares. Images were acquired with the 10× objective. The power of pump and Stokes beams at the sample was 20 and 30 mW, respectively. Length bar = 200 μm. (From Slipchenko, M.N., Chen, H., Ely, D.R. et al., Vibrational imaging of tablets by epi-detected stimulated Raman scattering microscopy, *Analyst*, 135, 2613–2619, 2010. Reproduced by permission of The Royal Society of Chemistry.)

chemical components demonstrated different distribution and amount of main excipients. For example, tablet from Pfizer contained the same amount of cellulose and dibasic in the form of small well-mixed particles. The tablet from Greenstone had more cellulose and less disbasic particles of about the same size as in Pfizer formulation. The tablet from Ethex had larger particles of main excipients compared to Pfizer tablet. Cellulose and dibasic particles were poorly mixed in the tablet from Upsher-Smith. In addition, SRS imaging provides high speed and high quality data which can be used for statistical analysis of particle sizes and distribution patterns in solid dosage forms.

These results demonstrated the capability of SRS microscopy in high-speed and high-resolution evaluation of API and excipients in the pharmaceutical tablets in terms of the particle size, structural integrity, and homogeneity. SRS microscopy with the submicron resolution and high imaging speed provides a potential technique for visualizing low-dose blend uniformity and dynamical studies of material degradation and drug release.

### 23.3.3 Drug Penetration through Skin

Efficient drug delivery to the skin is essential for the treatment of major dermatology disease, such as eczema, psoriasis, and acne. A profound understanding of how drug transports, including the rate and extent of absorption into and through the skin barrier, is desired to design the delivery system. Currently, adhesive tape-stripping is a widely used method to produce the drug's concentration profile across the membrane, in which the outermost layer of skin (the stratum corneum) is progressively removed and chemically analyzed (Herkenne et al. 2008). However, the measurement lacks sufficient spatiotemporal resolution. Most recently, Xie group demonstrated SRS microscopy as an ideal technique enabling label-free, nondestructive, real-time measurements in skin with high spatial 3D resolution (Saar et al. 2011).

In this work, two nonsteroidal anti-inflammatory drugs, ibuprofen and ketoprofen, were analyzed by a ps SRS system. Ketoprofen was dissolved in deuterated propylene glycol (PG) and deuterated ibuprofen was dissolved in PG. The drug solution was applied *ex vivo* to an excised piece of mouse ear for SRS imaging. Ketoprofen had a strong vibrational resonance from aromatic CH bond stretching at $\sim$1599 $cm^{-1}$, while deuterated ibuprofen and PG created unique vibrational resonances of CD bond at $\sim$2120 $cm^{-1}$. Both resonances were well distinguished from the fingerprint emanating from skin lipids. With these identified peaks, the penetration of drugs and PG through skin, especially the stratum corneum, was measured overtime by SRS imaging. With a high spatial 3D resolution, SRS images showed both drugs penetrate via the intercellular lipids of the stratum corneum and through the hair shafts. Based on these images, the concentration profile of drugs and PG at a different depth of skin overtime and chemical uptake kinetics across the stratum corneum were quantitatively analyzed. The concentration profiles showed a monotonical decrease from skin surface into the barrier for PG and drugs, while drugs exhibited a slower rate, consistent with the results obtained by more invasive and labor-intensive tape-stripping method. Because the cosolvent penetrated faster than the drug, the concentration of drug increased above its saturation solubility. The precipitation of drug crystals (ibuprofen) on the skin surface was directly visualized at the CD stretching frequency less than 30 min post-application. The 3D SRS imaging provided a new and direct proof of the so-called metamorphosis (Santos et al. 2011) of a topical drug formulation post-application to the skin.

In addition, the 3D imaging capability provided novel mechanistic insight into the topical drug delivery process, which is impossible for Raman or tape-stripping measurements. By SRS imaging and analyzing the penetration profile via hair shaft and via stratum corneum intracellular lipids overtime, PG penetration via the hair shaft was found to be rapid and attain steady-state before the first measurement at 26 min post-application of formulation. In contrast, the transport through the stratum corneum intercellular lipids was slower and increased over time during the 2 h experiment period, representing a non-steady state approaching to a constant flux. The data provided a direct experimental evidence of the idea deduced by Scheuplein 40 years ago (Scheuplein 1965) that the initial drug molecules crossing the skin came through low-resistance, shunt pathways (e.g., hair shafts) of limited capacity; the parallel but slower, transport across the bulk of the stratum corneum eventually overwhelms the transient pathways and completely dominates at steady state.

In conclusion, SRS microscopy revealed previously invisible features of topical drug transport including the 3D penetration profiles and uptake kinetics of ketoprofen and ibuprofen, distinct penetration pathways and different transport rates, and direct observation of drug crystals on the skin surface. SRS microscopy provided a nondestructive analysis of dermato-pharamacokinetics and real-time visualization of the drug transport process with a high 3D spatial resolution. In the future, the improved image acquisition speed, signal processing and analysis, especially in terms of defining more precisely the location of the skin surface, will allow the extraction of further quantitative information. The technology has great potential to be applied *in vivo* in models more relevant than the mouse skin in this work to understand and optimize formulations and delivery system for topical and transdermal drug.

## 23.4 Summary

With chemical selectivity and high imaging speed, and high 3D spatial resolution, CRS microscopy is powerful for characterization of drug delivery systems. CRS microscopy has shown great potential in tracking cellular uptake of nanocarriers, mapping drug distribution in tissues, and monitoring drug release from carriers such as polymer films and solid tablets. Continuing efforts in the development of hyperspectral CRS imaging will be critical for quantitative mapping of drug molecules in cells and tissues.

## References


Aaltonen, J., P. Heinänen, L. Peltonen et al. 2006. In situ measurement of solvent-mediated phase transformations during dissolution testing. *J. Pharm. Sci.* 95 (12): 2730–2737.

Belu, A.M., M.C. Davies, J.M. Newton, and N. Patel. 2000. Tof-sims characterization and imaging of controlled-release drug delivery systems. *Anal. Chem.* 72 (22): 5625–5638.

Chen, H., S. Kim, L. Li et al. 2008. Release of hydrophobic molecules from polymer micelles into cell membranes revealed by Forster resonance energy transfer imaging. *Proc. Natl Acad. Sci. USA* 105 (18): 6596–6601.

Csiszar, A., N. Hersch, S. Dieluweit et al. 2010. Novel fusogenic liposomes for fluorescent cell labeling and membrane modification. *Bioconj. Chem.* 21 (3): 537–543.

Debnath, S., P. Predecki, and R. Suryanarayanan. 2004. Use of glancing angle x-ray powder diffractometry to depth-profile phase transformations during dissolution of indomethacin and theophylline tablets. *Pharmacol. Res.* 21 (1): 149–159.

Herkenne, C., I. Alberti, A. Naik et al. 2008. In vivo methods for the assessment of topical drug bioavailability. *Pharm. Res.* 25 (1): 87–103.

Jarvis, L.M. 2009. Delivering the promise. *C&E News J.* 87: 18–27.

Kang, E., J. Robinson, K. Park, and J.-X. Cheng. 2007. Paclitaxel distribution in poly(ethylene glycol)/poly(lactide-co-glycolic acid) blends and its release visualized by coherent anti-Stokes Raman scattering microscopy. *J. Control Release* 122 (3): 261–268.

Kang, E., H. Wang, I.K. Kwon et al. 2006. In situ visualization of paclitaxel distribution and release by coherent anti-Stokes Raman scattering microscopy. *Anal. Chem.* 78 (23): 8036–8043.

Klueva, O., O. Olkhovyk, and R.J. Priore. 2010. Pharma applications of Raman chemical imaging. *Innovations in Pharmaceutical Technology*, pp. 34–37.

Levin, I. and R. Bhargava. 2005. Fourier transform infrared vibrational spectroscopic imaging: Integrating microscopy and molecular recognition. *Annu. Rev. Phys. Chem.* 56: 429–474.

Panyam, J. and V. Labhasetwar. 2003. Dynamics of endocytosis and exocytosis of poly(D,L-lactide-co-glycolide) nanoparticles in vascular smooth muscle cells. *Pharmacol. Res.* 20 (2): 212–220.

Pietzonka, P., B. Rothen-Rutishauser, P. Langguth et al. 2002. Transfer of lipophilic markers from PLGA and polystyrene nanoparticles to caco-2 monolayers mimics particle uptake. *Pharmacol. Res.* 19 (5): 595–601.

Ranade, V.V. 2010. Pharmaceutical applications of Raman spectroscopy. *Am. J. Ther.* 17 (1): 121.

Reich, G. 2005. Near-infrared spectroscopy and imaging: Basic principles and pharmaceutical applications. *Adv. Drug Deliv. Rev.* 57 (8): 1109–1143.

Saar, B.G., L.R. Contreras-Rojas, X.S. Xie, and R.H. Guy. 2011. Imaging drug delivery to skin with stimulated Raman scattering microscopy. *Mol. Pharm.* 8 (3): 969–975.

Sahoo, S.K., J. Panyam, S. Prabha, and V. Labhasetwar. 2002. Residual polyvinyl alcohol associated with poly (D,L-lactide-co-glycolide) nanoparticles affects their physical properties and cellular uptake. *J. Control Release* 82 (1): 105–114.

Sanhai, W.R., J.H. Sakamoto, R. Canady, and M. Ferrari. 2008. Seven challenges for nanomedicine. *Nat. Nanotechnol.* 3: 242–244.

Santos, P., A.C. Watkinson, J. Hadgraft, and M.E. Lane. 2011. Enhanced permeation of fentanyl from supersaturated solutions in a model membrane. *Int. J. Pharm.* 407 (1–2): 72–77.

Scheuplein, R.J. 1965. Mechanism of percutaneous adsorption. I. Routes of penetration and the influence of solubility. *J. Investig. Dermatol.* 45: 334–336.

Slipchenko, M.N., H. Chen, D.R. Ely et al. 2010. Vibrational imaging of tablets by epi-detected stimulated Raman scattering microscopy. *Analyst* 135: 2613–2619.

Slobodan, S. 2007. An in-depth analysis of Raman and near-infrared chemical images of common pharmaceutical tablets. *Appl. Spectrosc.* 61 (3): 239–250.

Szeto, H.H., P.W. Schiller, K. Zhao, and G. Luo. 2005. Fluorescent dyes alter intracellular targeting and function of cell-penetrating tetrapeptides. *FASEB J.* 19 (1): 118–120.

Tong, L., Y. Lu, R.J. Lee, and J.-X. Cheng. 2007. Imaging receptor-mediated endocytosis with a polymeric nanoparticle-based coherent anti-Stokes Raman scattering probe. *J. Phys. Chem. B* 111 (33): 9980–9985.

Torchilin, V.P. 2005. Recent advances with liposomes as pharmaceutical carriers. *Nat. Rev. Drug Discov.* 4 (2): 145–160.

Vicari, L., T. Musumeci, I. Giannone et al. 2008. Paclitaxel loading in PLGA nanospheres affected the in vitro drug cell accumulation and antiproliferative activity. *BMC Cancer* 8 (1): 212.

Windbergs, M., M. Jurna, H.L. Offerhaus et al. 2009. Chemical imaging of oral solid dosage forms and changes upon dissolution using coherent anti-Stokes Raman scattering microscopy. *Anal. Chem.* 81 (6): 2085–2091.

Wu, S. and N. Mcmillan. 2009. Lipidic systems for in vivo siRNA delivery. *The AAPS J.* 11 (4): 639–652.

Xu, P., E. Gullotti, L. Tong et al. 2008. Intracellular drug delivery by poly(lactic-co-glycolic acid) nanoparticles, revisited. *Mol. Pharmcol.* 6 (1): 190–201.

Zeitler, J.A., P.F. Taday, D.A. Newnham et al. 2007. Terahertz pulsed spectroscopy and imaging in the pharmaceutical setting—A review. *J. Pharm. Pharmacol.* 59 (2): 209–223.

# 24. Applications of Stimulated Raman Scattering Microscopy

Christian Freudiger, Daniel A. Orringer, and Xiaoliang Sunney Xie



As discussed in Chapter 4, stimulated Raman scattering (SRS) microscopy allows fast, three-dimensional imaging with chemical contrast based on intrinsic spectroscopic properties of the sample (Ploetz et al. 2007, Freudiger et al. 2008, Ozeki et al. 2009, Nandakumar et al. 2009, Saar et al. 2010a). It has superseded the older coherent Raman microscopy technique, coherent anti-Stokes Raman scattering (CARS) in that

- SRS is free from non-resonant background
- SRS has much improved sensitivity
- SRS does not suffer from spectral distortions
- SRS signal is linear in the concentration of the target species
- SRS does not have imaging artifacts due to phase-matching

This chapter summarizes important applications of SRS microscopy that capitalize on both high sensitivity and straightforward quantification of SRS. While CARS microscopy has often been criticized as being mainly useful for lipid imaging, SRS has allowed us to image a variety of biomolecules. Table 24.1 summarizes the growing number of molecular species that have been imaged with SRS microscopy thus far.

*Coherent Raman Scattering Microscopy*. Edited by Ji-Xin Cheng and X. Sunney Xie © 2013 CRC Press/Taylor & Francis Group, LLC. ISBN: 978-1-4398-6765-5.

**Table 24.1** Biomolecules Imaged with SRS Microscopy

| Mode | Raman Shift ($cm^{-1}$) | Compound | Reference |
|---|---|---|---|
| O–H | 3250 | Water | Saar et al. (2010a) and Roeffaers et al. (2010) |
| Aromatic C–H | 3060 | Phenylalanine | Roeffaers et al. (2010) |
| Alkene C–H | 3015 | Unsaturated lipids | Freudiger et al. (2008) and Roeffaers et al. (2010) |
| C–$H_3$ | 2950 | Proteins, lipids | Freudiger et al. (2008), Ozeki et al. (2009), Slipchenko et al. (2009), and Freudiger et al. (2011) |
| C–$H_2$ (asym) | 2885 | (Gel phase) lipids | Freudiger et al. (2011) |
| C–$H_2$ (sym) | 2845 | Lipids | Freudiger et al. (2008), Slipchenko et al. (2009), and Freudiger et al. (2011) |
| C–$D_3$ | 2120 | Deuterium labels | Saar et al. (2010a) |
| Amid I | 1660 | Protein | Nandakumar et al. (2009) and Slipchenko et al. (2009) |
| C=C | 1660 | Unsaturated lipids | Nandakumar et al. (2009) and Slipchenko et al. (2009) |
| Aromatic ring | 1600 | Lignin | Saar et al. (2010b) |
| Conjugated C=C | 1600 | Retinol, retinoic acid | Freudiger et al. (2008) |
| C–O | 1100 | Cellulose | Saar et al. (2010b) |
| O–P–O | 1090 | Nucleic acid | Zhang et al. (2011) |
| Phenyl ring | 1005 | Phenylalanine | Roeffaers et al. (2010) |
| P–$O_4^{3-}$ | 985 | Inorganic phosphate | Slipchenko et al. (2010) |
| O–P–O | 790 | Nucleic acid | Zhang et al. (2011) |
| S=O | 670 | DMSO | Freudiger et al. (2008) |

*Source:* Compiled from multiple sources.

# 24.1 Applications in Biomedical Research

## 24.1.1 Metabolism and Lipid Storage

Obesity is a major risk factor for chronic diseases, such as type 2 diabetes, cardiovascular disease, and hypertension. In order to prevent and treat obesity and its associated metabolic disorders, it is necessary to understand the regulatory mechanisms of fat accumulation and distribution at both cellular and organism levels. Although lipid-staining dyes exist, they do not work reliably (Hellerer et al. 2007, Wang et al. 2011) and they often cannot distinguish different types of lipids. In this section, we discuss how SRS imaging can be applied to study lipid metabolism in cells and organisms.

Raman spectra of different types of fatty acids are distinct (Figure 24.1A). While all fatty acids are visible based on the $CH_2$-stretching vibration at 2845 $cm^{-1}$, unsaturated

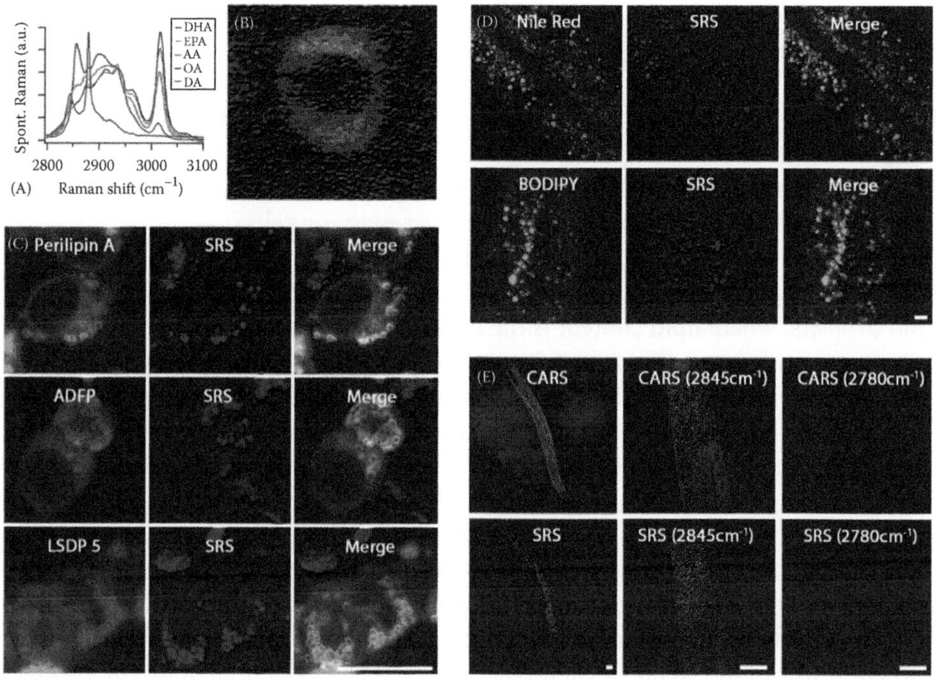


**FIGURE 24.1** Imaging and quantifying lipids with SRS microscopy. (A) Spontaneous Raman spectra of docosahexaenoic acid (DHA, with six C=C bonds), eicosapentaenoic acid (EPA, with five C=C bonds), arachidonic acid (AA, with four C=C bonds), oleic acid (OA, with a single C=C bond), and docosanoic acid (DA, saturated). The strong Raman peak around 3015 $cm^{-1}$ is characteristic of unsaturated fatty acids. The spectrum of DA is distinct from the others as it occurs in gel phase rather than liquid phase at room temperature. (B) SRS image of a giant unilamellar vesicle (GUV) to demonstrate high sensitivity of SRS for a single lipid bilayer. (C) Comparison of SRS lipid images at 2845 $cm^{-1}$ with fluorescences images of YFP labeled PAT family proteins in living cells. In the merged images, the proteins are seen to coat the lipid droplets as imaged with SRS microscopy. Scale: 10 μm. (D) Comparison of SRS lipid images at 2845 $cm^{-1}$ with lipid staining dyes Nile Red and BODIPY in *C. elegans*. Only partial overlap is seen in the merged images, suggesting that lipid dyes are not quantitative in larger organisms. Scale: 20 μm. (E) Comparison of SRS and CARS images at various magnifications and on (left and center column) and off resonance (right column) with the CH-stretching vibration at 2845 $cm^{-1}$. Scale: 50 μm. (Modified from Wang, M.C., et al., *Nature Methods*, 8, 135–138, 2011. http://www.nature.com/nmeth/journal/v8/n2/abs/nmeth.1556.html)

fatty acids exhibit a unique Raman band at 3015 $cm^{-1}$, attributable to the stretching mode of =C–H bond associated with C=C double bonds (Heinrich et al. 2008). The relative intensity of this 3015 $cm^{-1}$ mode is approximately proportional to the number of C=C double bonds in the lipid molecule. This isolated band can be selectively probed with narrowband SRS imaging (Freudiger et al. 2008). With multiplexed multi-band SRS imaging (Freudiger et al. 2011), it is also possible to probe the spectral shape of the complete CH-stretching region. In doing so we were able to distinguish between gel- and solid-phase lipids, and quantify cholesterol in the presence of other lipids. In summary, SRS microscopy is a powerful tool to probe various lipid types with high temporal and three-dimensional spatial resolution.

Figure 24.1B shows a lipid image of a giant unilamellar vesicle (GUV). This demonstrates that the sensitivity of fast imaging with SRS is sufficient to detect a single lipid bilayer based on intrinsic chemical contrast. Rather than a complete circle, SRS shows an inhomogeneous signal distribution because of a polarization effect excitation lasers with respect to the orientation of the C–C fatty acid chains, which are radially aligned at the top and bottom of the and tangentially at the sides. This effect can be exploited for orientation sensitive measurements or minimized by using circularly polarized beams.

To understand the SRS image contrast (Wang et al. 2011), we compare SRS images at 2845 $cm^{-1}$ with fluorescence microscopy images in living cells (Figure 24.1C). We tagged YFP to the three PAT family proteins Perilipin A, ADFP, and LSDP5, which are known to be lipid droplet associated proteins. Figure 24.1C confirms that SRS indeed highlights lipid droplets, where lipid content is the highest (Wang et al. 2011).

Figure 24.1D shows the comparison of SRS images with the two lipid staining dyes Nile Red and BODIPY. While these dyes work reasonably well in living cells, they are known for artifacts in more complex organism such as *Caenorhabditis elegans*, a common multicellular model organism for lipid biology (Mullaney and Ashrafi 2009). With SRS, lipid droplets can be observed throughout the entire organism. The signal is the strongest in the intestine, the primary site for lipid storage. Weaker signal originates from the hypodermis, oocytes, and early stage embryos (Wang et al. 2011). Nile Red fluorescence in *C. elegans* does not overlap with the SRS lipid signal (Figure 24.1D, top row) and localizes with granules of the gut. We further found that BODIPY, which is known to work more reliably than Nile Red in cells, highlights both gut granules and lipid droplets (Figure 24.1D, bottom row). This suggests that existing dyes are not suitable for imaging and quantifying lipid concentrations in textit *C. elegans* (Hellerer et al. 2007, Wang et al. 2011).

Finally we also compared the performance of SRS and the older technique CARS (Wang et al. 2011). While quantifying lipids based on the morphology of lipid droplets is possible with CARS (Hellerer et al. 2007), SRS allows for direct quantifications of signal levels due to the lack of the non-resonant background, a linear dependence of the signal on the concentration, and no spatial image artifacts due to phase-matching (see Chapter 4). Figure 24.1E shows the comparison of CARS and SRS image at different length scales (left and center row) and different Raman shifts (center and right row). The unspecific non-resonant background is clearly visible in the images and does not disappear when the difference frequency is tuned off the lipid vibration. Specifically, strong non-resonant CARS signal is generated from the proteins of the dermis, which is not observed in SRS (Figure 24.1E, center row). As such, SRS microscopy is the ideal candidate for imaging and quantifying lipids in living organisms (Wang et al. 2011).

We used SRS microscopy to perform an RNA interference screen to identify genes which regulate the total fat accumulation in *C. elegans* (Wang et al. 2011). Such screening is possible because the SRS signal from lipids correlates very well with traditional biochemical assays. Figure 24.2A shows the SRS lipid signal in wild-type *C. elegans* as well as two genotypes which are known to have up- (daf-2 mutant) and down-regulated lipid storage (lipase transgenic strain), respectively. In comparison with thin-layer/gas chromatography from 5000 worms (Figure 24.2B), the relative SRS signal from these mutants correlates well. The major advantage of SRS is that signal can be measured *in vivo* and quantified from 3 to 5 single worms, which has allowed us to perform RNAi screening of 272 genes (Wang et al. 2011). This screen identified nine genetic regulators of fat storage, whose inactivation increased fat content more than 25% (Figure 24.2C).

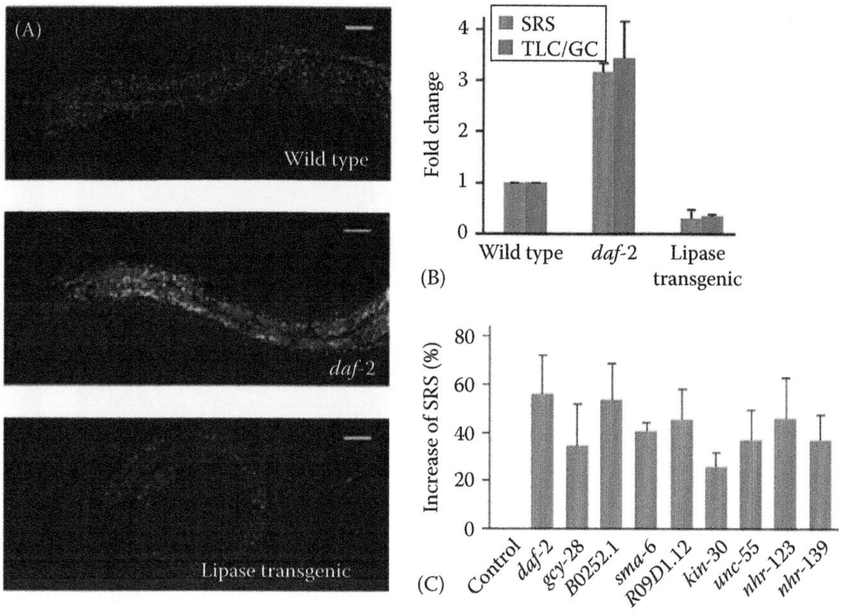


**FIGURE 24.2** RNAi Screen for lipid storage regulating genes. (A) Low resolution SRS images of *C. elegans* at 2845 cm$^{-1}$ allow quantification of the absolute lipid content of a worm. Images are from a wild type, a daf-2 mutant, and a lipase transgenic worm, respectively. (B) Results agree well with classical biochemistry methods (thin layer chromatography [TLC] coupled with gas chromatography), but can be acquired from a single worm rather than thousands for the classical methods. (C) This has allowed us to conduct an RNAi screen, which has identified nine previously unknown genes that regulate lipid storage. (Modified from Wang, M.C., et al., *Nature Methods*, 8, 135–138, 2011. http://www.nature.com/nmeth/journal/v8/n2/abs/nmeth.1556.html)

## 24.1.2 Biomass Conversion

Research into alternative energy has experienced dramatic growth in recent years, which was motivated by both the environmental impact of current fossil fuels, and by the unstable and uncertain sources of oil and natural gas (Herrera 2006). Under ideal conditions, currently unused plant materials, such as agricultural residues, forestry wastes, and energy crops, can be broken down by a series of chemical, enzymatic, and/or microbiological processes into ethanol or other biofuel sources (Kumar et al. 2009). The major challenge to be overcome in the widespread adoption of many biofuels is that biomass is intrinsically recalcitrant, making conversion into usable fuels inefficient.

Saar et al. used narrowband SRS microscopy to image the distribution of the two major components of plant samples: lignin (Figure 24.3A), which is responsible for recalcitrance; and polysaccharides (Figure 24.3B), which can be converted into fuels. These species have isolated vibrations at 1600 and 1100 cm$^{-1}$, respectively (Saar et al. 2010b).

Taking advantage of the high signal levels in SRS, images of plant materials could be recorded with contrast based on the Raman spectrum, but with orders of magnitude higher temporal resolution. Whereas previous lignin and cellulose images recorded with spontaneous Raman spectroscopy were obtained after hours, SRS images could be obtained in less than 1 min/frame. This allowed Saar et al. to record chemical movies of a degradation reaction in plant tissue as it occurred, and to process those movies

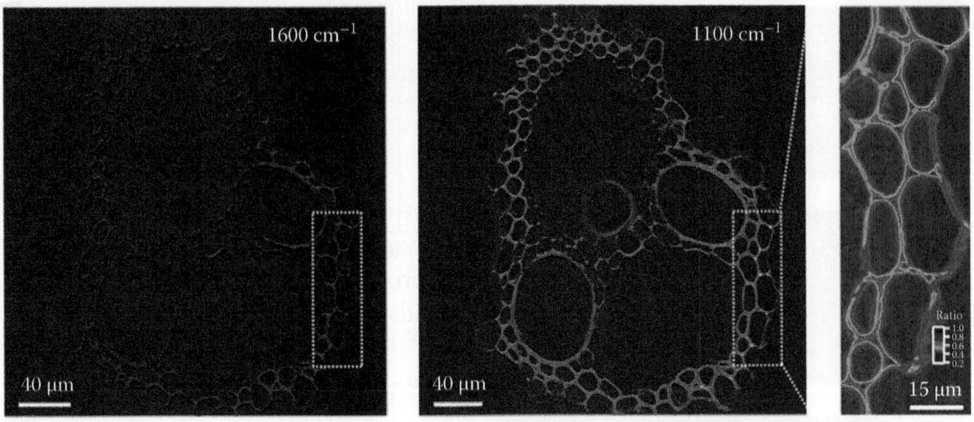


**FIGURE 24.3** Imaging of lignin and cellulose with SRS microscopy. (A) SRS image of a vascular bundle in a sample of corn stover at 1600 $cm^{-1}$ showing the lignin distribution. (B) SRS image of the same vascular bundle as in (A) showing the cellulose distribution at 1100 $cm^{-1}$. These images can be acquired simultaneously using the two-color SRS imaging. (C) Ratio of the lignin divided by the cellulose signal at higher magnification, obtained from the region surrounded by the dotted line in (A) and (B). This image highlights areas of particularly high lignin content, which appears primarily at the cell corners. (From Saar, B.G. et al., *Angew. Chem.*, 122, 5608–5611, 2010b. With permission.)

to obtain a spatially resolved reaction rate map, information that was not accessible to spontaneous Raman scattering because of its lower temporal resolution. Based on imaging, optimized lignin degradation procedures can be developed in subsequent studies.

### 24.1.3 Label-Free Cell and Tissue Imaging

One important application of SRS microscopy is label-free imaging of cells and tissue, because many stains are impossible to apply *in vivo*, or are toxic when applied. Other label-free optical techniques, such as confocal reflection (CR), two-photon excited autofluorescence (TPAF), second harmonic generation (SHG), third harmonic generation (THG) microscopy, and optical coherence tomography (OCT), have been suggested for this purpose (Tearney et al. 1997, Tai et al. 2005, Konig et al. 2007, Master and So 2009), but either do not have chemical specificity (CR, THG and OCT), or are limited to a few specific chemical species (TPAF and SHG) (Campagnola et al. 2002, Zipfel et al. 2003, Li et al. 2010).

SRS microscopy can highlight all structural components of tissue such as lipids, protein, and water based on $CH_2$–, $CH_3$– and OH-vibrations (Freudiger et al. 2008). We have also demonstrated *in vivo* skin imaging in mice and even humans (Saar et al. 2010a).

A challenge in tissue imaging was mapping of distributions of DNA based on the specific O–P–O-vibrations of the DNA backbone. Figure 24.4 demonstrates that DNA can be imaged specifically with SRS microscopy (Zhang et al. 2011). These results have inspired us to evaluate SRS microscopy as a tool for medical diagnostics of the central nervous system, where tissue biopsy carries major risks and label-free imaging is a necessity. Details will be discussed in a later section of this chapter.

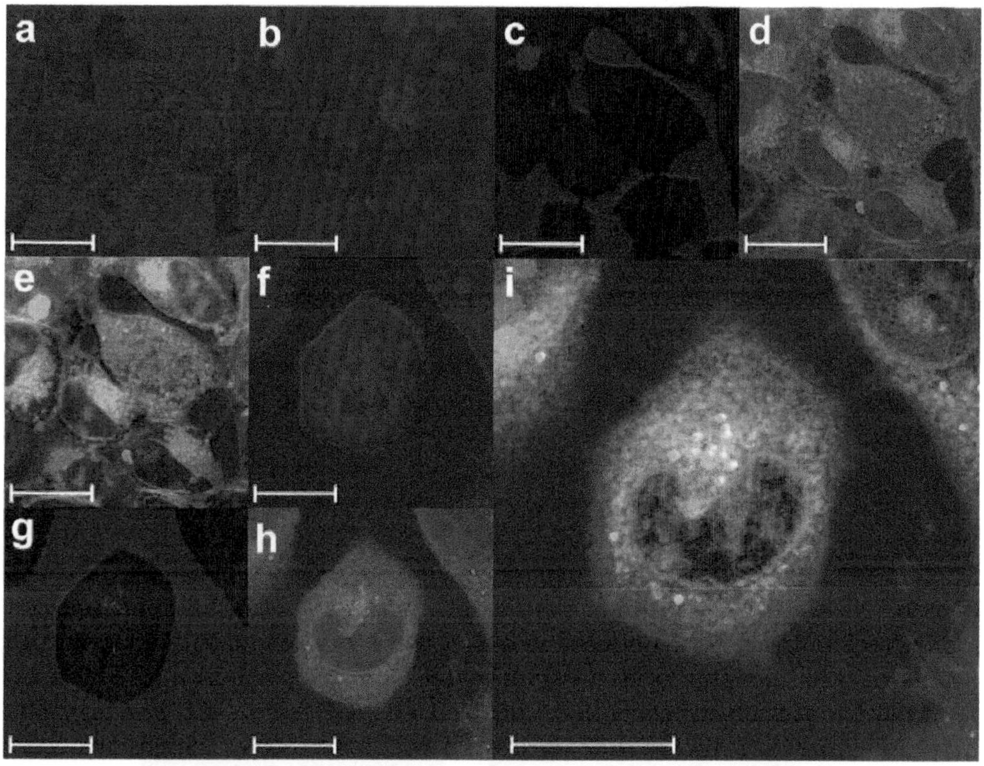


FIGURE 24.4 Label-free imaging of DNA. (A–E) SRS images of human cells (HEK-293) at (A) 790 $cm^{-1}$, (B) 1085 $cm^{-1}$, (C) 1000 $cm^{-1}$, (D) 2845 $cm^{-1}$ and (E) multicolor overlay of (A), (C), (D). (F–I) MCF-7 cells imaged at (F) 1085 $cm^{-1}$, (G) 1650 $cm^{-1}$, (H) 2845 $cm^{-1}$, and (I) overlay of (F–H). Scale: 20 μm. (From Zhang, X. et al., *ChemPhysChem*, 13(4):1054, 2012.)

## 24.1.4 Trans-Dermal Drug Delivery

Another important application of SRS microscopy is imaging drug delivery into skin. This has the potential to aid in developing formulations for topically applied drugs that enhance active ingredient penetration kinetics. Fluorescent labeling techniques cannot be used for this purpose because the labels themselves are usually much larger than small drug molecules and may thus perturb the transport properties under study. Confocal spontaneous Raman micro-spectroscopy can be used to obtain longitudinal penetration profiles without perturbative labels, but the lateral distribution is not available due to the long pixel dwell times (Caspers et al. 2003). Additionally, information from label-free tissue imaging can be used to show the distribution of drugs with respect to important tissue structures, such as hair ducts or the boundary between the *stratum corneum* and the viable epidermis.

This section shows the mapping of two compounds, dimethyl sulfoxide (DMSO), a skin penetration enhancer (Caspers et al. 2003), and retinoic acid (RA), which is used to treat acne, wrinkles, photo-aging, and acute promyelocytic leukemia (Van De Kerkhof et al. 2006). Figure 24.5A shows the Raman spectra of the compounds with vibrational resonances at 670 $cm^{-1}$ (DMSO) and 1570 $cm^{-1}$ (retinoic acid) that are isolated from typical skin spectra.

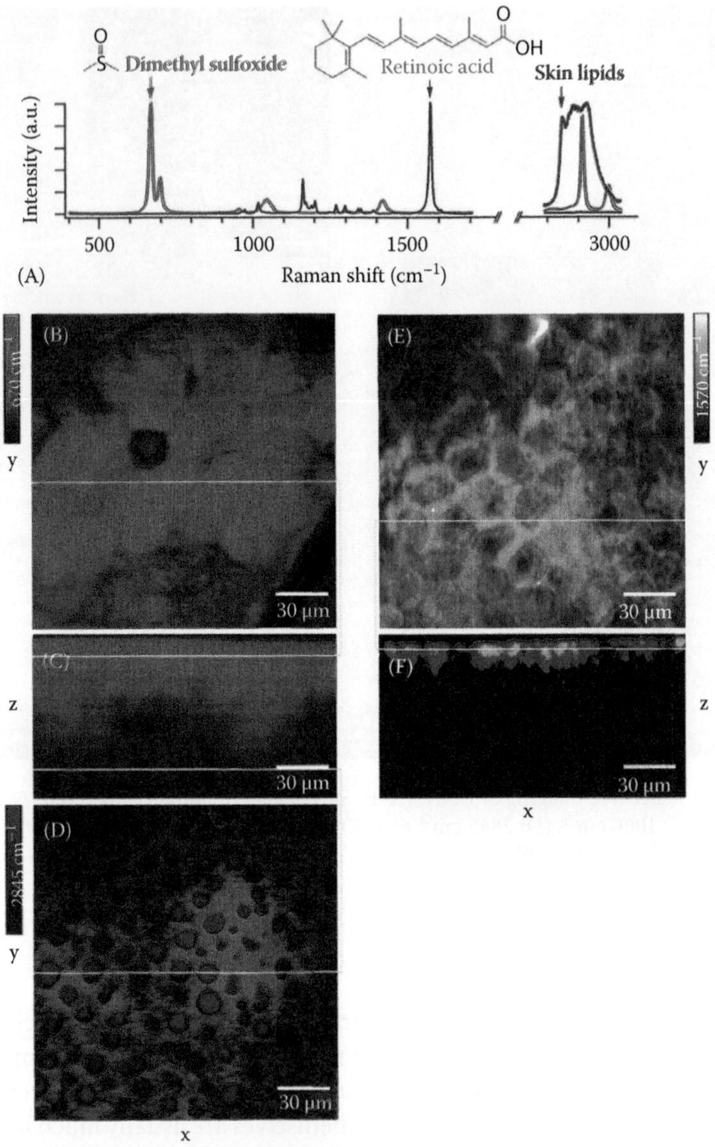


**FIGURE 24.5** Monitoring drug delivery into fresh murine skin by SRS microscopy. (A) Raman spectra of dimethyl sulfoxide (DMSO, green), retinoic acid (RA, blue), and typical lipids in murine skin (red). (B) Top view of the penetrated DMSO (green) in the *stratum corneum* imaged at 670 cm$^{-1}$ with SRS. (C) SRS DMSO depth profile through the line in (B). (D) Simultaneous two-color SRL image of DMSO (670 cm$^{-1}$, green) and lipid (2845 cm$^{-1}$, red) in the subcutaneous fat layer at a depth of ~65 μm through the lower line in (C). (E) Top view of the penetrated RA (blue) in the *stratum corneum* imaged at 1570 cm$^{-1}$ with SRS. (F) SRS RA depth profile through the line in (E). (From Saar, B.C. et al., *Sci. Mag.*, 330(6009), 1368, 2010.)

As a hydrophilic molecule, DMSO penetrates the skin via the protein phase and the DMSO image in the *stratum corneum* (Figure 24.5B) shows inverse contrast compared to the lipid image. A depth profile shows detectable DMSO over more than 60 μm (Figure 24.5C). The hydrophilic interaction with the tissue is confirmed in the subcutaneous fat layer. Simultaneous two-color imaging of lipid and DMSO allows us to show that the DMSO is insoluble in the lipid structures (Figure 24.5D). In contrast, retinoic acid, which is a hydrophobic molecule, penetrates via the lipid-rich intercellular spaces in the epidermis (Figure 24.5E and F). We found that penetration could be optimized by the use of a commercially available sonication device that is marketed for transdermal drug delivery in human patients (SonoPrep, Sontra Medical Corp., Franklin, MA).

For the initial study (Freudiger et al. 2008), mouse ear skin from white, wild-type mice was used *ex vivo*. With the development of *in vivo* SRS microscopy (Saar et al. 2010a) we repeated the retinol delivery experiment in living mice (Figure 24.6). In addition to the previously observed penetration through the lipid phase of the *stratum corneum ex vivo, in vivo* we observed that the drug also penetrates via the hair shaft into the sebaceous gland, a third hypothesized penetration route into skin (Prausnitz et al. 2004, Lademann et al. 2008). Averaging the three-dimensional data partially shown in Figure 24.6A through I, one can see the localization of retinol surrounding the hair and in the top of the sebaceous gland (Figure 24.6L). In the previous *ex vivo* work this pathway was not observed (Freudiger et al. 2008), indicating the importance of *in vivo* studies because the transport properties of small molecules can be affected by the skin temperature, moisture content, and other factors.

We were also able to acquire the first drug delivery study in a human volunteer (X.S.X) (Saar et al. 2010a). In this study, DMSO is also found to accumulate in the area surrounding the hair, though it does not completely penetrate into the hair itself. We used a general strategy for imaging of drug delivery: the hydrogen atoms in the drug molecule were replaced with the heavier isotope deuterium (D). This generated an isolated CD-stretching vibration around 2100 cm$^{-1}$ which falls into the "silent region" of Raman spectra of biological species and can thus be imaged with high sensitivity and specificity. This can be applied to any drug in which C–H bonds can be transformed into C–D groups, which generally has a minimal effect on the chemical properties of the molecule.

## 24.2 Application to Cancer Surgery and Diagnostics

### 24.2.1 Need for Optical Imaging in Cancer Surgery

Whenever possible, the treatment of solid tissue cancers begins with surgical resection. When tumor burden causes organ dysfunction, resection may be lifesaving. In addition, maximal cytoreduction diminishes the chance of recurrence, improves the efficacy of adjuvant therapies, and may ultimately prevent treatment failure.

Achieving maximal surgical resection relies on an accurate definition of tumor location and boundaries. Surgeons rely on cues such as color, texture, and vascularity to define the tumor margins. However, the gross borders of a tumor are often exceedingly difficult to judge macroscopically. In breast cancer surgery, achieving local tumor control through surgery has a proven survival benefit (Clarke et al. 2005). A variety of techniques have been developed to assist in the localization of tumors and definition of tumor margins. In the case of deep breast lesions, a wire, placed within an abnormality diagnosed on mammography, is used to guide surgeons toward a lesion. Because there are no reliable visual cues on which to base the borders of resection, the specimen is removed and subsequently then imaged with x-rays and traditional histopathologic methods to confirm whether the lesion has been removed. The process of wire-localized surgical biopsy is uncomfortable for the patient and often inefficient. While identifying cancer-containing breast

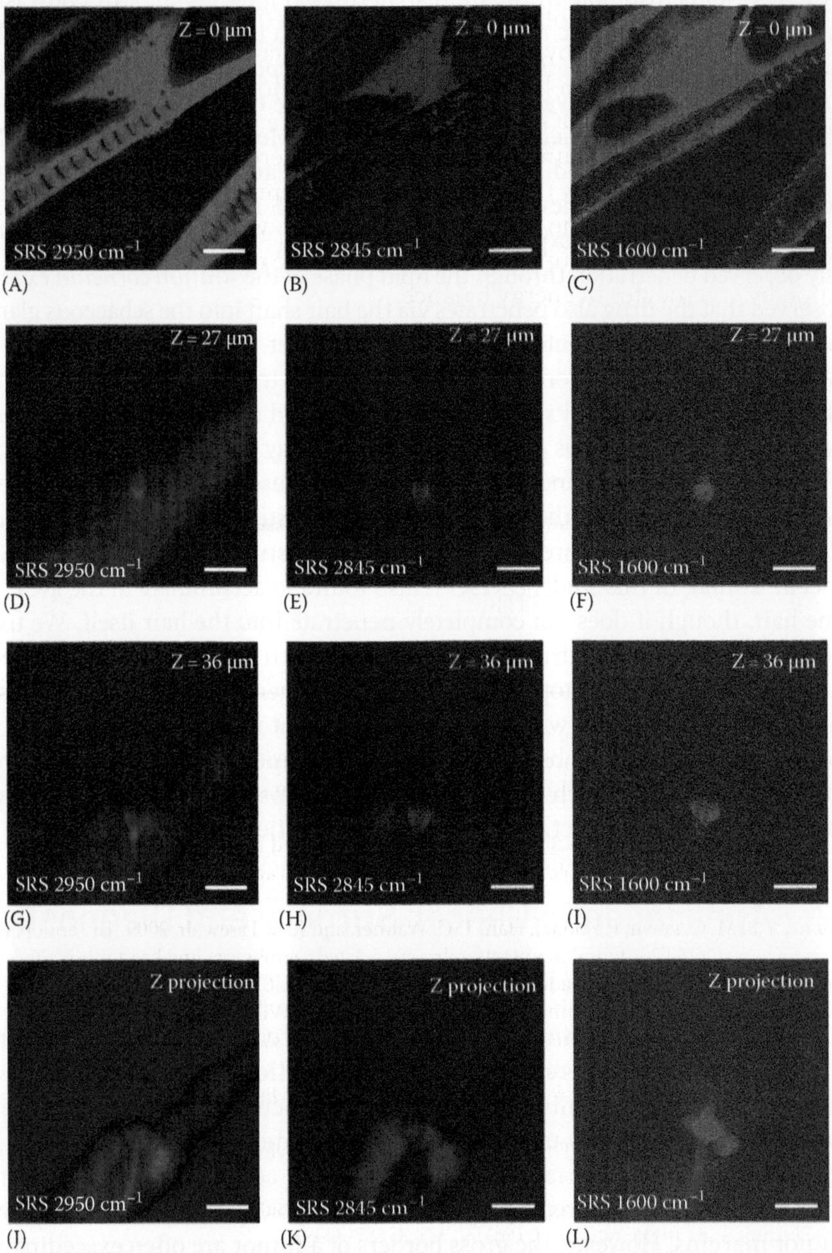


**FIGURE 24.6** *In vivo* imaging of drug penetration of retinol. (A–C) SRS images of hair and *stratum corneum* in the ear of a living mouse. (D–F) Images of a hair shaft within the viable epidermis. (G–I) Images of a sebaceous gland within the viable epidermis. (J–L) Depth projection of three-dimensional image stacks of the viable epidermis. Images are acquired at the indicated Raman shifts. Contrast coming primarily from protein (2950 $cm^{-1}$) and lipid (2845 $cm^{-1}$) shows the morphology of the skin with all its structural elements and sub-cellular resolution (see nuclei in D and K). Hairs are visible as solid structures in the protein image (D) and surrounded by oil secreted from the sebaceous glands in the lipid image (E). We are able to visualize with SRS that drug penetration of the topically applied retinol (C) occurs along the hair shaft (F, I, and L). Images are collected in transmission with 37 ms/frame acquisition speed and 512 × 512 pixel sampling. Scale: 25 μm.

tissue on a macroscopic level is often quite difficult, differentiating breast carcinoma from normal breast tissue is relatively straightforward on a microscopic level. If it were possible to microscopically examine a region corresponding to mammographic abnormality within the surgical field, surgical protocol for ensuring the best possible resection could be streamlined and become more uniformly effective in reducing tumor burden.

Brain tumor margins, even when they include the cortical surface, can also be impossible to distinguish by the naked eye. Yet, the degree of cytoreduction afforded by surgery is correlated closely with the patient's outcome (Berger 2010). Neurosurgeons integrate preoperative radiographic images, three-dimensional stereotactic navigation systems, intraoperative neurologic examination (during an awake craniotomy), and knowledge of the safe neuroanatomical corridors to determine a surgical strategy to safely maximize cytoreduction. Yet, an objective endpoint for determining when a resection is complete does not exist. Surgeons may leave regions of tumor behind at the time of surgery due to their inability to correctly identify tissue in the resection cavity as cancerous. They may also leave residual tumor behind after a resection if they feel important brain structures are at risk of being damaged. Conversely, an overly aggressive resection may damage functional areas of the brain resulting in neurologic deficit. Both under-resection and over-resection could be avoided if a technology was developed to enable surgeons to objectively define tumor margins. A microscopic examination of the surgical resection cavity that could be applied during an operation would allow surgeons to objectively define tumor margins by confirming the presence or absence of tumor cells.

Furthermore, microscopy techniques that enable the evaluation of tissue histology within the surgical field would have a major impact on the workflow and accuracy in the setting of cancer operations where specimens are taken to ensure negative histopathologic margins. The current surgical strategy for maximizing the extent of resection for many cancers, including breast (Sabel et al. 2009), skin (Veronesi et al. 1988), and prostate cancer (Valicenti et al. 2004), is often highly dependent on achieving negative surgical margins. Negative surgical margins are confirmed using traditional staining and light microscopy to evaluate the presence or absence of tumor cells in a given specimen. The process of procuring and preparing a specimen for margin analysis can be time-consuming and cumbersome to the operative flow. It is common for surgeons to wait up to 30 min in a busy surgical center for a biopsy result to become available. Often, intraoperative decisions are delayed during the waiting period for biopsy results. By bringing the capacity for histopathologic diagnosis to the surgical field, coherent Raman scattering techniques might dramatically reduce the logistical pitfalls that are commonplace in cancer operations where negative tissue margins must be achieved. By facilitating the process of tissue margin analysis, SRS might enable virtual biopsy of a greater number of specimens. Through improved analysis of tissue margins, SRS might help to reduce the risk of a false negative margin that could predispose to cancer recurrence.

### 24.2.2 Optical Biopsy with Coherent Raman Imaging

Coherent Raman imaging (CRI), comprising SRS and CARS microscopy, is an ideal candidate to solve this problem, because it intrinsically has the two important features:

1. Chemical contrast based on intrinsic vibrational properties of the molecules in the tissue
2. Optical three-dimensional sectioning based on nonlinear excitation

Thus CRI can offer many of the benefits of traditional pathology, but because of its unique properties it does not require the addition of dyes or stains to obtain contrast and also does not require physical sectioning of the tissue to obtain high quality images.

These properties have inspired much work in imaging diseases with CARS (Evans et al. 2007, Begin et al. 2009, Chowdary et al. 2010). Signal detection in the backward (epi) direction (Cheng et al. 2001, Saar et al. 2010a), imaging speeds up to video-rate (Evans et al. 2005, Saar et al. 2010a), as well as bio-compatibility of laser excitation intensities (Fu et al. 2006, Nan et al. 2006) have been demonstrated, allowing *in vivo* CRI (Evans et al. 2005, Fu et al. 2007a,b, 2008, Henry et al. 2009, Saar et al. 2010a), even in humans (Saar et al. 2010a).

Other optical imaging techniques have been suggested for the same purpose (Tearney et al. 1997, Tai et al. 2005, Konig et al. 2007, Master and So 2009). While they do offer three-dimensional imaging, confocal reflection (CR), third harmonic generation (THG) microscopy, and optical coherence tomography (OCT) lack the chemically specific contrast of pathology dyes and thus do not allow for multi-color imaging. Two-photon excited auto-fluorescence (TPAF) and second harmonic generation (SHG) have chemical specificity, but the signal originates from a few, specific molecules (e.g., riboflavin, NADH, tryptophan and collagen) (Campagnola et al. 2002, Zipfel et al. 2003, Li et al. 2010). In contrast, CRI allows the imaging of the main constituents of tissue (e.g., lipids, protein, water, blood, and DNA). As such, TPAF and SHG techniques are more similar to specific functional stains in traditional histopathology, which can be used after abnormalities have been recognized with fundamental morphological stains, such as hematoxylin and eosin (H&E). CRI is thus the best candidate to provide the foundation for versatile *in situ* tissue diagnosis similar to the "workhorse" stains in histopathology.

### 24.2.3 *In Vivo* Brain Tumor Imaging in Cranial Window Model

To demonstrate that CRI can indeed be utilized to image tissue *in vivo* on a cellular level, we imaged a brain tumor through cranial window. CARS imaging was used to differentiate cancerous tissue from adjacent normal tissue based on cellularity, cellular organization, and vascularity. In the implanted glioblastoma model, CARS imaging can define the precise cellular margin of the tumor-brain interface.

### 24.2.4 Comparison of SRS and CARS Imaging

While the tumor margin can be identified well with CARS imaging in Figure 24.7, the contrast does not reach the level of an H&E-stained section. We wanted to evaluate whether the contrast advantages of SRS compared to CARS are important for medical imaging. Figure 24.8 shows the comparison of CARS and SRS imaging of a mouse glioblastoma model *ex vivo* at both the $CH_2$-stretching vibration of lipids (2845 $cm^{-1}$) and the $CH_3$-stretching vibration, arising mainly from proteins (2950 $cm^{-1}$). The $CH_2$ SRS image shows the membrane and cytoplasm of the tumor cells, as well as residual myelinated axons (circles) in the tumor. The nuclei are completely dark due to the low concentration of lipids in the nucleus. While the $CH_3$ SRS image generally appears similar to the $CH_2$ SRS image, because lipids also have resonant $CH_3$ oscillators, nuclei show higher signal levels due to nuclear protein. The $CH_3$–$CH_2$ difference image can thus be used to generate a vibrational counterstain in a multi-color imaging as shown in Figure 24.9.

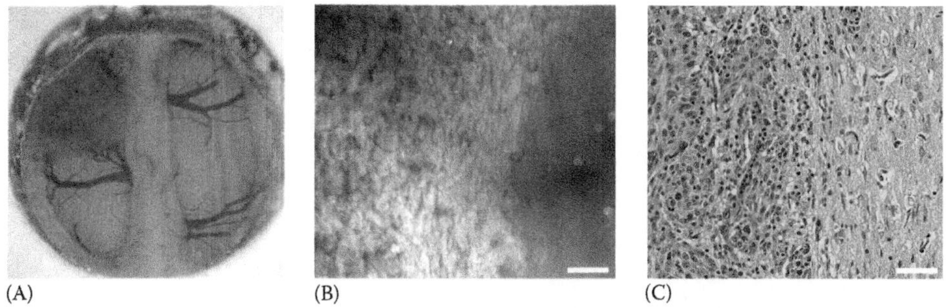


**FIGURE 24.7** Imaging of the brain tumor margin *in vivo*. (A) Photograph of the cranial window in glioblastoma model in a rat. The gross outline of the tumor can be seen based on a changed vascularity in the upper left quadrant. However, the exact tumor margin cannot be identified by the naked eye. (B) CARS image of the tumor margin acquired *in vivo* through the cranial window. Work is in progress to design and build an epi-SRS detector with high enough working distance to extend these studies to SRS. (From Saar, B.G. et al., *Science*, 330, 1368–1370, 2010a.) Images were acquired with 1 s/frame. The tumor is visible based on changes in the cellular morphology and tissue architecture on the left side. (C) Comparison with histopathology with H&E from animals with the same tumor type. Scale: 50 μm.

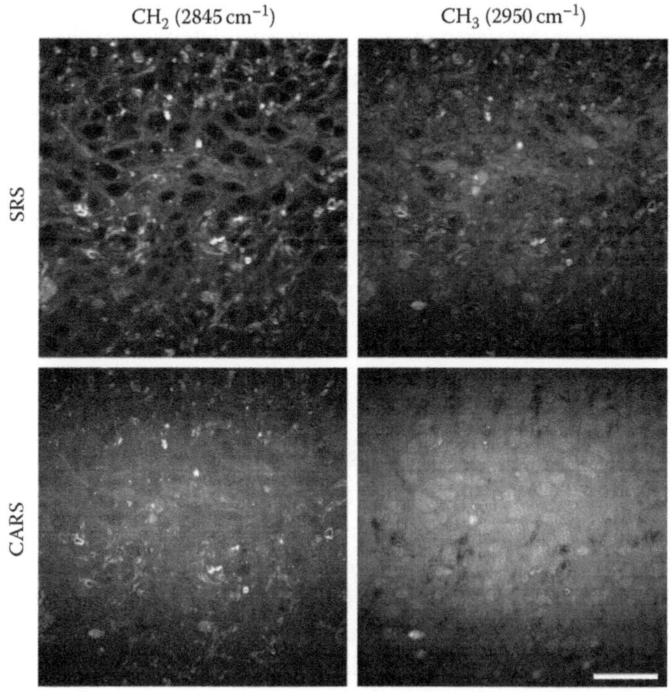


**FIGURE 24.8** Comparison of SRS and CARS imaging of the same region in a primary brain tumor model. SRS and CARS images with $CH_2$ contrast show high contrast for lipids. Compared to CARS, SRS is free from the non-resonant background, so the lipid-poor nuclei appear completely dark, resulting in a high nuclear-to-cytoplasmic contrast in SRS, which is similar to conventional H&E images. In the SRS image with $CH_3$ contrast, the nuclei show structure due to protein. This explains the origin of the non-resonant contrast inside the nuclei in the CARS image with $CH_2$ contrast. In the CARS image with $CH_3$ contrast, nuclei appear with positive contrast due to the spectral distortions observed in CARS. In summary, this figure shows that in both SRS and CARS, multi-color $CH_2$–$CH_3$ imaging allows us to specifically image cytoplasm and nuclei. However, the non-resonant background in CARS makes image interpretation and processing more straightforward with SRS. Scale bar: 25 μm.

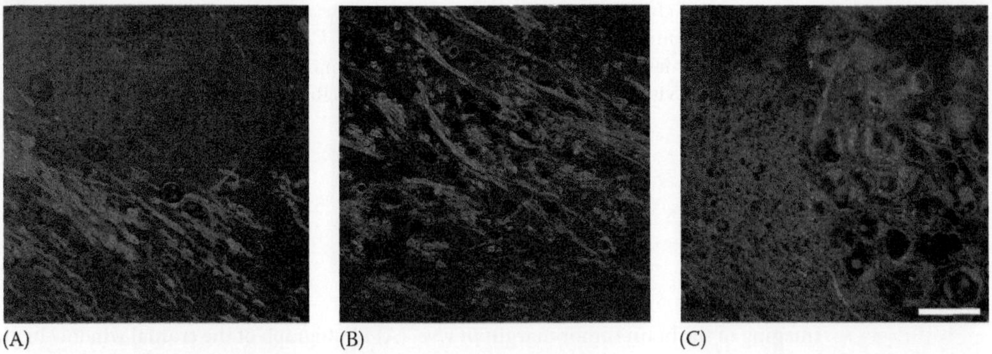

FIGURE 24.9 Differential diagnosis of brain tumors with SRS imaging. Multi-color images from (A) benign brain tissue, (B) an invasive glioblastoma, and (C) a breast-cancer metastasis in the brain. Green shows the $CH_2$ SRS image and blue shows the $CH_3$–$CH_2$ difference image. Malignant and benign tissue can be distinguished based on increased cellularity in the tumor. Primary and secondary tumors can be distinguished based on the invasion pattern at the tumor margin. Scale bar: 25 μm.

In contrast to the $CH_2$ SRS image, the CARS image does not show a high negative contrast for nuclei, which are filled in with the non-resonant background signal from nuclear protein. In the $CH_3$ CARS image, nuclei show up with positive contrast while lipid species are dim. This is because of the spectral distortions in CARS (Freudiger et al. 2008), which cause a dip of the $CH_2$ peak (Evans et al. 2007). Thus, while CARS imaging shows similar spectroscopic information in the CH-region as SRS, it has two limitations:

1. The $CH_2$ CARS image does not have strong nuclear-to-cytoplasmic contrast, the major feature of the H&E stain, due to the non-resonant background
2. Generation of a vibrational counterstain based on $CH_3$–$CH_2$ difference imaging is complicated by spectral distortions

## 24.2.5 SRS Multi-Color Imaging of the Tumor Margin

As discussed in the previous section, SRS imaging allows us to image tissue with high contrast and with multiple "colors" (chemical species). Figure 24.9 shows such multi-color images for benign brain tissue, an invasive glioblastoma, and a breast-cancer metastasis. The green color channel is generated from the $CH_2$ SRS image and the blue channel from the $CH_3$–$CH_2$ difference image. Individual myelinated axons and neuropil show up in green, and protein-rich nuclear features with sub-nuclear morphology in blue. This highlights the great level of detail that can be visualized with SRS, with the additional advantage that different colors encode for functional components in a manner that is very similar to H&E staining.

Due to the similar appearance to H&E, the same diagnostics features can be utilized by pathologists. The benign tissue (Figure 24.9A) can be distinguished from both types of tumor (Figure 24.9B and C), due to a lower density of nuclei, a characteristic known as "cellularity." Further, it is possible to distinguish primary brain tumors (Figure 24.9B) from metastasis (Figure 24.9C) based on the invasion pattern at the tumor margin. While primary tumors invade along the tracts of axons, metastases generally form a sharper interface with brain parenchyma due to their epithelioid character.

In conclusion, these first proof-of-principle experiments with SRS have shown that tissue can be imaged on the cellular level with excellent clarity. Multi-color SRS images highlight the same diagnostic features as traditional histopathology with H&E, but can be acquired without fixation, slicing, and staining. While more clinical studies are required to confirm that SRS can yield histopathologic diagnoses with the same accuracy as traditional light microscopy techniques, this original work provides us with the motivation to pursue clinical translation of SRS imaging.

## 24.3 Conclusion

With the recent advances in SRS microscopy, the imaging speed and sensitivity of coherent Raman scattering microscopy have been approaching theoretical limits. SRS is also in quantitative agreement with other analytical techniques and robust even in complex biological samples.

In the future, the research focus in the field is likely to shift from technical development toward applications. This chapter summarizes the exciting applications in biology and medicine investigated to date. They include studies on pharmacokinetics and lipid metabolism, as well as stain-free histopathology of cancer. These examples were chosen as they take full advantage of the unique properties of SRS microscopy compared to other imaging modalities such as fluorescence or confocal Raman microscopy, because fast label-free imaging in three dimensions is critical. For the imaging of drugs and metabolites, attachment of fluorophores to the small target molecules would drastically alter their properties. For medical imaging, high imaging speed is important to map large areas in close to real-time and intrinsic three-dimensional sectioning circumvents the need for physical sectioning of the tissue. Compared to approaches based on targeted molecular probes, SRS has chemical contrast based on intrinsic contrast of the tissue and has no complications associated with toxicity and uniform delivery of the contrast agents.

## References


Begin, S., E. Belanger, S. Laffray, R. Vallee, and D. Cote. In vivo optical monitoring of tissue pathologies and diseases with vibrational contrast. *Journal of Biophotonics*, 2(11):632–642, 2009.

Berger, M.S. Defining and achieving excellence in surgical neuro-oncology. *Clinical Neurosurgery*, 57(2):10–14, 2010.

Campagnola, P.J., A.C. Millard, M. Terasaki, P.E. Hoppe, C.J. Malone, and W.A. Mohler. Three-dimensional high-resolution second-harmonic generation imaging of endogenous structural proteins in biological tissues. *Biophysical Journal*, 82(1):493–508, 2002.

Caspers, P.J., G.W. Lucassen, and G.J. Puppels. Combined in vivo confocal Raman spectroscopy and confocal microscopy of human skin. *Biophysical Journal*, 85(1):572–580, 2003.

Cheng, J., A. Volkmer, L.D. Book, and X.S. Xie. An epi-detected coherent anti-Stokes Raman scattering (E-CARS) microscope with high spectral resolution and high sensitivity. *Journal of Physical Chemistry B*, 105(7):1277–1280, 2001.

Chowdary, P.D., Z. Jiang, E.J. Chaney, W.A. Benalcazar, D.L. Marks, M. Gruebele, and S.A. Boppart. Molecular histopathology by spectrally reconstructed nonlinear interferometric vibrational imaging. *Cancer Research*, 70:9562–9569, 2010.

Clarke, M., R. Collins, S. Darby, C. Davies, P. Elphinstone, E. Evans, J. Godwin et al., Effects of radiotherapy and of differences in the extent of surgery for early breast cancer on local recurrence and 15-year survival: An overview of the randomised trials. *Lancet*, 366(9503):2087, 2005.

Evans, C.L., E.O. Potma, M. Puoris'haag, D. Côté, C.P. Lin, and X.S. Xie. Chemical imaging of tissue in vivo with video-rate coherent anti-Stokes Raman scattering microscopy. *Proceedings of the National Academy of Sciences of the United States of America*, 102(46):16807, 2005.

Evans, C.L., X. Xu, S. Kesari, X.S. Xie, S.T.C. Wong, and G.S. Young. Chemically-selective imaging of brain structures with CARS microscopy. *Optics Express*, 15(19):12076–12087, 2007.

Freudiger, C.W., W. Min, G.R. Holtom, B. Xu, Dantus M., and X.S. Xie. Spectral imaging by stimulated Raman scattering. *Nature Photonics*, 2011a, http://www.nature.com/nphoton/journal/v5/n2/full/nphoton.2010.294.html

Freudiger, C.W., W. Min, B.G. Saar, S. Lu, G.R. Holtom, C. He, J.C. Tsai, J.X. Kang, and X.S. Xie. Label-free biomedical imaging with high sensitivity by stimulated Raman scattering microscopy. *Science*, 322(5909):1857, 2008.

Freudiger, C.W., B.G. Saar, R. Pfannl, Q. Zeng, L. Ottoboni, W. Ying, R.D. Folkerth et al., Stain-free histopathology with stimulated Raman scattering microscopy. in submission, 2011b. Accepted at Laboratory Investigation in 2012.

Fu, Y., T.B. Huff, H.W. Wang, H. Wang, and J.X. Cheng. Ex vivo and in vivo imaging of myelin fibers in mouse brain by coherent anti-Stokes Raman scattering microscopy. *Optics Express*, 16(24):19396, 2008.

Fu, Y., H. Wang, T.B. Huff, R. Shi, and J.X. Cheng. Coherent anti-Stokes Raman scattering imaging of myelin degradation reveals a calcium-dependent pathway in lyso-PtdCho-induced demyelination. *Journal of Neuroscience Research*, 85(13):2870–2881, 2007a.

Fu, Y., H. Wang, R. Shi, and J.X. Cheng. Characterization of photodamage in coherent anti-Stokes Raman scattering microscopy. *Optics Express*, 14(9):3942–3951, 2006.

Fu, D., T. Ye, T.E. Matthews, B.J. Chen, G. Yurtserver, and W.S. Warren. High-resolution in vivo imaging of blood vessels without labeling. *Optics Letters*, 32:2641–2643, 2007b.

Heinrich, C., A. Hofer, A. Ritsch, C. Ciardi, S. Bernet, and M. Ritsch-Marte. Selective imaging of saturated and unsaturated lipids by wide-field CARS-microscopy. *Optics Express*, 16(4):2699–2708, 2008.

Hellerer, T., C. Axang, C. Brackmann, P. Hillertz, M. Pilon, and A. Enejder. Monitoring of lipid storage in *Caenorhabditis elegans* using coherent anti-Stokes Raman scattering (CARS) microscopy. *Proceedings of the National Academy of Sciences*, 104(37):14658, 2007.

Henry, F.P., D. Cote, M.A. Randolph, E.A.Z. Rust, R.W. Redmond, I.E. Kochevar, C.P. Lin, and J.M. Winograd. Real-time in vivo assessment of the nerve microenvironment with coherent anti-Stokes Raman scattering microscopy. *Plastic and Reconstructive Surgery*, 123(2S):123S, 2009.

Herrera, S. Bonkers about biofuels. *Nature Biotechnology*, 24(7):755–760, 2006.

Konig, K., A. Ehlers, I. Riemann, S. Schenkl, R. Buckle, and M. Kaatz. Clinical two-photon microendoscopy. *Microscopy Research and Technique*, 70(5):398–402, 2007.

Kumar, P., D.M. Barrett, M.J. Delwiche, and P. Stroeve. Methods for pretreatment of lignocellulosic biomass for efficient hydrolysis and biofuel production. *Industrial and Engineering Chemistry Research*, 48(8):3713–3729, 2009.

Lademann, J., F. Knorr, H. Richter, U. Blume-Peytavi, A. Vogt, C. Antoniou, W. Sterry, and A. Patzelt. Hair follicles—An efficient storage and penetration pathway for topically applied substances. *Skin Pharmacology and Physiology*, 21(3):150–155, 2008.

Li, C., R.K. Pastila, C. Pitsillides, J.M. Runnels, M. Puoris'haag, D. Cote, and C.P. Lin. Imaging leukocyte trafficking in vivo with two-photon-excited endogenous tryptophan fluorescence. *Optics Express*, 18(2):988–999, 2010.

Master, B.R. and P.T.C. So. *Biomedical Nonlinear Optical Microscopy*. Oxford University Press, Oxford, U.K., 2009.

Mullaney, B.C. and K. Ashrafi. *C. elegans* fat storage and metabolic regulation. *Biochimica et Biophysica Acta (BBA)—Molecular and Cell Biology of Lipids*, 1791(6):474–478, 2009.

Nan, X., E.O. Potma, and X.S. Xie. Nonperturbative chemical imaging of organelle transport in living cells with coherent anti-Stokes Raman scattering microscopy. *Biophysical Journal*, 91(2):728–735, 2006.

Nandakumar, P., A. Kovalev, and A. Volkmer. Vibrational imaging based on stimulated Raman scattering microscopy. *New Journal of Physics*, 11:033026, 2009.

Ozeki, Y., F. Dake, S. Kajiyama, K. Fukui, and K. Itoh. Analysis and experimental assessment of the sensitivity of stimulated Raman scattering microscopy. *Optics Express*, 17(5):3651–3658, 2009.

Ploetz, E., S. Laimgruber, S. Berner, W. Zinth, and P. Gilch. Femtosecond stimulated Raman microscopy. *Applied Physics B: Lasers and Optics*, 87(3):389–393, 2007.

Prausnitz, M.R., S. Mitragotri, and R. Langer. Current status and future potential of transdermal drug delivery. *Nature Reviews Drug Discovery*, 3(2):115–124, 2004.

Roeffaers, M.B.J., X. Zhang, C.W. Freudiger, B.G. Saar, M. Ruijven, G. Dalen, C. Xiao, and X.S. Xie. Label-free imaging of biomolecules in food products using stimulated Raman microscopy. *Journal of Biomedical Optics*, accepted, 2010, http://www.ncbi.nlm.nih.gov/pubmed/21473164

Saar, B.G., C.W. Freudiger, C.M. Stanely, G.R. Holtom, and X.S. Xie. Video-rate molecular imaging in vivo with stimulated Raman scattering. *Science*, 330:1368–1370, 2010a, http://www.sciencemag.org/content/330/6009/1368.short

Saar, B.G., Y. Zeng, C.W. Freudiger, Y.S. Liu, M.E. Himmel, X.S. Xie, and S.Y. Ding. Label-free, real-time monitoring of biomass processing with stimulated Raman scattering microscopy. *Angewandte Chemie*, 122:5608–5611, 2010b, http://onlinelibrary.wiley.com/doi/10.1002/anie.201000900/abstract?deniedAcessCustomisedMessage=&userIsAuthenticated=false

Sabel, M.S., K. Rogers, K. Griffith, R. Jagsi, C.G. Kleer, K.A. Diehl, T.M. Breslin, V.M. Cimmino, A.E. Chang, and L.A. Newman. Residual disease after re-excision lumpectomy for close margins. *Journal of Surgical Oncology*, 99(2):99–103, 2009.

Slipchenko, M.N., H. Chen, D.R. Ely, Y. Jung, M.T. Carvajal, and J.X. Cheng. Vibrational imaging of tablets by epi-detected stimulated Raman scattering microscopy. *The Analyst*, 2010, http://pubs.rsc.org/en/Content/ArticleLanding/2010/AN/c0an00252f

Slipchenko, M.N., T.T. Le, H. Chen, and J.X. Cheng. High-speed vibrational imaging and spectral analysis of lipid bodies by compound Raman microscopy. *The Journal of Physical Chemistry B*, 113(21):7681–7686, 2009.

Tai, S.P., T.H. Tsai, W.J. Lee, D.B. Shieh, Y.H. Liao, H.Y. Huang, K. Zhang, H.L. Liu, and C.K. Sun. Optical biopsy of fixed human skin with backward-collected optical harmonics signals. *Optics Express*, 13(20):8231–8242, 2005.

Tearney, G.J., M.E. Brezinski, B.E. Bouma, S.A. Boppart, C. Pitris, J.F. Southern, and J.G. Fujimoto. In vivo endoscopic optical biopsy with optical coherence tomography. *Science*, 276(5321):2037, 1997.

Valicenti, R.K., I. Chervoneva, and L.G. Gomella. Importance of margin extent as a predictor of outcome after adjuvant radiotherapy for Gleason score 7 pT3N0 prostate cancer. *International Journal of Radiation Oncology, Biology, and Physics*, 58(4):1093–1097, 2004.

Van De Kerkhof, P.C.M., M.M. Kleinpenning, E.M.G.J. De Jong, M.J.P. Gerritsen, R.J. Van Dooren-Greebe, and H.A.C. Alkemade. Current and future treatment options for acne. *Journal of Dermatological Treatment*, 17(4):198–204, 2006.

Veronesi, U., N. Cascinelli, J. Adamus, C. Balch, D. Bandiera, A. Barchuk, R. Bufalino, P. Craig, J. De Marsillac, J.C. Durand, and others. Thin stage I primary cutaneous malignant melanoma. *New England Journal of Medicine*, 318(18):1159–1162, 1988, http://www.nejm.org/doi/pdf/10.1056/NEJM198805053181804

Wang, M.C., W. Min, C.W. Freudiger, G. Ruvkun, and X.S. Xie. RNA interference screening for fat regulatory genes with stimulated Raman scattering microscopy. *Nature Methods*, 8:135–138, 2011, http://www.nature.com/nmeth/journal/v8/n2/abs/nmeth.1556.html

Zhang, X., M.B.J. Roeffaers, S. Basu, B.G. Saar, C.W. Freudiger, and X.S. Xie. Imaging of DNA structures with stimulated Raman scattering microscopy. in submission, 2011, http://onlinelibrary.wiley.com/doi/10.1002/cphc.201100890/abstract

Zhang, X., M.B.J. Roeffaers, S. Basu, J.R. Daniele, D. Fu1, C.W. Freudiger, G.R. Holtom, and X.S. Xie. Label-free live-cell imaging of nucleic acids using stimulated Raman scattering microscopy. *ChemPhysChem*, 13(4):1054–1059, 2012.

Zipfel, W.R., R.M. Williams, R. Christie, A.Y. Nikitin, B.T. Hyman, and W.W. Webb. Live tissue intrinsic emission microscopy using multiphoton-excited native fluorescence and second harmonic generation. *Proceedings of the National Academy of Sciences of the United States of America*, 100(12):7075, 2003.

# 25. Applications of Coherent Anti-Stokes Raman Spectroscopy Imaging to Cardiovascular Diseases

**Han-Wei Wang, Michael Sturek, and Ji-Xin Cheng**



## 25.1 Introduction

Cardiovascular diseases (CVDs), which include atherosclerosis, heart disease, and stroke, have reached epidemic proportions (Naghavi et al. 2003). As the prevalence of this disease increases in developed and developing countries, particularly in the western world, there is an ever-growing demand for CVD research (Yusuf et al. 2001). For nearly the last 100 years, CVD has been the number one cause of death in the United States (AHA 2007). The number of American adults (age 18 and over) who suffer from one or more types of CVDs amounts to over 80 million (AHA 2010). Because of the prevalence of CVD, the direct and indirect cost associated with this disease is estimated to be over 430 billion dollars in 2007 in the United States.

Atherosclerosis, which is the major form of CVD, is a progression of arterial lesions caused by numerous risk factors, including hyperlipidemia, diabetes, and obesity (Libby 2002). Lipoproteins infiltrate the artery walls to form the fatty streak as the precursor of the lesion (Stary et al. 1994). Macrophages, which are differentiated from the extravasated monocytes, take up lipoproteins and become lipid-laden foam cells to contribute

*Coherent Raman Scattering Microscopy*. Edited by Ji-Xin Cheng and X. Sunney Xie © 2013 CRC Press/ Taylor & Francis Group, LLC. ISBN: 978-1-4398-6765-5.

substantially to atherosclerosis progression (Lusis 2000). The accumulation of foamy-like cells, dead cells, and fibrous matrix, as well as intracellular and extracellular lipid contents, comprises a necrotic core in more developed lesions. Fibrous tissues, calcified components, or fissure-induced hematoma are all present in the late lesion stage (Stary et al. 1992). Artery stenosis caused by the formation of a stable plaque or thrombus clot following the rupture of unstable plaque can cause chronic and acute syndromes, such as ischemia, angina pectoris, and acute myocardial infarction (Naghavi et al. 2003).

Several methods have been utilized for the study of atherosclerosis, including a variety of imaging techniques (Vallabhajosula and Fuster 1997, Fayad and Fuster 2001, Zandvoort et al. 2004, Le et al. 2007), catheter intervention (Kern 1995, Dauerman et al. 1998), electrocardiography, and genomic and proteomic methods (Eriksson et al. 2001, Osawa et al. 2002, Hofmann et al. 2004). Imaging techniques have been particularly valuable for laboratory research and clinical diagnosis. The ability to visualize plaques facilitates atherosclerosis studies on structure and pathobiology and guides therapeutic interventions, such as balloon angioplasty and stent placement.

Current imaging methods, such as x-ray angiography, magnetic resonance imaging (MRI), intravascular ultrasound (IVUS), computed tomography (CT), and optical coherence tomography (OCT), allow exquisite delineation of advanced lesions (Fayad and Fuster 2001, Choudhury et al. 2004, Jaffer et al. 2007, Bluemke et al. 2008, Kume et al. 2008, Sanz and Fayad 2008); however, these techniques have not yet reached optimal spatial resolution, nor have they enabled accurate identification of the lesion composition. As a gold standard, histology is phenomenal in biopsy studies but not feasible for live tissue imaging. With technical advances, fluorescence microscopy (including one-photon and two-photon methods) has been used for vascular and atherosclerosis studies (Amirbekian et al. 2007, Megens et al. 2007a,b) and also has been used to identify cellular and molecular composition *ex vivo* and *in vivo* with labeling (Zandvoort et al. 2004, Douma et al. 2007, Yu et al. 2007). In view of the limitations of labeling, including photobleaching, extra incubation period, and/or limited circulation lifetime, it is intriguing to explore label-free imaging methods, such as coherent anti-Stokes Raman spectroscopy (CARS) microscopy, for atherosclerosis studies.

## 25.2 Arterial Structures Imaged by Nonlinear Optical Microscopy

In general, artery structure can be roughly divided into three layers. The innermost layer from the lumen is the tunica intima, which comprises endothelium and internal elastic lamina. The second layer is the tunica media, which is mainly composed of elastic lamella and smooth muscle cells. The outermost layer is adventitia, which is mainly collagen and elastin (Cliff 1976, Stary et al. 1992). The roles of endothelial cells, such as selective permeability, thromboresistance, vasomotor tone, and regulation of immune and inflammation responses, are crucial for normal artery function as well as facilitating the onset of atherosclerotic lesions (Eckardstein 2004). Dysfunction and denudation of the endothelium could initiate the process of lipid infiltration, leukocyte adhesion, and the onset of atheroma (Ross and Glomset 1973, Bonetti et al. 2003). The functions of the smooth muscle cells include vessel contraction, connective tissue synthesis, and lipid metabolism (Bierman and Albers 1975, 1977, Eckardstein 2004). The migration and proliferation of smooth muscle cells in the transitional stage of atherosclerosis (Ross and Glomset 1973) and the synthesis of connective tissues secreted by smooth muscle cells in the late stage (Ross 1986) dominate the progression of plaque lesions. At present, the classification of atherosclerosis lesion states is made largely with imaging approaches (Ross and Klebanoff 1971, Stary et al. 1994, Fayad and Fuster 2001). Applying nonlinear optical (NLO) microscopy and CARS-based multimodal NLO microscopy for atherosclerosis will be discussed in the next section.

To visualize arterial structures without molecular labels, two-photon epifluorescence (TPEF) and second-harmonic generation (SHG) microscopy have been used. The high sensitivity of SHG microscopy to non-centrosymmetric structures has enabled visualization of collagen fibrils of the arterial wall (Zipfel et al. 2003, Zoumi et al. 2004). Sum frequency generation (SFG) (Shen 1989), which also derives its signal from non-centrosymmetric molecules but at the sum frequency of two excitation sources, offers imaging capability similar to that of SHG for biological tissues (Fu et al. 2007). Relying on intrinsic autofluorescence, TPEF microscopy has been applied to visualize elastin fibers (Zipfel et al. 2003, Boulesteix et al. 2005, Konig et al. 2005). Moreover, TPEF microscopy has been used to image fluorescently labeled endothelial cells, smooth muscle cells, and macrophages (Zandvoort et al. 2004). However, imaging arterial components with fluorescent labels faces tremendous technical challenges, including nonspecific binding and inefficient diffusion into arterial wall. It is known that endothelial cells and smooth muscle cells play leading roles in normal arterial function (Ross and Klebanoff 1971, Osawa et al. 2002, Dejana 2004) as well as in arterial disease (Ross 1986, Eriksson et al. 2001, Libby 2002). Therefore, label-free visualization of endothelial cells and smooth muscle cells would be a major improvement for investigating the early stages of CVDs.

Label-free visualization of significant arterial components, including arterial cells, has been reported using a microscope that integrated CARS, SFG, and TPEF imaging modalities on the same platform (Wang et al. 2008). In this work, spectrally resolved CARS, SFG, and TPEF signals were generated simultaneously from the same arterial sample by two synchronized picosecond lasers (Figure 25.1). This was the first demonstration that CARS imaging based on the $CH_2$ vibrations allowed visualization of endothelial cells, smooth muscle cells, elastin, and collagen fibrils of the arterial wall (Figure 25.2). The CH-rich residues of lysine in the cross-linking structures of elastin and collagen contribute to relatively weak contrast in CARS imaging. This work also measured, for the first time, the CARS spectra of elastin and collagen, which are protein fibrils, in arterial walls. Two CARS peaks, blue shifted probably by carbonyl groups, at 2870 and 2930 $cm^{-1}$ were assigned to symmetric and asymmetric, respectively. In addition, the polarization independence of the CARS signal from elastin and collagen is probably due to disordered orientation of $CH_2$-rich residues of lysines in the fibrils' cross-linking region. The demonstrated label-free imaging capability of multimodal NLO microscopy suggests its potential application to the studies of atherosclerosis *in vivo*.

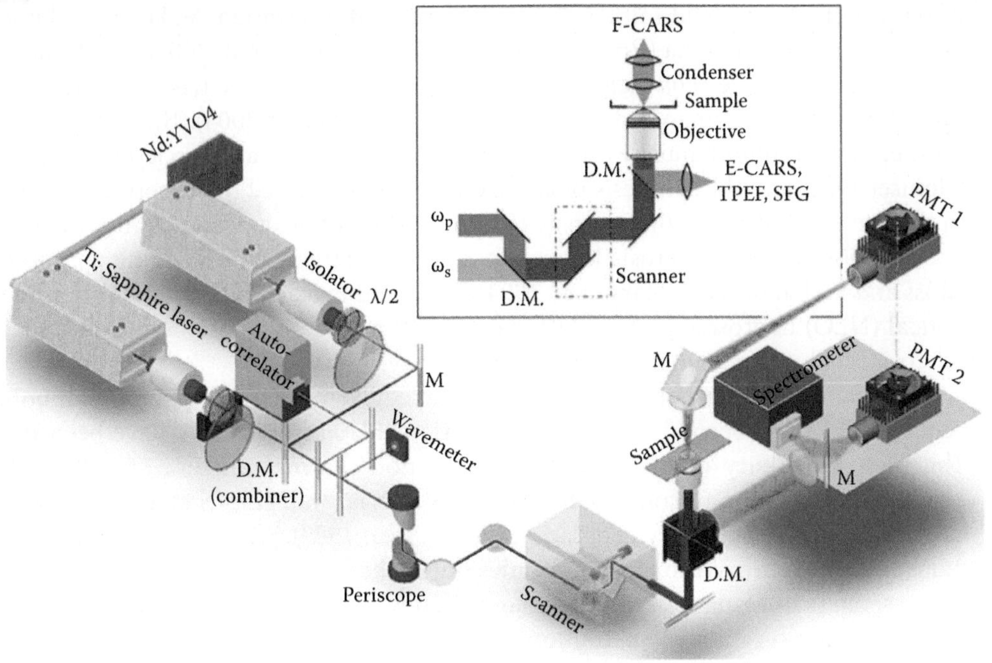


**FIGURE 25.1** A CARS-based multimodal NLO microscope. The system comprises picosecond lasers, colinear beam geometry, a laser-scanning microscope, and detectors. D.M., dichroic mirror; PMT, photomultiplier tube; M, mirror; $\omega_p$, pump beam; $\omega_s$, Stoke beam.

## 25.3 Atherosclerosis in Animal Models Imaged by Coherent Anti-Stokes Raman Spectroscopy-Based Multimodal NLO Microscopy

Atherosclerosis is a complex process, which has been a leading contributor to morbidity and mortality in the United States, and it has been on the rise globally (Yusuf et al. 2001). The statistics that account for the rise in incidence consequently call for new imaging techniques to advance the research and diagnosis of atherosclerosis. It is known that the atherosclerotic complications are not only dependent on stenosis or lumen occlusion, but also on the vulnerability to plaque rupture. Vulnerable plaque is characterized histologically by a thin fibrous cap over a soft, lipid-laden core (Stary 2000, Virmani et al. 2000). Therefore, diagnosis of atherosclerotic plaques should include both anatomical features for stenosis evaluation and molecular composition for vulnerability evaluation (Virmani et al. 2000, Naghavi et al. 2003). However, existing diagnostic tools for atherosclerosis are not capable of offering sufficient information about lesion composition to enable definitive diagnosis of vulnerable plaque in living specimens.

Several studies have highlighted the capability of CARS-based multimodal NLO microscopy for atherosclerosis research to visualize anatomical features and to analyze molecular composition of the lesions. Le et al. visualized for the first time, using CARS microscopy on a picosecond-laser platform with TPEF and SHG microscopy on a femtosecond-laser scheme, atherosclerotic plaques (Le et al. 2007) from Ossabaw pigs that are genetically predisposed to obesity, metabolic syndrome, and subsequent development of atherosclerosis

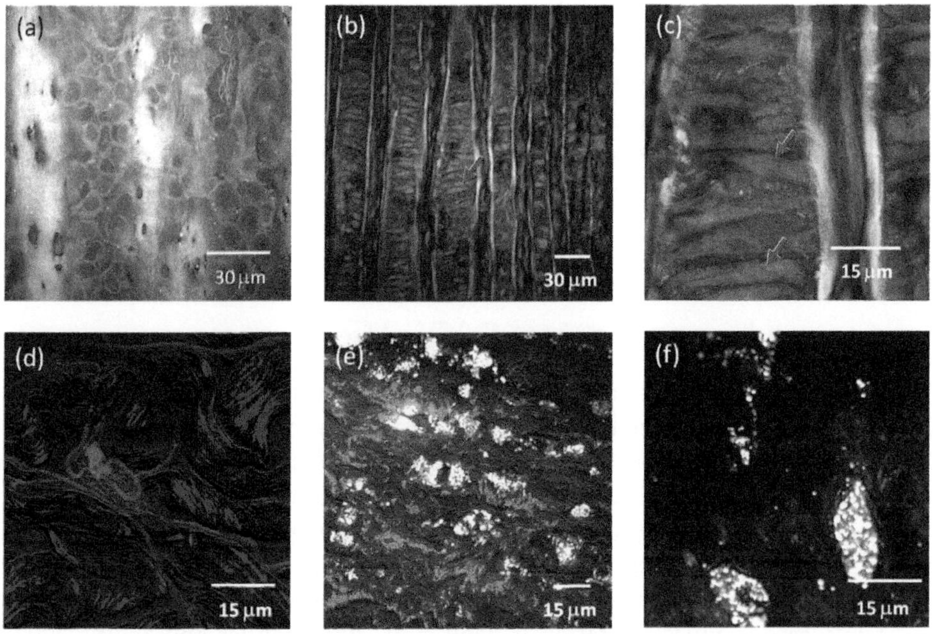


**FIGURE 25.2** Arterial cells and structures visualized by CARS-based multimodal NLO microscopy. (a) CARS imaging of endothelial cells from the luminal side of the artery at a depth adjacent to the internal elastic lamina. (b) Rod-shaped smooth muscle cells (as indicated by the yellow arrow) in layered media of an artery. The artery was sliced open and imaged from the luminal side. (c) A zoomed-in CARS image at the center of (b) co-localized with the SFG modality. The light-blue-colored SFG signal shows the extracellular fibrils around the smooth muscle cells (yellow arrows). (d) Collagen (blue) and elastic (green) fibrils in the adventitia of an artery. (e) Foam cells (bright white) and extracellular collagen (blue) in an atherosclerotic artery from a pig with metabolic syndrome. (f) Lipid-laden foam cells specifically visualized by CARS with submicron resolution. Intracellular vesicles of lipids (bright white) are clearly shown. Gray for CH-rich features by CARS; blue for collagen by SFG; green for elastin autofluorescence around 520 nm by TPEF. (a–d) Images of carotid arteries from lean Yorkshires. (e, f) Images of an iliac artery from Ossabaw swine model of metabolic syndrome.

(Sturek et al. 2007, Lee et al. 2009). In this pioneering study, Le et al. showed that a plaque can be detected with the multimodal NLO imaging. Several lesion characteristics, such as disorganized collagen fibrils in media, expression of collagen fibrils in thickened intima, and foamy cells and extracellular lipid deposits, directly revealed the location and composition of atherosclerotic plaques. Hence, Le et al. suggested that multimodal imaging can significantly improve the sensitivity and accuracy of plaque detection. Employing only picosecond lasers for the CARS, TPEF, and SFG imaging, Wang et al. further explored imaging capabilities of a CARS-based multimodal microscopy imaging of atherosclerosis. The study first demonstrated the imaging capability of identifying different atherosclerotic types (Wang et al. 2009) based on the scheme of atherosclerosis classification suggested by the American Heart Association (AHA) and modified by other authorities (Stary et al. 1994, 1995, Stary 2000, Virmani et al. 2000). Furthermore, Wang et al. (2009) showed that the multimodal approach employing CARS and SFG signals allows quantitation of collagen and lipid content in lesions from early to advanced stages.

More studies have shown increasing interest in using CARS microscopy for the studies of atherosclerotic lesions in animal models. Lim et al. utilized multimodal NLO microscopy to quantitatively measure the impact of a high-fat, high-cholesterol Western diet on the composition of atherosclerotic plaques in ApoE-deficient mice (Lim et al. 2010). Using a photonic crystal fiber-based CARS microscopy, the capability of the multimodal imaging was demonstrated using the alternative femtosecond system to study lesion development in myocardial infarction-prone, hyperlipidemic rabbits (Mostaço-Guidolin et al. 2010).

## 25.4 Atherosclerotic Plaques Studied by Multiplex and Hyperspectral CARS Microscopy

Determining the composition of lipids or lipid structures in an image of atherosclerotic plaque allows further understanding of the physiological role of lipids on the progression of the disease. A CARS spectrum contains the compositional information for chemical assessment, but it is not accessible in CARS imaging at a single Raman band. For this purpose, multiplex CARS and hyperspectral CARS have been developed to fill the niche.

Multiplex CARS utilizes a narrowband and a broadband excitation field to achieve simultaneous acquisition of a vibrational spectrum. It was first demonstrated with pico- and femto-second pulse excitation (Cheng et al. 2002, Müller and Schins 2002). The super continuum from a photonic crystal fiber has been extensively used for multiplex CARS development (Paulsen et al. 2003, Kee and Cicerone 2004, Kano and Hamaguchi 2005, Petrov and Yakovlev 2005). The multiplex CARS imaging is capable of observing spatially resolved chemical information by taking full vibrational spectra at each sample point within tens of milliseconds. Although the multiplex CARS imaging speed is not yet sufficient for the study of highly dynamic systems in millisecond physiological time-frames, it can be applied to the study of atherosclerosis at the tissue level in static conditions. Kim et al. employed multiplex CARS microscopy to study the chemical profiles of lipids of different morphologies presenting in atherosclerotic plaques in the ApoE-deficient mouse model. The study also demonstrated the potential of investigating the changes in the chemical profiles of atherosclerotic lipids in response to statin drug treatment (Kim et al. 2010).

Hyperspectral CARS imaging is another way to obtain spectrally resolved molecular maps. This method has been demonstrated recently by sequentially acquiring CARS images at a series of beating frequencies. The range of vibrational frequencies was achieved by spectral tuning of an optical parametric oscillator; thus, the CARS spectrum of each pixel can be obtained. Lim et al. have employed hyperspectral CARS microscopy in combination with principal component analysis for multivariate analysis to characterize the chemical composition of lipids in atherosclerotic plaque of varying morphology (Lim et al. 2011). Principal component analysis enables multidimensional datasets to be expressed in terms of fewer variables, thereby allowing the verification of small spectral differences imbedded in Raman spectra (Chan et al. 2006). The hyperspectral CARS imaging with the multivariate analysis allowed differentiation of cholesterol crystals from aliphatic lipid within atherosclerotic plaques (Lim et al. 2011). Although the average acquisition time for one hyperspectral image with the spectral data stack was reported to be about 20 min, automation of the laser tuning will highly improve the speed of the hyperspectral CARS imaging.

## 25.5 Summary and Discussion

The label-free imaging capability and chemical information of CARS microscopy, with additional functions provided by other NLO modalities, has allowed in-depth studies of atherosclerosis at tissue level. Applications of CARS-based multimodal imaging to *in vivo* atherosclerosis characterization highly depend on successful development of CARS endoscopy. Although *in vivo* imaging of atherosclerotic plaque in small animals, for example, ApoE-deficient mice, can be performed under NLO microscopy with invasive surgery (Figure 25.3), it is currently not possible to conduct studies in large animals with catheter-based systems used clinically in humans. The preliminary concept of CARS endoscopy (Légaré et al. 2006) and a portable CARS imaging device (Balu et al. 2010) have been demonstrated; nevertheless, a CARS intravascular catheter has not yet been realized, mainly because of the difficulties in tackling the miniaturization and perfecting the efficiency of the four-wave mixing process. Considering the tissue penetration depth of *ca.* 100 μm in CARS microscopy, atherosclerosis imaging in large animals has to be conducted using an intravascular catheter and imaging from the lumen side. An ideal CARS endoscopy device should also be bundled with other catheterization imaging methods, such as IVUS. Several advances have been reported in multiphoton endoscopy (Flusberg et al. 2005, Fu and Gu 2007) and in NLO contrast-enhanced OCT for *in situ* imaging (Bredfeldt et al. 2005, Su et al. 2007, Tang et al. 2007). In addition, an endoscope combining NLO imaging and Fourier domain OCT has recently been demonstrated on heart tissues with the use of a fiber-based femtosecond laser (Liu 2011). With the advances in CARS and in other coherent Raman scattering (CRS) processes such as stimulated Raman scattering (Freudiger et al. 2008, Saar et al. 2010), one can see the eventual reality of a CRS-based NLO intravascular catheter to facilitate *in vivo* studies and diagnosis in the years to come.

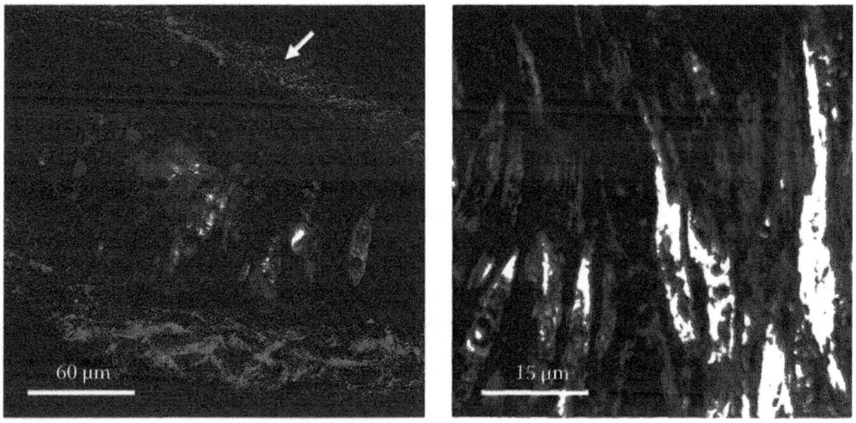


**FIGURE 25.3** *In vivo* imaging of an atherosclerotic femoral artery of an ApoE-deficient mouse (high-fat diet group, 37 weeks old). The left image shows the lipid-rich smooth muscle cells in the media of the mouse artery visualized by CARS imaging. The right image shows a zoomed-in image at deeper layer of the same artery. The arrow shows the collagen fibrils of the artery wall.

## References

AHA. 2007. Heart disease and stroke statistics—2007 update. Dallas, TX.

AHA. 2010. Heart disease and stroke statistics—2010 update. Dallas, TX.

Amirbekian, V., Lipinski, M. J., Briley-Saebo, K. C. et al. 2007. Detecting and assessing macrophages in vivo to evaluate atherosclerosis noninvasively using molecular MRI. *Proc. Natl. Acad. Sci. USA* 104: 961–966.

Balu, M., Liu, G., Chen, Z., Tromberg, B. J. and Potma, E. O. 2010. Fiber delivered probe for efficient CARS imaging of tissues. *Opt. Express* 18: 2380–2388.

Bierman, E. L. and Albers, J. J. 1975. Lipoprotein uptake by cultured human arterial smooth muscle cells. *Biochim. Biophys. Acta (BBA)—Lipids Lipid Metabol.* 388: 198–202.

Bierman, E. L. and Albers, J. 1977. Regulation of low density lipoprotein receptor activity by cultured human arterial smooth muscle cells. *Biochim. Biophys. Acta (BBA)—Lipids Lipid Metabol.* 488: 152–160.

Bluemke, D. A., Achenbach, S., Budoff, M. et al. 2008. Noninvasive coronary artery imaging: Magnetic resonance angiography and multidetector computed tomography angiography: A scientific statement from the American Heart Association Committee on cardiovascular imaging and intervention of the council on cardiovascular radiology and intervention, and the councils on clinical cardiology and cardiovascular disease in the young. *Circulation* 118: 586–606.

Bonetti, P. O., Lerman, L. O., and Lerman, A. 2003. Endothelial dysfunction: A marker of atherosclerotic risk. *Arterioscler. Thromb. Vasc. Biol.* 23: 168–175.

Boulesteix, T., Pena, A. M., Pages, N. et al. 2005. Micrometer scale ex vivo multiphoton imaging of unstained artery wall structure. *Cytometry* 69: 20–26.

Bredfeldt, J. S., Vinegoni, C., Marks, D. L., and Boppart, S. A. 2005. Molecularly sensitive optical coherence tomography. *Opt. Lett.* 30: 495–497.

Chan, J. W., Taylor, D. S., Zwerdling, T. et al. 2006. Micro-Raman spectroscopy detects individual neoplastic and normal hematopoietic cells. *Biophys. J.* 90: 648–656.

Cheng, J. X., Volkmer, A., Book, L. D., and Xie, X. S. 2002. Multiplex coherent anti-Stokes Raman scattering microspectroscopy and study of lipid vesicles. *J. Phys. Chem.* 106: 8493–8498.

Choudhury, R. P., Fuster, V., and Fayad, Z. A. 2004. Molecular, cellular and functional imaging of atherothrombosis. *Nat. Rev. Drug. Discov.* 3: 913–925.

Cliff, W. J. 1976. *Blood Vessels*. New York: Cambridge University Press.

Dauerman, H. L., Higgins, P. J., Sparano, A. M. et al. 1998. Mechanical debulking versus balloon angioplasty for the treatment of true bifurcation lesions. *J. Am. Coll. Cardiol.* 32: 1845–1852.

Dejana, E. 2004. Endothelial cell–cell junctions: Happy together. *Nat. Rev. Mol. Cell Biol.* 5: 261–270.

Douma, K., Megens, R. T. A., Reitsma, S. et al. 2007. Two-photon lifetime imaging of fluorescent probes in intact blood vessels: A window to sub-cellular structural information and binding status. *Microsc. Res. Tech.* 70: 467–475.

Eckardstein, A. V. 2004. *Atherosclerosis: Diet and Drugs*. New York: Springer.

Eriksson, E. E., Xie, X., Werr, J., Thoren, P., and Lindbom, L. 2001. Direct viewing of atherosclerosis in vivo: Plaque invasion by leukocytes is initiated by the endothelial selectins. *FASEB J.* 15: 1149–1157.

Fayad, Z. A. and Fuster, V. 2001. Clinical imaging of the high-risk or vulnerable atherosclerotic plaque. *Circ. Res.* 89: 305–316.

Flusberg, B. A., Cocker, E. D., Piyawattanametha, W. et al. 2005. Fiber-optic fluorescence imaging. *Nat. Methods* 2: 941–950.

Freudiger, C. W., Min, W., Saar, B. G. et al. 2008. Label-free biomedical imaging with high sensitivity by stimulated Raman scattering microscopy. *Science* 322: 1857–1861.

Fu, L. and Gu, M. 2007. Fibre-optic nonlinear optical microscopy and endoscopy. *J. Microsc.* 226: 195–206.

Fu, Y., Wang, H., Shi, R., and Cheng, J. X. 2007. Second harmonic and sum frequency generation imaging of fibrous astroglial filaments in ex vivo spinal tissues. *Biophys. J.* 92: 3251–3259.

Hofmann, C. S., Sullivan, C. P., Jiang, H.-Y. et al. 2004. B-Myb represses vascular smooth muscle cell collagen gene expression and inhibits neointima formation after arterial injury. *Arterioscler. Thromb. Vasc. Biol.* 24: 1608–1613.

Jaffer, F. A., Libby, P., and Weissleder, R. 2007. Molecular imaging of cardiovascular disease. *Circulation* 116: 1052–1061.

Kano, H. and Hamaguchi, H. 2005. Vibrationally resonant imaging of a single living cell by supercontinuum-based multiplex coherent anti-Stokes Raman scattering microspectroscopy. *Opt. Express* 13: 1322–1327.

Kee, T. W. and Cicerone, M. T. 2004. Simple approach to one-laser, broadband coherent anti-Stokes Raman scattering microscopy. *Opt. Lett.* 29: 2701–2703.

Kern, M. J. 1995. *The Caridac Catheterization Handbook*. St. Louis, MO: Mosby.

Kim, S.-H., Lee, E.-S., Lee, J. Y. et al. 2010. Multiplex coherent anti-Stokes Raman spectroscopy images intact atheromatous lesions and concomitantly identifies distinct chemical profiles of atherosclerotic lipids. *Circ. Res.* 106: 1332–1341.

Konig, K., Schenke-layland, K., Riemann, I., and Stock, U. A. 2005. Multiphoton autofluorescence imaging of intratissue elastic fibers. *Biomaterials* 26: 495–500.

Kume, T., Okura, H., Kawamoto, T. et al. 2008. Fibrin clot visualized by optical coherence tomography. *Circulation* 118: 426–427.

Le, T. T., Langohr, I. M., Locker, M. J., Sturek, M., and Cheng, J. X. 2007. Label-free molecular imaging of atherosclerotic lesions using multimodal nonlinear optical microscopy. *J. Biomed. Opt.* 12(5): 054007.

Lee, L., Alloosh, M., Saxena, R. et al. 2009. Nutritional model of steatohepatitis and metabolic syndrome in the Ossabaw miniature swine. *Hepatology* 50: 56–67.

Légaré, F., Evans, C. L., Ganikhanov, F., and Xie, X. S. 2006. Towards CARS endoscopy. *Opt. Express* 14: 4427–4432.

Libby, P. 2002. Inflammation in atherosclerosis. *Nature* 420: 868–874.

Lim, R. S., Kratzer, A., Barry, N. P. et al. 2010. Multimodal CARS microscopy determination of the impact of diet on macrophage infiltration and lipid accumulation on plaque formation in ApoE-deficient mice. *J. Lipid Res.* 51: 1729–1737.

Lim, R. S., Suhalim, J. L., Miyazaki-anzai, S. et al. 2011. Identification of cholesterol crystals in plaques of atherosclerotic mice using hyperspectral CARS imaging. *J. Lipid Res.* 52: 2177–2186.

Liu, G. 2011. Fiber-based combined optical coherence and multiphoton endomicroscopy. *J. Biomed. Opt.* 16: 036010.

Lusis, A. J. 2000. Atherosclerosis. *Nature* 407: 233–241.

Megens, R. T. A., Egbrink, M. G. O., Cleutjens, J. P. et al. 2007a. Imaging collagen in intact viable healthy and atherosclerotic arteries using fluorescently labeled CNA35 and two-photon laser scanning microscopy. *Mol. Imaging* 6: 247–260.

Megens, R. T. A., Reitsma, S., Schiffers, P. H. M. et al. 2007b. Two-photon microscopy of vital murine elastic and muscular arteries. *J. Vasc. Res.* 44: 87–98.

Mostaço-guidolin, L. B., Sowa, M. G., Ridsdale, A. et al. 2010. Differentiating atherosclerotic plaque burden in arterial tissues using femtosecond CARS-based multimodal nonlinear optical imaging. *Biomed. Opt. Express* 1: 59–73.

Müller, M. and Schins, J. M. 2002. Imaging the thermodynamic state of lipid membranes with multiplex CARS microscopy. *J. Phys. Chem. B* 106: 3715–3723.

Naghavi, M., Libby, P., Falk, E. et al. 2003. From vulnerable plaque to vulnerable patient: A call for new definitions and risk assessment strategies. Part I. *Circulation* 108: 1664–1672.

Osawa, M., Masuda, M., Kusano, K., and Fujiwara, K. 2002. Evidence for a role of platelet endothelial cell adhesion molecule-1 in endothelial cell mechanosignal transduction: Is it a mechanoresponsive molecule? *J. Cell Biol.* 158: 773–785.

Paulsen, H. N., Hilligsoe, K. M., Thogersen, J., Keiding, S. R., and Larsen, J. J. 2003. Coherent anti-Stokes Raman scattering microscopy with a photonic crystal fiber based light source. *Opt. Lett.* 28: 1123–1125.

Petrov, G. I. and Yakovlev, V. V. 2005. Enhancing red-shifted white-light continuum generation in optical fibers for applications in nonlinear Raman microscopy. *Opt. Express* 13: 1299–1306.

Ross, R. 1986. The pathogenesis of atherosclerosis—An update. *N. Engl. J. Med.* 314: 488–500.

Ross, R. and Glomset, J. A. 1973. Atherosclerosis and the arterial smooth muscle cell: Proliferation of smooth muscle is a key event in the genesis of the lesions of atherosclerosis. *Science* 180: 1332–1339.

Ross, R. and Klebanoff, S. J. 1971. In vivo synthesis of connective tissue proteins. *J. Cell Biol.* 50: 159–171.

Saar, B. G., Freudiger, C. W., Reichman, J. et al. 2010. Video-rate molecular imaging in vivo with stimulated Raman scattering. *Science* 330: 1368–1370.

Sanz, J. and Fayad, Z. A. 2008. Imaging of atherosclerotic cardiovascular disease. *Nature* 451: 953–957.

Shen, Y. R. 1989. Surface properties probed by SHG and SFG. *Nature* 337: 519–525.

Stary, H. C. 2000. Natural history and histological classification of atherosclerotic lesions: An update. *Arterioscler. Thromb. Vasc. Biol.* 20: 1177–1178.

Stary, H. C., Blankenhorn, D. H., Chandler, A. B. et al. 1992. A definition of the intima of human arteries and of its atherosclerosis-prone regions: A report from the Committee on Vascular Lesions of the Council on Arteriosclerosis, AHA. *Circulation* 85: 391–405.

Stary, H. C., Chandler, A. B., Dinsmore, R. E. et al. 1994. A definition of initial, fatty streak, and intermediate lesions of atherosclerosis: A report from the committee on vascular lesions of the council on arteriosclerosis, AHA. *Circulation* 14: 840–856.

Stary, H. C., Chandler, A. B., Dinsmore, R. E. et al. 1995. A definition of advanced types of atherosclerotic lesions and a histological classification of atherosclerosis: A report from the committee on vascular lesions of the council on arteriosclerosis, AHA. *Circulation* 92: 1355–1374.

Sturek, M., Alloosh, M., Wenzel, J. et al. 2007. Ossabaw Island miniature swine: Cardiometabolic syndrome assessment, in *Swine in the Laboratory: Surgery, Anesthesia, Imaging, and Experimental Techniques*, M. M. Swindle (ed.), Chapter 18. Boca Raton, FL: CRC Press.

Su, J., Tomov, I. V., Jiang, Y., and Chen, Z. 2007. High-resolution frequency-domain second-harmonic optical coherence tomography. *Appl. Opt.* 46: 1770–1775.

Tang, S., Sun, C.-H., Krasieva, T. B., Chen, Z., and Tromberg, B. J. 2007. Imaging subcellular scattering contrast by using combined optical coherence and multiphoton. *Opt. Lett.* 32: 503–505.

Vallabhajosula, S. and Fuster, V. 1997. Atherosclerosis: Imaging techniques and the evolving role of nuclear medicine. *J. Nucl. Med.* 38: 1788–1796.

Virmani, R., Kolodgie, F. D., Burke, A. P., Farb, A., and Schwartz, S. M. 2000. Lessons from sudden coronary death: A comprehensive morphological classification scheme for atherosclerotic lesions. *Arterioscler. Thromb. Vasc. Biol.* 20: 1262–1275.

Wang, H.-W., Langohr, I. M., Sturek, M., and Cheng, J.-X. 2009. Imaging and quantitative analysis of atherosclerotic lesions by CARS-based multimodal nonlinear optical microscopy. *Arterioscler. Thromb. Vasc. Biol.* 29: 1342–1348.

Wang, H.-W., Le, T. T., and Cheng, J.-X. 2008. Label-free imaging of arterial cells and extracellular matrix using a multimodal nonlinear optical microscope. *Opt. Commun.* 281: 1813–1822.

Yu, W., Braz, J. C., Dutton, A. M., Prusakov, P., and Rekhter, M. 2007. In vivo imaging of atherosclerotic plaques in apolipoprotein E deficient mice using nonlinear microscopy. *J. Biomed. Opt.* 12: 054008.

Yusuf, S., Reddy, S., Ounpuu, S., and Anand, S. 2001. Global burden of cardiovascular diseases part I: General considerations, the epidemiologic transition, risk factors and impact of urbanization. *Circulation* 104: 2746–2753.

Zandvoort, M., Engels, W., Douma, K. et al. 2004. TP microscopy for imaging of the vascular wall: A proof of concept study. *J. Vasc. Res.* 41: 54–63.

Zipfel, W. R., Williams, R. M., Christie, R. et al. 2003. Live tissue intrinsic emission microscopy using multiphoton-excited native fluorescence and second harmonic generation. *Proc. Natl Acad. Sci. USA* 100: 7075–080.

Zoumi, A., Lu, X., Kassab, G. S., and Tromberg, B. J. 2004. Imaging coronary artery microstructure using SHG and TPEF. *Biophys. J.* 87: 2778–2786.

# 26. Applications of CARS Microscopy to Tissue Engineering

**Annika Enejder and Christian Brackmann**



## 26.1 Overview

With this overview, we highlight the potential of CARS microscopy for visualization of the soft and living components forming the building blocks of replacement tissues: fully functional cells—either pre-cultured or in-growing cells from the host tissue—arranged in tissue-like hierarchical structures supported by natural or synthetic polymers. By probing vibrations characteristic for the cell membrane and the polymer, respectively, unique insights can be gained in (1) if and how the cells establish contact with the supporting scaffold; (2) whether the cells proliferate, differentiate, and function as native cells; and (3) how the architecture of the scaffold determines and can control the distribution and alignment of cells. The benefits of CARS microscopy here come to their best use; with minimal sample preparation, cell adhesion, proliferation, and integration can be studied in real time and under biologically realistic conditions. In addition, the full three-dimensional arrangement of the cells and the scaffold polymers can be visualized down to sub-micron level, also in thicker samples. With these unique capabilities, CARS microscopy has the potential to become an important instrument within tissue engineering.

*Coherent Raman Scattering Microscopy*. Edited by Ji-Xin Cheng and X. Sunney Xie © 2013 CRC Press/Taylor & Francis Group, LLC. ISBN: 978-1-4398-6765-5.

## 26.2 Introduction to Tissue Engineering

Tissue engineering and regenerative medicine seeks for fully functional alternatives to autologous implants for the replacement of damaged tissues or organs with reduced function. Complete recovery from a multitude of deadly and debilitating conditions is thereby foreseen, such as myocardial infarction, spinal injury, osteoarthritis, osteoporosis, diabetes, liver cirrhosis, and retinopathy, presently requiring daily medication or treatment for survival. An increasing need for replacement tissues is expected in the coming decades due to the aging population, wishing for a healthy and active life despite injuries and reduced function of some organs. Already in the year 2000, the lives of over 20 million patients were sustained or significantly improved by functional organ replacement and since then it has been estimated that the impacted population has grown at over 10% per year (Lysaght and O'Loughlin, 2000). In order to meet this increasing need, large efforts are presently made to design innovative, biocompatible materials that can support the regeneration of fully functional cells and organize them into tissue-like architectures. Artificial tissues are in general composed of a soft polymer matrix guiding and supporting the adhesion and integration of cells, either by seeding the scaffold with cells *in vitro*—sampled from functional parts of the host organ—or by promoting natural in-growth of cells from the surrounding tissue following implantation. For an optimal design of these innovative and complex constructs, deep and fundamental insights into the function and organization of cells into tissues are needed. Tissues consist of the most fascinating materials with an amazing degree of variation in structure and functionality, each optimized for its specific task through million years of evolution. In humans the astonishing number of 411 different cell types has been identified; there are, for example, epithelial cells forming smooth cell layers, assuring low-friction flow and efficient diffusion of gases, muscle cells containing long chains of contractile units, osteoblasts manufacturing inorganic bone minerals, and neurons possessing a branched tree of filaments transferring signals as electrochemical pulses (Vickaryous and Hall, 2006). The cells are organized as tissues by the extracellular matrix (ECM) composed of three major components (Lodish et al., 2008): (1) insoluble collagen and elastin fibers, which provide strength and resilience to the tissue; (2) adhesive matrix proteins (fibronectin, laminin), which anchor the cells to the ECM fibers; and (3) polysaccharides (glycosaminoglycans) linking the fibers and other ECM proteins into higher-order structural ECM modules and gel-like extracellular fillers (proteoglycans). These components can be assembled in an endless number of ways, and depending on their relative amounts and spatial arrangements, matrices with a wide range of properties are formed. Gel-like proteoglycans without any fibrous components form a plastic, protective, and insulating environment optimal for neurons. A highly organized collagen meshwork strengthened by calcium phosphate (hydroxyapatite) gives bone tissue its unique properties of rigidity and resilience with a tensile strength corresponding to that of steel (Levental et al., 2007), whereas a matrix with high contents of elastin fibers such as in arteries enables a repeated expansion/relaxation of more than a billion times during life without any signs of degradation (Daamen et al., 2007).

The challenge in the research field of tissue engineering is to create implants mimicking these highly sophisticated materials. Rather than attempting to recreate the complexity of entire organs *ex vivo*, a more realistic strategy is to develop synthetic matrices of polymers that are able to establish key interactions with cells and thereby unlock their innate powers of regeneration and overall organization. The chemical and micromechanical composition of the matrix has here been shown to play a crucial role, determining which cell lineage is developed, as well as cell spreading and proliferation (Engler et al., 2006). This is not so surprising considering that the ECM communicates directly with the intracellular signaling system via integrins and indirectly by controlled release of growth factors and other hormones, being stored in the proteoglycan filler (Berrier and Yamada, 2007; Hynes, 2009). Thus, one of the primary targets within the field is to construct biocompatible scaffold materials with macromolecular and morphological compositions that promote natural cell adhesion and regeneration. Supported by the recent advancements within the bio- and nanotechnologies, highly sophisticated scaffold materials emerge with tailored properties similar to those of native ECMs. A wide range of fibrous networks are constructed from decellularized tissues (Gilbert et al., 2006), as well as from self-assembled and electrospun bio- or synthetic polymers such as amyloid fibers, different polypeptides, collagens, elastin, fibrinogen, poly(glycolic acid) (PGA), and poly(lactic acid) (PLA) (Barnes et al., 2007; Zhang, 2003). Different hydrogels of ECM polysaccharides and correspondents can be shaped into porous scaffolds (Geckil et al., 2010) as well as from cellulose fibrils woven by bacteria (Czaja et al., 2007). Proteins with controlled higher-order structures and domains with different functionalities can be genetically encoded in DNA plasmids (Sengupta and Heilshorn, 2010). Laser microablation of polymers has been applied to construct repetitive, well-defined shapes such as honeycomb scaffolds, mimicking the ECM of heart tissue (Engelmayr et al., 2008). Biological, chemical, electrical, thermal, and mechanical properties have been further fine-tuned by forming composites of different fibers (Brown and Laborie, 2007; Heydarkhan-Hagvall et al., 2008; Luo et al., 2008) or together with, for example, ceramic materials (Thomas et al., 2007) and carbon nanotubes (Yan et al., 2008). Altogether, there are an impressive number of advanced technologies to produce artificial tissues available, and with the exciting recent innovations made within the bio- and nanotechnologies, proteomics, and stem cell biology, highly sophisticated materials are to be expected in the near future.

In order to assess the capability of these material as artificial ECMs, three-dimensional visual information is needed (1) on the scaffold morphology, typically composed of a hierarchical arrangement of nano- to microscale components, (2) with chemical specificity, so that the relative organization of individual building blocks in multi-component scaffolds can be distinguished and related to micromechanical properties, and not the least (3) on the interaction with different kinds of cells. Clearly, tissue engineering materials challenge present imaging technology from many aspects. They are comparatively thick samples requiring optical sectioning and a large probe depth. They are soft, living materials, preferably investigated without labeling or further sample preparation. In order to fine-tune the micromechanical properties, their chemical composition becomes increasingly complex, involving many different units in a broad size range spanning 100 nm to 100 μm. This requires the combination of sub-micron resolution and large field-of-view imaging. None of the microscopy techniques routinely applied within the field of tissue engineering (electron microscopy, fluorescence microscopy, atomic force microscopy, FTIR microscopy) fulfills these requirements. CARS microscopy, either alone or in combination with other nonlinear microscopy techniques, possesses all these features as will be illustrated in this overview. It has therefore the potential to become an important instrument within the dynamic research fields of tissue engineering and regenerative medicine.

## 26.3 CARS Microscopy of Tissue Scaffolds

The fundamental building block of the tissue scaffold is one or several polymers of either natural origin such as polypeptides and polysaccharides or different synthetic polymers such as PGA and PLA, hierarchically arranged as fibrous networks, foams, and fibrous or non-fibrous hydrogels (Lutolf and Hubbell, 2005). All these polymers exhibit high contents of carbon–hydrogen (C–H) bonds and consequently strong vibrational modes in the C–H stretch region ~2800–3000 $cm^{-1}$, which can readily be probed in CARS microscopy. Based on a long history of Raman spectroscopy, focusing on the so-called fingerprint region below ~1700 $cm^{-1}$, spectroscopic data in the C–H stretch region are scarce for scaffold materials. This makes it difficult to find relevant parameters for CARS microscopy in the scientific literature and motivates a compilation of the vibrational modes in this region for the most important scaffold materials being used within tissue engineering, as presented in Table 26.1. While most natural polymers exhibit vibrations at ~2885 and ~2940 $cm^{-1}$ (asymmetric $CH_2$ stretch vibrations), polypeptides and

**Table 26.1** Raman-Active Vibrations in the C–H Stretch Region of Important Tissue Scaffold Polymers

| | References | 2860–2989 $cm^{-1}$ | 2905–2916 $cm^{-1}$ | 2930–2943 $cm^{-1}$ | 2954–2965 $cm^{-1}$ | 2978–2993 $cm^{-1}$ |
|---|---|---|---|---|---|---|
| ***Polypeptides*** | | | | | | |
| Collagen | Edwards et al. (1997) | 2884 | | 2940 | | 2978 |
| Elastin | Green et al. (2008) | 2884 | | 2940 | | 2978 |
| ***Polysaccharides*** | | | | | | |
| Glycosamino-glycans | Ellis et al. (2009) | | 2905 | 2941 | | |
| Cellulose I (native) | Wiley and Atalla (1987) | 2885 | | 2941 | 2965 | |
| Cellulose II (synthetic) | Fischer et al. (2005) | 2889 | | 2943 | 2963 | |
| Chitin | Focher et al. (1992) | 2874 | | 2932 | 2963 | |
| ***Synthetic polymers*** | | | | | | |
| Poly glycolic acid (PGA) | Kister et al. (1997) | | | | 2954 | 2988 |
| Poly lactic acid (PLA) | Kister et al. (2000) | 2874 | | 2940 | | 2993 |
| Polyurethanes (PU) | Roohpour et al. (2009) | 2860 | | 2930 | | |
| Polycaprolactone (PLC) | Kister et al. (2000) | 2866 | 2916 | | | |

polysaccharides can be distinguished by their vibrations at 2978 cm$^{-1}$ (symmetric $CH_3$ stretch vibration) (Edwards et al., 1997) and 2965 cm$^{-1}$ (CH and $CH_2$ stretch vibrations) (Wiley and Atalla, 1987), respectively. Glycosaminoglycans lack the 2885 cm$^{-1}$ vibration but instead exhibit a characteristic vibration at 2905 cm$^{-1}$ (Ellis et al., 2009), which could form the basis for chemically selective imaging. Synthetic polymers show a more individual composition of vibrations; for instance PGA exhibits vibrations in the high-frequency region only (2954–2988 cm$^{-1}$) (Kister et al., 1997), whereas the two vibrations of PLC can be found at significantly lower frequencies (2866–2916 cm$^{-1}$) (Kister et al., 2000). Also the individual polymer units in a blend of PGA and PLA can be visualized by dual-vibration CARS microscopy probing the frequencies 2954 and 2874 cm$^{-1}$, respectively. From this we conclude that there are distinct spectral differences also in the C–H stretch region, which enables chemically specific imaging of multi-component scaffolds and integrating native ECM components. This should be further supported by the introduction of fiber-based white-light laser sources or fast wavelength scanning for CARS microspectroscopy, enabling monitoring of multiple components in the complex composite materials emerging within the field by characterizing the entire C–H stretch region in great detail.

CARS spectra of a few natural and synthetic tissue scaffold polymers are displayed in Figure 26.1, illustrating that their characteristic spectral shapes reported in the Raman literature remain, despite the influence of the unspecific nonresonant background

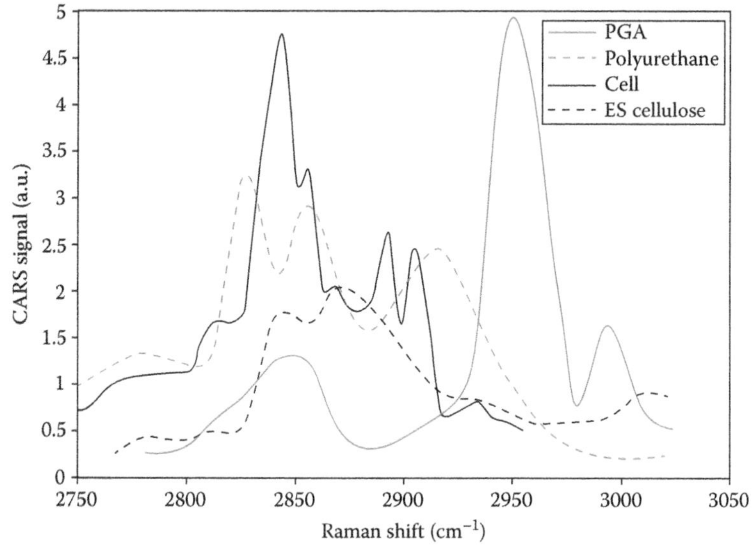


**FIGURE 26.1** CARS microscopy spectra in the C–H vibration region of important tissue scaffold materials: PGA (gray line), polyurethane (dashed gray line), and electrospun cellulose (dashed black line). All materials exhibit characteristic spectral shapes, which allow them to be distinguished in chemically complex and composite environments. Vibrational resonances for further scaffold materials are listed in Table 26.1. For comparison, a CARS spectrum of a fundamental cell model, the yeast cell, is shown (black line). The primary vibration utilized to visualize cells, 2845 cm$^{-1}$ (characteristic for lipid-rich cellular components such as the membrane and lipid stores), can here be identified as the most prominent peak, clearly distinctive from any of the resonances of the scaffold materials. This enables chemically specific imaging of the bio-functional units of artificial tissues simultaneously with its structural components and thus detailed investigations of cell–scaffold interaction mechanisms.

generated by all molecules in the probe volume. In general, the impact of the nonresonant background can be observed as a spectrally independent off-set to the resonant CARS signal together with a dispersive shift of the vibrational peak to lower Raman shifts, originating from the mixing between the resonant and nonresonant parts of the CARS signal (Lotem et al., 1976). The spectra in Figure 26.1 indicate a relatively small nonresonant contribution compared to the resonant part, probably as a result of the strong resonances of the different C–H bonds present at large concentrations in these polymers.

In this context, a comparison with a typical CARS spectrum of cells in the corresponding vibrational region is relevant, in order to identify vibrations where cellular components selectively can be visualized and separated from the scaffold fibers. Only then can detailed and reliable information on the contact established between the cells and the scaffold be extracted from the images. Cellular components that give the strongest resonant CARS signal and best image contrast when probing the C–H vibrational region are intracellular lipid stores and cell membranes. An average CARS spectrum of these components in one of the most widely used biological cell models—the yeast cell—is exemplified in Figure 26.1, showing a signature typical for lipids significantly different from that of the scaffold materials. Figure 26.1 suggests that optimal chemical contrast for selective imaging of cells interacting with a polyurethane scaffold is achieved by dual-vibration CARS microscopy probing ~2845 $cm^{-1}$ (cells) and ~2917 $cm^{-1}$ (scaffold). The CH-stretch vibration at ~2845 $cm^{-1}$ is also favorable to monitor cells in scaffolds from electrospun cellulose and PGA, but here combined with the characteristic vibrations at 2872 (cellulose) and 2948 $cm^{-1}$ (PGA), respectively, high-lightening the scaffolds. Compared to the listed Raman vibrations found in the literature (Table 26.1), optimal frequencies for CARS are slightly lower as a result of the previously discussed dispersive effect due to the mixing with the nonresonant background (Lotem et al., 1976).

Figure 26.2 shows a collection of CARS microscopy images of different tissue scaffolds. In Figure 26.2A, a dense web of cellulose fibers bioengineered by the bacteria *Gluconacetobacter xylinus* is visualized by CARS microscopy. These fibers—just as the more abundant plant cellulose fibers—consist of unidirectionally aligned glycan chains, interlocked by hydrogen bonds so that crystalline subunits are produced (Cellulose I). However, the microbial cellulose has a unique collagen-like structure, significantly different from that of plant cellulose fibers. It has a hierarchical structure where the smallest unit consists of nine parallel polyglycan chains forming left-handed triple helices co-assembled into a twisted protofibril (Ross et al., 1991). Multiple protofibrils are bundled into 4–7 nm sized crystalline microfibrils, which in turn form ribbons with a cross section of ~7 × 70–145 nm and finally bands with a width of ~500 nm (Klemm et al., 2005). This hierarchical arrangement forms an extensive surface area for the binding of water molecules. Dielectric spectroscopy confirms that each microfibril is surrounded by a water sheet (Gelin et al., 2007), which compiled as a cellulose matrix results in a hydrogel-like material consisting of $99\%_{wt}$ water. Due to the high water content relative to cellulose, CARS microscopy probing the CH-vibration region is infeasible for this material since the insignificant resonant signal from the cellulose microfibrils is overwhelmed by the nonresonant contribution from all water molecules, resulting in a featureless CARS microscopy image (data not shown). Instead, these cellulose ribbons characterized by high amounts of bound water can be outlined by probing the broad OH vibration

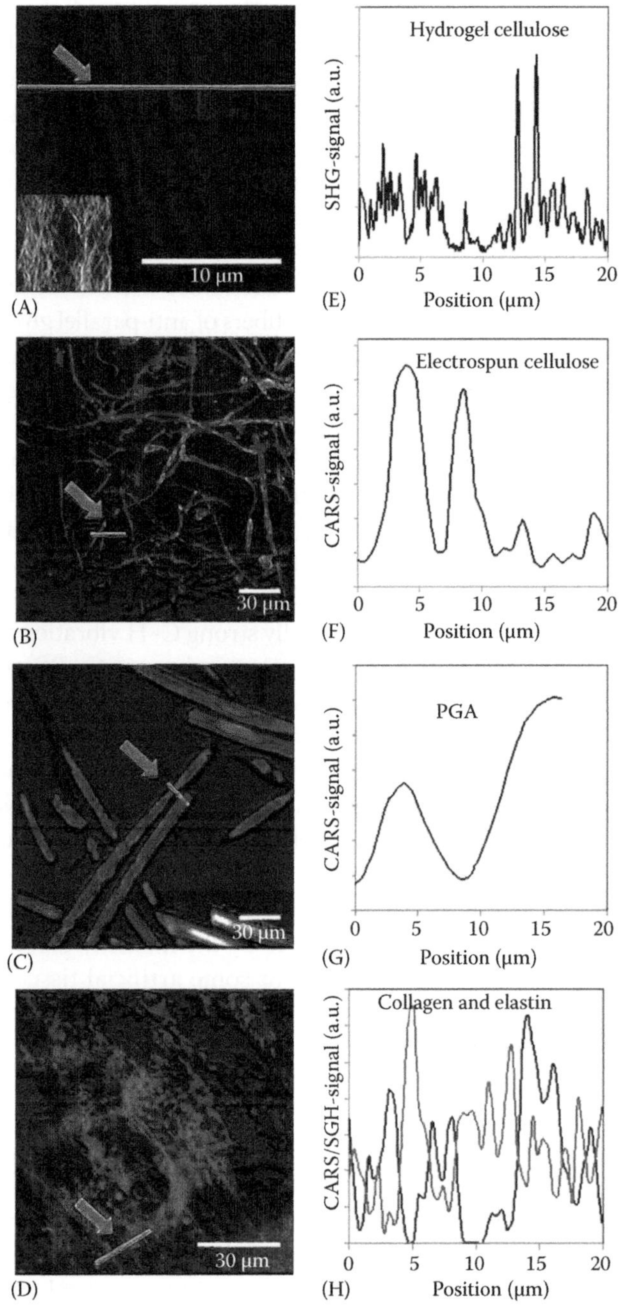


**FIGURE 26.2** CARS microscopy images of native and artificial tissue scaffolds: (A) hydrogel cellulose woven by the bacteria *Gluconacetobacter xylinus*, probed at the 3220 cm$^{-1}$ OH vibration, characteristic for water and cellulose (the inset is the corresponding SHG microscopy image, showing the cellulose fibers with better contrast); (B) compact electrospun cellulose fibers with low water content, better probed at the C–H vibration ~2972 cm$^{-1}$; (C) PGA probed at the strong C–H vibration ~2950 cm$^{-1}$; and (D) arterial tissue from sheep showing the distribution of elastin (magenta, CARS microscopy at 2845 cm$^{-1}$) and collagen (blue, SHG microscopy). Note the different length scales used in the images. Intensity profiles along the lines (yellow) indicated by arrows are plotted in (E–H) on the same length scale (0–20 μm) in order to highlight the vastly different fiber dimensions and for comparison with typical dimensions of native ECM components (H).

(here the flank ~3220 $cm^{-1}$), distinctive for both water and cellulose; see Figure 26.2A. However, as shown in the inset of Figure 26.2B hydrogel cellulose is better visualized by second harmonic generation (SHG) microscopy, as a result of the collagen-like twisted arrangement of the microfibrils rendering them SHG active. Cellulose scaffolds can also be formed by electrospinning a solution of plant cellulose, with the advantage that fiber dimensions and scaffold densities can be precisely controlled and optimized to the preferences of the host tissue and cells. Electrospinning requires the cellulose to be chemically dissolved and regenerated back to pure cellulose by hydrolysis (Frey, 2008). The parallel alignment of the glycan chains found in native cellulose and the hierarchical composition of the cellulose fibers is then lost, and compact fibers of anti-parallel glycan chains (Cellulose II) are instead formed. While this macromolecular arrangement is unfavorable for SHG microscopy for symmetry reasons, the compact fibers of pure cellulose formed in the electrospinning post-processing generate a strong resonant CARS signal at ~2872 $cm^{-1}$ with little interference from the nonresonant background of microfibril-bound water. A CARS microscopy image of the morphology of an electrospun cellulose scaffold is exemplified in Figure 26.2B. The fibers of synthetic polymers used for tissue scaffolds are also typically composed of compact matter, making them readily detectable by CARS microscopy. CARS microscopy of PGA is particularly favorable, since it in addition to the compact polymer arrangement exhibits an exceptionally strong C–H vibration at ~2950 $cm^{-1}$ (compare CARS spectrum in Figure 26.1). Figure 26.2C shows a CARS microscopy image of a PGA tissue scaffold with large inter-fiber distance, favorable for three-dimensional cell integration. The images of the different scaffold materials (Figure 26.2A through C) and the corresponding intensity profiles in Figure 26.2E through G illustrate that vastly different morphologies can be manufactured. As indicated in the profile plot in Figure 26.2E, the hydrogel cellulose consists of a dense network of 100–500 nm sized bundles, resulting in a high tensile strength, though with little space for deep cell integration (Bäckdahl et al., 2006). Scaffolds with thicker and less densely distributed cellulose fibers can be manufactured by electrospinning, as indicated by the intensity plot in Figure 26.2F. This promotes deeper cell integration, favorable for some artificial tissue constructs. In general, the electrospun cellulose scaffold has the advantage that it can be shaped in different designs with fiber diameters in the range of 0.3–3 μm, aligned or randomly oriented at different densities (Frey, 2008), in order to meet cell-specific requirements on the extracellular environment for optimal cell integration. A scaffold with even larger space between multi-micron-sized fibers can be achieved by the use of the artificial polymer PGA, as illustrated by the CARS microscopy image in Figure 26.2C and the corresponding intensity profile plot in Figure 26.2G. This architecture promotes not only integration of larger cells, but also *in situ* vascularization—a crucial issue for long-term survival of cells in a three-dimensional tissue scaffold. For comparison, the structural arrangement of native ECM in arterial tissue is depicted in Figure 26.2D with the corresponding intensity profiles in Figure 26.2H. A fine web of sub-micron-sized elastin fibers is here complemented by mechanically strong, thicker collagen bundles, forming a highly tensile scaffold able to accommodate to the pulsed blood flow while still being strong enough to sustain the high pressures generated by the beating heart muscle. The two images were collected by simultaneous CARS (2845 $cm^{-1}$) and SHG microscopy of the elastin and collagen network in the tissue, respectively. This highlights the need for the design of more advanced, multi-component scaffolds in order to mimic the complex function and material properties of

ECM in native tissue. Thus, a wide range of multi-component and multi-functional artificial tissue scaffolds are presently being suggested (Kwon and Matsuda, 2005; Li et al., 2006; Brown and Laborie, 2007; Thomas et al., 2007; Heydarkhan-Hagvall et al., 2008; Luo et al., 2008). This signifies a new era in the development of tissue scaffolds, from production of materials with advantageous properties at a macroscopic scale (e.g., high mechanical strength and elasticity) to customized assembling of bioengineered or native ECM molecules forming microenvironments chemically and morphologically favorable for the adhesion and proliferation of specific cell lines. With this paradigm shift, new techniques for microscale characterization are needed in the field of tissue engineering. Considering the unique capabilities of CARS microscopy with chemically specific imaging of soft matter, enabling label-free, three-dimensional characterization at sub-micron resolution of the individual components in artificial tissue scaffolds, it has the potential to become an important tool within the field.

## 26.4 CARS Microscopy of Cell Adhesion

While the scaffold sets the macroscopic material properties of the artificial tissue such as elasticity and mechanical strength, it also determines the microscopic morphology and surface chemistry crucial for close and natural interaction with living cells. Cell interaction takes place by the formation of adhesion complexes, which rather than fix, static contact points can be considered as dynamic extracellular sensing organs. The adhesion complexes consist of local assemblies of transmembrane receptors in the cell membrane with integrins being the most important category. The integrins are composed of a large extracellular domain that binds to proteins (e.g., fibronectin, vitronectin, and laminin) in the ECM/scaffold and a short cytoplasmic domain that links to and reorganizes the actin cytoskeleton (Hynes, 2002). The interaction with artificial scaffolds usually takes place through adsorbed proteins from the culture medium (Nikolovski and Mooney, 2000) or active surface modification with ECM components (Sreejalekshmi and Nair, 2011). The adhesion complexes provide the cell with important information on the local microenvironment, integrated through different cellular signaling networks, and eventually generate responses that control fundamental functions such as cell survival, proliferation, migration, and differentiation. Thus, the adhesion complexes represent major molecular hubs, where mechanical forces from the surroundings and biochemical signals converge. The spatial/temporal distribution, traction forces, and lifetime of the adhesion complexes are primarily set and can be controlled by the properties of the scaffold: (1) its surface chemistry determining the density and kinds of surface-adsorbed proteins; (2) the micromechanical properties such as rigidity and friction; and (3) the micro-geometry: two-dimensional versus three-dimensional substrates, aligned or randomly oriented scaffold components, porous or compact matrices, etc. (Vogel and Sheetz, 2006). Based on the spatial/temporal/affinity characteristics of the adhesion complexes, they can be classified into three main groups: focal complexes, focal adhesions, and fibrillar adhesions (Geiger et al., 2001). Focal complexes are transient dot-like structures (~1 μm) present at the edges of lamellipodia or filopodia, that is, cell extensions formed during cell migration. Thus, they can usually be found immediately behind the leading edge of spreading or migrating cells. Focal adhesions are flat, elongated structures (2–5 μm) located at the periphery of cells, forming strong and mature contact points. Finally, fibrillar adhesions are highly stable complexes enriched in the central region of the cell primarily as elongated complexes of sizes up to 10 μm, but also as dots (1 μm).

Over the last decades, the molecular diversity of these adhesions and their roles in cell migration and function in a matrix of ECM components have been extensively studied, in particular by immunostaining the receptor proteins and the actin filament, visualizing them by fluorescence microscopy (Worth and Parsons, 2008). Thus, microscopy-based research has a central role in the characterization and understanding of these mechanisms. The interest for new technologies is growing with higher requirements on resolution and long-term live-cell measurements, and with an increasing awareness for the difficulties and uncertainties associated with the use of fluorescent marker molecules. In particular, labeling of cells in a three-dimensional matrix requires special protocols and extensive experience to accommodate the long time required for antibodies to diffuse due to the limited permeability. Still, it frequently results in inhomogeneous staining, which then is misinterpreted as the establishment of few adhesion points. Thus, there is a great desire within the field for label-free microscopy techniques in order to study cell adhesion in three-dimensional tissue scaffolds avoiding these uncertainties. Clearly, CARS microscopy has here an important task to fulfill. Though CARS microscopy is not able to visualize individual transmembrane proteins, it can provide important complementary information on adhesion-associated shape and density changes in the cell membrane by probing the CH-vibration at 2845 $cm^{-1}$. Images in the present work show that this kind of information reveals which adhesion mechanisms are involved. Furthermore, few reports exist on the role of the cell membrane in conjunction with cell adhesion on tissue scaffolds. Considering the substantial risk that bulky labeling molecules alter the physical and dynamic properties of the proteins and the lipid bilayer of the membrane, it is questionable whether a representative picture is given by fluorescence microscopy. With the unique capabilities of CARS microscopy for three-dimensional, label-free, live-cell imaging, it has a strong potential to shine new light on the *dynamics* of cell–matrix adhesion process in realistic three-dimensional scaffolds, including on the role of the cell membrane.

The potential of CARS microscopy for visualization of cell adhesion is exemplified in Figures 26.3 and 26.4, showing different kinds of adhesion mechanisms. In Figure 26.3, a series of CARS/SHG co-localization images of a detail of an artificial blood vessel is shown. The blood vessel, composed of a cellulose scaffold, is explanted following a biocompatibility study in rat. The cellulose fibers of the scaffold are visualized by SHG microscopy (excitation at 817 nm and detection at 409 nm) and color coded in blue. The host-tissue cells (here lipid-rich adipocytes) probed at the 2845 $cm^{-1}$ vibration by CARS microscopy (excitation at 817 nm in combination with 1064 nm and detection at 663 nm) appear in yellow. Figure 26.3C is a close-up of the sub-area marked in Figure 26.3B, which in turn is a close-up of the marked area in Figure 26.3A. Already from Figure 26.3A it is clear that the cell has formed a tight contact with the cellulose scaffold, and in Figure 26.3B localized bands of cellulose fibers aligned across the contact border between the cell and the scaffold can be distinguished. These bundles most likely represent external parts of the adhesion points, in turn linked to localized assemblies of transmembrane proteins at the corresponding site underneath in the cell membrane. No signs of intensity variations or shape changes can be observed in the CARS microscopy images of the cell. However, the CARS images must in this particular case be interpreted with care, since the dominant part of the CARS signal most likely originates

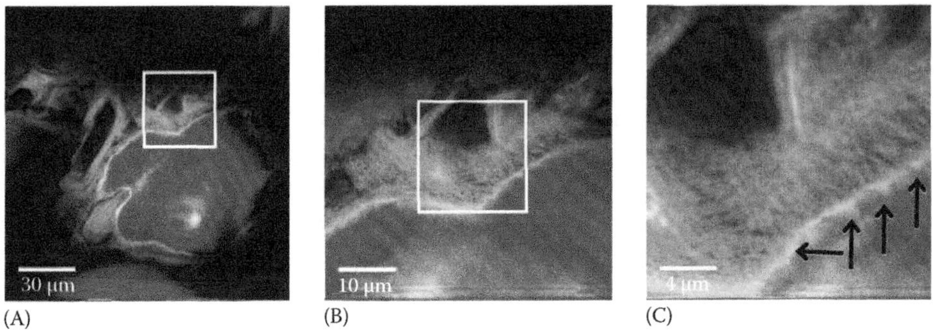


**FIGURE 26.3** Cells (adipocytes shown in yellow, CARS microscopy at 2845 cm$^{-1}$) integrated from the host tissue into the cellulose scaffold (blue, SHG microscopy) of an artificial blood vessel previously implanted in rat. (B) is a close-up of the area marked in (A), revealing aligned fibers across the cell–scaffold contact border. By zooming in further in (C) (corresponding to marked area in B), the extracellular parts of individual adhesion complexes can be identified as ~2 μm long and ~1 μm wide cellulose bundles, highlighted by arrows. No filopodia can be distinguished, which together with the size and location (the periphery of the cell) of the adhesion points indicates that they can be categorized as focal adhesions.

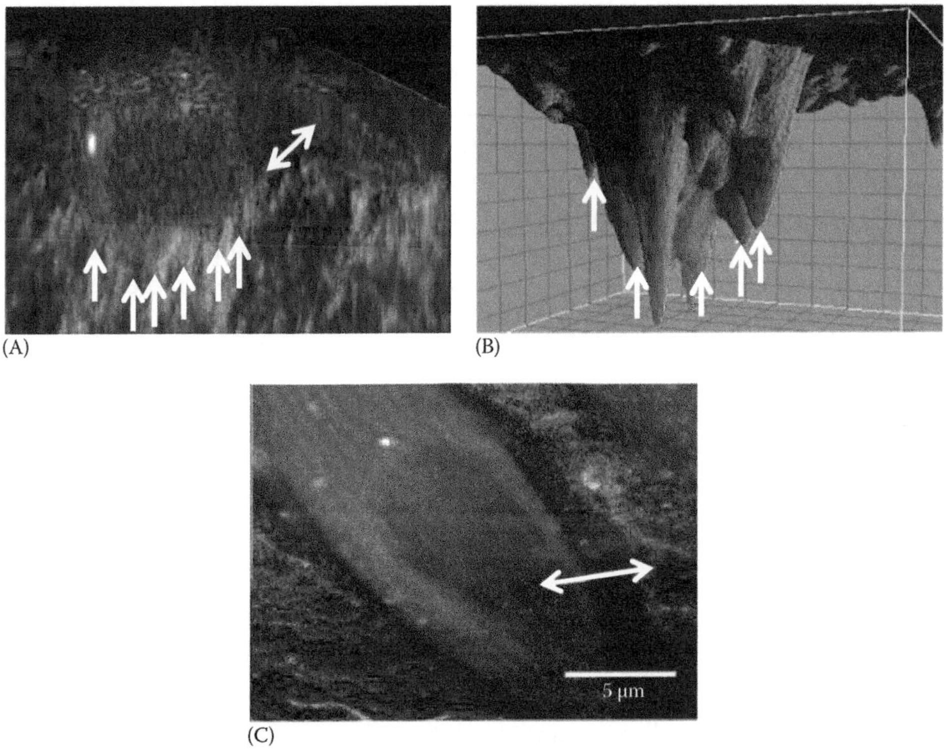


**FIGURE 26.4** Smooth muscle cells (SMCs in orange, CARS microscopy at 2845 cm$^{-1}$) seeded on a cellulose scaffold (blue, SHG microscopy). (A) Volume image showing a cross section of a single SMC and individual adhesion complexes formed between the cell and scaffold fibers, as indicated by the arrows underneath the cell. The cell reaches out to the scaffold fibers by the formation of filopodia, which suggests that adhesion points are focal complexes. (B) The entire distribution of filopodia under the cell showed as a CARS intensity surface plot. (C) The central plane (perpendicular to the cross section shown in (A)) of the z-stack forming the volume image (A) reveals that connections are also formed with neighboring cells as indicated by the arrow. This connection is also pointed out in (A).

from the periphery of the dense lipid store filling up the entire cell volume of the adipocyte. Small variations in the cell membrane structure would here be difficult to identify. In the close-up (Figure 26.3C), the extracellular parts of individual adhesion complexes can be distinguished as ~2 μm long and ~1 μm wide cellulose bundles, highlighted by arrows. Their size (1–2 μm), shape (elongated), and location (the periphery of the cell) all suggest that these adhesion points can be categorized as focal adhesions. This indicates that cellulose scaffolds enable anchorage in the host tissue by the formation of natural adhesion complexes with surrounding cells *in situ*.

Similar observations were made by CARS microscopy also when cellulose scaffolds were seeded with cells *in vitro* (Brackmann et al., 2011a). With the aim to develop functional artificial blood vessels capable of pulsatile dilatations and contractions for natural blood flow and reduced risk for clot formation, cellulose scaffolds were seeded with smooth muscle cells (SMCs) and studied by simultaneous CARS (2845 $cm^{-1}$, cells) and SHG (409 nm, cellulose scaffold) microscopy. Close-up images reveal how SMCs form tight contact with individual cellulose bundles at the periphery of the cells in a similar way as the adipocytes in Figure 26.3, that is, they establish focal adhesions with the scaffold. In addition, since the *in vitro* growth of the SMCs allows collection of images throughout the entire cell integration process—from the initial cell adsorption to the migration of well-established cells—the more transient (focal complexes) as well as the long-term (fibrillar adhesions) contact points can be visualized. As a clear proof for the formation of focal complexes, several CARS microscopy images reveal micrometer-sized cell extensions (filopodia) connecting to cellulose bundles at the distal ends. Examples of these transient connections can be observed in the CARS volume rendering in Figure 26.4A, showing a cross section (side view) of an SMC on a cellulose scaffold. The adhesion points are here pointed out by arrows. By plotting the CARS intensity distribution in the cell membrane plane as a surface plot, a clear representation of all filopodia underneath the cell is provided (Figure 26.4B). In addition, Figure 26.4C highlights that CARS microscopy can also provide important information on cell–cell interactions. Figure 26.4C shows the central plane through the cell in Figure 26.4A, revealing that a connection to the neighboring cell to the right has been established. For clarity, the corresponding site has been indicated also in Figure 26.4A. Finally, examples of the appearance of the most stable and long-term kind of adhesions are given for a collection of SMCs in Figure 26.5A. Already after 4 days of growth, multi-micron-sized, dense local collections of cellulose can be observed underneath the central region of the cells. This is indeed representative of fibrillar adhesions, providing high stability.

Information on the type and spatiotemporal distribution of adhesion foci that cells form in their interaction with artificial tissue scaffolds is of highest interest in tissue engineering, as it determines important properties such as to which extent the scaffold supports natural cell regeneration, differentiation, and tissue-like organization. Numerous studies exist on how modifications of the scaffold micromechanical properties, morphology, and surface chemistry enhance cell proliferation and migration (Vogel and Sheetz, 2006; Sreejalekshmi and Nair, 2011). However, a characteristic of these studies is that the effects of the material modifications are assessed as cell adhesion on two-dimensional substrates rather than within three-dimensional scaffolds, quantified by automatic colorimetric methods (cell attachment evaluated as optical density)

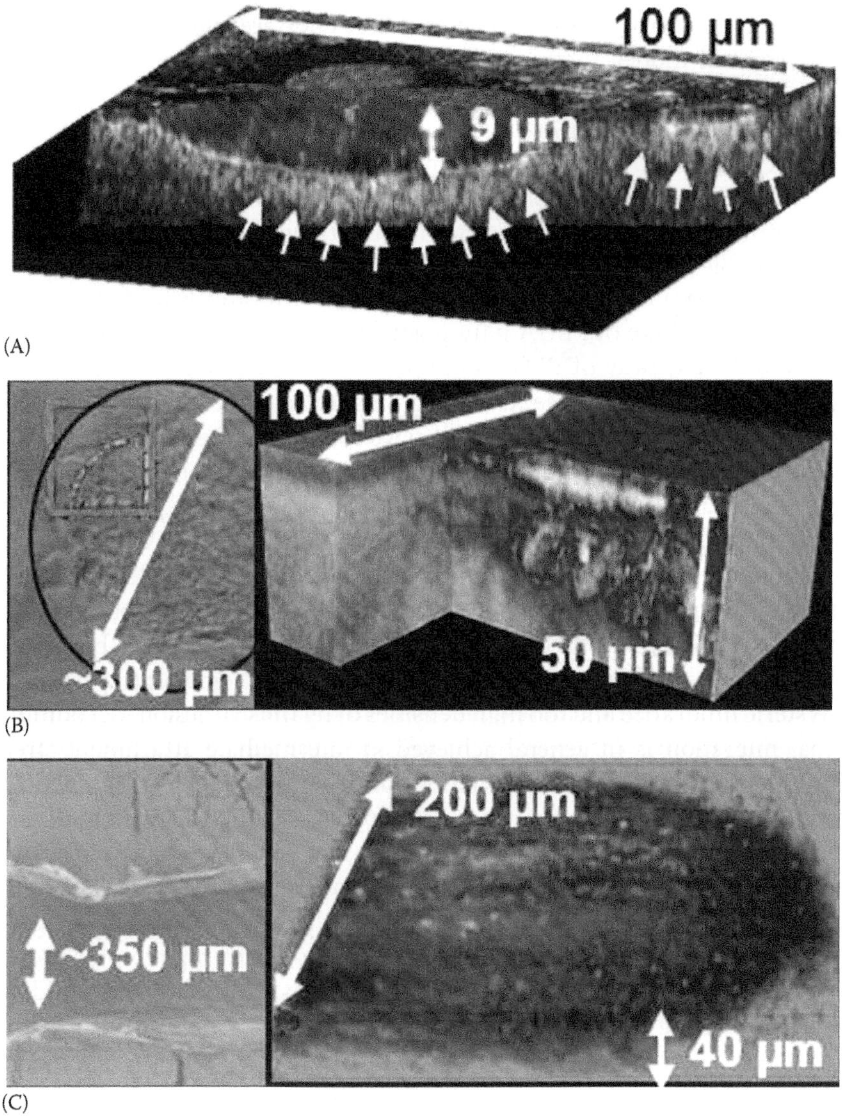


**FIGURE 26.5** CARS microscopy images (2845 $cm^{-1}$) of cell integration in scaffolds with different architectures. (A) SMCs (orange) grown 4 days on a *compact* cellulose scaffold (SHG microscopy, cyan) have integrated one to two cell layers into the scaffold, remodeled it, and formed multi-micron-sized, dense local collections of cellulose underneath the central region of the cells. (B) Osteoprogenitor cells (orange; background—low intensity: magenta) seeded on a *porous* cellulose scaffold with pore diameters of 300–500 μm, one of which is depicted by brightfield microscopy to the left. The area corresponding to the CARS volume image to the right is marked by a square and the periphery of the pore is highlighted by a black line. Cells fill up the pores, here indicated by a dashed orange line, also deeper into the scaffold. (C) Toward efficient load-bearing artificial tissues: guided migration of fibroblast cells (orange) in a cellulose scaffold with ~350 μm sized *channels*, a cross section of which is depicted by electron microscopy to the left. The alignment of the cells within the channels can be noted in the volume image to the right, also showing aligned collagen fibers (blue, SHG microscopy) generated by the fully functional cells. Thus, the peripherally located cells seem to "transmit" their alignment with the channel walls to the centrally located cells by cell–cell interactions.

or at best assessed by low-resolution fluorescence microscopy (adhesion contact area). Since micromechanical and micro-geometrical cues are significantly different in the two-dimensional model, these studies do not provide a representative picture of cell adhesion mechanisms or response in a three-dimensional scaffold (Friedl and Brocker, 2000; Pedersen and Swartz, 2005). The series of images in Figures 26.3 through 26.5 proves that CARS microscopy is able to not only generate unique visualizations of cell–matrix and cell–cell interaction mechanisms in a realistic three-dimensional tissue scaffold setting, but also provide important qualitative and quantitative information on adhesion complex types, spatial distribution, and density as well as dynamics.

## 26.5 CARS Microscopy of Cell Migration

Cell migration has primarily been studied on surfaces, where it begins with the extension of a protrusion, the formation of adhesive complexes, the generation of stress in the cytoskeleton followed by movement of the cell body, and finally the detachment of the trailing edge (Lauffenburger and Horwitz, 1996). Thus, cell migration takes place as a dynamic equilibrium between forces attaching the cell to the ECM/scaffold surface and forces trying to release the cell, which in turn depend on a complex interplay between the mechanical, geometrical, and chemical properties of the surrounding microenvironment on one hand and of the cell on the other. This is a delicate balance. Compact materials with high stiffness favor cell migration, though only until preventing factors such as steric hindrance and too high densities of ligands/adhesion foci dominate. Thus, maximal migration is in general achieved at intermediate attachment strength; low motility is observed both for combinations of scaffolds and cells that together generate too high adhesiveness since the cytoskeletal forces then are too weak to disrupt attachments and for combinations that result in too low adhesiveness since the cytoskeletal forces then cannot be transferred effectively enough (Dimilla et al., 1991). Theoretical models suggest that in order to achieve high cellular integration for a category of cells, each with a critical number and turnover of adhesion-forming receptor proteins, there exists an optimal value of stiffness for efficient force transfer that should be considered when designing tissue scaffolds (Zaman et al., 2005). Consequently, when selecting materials and designing the morphology of a tissue scaffold for a specific cell category, the tendency to express adhesion receptor proteins of the cell as well as the affinity to and organization of the cytoskeleton must carefully be considered for optimal cell integration. For instance, large spindle-shaped cells, that is, fibroblasts, endothelial cells, and many tumor cells, exhibit high integrin expression and strong cytoskeletal contractility, whereas smaller cells such as T-lymphocytes and dendritic cells express low levels of integrins. To this, the geometrical properties of the scaffold must be taken into account: the microscale roughness causing friction, compact or porous morphologies, random or aligned arrangements of scaffold components, etc. In particular, the cell response to distinct curvatures in the scaffold is of increasing interest, with the prospects of actively guiding and controlling cell migration (Luo and Shoichet, 2004; Yang et al., 2005; Badami et al., 2006). Cells orient themselves in such a way as to experience minimal distortion of the cytoskeleton (Schwarz and Bischofs, 2005), which results in contact guidance, that is, alignment and migration along fibers or on grooved surfaces. An improved understanding of these mechanisms is needed to achieve directed cell migration and promoted in-growth by intelligent design of the scaffold architecture with respect to fiber size, orientation, and surface topology (Luo and Shoichet, 2004; Yang et al., 2005; Badami et al., 2006).

Present understanding of cell integration in scaffolds is primarily based on studies of cell migration on two-dimensional surfaces, due to the limited capability of standard microscopy techniques to probe deep, particularly in these highly light-scattering materials, and due to difficulties with inhomogeneous fluorescent staining in compact, three-dimensional matrices. However, cells in a three-dimensional matrix experience a completely different environment and consequently respond differently than when they are attached to a two-dimensional surface (Friedl and Brocker, 2000; Pedersen and Swartz, 2005). An important difference is that they face spatial barriers and matrix friction when propagating through the matrix, which in turn forces them to adapt their shape and enzymatically degrade or reorganize the scaffold components. Interesting is also that fewer discrete focal contacts and intracellular stress fibers can be observed for cells migrating in three-dimensional scaffolds and that they seem to take a different role in the migration process than in conjunction with migration on surfaces. Instead of bringing the cell forward like a rolling tank, the cytoskeletal contractility is primarily utilized to clear away scaffold components, reducing the steric resistance and thereby indirectly moving the cell forward. There even exists a migration scheme largely independent of adhesion complexes for smaller cells migrating through a three-dimensional matrix, the so-called amoeboid mode of migration, where movement is achieved through the formation of pseudopods in matrix pores, pulling the cell body through the pore via cell shape changes (Friedl et al., 1998). Clearly, to fully understand the underlying mechanisms by which cells migrate in tissue scaffolds, it is necessary to study the movement of cells in three-dimensional environments.

Figure 26.5 illustrates the capabilities of CARS microscopy to provide high-quality, three-dimensional visual overviews of the cell integration process in tissue scaffolds of different characters at depths down to ~50–100 μm, providing relevant data such as cell integration depth and degree of cell alignment. In addition, an in-depth understanding of the underlying mechanism can be acquired by zooming in on individual cells in order to assess their adhesion patterns and interaction with single scaffold components at a sub-micron resolution. In Figure 26.5A, the migration of smooth muscles cells (CARS microscopy at 2845 $cm^{-1}$, color coded in orange) can be observed in a tissue scaffold of a compact arrangement of cellulose fibers (SHG microscopy at 409 nm, color coded in cyan). The cells have achieved an integration depth corresponding to one to two cell layers and formed a region with increased density of cellulose fibers immediately below the cell collection. This indicates that the migration of SMCs in three-dimensional scaffolds takes place by utilizing the cytoskeletal forces to reshape the scaffold matrix according to the "Fibroblast-like" migration scheme. This is reasonable since both cell categories host densely packed transmembrane proteins and well-developed intracellular stress filaments (Friedl and Brocker, 2000). Regions with increased cellulose fiber densities are not only observed below the cells but also beside and, for cells submerged into the scaffold, even above the cells (manuscript in preparation, Brackmann et al.). This confirms that it is not merely a passive compression effect but indeed involves the transmission of cytoskeletal forces through the adhesion complexes distributed in the entire cell membrane. In order to achieve deeper cell integration, different scaffold morphologies are required. By synthesizing the hydrogel cellulose scaffold in the presence of paraffin wax microspheres with diameters in the size range of 300–500 μm, which subsequently were leached out, a porous tissue scaffold was designed as a potential supporting material for osseous tissue (Backdahl et al., 2008). Consequently, cell integration of osteoprogenitor cells was investigated (Brackmann et al., 2011b). A bright-field microscopy image of a single pore is shown in Figure 26.5B, also indicating the area visualized by CARS microscopy (2845 $cm^{-1}$) displayed to the right. The high-intensity regions (orange) (low intensity: magenta) map the distribution of osteoprogenitor cells, showing a dense collection of thriving cells. From these images, we conclude that the majority of cells were located in and readily filling up the pores. Clearly, significantly larger cell-integration depths were achieved with a microporous scaffold morphology. Furthermore, to investigate the ability to guide the cell migration, integration of fibroblast cells and collagen generation was studied in a cellulose scaffold with molded channels with a diameter of ~500 μm (Martinez et al., 2011). An electron microscopy image of a channel cross section is shown in Figure 26.5C, and the distributions of cells (orange) and generated collagen fibers (blue) in the channel are shown to the right as an overlay CARS (2845 $cm^{-1}$) and SHG (409 nm) microscopy image (volume rendering). From the CARS image, it can be concluded that the channels were completely filled with cells, nicely aligned along the central axis of the channel. In addition, the SHG microscopy image reveals that the cells readily produce native extracellular components and collagen fibers, also oriented along the channels. Interestingly, this indicates that the contact guidance of the peripherally located cells along the channel walls is "transmitted" also to the centrally located cells by cell–cell interactions. This opens up for the synthesis of an efficient load-bearing tissue scaffold, for instance as a replacement knee meniscus. Considering that surgical treatments of meniscal lesions are the most common procedures in the orthopedic field, this construct is likely to attract high commercial interest.

Altogether, this collection of images (Figures 26.3 through 26.5) illustrates that CARS microscopy is able not only to generate quite unique visualizations of cell–scaffold and cell–cell interaction mechanisms of high relevance within tissue engineering, but also to provide important qualitative and quantitative information such as migration depth, adhesion type, size, shape, orientation, and dynamics.

## 26.6 CARS Microscopy of Cell Proliferation and Differentiation

The ability of stem cells to self-renew and differentiate to all kinds of specialized cell types has been prophesized to become the universal solution of incorporating biological function in artificial tissues. However, major obstacles have been identified: (1) stem cells are difficult to differentiate *in vitro* and their specialized function is soon lost also in the host tissue; (2) they form cell layers—few preserved three-dimensional tissue-like architectures have been reported; and (3) they are easily damaged and have difficulties to survive the transplantation procedure. Thus, one of the primary conclusions derived from the efforts in stem-cell research for tissue engineering made in the past years is that more knowledge is needed to be able to design host environments that support long-term survival and function of cells for successful, therapeutic results. Fundamental is that the scaffold must be able to support natural cell function as in the native tissue, including cell proliferation and differentiation. Hence, it must be able to host the multitude of molecules such as growth factors (Babensee et al., 2000), which activate important intracellular signal transduction pathways that direct cell survival, proliferation, and differentiation, but also to have a morphology that ensures sufficient access to nutrients and oxygen. The biomechanical architecture of the scaffold has also *per se* a major impact on cell function, which can be utilized. Several studies on stem cell differentiation show that by tailoring the stiffness or elasticity of the scaffold, the development of the stem cells can be controlled. Native mesenchymal stem cells, initially small and round, develop increasingly branched, spindle, and polygonal shapes, respectively, when grown on matrices with an elasticity corresponding to that of brain, muscle, and bone tissues (Engler et al., 2006). Thus, stem cells commit to neurogenic lineages in softer scaffolds and to osteogenic lineages in rigid materials.

Crucial for the evaluation of the biological function of different artificial-tissue constructs is to visually be able to follow cell proliferation and differentiation all the way to three-dimensional tissue-like structures. Techniques such as fluorescence microscopy, electron microscopy, and AFM all involve harsh sample preparation or thin sectioning, making it impossible to study these soft materials with living cells. The early phases of cell differentiation can be studied by fluorescence microscopy, but as soon as the cells form larger collections, it is difficult to get the fluorophores evenly distributed in the scaffold and in the tissues. Thus, there is an urgent need for a noninvasive microscopy technique for tomographic, label-free imaging of soft, living samples. CARS microscopy fulfills all these requirements and has a strong potential to become an important instrument in the development of biologically functional artificial tissues.

Parameters that are of interest to assess with respect to cell proliferation and differentiation include (1) the impact of the molecular composition, the micromechanical properties, and the architecture of the scaffold on the area and volume of cell colonies; (2) cell morphology; and (3) amounts of native ECM components generated. Figure 26.6 shows a set of CARS microscopy images (2845 $cm^{-1}$) of osteoprogenitor cells seeded in a cellulose scaffold with a porous architecture following 6 days of growth (Brackmann et al., 2011b). In some pores, merely isolated cells can be observed, still migrating and establishing contacts through branched outgrowths as exemplified in Figure 26.6A. In other pores, the cells have started to proliferate, resulting in smaller colonies (Figure 26.6B) and even dense populations (Figure 26.6C). As a simple measure of cell proliferation, the area or volume of cells relative to total scaffold area/volume can readily be quantified from the CARS microscopy images, here 13% (Figure 26.6A), 31% (Figure 26.6B), and 46% (Figure 26.6C). From this, we conclude that the porous scaffold environment supports cell proliferation splendidly, but the degree of proliferation is dependent on the local microenvironment of the cells in the pores. Also the cell differentiation process can be followed by CARS microscopy. This is nicely illustrated by Cheng and colleagues, who have studied the growth of dorsal root ganglia versus time in a hydrogel matrix (Conovaloff et al., 2009). Unique images of the daily progression of neurite growth were collected (Figure 26.7A and B, from Conovaloff et al. [2009]), from which the

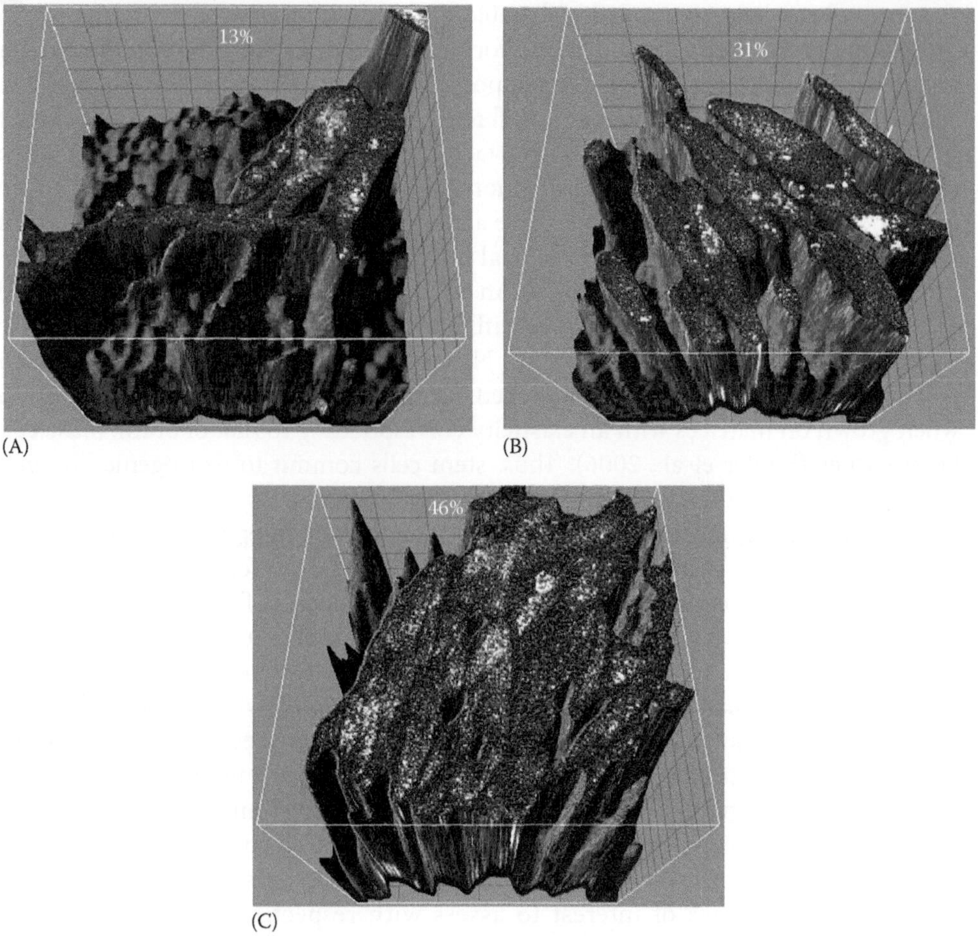


**FIGURE 26.6** The proliferation of osteoprogenitor cells (orange) in a cellulose scaffold with a porous architecture can be monitored and quantified by CARS microscopy ($2845\,cm^{-1}$). In some pores (A), merely a few isolated cells can be found, still seeking contacts with branched outgrowths. In other pores, small cell colonies (B) are formed and even dense populations (C) filling the entire pore volume. Numbers shown are the relative pore area filled with cells, as a simple measure of the progression of the proliferation. This illustrates that the proliferation rate is dependent on the local pore environment.

outgrowth pattern, neurite length, and density could be followed in great detail, all important parameters to assess the level of cell differentiation and specialized function. Figure 26.7C and D illustrates the cell differentiation process of SMCs seeded on cellulose scaffolds. Four days after seeding, the cells have started to proliferate and form small colonies well integrated in the scaffold (Figure 26.7C). However, they all show a round shape, characteristic for immature cells. After 28 days, the cell morphology is of a completely different character (Figure 26.7D). The cells have here differentiated into an elongated and aligned shape. In addition, from the architecture of the cellulose scaffold exhibiting repetitive, band-like structures, it is clear that the cells have developed the functionality of mature muscle cells of contraction and reshape the artificial ECM into band-like structures.

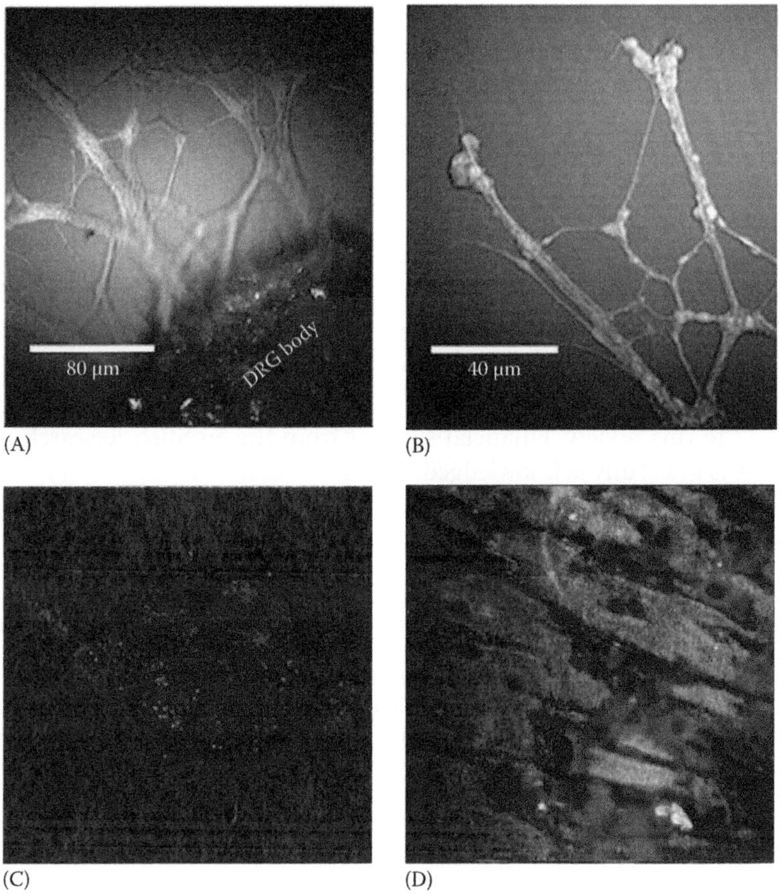


FIGURE 26.7 CARS microscopy images (2845 cm$^{-1}$) of cells differentiating from an immature state (A and C) to a cell line with fully established specialized function (B and D). In (A) and (B), neurite outgrowth from dorsal root ganglia in a hydrogel matrix can be followed. (From Conovaloff, A. et al., *Organogenesis*, 5, 231, 2009. With permission.) The differentiation of SMCs (orange) in a cellulose scaffold (blue, SHG microscopy) can clearly be recognized by comparing cell shape and alignment in images collected 4 (C) and 28 (D) days after the cell seeding. Note the repetitive, band-like structures of the surrounding scaffold indicating that the cells have developed the functionality of mature muscle cells to contract and transfer their internal forces to the ECM.

## 26.7 Summary

CARS microscopy is able to visualize and characterize native and artificial tissue scaffolds at high spatial resolution in three dimensions. Furthermore, images showing minute changes in the cell membrane that occur in conjunction with cell adhesion as well as time-lapse image series of cell migration patterns, cell proliferation, and differentiation can be acquired. Samples studied—all fragile arrangements of biopolymers and living cells—are unlikely to sustain the conventional sample preparation procedures for microscopy without structural and biochemical changes and only with great uncertainty can be labeled. Hence, CARS microscopy has the potential to provide a more realistic picture of the mechanisms behind cell–scaffold and cell–cell interaction processes than presently being extrapolated from studies on two-dimensional surfaces and cells with fluorophores attached to their transmembrane proteins and cytoskeleton. This is crucial to be able to take the challenges presently emerging within tissue engineering, including targeted design of adhesion molecules for specific cell types as needed for guided tissue regeneration and synthesis of materials exhibiting micromechanical responsiveness corresponding to that of living tissues. Hence, CARS microscopy has the prospects of becoming a highly useful and eventually even a standard instrument within tissue engineering.

## Acknowledgments

This work would not have been possible without the long-term, fruitful collaboration with Professor P. Gatenholm, Bipolymer Science, Chalmers University of Technology, Sweden. We would like to thank Professor J.-X. Cheng et al. for contributing images in Figure 26.7 to this review. Financial support from the Swedish Research Council and VINNOVA is gratefully acknowledged.

## References

Babensee, J. E., L. V. McIntire, and A. G. Mikos. 2000. Growth factor delivery for tissue engineering. *Pharmaceutical Research* 17: 497–504.

Backdahl, H., M. Esguerra, D. Delbro, B. Risberg, and P. Gatenholm. 2008. Engineering microporosity in bacterial cellulose scaffolds. *Journal of Tissue Engineering and Regenerative Medicine* 2: 320–330.

Bäckdahl, H., G. Helenius, A. Bodin, U. Nannmark, B. R. Johansson, B. Risberg, and P. Gatenholm. 2006. Mechanical properties of bacterial cellulose and interactions with smooth muscle cells. *Biomaterials* 27: 2141–2149.

Badami, A. S., M. R. Kreke, M. S. Thompson, J. S. Riffle, and A. S. Goldstein. 2006. Effect of fiber diameter on spreading, proliferation, and differentiation of osteoblastic cells on electrospun poly(lactic acid) substrates. *Biomaterials* 27: 596–606.

Barnes, C. P., S. A. Sell, E. D. Boland, D. G. Simpson, and G. L. Bowlin. 2007. Nanofiber technology: Designing the next generation of tissue engineering scaffolds. *Advanced Drug Delivery Reviews* 59: 1413–1433.

Berrier, A. L. and K. M. Yamada. 2007. Cell-matrix adhesion. *Journal of Cellular Physiology* 213: 565–573.

Brackmann, C., M. Esguerra, D. Olausson, D. Delbro, A. Krettek, P. Gatenholm, and A. Enejder. 2011a. Coherent anti-Stokes Raman scattering microscopy of human smooth muscle cells in bioengineered tissue scaffolds. *Journal of Biomedical Optics* 16: 021115.

Brackmann, C., M. Zaborowska, J. Sundberg, P. Gatenholm, and A. Enejder. 2012. In situ imaging of collagen synthesis by osteoprogenitor cells in microporous bacterial cellulose scaffolds. *Tissue Engineering C* 18: 227–234.

Brown, E. E. and M. P. G. Laborie. 2007. Bioengineering bacterial cellulose/poly(ethylene oxide) nanocomposites. *Biomacromolecules* 8: 3074–3081.

Conovaloff, A., H. W. Wang, J. X. Cheng, and A. Panitch. 2009. Imaging growth of neurites in conditioned hydrogel by coherent anti-Stokes Raman scattering microscopy. *Organogenesis* 5: 231–237.

Czaja, W. K., D. J. Young, M. Kawecki, and R. M. Brown. 2007. The future prospects of microbial cellulose in biomedical applications. *Biomacromolecules* 8: 1–12.

Daamen, W. F., J. H. Veerkamp, J. C. M. van Hest, and T. H. van Kuppevelt. 2007. Elastin as a biomaterial for tissue engineering. *Biomaterials* 28: 4378–4398.

Dimilla, P. A., K. Barbee, and D. A. Lauffenburger. 1991. Mathematical-model for the effects of adhesion and mechanics on cell-migration speed. *Biophysical Journal* 60: 15–37.

Edwards, H. G. M., D. W. Farwell, J. M. Holder, and E. E. Lawson. 1997. Fourier-transform Raman spectroscopy of ivory.2. Spectroscopic analysis and assignments. *Journal of Molecular Structure* 435: 49–58.

Ellis, R., E. Green, and C. P. Winlove. 2009. Structural analysis of glycosaminoglycans and proteoglycans by means of Raman microspectrometry. *Connective Tissue Research* 50: 29–36.

Engelmayr, G. C., M. Y. Cheng, C. J. Bettinger, J. T. Borenstein, R. Langer, and L. E. Freed. 2008. Accordion-like honeycombs for tissue engineering of cardiac anisotropy. *Nature Materials* 7: 1003–1010.

Engler, A. J., S. Sen, H. L. Sweeney, and D. E. Discher. 2006. Matrix elasticity directs stem cell lineage specification. *Cell* 126: 677–689.

Fischer, S., K. Schenzel, K. Fischer, and W. Diepenbrock. 2005. Applications of FT Raman spectroscopy and micro spectroscopy characterizing cellulose and cellulosic biomaterials. *Macromolecular Symposia* 223: 41–56.

Focher, B., A. Naggi, G. Torri, A. Cosani, and M. Terbojevich. 1992. Structural differences between chitin polymorphs and their precipitates from solutions—Evidence from CP-MAS C-13-NMR, FT-IR AND FT-Raman spectroscopy. *Carbohydrate Polymers* 17: 97–102.

Frey, M. W. 2008. Electrospinning cellulose and cellulose derivatives. *Polymer Reviews* 48: 378–391.

Friedl, P. and E. B. Brocker. 2000. The biology of cell locomotion within three-dimensional extracellular matrix. *Cellular and Molecular Life Sciences* 57: 41–64.

Friedl, P., K. S. Zanker, and E. B. Brocker. 1998. Cell migration strategies in 3-D extracellular matrix: Differences in morphology, cell matrix interactions, and integrin function. *Microscopy Research and Technique* 43: 369–378.

Geckil, H., F. Xu, X. H. Zhang, S. Moon, and U. Demirci. 2010. Engineering hydrogels as extracellular matrix mimics. *Nanomedicine* 5: 469–484.

Geiger, B., A. Bershadsky, R. Pankov, and K. M. Yamada. 2001. Transmembrane extracellular matrix-cytoskeleton crosstalk. *Nature Reviews Molecular Cell Biology* 2: 793–805.

Gelin, K., A. Bodin, P. Gatenholm, A. Mihranyan, K. Edwards, and M. Stromme. 2007. Characterization of water in bacterial cellulose using dielectric spectroscopy and electron microscopy. *Polymer* 48: 7623–7631.

Gilbert, T. W., T. L. Sellaro, and S. F. Badylak. 2006. Decellularization of tissues and organs. *Biomaterials* 27: 3675–3683.

Green, E., R. Ellis, and P. Winlove. 2008. The molecular structure and physical properties of elastin fibers as revealed by Raman microspectroscopy. *Biopolymers* 89: 931–940.

Heydarkhan-Hagvall, S., K. Schenke-Layland, A. P. Dhanasopon, F. Rofail, H. Smith, B. M. Wu, R. Shemin et al. 2008. Three-dimensional electrospun ECM-based hybrid scaffolds for cardiovascular tissue engineering. *Biomaterials* 29: 2907–2914.

Hynes, R. O. 2002. Integrins: Bidirectional, allosteric signaling machines. *Cell* 110: 673–687.

Hynes, R. O. 2009. The extracellular matrix: Not just pretty fibrils. *Science* 326: 1216–1219.

Kister, G., G. Cassanas, M. Bergounhon, D. Hoarau, and M. Vert. 2000. Structural characterization and hydrolytic degradation of solid copolymers of D,L-lactide-co-epsilon-caprolactone by Raman spectroscopy. *Polymer* 41: 925–932.

Kister, G., G. Cassanas, and M. Vert. 1997. Morphology of poly(glycolic acid) by IR and Raman spectroscopies. *Spectrochimica Acta Part A—Molecular and Biomolecular Spectroscopy* 53: 1399–1403.

Klemm, D., B. Heublein, H. P. Fink, and A. Bohn. 2005. Cellulose: Fascinating biopolymer and sustainable raw material. *Angewandte Chemie-International Edition* 44: 3358–3393.

Kwon, I. K. and T. Matsuda. 2005. Co-electrospun nanofiber fabrics of poly(L-lactide-co-epsilon-caprolactone) with type I collagen or heparin. *Biomacromolecules* 6: 2096–2105.

Lauffenburger, D. A. and A. F. Horwitz. 1996. Cell migration: A physically integrated molecular process. *Cell* 84: 359–369.

Levental, I., P. C. Georges, and P. A. Janmey. 2007. Soft biological materials and their impact on cell function. *Soft Matter* 3: 299–306.

Li, M., M. J. Mondrinos, X. Chen, M. R. Gandhi, F. K. Ko, and P. I. Lelkes. 2006. Co-electrospun poly(lactide-co-glycolide), gelatin, and elastin blends for tissue engineering scaffolds. *Journal of Biomedical Materials Research Part A* 79A: 963–973.

Lodish, H., A. Berk, C. A. Kaiser, M. Krieger, M. P. Scott, A. Bretscher, H. Ploegh et al. 2008. *Molecular Cell Biology*. New York: W. H. Freeman and Company.

Lotem, H., R. T. Lynch, and N. Bloembergen. 1976. Interference between Raman resonances in 4-wave difference mixing. *Physical Review A* 14: 1748–1755.

Luo, Y. and M. S. Shoichet. 2004. A photolabile hydrogel for guided three-dimensional cell growth and migration. *Nature Materials* 3: 249–253.

Luo, H. L., G. Y. Xiong, Y. Huang, F. He, W. Wang, and Y. Z. Wan. 2008. Preparation and characterization of a novel COL/BC composite for potential tissue engineering scaffolds. *Materials Chemistry and Physics* 110: 193–196.

Lutolf, M. P. and J. A. Hubbell. 2005. Synthetic biomaterials as instructive extracellular microenvironments for morphogenesis in tissue engineering. *Nature Biotechnology* 23: 47–55.

Lysaght, M. J. and J. A. O'Loughlin. 2000. Demographic scope and economic magnitude of contemporary organ replacement therapies. *ASAIO Journal* 46: 515–521.

Martinez, H., C. Brackmann, A. Enejder, and P. Gatenholm. 2012. Mechanical stimulation of fibroblasts in micro-channeled bacterial cellulose scaffolds enhances production of oriented collagen fibers. *Journal of Biomedical Materials Research: Part A* 100: 948–957.

Nikolovski, J. and D. J. Mooney. 2000. Smooth muscle cell adhesion to tissue engineering scaffolds. *Biomaterials* 21: 2025–2032.

Pedersen, J. A. and M. A. Swartz. 2005. Mechanobiology in the third dimension. *Annals of Biomedical Engineering* 33: 1469–1490.

Roohpour, N., J. M. Wasikiewicz, D. Paul, P. Vadgama, and I. U. Rehman. 2009. Synthesis and characterisation of enhanced barrier polyurethane for encapsulation of implantable medical devices. *Journal of Materials Science-Materials in Medicine* 20: 1803–1814.

Ross, P., R. Mayer, and M. Benziman. 1991. Cellulose biosynthesis and function in bacteria. *Microbiological Reviews* 55: 35–58.

Schwarz, U. S. and I. B. Bischofs. 2005. Physical determinants of cell organization in soft media. *Medical Engineering and Physics* 27: 763–772.

Sengupta, D. and S. C. Heilshorn. 2010. Protein-engineered biomaterials: Highly tunable tissue engineering scaffolds. *Tissue Engineering Part B—Reviews* 16: 285–293.

Sreejalekshmi, K. G. and P. D. Nair. 2011. Biomimeticity in tissue engineering scaffolds through synthetic peptide modifications—Altering chemistry for enhanced biological response. *Journal of Biomedical Materials Research Part A* 96A: 477–491.

Thomas, V., D. R. Dean, M. V. Jose, B. Mathew, S. Chowdhury, and Y. K. Vohra. 2007. Nanostructured biocomposite scaffolds based on collagen coelectrospun with nanohydroxyapatite. *Biomacromolecules* 8: 631–637.

Vickaryous, M. K. and B. K. Hall. 2006. Human cell type diversity, evolution, development, and classification with special reference to cells derived from the neural crest. *Biological Reviews* 81: 425–455.

Vogel, V. and M. Sheetz. 2006. Local force and geometry sensing regulate cell functions. *Nature Reviews Molecular Cell Biology* 7: 265–275.

Wiley, J. H. and R. H. Atalla. 1987. Band assignments in the Raman-spectra of celluloses. *Carbohydrate Research* 160: 113–129.

Worth, D. C. and M. Parsons. 2008. Adhesion dynamics: Mechanisms and measurements. *International Journal of Biochemistry and Cell Biology* 40: 2397–2409.

Yan, Z. Y., S. Y. Chen, H. P. Wang, B. Wang, C. S. Wang, and J. M. Jiang. 2008. Cellulose synthesized by *Acetobacter xylinum* in the presence of multi-walled carbon nanotubes. *Carbohydrate Research* 343: 73–80.

Yang, F., R. Murugan, S. Wang, and S. Ramakrishna. 2005. Electrospinning of nano/micro scale poly(L-lactic acid) aligned fibers and their potential in neural tissue engineering. *Biomaterials* 26: 2603–2610.

Zaman, M. H., R. D. Kamm, P. Matsudaira, and D. A. Lauffenburger. 2005. Computational model for cell migration in three-dimensional matrices. *Biophysical Journal* 89: 1389–1397.

Zhang, S. G. 2003. Fabrication of novel biomaterials through molecular self-assembly. *Nature Biotechnology* 21: 1171–1178.

# 27. Dietary Fat Absorption Visualized by CARS Microscopy

Kimberly K. Buhman



## 27.1 Introduction

### 27.1.1 Dietary Fat

The chemical structure of dietary fat is a triacylglycerol (TAG). TAG contains a glycerol backbone and three fatty acids joined via ester bonds. Fatty acids are carboxylic acids with hydrocarbon tails, which vary in chain length and saturation. Fat (or TAG) is the most energy dense of the three major energy-yielding nutrients present in human diets. In addition, it provides essential structural and signaling functions for cells in normal physiology and aids in the absorption of fat soluble vitamins. When present in excess amounts in the diet (>35% fat), fat contributes to an imbalance in energy due to increased energy intake and thus contributes to the problems of overweight and obesity [31–33,35,52]. In addition, excess dietary fat contributes to elevated blood lipid concentrations particularly at short times after meals. Elevated blood lipid levels in fasting and postprandial conditions are risk factors for cardiovascular disease [19,29,41].

*Coherent Raman Scattering Microscopy*. Edited by Ji-Xin Cheng and X. Sunney Xie © 2013 CRC Press/Taylor & Francis Group, LLC. ISBN: 978-1-4398-6765-5.

### 27.1.2 Dietary Fat Absorption

The absorption of dietary fat is a nearly complete process with greater than 95% of dietary fat consumed being absorbed by the small intestine whether a low- or high-fat diet is consumed [46]. Quantitatively, the amount of dietary fat absorbed is determined by quantifying the difference between the amount of dietary fat consumed and the amount of fat excreted in feces over several days. This process, however, is not only regulated over several days but also in response to each meal containing dietary fat.

The bulk of dietary fat absorption occurs in the jejunum, or midsection, of the small intestine and includes multiple steps (Figure 27.1). First, dietary fat in the form of TAG is digested or hydrolyzed by pancreatic lipase, generating two, free fatty acids (FFAs) and a monoacylglycerol (MAG) per TAG molecule. These products are emulsified with the help of phospholipids and bile acids present in bile to form micelles and then transported into the absorptive cells of the small intestine, enterocytes. Within enterocytes, the re-esterification of fatty acyl CoAs and MAG to TAG is primarily mediated by the acylcoenzyme A:monoacylglycerol transferase (MGAT) pathway for TAG biosynthesis [20]. MGAT catalyzes a reaction between fatty acyl CoA and MAG to make diacylglycerol (DAG). Acylcoenzyme A:diacylglycerol transferase (DGAT) catalyzes a reaction between fatty acyl CoA and DAG to make TAG. The major MGAT enzyme in mouse enterocytes is MGAT2 [57]. Two DGAT enzymes are present in enterocytes of mice, including DGAT1 and DGAT2 [5]. The TAGs are then incorporated into the core of specialized lipoproteins, chylomicrons (CMs), which have the capacity for carrying large amounts of TAG, a neutral lipid, in the aqueous environment of blood. CMs are secreted from enterocytes via the lymphatic system and thoracic duct into circulation and delivered to tissues where they are either stored or oxidized based on physiological need [21,26,45]. Under high-fat dietary challenges, TAG is also found packaged in enterocytes in cytoplasmic lipid droplets (CLDs) [59]. We recently demonstrated that this storage pool of TAG in enterocytes expands and depletes relative to the fed-fasted state and is present whether mice are acutely or chronically challenged by high levels of dietary

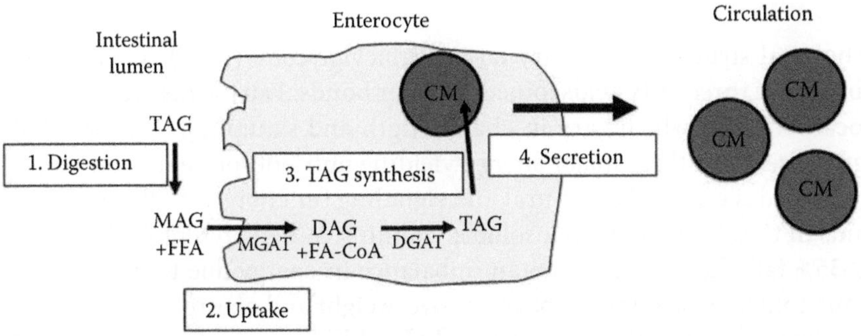


**FIGURE 27.1** Dietary fat absorption. 1. Digestion. Dietary fat in the form of TAG is first hydrolyzed in the intestinal lumen to FFA and MAG via pancreatic lipase. 2. Uptake. These digestive products are emulsified with the help of phospholipids and bile acids from bile and are transported into the absorptive cell of the intestine, enterocytes. 3. TAG synthesis. The digestive products are re-esterified to make DAG and then TAG primarily via the monoacylglycerol acyltransferase (MGAT) pathway. 4. Secretion. The TAG is packaged into the core of a CM particle for secretion into circulation and delivery to tissues for storage and/or oxidation based on physiological need.

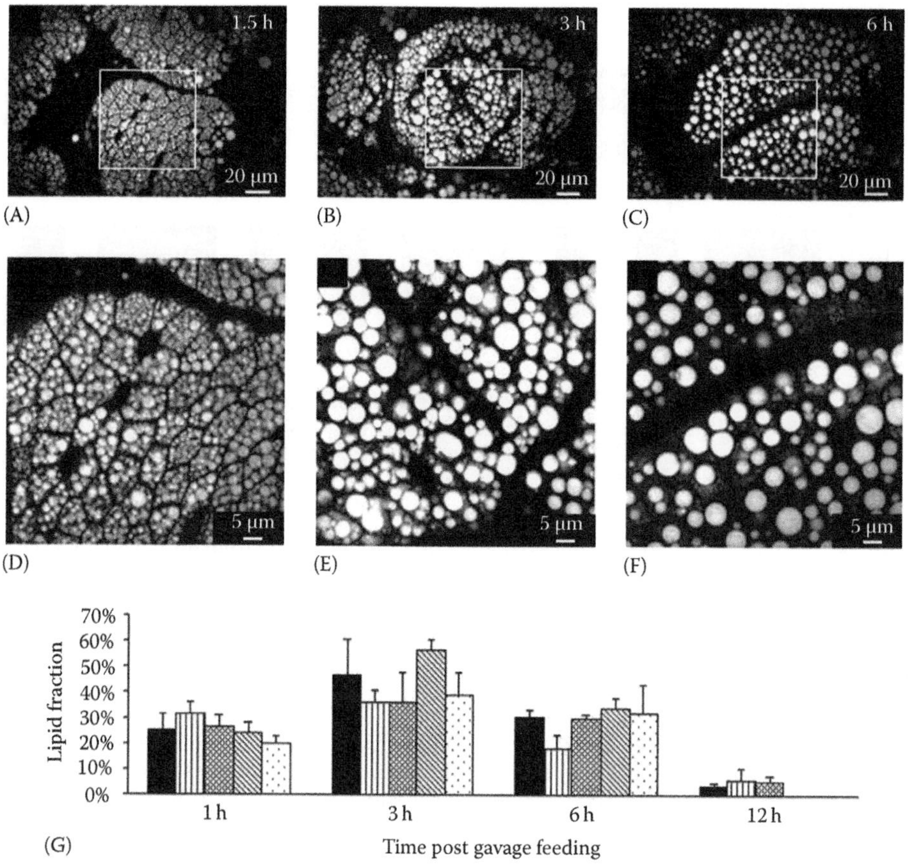


**FIGURE 27.2** *In vivo* CARS imaging demonstrates dynamic accumulation and depletion of TAG in CLDs during the process of dietary fat absorption. (A, D) At 1 h post-gavage of 300 μL olive oil, enterocytes were filled with small CLDs with an average diameter of 2.7 μm. (B, E) At 3 h post-gavage, larger CLDs with an average diameter of 8.7 μm were observed. (C, F) At 6 h post-gavage, compared to the level at 3 h post-gavage the CLD accumulation decreased with an average diameter of 5.0 μm. (D–F) are zoom-in views of (A–C), respectively. (G) Variation of the amount of CLDs based on area in enterocytes with different times postconsumption of dietary fat. Data were obtained from Image-J analysis of *ex vivo* CARS images of upper jejunum of small intestine tissues (each bar represents an individual mouse). (From Zhu, J. et al., *J. Lipid Res.*, 50, 1080, 2009. With permission. © 2009 the American Society for Biochemistry and Molecular Biology.)

fat (Figures 27.2 and 27.3) [24,59]. In addition, TAG stored in the intestine has recently been shown to be the source of a rapid triglyceridemic response in response to a second meal in humans [9,34]. Therefore, regulation of this pool of TAG is likely important in determining postprandial blood TAG concentrations and assimilation of dietary fat.

### 27.1.3 Challenges in Imaging Dietary Fat Absorption

Multiple physiological and technical challenges exist for imaging dietary fat absorption. The process of dietary fat absorption is a dynamic process that occurs in the small intestine—an internal organ. Imaging of this process in real time is desired; however, previously available techniques for imaging lipids required tissue fixation and staining

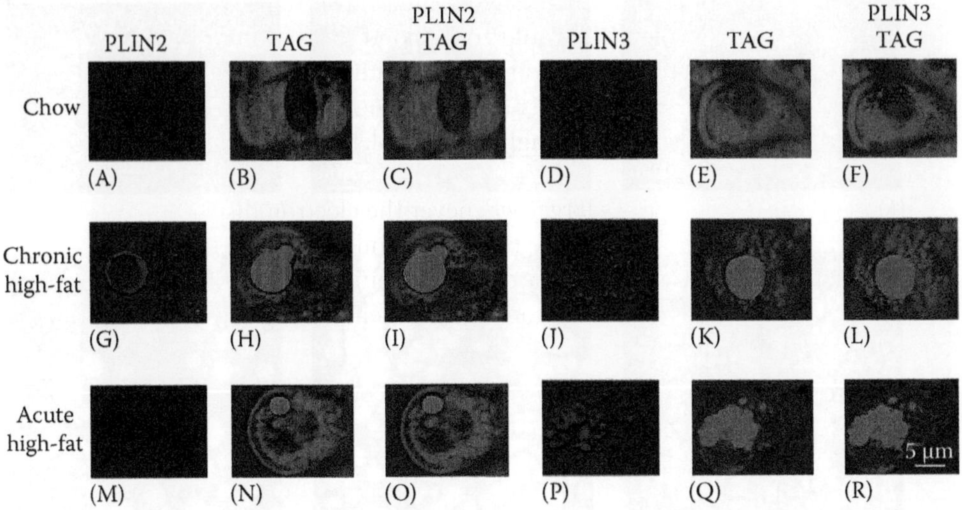


**FIGURE 27.3** Cellular localization of PLIN2 and PLIN3 in enterocytes from mice acutely and chronically challenged by dietary fat. Enterocytes were isolated from mice fed chow, chronic high fat, or acute high fat challenged and imaged by CARS microscopy for triglyceride (green) or by immunocytochemistry for PLIN2 or PLIN3 (red). Merged CARS and fluorescence images help identify cellular localization. Chow and chronic high-fat-fed mice were fasted for 2 h at the beginning of the light cycle before sample collection (A–L). Acute challenged mice were fasted for 4 h at the beginning of the light cycle, given 300 μL of olive oil via gavage, and samples collected 3 h post-gavage (M–R). (Modified from *Biochim. Biophys. Acta (Cellular Biology of Lipids)*, 1791, Lee, B., Zhu, J., Wolins, N.E., Cheng, J.X., and Buhman, K.K., Differential association of adipophilin and TIP47 proteins with cytoplasmic lipid droplets in mouse enterocytes during dietary fat absorption, 1173–1180. Copyright 2009, with permission from Elsevier.)

of tissue samples isolated at specific times. Because tissue biopsies from the small intestine are quite invasive, this requires animals to be euthanized at each time point. This results in large animal numbers per study and also prevents real-time imaging. In addition, tissue fixation is also limiting due to the loss of lipid from the sample in the fixation process. Furthermore, stains available such as Nile red, oil red o, and osmium tetroxide have limited molecular specificity. Finally, because fluorescent probes are often of a similar size as a lipid molecule, fluorescently labeled lipids do not always behave in terms of function or transport as do unlabeled lipids.

## 27.2 Methods

### 27.2.1 Coherent Anti-Stokes Raman Scattering Microscopy

Coherent anti-Stokes Raman scattering (CARS) microscopy generates images of cells and tissues with chemical specificity without the need for fixation or outside labeling [11,12]. In addition, the images have three-dimensional, submicron spatial resolution similar to images generated from multiphoton microscopes. Also, the acquisition rate of the images is fast. Most of the current applications for CARS microscopy monitor a phenotypic response of TAG in cells and tissues to chemical treatments or biological modifications. Raman scattering signals are present for many types of molecules in cells; however, the strongest signal arises from the C–H bond stretches found in lipids such as fatty acids. For CARS imaging of fatty acids, the pump laser at 790 nm (12658 $cm^{-1}$)

and the Stokes laser at 1018 nm (9823 $cm^{-1}$) provide a wavenumber difference centered at 2840 $cm^{-1}$. Because the chemical structure of dietary fat is a TAG, containing three fatty acids, this imaging technique is ideal for imaging the process of dietary fat absorption. CARS microscopy is capable of providing similar results as histological staining and biochemical analysis of the small intestine; however, it is even better due to avoiding tissue fixation and the use of non-specific reagents.

### 27.2.2 CARS Microscopy in Real Time

In addition to its label-free chemical specificity, CARS microscopy is a nondestructive method and can therefore be used for real-time imaging in live cells and animals. The penetration depth of CARS microscopy is approximately only 100 μm, allowing real-time imaging of individual cells and surface structures such as skin [13]. The absorptive cells of the small intestine, enterocytes, are present on the surface facing the small intestine lumen. Imaging these cells is currently possible in *ex vivo* samples and *in vivo* in mice with a surgical procedure. Using a surgical procedure to open the small intestine in anesthetized mice has allowed the dynamic process of dietary fat absorption taking place within enterocytes to be visualized by CARS microscopy in real time for up to 6 h [59]. A CARS endoscope would provide access to deeper tissues *in vivo*, such as the lining of the gastrointestinal tract, thus providing additional potential applications in clinical diagnostics.

### 27.2.3 CARS Microscopy and Two-Photon Excited Fluorescence Microscopy

Another advantage of CARS microscopy is its ability to be combined with other two-photon imaging modalities such as two-photon excited fluorescence microscopy (TPEF) [10]. The use of a femtosecond laser allows for the excitation source to induce both CARS and TPEF signals at the same time. This allows both CARS and TPEF to be present in one, laser-scanning microscope platform. Combining lipid imaging by CARS and immunohistochemistry of proteins via TPEF has allowed for investigation of cellular localization and regulation of specific proteins involved in CLD biology and TAG metabolism in enterocytes in response to dietary fat [24] (Figure 27.3).

## 27.3 New Findings

### 27.3.1 Dynamic Presence of Cytoplasmic Lipid Droplets in Enterocytes

Recently, CLDs have gained significant attention as a bona fide and active cellular organelle [14,50]. TAG-rich CLDs are present in enterocytes in response to an oral TAG challenge in many species, including *Caenorhabditis elegans*, pythons, mice, rats, and humans [59]. Using CARS microscopy, we demonstrated that after an oral TAG challenge, TAG-rich CLDs increase in size and amount in enterocytes up to 3 h post-challenge and decrease in size and amount until they are absent by 12 h post-challenge in mice (Figure 27.2) [59]. In addition, when dietary fat is fed chronically, TAG-rich CLDs are also present in enterocytes of mice in the fed state (Figure 27.3E and N) [59]. It is unknown how different amounts of dietary fat, types of dietary fat, and health problems such as obesity and diabetes regulate trafficking of dietary fat through this cellular organelle. It is also unknown how factors known to regulate the synthesis and breakdown of CLDs in other cell types regulate this process in enterocytes. These factors include enzymes in TAG biosynthesis [58], CLD-associated proteins [2], and lipolytic enzymes [22]. Specific proteins and enzymes in these pathways and their interaction with CLDs most likely determine the fate of dietary fat in enterocytes and ultimately regulate trafficking of dietary fat to specific places and for specific purposes such as storage, building, and energy production.

### 27.3.2 Proteins Associated with Cytoplasmic Lipid Droplets in Enterocytes

Multiple proteins have been found in association with CLDs in various cell types in proteomic studies [1,4,8,16,25,38,47,49,51,54]. The composition of the proteins present on CLDs is complex and dynamic and regulates their anabolism and catabolism. The perilipin (PLIN) family is a family of CLD-associated proteins that has been studied recently. This family is encoded by five genes that share an amino terminal, highly conserved PAT domain, that defines the family, but are expressed in a tissue-specific manner [2,3]. The PLIN family includes PLIN1, PLIN2 (adipophilin, adipose differentiation-related protein, ADRP, and ADFP), PLIN3 (tail-interacting protein 47, TIP47), PLIN4 (S3–12), and PLIN5 (OXPAT, MLDP, PAT-1, and LSDP5). Their characteristic tissue expression patterns, cellular localization, and binding specificities suggest that they each have unique roles in the synthesis, structure, and catabolism of CLDs [53]. Overexpression of PLIN family members drives TAG storage, and specific members of the family drive CLD size, fatty acid oxidation, and lipoprotein synthesis and secretion, and regulates lipolysis of TAG in CLD stores in various cell types [2]. Expression of PLIN family members on CLDs forms the regulatory border between the TAG stored in the CLD and the cellular machinery that manages the metabolism of the TAG. Regulation of these processes is most likely mediated by the recruitment of other proteins and enzymes involved in these pathways to the CLD. For example, PLINs on CLDs have been shown to interact with lipolytic enzymes such as hormone-sensitive lipase and adipocyte TAG lipase-activating protein, CGI-58 [44,56]. In addition, DGAT2 was identified in association with CLDs and overexpression of DGAT2 promotes CLD association with mitochondria [28,42].

PLIN1 and PLIN2 are constitutively expressed proteins; however, they are found in cells associated with CLDs only when CLDs are present and are degraded through a proteasome-dependent pathway when CLDs are absent [27,55]. PLIN3, PLIN4, and PLIN5 are exchangeable CLD proteins as they localize to the cytoplasm when TAG stores in CLDs are not present and rapidly associate with newly synthesized TAG in the formation of CLDs. Interestingly, PLIN3 was found to be most abundantly expressed (10 times the median for all tissues) in mouse small intestine compared to 44 other mouse tissues in a high-throughput gene expression analysis [43]. We found that PLIN2 and PLIN3 are the only PLIN family members present in intestinal mucosa [24].

We identified PLIN2 and PLIN3 associated with CLDs in enterocytes under two different dietary fat challenge conditions where TAG accumulates in CLDs in enterocytes (Figure 27.3) [24]. In one condition, low-fat, chow-fed mice were given olive oil by oral gavage; enterocytes were collected at 2 h post-gavage and analyzed. We found PLIN3 present in enterocytes and localized to CLDs but did not identify PLIN2 in enterocytes under this condition. In the second condition, mice were fed a high-fat diet for 2 weeks, withheld food for 2 h at the beginning of the light cycle, and analyzed. We found PLIN2 present and associated with CLDs in enterocytes. PLIN3, on the other hand, was present in the cytoplasm, but not associated with CLDs. These two experimental conditions differ in the composition of the high-fat dietary challenge, the timing relative to the high-fat dietary challenge, and the acute versus chronic nature of the high-fat dietary challenge. In the first condition, the high-fat dietary challenge included only fat, one time, and was well controlled for analysis of time post-challenge. In the second condition, the high-fat dietary challenge included fat, carbohydrate, and protein; was present daily for 2 weeks; and was not well controlled for analysis of time post-challenge. One interpretation of the earlier results is that PLIN2 and PLIN3 differentially associate with CLDs in enterocytes in response to the composition of the high-fat challenge with PLIN2 being present on lipid droplets when carbohydrates and proteins are also included in the challenge. A second interpretation of these results is that PLIN3 associates with "new" CLDs forming shortly after fat is consumed and PLIN2 associates with "old" CLDs that have been stored for longer time periods in enterocytes. A third interpretation of these results is that PLIN2 and PLIN3 differentially associate with CLDs in enterocytes in response to acute or chronic high-fat dietary challenges as the condition where PLIN2 was found associated was after the high-fat dietary challenge was present every day for 2 weeks. Future studies using CARS microscopy and TPEF will help to elucidate the role of PLIN2 and PLIN3 in intestinal TAG metabolism.

### 27.3.3 Functions of Enzymes in TAG Biosynthesis in Enterocytes

Although DGAT1 and DGAT2 are both expressed in most cell types; their relative levels in different cell types vary. DGAT2 is highest in liver and adipose tissue, whereas DGAT1 is highest in small intestine [6,7]. In the absence of DGAT1, mice are resistant to diet-induced obesity and have less body fat [40]. Although fasting blood TAG concentrations are normal, their postprandial triglyceridemic response is blunted [5,23]. Interestingly, the only cell type identified that has more TAG in DGAT1-deficient mice than wild-type mice when fed a high-fat diet is the enterocyte during the process of dietary fat absorption [5,23]. DGAT1-deficient mice do not have a defect in quantitative dietary fat absorption [5,23]. To determine whether the loss of DGAT1 in the intestine contributes to their resistance to diet-induced obesity, we generated mice with intestine-specific overexpression of DGAT1 ($Dgat1^{Int20X}$) and crossed them with DGAT1-deficient mice ($Dgat1^{-/-}$) to generate mice with expression of DGAT1 only in the intestine ($Dgat1^{IntONLY}$) [23]. Using CARS microscopy, we found that TAG accumulated in CLDs in enterocytes of WT mice in response to a high-fat challenge was not present in mice with intestine-specific overexpression of DGAT1 (Figure 27.4A). We also found using CARS microscopy that the abnormal accumulation of TAG in enterocytes of DGAT1-deficient mice was corrected (Figure 27.4A) [23]. In addition, the resistance to diet-induced obesity and hepatic steatosis were no longer present in mice with intestine only expression of DGAT1 [23]. These results support the important role of regulation of TAG storage and secretion by enterocytes in energy balance.

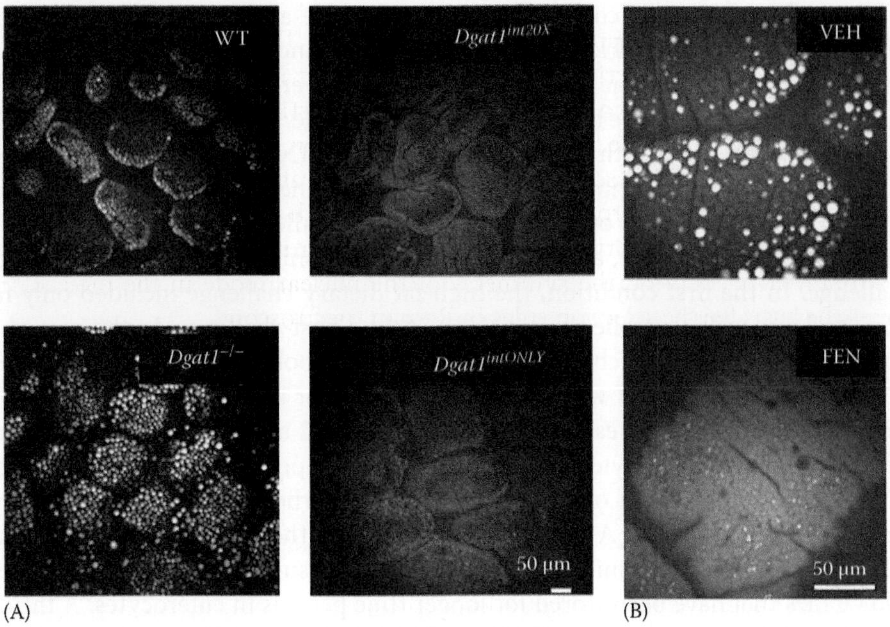


**FIGURE 27.4** (A) Abrogated TG storage in enterocytes of mice with increased *Dgat1* mRNA abundance. Representative CARS images of TG storage in CLDs in enterocytes representing upper jejunum of female WT, *Dgat1*$^{-/-}$, *Dgat1*$^{Int20X}$, and *Dgat1*$^{IntONLY}$ mice fed a HF diet for 9 weeks. These mice were euthanized after fasting for 2h. (From Lee, B. et al., *J. Lipid Res.*, 51, 1770, 2010. With permission. © 2010, the American Society for Biochemistry and Molecular Biology.) (B) Fenofibrate decreases TG storage in enterocytes of HF-fed mice. C57BL/6, male mice were fed a HF diet for 2 weeks and then continued on the HF diet in combination with daily oral administration of fenofibrate for 5 days. Mice were fasted for 2h before euthanasia. Representative images of TG storage in CLDs in enterocytes in the upper jejunum of vehicle (VEH)- and fenofibrate (FEN)-treated mice, using 20× objective at 1× and 3× magnification. (Modified from *Biochim. Biophys. Acta* (*Cellular Biology of Lipids*), 1811, Uchida, A., Slipchenko, M.N., Cheng, J.X., and Buhman, K.K., Fenofibrate, a peroxisome proliferator-activated receptor alpha agonist, alters triglyceride metabolism in enterocytes of mice, 170–176. Copyright 2011, with permission from Elsevier.)

### 27.3.4 Regulation of Enterocyte TAG Metabolism by Hypolipidemic Drugs

Fenofibrate, a drug in the fibrate class of amphipathic carboxylic acids, has multiple blood lipid modifying actions, which are beneficial to the prevention of atherosclerosis [18,36]. One of its strongest actions is on lowering fasting and postprandial blood TAG concentrations [37]. Fenofibrate binds and activates the transcription factor, peroxisome proliferator–activated receptor alpha (PPARα). PPARα regulates genes involved in TAG metabolism, including uptake, trafficking, and oxidation. The major mechanism for its hypotriglyceridemic action is thought to be improved clearance of TAG from circulation [15,17,18,30,39]. Blood TAG concentrations are always a balance of input and clearance, and therefore, we investigated the effects of fenofibrate on the input of dietary TAG from enterocytes. We found that high-fat-fed mice treated with fenofibrate have significantly less TAG storage in enterocytes by CARS microscopy [48] (Figure 27.4B). This result is consistent with a direct action of fenofibrate on enterocyte TAG metabolism.

## 27.4 Summary

Dietary fat absorption is an important process for maintaining health and has the potential to be altered for treating diseases such as obesity and cardiovascular disease. Further understanding of how dietary fat is digested, taken up, metabolized, and secreted from the absorptive cells of the small intestine, enterocytes, is critical for the development of dietary recommendations for health and treatments for disease. CARS microscopy has already proven to be an effective tool for further understanding the process of dietary fat absorption, including imaging this process in real time, discovering the role of proteins associated with CLDs and in TAG biosynthesis in this process, and understanding the effects of hypolipidemic drugs on this process.

## References

1. Beller, M., D. Riedel, L. Jansch, G. Dieterich, J. Wehland, H. Jackle, and R. P. Kuhnlein. 2006. Characterization of the Drosophila lipid droplet subproteome. *Mol. Cell. Proteomics* 5: 1082–1094.
2. Bickel, P. E., J. T. Tansey, and M. A. Welte. 2009. PAT proteins, an ancient family of lipid droplet proteins that regulate cellular lipid stores. *Biochim. Biophys. Acta* 1791: 419–440.
3. Brasaemle, D. L. 2007. Thematic review series: Adipocyte biology. The perilipin family of structural lipid droplet proteins: Stabilization of lipid droplets and control of lipolysis. *J. Lipid Res.* 48: 2547–2559.
4. Brasaemle, D. L., G. Dolios, L. Shapiro, and R. Wang. 2004. Proteomic analysis of proteins associated with lipid droplets of basal and lipolytically stimulated 3T3-L1 adipocytes. *J. Biol. Chem.* 279: 46835–46842.
5. Buhman, K. K., S. J. Smith, S. J. Stone, J. J. Repa, J. S. Wong, F. F. Knapp, B. J. Burri, R. L. Hamilton, N. A. Abumrad, and R. V. Farese. 2002. DGAT1 is not essential for intestinal triacylglycerol absorption or chylomicron synthesis. *J. Biol. Chem.* 277: 25474–25479.
6. Cases, S., S. J. Smith, Y. W. Zheng, H. M. Myers, S. R. Lear, E. Sande, S. Novak et al. 1998. Identification of a gene encoding an acyl CoA: Diacylglycerol acyltransferase, a key enzyme in triacylglycerol synthesis. *Proc. Natl Acad. Sci. USA* 95: 13018–13023.
7. Cases, S., S. J. Stone, P. Zhou, E. Yen, B. Tow, K. D. Lardizabal, T. Voelker, and R. V. Farese. 2001. Cloning of DGAT2, a second mammalian diacylglycerol acyltransferase, and related family members. *J. Biol. Chem.* 276: 38870–38876.
8. Cermelli, S., Y. Guo, S. P. Gross, and M. A. Welte. 2006. The lipid-droplet proteome reveals that droplets are a protein-storage depot. *Curr. Biol.* 16: 1783–1795.
9. Chavez-Jauregui, R. N., R. D. Mattes, and E. J. Parks. 2010. Dynamics of fat absorption and effect of sham feeding on postprandial lipema. *Gastroenterology*. 139(5): 1538–1548.
10. Chen, H., H. Wang, M. N. Slipchenko, Y. Jung, Y. Shi, J. Zhu, K. K. Buhman, and J. X. Cheng. 2009. A multimodal platform for nonlinear optical microscopy and microspectroscopy. *Opt. Express* 17: 1282–1290.
11. Cheng, J. X. 2007. Coherent anti-Stokes Raman scattering microscopy. *Appl. Spectrosc.* 61: 197–208.
12. Cheng, J. X. and X. S. Xie. 2004. Coherent anti-Stokes Raman scattering microscopy: Instrumentation, theory, and applications. *J. Phys. Chem. B* 108: 827–840.
13. Evans, C. L., E. O. Potma, M. Puoris'haag, D. Cote, C. P. Lin, and X. S. Xie. 2005. Chemical imaging of tissue in vivo with video-rate coherent anti-Stokes Raman scattering microscopy. *Proc. Natl Acad. Sci. USA* 102: 16807–16812.
14. Farese, R. V., Jr. and T. C. Walther. 2009. Lipid droplets finally get a little R-E-S-P-E-C-T. *Cell* 139: 855–860.
15. Ferreira, A. V., G. G. Parreira, A. Green, and L. M. Botion. 2006. Effects of fenofibrate on lipid metabolism in adipose tissue of rats. *Metabolism* 55: 731–735.
16. Fujimoto, Y., H. Itabe, J. Sakai, M. Makita, J. Noda, M. Mori, Y. Higashi, S. Kojima, and T. Takano. 2004. Identification of major proteins in the lipid droplet-enriched fraction isolated from the human hepatocyte cell line HuH7. *Biochim. Biophys. Acta* 1644: 47–59.

17. Hahn, S. E. and D. M. Goldberg. 1992. Modulation of lipoprotein production in Hep G2 cells by fenofibrate and clofibrate. *Biochem. Pharmacol.* 43: 625–633.
18. Heller, F. and C. Harvengt. 1983. Effects of clofibrate, bezafibrate, fenofibrate and probucol on plasma lipolytic enzymes in normolipaemic subjects. *Eur. J. Clin. Pharmacol.* 25: 57–63.
19. Kannel, W. B. and R. S. Vasan. 2009. Triglycerides as vascular risk factors: New epidemiologic insights. *Curr. Opin. Cardiol.* 24: 345–350.
20. Kayden, H. J., J. R. Senior, and F. H. Mattson. 1967. The monoglyceride pathway of fat absorption in man. *J. Clin. Invest.* 46: 1695–1703.
21. Kindel, T., D. M. Lee, and P. Tso. 2010. The mechanism of the formation and secretion of chylomicrons. *Atheroscler. Suppl.* 11: 11–16.
22. Lass, A., R. Zimmermann, M. Oberer, and R. Zechner. 2011. Lipolysis—A highly regulated multi-enzyme complex mediates the catabolism of cellular fat stores. *Prog. Lipid Res.* 50: 14–27.
23. Lee, B., A. M. Fast, J. Zhu, J. X. Cheng, and K. K. Buhman. 2010. Intestine-specific expression of acyl CoA:diacylglycerol acyltransferase 1 reverses resistance to diet-induced hepatic steatosis and obesity in Dgat1−/− mice. *J. Lipid Res.* 51: 1770–1780.
24. Lee, B., J. Zhu, N. E. Wolins, J. X. Cheng, and K. K. Buhman. 2009. Differential association of adipophilin and TIP47 proteins with cytoplasmic lipid droplets in mouse enterocytes during dietary fat absorption. *Biochim. Biophys. Acta* 1791: 1173–1180.
25. Liu, P., Y. Ying, Y. Zhao, D. I. Mundy, M. Zhu, and R. G. Anderson. 2004. Chinese hamster ovary K2 cell lipid droplets appear to be metabolic organelles involved in membrane traffic. *J. Biol. Chem.* 279: 3787–3792.
26. Mansbach, C. M. and F. Gorelick. 2007. Development and physiological regulation of intestinal lipid absorption. II. Dietary lipid absorption, complex lipid synthesis, and the intracellular packaging and secretion of chylomicrons. *Am. J. Physiol. Gastrointest. Liver Physiol.* 293: G645–G650.
27. Masuda, Y., H. Itabe, M. Odaki, K. Hama, Y. Fujimoto, M. Mori, N. Sasabe, J. Aoki, H. Arai, and T. Takano. 2006. ADRP/adipophilin is degraded through the proteasome-dependent pathway during regression of lipid-storing cells. *J. Lipid Res.* 47: 87–98.
28. McFie, P. J., S. L. Banman, S. Kary, and S. J. Stone. 2011. Murine diacylglycerol acyltransferase-2 (DGAT2) can catalyze triacylglycerol synthesis and promote lipid droplet formation independent of its localization to the endoplasmic reticulum. *J. Biol. Chem.* 286(32): 28235–28246.
29. Onat, A., I. Sari, M. Yazici, G. Can, G. Hergenc, and G. S. Avci. 2006. Plasma triglycerides, an independent predictor of cardiovascular disease in men: A prospective study based on a population with prevalent metabolic syndrome. *Int. J. Cardiol.* 108: 89–95.
30. Oosterveer, M. H., A. Grefhorst, T. H. van Dijk, R. Havinga, B. Staels, F. Kuipers, A. K. Groen, and D. J. Reijngoud. 2009. Fenofibrate simultaneously induces hepatic fatty acid oxidation, synthesis, and elongation in mice. *J. Biol. Chem.* 284: 34036–34044.
31. Prentice, A. M. 1998. Manipulation of dietary fat and energy density and subsequent effects on substrate flux and food intake. *Am. J. Clin. Nutr.* 67: 535S–541S.
32. Ramirez, I. and M. I. Friedman. 1990. Dietary hyperphagia in rats: Role of fat, carbohydrate, and energy content. *Physiol. Behav.* 47: 1157–1163.
33. Ravussin, E. and P. A. Tataranni. 1997. Dietary fat and human obesity. *J. Am. Diet. Assoc.* 97: S42–S46.
34. Robertson, M. D., M. Parkes, B. F. Warren, D. J. Ferguson, K. G. Jackson, D. P. Jewell, and K. N. Frayn. 2003. Mobilisation of enterocyte fat stores by oral glucose in humans. *Gut* 52: 834–839.
35. Rolls, B. J. 1995. Carbohydrates, fats, and satiety. *Am. J. Clin. Nutr.* 61: 960S–967S.
36. Rosenson, R. S., D. A. Wolff, A. L. Huskin, I. B. Helenowski, and A. W. Rademaker. 2007. Fenofibrate therapy ameliorates fasting and postprandial lipoproteinemia, oxidative stress, and the inflammatory response in subjects with hypertriglyceridemia and the metabolic syndrome. *Diabetes Care* 30: 1945–1951.
37. Sandoval, J. C., Y. Nakagawa-Toyama, D. Masuda, Y. Tochino, H. Nakaoka, R. Kawase, M. Yuasa-Kawase et al. 2010. Fenofibrate reduces postprandial hypertriglyceridemia in CD36 knockout mice. *J. Atheroscler. Thromb.* 17: 610–618.
38. Sato, S., M. Fukasawa, Y. Yamakawa, T. Natsume, T. Suzuki, I. Shoji, H. Aizaki, T. Miyamura, and M. Nishijima. 2006. Proteomic profiling of lipid droplet proteins in hepatoma cell lines expressing hepatitis C virus core protein. *J. Biochem.* 139: 921–930.
39. Schoonjans, K., M. Watanabe, H. Suzuki, A. Mahfoudi, G. Krey, W. Wahli, P. Grimaldi, B. Staels, T. Yamamoto, and J. Auwerx. 1995. Induction of the acyl-coenzyme A synthetase gene by fibrates and fatty acids is mediated by a peroxisome proliferator response element in the C promoter. *J. Biol. Chem.* 270: 19269–19276.

40. Smith, S. J., S. Cases, D. R. Jensen, H. C. Chen, E. Sande, B. Tow, D. A. Sanan, J. Raber, R. H. Eckel, and R. V. Farese. 2000. Obesity resistance and multiple mechanisms of triglyceride synthesis in mice lacking Dgat. *Nat. Genet.* 25: 87–90.
41. Stalenhoef, A. F. and G. J. de. 2008. Association of fasting and nonfasting serum triglycerides with cardiovascular disease and the role of remnant-like lipoproteins and small dense LDL. *Curr. Opin. Lipidol.* 19: 355–361.
42. Stone, S. J., M. C. Levin, P. Zhou, J. Han, T. C. Walther, and R. V. Farese, Jr. 2009. The endoplasmic reticulum enzyme DGAT2 is found in mitochondria-associated membranes and has a mitochondrial targeting signal that promotes its association with mitochondria. *J. Biol. Chem.* 284: 5352–5361.
43. Su, A. I., M. P. Cooke, K. A. Ching, Y. Hakak, J. R. Walker, T. Wiltshire, A. P. Orth et al. 2002. Large-scale analysis of the human and mouse transcriptomes. *Proc. Natl Acad. Sci. USA* 99: 4465–4470.
44. Subramanian, V., A. Rothenberg, C. Gomez, A. W. Cohen, A. Garcia, S. Bhattacharyya, L. Shapiro et al. 2004. Perilipin A mediates the reversible binding of CGI-58 to lipid droplets in 3T3-L1 adipocytes. *J. Biol. Chem.* 279: 42062–42071.
45. Tso, P. and J. A. Balint. 1986. Formation and transport of chylomicrons by enterocytes to the lymphatics. *Am. J. Physiol.* 250: G715–G726.
46. Tso, P., J. A. Balint, M. B. Bishop, and J. B. Rodgers. 1981. Acute inhibition of intestinal lipid transport by Pluronic L-81 in the rat. *Am. J. Physiol.* 241: G487–G497.
47. Turro, S., M. Ingelmo-Torres, J. M. Estanyol, F. Tebar, M. A. Fernandez, C. V. Albor, K. Gaus, T. Grewal, C. Enrich, and A. Pol. 2006. Identification and characterization of associated with lipid droplet protein 1: A novel membrane-associated protein that resides on hepatic lipid droplets. *Traffic* 7: 1254–1269.
48. Uchida, A., M. N. Slipchenko, J. X. Cheng, and K. K. Buhman. 2011. Fenofibrate, a peroxisome proliferator-activated receptor alpha agonist, alters triglyceride metabolism in enterocytes of mice. *Biochim. Biophys. Acta* 1811: 170–176.
49. Umlauf, E., E. Csaszar, M. Moertelmaier, G. J. Schuetz, R. G. Parton, and R. Prohaska. 2004. Association of stomatin with lipid bodies. *J. Biol. Chem.* 279: 23699–23709.
50. Walther, T. C. and R. V. Farese, Jr. 2009. The life of lipid droplets. *Biochim. Biophys. Acta* 1791: 459–466.
51. Wan, H. C., R. C. Melo, Z. Jin, A. M. Dvorak, and P. F. Weller. 2007. Roles and origins of leukocyte lipid bodies: Proteomic and ultrastructural studies. *FASEB J.* 21: 167–178.
52. Warwick, Z. S. and S. S. Schiffman. 1992. Role of dietary fat in calorie intake and weight gain. *Neurosci. Biobehav. Rev.* 16: 585–596.
53. Wolins, N. E., B. K. Quaynor, J. R. Skinner, A. Tzekov, M. A. Croce, M. C. Gropler, V. Varma et al. 2006. OXPAT/PAT-1 is a PPAR-induced lipid droplet protein that promotes fatty acid utilization. *Diabetes* 55: 3418–3428.
54. Wu, C. C., K. E. Howell, M. C. Neville, J. R. Yates, III, and J. L. McManaman. 2000. Proteomics reveal a link between the endoplasmic reticulum and lipid secretory mechanisms in mammary epithelial cells. *Electrophoresis* 21: 3470–3482.
55. Xu, G., C. Sztalryd, X. Lu, J. T. Tansey, J. Gan, H. Dorward, A. R. Kimmel, and C. Londos. 2005. Post-translational regulation of adipose differentiation-related protein by the ubiquitin/proteasome pathway. *J. Biol. Chem.* 280: 42841–42847.
56. Yamaguchi, T., N. Omatsu, S. Matsushita, and T. Osumi. 2004. CGI-58 interacts with perilipin and is localized to lipid droplets. Possible involvement of CGI-58 mislocalization in Chanarin-Dorfman syndrome. *J. Biol. Chem.* 279: 30490–30497.
57. Yen, C. L. and R. V. Farese, Jr. 2003. MGAT2, a monoacylglycerol acyltransferase expressed in the small intestine. *J. Biol. Chem.* 278: 18532–18537.
58. Yen, C. L., S. J. Stone, S. Koliwad, C. Harris, and R. V. Farese, Jr. 2008. Thematic review series: Glycerolipids. DGAT enzymes and triacylglycerol biosynthesis. *J. Lipid Res.* 49: 2283–2301.
59. Zhu, J., B. Lee, K. K. Buhman, and J. X. Cheng. 2009. A dynamic, cytoplasmic triacylglycerol pool in enterocytes revealed by ex vivo and in vivo coherent anti-Stokes Raman scattering imaging. *J. Lipid Res.* 50: 1080–1089.

# Index

## L



## M

## T



## V

Printed and bound by CPI Group (UK) Ltd, Croydon, CR0 4YY
25/10/2024
01779335-0001